AF477870

ASTROCHEMISTRY:
RECENT SUCCESSES AND
CURRENT CHALLENGES

IAU SYMPOSIUM No. 231

COVER ILLUSTRATION:

This view was taken looking out over the Pacific Ocean, on Ocean View Boulevard, near Asilomar, CA, on a windy day in June 2004.

Above the horizon, a few of the recent successes and current challenges in astrochemistry are shown. The detectability and identification of organic molecules in space are symbolized by (left to right): glycine, ethylene glycol, benzene, glycolaldehyde, acetic acid, vinyl alcohol and pyrimidine. The amazing detection of ND_3, the model reaction scheme leading to deuterated forms of H_3^+, and the subsequent detection of D_2H^+, all recent highlights of deuterium fractionation chemistry, are also illustrated. Extragalactic astrochemistry is represented by the detection of CO in a quasar host galaxy at redshift 6.42. Further successes and challenges will be provided by new and forthcoming observing facilities. Among these are (left to right): JWST, Herschel, ALMA and Spitzer.

Picture: courtesy of Andrew Markwick-Kemper.

INTERNATIONAL ASTRONOMICAL UNION

UNION ASTRONOMIQUE INTERNATIONALE

ASTROCHEMISTRY: RECENT SUCCESSES AND CURRENT CHALLENGES

PROCEEDINGS OF THE 231st SYMPOSIUM OF THE
INTERNATIONAL ASTRONOMICAL UNION
HELD IN PACIFIC GROVE, CALIFORNIA, USA
AUGUST 29–SEPTEMBER 2, 2005

Edited by

DARIUSZ C. LIS
California Institute of Technology, Pasadena, CA, USA

GEOFFREY A. BLAKE
California Institute of Technology, Pasadena, CA, USA

and

ERIC HERBST
Ohio State University, Columbus, OH, USA

CAMBRIDGE UNIVERSITY PRESS
The Edinburgh Building, Cambridge CB2 2RU, UK
40 West 20th Street, New York, NY 10011–4211, USA
477 Williamstown Road, Port Melbourne, VIC 3207, Australia
Ruiz de Alarcón 13, 28014 Madrid, Spain
Dock House, The Waterfront, Cape Town 8001, South Africa

First published 2006

Printed in the United Kingdom at the University Press, Cambridge

Typeset in System LaTeX 2_ε

A catalogue record for this book is available from the British Library

Library of Congress Cataloguing in Publication data

ISBN-13 978 0521 85202 9 hardback
ISBN-10 0521 85202 1 hardback
ISSN 1743–9213

Table of Contents

Section A. Star Formation
Chair: Ewine van Dishoeck

Section B. Basic Processes I
Chair: John Maier

Section F. Extragalactic Molecules
Chair: Thomas Phillips

Section G. Special Session
Chair: Gary Melnick

Section H. Formation of Molecular Hydrogen
Chair: Eric Herbst

Section I. Circumstellar Disks
Chair: Yuri Aikawa

Section J. Basic Processes II
Chair: Helen Fraser

Section K. Solar System Connection
Chair: Dariusz Lis

Section L. Evolved Stars
Chair: Al Glassgold

Ending the Symposium

Report on the Panel Discussion

From left to right: Frances Townes, Charles Townes,
and Ewine van Dishoeck.
Photo: G. Shaw

Preface

The latest in a series of IAU-sponsored symposia on astrochemistry, IAU Symposium 231, "Astrochemistry Throughout the Universe: Recent Successes and Current Challenges," was held from 29 August to 2 September 2005 at the lovely and tranquil setting provided by the Asilomar Conference Grounds, located in Pacific Grove, California within a short distance of the Pacific Ocean. Three hundred participants from 26 countries managed to discuss and learn many new and exciting aspects of their scientific fields without ignoring the attractions of the nearby beach. An evening alfresco banquet with a barbeque and bonfire and an excursion to the Monterey Bay Aquarium provided times to relax. The Local Organizing Committee, chaired by Tom Phillips, who was aided by Darek Lis, Geoffrey Blake, Andrew Markwick-Kemper, Susanna Widicus Weaver, Mary Ellen Barba, Jake Llamas, and others, provided an efficient and smooth operation that left all participants pleased. By the end of the week it had become clear that, true to the title of the symposium, both recent successes and current challenges exist in abundance.

The symposium was organized by the IAU Working Group on Astrochemistry, under the sponsorship of IAU Commission 34 (IAU Division VI), with co-sponsorship provided by Commissions 14, 15, 16, 28, and 40. The symposium was financed in large part through the support of the International Astronomical Union, California Institute of Technology, Jet Propulsion Laboratory, National Radio Astronomy Observatory, Spitzer Science Center, and Infared Processing and Analysis Center. To all of these organizations, we owe a large debt of thanks. This meeting followed a path provided by earlier distinguished conferences. Starting with the fondly remembered meeting at Mont Tremblant, Canada in 1979, meetings with sponsorship similar to this one were held in India in 1985 (IAU Symposium 120), Brazil in 1991 (IAU Symposium 150), the Netherlands in 1996 (IAU Symposium 178), and South Korea in 1999 (IAU Symposium 197). Each meeting has been larger than its predecessor, testifying to the growth of astrochemistry. This particular meeting was both similar to and different from its predecessors. The major similarity was the diversity of scientific disciplines, with astronomy, chemistry, physics, and even geology represented, while the major difference was the accent on youthful speakers, who provided a much needed sense of vigor.

The scientific program of the symposium consisted of 58 oral presentations and 217 posters. The talks were divided into reviews, normal invited talks, and special hot-topic presentations chosen mainly by the SOC from poster submissions. The posters were presented in three separate sessions of at least two hours each, and indicated the strong diversity and many recent successes of the field. Since the molecular universe now covers so many different sources and topics, the wealth of posters was almost overwhelming in its scope. For those of us who failed to come to grips with more than a small fraction of these fine presentations, the posters will be stored in their entirety on the symposium web site (`http://asilomar.caltech.edu/`) if their authors choose to upload them.

The oral program started with an introduction to the topic of molecules in space by the Nobel Laureate Charles H. Townes, whose team was the first to detect a polyatomic molecule (ammonia) in space. The first session was devoted to the topic of star formation, a field of intense current interest in which the role played by molecules as probes is pivotal. The field of single star formation of low-mass stars was emphasized with two newly discovered evolutionary stages—the pre-stellar core and the hot corino—explained. Pre-stellar cores undergo a nearly isothermal collapse and form a dense and cold condensation at their centers, with the density in the range $10^6 - 10^7$ cm^{-3}. Most species heavier than hydrogen and helium are depleted from the gas in this central condensation, and this depletion produces a very strong deuterium isotopic fractionation leading to the formation of relatively high abundances of multiply deuterated species. Following the pre-stellar stage, the collapse becomes adiabatic and the central condensation heats up and collapses to form a star. In addition to winds and shocks, the protostellar stage in low-mass star formation is now known to contain hot corinos, which are the equivalent of the hot cores surrounding young stellar objects of high mass, although the corinos are smaller, are not as warm, and may not contain as many hydrogen-rich organic molecules. These species arise in part from a warm gas-phase chemistry that starts with evaporated mantles of interstellar dust particles produced by increasing temperatures. This warm-up period is not as simple as previously assumed, since desorption of dirty ices is a rather complex affair. Although much of the discussion was on low-mass star formation, the high-mass case was also considered by speakers. Detailed models including chemistry and a variety of physical processes such as radiative transfer and hydrodynamic collapse were discussed.

Once low-mass and possibly high-mass stars are formed, they are often surrounded by circumstellar disks, which are of unique importance because they can be precursors to systems of planets. Both observations and chemical models of these objects were highlighted. Needless to say, astrochemistry is concerned with more than stellar and planetary formation. Indeed, some of the most interesting chemistry occurs in the envelopes of late-type stars and in evolved objects such as the Red Rectangle, especially under carbon-rich conditions. These topics were also presented.

Star formation cannot be comprehended without an understanding of energetic interfaces, and this topic was explored in the context of shocks, photon-dominated regions (PDR's), and X-ray-dominated regions (XDR's). The Horsehead Nebula, a particularly well-studied PDR, contains large abundances of hydrocarbons fairly close to its edge, and an explanation of these abundances still represents a challenge to modellers. The discovery of a new interstellar molecule—CF^+—at a perfectly explainable abundance showed that simple chemical reactions can explain the abundance of this species.

Progress in astrochemistry relies to a great extent on our knowledge of basic chemical and physical processes, so the symposium continued a tradition of sessions on basic processes consisting of presentations on both experimental and theoretical methods. The first of the two general sessions started with a review on what is known and what is needed to be known and continued with talks on molecular spectroscopy, inelastic collisions, low temperature reactive collisions, deuterium fractionation, and dissociative recombination. It was brought out that laboratory evidence indicates strongly that methanol cannot be synthesized in the gas phase and must be produced on grain surfaces and then ejected into the gas in even the coldest regions. But if methanol is so ejected, what about other mantle species? And if all mantle species are desorbed at small but noticeable amounts into the gas, can the use of purely gas-phase models, hitherto rather successful, be continued? Clearly, the challenge of including surface chemistry and gas-surface interactions is a major one. The second session consisted of experimental talks on surface chemistry and desorption, followed by a talk by an ab initio quantum chemist—on the topic of photochemistry on ice surfaces. The session was concluded with talks on PAH identification and spectroscopy and on silicate phases in dust grains. A third session—on the details of molecular hydrogen formation on grain surfaces—showed that powerful experimental methods are now being brought to bear on a process that is probably the most basic and important chemical reaction in the universe. Unfortunately, the data and its interpretation by various investigators are not yet in harmonious agreement.

Presenting a major challenge, diffuse interstellar clouds seem to become more complex each year, and their complexity was fully exposed if not fully understood. Not only are these regions the sites of the still unexplained diffuse interstellar bands and the still mysterious ion CH^+, they are now known to contain polyatomic molecules at hard-to-imagine abundances, including amounts of the simplest polyatomic species—H_3^+—explainable only in terms of a high ionization rate possibly caused by low-energy cosmic rays. The only model presented of a classical diffuse cloud contains three phases: a cool low-density region with high-density nuggets and a shock.

Part of the appeal of astrochemistry lies in the probing of the extent of molecular complexity in assorted sources, and the subject of complex molecules warranted a session of talks on the difficulties of their unambiguous detection despite some past and recent successes using the GBT, and on their possible formation in the gas, via thermal reactions on grain surfaces, and via high-energy surface processes initiated by ionic bombardment.

To balance the accent on youth throughout the oral program, a panel discussion on challenges to the field involving some illustrious but mature astrochemists was organized and proved to be quite entertaining. Two issues seemed to dominate the discussion: the continuing challenge posed by our lack of knowledge of surface processes and the issue of complexity in chemical models. This latter issue appeared to excite the most controversy, with the mature scientists split between coming to grips with if not exactly welcoming complexity in chemical models and preferring back-of-the-envelope estimations.

A special session on results obtained with space-based and airborne telescopes was held. We learned about a possible discovery of interstellar molecular oxygen with the Odin satellite, and heard a summary of the many exciting discoveries of SWAS. New astrochemical results from the Spitzer c2d project, and a teasing presentation of the future glories of SOFIA ended the session.

Although much of astrochemistry deals with interstellar clouds in our own Galaxy, the field is rapidly expanding its region of interest to include nearby galaxies, objects at high redshift, and even the formation of early structures. Talks on each of these subjects were given showing that successes, although less detailed than in the Milky Way, exist.

The links among astrochemistry, the solar system, and extra-solar planets were explored in a pivotal session. The chemistry of extra-solar planets is a field still in its infancy, but with immense potential to overcome current challenges. Objects much closer to home still retain some mysteries: as we learned, comets are chemical factories not fully understood even if pristine by solar system standards, meteorites show complex organic chemistry but much processing, and interplanetary dust particles are amazingly complex given their small sizes.

The oral program was concluded with a fine summary by Alex Dalgarno, whose soft-spoken but incisive commentary never fails to hold the listener's attention. All of the attendees look forward to the next symposium in this fine series, to be held in approximately five-years' time at a venue in southern Europe. By then or shortly thereafter, a new generation of telescopes, including the Herschel Space Observatory and ALMA, will begin to bring new and unimagined advances providing both excitement and difficulty to those of us who must understand them.

Through reading the written contributions in this volume, graciously provided by the speakers at the symposium under severe time pressures, the reader can gain both a sense of the scope of astrochemistry in the early twenty-first century and its immense possibilities for future growth. We hope that this volume will both interest the reader in astrochemistry and help to bring new practitioners to the field.

Eric Herbst, Secretary SOC,
Columbus, OH, October 6, 2005

THE ORGANIZING COMMITTEE

Scientific

L. J. Allamandola (USA)
J. H. Black (Sweden)
G. A. Blake (USA)
P. Caselli (Italy)
E. F. van Dishoeck (Chair, Netherlands)
P. Ehrenfreund (Netherlands)
G. Garay (Chile)
M. Guélin (France)
C. Henkel (Germany)
E. Herbst (Secretary, USA)

U. G. Jørgensen (Denmark)
J. P. Maier (Switzerland)
K. M. Menten (Germany)
T. J. Millar (United Kingdom)
Y. C. Minh (South Korea)
M. Ohishi (Japan)
A. C. Raga (Mexico)
J. Rawlings (United Kingdom)
B. Rowe (France)
J. Yang (China)

Local

M. E. Barba
G. A. Blake
D. C. Lis
J. Llamas

A. J. Markwick-Kemper
S. McCurdy
T. G. Phillips (Chair)
S. Widicus Weaver

Acknowledgments

The symposium was sponsored by the IAU Commissions No. 34 (Interstellar Matter), No. 14 (Atomic and Molecular Data), No. 15 (Comets and Minor Planets), No. 16 (Physical Studies of Planets and Satellites), No. 28 (Galaxies), and No. 40 (Radio Astronomy).

The Local Organizing Committee operated under the auspices of the
California Institute of Technology.

Funding by
California Institute of Technology,
International Astronomical Union,
Infrared Processing and Analysis Center,
Jet Propulsion Laboratory,
Spitzer Science Center,
and
National Radio Astronomy Observatory
is gratefully acknowledged.

Editors' Note

This volume contains the review and invited papers presented at the IAU Symposium 231. A complete list of contributed poster papers is included at the end of this volume and the posters, together with many PowerPoint presentations of invited talks, are available on the Symposium web site (`http://asilomar.caltech.edu/`).

Dariusz C. Lis, Geoffrey A. Blake, and Eric Herbst
December 5, 2005

Participants

Susanne **Aalto**, Onsala Space Observatory, Sweden — susanne@oso.chalmers.se
Nicholas **Abel**, University of Kentucky, USA — npabel2@uky.edu
Tom **Abel**, Stanford University, USA — tabel@stanford.edu
Kinsuk **Acharyya**, Center for Space Physics, India — acharyya@strw.leidenuniv.nl
Marcelino **Agúndez**, Instituto de Estructure de la Materia, Spain — marce@damir.iem.csic.es
Yuri **Aikawa**, Kobe University, Japan — aikawa@kobe-u.ac.jp
Isabel **Aleman**, Universidade de São Paulo, Brazil — isabel@astro.iag.usp.br
Ashraf **Ali**, NASA - Goddard Space Flight Center, USA — xr2aa@lepvax.gsfc.nasa.gov
Louis **Allamandola**, NASA - Ames Research Center, USA — lallamandola@mail.arc.nasa.gov
Stefan **Andersson**, Leiden Observatory, Netherlands — s.andersson@chem.leidenuniv.nl
Yiping **Ao**, Purple Mountain Observatory, China — ypao@pmo.ac.cn
Aldo **Apponi**, University of Arizona, USA — aapponi@virgo.as.arizona.edu
Héctor **Arce**, American Museum of Natural History, USA — harce@amnh.org
Marc **Audard**, Columbia University, USA — audard@astro.columbia.edu
Jean-Pierre **Aycard**, Université de Provence, France — jean-pierre.aycard@up.univ-mrs.fr
Mary Ellen **Barba**, Caltech/IPAC, USA — meb@ipac.caltech.edu
Peter **Barnes**, University of Sidney, Australia — peterb@physics.usyd.edu.au
Arnd **Baurichter**, University of Southern Denmark, Denmark — arnd@fysik.sdu.dk
Eric **Becklin**, UCLA, USA — becklin@astro.ucla.edu
Thomas **Bell**, University College London, United Kingdom — tab@star.ucl.ac.uk
Christopher **Bennett**, University of Hawaii, USA — cjbennet@hawaii.edu
Frank **Bensch**, RAIUB University Bonn, Germany — fbensch@astro.uni-bonn.de
Astrid **Bergeat**, CNRS - Université Bordeaux I, France — a.bergeat@lpcm.u-bordeaux1.fr
Edwin **Bergin**, University of Michigan, USA — ebergin@umich.edu
Johannes **Berndt**, Ruhr University Bochum, Germany — jb@ep2.rub.de
Max **Bernstein**, NASA - Ames Research Center, USA — mbernstein@mail.arc.nasa.gov
Coralie **Berteloite**, Université de Rennes 1, France — coralie.berteloite@univ-rennes1.fr
Nicholas **Betts**, University of Colorado, USA — Nicholas.Betts@colorado.edu
Ludovic **Biennier**, CBRS - Université Rennes, France — ludovic.biennier@univ-rennes1.fr
Ofer **Biham**, Hebrew University, Israel — biham@phys.huji.ac.il
Suzanne **Bisschop**, Leiden Observatory, Netherlands — bisschop@strw.leidenuniv.nl
John **Black**, Onsala Space Observatory, Sweden — jblack@oso.chalmers.se
Geoffrey **Blake**, Caltech, USA — gab@gps.caltech.edu
Heloisa **Boechat**, Universidade Federal do Rio de Janeiro, Brazil — heloisa@ov.ufrj.br
Adwin **Boogert**, Caltech, USA — acab@astro.caltech.edu
Catherine **Boone**, University of Colorado, USA — catherib@colorado.edu
Oliver **Botta**, International Space Science Institute, Switzerland — oliver.botta@issi.unibe.ch
Sandrine **Bottinelli**, University of Hawaii, USA — sandrine@ifa.hawaii.edu
Janet **Bowey**, University College London, United Kingdom — jeb@star.ucl.ac.uk
Rogier **Braakman**, Caltech, USA — rogier@caltech.edu
Philippe **Bréchignac**, University of Paris-Sud, France — Philippe.Brechignac@ppm.u-psud.fr
Christian **Brinch**, Leiden Observatory, Netherlands — brinch@strw.leidenuniv.nl
Crystal **Brogan**, University of Hawaii, USA — cbrogan@ifa.hawaii.edu
Joanna **Brown**, Caltech, USA — jmb@astro.caltech.edu
Don **Brownlee**, University of Washington, USA — brownlee@astro.washington.edu
Natacha **Bruneleau**, CESR-CNRS, France — natacha.bruneleau@cesr.fr
Sandra **Brünken**, Harvard-Smithsonian Center for Astrophysics, USA — bruenken@ph1.uni-koeln.de
Harold **Butner**, Joint Astronomy Center, USA — h.butner@jach.hawaii.edu
Jan **Cami**, NASA - Ames Research Center, USA — jcami@mail.arc.nasa.gov
Andre **Canosa**, CNRS - Université de Rennes 1, France — andre.canosa@univ-rennes1.fr
Paola **Caselli**, INAF-Osservatorio Astrofisico di Arcetri, Italy — caselli@arcetri.astro.it
Emmanuel **Caux**, Centre d'Etude Spatiale des Rayonnements, France — caux@cesr.fr
Stephanie **Cazaux**, INAF-Osservatorio Astrofisico di Arcetri, Italy — cazaux@arcetri.astro.it
Christine **Cecala**, University of Illinois at Indiana-Champaign, USA — ccecala2@uiuc.edu
Cecilia **Ceccarelli**, Observatoire de Grenoble, France — Cecilia.Ceccarelli@obs.ujf-grenoble.fr
Jose **Cernicharo**, DAMIR - IEM - CSIC, Spain — cerni@damir.iem.csic.es
Qiang **Chang**, Ohio State University, USA — changq@pacific.mps.ohio-state.edu
Steven **Charnley**, NASA - Ames Research Center, USA — charnley@dusty.arc.nasa.gov
Thierry **Chiavassa**, Université de Provence, France — thierry.chiavassa@up.univ-mrs.fr
Chris **Churchill**, New Mexico State University, USA — cwc@nmsu.edu
Claudia **Comito**, MPIfR, Germany — ccomito@mpifr-bonn.mpg.de
Nick **Cox**, University of Amsterdam, Netherlands — ncox@science.uva.nl
Pierre **Cox**, IRAM, France — cox@iram.fr
Susan **Creighan**, University College London, United Kingdom — s.creighan@ucl.ac.uk
Dale **Cruikshank**, NASA - Ames Research Center, USA — dcruikshank@mail.arc.nasa.gov
Hermina **Cuppen**, Ohio State University, USA — cuppen@mps.ohio-state.edu
Alexander **Dalgarno**, Harvard-Smithsonian Center for Astrophysics, USA — adalgarno@cfa.harvard.edu
Fabien **Daniel**, DAMIR, Instituto de Estructura de la Materia, CSIC, Spain — daniel@damir.iem.csic.es
Emmanuel **Dartois**, Institut d'Astrophysique Spatiale, France — emmanuel.dartois@ias.u-psud.fr
Michael **Davis**, Open University, United Kingdom — M.P.Davis@open.ac.uk
Anita **Dawes**, Open University, United Kingdom — A.Dawes@open.ac.uk
Thijs **de Graauw**, SRON/Leiden Observatory, Netherlands — thijsdg@sron.rug.nl
Dominique **Deboffle**, CNRS - IAS, France — dominique.deboffle@ias.u-psud .fr
Didier **Despois**, Observatoire de Bordeaux, France — despois@obs.u-bordeaux1.fr
Steven **Doty**, Denison University, USA — doty@denison.edu
Meredith **Drosback**, University of Colorado, USA — meredith.drosback@colorado.edu
Marie-Lise **Dubernet**, LERMA - Observatoire de Paris, France — marie-lise.dubernet@obspm.fr
Anne **Dutrey**, Observatoire de Bordeaux, France — Anne.Dutrey@obs.u-bordeaux1.fr
Yves **Ellinger**, CNRS, France — ellinger@mnhn.fr
Torsten **Elwert**, University of Kentucky, USA — elwert@pa.uky.edu
Martin **Emprechtinger**, Universität Köln, Germany — emprecht@ph1.uni-koeln.de
Neal **Evans**, University of Texas at Austin, USA — nje@astro.as.utexas.edu
Edith **Falgarone**, ENS - Observatoire de Paris, France — edith@lra.ens.fr
Paul **Feldman**, Herzberg Institutue of Astrophysics, Canada — paul.feldman@nrc.gc.ca
Gary **Ferland**, University of Kentucky, USA — gary@pa.uky.edu
Michel **Fich**, University of Waterloo, Canada — fich@uwaterloo.ca
Jean-Hugues **Fillion**, Observatoire de Paris - Univ. de Cergy-Pontoise, France — jean-hugues.fillion@obspm.fr
José Pablo **Fonfría Expósito**, DAMIR - Instituto de Estructura de la Materia - CSIC, Spain — pablo@damir.iem.csic.es
William **Forrest**, University of Rochester, USA — forrest@pas.rochester.edu

Helen **Fraser**, University of Strathclyde, United Kingdom — h.fraser@phys.strath.ac.uk
Rachel **Friesen**, University of Victoria, Canada — rfriesen@uvastro.phys.uvic.ca
Guido **Fuchs**, Leiden Observatory, Netherlands — fuchs@strw.leidenuniv.nl
Gazinur **Galazutdinov**, Korea Astronomy and Space Science Institute, South Korea — gala@boao.re.kr
Robin **Garrod**, Ohio State University, USA — rgarrod@mps.ohio-state.edu
Thomas **Geballe**, Gemini Observatory, USA — tgeballe@gemini.edu
Vincent **Geers**, Leiden Observatory, Netherlands — vcgeers@strw.leidenuniv.nl
Wolf **Geppert**, Stokholm University, Sweden — wgeppert@hotmail.com
Maryvonne **Gerin**, LERMA, France — gerin@lra.ens.fr
Thomas **Giesen**, Universität Köln, Germany — giesen@ph1.uni-koeln.de
Al **Glassgold**, UC Berkeley, USA — aglassgold@astron.berkeley.edu
Simon **Glover**, American Museum of Natural History, USA — scog@amnh.org
Javier **Goicoechea**, ENS - Observatoire de Paris, France — javier@lra.ens.fr
Donald **Goldsmith**, Interstellar Media, USA — dgsmith2@ix.netcom.com
Fabien **Goulay**, UC Berkeley, USA — fgoulay@lbl.gov
Simon **Green**, University of Nottingham, United Kingdom — pcxsdg1@nottingham.ac.uk
Faustine **Grossemy**, IAS - Université Paris XI, France — faustine.grossemy@ias.u-psud.fr
Michel **Guelin**, IRAM, France — guelin@iram.fr
Stéphane **Guilloteau**, Observatore de Bordeaux, France — guilloteau@obs.u-bordeaux1.fr
DeWayne **Halfen**, University of Arizona, USA — halfendt@as.arizona.edu
George **Helou**, Caltech/IPAC, USA — helou@ipac.caltech.edu
Louis **d'Hendecourt**, CNRS, France — ldh@ias.fr
Thomas **Henning**, MPIfA, Germany — henning@mpia.de
Eric **Herbst**, Ohio State University, USA — herbst@mps.ohio-state.edu
Hiroshi **Hidaka**, Hokkaido University, Japan — hidaka@neko.lowtem.hokudai.ac.jp
Carolin **Hieret**, MPIfR, Germany — chieret@mpifr-bonn.mpg.de
Pierre **Hily-Blant**, IRAM, France — hilyblan@iram.fr
Tomoya **Hirota**, National Astronomical Observatory of Japan, Japan — tomoya.hirota@nao.ac.jp
Åke **Hjalmarson**, Onsala Space Observatory, Sweden — hjalmar@oso.chalmers.se
Michiel **Hogerheijde**, Leiden Observatory, Netherlands — michiel@strw.leidenuniv.nl
Jan **Hollis**, NASA - Goddard Space Flight Center, USA — mhollis@milkyway.gsfc.nasa.gov
Philip **Holtom**, Open University, United Kingdom — P.Holtom@open.ac.uk
Liv **Hornekær**, Aarhus University, Denmark — liv@phys.au.dk
Martin **Houde**, University of Western Ontario, Canada — houde@astro.uwo.ca
James **Hough**, University of Hertfordshire, United Kingdom — jhh@star.herts.ac.uk
Douglas **Hudgins**, NASA - Ames Research Center, USA — dhudgins@mail.arc.nasa.gov
Friedrich **Huisken**, University of Jena - MPIfA, Germany — friedrich.huisken@uni-jena.de
Todd **Hunter**, Harvard-Smithsonian Center for Astrophysics, USA — thunter@cfa.harvard.edu
William **Irvine**, University of Massachusetts, USA — irvine@astro.umass.edu
Izaskun **Jimenez-Serra**, DAMIR - Instituto de Estructure de la Materia - CSIC, Spain — izaskun@damir.iem.csic.es
Christine **Joblin**, CESR - CNRS, France — christine.joblin@cesr.fr
Doug **Johnstone**, Herzberg Instute of Astrophysics, Canada — doug.johnstone@nrc-cnrc.gc.ca
Anthony **Jones**, Institut d'Astrophysique Spatiale, France — Anthony.Jones@ias.u-psud.fr
Bastiaan **Jonkheid**, Leiden Observatory, Netherlands — jonkheid@strw.leidenuniv.nl
Jes **Jørgensen**, Harvard-Smithsonian Center for Astrophysics, USA — jjorgensen@cfa.harvard.edu
Kay **Justtanont**, Stockholm Observatory, Sweden — kay@astro.su.se
Mika **Juvela**, Helsinki University Observatory, Finland — mika.juvela@helsinki.fi
Inga **Kamp**, STScI, USA — kamp@stsci.edu
Michael **Kaufman**, San Jose State University - NASA Ames, USA — kaufman@ism.arc.nasa.gov
Akiko **Kawamura**, Nagoya University, Japan — kawamura@a.phys.nagoya-u.ac.jp
Matthew **Kelley**, Caltech, USA — mjkolloy@its.caltech.edu
Tom **Kerr**, Joint Astronomy Center, USA — tkerr@jach.hawaii.edu
Helen **Kirk**, University of Victoria, Canada — hkirk@uvastro.phys.uvic.ca
Claudia **Knez**, University of Texas at Austin, USA — claudia@astro.as.utexas.edu
Eva **Kovacevic**, Ruhr Universität Bohum, Germany — ek@ep2.rub.de
Carsten **Kramer**, Universität Köln, Germany — kramer@ph1.uni-koeln.de
Holger **Kreckel**, MPIfNP, Germany — holger.kreckel@mpi-hd.mpg.de
Jacek **Krełowski**, Nicolaus Copernicus University, Poland — jacek@astri.uni.torun.pl
Geert-Jan **Kroes**, Leiden University, Netherlands — g.j.kroes@chem.leidenuniv.nl
Yi-Jehng **Kuan**, National Taiwan Normal University, Taiwan — kuan@sgrb2.geos.ntnu.edu.tw
Stan **Kurtz**, UNAM, Mexico — s.kurtz@astrosmo.unam.mx
Fred **Lahuis**, Leiden Observatory, Netherlands — f.lahuis@sron.rug.nl
William **Langer**, NASA - JPL, USA — william.langer@jpl.nasa.gov
William **Latter**, NASA HSC/Caltech, USA — latter@ipac.caltech.edu
Brandon **Lawton**, New Mexico State University, USA — blawton@nmsu.edu
Franck **Le Petit**, Onsala Space University, Sweden — Franck.LePetit@obspm.fr
Sébastien **Le Picard**, Université de Rennes 1, France — sebastien.le-picard@univ-rennes1.fr
Jeong-Eun **Lee**, ORNL - University of Kentucky, USA — jelee@astro.as.utexas.edu
Teck-Ghee **Lee**, University of Texas at Austin, USA — leetg@ornl.gov
Bertrand **Lefloch**, Observatoire de Grenoble, France — lefloch@obs.ujf-grenoble.fr
Silvia **Leurini**, MPIfR, Germany — sleurini@mpifr-bonn.mpg.de
Chris **Lintott**, University College London, United Kingdom — cjl@star.ucl.ac.uk
Darek **Lis**, Caltech, USA — dcl@submm.caltech.edu
René **Liseau**, Stockholm University, Sweden — rene@astro.su.se
Harvey **Liszt**, NRAO, USA — hliszt@nrao.edu
Sheng-Yuan **Liu**, ASIAA, Taiwan — syliu@asiaa.sinica.edu.tw
Fred **Lo**, NRAO, USA — flo@nrao.edu
Suzanne **Madden**, CEA - Saclay Servic d'Astrophysique, France — smadden@cea.fr
John **Maier**, University of Basel, Switzerland — j.p.maier@unibas.ch
Giuliano **Malloci**, INAN - Osservatorio Astronomico di Cagliari, Italy — gmalloci@ca.astro.it
Margot **Mandy**, University of Northern British Columbia, Canada — mandy@unbc.ca
Sébastien **Maret**, University of Michigan, USA — smaret@umich.edu
Andrew **Markwick-Kemper**, University of Virginia, USA — andrewjmk@gmail.com
Ciska **Markwick-Kemper**, University of Virginia, USA — fk2n@virginia.edu
Jesus **Martin-Pintado**, DAMIR - Inst. de Estructura de la Materia - CSIC, Spain — jmartin.pintado@iem.cfmac.csic.es
Brenda **Matthews**, Herzberg Institute of Astrophysics, Canada — brenda.matthews@nrc-cnrc.gc.ca
Ben **McCall**, University of Illinois at Urbana-Champaign, USA — bjmccall@uiuc.edu
June **McCombie**, University of Nottingham, United Kingdom — june.mccombie@nottingham.ac.uk
Martin **McCoustra**, University of Nottingham, United Kingdom — martin.mccoustra@nottingham.ac.uk
Donald **McNaughton**, Monash University, Australia — don.mcnaughton@sci.monash.edu.au
Gwendolyn **Meeus**, AIP, Germany — gwen@aip.de
David **Meier**, University of Illinois - NRAO, USA — meierd@astro.uiuc.edu

Rowin **Meijerink**, Leiden Observatory, Netherlands — meijerin@strw.leidenuniv.nl
Gary **Melnick**, Harvard-Smithsonian Center for Astrophysics,, USA — gmelnick@cfa.harvard.edu
Bruno **Merín**, Leiden Observatory, Netherlands — merin@strw.leidenuniv.nl
Stefanie **Milam**, University of Arizona, USA — stemil@as.arizona.edu
Tom **Millar**, University of Manchester, United Kingdom — Tom.Millar@manchester.ac.uk
George **Mitchell**, Saint Mary's University, Canada — gmitchell@ap.stmarys.ca
Marla **Moore**, NASA - Goddard Space Flight Center, USA — ummhm@lepvax.gsfc.nasa.gov
Oscar **Morata**, Ohio State University, USA — omorata@mps.ohio-state.edu
Mark **Morris**, UCLA, USA — morris@astro.ucla.edu
Gloria **Moyano**, Australian National University, Australia — moyano@rsc.anu.edu.au
Robin **Mukerji**, Open University, United Kingdom — r.mukerji@open.ac.uk
Giacomo **Mulas**, INAF - Osservatorio Astronomico di Cagliari, Italy — gmulas@ca.astro.it
Holger **Müller**, Universität Köln - MPIfR, Germany — hspm@ph1.uni-koeln.de
Akihiro **Nagaoka**, Hokkaido University, Japan — nagaoka@neko.lowtem.hokudai.ac.jp
Joan **Najita**, NOAO, USA — najita@noao.edu
David **Neufeld**, Johns Hopkins University, USA — neufeld@pha.jhu.edu
Roisin **Ni Chuimin**, University of Manchester, United Kingdom — roisin.nichuimin@nuigalway.ie
Jarosław **Nirski**, Nicolaus Copernicus University, Poland — nirski@astri.uni.torun.pl
Hideko **Nomura**, Kobe University, Japan — hnomura@kobe-u.ac.jp
Dana **Nuccitelli**, UC Davis, USA — dana1981@yahoo.com
Masatoshi **Ohishi**, National Astronomical Observatory of Japan, Japan — masatoshi.ohishi@nao.ac.jp
Cristina **Oliveira**, Johns Hopkins University, USA — oliveira@pha.jhu.edu
A.O. Henrik **Olofsson**, Stockholm University, Sweden — henrik@oso.chalmers.se
Hans **Olofsson**, Onsala Space Observatory, Sweden — hans@astro.su.se
Evert **Olsson**, Onsala Space Observatory, Sweden — evert@oso.chalmers.se
Laurent **Pagani**, Observatoire de Paris, France — laurent.pagani@obspm.fr
Maria Elisabetta **Palumbo**, INAF - Osservatorio Astrofisico di Catania, Italy — mepalumbo@ct.astro.it
Padelis **Papadopoulos**, ETH Zurich, Switzerland — papadop@phys.ethz.ch
Juan **Pardo**, CSIC, Spain — pardo@damir.iem.csic.es
Berengere **Parise**, MPIfR, Germany — bparise@mpifr-bonn.mpg.de
In Hee **Park**, Ohio State University, USA — ihpark@mps.ohio-state.edu
Amit **Pathak**, Gorakhpur University, India — amitpathak1234@rediffmail.com
Francoise **Pauzat**, CNRS, France — pauzat@mnhn.fr
Els **Peeters**, NASA - Ames Research Center, USA — epeeters@mail.arc.nasa.gov
Yvonne **Pendleton**, NASA - Ames Research Center, USA — ypendleton@mail.arc.nasa.gov
Ruisheng **Peng**, Caltech Submillimeter Observatory, USA — peng@submm.caltech.edu
Carina **Persson**, Onsala Space Observatory, Sweden — carina@oso.chalmers.se
Thomas **Phillips**, Caltech, USA — phillips@submm.caltech.edu
Sergio **Pilling**, UFRJ, Brazil — sergiopilling@yahoo.com.br
Jorge Luis **Pineda Galvez**, RAIUB University Bonn, Germany — jopineda@astro.uni-bonn.de
Valerio **Pirronello**, DMFCI - Università di Catania, Italy — pirronello@dmfci.unict.it
Rene **Plume**, University of Calgary, Canada — plume@ism.ucalgary.ca
Edward **Polehampton**, MPIfR, Germany — epoleham@mpifr-bonn.mpg.de
Klaus **Pontoppidan**, Leiden Observatory, Netherlands — pontoppi@strw.leidenuniv.nl
Marta **Pułecka**, Nicolaus Copernicus Astronomical Center, Poland — mara@ncac.torun.pl
Denis **Puy**, Observatoire de Geneve, Switzerland — Denis.Puy@obs.unige.ch
Chunhua **Qi**, Smithsonian Astrophysical Observatory, USA — cqi@cfa.harvard.edu
Mercedes **Ramos-Lerate**, University College London, United Kingdom — M.R.Lerate@rl.ac.uk
Sofia **Ramstedt**, Stockholm Observatory, Sweden — sofia@astro.su.se
Jonathan **Rawlings**, University College London, England — jcr@star.ucl.ac.uk
Anthony **Remijan**, NASA - Goddard Space Flight Center, USA — aremijan@pop900.gsfc.nasa.gov
Miguel Angel **Requena Torres**, DAMIR - Inst. de Estructura de la Materia - CSIC, Spain — requena@damir.iem.csic.es
Jeonghee **Rho**, Spitzer Science Center, USA — rho@ipac.caltech.edu
Matthew **Richter**, UC Davis, USA — richter@physics.ucdavis.edu
Marianna **Ridderstad**, Helsinki University Observatory, Finland — marianna@astro.helsinki.fi
Isabelle **Ristorcelli**, CESR, France — Isabelle.Ristorcelli@cesr.fr
Helen **Roberts**, University of Manchester, United Kingdom — helen.h.roberts@manchester.ac.uk
Steve **Rodgers**, NASA - Ames Research Center, USA — rodgers@dusty.arc.nasa.gov
Arturo **Rodríguez-Franco**, DAMIR - Inst. de Estructure de la Materia - CSIC, Spain — arturo@damir.iem.csic.es
Markus **Röllig**, Universität Köln, Germany — roellig@ph1.uni-koeln.de
Evelyne **Roueff**, Observatoire de Paris, France — evelyne.roueff@obspm.fr
Monica **Rubio**, Universidad de Chile, Chile — mrubio@das.uchile.cl
Paul **Ruffle**, University of Manchester, United Kingdom — paul.ruffle@postgrad.manchester.ac.uk
Itsuki **Sakon**, University of Tokyo, Japan — isakon@astron.s.u-tokyo.ac.jp
Farid **Salama**, NASA - Ames Research Center, USA — Farid.Salama@nasa.gov
Peter **Sarre**, University of Nottingham, United Kingdom — Peter.Sarre@Nottingham.ac.uk
Daniel Wolf **Savin**, Columbia Astrophysics Laboratory, USA — savin@astro.columbia.edu
Peter **Schilke**, MPIfR, Germany — schilke@mpifr-bonn.mpg.de
Stephan **Schlemmer**, Universität Köln, Germany — schlemmer@ph1.uni-koeln.de
Fredrik **Schöier**, Stockholm Observatory, Sweden — fredrik@astro.su.se
Katharina **Schreyer**, AAIU Jena, Germany — martin@astro.uni-jena.de
Sara **Seager**, Carnegie Institution of Washington, USA — seager@dtm.ciw.edu
Dmitry **Semenov**, MPIfA, Germany — semenov@mpia.de
Osama **Shalabiea**, Cairo University, Egypt — shalabiea@yahoo.com
Gargi **Shaw**, University of Kentucky, USA — gshaw@pa.uky.edu
Valery **Shematovich**, Russian Academy of Sciences, Russian Federation — shematov@inasan.rssi.ru
Ian **Sims**, Université de Rennes 1, France — ian.sims@univ-rennes1.fr
Arfon **Smith**, University of Nottingham, United Kingdom — pcxams@nottingham.ac.uk
Mark **Smith**, University of Arizona, USA — msmith@u.arizona.edu
Ronald **Snell**, University of Massachusetts, USA — snell@astro.umass.edu
Theodore **Snow**, University of Colorado, USA — tsnow@casa.colorado.edu
In-Ok **Song**, Sejong University, South Korea — pcxios@nottingham.ac.uk
Paule **Sonnentrucker**, Johns Hopkins University, USA — sonnentr@pha.jhu.edu
Marco **Spaans**, Kapteyn Astronomical Institutue, Netherlands — spaans@astro.rug.nl
Henrik **Spoon**, Cornell University, USA — spoon@astro.cornell.edu
Phillips **Stancil**, University of Georgia, USA — stancil@physast.uga.edu
Amiel **Sternberg**, Tel Aviv University, Israel — amiel@wise.tau.ac.il
Vladimir **Strelnitski**, Maria Mitchell Observatory, USA — vladimir@mmo.org
Mario **Tafalla**, Observatorio Astronomico Nacional, Spain — m.tafalla@oan.es
Junko **Takahashi**, Meiji Gahuin University, Japan — juntaka@law.meijigakuin.ac.jp
Dahbia **Talbi**, CNRS, France — talbi@mnhn.fr
Xiaofeng **Tan**, NASA - Ames Research Center, USA — xtan@mail.arc.nasa.gov

Achim **Tappe**, Caltech/IPAC - SSC, United Stats — tappe@ipac.caltech.edu
Belén **Tercero**, DAMIR - Instituto de Estructura se la Materia - CSIC, Spain — belen@damir.iem.csic.es
David **Teyssier**, DAMIR - CSIC / SRON, Spain — teyssier@sron.rug.nl
Patrice **Theulé**, Harvard Smithsonian Center for Astrophysics, USA — ptheule@cfa.harvard.edu
Sven **Thorwirth**, MPIfR, Germany — sthorwirth@mpifr-bonn.mpg.de
Alexander **Tielens**, Kapteyn Astronomical Institute, Netherlands — tielens@astro.rug.nl
Radmila **Topalovic**, University of Nottingham, United Kingdom — pcxrt@nottingham.ac.uk
Carmen **Tornow**, Institute of Planetary Research, Germany — carmen.tornow@dlr.de
Charles **Townes**, UC Berkeley, USA — cht@ssl.berkeley.edu
Ewine **van Dishoeck**, Leiden Observatory, Netherlands — ewine@strw.leidenuniv.nl
Tim **van Kempen**, Leiden Observatory, Netherlands — kempen@strw.leidenuniv.nl
Charlotte **Vastel**, Caltech, USA — vastel@submn.caltech.edu
Gianfranco **Vidali**, Syracuse University, USA — gvidali@syr.edu
Uma **Vijh**, University of Toledo, USA — uvijh@astro.utoledo.edu
Serena **Viti**, University College London, United Kingdom — sv@star.ucl.ac.uk
Valentine **Wakelam**, Ohio State University, USA — wakelam@mps.ohio-state.edu
Malcolm **Walmsley**, Arcetri Observatory, Italy — walmsley@arcetri.astro.it
Adam **Walters**, CESR - Toulouse University, France — adam.walters@cesr.fr
Naoki **Watanabe**, Hokkaido University, Japan — watanabe@lowtem.hokudai.ac.jp
William **Whyatt**, University College London, United Kingdom — wwhyatt@star.ucl.ac.uk
Susanna **Widicus Weaver**, University of Illinois, USA — swidicus@caltech.edu
Dmitri **Wiebe**, Russian Academy of Sciences, Russian Federation — dwiebe@inasan.ru
Karen **Willacy**, NASA - JPL, USA — karen.willacy@jpl.nasa.gov
David **Williams**, University College London, United Kingdom — daw@star.ucl.ac.uk
Christine **Wilson**, McMaster University - SMA, Canada — wilson@physics.mcmaster.ca
Eva **Wirström**, Onsala Space Observatory, Sweden — eva@oso.chalmers.se
Adolf **Witt**, University of Toledo, USA — awitt@dusty.astro.utoledo.edu
Mark **Wolfire**, University of Maryland, USA — mwolfire@astro.umd.edu
Angela **Wolff**, University College London, United Kingdom — a.wolff@ucl.ac.uk
Paul **Woods**, NASA - JPL, USA — pwoods@thisvi.jpl.nasa.gov
David **Woon**, Molecular Research Institute, USA — woon@purisima.molres.org
Alwyn **Wootten**, NRAO, USA — awootten@nrao.edu
Hiroshige **Yoshida**, Caltech Submillimeter Observatory, USA — hiro@submm.caltech.edu
Igor **Zinchenko**, Russian Academy of Sciences, Russian Federation — zin@appl.sci-nnov.ru

Astrochemistry: Recent Successes and Current Challenges
Proceedings IAU Symposium No. 231, 2005
D.C. Lis, G.A. Blake & E. Herbst, eds.

© 2006 International Astronomical Union
doi:10.1017/S1743921306006995

Observations of Low-Mass Protostars: Cold Envelopes and Hot Corinos

Cecilia Ceccarelli

Laboratoire d'Astrophysique de Grenoble,
414 rue de la Piscine BP 53, 38041 Grenoble, France
email: Cecilia.Ceccarelli@obs.ujf-grenoble.fr

Abstract. Recent years have seen substantial progresses in our understanding of solar type protostellar structure, and particularly of the chemical structure of the protostellar envelopes. On the one hand, the cold outer regions keep intact the memory of the previous pre-collapse phase, when the dust is so cold and dense that almost all molecules freeze out onto the dust grain mantles. The gas-phase chemical composition undergoes dramatic changes, the most spectacular aspect of which is the huge increase of the molecular deuteration degree, which can reach 13 orders of magnitude with respect to the elemental D/H ratio. On the other hand, in the innermost regions of the envelope – the so-called hot corinos – the grain mantles evaporate when the dust temperatures exceed 100 K, injecting into the gas phase hydrogenated molecules, such as formaldehyde and methanol. Those molecules probably undergo chemical reactions that form more complex organic molecules, which have now also been observed in low-mass hot corinos. In this contribution, I review what we have and have not recently understood concerning both the cold envelopes and the hot corinos of solar-type protostars.

Keywords. astrochemistry — stars: formation

---

1. Introduction

Low-mass stars form from cold and dense condensations inside molecular clouds. The stages just before the collapse sets in are likely represented by the so-called pre-stellar cores (see Tafalla's and Shematovich's contributions in this volume). Once the collapse takes over, a protostar is formed: the central object – the future star – grows in mass by accreting material from the surrounding envelope, which, at the beginning, completely obscures it. This phase is likely represented by so-called Class 0 protostars: low-luminosity ($\leqslant 50\ L_\odot$) sources that emit most of their energy in the sub-millimeter wavelength range (André *et al.* 2000). The accretion of the central object – namely the infall motion of the envelope material – is accompanied by the spectacular phenomenon of outflow, which re-injects into the interstellar medium up to one-third of the infalling material. As time passes, the protostellar envelope is progressively dispersed: the protostar is believed to become a Class I source and then a pre-main-sequence star surrounded by a proto-planetary disk, which eventually may form comets, meteorites and planets (see Blake's and Kamp's contributions in this volume).

One of the major and more fascinating questions linked to the process of the formation of a star and its planetary system, especially if similar to our own solar system, is: *What is the chemical budget acquired during the protostellar phase and inherited by the forming planets?* Answering this question should shed some light on our own origins. This ultimate question splits into several smaller questions: What molecules are formed during the protostellar phase? Do molecules exist that are formed prevalently during the protostellar phase, and that are therefore a hallmark of this period? What is their fate? Do they condense onto the grain mantles during the proto-planetary phase? Are they incorporated into the planetesimals that eventually form comets, meteorites and planets? Are these pristine molecules released into the nascent planetary atmosphere during the early, intense comet bombardment? And what is the ultimate molecular complexity reached during star formation?

Since molecules formed during the protostellar phase might potentially be found at a later stage, even in planets, knowing the chemical composition of the protostellar envelope is therefore of paramount importance. This contribution deals with what we have learned in the last few years about the chemical composition of protostars in their first stage, the Class 0 phase. Although these sources are characterized by and even classified because of their cold appearance – the continuum emission of the cold ($\sim$ 30 K) dust of the external regions of the envelope – it is now recognized that their envelopes have important temperature gradients ($\S$ 2). The cold outer shells hide inner regions where the temperature reaches the sublimation temperature of the grain mantles ($\sim$ 90 $-$ 100 K). From a chemical point of view, this divides the protostellar envelopes into two distinct regions: the *outer envelope* ($\S$ 3), where the chemistry is typical of cold and dense gas, and the inner envelope, where the chemistry is dominated by the mantle sublimation in warm gas. In $\S$ 4, we will show that these inner regions share many of the characteristics of the hot cores discovered in high-mass protostars more than a decade ago (Kurtz *et al.* 2000). Even though they are not just a scaled version of the hot cores, the inner regions of low-mass protostars are now known as *hot corinos* (Ceccarelli 2004b).

2. The Physical and Chemical Structure of Class 0 Protostars

Class 0 sources are defined as low-luminosity sources, with a proven central object and centrally condensed envelope, whose luminosity in sub-millimeter to millimeter wavelengths, L_{smm}, is an important fraction of the bolometric luminosity, L_{bol}: L_{smm}/L_{bol} $\geqslant$0.5% (André *et al.* 2000). This phenomenological classification is meant to correspond to a physical state: the mass of the inner hidden object is lower than the mass of the surrounding envelope. In the literature, other criteria have also been proposed and used, as, for example, the bolometric temperature (Chen *et al.* 1995). The goal of all the proposed criteria has been to identify the properties of protostars in the so-called "main accretion phase", i.e. the phase when the future star acquires the bulk of the mass. At present, fewer than 100 protostars have been counted and, to my knowledge, the most up-to-date database of Class 0 sources has been published by Froebrich (2005).

Evidently, the presence of a central object and a heating source implies the existence of a temperature gradient across the envelope. The first theoretical modeling of a spherical envelope undergoing collapse, and whose initial conditions are those of the singular isotherm sphere (SIS), as theorized by Shu and collaborators in the seventies (Shu 1977), was published by Ceccarelli *et al.* (1996). The model solves in a self-consistent way the gas thermal balance, the radiative transfer and the chemical composition across the envelope as functions of time. Successively, models with larger chemical networks have been developed (Rodgers & Charnley 2003; Doty *et al.* 2004; Lee *et al.* 2004) and with inner photodissociation regions (PDRs) or X-ray-dominated regions (XDRs) (Stauber *et al.* 2004, 2005). Figure 1 shows the predicted physical structure of the envelope of IRAS 16293-2422, the prototype of Class 0 sources. The gas and dust temperature as well as the density profiles have been derived by modeling molecular line observations (Ceccarelli *et al.* 2000a,b).

The gas temperature tracks very closely the dust temperature, except in the very outer and inner regions. In the outer region ($\geqslant$ 400 AU), the CO lines are optically thin and, consequently, the CO line emission is particularly efficient in cooling the gas, which

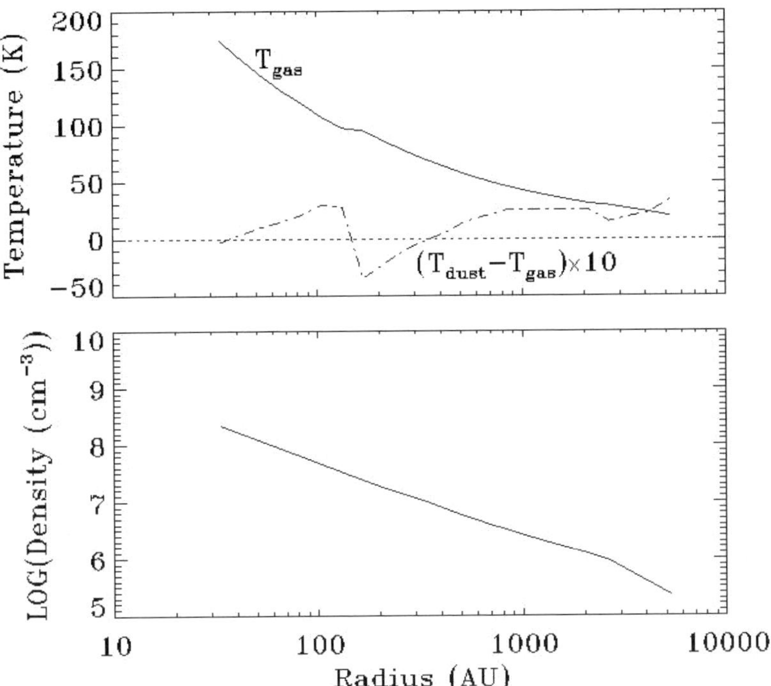

Figure 1. The temperature (upper panel) and density (lower panel) profile of the envelope surrounding the Class 0 protostar IRAS 16293-2422 (Figure from Ceccarelli *et al.* 2000a).

becomes cooler than the dust. On the contrary, in the inner regions between about 150 and 400 AU, the gas is warmer than the dust, because the CO and O lines (the major gas coolants in that region) are optically thick and, therefore, not very efficient in cooling the gas. At about 150 AU, the grain mantles sublimate injecting into the gas phase a large quantity of water molecules. However, the water molecules become more a heating agent than a cooling one, because H_2O molecules are pumped radiatively by the NIR photons of the warm dust and de-excited by collisions, giving rise to a net heating rate. A similar analysis has been carried out for another Class 0 source, NGC 1333 IRAS 4 (Maret *et al.* 2002). Basically, the same behavior observed in IRAS 16293-2422 applies to this protostar too.

Another approach used in the literature is that of deriving the dust temperature and density of the protostellar envelopes by modeling the continuum emission, specifically the spectral energy distribution (SED) and the spatial distribution (maps) simultaneously (*e.g.*, Jorgensen *et al.* 2002). In this case the dust density is assumed to follow a power law, and the gas temperature to be identical to the dust temperature. The two methods give approximatively the same results, which likely just means that the approximations used are valid when considering the uncertainties associated with the observational data. Moreover, this also means that one can rely, to a first approximation, on the predictions obtained by these models.

In the context of this review, the important result of these models is the prediction of regions with dust temperatures larger than the sublimation temperature of the grain mantles (~ 100 K). This prediction has some very immediate consequences: in this region grain mantles evaporate, and, therefore, the abundance of the mantle components should increase with respect to the abundance in the cold ($T_{dust} \leqslant 100$ K) envelope. Water is the major component of the ice mantles, but unfortunately water lines are not observable

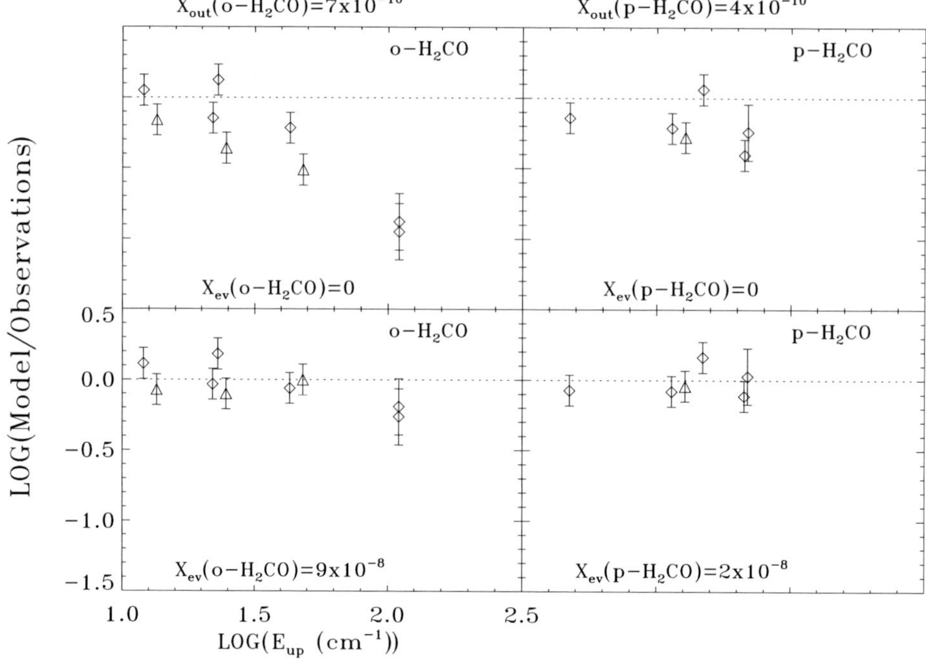

Figure 2. Ratio between the observed and predicted line fluxes as a function of the upper level energy for a model with the parameters described in the text (*bottom panels*) and for a model with no evaporation in the inner region (*top panels*). The diamonds represent the $H_2^{12}CO$ lines while the triangles the $H_2^{13}CO$ lines. Note that the logarithm of the upper level energies of the $H_2^{13}CO$ lines has been shifted by 0.05 to make the points distinguishable in the plot (Figure from Ceccarelli *et al.* 2000b).

with ground-based telescopes. The spectrometer LWS aboard ISO provided observations of water lines towards a few Class 0 sources (Ceccarelli *et al.* 1999), but with a relatively poor spatial and spectral resolution, so that the interpretation of the data is not unique. Nonetheless, if the water line spectra are interpreted as emitted in the envelope, allowing a jump in the water abundance in the $T_{dust} \geqslant 100$ K region, the agreement among the model predictions and the observed line fluxes is rather good. The analysis has been carried out for two cases: IRAS 16293-2422 and NGC 1333 IRAS 4 (Ceccarelli *et al.* 2000a; Maret *et al.* 2002). In both cases, the observed data are reproduced if the water abundance is $\sim 3 \times 10^{-7}$ in the outer envelope, and jumps by about a factor of 10 in the inner region. Formaldehyde is also an important grain-mantle component and this molecule has the advantage that its transitions lie in the millimeter to sub-millimeter wavelength range and are, therefore, observable with ground-based telescopes. These observations have a much better spatial and spectral resolution than ISO-LWS and provide stronger constraints on the models. Figure 2 shows the ratio between the theoretical predictions and the observed line fluxes towards IRAS 16293-2422 for two cases: with a H_2CO abundance constant across the envelope, and with the H_2CO abundance profile following a jump function (Ceccarelli *et al.* 2000b). The constant abundance model fails to reproduce the observed H_2CO spectrum, because the transitions with high ($\geqslant 70$ K) upper energies are underestimated by one order of magnitude. The abundance jump model instead reproduces the observations quite well. *The presence of a jump in the abundance of formaldehyde lends strong support to the hypothesis of the mantle ice sublimation.*

A similar analysis, in terms of abundance jump models, has been carried out with several other molecules in IRAS 16293-2422 (Schöier *et al.* 2002). Molecules, like HC_3N,

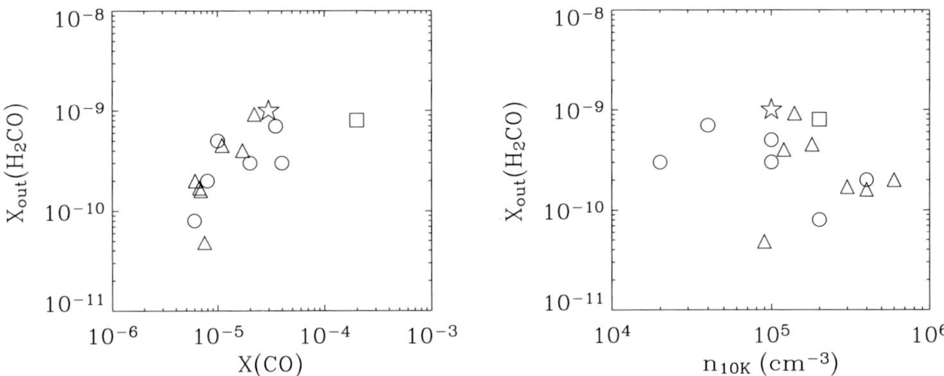

Figure 3. The abundances of H_2CO and CO in the outer envelopes of Class 0 sources and in pre-stellar cores (*left panel*) and the H_2CO abundance versus the density where $T_{\rm dust} = 10$ K (*right panel*). Circles represent Class 0 sources, the star represents IRAS 16293-2422, the square represents VLA1623 – the only targeted source where no H_2CO jump could be detected – and the triangles represent the pre-stellar cores of the Bacmann *et al.* (2002) sample (Figure from Maret *et al.* 2004).

CH_3CN, CH_3OH, SO and SO_2, which are thought to be frozen out onto the grain mantles, display jumps in their abundance by more than one order of magnitude. Also, this work confirms the basic prediction of a region, where the grain mantles sublimate.

All these studies refer to the case of IRAS 16293-2422. The question arises whether IRAS 16293-2422 is representative of Class 0 sources, or it is a peculiar object in the astronomical zoo. A survey of the H_2CO line emission towards a bit less than a dozen Class 0 sources has shown that all the targeted sources, except VLA16293, have a region where the formaldehyde abundance jumps by more than one order of magnitude (Maret *et al.* 2004). Several of these sources also show jumps in the methanol abundance (Maret *et al.* 2005). The sizes of the jump regions range from 10 to 150 AU (see Table 1), and are, therefore, comparable to the size of the solar system. Both the estimates of the abundances and the warm region sizes suffer a relatively large uncertainty, for they are derived by the multi-frequency analysis of *single-dish* measurements, which encompass regions between 1000 and 5000 AU. Finally, an additional uncertainty is also connected to the details of the model used to interpret the data (Jorgensen *et al.* 2005b). In the following, I will review the characteristics of the outer and inner regions respectively, giving further arguments in favor of this structure.

3. The Cold Outer Envelopes

The most outstanding characteristic of the outer envelopes is that, from the chemical point of view, they do not seem to be touched by the collapse. They keep intact the memory of the pre-collapse phase, for they are chemically identical to pre-stellar cores. This similarity is based on two key properties of the outer envelopes and pre-stellar cores: (i) the large depletion of CO and the heavy-element-bearing molecules; and (ii) the strong enhancement of molecular deuteration.

Figure 3 shows the abundance of H_2CO and CO in the outer envelopes of Class 0 sources (Maret *et al.* 2004) and in a sample of pre-stellar cores (Bacmann *et al.* 2002).

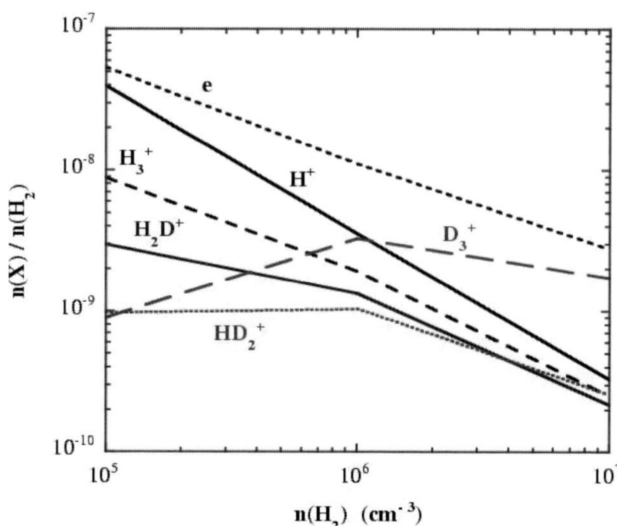

Figure 4. Abundances of major ions and electrons in the standard model (dust radius = 0.1 μm, T = 10 K, $\zeta = 3 \times 10^{-17}$ s^{-1}) as functions of the density of molecular hydrogen (Figure from Walmsley *et al.* 2004).

First, CO and H$_2$CO are highly depleted, up to a factor almost 100. This is due to the condensation of these molecules on the grain mantles. Second, the two samples are practically indistinguishable on this plot, suggesting that indeed the pre-stellar cores are likely the precursors of Class 0 sources. The other amazing aspect of Figure 3 is the tight linear correlation between the CO and H$_2$CO abundances. In principle, the binding energies of CO and H$_2$CO are different, and H$_2$CO should freeze-out onto the grain mantles at lower densities than CO. The observed correlation suggests a common chemical origin for the two species.

The second key aspect pre-stellar cores and Class 0 outer envelopes share is the strongly enhanced molecular deuteration. When considering the processes at the basis of molecular deuteration, this is not surprising, but actually is a tautology with respect to the first aspect. It is in fact the diminishing of the CO abundance in the gas phase that causes the extreme molecular deuteration observed in both classes of objects. In fact, the deuterium atoms are transmitted to the molecules by proton exchange with H$_2$D$^+$, which is formed by the reaction of H$_3^+$ with HD. Under standard conditions, H$_3^+$ is destroyed by the reaction with CO (the most abundant molecule after H$_2$) and, to a much smaller extent, by electron recombination and reaction with HD. However, in CO-depleted and low-ionization gas, the major destruction route of H$_3^+$ becomes the reaction with HD which forms H$_2$D$^+$. As a consequence, H$_2$D$^+$ can become even more abundant than H$_3^+$. Actually, in extreme CO-depletion conditions, H$_2$D$^+$/H$_3^+$ reaches the value of 2, and the most abundant ion can become D$_3^+$, the final product of the chain that forms H$_2$D$^+$, HD$_2^+$ and D$_3^+$, as shown in Figure 4 (Roberts *et al.* 2003; Flower *et al.* 2004; Ceccarelli & Dominik 2005).

The new classes of models for molecular deuteration mentioned above were triggered by the numerous observations of dramatically enhanced D/H ratios, proved by the detection of several doubly and triply deuterated molecules. When considering that the cosmic elemental abundance of deuterium is $\sim 10^{-5}$ times the hydrogen abundance, the adjective "dramatic" is barely an exaggeration. Although doubly deuterated formaldehyde had been detected fifteen years ago in Orion by Turner (1990), the measured abundance ($\sim 0.2\%$ of H_2CO) did not seem to draw particular attention (probably because Orion has always been considered a peculiar region). The discovery that D_2CO has an abundance about 5% that of H_2CO in a low-mass protostar, IRAS 16293-2422 (Ceccarelli *et al.* 1998), on the contrary, started a flurry of activity in this field, and opened the quest for multiply deuterated molecules. In a few years, the following deuterated species have been detected (see Ceccarelli 2004a for a recent review of the subject): doubly and triply deuterated ammonia (Roueff *et al.* 2000; Loinard *et al.* 2001; Lis *et al.* 2002; van der Tak *et al.* 2002), doubly deuterated hydrogen sulfide (Vastel *et al.* 2003), and doubly and triply deuterated methanol (Parise *et al.* 2002, 2004). In addition, the following detailed studies of doubly deuterated formaldehyde have been carried out:

- Towards IRAS 16293-2422, to characterize and better constrain the exact physical conditions of the deuterated gas (Loinard *et al.* 2000);
- Towards IRAS 16293-2422 to measure the deuteration ratio across the envelope, with results larger at ~ 2000 AU from the center, i.e. in the outer envelope (Ceccarelli *et al.* 2001);
- In a sample of protostars, to prove that IRAS 16293-2422 is not a peculiarity but representative of Class 0 sources with respect to the observed degree of formaldehyde deuteration (Loinard *et al.* 2002);
- In a sample of pre-stellar cores, to prove that the molecular deuteration observed in the Class 0 sources sets in during the pre-collapse phase (Bacmann *et al.* 2003).

All these observational studies helped to elucidate the mechanisms underlying the observed molecular deuteration (see Roberts' contribution in this volume). There exist two main routes, depending on the molecule: in the gas phase (Roberts & Millar 2000) or on the grain surfaces (Tielens 1983). Likely, ammonia is a gas-phase product (Roueff *et al.* 2000), whereas methanol and formaldehyde are grain-surface products (Ceccarelli *et al.* 2001; Parise *et al.* 2002). Whatever the mechanism, at the heart of the two routes is the enhancement of the deuteration of H_3^+. In the first case, gas-phase chemistry, the deuterated forms of H_3^+ directly transmit the deuteration to the other molecules (see above). In the second case, grain-surface chemistry, the deuteration is mostly due to the accreting gas at the moment of mantle formation, which, in turns, depends on the deuterated forms of H_3^+. In this respect, a key detection was that of H_2D^+ towards the pre-stellar core L 1544 (Caselli *et al.* 2003). This key ion had been searched for decades without success toward massive star forming regions, and it was finally detected with an abundance comparable to that of H_3^+ in this little and cold condensation. This observation was likely the last step that led to the new class of chemical models described at the beginning of this paragraph. Finally, the detection of HD_2^+ was one important confirmation of the proposed new schemes (Vastel *et al.* 2004).

Although our comprehension of the mechanisms leading to molecular deuteration in the interstellar medium has undoubtedly improved in these last few years, there still remain a number of unresolved questions. First, what is the relative role of gas-phase versus grain-surface chemistry? It is not completely clear yet what molecules are formed in one or the other way, and to what extent. This comes along with the fact that the models still have difficulties in reproducing all the details of the observations. Figure 5 illustrates an example linked to formaldehyde and methanol formation. Both molecules are thought to be formed on grain surfaces via successive hydrogenation of CO (Tielens & Hagen 1982; Charnley *et al.* 1997). Based on this theory, the deuteration ratios in the two molecules should be approximatively similar, because both molecules form almost simultaneously when CO condenses out onto the grain mantles. However, observations of the various forms of the two molecules do not fully agree with this picture, as shown

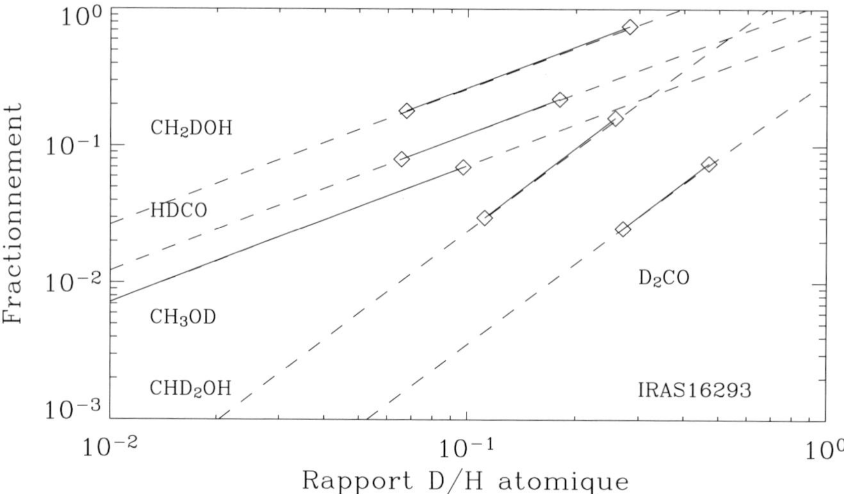

Figure 5. Comparison between the observed fractionations towards IRAS 16293-2422 (solid lines) and the predictions of the grain model (dashed lines) by Stantcheva & Herbst (2003). The data are from Parise *et al.* (2002) and Parise *et al.* (2004) and the Figure has been kindly provided by B. Parise.

in Figure 5. It refers to the case of IRAS 16293-2422, but a survey of several Class 0 sources confirms that similar plots can be drawn for the other surveyed sources (Parise *et al.*, in preparation).

There is a slight discrepancy between the value of the required D/H in accreting atomic gas, when considering the various deuterated species. D_2CO requires systematically larger atomic D/H ratios than HDCO and any other form of deuterated methanol. The CH_3OD/CH_3OH ratio is systematically much lower than that of the other forms of methanol. At this point it is still not completely clear whether this is because of gas-phase reactions taking place after these molecules are injected into the gas phase after the mantle sublimation (Parise *et al.* 2002; Osamura *et al.* 2004).

Another puzzle is represented by deuterated water. Searches for solid HDO have been negative so far, giving some stringent upper limits to the water deuteration. In particular, the value of $\leqslant 2\%$ (Dartois *et al.* 2003; Parise *et al.* 2003) is considerably less than the deuteration observed in the other molecules. However, observations of solid H_2O and HDO may be "contaminated" by the contribution of the molecular cloud in front of the observed protostar (see for example the discussion in Boogert *et al.* 2002). A more stringent constraint is set by observations of gas-phase HDO in the outer envelope and hot corino, where the icy mantles sublimate. Observations towards IRAS 16293-2422 set some stringent constraint to the deuteration of water: in the sublimated ices it is $\sim 3\%$, whereas in the cold envelope it is $\leqslant 0.2\%$ (Parise *et al.* 2005). Even considering a water abundance in the ices of 10^{-4}, the measured HDO/H_2 of 3% is still low with respect to the available elemental deuterium, by about a factor 5. Therefore, water deuteration seems to follow a different route than formaldehyde, methanol, and the other molecules.

Table 1. Properties of the inner envelopes of Class 0 sources.

Source	Radius (AU)	Density (cm^{-3})	H_2CO Outer	H_2CO Inner	CH_3OH Outer	CH_3OH Inner
NGC 1333-IRAS 4A	53	2×10^9	2×10^{-10}	2×10^{-8}	7×10^{-10}	$< 1 \times 10^{-8}$
NGC 1333-IRAS 4B	27	2×10^8	5×10^{-10}	3×10^{-6}	2×10^{-9}	7×10^{-7}
NGC 1333-IRAS 2	47	3×10^8	3×10^{-10}	2×10^{-7}	1×10^{-9}	3×10^{-7}
L 1448-MM	20	2×10^8	7×10^{-10}	6×10^{-7}	2×10^{-9}	5×10^{-7}
L 1448-N	20	1×10^8	3×10^{-10}	1×10^{-6}	7×10^{-10}	$< 4 \times 10^{-7}$
L 1157-MM	40	8×10^8	8×10^{-11}	1×10^{-8}	3×10^{-10}	$< 3 \times 10^{-8}$
IRAS 16293-2422	2×10^8	133	1×10^{-9}	1×10^{-7}	1×10^{-9}	1×10^{-7}

Notes: The second column reports the radius, at which the dust temperature exceeds the grain-mantle sublimation temperature (~ 100 K). The following columns report the formaldehyde and methanol abundances in the outer and inner envelope respectively (from Maret *et al.* 2004, 2005).

4. The Hot Corinos

Table 1 summarizes some important known properties of the inner regions, the hot corinos.

Hot corinos have sizes comparable to the sizes of the solar system, with radii smaller than about 50 AU, with the notable exception of IRAS 16293-2422, which, on this very particular aspect is not representative of the other Class 0 sources. The density is relatively large, larger than 10^8 cm^{-3} in all studied hot corinos. The abundances of formaldehyde and methanol jump by more than one order of magnitude with respect to the abundances in the outer envelopes, supporting the theory that grain mantles sublime in these regions.

A key question, and one relevant for the chemical structure of the Class 0 sources, is what molecules are found in the hot corinos and, in particular, whether organic complex molecules are formed, in analogy with massive hot cores. More generally, are hot corinos scaled versions of hot cores, or does a peculiar chemistry take place? And how reliable are the hot corino sizes derived from the analysis of the single-dish data?

Once again, the benchmark for these studies has been IRAS 16293-2422. In the hot corino of this source, complex organic molecules were detected for the first time (Cazaux *et al.* 2003), and this hot corino was also the first to be imaged (Kuan *et al.* 2004; Bottinelli *et al.* 2004b).

Figure 6 shows the extremely rich spectrum of IRAS 16293-2422 at 98 GHz, and the lines from identified complex molecules. In practice, all the complex molecules typical of massive hot cores searched for were detected, with abundances similar to, if not higher than, those measured in the hot cores (Cazaux *et al.* 2003), including O and N-bearing molecules such as formic acid, HCOOH, acetaldehyde, CH_3CHO, methyl formate, CH_3OCHO, dimethyl ether, CH_3OCH_3, acetic acid, CH_3COOH, methyl cyanide, CH_3CN, ethyl cyanide, C_2H_5CN and propyne, CH_3CCH. Two years after the first detection in IRAS 16293-2422, complex organic molecules have also been detected towards the other Class 0 sources NGC 1333 IRAS 4A and B, and NGC 1333 IRAS 2 (Bottinelli *et al.* 2004a; Jorgensen *et al.* 2005a; Bottinelli *et al.* 2005b).

The number of hot corinos, where complex organic molecules have been detected is evidently much too small to draw any conclusions based on statistical considerations (see Table 2). However, a first conclusion is that complex organic molecules seem to be a rather common product in hot corinos. In this aspect, IRAS 16293-2422 does not seem peculiar, and may be representative of the Class 0 sources. However, there is a relatively

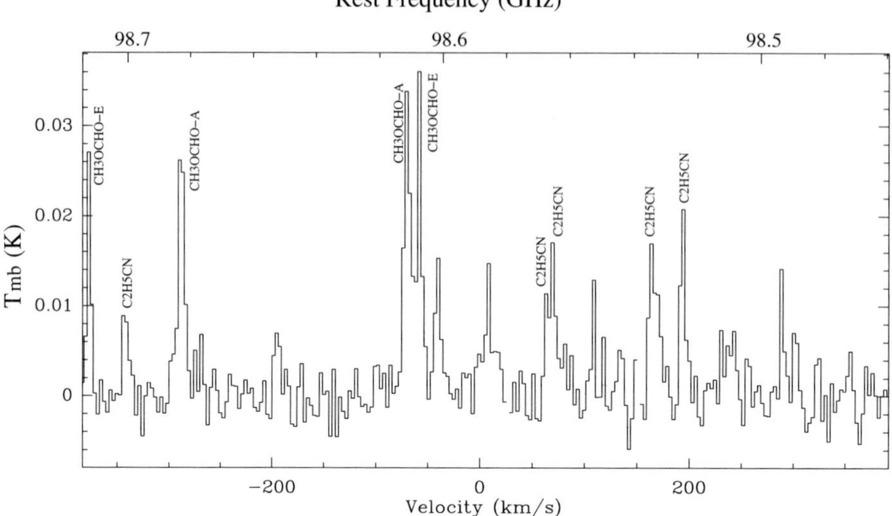

Figure 6. The spectrum at 98 GHz of IRAS 16293-2422 (Cazaux *et al.* 2003).

Table 2. Abundances of the complex organic molecules observed in Class 0 sources.[†]

Molecule	IRAS 16293	IRAS 4A	IRAS 4B	IRAS 2A
H_2CO	1×10^{-7}	2×10^{-8}	3×10^{-6}	2×10^{-7}
CH_3OH	1×10^{-7}	$< 1 \times 10^{-8}$	3×10^{-6}	2×10^{-7}
$HCOOCH_3$-A	1.7×10^{-7}	3.4×10^{-8}	1.1×10^{-6}	$< 6.7 \times 10^{-7}$
$HCOOH$	6.2×10^{-8}	4.6×10^{-9}	$< 1.0 \times 10^{-6}$	$< 1.2 \times 10^{-7}$
CH_3OCH_3	2.4×10^{-7}	$< 2.8 \times 10^{-8}$	$< 1.2 \times 10^{-6}$	3.0×10^{-8}
CH_3CN	1.0×10^{-8}	1.6×10^{-9}	9.5×10^{-8}	8.7×10^{-9}
C_2H_5CN	1.2×10^{-8}	$< 1.2 \times 10^{-9}$	$< 7.5 \times 10^{-7}$	$< 1.0 \times 10^{-7}$

[†]Bottinelli *et al.* (2005b)

large spread in the derived abundances of the detected organic complex molecules. At this point it is not clear what causes this spread, whether it is the age of the hot corino, the mantle composition, or something else.

So far, results based on single-dish observations have been discussed. As previously remarked, these observations encompass regions of at least 1000 AU, whereas the predicted sizes of the hot corinos do not exceed 150 AU (Table 1). It is therefore more than legitimate to verify these predictions with high-spatial resolution, interferometric observations.

The first images of a hot corino have been obtained in the direction of IRAS 16293-2422 with the interferometers SMA (Kuan *et al.* 2004; Chandler *et al.* 2005) and PdB (Bottinelli *et al.* 2004b). Figure 7 shows the images obtained at PdB in the 1.3 mm continuum and in one methyl format line (Bottinelli *et al.* 2004b). Here are a few striking remarks:

• The line emission is concentrated in two spots, centered on the two objects forming the proto-binary system. The whole line flux detected with the single dish is recovered in the two spots, which means that all the single-dish emission originates in those two spots.

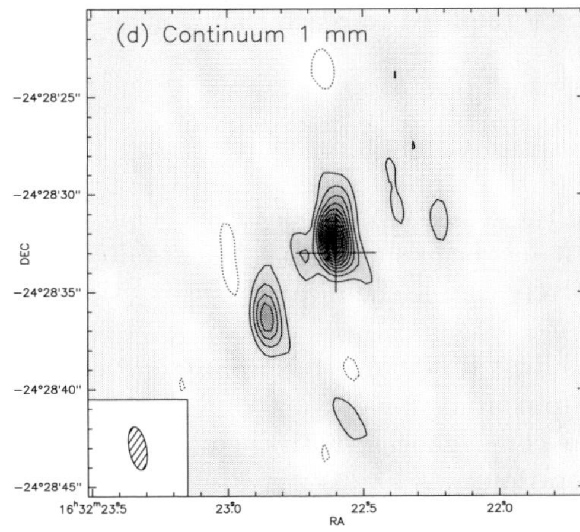

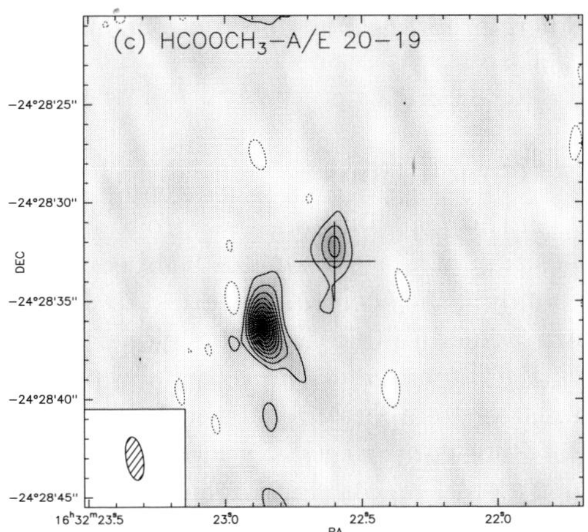

Figure 7. PdB images of IRAS 16293-2422 (Bottinelli *et al.* 2004b): the 1.3 mm continuum (*left panel*) and a line of methyl formate (*right panel*).

- No signs of outflowing material are associated with the methyl formate emission: the emission is compact and the line profile is relatively narrow.
- The brightest – and resolved – spot has a radius of $\sim$ 150 AU, exceedingly close to the "old" predictions, based on the low-resolution water-line spectrum by ISO-LWS (Ceccarelli *et al.* 2000a).
- The two objects of the proto-binary system forming IRAS 16293-2422 do not appear to have the same characteristics. The north source (B in the literature) is bright in the 1.3 mm continuum, but barely detected in the methyl formate line. On the contrary, the south source shines in the molecular line and is barely detected in the continuum.

PdB images of NGC 1333 IRAS 4 have recently been obtained by Bottinelli *et al.* (2005a). As in the case of IRAS 16293-2422, IRAS 4A is a binary system, and the observed ethyl cyanide line emission is concentrated in two spots, centered on the two proto-binary system objects. Again, the two objects do not have the same appearance, with the strength in the continuum and line emission once again reversed.

It is to be emphasized, however, that in both cases, the fact that the two binary objects have a different appearance in the line emission does not necessarily imply that the abundance in the two hot corinos is different. In fact, the weakest hot corino in both binary systems is unresolved and the line likely optically thick (see the discussion in Bottinelli *et al.* 2004b). Evidently, high-resolution observations of organic molecules are a valuable tool to study the chemistry of the hot corinos, and how it depends on the physical conditions, rather than the mantle composition and age, which presumably are the same for the two objects of the binary system. In this respect, it is worth mentioning the potential of the sulfur-bearing molecules in estimating the age of the hot corinos. For many years, S-bearing molecules have been thought to be potential "chemical clocks" (Charnley 1997; Hatchell *et al.* 1999; Wakelam *et al.* 2004a). Recently, Wakelam *et al.* (2004a) have shown that although it is a very delicate exercise, which requires sophisticated modeling, measurements of the SO, SO_2, H_2S and CS can indeed give useful constraints on the hot corino age (see also Wakelam *et al.* 2004b, 2005). The method applied to IRAS 16293-2422 gives an estimate of the hot corino age of about 3000 yr (Wakelam *et al.* 2004a). Note that this is the age of the hot corino and not the age of the protostar, though. In other words, this is the time since the ice mantles sublimated in the 150 AU radius region, which depends on the density and luminosity of the source.

Therefore, this time does not account for the time required to reach that luminosity and density.

5. Conclusions

In this contribution, new progress reached in these last years about the chemical composition of Class 0 sources has been reported. It has been shown that the envelopes surrounding the accreting future stars have a relatively simple chemical structure. Molecules are formed in two distinct regions:

(i) The *outer envelope*, defined by where the dust temperature is lower than the sublimation temperature of the grain mantles. From a chemical point of view, the outer envelopes are practically identical to pre-stellar cores, sharing with them a large molecular depletion and an extreme molecular deuteration.

(ii) The *hot corino*, where the chemistry is dominated by sublimation of the grain mantles. These regions are dense ($\sim 10^8$ cm^{-3}), warm ($\geqslant 100$ K) and compact ($\leqslant 150$ AU in radius). Their major characteristic is an enrichment in complex organic molecules.

While several studies have appeared in the literature on cold envelope chemistry, much less is known about the hot corinos, and their nature is still debated (as is evident by the questions asked during the presentation of this review). I expect that substantial progress will be obtained in the incoming years, thanks to the exploitation of available instruments, the advent of new instruments like ALMA, as well as the development of new chemical models for the hot corinos.

Acknowledgements

I would like to acknowledge and thank the enthusiastic support of the WAGOS group, which did much of the work presented in this review, specifically A. Bacmann, S. Bottinelli, A. Castets, E. Caux, S. Cazaux, B. Lefloch, S. Maret, B. Parise, L. Pagani, X. Tielens, C. Vastel and V. Wakelam. I also acknowledge the financial support of the French National Program entitled "Physique et Chimie du Milieu Interstellaire".

References

André, P., Ward-Thompson, D., & Barsony, M. 2000, *Protostars and Planets IV*, 59

Bacmann, A., Lefloch, B., Ceccarelli, C., Castets, A., Steinacker, J., & Loinard, L. 2002, *A&A* 389, L6

Bacmann, A., Lefloch, B., Ceccarelli, C., Steinacker, J., Castets, A., & Loinard, L. 2003, *Ap. J.* 585, L55

Boogert, A.C.A., Hogerheijde, M.R., Ceccarelli, C., Tielens, A.G.G.M., van Dishoeck, E.F., Blake, G.A., Latter, W.B., & Motte, F. 2002, *Ap. J.* 570, 708

Bottinelli, S., Ceccarelli, C., Lefloch, B., Williams, J.P., Castets, A., Caux, E., Cazaux, S., Maret, S., Parise, B., & Tielens, A.G.G.M. 2004a, *Ap. J.* 615, 354

Bottinelli, S., Ceccarelli, C., Neri, R., & Williams, J.P. 2005a, *Ap. J.*, in preparation

Bottinelli, S., Ceccarelli, C., Neri, R., Williams, J.P., Caux, E., Cazaux, S., Lefloch, B., Maret, S., & Tielens, A.G.G.M. 2004b, *Ap. J.* 617, L69

Bottinelli, S., Ceccarelli, C., Williams, J.P., & Lefloch, B. 2005b, *Ap. J.*, submitted

Caselli, P., van der Tak, F.F.S., Ceccarelli, C., & Bacmann, A. 2003, *A&A* 403, L37

Cazaux, S., Tielens, A.G.G.M., Ceccarelli, C., Castets, A., Wakelam, V., Caux, E., Parise, B., & Teyssier, D. 2003, *Ap. J.* 593, L51

Ceccarelli, C. 2004a, in The Dense Interstellar Medium in Galaxies, ed. S. Pfalzner *et al.* (Berlin: Springer Verlag), 473

Ceccarelli, C. 2004b, in Star Formation in the Interstellar Medium, ed. D. Johnstone *et al.* (ASP Conference Series), 323, 195

Ceccarelli, C., Castets, A., Caux, E., Hollenbach, D., Loinard, L., Molinari, S., & Tielens, A.G.G.M. 2000a, *A&A* 355, 1129

Ceccarelli, C., Castets, A., Loinard, L., Caux, E., & Tielens, A.G.G.M. 1998, *A&A* 338, L43

Ceccarelli, C., Caux, E., Loinard, L., Castets, A., Tielens, A.G.G.M., Molinari, S., Liseau, R., Saraceno, P., Smith, H., & White, G. 1999, *A&A* 342, L21

Ceccarelli, C., & Dominik, C. 2005, *A&A* 440, 583

Ceccarelli, C., Hollenbach, D.J., & Tielens, A.G.G.M. 1996, *Ap. J.* 471, 400

Ceccarelli, C., Loinard, L., Castets, A., Tielens, A.G.G.M., & Caux, E. 2000b, *A&A* 357, L9

Ceccarelli, C., Loinard, L., Castets, A., Tielens, A.G.G.M., Caux, E., Lefloch, B., & Vastel, C. 2001, *A&A* 372, 998

Chandler, C.J., Brogan, C.L., Shirley, Y.L., & Loinard, L. 2005, *Ap. J.* 632, 371

Charnley, S.B. 1997, *Ap. J.* 481, 396

Charnley, S.B., Tielens, A.G.G.M., & Rodgers, S.D. 1997, *Ap. J.* 482, L203

Chen, H., Myers, P.C., Ladd, E.F., & Wood, D.O.S. 1995, *Ap. J.* 445, 377

Dartois, E., Thi, W.-F., Geballe, T.R., Deboffle, D., d'Hendecourt, L., & van Dishoeck, E. 2003, *A&A* 399, 1009

Doty, S.D., Schoier, F.L., & van Dishoeck, E.F. 2004, *A&A* 418, 1021

Flower, D.R., Pineau des Forêts, G., & Walmsley, C.M. 2004, *A&A* 427, 887

Froebrich, D. 2005, *Ap. J. Suppl.* 156, 169

Hatchell, J., Roberts, H., & Millar, T.J. 1999, *A&A* 346, 227

Jorgensen, J.K., Bourke, T.L., Myers, P.C., Schöier, F.L., van Dishoeck, E.F., & Wilner, D.J. 2005a, *Ap. J.* 632, 973

Jorgensen, J.K., Schöier, F.L., & van Dishoeck, E.F. 2002, *A&A* 389, 908

– 2005b, *A&A* 437, 501

Kuan, Y.-J., Huang, H.-C., Charnley, S.B., Hirano, N., Takakuwa, S., Wilner, D.J., Liu, S.-Y., Ohashi, N., Bourke, T.L., Qi, C., & Zhang, Q. 2004, *Ap. J.* 616, L27

Kurtz, S., Cesaroni, R., Churchwell, E., Hofner, P., & Walmsley, C.M. 2000, *Protostars and Planets IV*, 299

Lee, J.-E., Bergin, E.A., & Evans, N.J. 2004, *Ap. J.* 617, 360

Lis, D.C., Roueff, E., Gerin, M., Phillips, T.G., Coudert, L.H., van der Tak, F.F.S., & Schilke, P. 2002, *Ap. J.* 571, L55

Loinard, L., Castets, A., Ceccarelli, C., Caux, E., & Tielens, A.G.G.M. 2001, *Ap. J.* 552, L163

Loinard, L., Castets, A., Ceccarelli, C., Lefloch, B., Benayoun, J.-J., Caux, E., Vastel, C., Dartois, E., & Tielens, A.G.G.M. 2002, *Planet. Space Sci.* 50, 1205

Loinard, L., Castets, A., Ceccarelli, C., Tielens, A.G.G.M., Faure, A., Caux, E., & Duvert, G. 2000, *A&A* 359, 1169

Maret, S., Ceccarelli, C., Caux, E., Tielens, A.G.G.M., & Castets, A. 2002, *A&A* 395, 573

Maret, S., Ceccarelli, C., Caux, E., Tielens, A.G.G.M., Jörgensen, J.K., van Dishoeck, E., Bacmann, A., Castets, A., Lefloch, B., Loinard, L., Parise, B., & Schöier, F.L. 2004, *A&A* 416, 577

Maret, S., Ceccarelli, C., Tielens, A.G.G.M., Caux, E., Lefloch, B., Faure, A., Castets, A., & Flower, D.R. 2005, *A&A*, in press

Osamura, Y., Roberts, H., & Herbst, E. 2004, *A&A* 421, 1101

Parise, B., Castets, A., Herbst, E., Caux, E., Ceccarelli, C., Mukhopadhyay, I., & Tielens, A.G.G.M. 2004, *A&A* 416, 159

Parise, B., Caux, E., Castets, A., Ceccarelli, C., Loinard, L., Tielens, A.G.G.M., Bacmann, A., Cazaux, S., Comito, C., Helmich, F., Kahane, C., Schilke, P., van Dishoeck, E., Wakelam, V., & Walters, A. 2005, *A&A* 431, 547

Parise, B., Ceccarelli, C., Tielens, A.G.G.M., Herbst, E., Lefloch, B., Caux, E., Castets, A., Mukhopadhyay, I., Pagani, L., & Loinard, L. 2002, *A&A* 393, L49

Parise, B., Simon, T., , Caux, E., Dartois, E., Ceccarelli, C., Rayner, J., & Tielens, A.G.G.M. 2003, *A&A*, 410, 897

Roberts, H., Herbst, E., & Millar, T.J. 2003, *Ap. J.* 591, L41

Roberts, H., & Millar, T.J. 2000, *A&A* 364, 780

Rodgers, S.D., & Charnley, S.B. 2003, *Ap. J.* 585, 355

Roueff, E., Tiné, S., Coudert, L.H., Pineau des Forêts, G., Falgarone, E., & Gerin, M. 2000, *A&A* 354, L63

Schöier, F.L., Jorgensen, J.K., van Dishoeck, E.F., & Blake, G.A. 2002, *A&A* 390, 1001

Shu, F.H. 1977, *Ap. J.* 214, 488

Stantcheva, T., & Herbst, E. 2003, *MNRAS* 340, 983

Stauber, P., Doty, S.D., van Dishoeck, E.F., & Benz, A.O. 2005, *A&A* 440, 949

Stauber, P., Doty, S.D., van Dishoeck, E.F., Jorgensen, J.K., & Benz, A.O. 2004, *A&A* 425, 577

Tielens, A.G.G.M. 1983, *A&A* 119, 177

Tielens, A.G.G.M., & Hagen, W. 1982, *A&A* 114, 245

Turner, B.E. 1990, *Ap. J.* 362, L29

van der Tak, F.F.S., Schilke, P., Müller, H.S.P., Lis, D.C., Phillips, T.G., Gerin, M., & Roueff, E. 2002, *A&A* 388, L53

Vastel, C., Phillips, T.G., Ceccarelli, C., & Pearson, J. 2003, *Ap. J.* 593, L97

Vastel, C., Phillips, T.G., & Yoshida, H. 2004, *Ap. J.* 606, L127

Wakelam, V., Caselli, P., Ceccarelli, C., Herbst, E., & Castets, A. 2004a, *A&A* 422, 159

Wakelam, V., Castets, A., Ceccarelli, C., Lefloch, B., Caux, E., & Pagani, L. 2004b, *A&A* 413, 609

Wakelam, V., Ceccarelli, C., Castets, A., Lefloch, B., Loinard, L., Faure, A., Schneider, N., & Benayoun, J.-J. 2005, *A&A* 437, 149

Walmsley, M., Flower, D., & des Forets, G.P. 2004, *A&A* 418, 1035

Discussion

RAWLINGS: Concerning the so-called "hot corinos": these objects are barely resolvable, so what evidence is there that they are the low-mass analogues of hot cores, rather than – say – outflow interfaces?

WOOTTEN: In the two binary objects, only one component shows hot corino aspects. In both, that same component is the only one powering a flow, seen on scales as small as 1/3 AU in 16293 in H_2O maser emission. Since we know there are active flow-driven chemistries occurring in the two hot-corino sources, is it is perhaps too early to rule out the contributions of outflows to enhancement of emissions from more complex molecules? To discern the roles played by outflow and by protostellar heating, we need more sensitivity and more resolution, such as will be provided by ALMA.

CECCARELLI: Although it is possible, in principle, that the observed jumps in the abundances of typical hot core molecules are caused by the interaction between the outflow and the envelope, several reasons are in favor of the dust sublimation region (hot corino) hypothesis: (1) This is a very simple and natural explanation, which is based on the basic prediction that the temperature of the envelope increases towards the center, coupled with the SED+SCUBA map observations fits. (2) The high-resolution maps of the complex organic molecules (Kuan *et al.* 2004; Bottinelli *et al.* 2004) do not show any hint of the outflow interaction, but agree with the point (1) predictions.

GLASSGOLD: If IRAS 16293 is not a "freak", why is the size of its hot inner region so much larger than other Class 0 sources?

CECCARELLI: The size of the hot inner region depends on the luminosity and density. IRAS 16293 is relatively bright (~ 15 $L_\odot$) and with a relatively low-mass envelope with respect to the other studied Class 0 sources. Whether this is because IRAS 16293 is younger or less massive than the other sources is not clear. The remarkable fact is that it is chemically representative of the other sources. This says to us that the initial conditions are, after all, not too different in Class 0 sources, based on both the deuteration degree and the measure of complex organic molecules.

GUÉLIN: You have compared in a table the fractional abundances of molecules in the hot and cold components of several "corinos". The hot-core component is, however, hardly or unresolved even with interferometers like the BdBI or SMA. How can you then compare the beam-averaged abundances in unresolved sources, which lie at different distances and presumably have different sizes?

CECCARELLI: The abundances have been derived by comparing the observations with the predictions of a model which takes into account the density and temperatures profiles of each studied source. The profiles have been derived from the SED and SCUBA maps of each source – based on this structure the model predicts the sizes where the dust temperature exceeds 100 K, the sublimation temperature of the water ices (in first approximation). Therefore, the derived abundances, although they are model dependent, are not derived by "simple" beam-averaged column densities.

NEUFELD: If I understand you correctly, you have assumed a value of 100 K for the temperature at which the "hot-core chemistry", as you call it, turns on. This reflects a theoretical presumption about what these regions are. Is it possible, in cases where you have large numbers of observed transitions, to treat the "turn-on" temperature as a *free* parameter that will be constrained by the observations?

CECCARELLI: Indeed, we have evidence that icy mantles start to sublimate at a lower temperature in IRAS 16293-2422, based on the H_2CO observations (see Ceccarelli *et al.* 2001), because of the large number of observed transitions. As regards the organic complex molecules, sign spots of the "hot core chemistry", the best way to constrain the sublimation mantle temperature is so far the interferometric observations obtained toward IRAS 16293-2422 (Kuan *et al.* 2004; Bottinelli *et al.* 2004). But, in principle, yes, a multi-frequency study with a large number of transition can also constrain the "turn-on" temperature, as in the case of H_2CO.

Photo: E. Herbst

Astrochemistry: Recent Successes and Current Challenges
Proceedings IAU Symposium No. 231, 2005
D.C. Lis, G.A. Blake & E. Herbst, eds.
© 2006 International Astronomical Union
doi:10.1017/S1743921306007009

Observations of Pre-Stellar Cores

M. Tafalla

Observatorio Astronómico Nacional, Alfonso XII 3, E-28014 Madrid, Spain
email: m.tafalla@oan.es

Abstract. Our understanding of the physical and chemical structure of pre-stellar cores, the simplest star-forming sites, has significantly improved since the last IAU Symposium on Astrochemistry (South Korea, 1999). Research done over these years has revealed that major molecular species like CO and CS systematically deplete onto dust grains in the interior of pre-stellar cores, while species like N_2H^+ and NH_3 survive in the gas phase and can usually be detected toward the core centers. Such a selective behavior of molecular species gives rise to a differentiated (onion-like) chemical composition, and manifests itself in molecular maps as a dichotomy between centrally peaked and ring-shaped distributions. From the point of view of star-formation studies, the identification of molecular inhomogeneities in cores helps to resolve past discrepancies between observations made using different tracers, and brings the possibility of self-consistent modelling of the core internal structure. Here I present recent work on determining the physical and chemical structure of two pre-stellar cores, L1498 and L1517B, using observations in a large number of molecules and Monte Carlo radiative transfer analysis. These two cores are typical examples of the pre-stellar core population, and their chemical composition is characterized by the presence of large 'freeze out holes' in most molecular species. In contrast with these chemically processed objects, a new population of chemically young cores has begun to emerge. The characteristics of its most extreme representative, L1521E, are briefly reviewed.

Keywords. astrochemistry — ISM: clouds — ISM: molecules — molecular processes — radiative transfer — radio lines: ISM — stars: formation

---

1. Pre-stellar Cores as Star-Forming Sites

Pre-stellar (or starless) cores are the simplest star-forming sites. They are isolated, lie nearby, and form one star (or one binary) at the time; they closely resemble the theorist's ideal of a star forming region. When observed in a molecular tracer like ammonia, a pre-stellar core appears as a centrally concentrated structure containing one or a few solar masses of material and having a typical size of about 0.1 pc (see Figure 1 and Benson & Myers 1989 for further global properties of cores).

Pre-stellar cores are the dominant star-forming sites in nearby molecular clouds like Taurus-Auriga, where stars like our Sun are currently forming in the so-called "isolated mode" (*e.g.*, Shu, Adams, & Lizano 1987). They are not, however, the dominant star-forming regions of our galaxy, as most stars in the Milky Way have formed in groups (*e.g.*, Adams & Myers 2001), and therefore must result from the collapse of more complex gas structures. Still, star formation in isolated pre-stellar cores seems to involve most of the physical elements that we associate with the birth of a typical low-mass star, like gravitational infall, disk formation, and bipolar outflow ejection. All these elements, in fact, were first identified in stars forming in isolated cores, and they can be studied with great detail in these simple environments.

The above reasons of simplicity make pre-stellar cores ideal sites to study the initial conditions of star-formation (*e.g.*, Ward-Thompson *et al.* 1999). Cores are also the most promising places to test the different competing models of star formation, in particular the (fast) turbulence driven scenario (Mac Low & Klessen 2004) and the (slow)

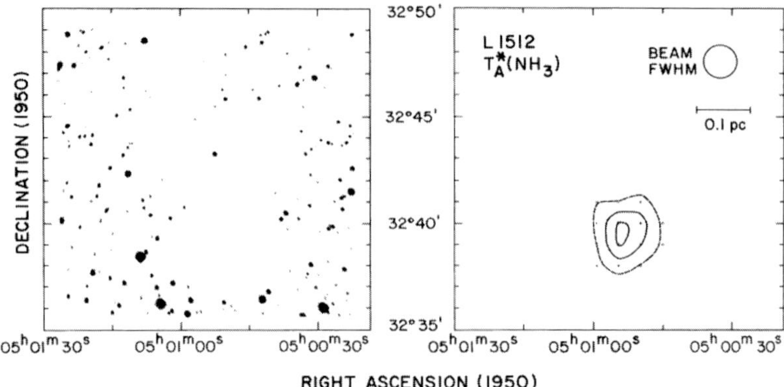

Figure 1. A representative pre-stellar core, L1512. *Left:* Optical image from the Palomar Sky Survey, where L1512 appears as a patch of obscuration against the background of stars. *Right:* Map of the NH_3 emission toward the same region showing L1512 as a centrally concentrated (slightly resolved) object. Figure from Benson & Myers (1989).

magnetically-mediated star formation models (Shu *et al.* 1987; Mouschovias & Ciolek 1999). By determining how pre-stellar cores contract out of the more diffuse ambient cloud and by what mechanism cores lose their gravitational support and start collapsing to form a star, we may be able to observationally distinguish between these two models.

2. Cores as Laboratories of ISM Chemistry

Pre-stellar cores are also some of the simplest chemical laboratories in the interstellar medium (ISM) because of their isolation, approximately spherical geometry, and low (10 K), close to constant gas temperature. Despite this simplicity, however, pre-stellar cores are not chemically homogeneous, and the ignorance of this fact has caused, in the past, serious difficulties when attempting to derive the core internal structure from observations. Early warnings of chemical inhomogeneities were the striking discrepancies found when mapping cores using different molecular tracers. Classic examples of this problem are the observations made using NH_3 and CS, two molecular species expected to trace similar gas conditions. As found by Zhou *et al.* (1989), maps made in NH_3 and CS present systematically different sizes (NH_3 maps are a factor of 2 smaller), different shapes, and often different peak positions. Resolution effects or radiative transfer complexities were soon found insufficient to explain these discrepancies.

A hint of a solution to the tracer discrepancies comes from the recent realization that molecules such as CO and CS are systematically depleted at the centers of cores due to the freeze out onto cold dust grains (Kuiper *et al.* 1996; Willacy *et al.* 1998; Kramer *et al.* 1999; Caselli *et al.* 1999). The identification of molecular freeze out in cores (long expected from theoretical grounds, see Watson & Salpeter 1972) has renewed the interest in the study of dense-core chemistry, and brings with it the promise of resolving the old tracer discrepancies with relatively simple chemical processes. From the point of view of star-formation studies, the identification of freeze out in some molecules is forcing a re-evaluation of the behavior of the different dense gas tracers under typical core conditions, as we depend on them to infer basic core properties like gas temperature and kinematics. Determining the chemical structure of cores has therefore become a necessary step in our attempt to understand how stars are born.

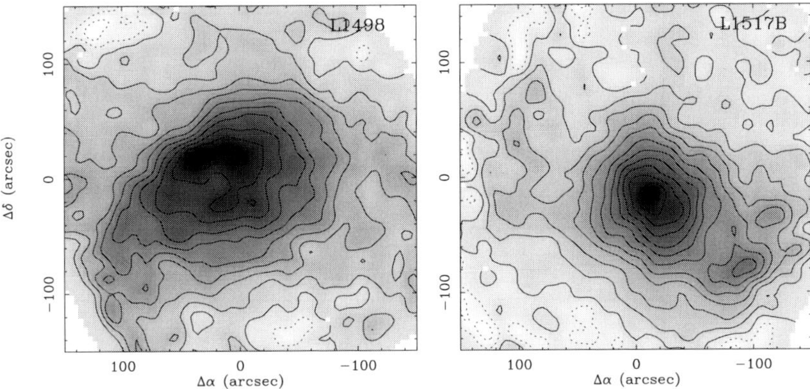

Figure 2. 1.2 mm continuum maps of L1498 (*left*) and L1517B (*right*). Note the central concentration and close-to-round shape of the emission. First contour and contour spacing are 2 mJy/11″-beam.

3. A Molecular Survey of L1498 and L1517B

To understand the chemical behavior of dense-gas tracers in star-forming regions, we have carried out a systematic molecular line survey of two pre-stellar cores in the Taurus-Auriga cloud complex, L1498 and L1517B (Tafalla, Myers, Caselli, & Walmsley 2004; Tafalla *et al.* 2005, in preparation). We selected these two cores (Fig. 2) for being isolated, close to round, and otherwise typical cores of their surrounding cloud, and we have observed them in the 1.2 mm continuum and in a large number of molecular lines from thirteen different species using the IRAM 30m, Effelsberg 100 m, and FCRAO 14 m radio telescopes.

The goal of this project is to model self-consistently all the observed emission in order to determine how the different species trace the core interior, and to provide a high quality set of molecular abundances for testing chemical models. We can divide the analysis of the observations in two steps. First, we determine the physical parameters of the cores by modelling their distributions of density, temperature, and gas kinematics assuming that the cores are spherically symmetric. Once these parameters have been fixed, the cores can be seen as laboratories of known physical properties, and the abundances of the different molecular species can be derived directly by fitting their observed emission.

3.1. *Physical Structure of L1498 and L1517B*

To derive the density profiles of L1498 and L1517B we rely on the dust continuum emission, as this is the most unbiased tracer of the core column density. We assume a 1.2 mm dust emissivity of 0.005 cm^2 g^{-1} and a dust temperature of 10 K, and we note that these parameters are the largest source of uncertainty of the whole analysis (see Tafalla *et al.* 2004 for further details). By fitting the radial profiles of 1.2 mm continuum emission, we find density profiles very close to those expected for isothermal spheres with central densities of 10^5 and 2×10^5 cm^{-3} for L1498 and L1517B, respectively (see Alves *et al.* 2001 and Evans *et al.* 2001 for similar fits to other pre-stellar cores).

To derive the gas temperature profile, we use the NH$_3$ emission, that we will see below traces well the inner core. From the combined analysis of the emission from the metastable J, K=1,1 and 2,2 levels, we derive constant temperature profiles for both cores with values close to 10 K.

Finally, we derive core turbulent profiles using the linewidth of NH$_3$(1,1) complemented with other species. We subtract the (constant) thermal component and find a constant

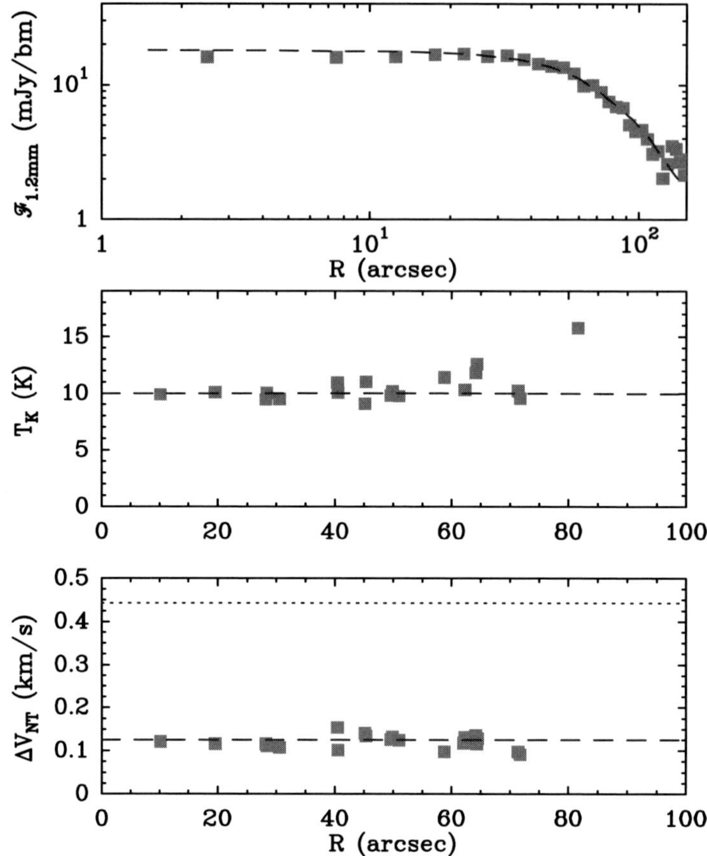

Figure 3. Determination of the physical parameters of L1498. *Top:* Radial profile of 1.2 mm continuum emission from the map of Figure 2 (squares) and prediction from the best fit density determination (dashed lines). *Middle:* NH_3-derived gas kinetic temperature estimate (squares) and constant 10 K fit (dashed lines). *Bottom:* Non-thermal linewidth component derived from $NH_3(1,1)$ spectra (squares) and constant component with FWHM = 0.125 $\mathrm{km\,s^{-1}}$ (dashed lines). The dotted line indicates the expected value for a sonic component.

turbulent component of less than $1/3^{rd}$ the speed of sound. Such a low level of turbulence would seem problematic for current turbulent models of star formation.

3.2. *Chemical Structure of L1498 and L1517B*

Once the physical structure of each core has been determined, the only free parameter left to fit the observed emission of a given species is the radial distribution of its abundance. To convert this distribution into a predicted line intensity, we use a Monte Carlo radiative transfer code that assumes spherical symmetry (Bernes 1979). For each observed transition, we require that the model fits both the radial distribution of integrated intensity (derived by averaging the data azimuthally) and the line spectrum observed toward the core center. When possible, we use the thin emission of a rare isotopologue (such as $C^{18}O$ and $C^{34}S$) to determine the abundance of the major species (in this case CO and CS). For most molecules we have observed two or more transitions, thus their abundance profile is over-determined by the data.

In a meeting like this one is important to emphasize that the analysis presented here can only be carried out if certain molecular parameters have been previously determined.

Among these parameters, the collision rates are critical because they regulate the excitation of the levels and therefore the relation between model abundance and predicted intensity. In fact, the availability of collision rates with H_2 or He was a main criterion for selecting the species observed in this survey. Another important molecular parameter needed to fit the narrow lines observed in L1498 and L1517B is the frequency of each transition. For our modelling, accuracies of 10 kHz are required, and although this level is now easily achieved in the laboratory, not all important molecular transitions have yet been measured to this precision; accurate rest frequencies of molecular ions are especially critical. For our work, we have used the most recent frequency determinations from the CDMS, Gottlieb and collaborators, Dore and collaborators, and the JPL catalog.

To illustrate the process of deriving abundance profiles from the data, we present in Figure 4 (top) a sample of integrated intensity maps for the L1498 core. The three panels in the left column show centrally concentrated distributions corresponding to the 1.2 mm continuum, $N_2H^+(1-0)$, and $NH_3(1,1)$. The rest of the maps, on the other hand, show ring-like distributions with a relative minimum at the dust peak. These ring-like distributions are quite fragmented and slightly different for each molecule, although there are systematic features, such as a brighter peak to the southeast.

When we convert the above maps into radial profiles of integrated intensity and attempt to fit them with different abundance profiles, we obtain the results shown in the bottom part of Figure 4. For each molecule, this figure shows two model predictions: a constant abundance model chosen to fit the emission in the outer core (dashed lines) and a best-fit model (solid lines). As the figure shows, the N_2H^+ emission is fit reasonably well by the constant abundance model, while the observed NH_3 emission is more centrally concentrated than predicted by a constant abundance model. This species, therefore, requires a significant enhancement of abundance toward the core center.

In agreement with the expectation from the ring-shaped maps, the rest of the molecules cannot be fit with a constant abundance model, as these models clearly overestimate the central intensity by a factor of 2 or more when forced to fit the outer core emission. Only using a sharp central abundance drop can both the outer and inner emission be simultaneously fit. For this reason, we have chosen for our best-fit models simple step functions with close to zero abundance toward the center. From the quality of the fit, we conclude that the data are consistent with a (close to) total absence of certain molecules at the core center.

Although not shown here for lack of space, the abundance results for L1517B are very similar to those for L1498 (a full account of the analysis will be presented in Tafalla *et al.* 2005, in preparation). The outer abundances for both cores are in fact rather close, and for most species they agree within a factor of 2; this suggests that both objects have contracted from gas having similar chemical compositions. The size of the central hole, on the other hand, is different in the two cores. L1517B presents significantly smaller central molecular holes (about 50% smaller than L1498), which may be related to the more concentrated gas distribution found in this core.

3.3. *Consequences for Dense Core Studies*

L1498 and L1517B seem in every aspect typical pre-stellar cores. More restricted studies of other cores by different authors, such as those presented in this meeting in the posters of Butner *et al.*, Buckle *et al.*, Friesen *et al.*, Zinchenko *et al.*, show molecular abundance patterns that are very similar to those found in L1498 and L1517B; it seems therefore natural to assume that the abundance profiles of L1498 and L1517B are representative of the pre-stellar core population as a whole (but see below for exceptions). Thus, despite

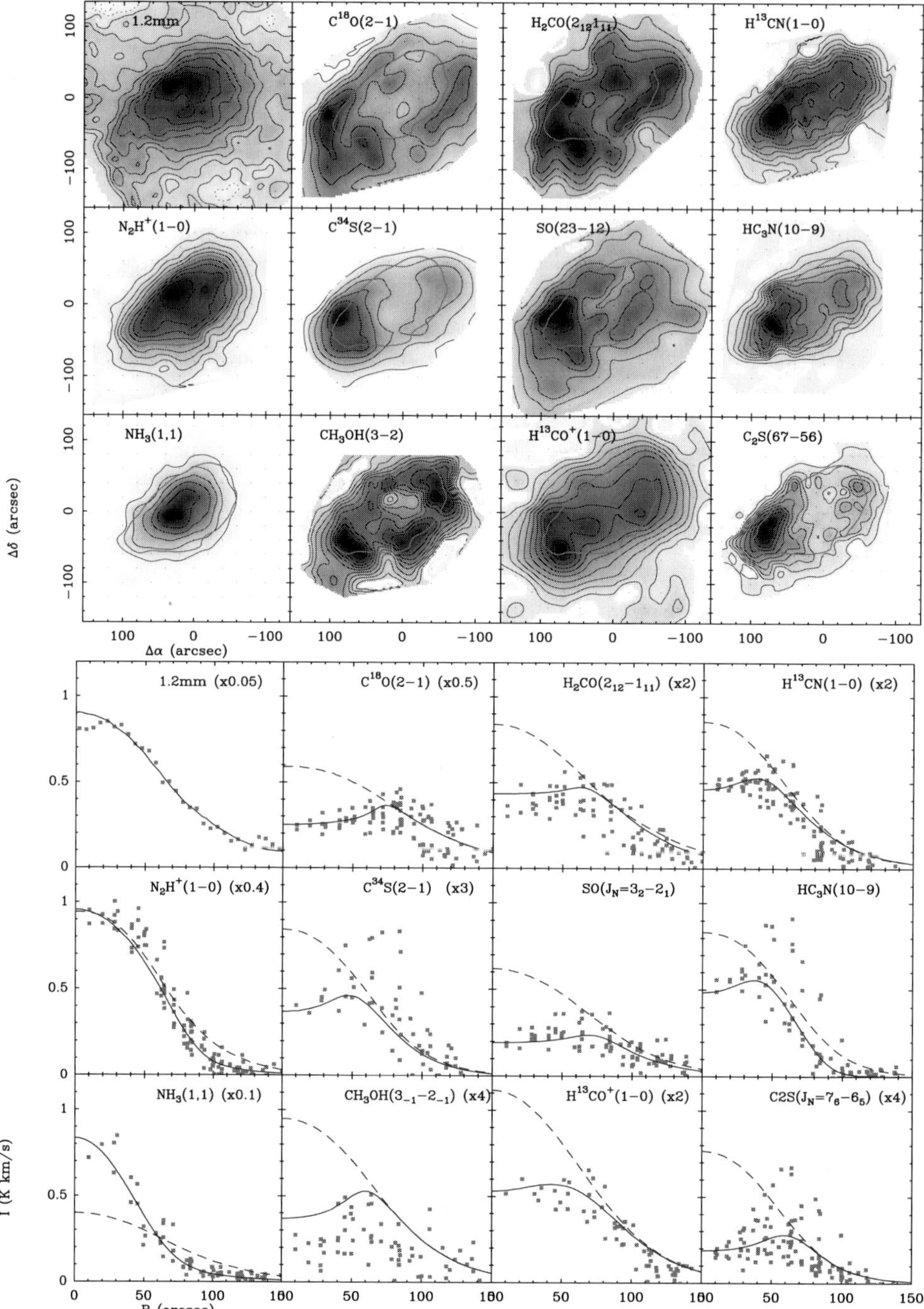

Figure 4. Partial results of the molecular survey toward L1498. *Top:* Maps of 1.2 mm continuum and molecular lines for a sample of transitions. The three panels in the leftmost column contain centrally concentrated emission maps, while the rest of the panels show ring-shaped distributions. *Bottom:* Radial profiles of emission derived from the maps in the top (squares) together with the emission predictions from two Monte Carlo models. The dashed lines are the predictions from constant abundance models set to fit the outer core emission, while the solid lines are predictions from the best-fit models.

their expected simplicity, the observations force us to conclude that pre-stellar cores must have a strongly differentiated chemical composition.

If chemical inhomogeneities are part of the initial conditions of star formation, they need to be considered seriously when sampling star-forming gas with molecular tracers. The bottom panels in Figure 4, in particular, show that many molecular maps of a core may reflect more its chemical composition than its physical structure, and illustrate the danger in attributing emission peaks of even thin lines like $C^{34}S(2\text{--}1)$ to actual core sub-structure. This analysis demonstrates that one needs to be careful when choosing dense gas tracers, and that N_2H^+ and NH_3 are two of the best choices. These species, together with their isotopologues and the H_2D^+ ion recently detected by Caselli *et al.* (2003) in the pre-stellar core L1544, are probably all the molecular tracers left to study the conditions in the inner core (even N_2H^+ may deplete at high densities, see Bergin *et al.* 2002 and Pagani *et al.* 2005). Posters in this meeting by Crapsi *et al.* and Vastel *et al.* show the use of these tracers to study the central conditions of pre-stellar cores.

The modelling of pre-stellar core chemistry is the subject of several contributions in this meeting (talks by Shematovich and Roberts, and posters by Aikawa *et al.*, Rawlings, Lee *et al.*, Walmsley *et al.*), so I refer to them for further details. The only point worth emphasizing here is that the unique behavior of species like N_2H^+ and NH_3 strongly suggests that their survival in the gas phase results from a low binding energy of N_2, as initially suggested by Bergin & Langer (1997) and Charnley (1997). However, similar binding energies for N_2 and CO have been recently measured by Öberg *et al.* (2005), a result that has been used to claim that other mechanisms, like a low sticking coefficient for molecular or atomic N, may be needed to explain the behavior of N_2H^+ and NH_3 (Flower *et al.* 2005). Further work in this topic is clearly needed if we are to understand and use to our advantage the peculiar chemistry of these two important dense gas tracers.

4. Searching for Young Pre-Stellar Cores

If we observe pre-stellar cores selected from NH_3 surveys (like those from Myers and collaborators), we find that molecular depletion is the norm without exception. This systematic trend suggests that NH_3-selected cores are significantly advanced in their process of gas contraction from the ambient cloud, and that there must exist a population of younger cores that are their precursors. Identifying such a population of cores is interesting not only because of their chemical properties, but because its members should constitute a missing link between cloud and core conditions, and they will therefore provide useful clues on the (mysterious) physical process that drives core condensation out of ambient material.

In the search for young pre-stellar cores, we can take advantage of their expected chemical properties. One of them is a lower degree of CO freeze out at the center, which should make a young pre-stellar core appear centrally concentrated in a $C^{18}O$ map. Another expected property is a low abundance of late-time species like N_2H^+ and NH_3, which will translate into weak emission in any line of these tracers. To relate these properties, we define the intensity ratio

$$R = \frac{I[C^{18}O(1-0)]}{I[N_2H^+(1-0)]},$$

where the intensities are evaluated at the core center. This ratio is an easily measurable quantity, as both lines lie in the 3 mm wavelength band and can therefore be observed with similar angular resolution. Young cores are expected to have relatively large central $C^{18}O(1\text{--}0)$ emission together with weak $N_2H^+(1\text{--}0)$ lines, so they are expected to have

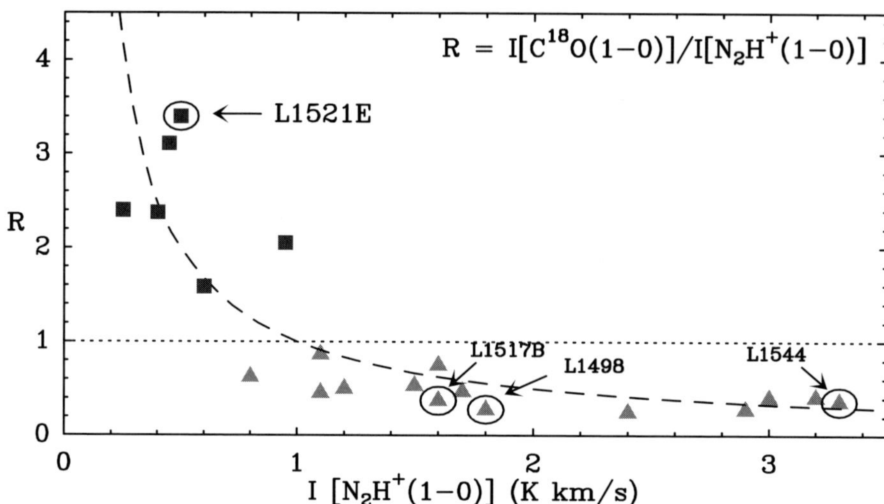

Figure 5. R $(=$I$[C^{18}O(1{-}0)]/$I$[N_2H^+(1{-}0)])$ as function of I$[N_2H^+(1{-}0)]$ for two samples of cores. The triangles are "classical" starless cores selected for their easily detectable NH_3 emission, while the squares are starless cores selected for their weak NH_3 emission. The dotted horizontal line marks the expected approximate boundary between cores with and without $C^{18}O$ freeze out, and the dashed line is the prediction from a toy model of core chemical evolution. Note how L1521E stands out among the young core candidates (R$>$ 1).

relatively large values of the R parameter. Conversely, old cores are expected to have relatively weak central $C^{18}O(1{-}0)$ emission and strong $N_2H^+(1{-}0)$ lines, so they should be characterized by relatively low values of R. The approximate boundary between these two behaviors can be calculated using the ratio predicted by our Monte Carlo model for cores like L1498 and L1517B assuming constant abundances for the two species. In this way, we find the expected separation between young and old cores near the value $R = 1$.

As mentioned before, searches for young cores using NH_3-selected candidates seem doomed to fail. This is illustrated by the triangles in Figure 5, which depict a survey of that type of objects carried out with the FCRAO telescope. In this figure, all selected objects lie below the $R = 1$ line, including (not surprisingly) L1498 and L1517B studied before. The objects from this survey probably span a range of ages, as suggested by the range of $N_2H^+(1{-}0)$ intensities that starts near 1 and ends past 3 with evolved cores like L1544 (*e.g.*, Crapsi *et al.* 2005). However, they seem to be missing the youngest cores.

To identify candidates as young starless cores we need to include cores having weak NH_3 emission, and this has been done selecting sources from the survey of Suzuki *et al.* (1992) and from a survey of the L1521 filament in Taurus using the FCRAO telescope. These cores finally fill the region in the plot expected for young cores, as they have low $N_2H^+(1{-}0)$ intensities together with large R values. Among these objects, L1521E has the largest R ratio ($=3.4$), and therefore appears as the best candidate for a young starless core. Previous suggestions of this core being extremely young have been made by Suzuki *et al.* (1992) and Hirota *et al.* (2002).

Given the unusual characteristics of L1521E, we have carried out a molecular survey of this core in a similar manner as we have studied L1498 and L1517B. From a preliminary analysis of these data, we conclude that it has no significant CO or CS central depletion (Tafalla & Santiago 2004), and that its N_2H^+ and NH_3 abundances are 8 and 20 times lower, respectively, than L1498 and L1517B. These characteristics truly classify L1521E as a chemically young pre-stellar core, and therefore suggest that this object has contracted from the ambient cloud to its observed state rather recently. Surprisingly, however, L1521E has a central density of 10^5 cm^{-3}, which is very similar to that found in L1498 and L1517B. This high density contradicts the expectation that a young core should be less dense, and suggests that L1521E may have contracted faster than the others (also Aikawa *et al.* 2005). Clearly, more work is needed to clarify the origin of this group of starless cores, and recent studies of similar systems are encouraging (Hirota *et al.* 2004, Morata *et al.* 2005). The poster by Hirota and Yamamoto in this meeting provides recent results on this topic.

Acknowledgements

I thank the organizers for their invitation and for a highly enjoyable and productive meeting. Part of the work presented here is the result of an ongoing collaboration with Joaquín Santiago, Phil Myers, Paola Caselli, Malcolm Walmsley, Claudia Comito, and Antonio Crapsi. I thank them for help and discussions on pre-stellar cores over the last several years.

References

Adams, F.C. & Myers, P.C. 2001, *Ap. J.* 553, 744

Aikawa, Y., Herbst, E., Roberts, H., & Caselli, P. 2005, *Ap. J.* 620, 330

Alves, J., Lada, C.J., & Lada, E.A. 2001, *Nature* 409, 159

Benson, P.J. & Myers, P.C. 1989, *Ap. J. Suppl.* 71, 89

Bergin, E.A. & Langer, W.D. 1997, *Ap. J.* 486, 316

Bergin, E.A., Alves, J., Huard, T., & Lada, C.J. 2002, *Ap. J.* 570, L101

Bernes, C. 1979, *A&A* 73, 67

Caselli, P., Walmsley, C.M., Tafalla, M., Dore, L. & Myers, P.C. 1999, *Ap. J.* 523, L165

Caselli, P., van der Tak, F.F.S., Ceccarelli, C., & Bacmann, A. 2003, *A&A* 403, L37

Charnley, S.B. 1997, *MNRAS* 291, 455

Crapsi, A., Caselli, P., Walmsley, C.M., Myers, P.C., Tafalla, M., Lee, C.W., & Bourke, T.L. 2005, *Ap. J.* 619, 379

Evans, N.J., II, Rawlings, J.M.C., Shirley, Y.L., & Mundy, L.G. 2001 *Ap. J.* 557, 193

Flower, D.R., Pineau Des Forêts, G., & Walmsley, C.M.2005 *A&A* 436, 933

Hirota, T., Ito, T., & Yamamoto, S. 2002 *Ap. J.* 565, 359

Hirota, T., Maezawa, H., & Yamamoto, S. 2004 *Ap. J.* 617, 399

Kramer, C., Alves, J., Lada, C.J., Lada, E.A., Sievers, A., Ungerechts & Walmsley, C.M. 1999, *A&A* 342, 257

Kuiper, T.B.H., Langer, W.D., & Velusamy, T. 1996, *Ap. J.* 468, 761

Mac Low, M.-M. & Klessen, R.S. 2004, *Rev. Mod. Phys.* 76, 125

Morata, O., Girart, J.M., & Estalella, R. 2005 *A&A* 435, 113

Mouschovias, T.C., & Ciolek, G.E. 1999, in C.J. Lada and N.D. Kylafis (eds.), *The Origin of Stars and Planetary Systems (Dordrecht: Kluwer)*, p. 305

Öberg, K.I., van Broekhuizen, F., Fraser, H.J., Bisschop, S.E., van Dishoeck, E.F., & Schlemmer, S. 2005, *A&A* 621, L33

Pagani, L., Pardo, J.-R., Apponi, A.J., Bacmann, A., & Cabrit, S. 2005, *A&A* 429, 181

Shu, F.H., Adams, F.C., & Lizano S. 1987, *ARAA* 25, 23

Suzuki, H., Yamamoto, S., Ohishi, M., Kaifu, N., Ishikawa, S.-I., Hirahara, Y., & Takano, S. 1992, *Ap. J.* 392, 551

Tafalla, M., Myers, P.C., Caselli, P., & Walmsley, C.M. 2004, *A&A* 416, 191

Tafalla, M. & Santiago, J. 2004, *A&A* 414, L53

Ward-Thompson, D., Motte, F., & André, P. 1999, *A&A* 305, 143

Watson, W.D. & Salpeter, E.E. 1972, *Ap. J.* 175, 659

Willacy, K., Langer, W.D., & Velusamy, T. 1998, *Ap. J.* 507, L171

Zhou, S., Wu, Y., Evans, N.J.II, Fuller, G.A., & Myers, P.C. 1989, *Ap. J.* 346, 168

Discussion

GEBALLE: You stated at the beginning of your talk that some pre-stellar cores display outflows. How is it possible without a star (or protostar) in the core and such objects being "pre-stellar" cores?

TAFALLA: I'm afraid there is some misunderstanding here. I mentioned that stars forming in cores like L1498 and L1517B will develop outflows, but of course they need to form a protostar first. As far as we know (and we have looked), these two cores are starless and therefore have no outflow.

JOHNSTONE: We should be careful using terms like "young" when the evolution of a core is unknown. How a core forms and evolves is still poorly known and perhaps we should be clearer in the terminology using expressions like "early differentiation" or the like. Still, the ratio relation clearly shows an age direction with lower R's being more evolved cores.

TAFALLA: I agree that we have to be careful connecting the chemical and physical evolution of cores. At this point, when we say "young" we usually mean having low chemical processing according to models. The main unknown connecting physical and chemical evolution is whether all cores start with similar chemical composition, so chemical differences can be interpreted as differences either in age or in speed of contraction.

PAPADOPOULOS: A note rather than a question. Quiescent pre-stellar cores are not a problem for the turbulent-driven star formation, since from the first proposition by Larson to the latest work by Krumholz and McKee, such quiescent cores/places in the ISM are the ones predicted to detach from the turbulent hierarchy and settle to a star-formation-capable phase with thermal energy dominating over the turbulent one.

TAFALLA: That is exactly why identifying young cores is so interesting for testing core contraction models. If cores form by the dissipation of turbulence, we should find a systematic decrease in linewidth with age.

FALGARONE: What is the contribution of the intrinsic bistability of steady-state chemistry in dense cores to the observed chemical differentiation?

TAFALLA: Not being an expert on chemical models, I am not the best person to answer this question. As far as I know, there is no analysis of bistability that takes into account molecular freeze out and dust chemistry. Probably someone should look into this problem.

HENNING: If you assume a sphere, you will get a sphere. How good is this assumption? Actually, we demonstrated that one better should do a 3D analysis

TAFALLA: We especially chose our cores for their simple geometry and close to round shape in the plane of the sky. Of course we cannot constrain the line of sight dimension, so we assume that it will not be too different from what we observe in the plane of the sky. By observing multiple cores, we expect to minimize the chance of being confused by objects having an anomalous line-of-sight dimension.

Astrochemistry: Recent Successes and Current Challenges
Proceedings IAU Symposium No. 231, 2005
D.C. Lis, G.A. Blake & E. Herbst, eds.

© 2006 International Astronomical Union
doi:10.1017/S1743921306007010

Modelling of Deuterium Chemistry in Star-Forming Regions

Helen Roberts

University of Manchester, School of Physics and Astronomy,
Sackville Street, Manchester, M60 1QD, UK
email: helen.h.roberts@manchester.ac.uk

Abstract. Several new multiply deuterated species have been detected over the past three years, including ND_3 (van der Tak *et al.* 2002; Lis *et al.* 2002), CHD_2OH, CD_3OH (Parise *et al.* 2002, 2004), D_2S (Vastel *et al.* 2003), HD_2^+ (Vastel *et al.* 2004) and D_2CS (Marcelino *et al.* 2005). In addition, mono-deuterated species have been observed with abundances $>10\%$ of their un-deuterated analogues (e.g. CH_2DOH observed by Parise *et al.* 2002; NH_2D observed by Saito *et al.* 2000 and Hatchell 2003). These are remarkable results, given that the underlying abundance of deuterium in the local interstellar medium (ISM) is $\sim 10^{-5}$ times lower than that of hydrogen (Linsky 1998; Sonneborn *et al.* 2000).

Such large enhancements in the abundances of deuterium-bearing molecules can either be due to gas-phase or to grain-surface fractionation. Grain-surface reactions are undoubtedly important in producing saturated species such as methanol, water, ammonia, and hydrogen sulphide. Water ice is observed to be abundant and ubiquitous throughout the ISM, and enhanced abundances of gas-phase NH_3, CH_3OH, H_2CO and H_2S (among others) are observed in warmer regions around protostars where grain mantles have evaporated.

Recent observational and theoretical evidence suggests that the deuterium fractionation in star-forming regions is set by gas-phase and grain-surface reactions during the cold, dense pre-protostellar phase. For species which form on grain surfaces via H atom addition to CO, N, O and S, the deuterium fractionation on grains comes from the relative amounts of atomic D and H which are accreting from the gas. The observations of deuterated methanol and D_2S require that the gas-phase atomic D/H ratio at the time the molecules formed was $\geqslant 0.1$.

This paper presents results from chemical models of the prestellar core phase of star formation, showing how this high atomic D/H ratio can be produced, and discusses how models can also be used to look at deuterium fractionation in the protostellar stages of star formation.

Keywords. ISM: abundances — ISM: molecules — molecular processes — stars: formation

1. Introduction

Observations of deuterated molecules are extremely useful probes of the physical conditions in astronomical objects. Abundances of deuterated molecular ions can be used to determine the ionisation fraction in dark clouds and prestellar cores, a parameter which regulates the collapse of magnetised gas, and, thus, is important in models of star formation. Where ices have evaporated and released the products of grain surface chemistry (around new protostars or in cometary comae), deuterium fractionation can provide valuable information on the temperatures under which the ices formed and how species have been processed within them. The underlying abundance of deuterium is an important cosmological parameter, but it cannot be directly determined outside of the local ISM; observations of deuterated molecules, however, can be made throughout the Galaxy, and beyond, and used to infer the underlying D/H ratio.

Table 1. A sample of the molecular D/H ratios in selected sources.

	TMC-1 CP	L134N	L1544	Barnard 1	NGC1333 IRAS4A	IRAS 16293–2422
DCO^+	0.02^1	0.18^1	0.04^{12}	—	0.01^{19}	0.01^{22}
N_2D^+	0.08^1	0.3^{10}	0.2^{12}	0.15^{10}	—	—
NH_2D	0.01^1	0.1^1	0.13^{13}	0.07^{16}	0.07^{13}	0.1^{22}
NHD_2	—	0.05^{11}	—	—	—	—
ND_3	—	—	—	0.001^{17}	0.001^{20}	—
DCN	0.023^2	0.05^5	—	—	—	—
DNC	0.015^3	—	0.03^{14}	—	—	—
HDO	—	—	—	—	—	0.03^{23}
HDCO	—	0.07^5	—	0.14^{18}	—	0.15^{22}
D_2CO	—	—	0.04^{15}	0.06^{18}	0.07^{21}	0.06–0.13^{24}
HDCS	0.02^4	—	—	0.3^{18}	—	—
D_2CS	—	—	—	0.1^{18}	—	—
CH_3OD	0.026^5	$<0.03^5$	—	—	—	0.015^{25}
CH_2DOH	—	—	—	—	—	0.3^{25}
CHD_2OH	—	—	—	—	—	0.06^{25}
CD_3OH	—	—	—	—	—	0.014^{25}
C_2D	0.01^6	—	—	—	—	—
C_4D	0.004^7	—	—	—	—	—
DC_3N	0.03–0.1^8	—	—	—	—	—
DC_5N	0.013^9	—	—	—	—	—

References: [1]Tiné *et al.* 2000; [2]Wootten 1987; [3]Guélin *et al.* 1982; [4]Minowa *et al.* 1997; [5]Turner 2001; [6]Millar *et al.* 1989; [7]Turner 1989; [8]Howe *et al.* 1994; [9]Macleod *et al.* 1981; [10]Gerin *et al.* 2001; [11]Roueff *et al.* 2000; [12]Caselli *et al.* 2002; [13]Shah & Wootten 2001; [14]Hirota *et al.* 2003; [15]Bacmann *et al.* 2002; [16]Saito *et al.* 2000; [17]Lis *et al.* 2002; [18]Marcelino *et al.* 2005; [19]Stark *et al.* 1999; [20]van der Tak *et al.* 2002; [21]Loinard *et al.* 2002; [22]van Dishoeck *et al.* 1995; [23]Parise *et al.* 2005; [24]Ceccarelli *et al.* 2001; [25]Parise *et al.* 2004.

These studies rely on our understanding of the chemical routes leading to molecular deuteration. Over the past few years, the observations of deuterated species with abundances many orders of magnitude higher than expected have led to renewed interest in the subject of deuterium fractionation and a re-examination of the chemical models.

Table 1 shows abundances of deuterated species relative to their completely hydrogenated analogues for a variety of interstellar and protostellar sources. TMC-1 is the archetypal dark cloud; L134N was also thought to be a dark cloud, but recent observations (Pagani *et al.* 2005) suggest it may have more in common with prestellar cores like L1544 and Barnard 1. NGC1333 IRAS4A and IRAS 16293-2422 are warmer, low-mass protostellar sources.

The paper is arranged as follows: §2 explains the basic deuterium fractionation processes and shows how molecular D/H ratios in cold clouds can be estimated, §3 presents results from prestellar core models, showing how high deuteration fractionation is achieved, while §4 discusses the ratios observed in protostellar sources.

Throughout, [X] denotes the fractional abundance of species X with respect to H_2, $N(X)$ is the column density of X, and 'molecular D/H ratio' refers to the ratio of a deuterated species abundance to that of its totally hydrogenated analogue.

2. Dark Clouds

The enhanced abundances of deuterated species in interstellar dark clouds such as TMC-1, which have temperatures of 10 K and H_2 densities of 10^4 cm^{-3}, are typically on the order of a few percent (see Table 1). It has long been understood (see e.g. Millar *et al.* 1989) that these are due to chemical fractionation. A full list of the fractionation reactions used in the models can be found in Roberts & Millar (2000) and Roberts *et al.* (2004), but a selection that will be referred to in this paper are listed below.

$$H_3^+ + HD \rightleftharpoons H_2D^+ + H_2 + 220K \tag{2.1}$$

$$H_2D^+ + HD \rightleftharpoons HD_2^+ + H_2 + 187K \tag{2.2}$$

$$HD_2^+ + HD \rightleftharpoons D_3^+ + H_2 + 234K \tag{2.3}$$

$$CH_3^+ + HD \rightleftharpoons CH_2D^+ + H_2 + 370K \tag{2.4}$$

$$CH_2D^+ + HD \rightleftharpoons CHD_2^+ + H_2 + 370K \tag{2.5}$$

Fractionation occurs because of the exothermicity of the forward reactions, so if the rate coefficient for reaction (x) is $k_{(x)}$, the rate for the reverse reaction is $k_{(x)}\,e^{-\Delta E/T}$, where ΔE is the energy barrier, in K, as listed for each reaction.

The molecular ion, H_3^+, plays a crucial role the ion-molecule chemistry which dominates in dark clouds (see e.g. Herbst 2000), while the bulk of the deuterium in the cloud is assumed to be incorporated into HD. As reaction (2.1) shows, these two species react efficiently to form H_2D^+ and H_2. Due to the energy barrier, however, the reverse reaction is very slow for $T \leqslant 30$ K so $[H_2D^+]$ becomes enhanced relative to $[H_3^+]$. This enhancement can be estimated via the equation:

$$\frac{[H_2D^+]}{[H_3^+]} = \frac{k_{(2.1)}}{k_{(2.1)}\exp(-220/T) + k_{(2.2)}[HD] + k_{CO}[CO] + k_{d.r.}[e^-]} \frac{[HD]}{[H_2]}. \tag{2.6}$$

As well as being destroyed by reactions (2.1) and (2.2), H_2D^+ can also react via proton transfer to neutral species (represented here by CO, as it is the most abundant of them) and by dissociative recombination (d.r.) with electrons.

In dark clouds, at 10 K, $k_{(2.1)}\exp(-220/T) \sim 10^{-18}$ cm^3 s^{-1}, while the fractional abundance of HD is 3×10^{-5} and $k_{(2.2)} = 8 \times 10^{-10}$ cm^3 s^{-1}, so these are minor channels. Destruction is dominated by CO ($[CO] \sim 10^{-4}$; $k_{CO} \sim 10^{-9}$ cm^3 s^{-1}) and electrons ($[e^-] \sim 2 \times 10^{-7}$; $k_{d.r.} \sim 3\times10^{-7}$ cm^3 s^{-1}), which gives an enhancement factor of $\sim 10^4$ and an H_2D^+/H_3^+ ratio of ~ 0.1. When H_3^+ and H_2D^+ then react the assumption is that, if there is more than one place for the D to end up in the products, the branching between the different channels will be statistical; e.g. for the reaction of CO with H_2D^+ there is twice as much chance to transfer a proton and form HCO^+, so $DCO^+/HCO^+ \sim \frac{1}{3}\, H_2D^+/H_3^+$. Similar approximations can be made for other molecules, explaining the fractionation of a few percent which is typically observed towards TMC-1.

3. Prestellar Cores

So how can ratios significantly higher than this be produced? In recent years submm cameras like SCUBA and SHARC have allowed observers to directly map the dust distribution in dense clouds and compare this to molecular abundances. In an increasing number of sources, key tracer molecules such as CO and CS are found to be heavily depleted at the dust peak. The best explanation for this is that the molecules are freezing onto the grain surfaces. Depletion of CO has been observed in L134N (Pagani *et al.* 2005) and L1544 (Caselli *et al.* 1999), sources which also have high molecular D/H ratios (see Table 1).

It has long been predicted that freeze-out of neutral species such as CO will enhance molecular D/H ratios (Brown & Millar 1989). This is clear from eq. (2.6): if [CO] is reduced, the enhancement factor increases. In prestellar cores the ionisation fraction is also expected to be lower than in dark clouds, as the higher densities mean collisions

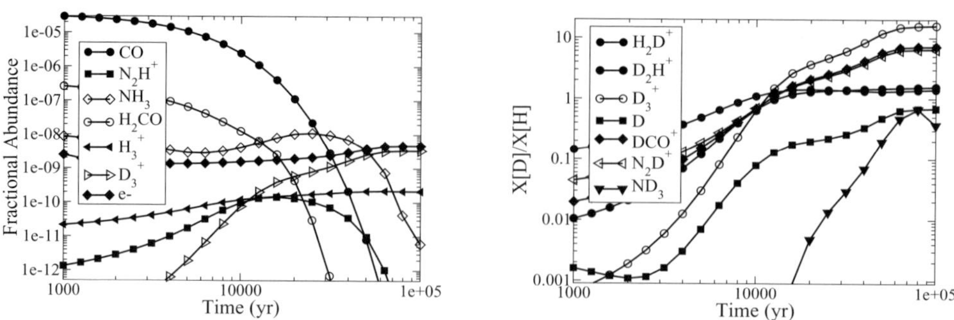

Figure 1. Predicted evolution of fractional abundances (*left*) and molecular D/H ratios (*right*) for selected species, predicted by an accretion model at $8\,\mathrm{K}$, $n(\mathrm{H}_2) = 10^6\,\mathrm{cm}^{-3}$.

leading to recombination reactions happen more frequently. Observations of molecular ions in L1544 suggest $[\mathrm{e}^-]{\sim}$a few times 10^{-9} (Caselli *et al.* 2002). The second term in the denominator in eq. (2.6), reaction (2.2), then becomes the dominant destruction term, and limits the enhancement in $\mathrm{H}_2\mathrm{D}^+/\mathrm{H}_3^+$. Reaction (2.2) forms HD_2^+, however, which can also pass on its fractionation to other neutral species (and more efficiently than $\mathrm{H}_2\mathrm{D}^+$ since, statistically, it will transfer a deuteron $\tfrac{2}{3}$ of the time), so one could write an equation for enhancement of $\mathrm{HD}_2^+/\mathrm{H}_3^+$ very similar to eq. (2.6). Once most of the CO has disappeared from the gas and with $[\mathrm{e}^-] \sim 3 \times 10^{-9}$ the enhancement is $\sim 10^5$: $[\mathrm{HD}_2^+] \sim [\mathrm{H}_2\mathrm{D}^+] \sim [\mathrm{H}_3^+]$.

Reaction (2.3) of HD_2^+ with HD forms D_3^+. Again, there is a barrier in the reverse direction, and, obviously, no further exchange with HD can occur, so in the prestellar cores where the CO and electron abundances are low, the $\mathrm{D}_3^+/\mathrm{H}_3^+$ ratio can become very large. In the limit where all the CO has frozen out:

$$\frac{\mathrm{D}_3^+}{\mathrm{H}_3^+} \sim \frac{k_{(2.3)}}{k_{d.r.}[\mathrm{e}^-]}\frac{\mathrm{HD}}{\mathrm{H}_2} \sim 20, \tag{3.1}$$

but even before all the neutral species freeze-out, the $\mathrm{D}_3^+/\mathrm{H}_3^+$ ratio is many times higher than the HD/H_2 ratio (and when D_3^+ reacts it always produces a deuterated product), leading to high fractionation in other species.

This is illustrated in Figure 1, which shows results from a chemical model which includes multiple deuteration, gas-phase reactions, and freeze-out of species onto grains (see Roberts *et al.* 2003, 2004 for full details). The temperature is 8 K and the H_2 density is 10^6 cm^{-3}. The molecular D/H ratios increase over time, while the abundances of CO and other heavy species decrease as they accrete onto the grains. It is assumed that H_2, HD and D_2 are formed on the grains and returned to the gas-phase, so at the end-point of the model only species composed entirely of H and D (along with He, He$^+$ and electrons) remain in the gas-phase. After 10^5 yr this model predicts that D_3^+ will be the dominant ion: $[\mathrm{D}_3^+] \sim [\mathrm{e}^-] \sim$ a few times 10^{-9}. It also agrees with our simplified estimates from eqs. (2.6) and (3.1) that $\mathrm{H}_2\mathrm{D}^+/\mathrm{H}_3^+ \sim \mathrm{HD}_2^+/\mathrm{H}_3^+ \sim 1$ and $\mathrm{D}_3^+/\mathrm{H}_3^+ \sim 20$.

In L1544, $\mathrm{H}_2\mathrm{D}^+$ and HD_2^+ have been detected (by Caselli *et al.* 2003 and Vastel *et al.* 2004, respectively), and their abundance determinations are in agreement with model predictions. The models suggest that D_3^+ is the best ion to trace the ionisation fraction of a prestellar core in the last stages of collapse. As D_3^+ is a symmetrical molecule, however, it does not have a sub-mm spectrum and, thus, cannot be observed in these regions. Determining the relative abundances of $\mathrm{H}_2\mathrm{D}^+$ and HD_2^+, and using a detailed

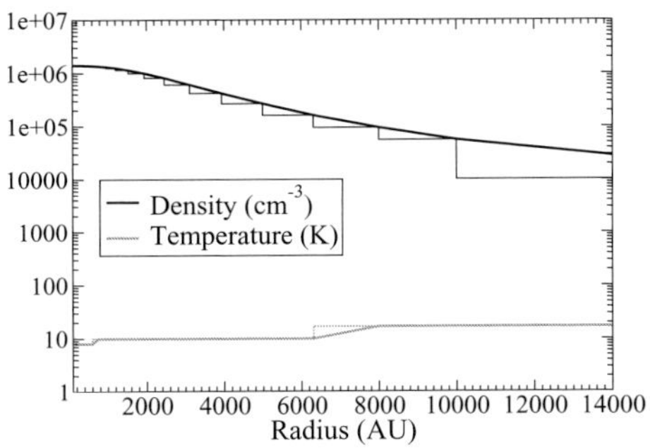
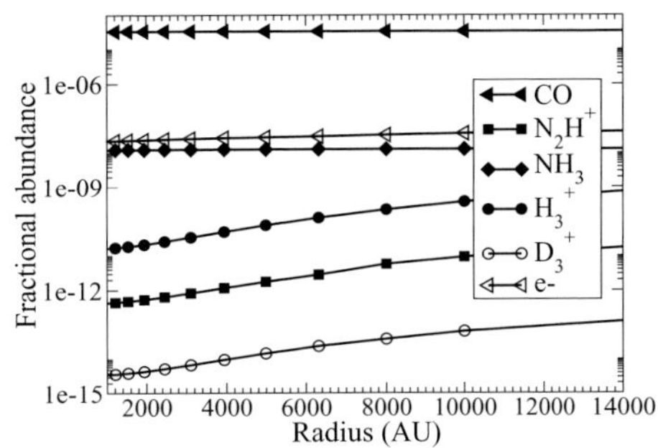

Figure 2. *Left:* Physical parameters of L1544 as a function of radius (Tafalla *et al.* 2002), along with the model fit. *Right:* Abundances of selected species as a function of radius predicted by the model of L1544 after 100 yr of freeze-out.

chemical model of the core centre, could prove useful in future attempts to determine $[e^-]$.

Another important prediction of the model is the atomic D/H ratio, which rises from 0.1 to $\sim$0.6 between 10^4 and 10^5 yr. As mentioned above, observations of molecular D/H ratios in protostellar sources for species which are assumed to have formed on the grains, requires that the atomic D/H ratio in the gas-phase at the time the species formed was $\geqslant$0.1. This is discussed further in §4.

3.1. *D$_2$CO in L1544: An Updated Comparison*

Another piece of observational evidence which links freeze-out with high deuteration is the study by Bacmann *et al.* (2002, 2003) of CO depletion and D$_2$CO fractionation in six prestellar cores. These authors found D$_2$CO/H$_2$CO ratios ranging between 0.01 and 0.1, with the largest D$_2$CO ratios from the sources which have the highest CO depletion. For L1544 their measured D$_2$CO/H$_2$CO ratio is 0.04 and the CO depletion factor (compared to the canonical value of 10^{-4} with respect to H$_2$) is 14.

The above model requires CO to be depleted by $\sim$40 to reproduce the observed D$_2$CO/H$_2$CO ratio. One has to remember, though, that the above model only represents physical conditions at the centre of a prestellar core. Bacmann *et al.* (2002) calculate an overall density and depletion factor for each prestellar core, but, in reality, these quantities increase towards the core centres. The high D$_2$CO/H$_2$CO ratios observed by Bacmann *et al.* may, therefore, originate in the inner regions of the prestellar cores, where the CO is heavily depleted, while the bulk of the CO emission is coming from a lower density, less depleted, envelope. In Roberts *et al.* (2004) the density structure of L1544 was approximated using a set of models at different densities, based on the density profile determined by Tafalla *et al.* (2002). Molecular abundances from each model were multiplied by the assumed path length and then summed to get a prediction for the column density through the centre of each core. This resulted in significantly better agreement: a predicted D$_2$CO/H$_2$CO ratio of 0.03 for an overall CO depletion factor of 20.

The chemical reaction network in the model has now been updated. Figure 2 shows the fit to the observed temperature and density profiles and the distribution of species as a function of radius at an early time point in the model. The CO abundance is initially assumed to be close to its canonical value and constant across the core. Figure 3 shows the predicted distribution of species after 10^4 yr and after 10^5 yr. After 10^4 yr the abundance of CO is starting to decline at the core centre, as the higher density means freeze-out occurs more rapidly. The abundance of D$_3^+$ is higher closer to the core centre, and is

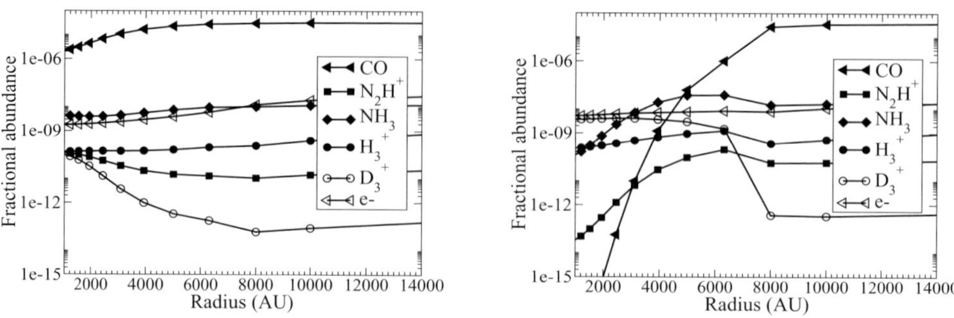

Figure 3. Abundances of selected species as a function of radius predicted by the model of L1544. *Left:* after 10^4 yr; *Right:* after 10^5 yr.

equal to that of H_3^+ within $\sim$1000 AU. Once 10^5 yr have passed the CO is completely frozen out at the core centre and $[D_3^+] \sim [e^-]$.

One of the major changes to the deuterium chemistry in this latest model is the way formaldehyde is fractionated. Roberts *et al.* (2004) pointed out that the level of fractionation might be overestimated by only considering the H_3CO^+ form of protonated formaldehyde (along with its deuterated analogues), rather than the lower energy H_2COH^+ isomer. If D_3^+ reacts with H_2CO to form H_2DCO^+, then H_2DCO^+ can recombine with electrons to produce HDCO. The HDCO also reacts with D_3^+, producing HD_2CO^+ and then D_2CO. Using this scheme, the fractionation in D_2CO will track that of D_3^+ and will, therefore, be enhanced in regions where CO is depleted.

Recent quantum chemical calculations by Osamura *et al.* (2005), however, suggest that the above route to D_2CO is very inefficient. Instead, when D_3^+ reacts with H_2CO, H_2COD^+ is formed, which then dissociates back to H_2CO, and similarly for HDCO. Instead of H_2CO being converted to D_2CO, D_2CO forms via the analogue of the reaction which forms the bulk of the H_2CO:

$$CHD_2 + O \longrightarrow D_2CO + H. \tag{3.2}$$

This does result in a some enhancement of the D_2CO/H_2CO ratio, since the CHD_2/CH_3 ratio is enhanced through the CHD_2^+ ion (reactions (2.4) and (2.5)). The energy barrier for destruction of deuterated CH_3^+ by H_2 is higher than for H_3^+, so this process is efficient up to temperatures of $\sim$50 K, but, as shown by Millar *et al.* (2000), the depletion of neutral species does not have the same dramatic effect on CH_3^+ fractionation that it does for H_3^+.

Instead of simply presenting column densities through the core centre, these latest model results are used to calculate beam-averaged column densities, which should allow a better comparison with observations. Averaging over a $10''$ beam, at the distance of L1544 (220 pc), for times between 10^4 and 10^5 yr, the column density of CO decreases from 10^{18} to 4×10^{17} cm^{-2}, while that of H_2D^+ is 2–5 $\times$ 10^{13} cm^{-2}, consistent with the observations of $N(CO) \sim 9 \times 10^{17}$ cm^{-2} (Bacmann *et al.* 2002) and $N(H_2D^+) = 5 \times 10^{13}$ cm^{-2} (Caselli *et al.* 2003). Figure 4 shows predictions for the deuterium fractionation in selected molecules, based on column density ratios. For H_2D^+ and HD_2^+, after a few times 10^4 yr, the ratios are in agreement with the one-point model and with the observations (i.e. $[HD_2^+] \sim [H_2D^+] \sim [H_3^+]$), but the highest predicted D_2CO/H_2CO ratio is now 0.001. It is worth noting, however, that $N(D_2CO)$ is a few times 10^{12} cm^{-2}, compared with the observed value of 9×10^{11} cm^{-2} (Bacmann *et al.* 2003), so the

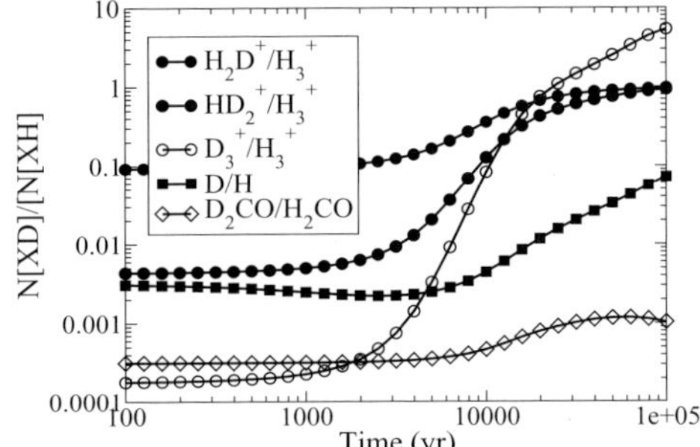

Figure 4. Molecular D/H ratios evolving over time, predicted for a core with the structure of L1544, that one would expect to observe with a $10''$ beam.

problem is not that the models don't produce enough D_2CO, rather that they produce too much H_2CO.

One possible solution could be the inclusion of surface chemistry in the models. The grains do not merely act as passive repositories for the species freezing out, they also catalyse chemical reactions. This has long been known as the only efficient way to produce H_2 in dark clouds, while the enhancement of H_2O, H_2CO, CH_3OH, NH_3 and H_2S observed in protostellar sources indicates that atoms and unsaturated molecules are readily hydrogenated. It is expected, therefore, that the relative abundances of H_2CO, $HDCO$, and D_2CO formed from the solid-state CO, will depend upon the relative amounts of H and D landing on the grains. With the high atomic D/H ratio predicted by the above models, it is possible to produce a high D_2CO/H_2CO ratio on the surface, and, although desorption processes at low temperatures are not well understood, there could be mechanisms for returning some of this formaldehyde to the gas-phase. There are two problems with this hypothesis, though. The first is that there is a significant abundance of H_2CO formed in the gas-phase, and unless larger amounts were formed on and desorbed from the surface, this would dilute the surface fractionation. The second is that if significant amounts of neutral species continuously desorb from the surface, they will react with H_3^+ and its analogues, competing with the deuterium fractionation reactions. This will reduce the molecular D/H ratios including, crucially, the *atomic* D/H ratio, meaning that the fractionation produced on the grain-surfaces may not then reach the level required.

4. Low-Mass Protostars

The regions surrounding low-mass protostars are more complex, containing multiple temperature and density components, and the chemistry may have been impacted by shocks from the protostellar outflow as well as thermal heating. Despite the higher temperatures, which would seem to preclude chemical fractionation via reactions (2.1)–(2.3), molecular D/H ratios can be very high. In fact, the highest ratios measured for formaldehyde, methanol and hydrogen sulphide are observed towards these regions. The thing which is immediately obvious about this list of species is they are typically those which do not form efficiently via ion-molecule chemistry or are seen to have abundance jumps close to the protostar (e.g. Maret *et al.* 2004; Parise *et al.* 2005), indicating that they have evaporated from grain surfaces. The DCO^+/HCO^+ ratios in these sources are typically lower than in the prestellar cores (see Table 1), supporting the assumption that, at higher temperatures, fractionation is not dominated by deuterated H_3^+.

It is now generally accepted that the deuterium fractionation in the protostellar cores was set during the cold, dense prestellar phase, when the atomic D/H ratio in the gas was very high. Methanol and hydrogen sulphide are probably the best molecules to use to probe deuterium fractionation processes on grain surfaces. Formaldehyde and ammonia, whose deuterated analogues have also been widely observed in protostellar sources, can form via both gas-phase and grain-surface chemistry so the resulting fractionation in the warm gas will depend on the proportions coming from each route.

Current protostellar models usually assume the initial grain-mantle composition, rather than calculating it via a chemical model. At t = 0 yr everything is frozen onto the grains, and the star is assumed to 'switch-on', causing either instantaneous or gradual heating which desorbs species from grains. If the model predicts that the molecular D/H ratios remain constant for a long time after evaporation, then observations can be used to directly determine the surface fractionation, and estimate the temperature at which the mantles formed. This has been done for HDS/H_2S near high-mass protostars (Hatchell *et al.* 1999). If, instead, the fractionation changes after evaporation (as appears to be the case for certain isotopomers of methanol) then the observed molecular D/H ratios can be used to estimate the time since mantle evaporation (Osamura *et al.* 2004).

5. Conclusions and Future Directions

Since the last Astrochemistry Symposium (IAUS 197, Korea, 1999) observations of enormous levels of deuterium fractionation have been made, proving a challenge for the models to explain (e.g. Loinard *et al.* 2001; Ceccarelli 2002; Parise *et al.* 2002). One of the major successes of the past three years, therefore, is the revision of the low-temperature chemical fractionation mechanism to include multiply deuterated forms of H_2D^+ (Lis *et al.* 2002; Phillips & Vastel 2003). This leads naturally to high molecular D/H ratios in low-temperature gas, but also goes a long way towards explaining the ratios in protostellar sources with the prediction of a very high atomic D/H ratio which lasts for $\sim 10^5$ yr (Roberts *et al.* 2003). It also suggests a potentially useful way to measure the ionisation fraction at the centre of a prestellar core in the very last stages before the star forms (Flower *et al.* 2004).

Obviously, there are improvements which still need to be made to the models before we can fully exploit the potential of molecular D/H ratios to trace conditions throughout the different stages of star-formation. Most of these uncertainties involve the gas-grain interaction and the grain-surface chemistry. To improve the models further we need to include desorption mechanisms at low-temperatures, and accurate binding energies for species on different surfaces are required. Many of the parameters needed for detailed surface chemistry models are also uncertain, although a lot of theoretical and experimental work is being carried out, and is described elsewhere in this volume.

For more accurate gas-phase fractionation models, we do need to confirm the rate coefficients for the fractionation reactions at 10 K, and for the dissociative recombination of H_3^+ and its deuterated analogues. Experiments to determine whether the branching ratios for dissociative recombination of deuterated ions are statistical, or if the heavier deuterium is more likely to be retained in the larger products, would also be extremely useful.

References

Bacmann, A., Lefloch, B., Ceccarelli, C., Castets, A., Steinacker, J., & Loinard, L. 2002, *A&A* 389, L6
Bacmann, A., Lefloch, B., Ceccarelli, C., Steinacker, J., Castets, A., & Loinard, L. 2003, *Ap. J.* 585, L55

Bell, M.B., Avery, L.W., Matthews, H.E., Feldman, P.A., Watson, J.K.G., Madden, S.C., & Irvine, W.M. 1988, *Ap. J.* 326, 924

Brown, P.B., & Millar, T.J. 1989, *MNRAS* 237, 661

Caselli, P., Walmsley, C.M., Tafalla, M., Dore, L., & Myers, P.C. 1999, *Ap. J.* 523, L165

Caselli, P., van der Tak, F.F.S., Ceccarelli, C., & Bacmann, A. 2003, *A&A* 403, L37

Caselli, P., Walmsley, C.M., Zucconi, A., Tafalla, M., Dore, L., & Myers, P.C. 2002, *Ap. J.* 565, 344

Ceccarelli, C. 2002, *P& SS* 50, 1267

Ceccarelli, C., Loinard, L., Castets, A., Tielens, A.G.G.M., Caux, E., Lefloch, B., & Vastel, C. 2001, *A&A* 372, 998

Flower, D.R., Pineau des Forêts, G., & Walmsley, C.M. 2004, *A&A* 427, 887

Gerin, M., Combes, F., Wlodarczak, G., Encrenaz, P., & Laurent, C. 1992, *A&A* 253, L29

Gerin, M., Pearson, J.C., Roueff, E., Falgarone, E., & Phillips, T.G. 2001, *Ap. J.* 551, L193

Guélin, M., Langer, W.D., & Wilson, R.W. 1982, *A&A* 107, 107

Hatchell, J. 2003, *A&A* 403, L25

Hatchell, J., Roberts, H., & Millar, T.J. 1999, *A&A* 346, 227

Herbst, E. 2000, *Phil. Trans. R. Soc. Lond. A.* 358, 2523

Hirota, T., Ikeda, M., & Yamamoto, S. 2003, *Ap. J.* 594, 859

Howe, D.A., Millar, T.J., Schilke, P., & Walmsley, C.M. 1994, *MNRAS* 267, 59

Linsky, J.L. 1998, *Space Sci. Rev.* 84, 285

Lis, D.C., Roueff, E., Gerin, M., Phillips, T.G., Coudert, L.H., van der Tak, F.F.S., & Schilke, P. 2002, *Ap. J.* 571, L55

Loinard, L., Castets, A., Ceccarelli, C., Caux, E., & Tielens, A.G.G.M. 2001, *Ap. J.* 552, L163

Loinard, L., Castets, A., Ceccarelli, C., *et al.* 2002, *P&SS*, 50, 1205

MacLeod, J.M., Avery, L.W., & Broten, N.W. 1981, *Ap. J.* 251, L33

Maret, S., Ceccarelli, C., Caux, E., *et al.* 2004, *A&A* 416, 577

Marcelino, N., Cernicharo, J., Roueff, E., Gerin, M., & Mauersberger, R. 2005, *Ap. J.* 620, 308

Millar, T.J., Bennett, A., & Herbst, E. 1989, *Ap. J.* 340, 906

Millar, T.J., Roberts, H., Markwick, A.J., & Charnley, S.B. 2000, *Phil. Trans. R. Soc. Lond. A.* 358, 2359

Minowa, H., Satake, M., Hirota, T., Yamamamoto, S., Ohishi, M., & Kaifu, N. 1997, *Ap. J.* 491, L63

Osamura, Y., Roberts, H., & Herbst, E. 2004, *A&A* 421, 1101

Osamura, Y., Roberts, H., & Herbst, E. 2005, *Ap. J.* 621, 348,

Pagani, L., Pardo, J.-R., Apponi, A.J., Bacmann, A., & Cabrit, S. 2005, *A&A* 429, 181

Parise, B., Ceccarelli, C., Tielens, A.G.G.M., *et al.* 2002, *A&A* 393, L49

Parise, B., Castets, A., Herbst, E., *et al.* 2004, *A&A* 416, 159

Parise, B., Caux, E., & Castets, A., *et al.* 2005, *A&A* 431, 547

Phillips, T.G., & Vastel, C. 2003, in Chemistry as a Diagnostic of Star Formation, ed. C.L. Curry & M. Fich, NRC Press, Ottowa, Canada, p. 3

Roberts, H., & Millar, T.J. 2000, *A&A* 361, 388

Roberts, H., Herbst, E., , & Millar, T.J. 2003, *Ap. J.* 591, L41

Roberts, H., Herbst, E., & Millar, T.J. 2004, *A&A* 424, 905

Roueff, E., Tiné, S., Coudert, L.H., Pineau des Forêts, G., Falgarone, E., & Gerin, M. 2000, *A&A* 354, L63

Saito, S., Ozeki, H., Ohishi, M., & Yamamoto, S. 2000, *Ap. J.* x535, 227

Shah, R.Y., & Wootten, A. 2001, *Ap. J.* 554, 933

Sonneborn, G., Tripp, T. M., Ferlet, R., Jenkins, E.B., Sofia, U.J., Vidal-Madjar, A., & Wozniak, P. 2000, *Ap. J.* 545, 277

Stark, R., van der Tak, F.F.S., & van Dishoeck, E.F. 1999, *Ap. J.* 521, L67

Stantcheva, T., & Herbst, E. 2003, *MNRAS* 340, 983

Tafalla, M., Myers, P.C., Caselli, P., Walmsley, C.M., & Comito, C. 2002, *Ap. J.* 569, 815

Tiné, S., Roueff, E., Falgarone, E., Gerin, M., & Pineau des Forêts, G. 2000, *A&A* 356, 1039

Turner, B.E. 1989, *Ap. J.* 347, L39

Turner, B.E. 2001, *ApJS* 136, 579

van der Tak, F.F.S., Schilke, P., Müller, H.S.P., Lis, D.C., Phillips, T.G., Gerin, M., & Roueff, E. 2002, *A&A* 388, L53

van Dishoeck, E.F., Blake, G.A., Jansen, D.J., & Groesbeck, T.D. 1995, *Ap. J.* 447, 760

Vastel, C., Phillips, T.G., Ceccarelli, C., & Pearson, J. 2003, *Ap. J.* 593, L97

Vastel, C., Phillips, T.G., & Yoshida, H. 2004, *Ap. J.* 606, L127

Wootten, A. 1987, in *Astrochemistry*, ed. M.S. Vardya S.P. Tarafdar (Reidel, Dordrecht) p. 311

Discussion

RAWLINGS: The potential surfaces (and the dissociative recombination rates) for the deuterated forms of H_3^+ are highly uncertain. How does this affect the modeled D_3^+, HD_2^+, H_2D^+ abundances?

ROBERTS: For H_2D^+/H_3^+ and HD_2^+/H_2D^+ the destruction is dominated by reaction of the deuterated ion with HD (as the electron fraction is so low at densities of 10^6 cm^{-3}), so the dissociative recombination rate does not have a major effect. For D_3^+/HD_2^+, however, recombination with electrons is its only destruction channel (in the last stage of depletion) so here, yes, that rate is an important parameter.

HERBST: (1) For protostellar cores, should we not distinguish between inner (warmer) and outer (cooler) regions? (2) How will models be affected by the new measurement that $CH_3OH_2^+ + e^-$ does not produce $CH_3OH + H$?

ROBERTS: (1) Yes. I am currently working on this. (2) If methanol is destroyed by dissociated recombination, then the selective fall in methanol isotopologues wih an –OD group may no longer occur. In that case we would have to look for alternate (possibly grain-surface) processes (see talk by Watanabe).

WALMSLEY: Do you assume "normal ISM grains" in your models? This can have a considerable effect on the degree of D fractionation.

ROBERTS: Yes, we only consider "normal grains". The nature of the grain surfaces is something we should consider in the future.

Astrochemistry: Recent Successes and Current Challenges
Proceedings IAU Symposium No. 231, 2005
D.C. Lis, G.A. Blake & E. Herbst, eds.

© 2006 International Astronomical Union
doi:10.1017/S1743921306007022

Self-Consistent Theoretical Models of Collapsing Pre-Stellar Cores

Valery I. Shematovich[1], Boris M. Shustov[1], Dmitri S. Wiebe[1], Yaroslav N. Pavlyuchenkov[1], and Zhi-Yun Li[2]

[1]Institute of Astronomy of the Russian Academy of Sciences, 48 Pyatnitskaya str., 119017 Moscow, Russia
email: shematov@inasan.rssi.ru
[2]Department of Astronomy, University of Virginia, P.O. Box 3818, Charlottesville, VA 22903, USA
email: zl4h@virginia.edu

Abstract. We present a coupled dynamical and chemical model for collapsing pre-stellar cores (Li *et al.* 2002; Shematovich *et al.* 2003a,b; Pavlyuchenkov *et al.* 2003). It treats the dynamics of thermally and magnetically supported cores in 1D, with an extended chemical network incorporated. The latest version of the model includes UV-irradiation of the core envelope. We have also developed a 2D Monte Carlo model of radiative transfer to compute molecular line profiles for comparison with observations.

The model allowed us to constrain evolutionary scenarios for collapsing pre-stellar cores, to calculate molecular line profiles from the spatial distribution of chemical species and the velocity field, and to characterize the chemical properties of dense cores.

We have determined line profiles along multiple lines of sight through a given pre-stellar core. This allowed us to compare model predictions with the observational maps of molecular lines available for L1544 and other well studied cores. The comparison of synthetic and observed line profile maps contributed to the understanding of the velocity field and pattern of chemical differentiation observed in individual cores.

Keywords. line: profiles — ISM: magnetic fields — ISM: molecules — stars: formation — ISM: individual (L1544)

1. Introduction

It is by now well known that the chemical composition of starless (pre-stellar) cores is highly inhomogeneous (e.g., M. Tafalla, this volume). Some (like N-bearing) molecules tend to be centrally peaked, while distributions of the others (like CO or S-bearing species) are reminiscent of concentric shells forming a so-called onion-like structure. One way to study this inhomogeneity is to fit the observed line profiles with theoretical ones computed using some *ad hoc* representation for the radial abundance profile.

The simplest function that mimics the behavior of molecular abundances in starless cores is a step function in which two abundance values are adopted. One of them is attributed to the core envelope, while the other is associated with the dense central region to take molecular freeze-out into account. Smoother representations are also possible, such as a power law or an exponential function. Observations are used to find the "best fit" parameters that provide the closest agreement between theoretical and observed abundances, column densities, line profiles, etc. This approach is utilized by Tafalla *et al.* (2002), Lee *et al.* (2003), Jørgensen *et al.* (2004), Doty *et al.* (2004), as well as in other studies.

Table 1. Timescales in pre-stellar cores

Process	Mechanism	Timescale (yr)	Timescale for $n = 10^4$ cm^{-3}
Chemistry	Cosmic-ray ionization[a,b]	$2.6 \times 10^9/n$	2.6×10^5
Freeze-out	Gas-grain collision	$3 \times 10^9/n$	3×10^5
Cooling	Radiative emissions	10^6	10^6
Collapse	Gravity	$2.4 \times 10^7/\sqrt{n}$	2.4×10^5
Ambipolar diffusion	Ion-neutral drift[c]	$4 \times 10^{13} x_e$	4×10^5

[a] assuming cosmic ray ionization rate $\zeta = 10^{-17}$ s^{-1};
[b] n is the number density of H nuclei;
[c] x_e is the fractional ionization.

Lee *et al.* (2005) have shown that the simplified representations with "best fit" parameters do reproduce the general abundance trends both in pre-stellar cores and at later stages, when a central heating source has already appeared. However, this technique is only able to shed some light on the current state of an object without giving any clues to its past evolution. The real way to understand the initial conditions of star formation and to elucidate the details of pre-stellar collapse is presented by dynamical models.

The simultaneous modeling of dynamical and chemical evolution is a complicated and computationally demanding task. Possible methods to simplify this task include either a simple chemistry on top of complex dynamics (e.g., Nelson & Langer 1999; Desch & Mouschovias 2001), or a complicated chemistry on top of simple dynamics. In the latter case the dynamical evolution of the core can be represented by a pre-defined dependence $n(r, t)$ (like a sequence of Bonnor-Ebert spheres), by a known dynamical solution (free-fall, Shu or Larson-Penston solution), or by the solution of Euler's equations in 1D. This approach is utilized by Rawlings *et al.* (1992), Bergin & Langer (1997), Aikawa *et al.* (2001, 2003, 2005), Lee *et al.* (2004) and by others. The important feature of this approach is that the dynamics is modeled separately and, thus, is decoupled from the chemistry.

Why does one care at all about simultaneous modeling of chemistry and dynamics? The reason is that molecular species are not just passive tracers of dynamical evolution. They influence the dynamics via two primary aspects: radiative cooling and ambipolar diffusion. The latter depends on the atomic and molecular ion density and allows the core to contract despite magnetic support. Timescales for various chemical and dynamical processes are summarized in Table 1. For a typical density in molecular clouds all these timescales are comparable. This implies that in models of early stages of pre-stellar evolution, chemistry and dynamics are to be modeled self-consistently, with dynamical parameters (density, temperature) entering the chemical equations and chemistry-dependent parameters (cooling function, ion-neutral drift efficiency) entering the dynamical equations.

The first attempt to account for the chemistry-dependent thermal balance in a hydrodynamical model was made by Gerola & Glassgold (1978). For a number of years, the self-consistent approach to modeling of starless cores has been developed by the Moscow astrochemistry group. In one series of models, we studied the chemical and dynamical evolution of a UV irradiated core—an evaporated gaseous globule (EGG, see Figure 1). In such models, the equations of hydrodynamics, chemical kinetics, and energy budget are integrated simultaneously, without imposing any critical pre-defined conditions (Shematovich *et al.* 1997). It has been shown that even a core immersed in the

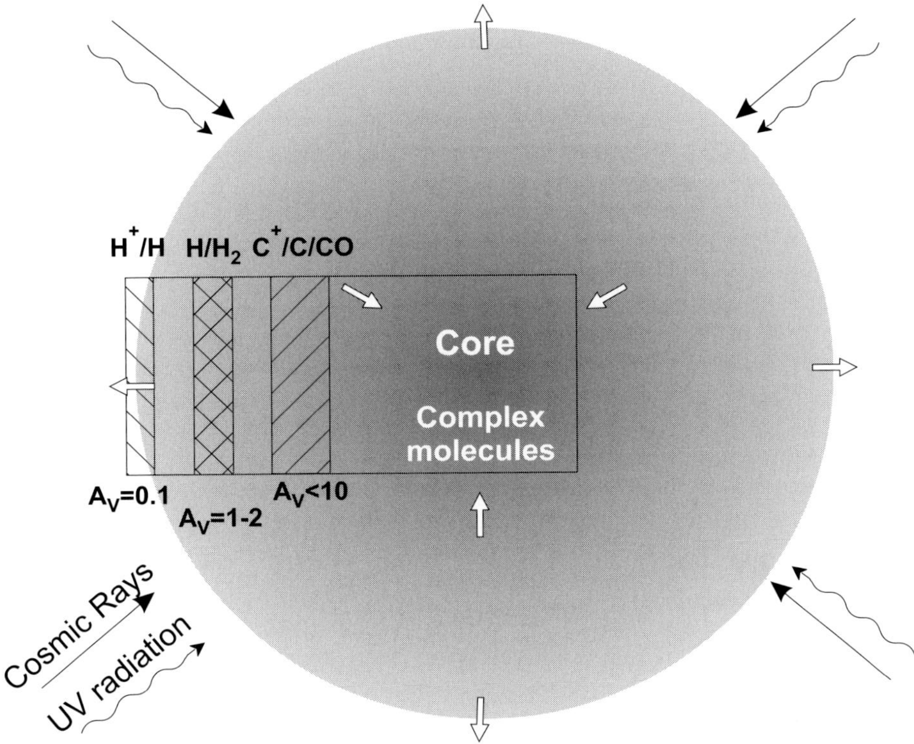

Figure 1. Sketch of a dense pre-stellar core.

diffuse interstellar UV field is separated into two dynamically distinct regions: a collapsing inner part and an expanding envelope. The relative mass fractions in these regions depend on the UV intensity and the initial cloud mass. For low-mass cores ($< 10\,M_\odot$) the photoevaporation time is short, and such cores are completely destroyed by diffuse UV radiation. More massive cores can survive weak external heating caused by absorbed UV photons. The core is subject to radiation-driven implosion which is much faster than the "normal" gravitational collapse. On the other hand, heating of the external layers and their evaporation cause the gradual decrease of the core mass. Between the collapsing core and its evaporating envelope a transition region is formed, which propagates inside the core, where the cold dark chemistry has dominated before. Icy mantles evaporate in this transition region, causing local enhancement of some species' abundances, especially, of water vapor.

In another series of models we investigated the contraction of a magnetized core driven by ambipolar diffusion (Li *et al.* 2002; Shematovich *et al.* 2003a,b). In this contribution, we present new results related to this model.

2. Model of a Magnetized Core

Our model is based on a 1D representation of the magnetic field, in which the magnetic pressure is retained in the model, while the magnetic tension is neglected. This model is by no means intended to reproduce the kinematic structure of a contracting core; in fact, in Pavlyuchenkov *et al.* (2003) we show that it does not. However, it does capture the main temporal features of a magnetized core evolution—namely, the quasi-static contraction at the stage when the magnetic support is important and a nearly free-fall collapse afterwards.

We assume that the core is isothermal at 10 K and consists of gas and dust that are well mixed. The equations that describe the evolution of a magnetized, self-gravitating cloud are (Li 1999)

$$\frac{\partial \rho}{\partial t} = -\frac{1}{r^2}\frac{\partial}{\partial r}(r^2 \rho v) \tag{2.1}$$

$$\frac{\partial v}{\partial t} = -v\frac{\partial v}{\partial r} - \frac{GM_r}{r^2} - \frac{1}{\rho}\frac{\partial}{\partial r}\left(\rho a^2 + \frac{B^2}{8\pi}\right) \tag{2.2}$$

$$v_{\mathrm{i}} - v = -\tau_{\mathrm{ff}}\nu_{\mathrm{ff}}\frac{1.4}{4\pi\rho}\frac{\partial}{\partial r}\left(\frac{B^2}{2}\right) \tag{2.3}$$

$$\frac{\partial B}{\partial t} = -\frac{1}{r}\frac{\partial}{\partial t}(rv_{\mathrm{i}}B). \tag{2.4}$$

Here a is the isothermal sound speed, M_r is the mass enclosed within a radius r, v and v_{i} are velocities of neutral and ionized species, τ_{ff} is the free-fall time for density ρ, and ν_{ff} is the magnetic coupling parameter. The latter value is the ratio of the local free-fall time to the magnetic field-neutral coupling timescale. It is this parameter that defines the connection between the chemical and dynamical evolution of the core. The ion density that is needed to compute ν_{ff} is evaluated with a time-dependent chemical model that includes approximately 2000 gas-phase reactions, involving 59 neutral and 83 ionized species from the UMIST 95 ratefile (Millar *et al.* 1997), as well as 53 grain-surface diffusive reactions (Hasegawa & Herbst 1993). Gas and dust phases are coupled through adsorption-desorption processes. Desorption energies are taken from Hasegawa & Herbst (1993) with updates from Aikawa *et al.* (2001). Adsorption is parameterized with the sticking probability s. The core mass is assumed to be $\sim 20 M_{\odot}$, with an initial density of $n_{\mathrm{H_2}} = 10^3$ cm^{-3}. The initial state of the core is magnetostatic. If there were no ambipolar diffusion, the core would stay in this state indefinitely long.

3. Chemical Differentiation in the Magnetized Pre-Stellar Cores

Slow contraction is naturally achieved in a strongly magnetized core, whose evolution is driven by ambipolar diffusion over several dynamic times (Li 1999; Ciolek & Basu 2000). The relatively long evolution time should leave a strong imprint on the core chemistry, since the chemical and dynamical timescales are comparable. In a magnetic cloud, the interplay between dynamics and chemistry is highly nonlinear: the density evolution affects the chemistry, which determines the abundances of charged particles, which in turn controls the rate of ambipolar diffusion that regulates the density evolution.

The coupled dynamical and chemical model was used to study the chemical appearance of pre-stellar cores (Shematovich *et al.* 2003a,b) with special attention to the effects of the strength of magnetic fields, initial chemical composition, sticking probability, adsorption energies, cosmic ray ionization rate, and cloud mass. The model results were compared with the data available on the well studied core L1544. The main conclusion from our previous studies is that the model, in which the cloud is magnetically supported for several million years before collapsing dynamically, provides a good overall fit to the observed column densities of various species in L1544. The fit is significantly worse for the non-magnetic model, in which the cloud collapses promptly. In particular, the radius of the ring in the distribution of sulfur-bearing molecules is much smaller than observed. Varying the model parameters, we have been able to build a model which reproduces observed column densities of CO, CS, CCS, HCO$^+$, NH$_3$, and N$_2$H$^+$ within a factor of 3.

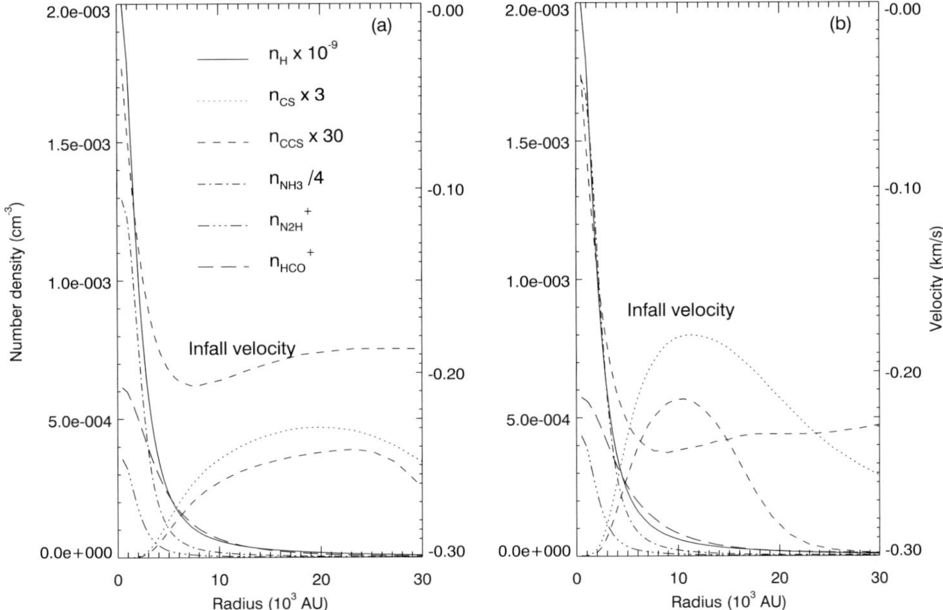

Figure 2. Radial profiles of the *volume* densities of the hydrogen nuclei (heavy solid line) and 5 selected molecular species, plotted together with the profile of the infall velocity (heavy dashed line) at a time when the H_2 central density reaches a value of 10^6 cm^{-3}. Note that the high-density central region has a relatively small infall speed, and is well traced by the species NH_3 and N_2H^+, and to a lesser extent, by HCO^+. The faster contracting inner envelope is, on the other hand, better traced by the centrally-depleted species CS and CCS. Some curves are shifted by various factors for clarity. (*a*) Standard model; (*b*) Same model but with ion sticking probability $s_i = 1$.

Our previous studies represent an important illustration of the influence that chemistry exerts on dynamics. Cosmic rays are the only noticeable source of ionization in these cores; thus, the fractional ionization is determined by the molecular chemistry, driven by cosmic rays. *Varying only parameters of the chemical model we change the contraction timescale by a factor of 3.* In the standard model the density grows from 10^3 to 10^6 cm^{-3} over 6.5 Myr. If we raise the sticking probability for ions from 0.3 to 1, the contraction accelerates due to the decreased abundance of ions, and the evolutionary time shortens to 4.9 Myr. If the surface chemistry is ignored in the model, the contraction lasts for 10.9 Myr. Finally, if ion recombination on grains is prohibited, the timescale increases up to 14.4 Myr.

In this paper, we emphasize the dependence of the developed model on the chemical exchange between the gas and dust phases. The standard chemical and dynamical model of the magnetized pre-stellar core of Shematovich *et al.* (2003a), having sticking probabilities for neutrals (s_n) and ions (s_i) equal to 0.3 for both neutrals and ions is analyzed together with the model in which $s_n = 0.3$ and $s_i = 1$. The higher value of the sticking probability for ions causes a stronger chemical coupling between the gas and dust, and corresponds to the case when dust grains are mainly negatively charged. The chemistry network of our model was updated by taking into account the new data on (i) rate coefficients and branching ratios for dissociative recombination of molecular ions—H_3^+, CH_3^+, H_3O^+, N_2H^+, NH_4^+, CH^+–CH_4^+, etc. (see discussion by Herbst 2005); and (ii) new and re-estimated rate coefficients of neutral-neutral reactions (Smith *et al.* 2004).

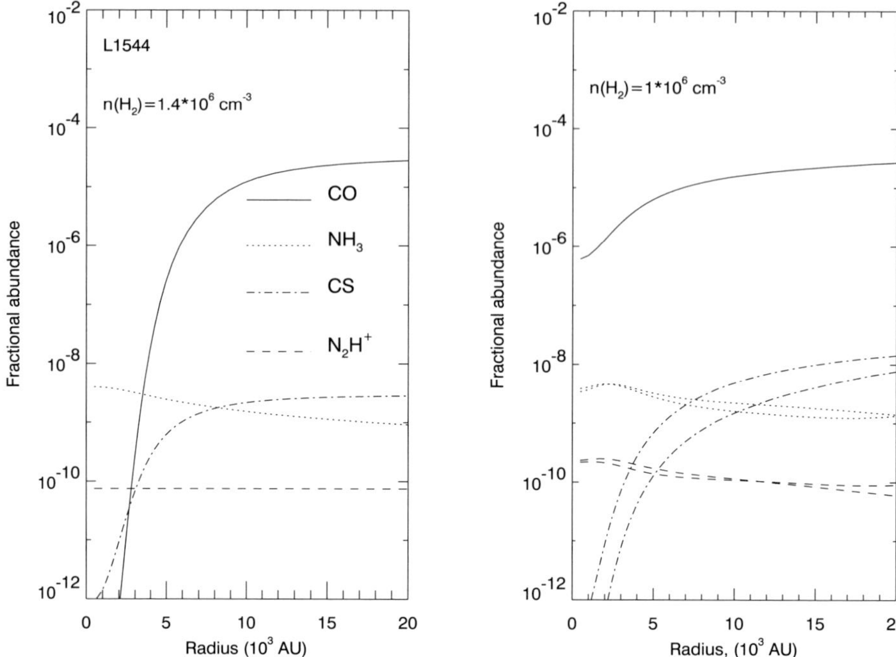

Figure 3. *Right:* Patterns of chemical differentiation in the core models with ion sticking coefficients $s_i = 0.3$ (thick lines) and 1 (thin lines). *Left:* "Best fit" model of line observations of the pre-stellar core L1544 (Tafalla *et al.* 2002).

Predictions of our standard model on chemistry are in agreement with observations (Tafalla *et al.* 2002; Tafalla, this volume). As can be seen from Figure 2, the following features are typical of pre-stellar cores approaching a central density of 10^5–10^6 cm^3 of hydrogen nuclei:

— There is a general trend for the fractional abundance to decrease toward the core center, with a few exceptions. This central depletion is particularly prominent for the sulphur-bearing molecules CS and CCS.

— The nitrogen-bearing species, such as NH_3, HCN and N_2H^+, show little depletion or even enhancement in the central region. Molecular nitrogen becomes one of the most abundant species in the gas phase.

— CO remains abundant in the outer envelope, but decreases by more than an order of magnitude in the very central region.

— The gas-phase water is strongly depleted, therefore it becomes a main constituent of the icy mantles of dust grains.

Our main conclusion is that the observed chemical differentiation can be explained as a natural consequence of the core evolution. The dense cores, formed from a more diffuse background (strongly magnetized or not) by gravitational contraction, will have a plateau-envelope density structure, and a distinctive velocity field: zero velocity at the center, nearly linear increase in the central plateau region, with a peak somewhere in the inner envelope. If the abundances of NH_3 and N_2H^+ are not depleted or even enhanced, they should preferentially trace the low-speed plateau region. Strongly depleted species, such as CS and CCS, on the other hand, should be more sensitive to the inner envelope, which has a faster infall motion (see Figure 2). The depletion pattern is not strongly changed when the sticking probabilities are high, $s_i = 1$. Nevertheless, in this case we have both higher abundances of sulphur-bearing molecules and higher infall velocity (Figure 2b). The chemical differentiation offers an exciting possibility of determining this distinctive velocity signature, and may directly probe the core growth.

For the set of models considered in Shematovich *et al.* (2003a,b), we find that the magnetic model with mixed initial metal abundances best reproduces the data on the velocity field and column densities of various species in L1544. Tafalla *et al.* (2002) modeled the data from L1544 carefully, and deduced fractional abundances for $C^{18}O$, CS, N_2H^+, and NH_3, as well as the hydrogen number density as functions of the radius from the center. A comparison between the patterns of chemical differentiation in the core models with ion sticking coefficients $s_i = 0.3$ (thick lines) and 1 (thin lines) and the "best fit" model of line observations of the pre-stellar core L1544 (Tafalla *et al.* 2002) is presented in Figure 3.

Radial distributions of $C^{18}O$, CS, N_2H^+, and NH_3 provide a more stringent test of model predictions than the integrated column densities used previously. From this comparison we would like to mention, that the density profile provided by the considered self-consistent models is close to that inferred by Tafalla *et al.* (2002). Next, the model appears to fit the inferred abundance distributions of all four species reasonably well, particularly for NH_3. The most notable discrepancy lies in $C^{18}O$ at the smallest radii (inside about 10^4 AU), although we note that the distribution in the innermost region is not well constrained observationally because of finite spatial resolution. The predicted N_2H^+ abundance increases gradually inward, which is different from the inferred constant distribution. The difference is relatively modest, however, within a factor of 2 or so. The predicted CS abundance is significantly higher than the inferred value at large radii (beyond a few times 10^4 AU), which may require a better treatment of the sulphur chemistry, which is somewhat uncertain.

4. Synthetic Line Profiles

The potential of the coupled dynamical and chemical model is fully realized by constructing and analyzing the synthetic line profiles at various core locations, because such model provides a simultaneous determination of the spatial distribution of a given chemical species *and* the velocity field, the two ingredients for line profile modeling. Line profiles of species such as CS and HCO^+ are routinely employed to deduce infall motions in dense cores. The fact that these species may be depleted poses a problem. Rawlings & Yates (2001) examined the line profiles in a collapsing core in some detail, and concluded that models assuming chemical uniformity or simple monotonic variations are unacceptable.

We have calculated the line profiles of $C^{18}O$ (1–0), CS (2–1), and HCO^+ (3–2) transitions for the models with ion sticking coefficient equal to 0.3 (solid lines) and 1 (dashed lines) with the radiative trasfer code URANIA (Pavlyuchenkov *et al.* 2003; Pavlyuchenkov & Shustov 2004), and have compared these spectra with observed line profiles for L1544 taken from Tafalla *et al.* (2002), Lee *et al.* (2001), and Lee *et al.* (2003), correspondingly. This comparison is shown in Figure 4. We see that the intensities of the calculated line profiles are slightly higher than the observed values and the lines are double-peaked with strong dips for CS (2–1), and HCO^+ (3–2) lines. These features are mainly caused by the relatively high infall speed in the model core and excess abundances in the outer envelope of the pre-stellar core. The synthetic line profiles calculated with the high ion sticking probability are shown by dashed lines. These profiles have a slightly higher intensity because of the shorter time for molecules to freeze out. The calculated $C^{18}O$ (1–0) line profile is still double-peaked compared with the observed line because of the

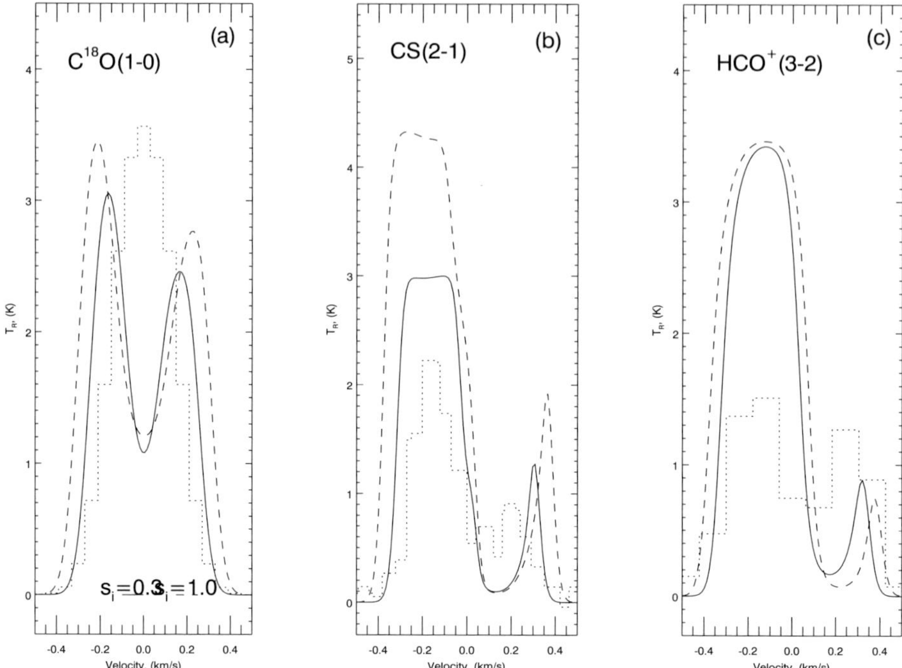

Figure 4. Comparison between observed and synthetic line profiles of C^{18}O (1–0), CS (2–1), and HCO$^+$ (3–2) transitions for models with ion sticking coefficients $s_i = 0.3$ (solid lines) and 1 (dashed lines). Observed molecular lines from the L1544 core are shown by dotted lines. (*a*) C^{18}O (1–0) line profile was taken from Tafalla *et al.* (2002); (*b*) CS (2–1) line profile was taken from Lee *et al.* (2002); (*c*) HCO$^+$ (3–2) line profile was taken from Lee *et al.* (2003).

relatively high CO abundance in the core envelope. These differences between modeled and observed line profiles are mainly caused by the high infall speeds obtained in the considered models and are a shortcoming of the 1D model of the dynamical and chemical evolution of magnetized starless cores. They can be overcome by reducing the infall speed to the level of half of the sound speed, and by reducing the CO abundance by exposing the outer core envelope to interstellar UV radiation.

There is an additional indication of the more complex velocity field in the L1544 core. In Figure 5, we compare the calculated and observed maps of the HCO$^+$ (3–2) transitions in this core (Gregersen & Evans 2000). It is seen that it is impossible to reproduce in the calculated map the changing ratio between blue and red wings in the HCO$^+$ (3–2) line profiles. This effect could be referred to the slow rotation of the L1544 core. Such a complicated velocity field can be hopefully reproduced by 2D or 3D versions of the self-consistent chemical and physical model of pre-stellar core evolution.

5. Conclusions

Self-consistent chemical and dynamical models seem to be very promising for investigation of the very early stages of star formation. Many features of pre-stellar cores are interpreted in a natural manner. The presented results suggest that the chemical differentiation observed in L1544 can be explained quantitatively in the context of the standard scenario for isolated low-mass core formation, involving ambipolar diffusion over several dynamic times. We plan to strengthen this conclusion by refining our model, and

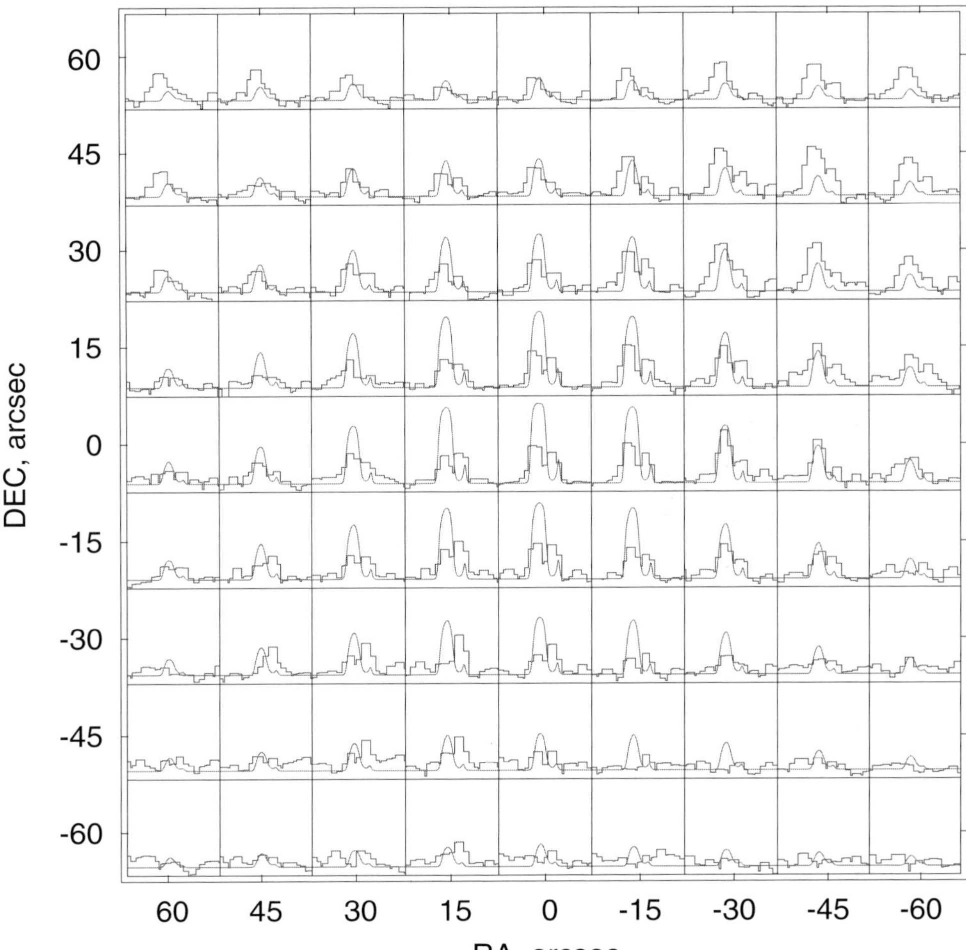

Figure 5. Map of the observed and synthetic line profiles of HCO$^+$ (3–2) transitions for the standard model. Observed HCO$^+$ (3–2) map for the L1544 core, taken from Gregersen & Evans (2000), is shown by histogram-like curves.

developing 2D and 3D versions of our self-consistent chemical and dynamical model needed to improve the interpretation of high-resolution observations.

Acknowledgements

The Moscow astrochemistry group is supported by RFBR grant number 04-02-16637. Work on star formation at the University of Virginia is supported in part by NSF grant AST-0307368.

References

Aikawa, Y., Herbst, E., Roberts, H., & Caselli, P. 2005, *Ap. J.* 620, 330
Aikawa, Y., Ohashi, N., & Herbst, E. 2003, *Ap. J.* 593, 906
Aikawa, Y., Ohashi, N., Inutsuka, Sh., Herbst, E., & Takakuwa, Sh. 2001, *Ap. J.* 552, 639
Bergin, E.A. & Langer, W.D. 1997, *Ap. J.* 486, 316

Ciolek, G.E. & Basu, S. 2000, *Ap. J.* 529, 925

Desch, S.J. & Mouschovias, T.Ch. 2001, *Ap. J.* 550, 314

Doty, S.D., Schöier, F.L., & van Dishoeck, E.F. 2004, *A&A* 418, 1021

Gerola, H. & Glassgold, A.E. 1978, *Ap. J. Suppl.* 37, 1

Gregersen, E.M., & Evans, N.J. 2000, *Ap. J.* 538, 260

Hasegawa, T.I., & Herbst, E. 1993, *MNRAS* 263, 589

Herbst, E. 2005, *J. Phys.: Conf. Ser.* 4, 17

Jørgensen, J.K., Schöier, F.L., & van Dishoeck, E.F. 2004, *A&A* 416, 603

Lee, C.-W., Myers, P.C., & Tafalla, M. 2001, *Ap. J. Suppl.* 136, 703

Lee, J.-E., Bergin, E.A., & Evans, N.J. 2004, *Ap. J.* 617, 360

Lee, J.-E., Evans, N.J., & Bergin, E.A. 2005, *Ap. J.*, in press (astro-ph/0506086)

Lee, J.-E., Evans, N.J., Shirley, Y.L., & Tatematsu, K. 2003, *Ap. J.* 583, 789

Li, Z.-Y. 1999, *Ap. J.*, 526, 806

Li, Z.-Y., Shematovich, V.I., Wiebe, D.S., & Shustov, B.M. 2002, *Ap. J.* 569, 792

Millar, T.J., Farquhar, P.R.A., & Willacy, K. 1997, *A& AS* 121, 139

Nelson, R.P. & Langer, W.D. 1999, *Ap. J.* 524, 923

Pavlyuchenkov, Ya.N. & Shustov, B.M. 2004, *Astronomy Reports* 48, 315

Pavlyuchenkov, Ya.N., Shustov, B.M., Shematovich, V.I., Wiebe, D.S., & Li, Z.-Y. 2003, *Astronomy Reports* 47, 176

Rawlings, J.M.C. & Yates, J.A. 2001, *MNRAS* 326, 1423

Rawlings, J.M.C., Hartquist, T.W., Menten, K.M., & Williams, D.A. 1999 *MNRAS* 255, 471

Shematovich, V.I., Wiebe, D.S., & Shustov, B.M. 1997, *MNRAS* 292, 601

Shematovich, V.I., Wiebe, D.S., Shustov, B.M., & Li, Z.-Y., 2003a, *Ap. J.* 588, 894

Shematovich, V.I., Wiebe, D.S., Shustov, B.M., Pavlyuchenkov, Ya.N., & Li, Z.-Y. 2003b, in *Chemistry as a diagnostic of star formation,* eds. C.L. Curry & M. Fich (Ottawa: NRC Research Press), p. 97

Shu, F. 1977, *Ap. J.* 214, 488

Smith, I.W.M., Herbst, E., & Chang, Q. 2004, *MNRAS* 350, 323

Tafalla, M., Myers, P.C., Caselli, P., Walmsley, C.M., & Comito, C. 2002, *Ap. J.* 569, 815

Discussion

AIKAWA: You showed that the model of Shu (1977) (without magnetic field) fit the observed line profiles better than your model with ambipolar diffusion. Usually people expect that a magnetized core collapses more slowly than that without magnetic field. Would you tell me why you obtained a faster infall velocity than Shu's model?

SHEMATOVICH: Our new self-similar model is a generalization of Shu-like solution to the pre-stellar phase, including an infall speed that is treated as a free parameter. It is set to a value that is about half of the sound speed for the particular example shown.

VITI: Regarding the problem with HCO^+ (3–2) line observations, have you included rotation in your models? Could rotation affect your fits?

SHEMATOVICH: Yes, the core rotation can improve the approach of the synthetic map to the observed one. As we are talking about self-consistent models, then the correct inclusion of core rotation into the model can be done at least in the frame of a 2D model of core dynamics.

Astrochemistry: Recent Successes and Current Challenges
Proceedings IAU Symposium No. 231, 2005
D.C. Lis, G.A. Blake & E. Herbst, eds.

© 2006 International Astronomical Union
doi:10.1017/S1743921306007034

The Young Massive Star Environment

Stan Kurtz

Centro de Radioastronomía y Astrofísica, UNAM Morelia, Michoacán 58089, México
email: s.kurtz@astrosmo.unam.mx

Abstract. Molecular cloud cores—whether stellar, non-stellar, proto-stellar, or pre-proto-stellar—have come to be recognized as key elements in the study of star formation. For both high- and low-mass stars, these cores hold key information for the final make-up of the stellar cluster and the physical processes by which the cluster and/or the individual stars form. In the case of massive stars, the chemical effects extend even beyond the molecular core, reaching into the photo-dissociated region produced by the harder spectrum of the massive stars. We discuss the directions in which these studies are taking us, with special attention given to those aspects which are unique to high-mass star-forming regions.

Keywords. masers — ISM: clouds — ISM: molecules — stars: formation — stars: pre–main-sequence

1. Introduction

Massive stars (larger than about 10 $M_\odot$) are intrinsically interesting objects for study. Even more, they are astronomically important objects to study for a variety of reasons. During their short main-sequence lifetimes they deposit large amounts of energy and momentum into the interstellar medium (ISM). In the process, they heat—and ultimately destroy—the molecular clouds in which they formed. By doing so, they affect the rate and composition of star formation within a galaxy. Of particular interest from a chemical standpoint, the ultraviolet photon flux from massive stars creates H II regions and Photon Dominated Regions (PDRs), producing ions and radicals and driving photo-chemical reactions. The energy from the stars promotes endothermic reactions and evaporates grain mantles, resulting in order-of-magnitude abundance increases. In their death as supernovae, massive stars seed the ISM with heavy elements, accelerate cosmic rays which then heat molecular clouds, producing free radicals in the process, and destroy dust grains in the strong supernova shocks.

One can characterize a "typical" massive star, say of O5 spectral type, as being about 40 $M_\odot$, 20 $R_\odot$, 40,000 K, and having a luminosity of around 3×10^5 $L_\odot$ and a main-sequence lifetime of about 1.3×10^6 yr. Perhaps more crucially, one can characterize massive star formation regions as being *complex*. The Orion massive star forming region, for example, contains not only the massive stars (and many low-mass stars) of the Trapezium cluster, but it also hosts compact and diffuse H II regions, a PDR, molecular outflows, jets, a hot molecular core, masers, proplyds, etc. When all these phenomena (some evolving on very short timescales) are combined within a volume of a few tenths of a cubic parsec, and placed at a typical distance of 5 kpc, it becomes exceedingly difficult to disentangle all of the physical processes involved.

Moreover, the fact that massive star formation typically occurs within dense condensations of molecular clouds, with visual extinctions of order a thousand, renders these regions completely inaccessible at optical, and often near-IR, wavelengths. The most

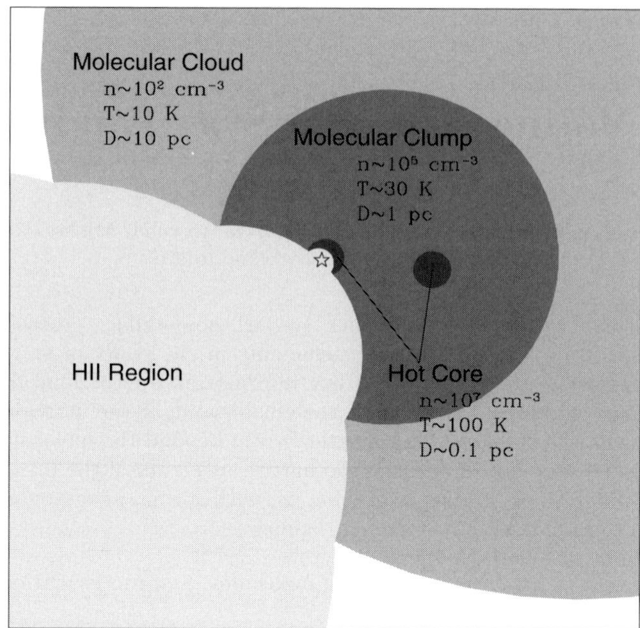

Figure 1. A schematic representation showing the hierarchy of structures within molecular clouds. Clouds have clumps, which in turn have cores. A massive star, forming within a hot molecular core, initially produces an ultracompact H II region, but may also give rise to much more extended ionized regions if the Lyman photon flux encounters the lower densities of the clump and the cloud. From Kim & Koo (2001).

common tracers are seen in the radio regime: compact H II regions, masers of various chemical species, and far-IR/sub-mm continuum sources.

2. The Star Formation Process

Low-mass star formation studies, not hampered by the great distances, high extinctions, and complex ecologies of high-mass star formation studies, have advanced significantly. Although there are still many details to be worked out (*e.g.*, Königl & Pudritz 2000) and there are still controversies about the pre-protostellar stage (*e.g.*, Vázquez *et al.* 2000), there is general agreement on the overall scheme of disk accretion by which low-mass stars form (*e.g.*, Shu *et al.* 1987).

There is much less agreement about the process of high-mass star formation. Probably the most accepted view is essentially a scaled-up version of low-mass star formation, of infall through an accretion disk. In this view, gravitational instabilities in a molecular cloud initiate collapse of the cloud material, which proceeds on the free-fall time scale of 10^5 yr. When sufficient material has fallen into the center of the instability, a hydrostatic core forms, which evolves on the Kelvin-Helmholtz time scale of 10^4 yr. This is qualitatively the same as for low-mass star formation, but on shorter timescales, due to the higher densities and energies. As the molecular density continues to increase, the collapsing core eventually becomes optically thick, and the energy released by the gravitational collapse begins to heat the core, producing the so-called Hot Molecular Core phase. Eventually, nuclear reactions begin, a star is born, and the resulting ultraviolet photon flux ionizes the surrounding material, producing an ultracompact (UC) H II region.

The various stages of this process are indicated schematically in Figure 1. The nomenclature in the literature is by no means uniform; here we adopt the usage that within the larger molecular cloud one encounters *clumps* of densities around 10^5 cm^{-3}, temperatures around 30 K, and sizes around 1 pc. Hierarchical fragmentation gives rise to *hot molecular cores* within the clump, that are even denser (10^7 cm^{-3}), hotter (> 100 K), and smaller (0.1 pc). In this model of star formation it is the formation of the protostars within the cores that heats them, leading to the hot molecular core phase. The massive star(s), once formed, will ionize the surrounding dense molecular gas, giving rise to an UC H II region. Depending on the HMC density (possibly with gradients), and its position with respect to the clump and cloud, the UC H II region can remain ultracomapct, or can evolve into a compact H II region, with or without extended, diffuse emission (Franco *et al.* 2006).

As noted above, there is certainly no universal agreement on the mechanism(s) of massive star formation. A rather different proposal from the infall and accretion disk scenario is that massive stars form by the merger, or coalescence, of lower-mass stars or protostellar objects (Bonnell *et al.* 1998; Bally & Zinnecker 2005). Also clear is that in some environments, different *modes* of massive star formation occur. An example is the "collect-and-collapse" model in which an expanding H II region sweeps up molecular material until it becomes gravitationally unstable, collapses, and forms a new massive star, near the border of the H II region. See Deharveng *et al.* (2005) for numerous examples. The existence of different modes of massive star formation seems likely; the existence of different mechanisms of massive star formation seems possible. Nevertheless, at present, the most popular model for the most common mode of massive star formation in the Milky Way is the scaled-up version of low-mass star formation. For the remainder of the present discussion we will presume this model.

3. UC H IIs and HMCs: The Passing of the Baton

Throughout most of the 1970s and 1980s, the chief tracers of massive star formation were the compact and ultracompact H II regions these stars formed soon after their birth. The bright radio continuum emission from these nebulae serves as an effective signpost of recent massive star formation. But naturally it was desired to push even further back toward the beginning, with some tracer of even earlier evolutionary states. In a series of papers at the beginning of the 1990s, Cesaroni, Walmsley & Churchwell very clearly passed the baton from UC H II regions to Hot Molecular Cores as this earlier tracer (Churchwell *et al.* 1990; Cesaroni *et al.* 1991; Cesaroni *et al.* 1992).

Throughout the early 1990s, much of the HMC work centered on characterizing the cores: their size, temperature, density, location, and clarifying if they were internally or externally heated. Shown in Figure 2 are several examples of this work. Both the ammonia and methyl cyanide molecules have proven convenient to probe the physical properties of these regions, determine their locations, and confirm that they are, in general, internally heated. This last point is worthy of mention, because for a number of years there was some discussion about whether these objects had an internal energy source or if they were heated by an external source, typically a nearby star. Although there almost certainly are examples of molecular clumps that do *not* contain a massive protostar, their prospects for aiding the study of high-mass star formation were less immediate, and they were deemed of lesser interest.

Much of the characterization of HMC properties was done by people whose primary interest was the star formation process. Thankfully, other people, with more chemical interests, were also involved, and numerous studies of the chemical processes within HMC have been made, including Charnley (1997), Doty & Neufeld (1997), Millar *et al.* (1997),

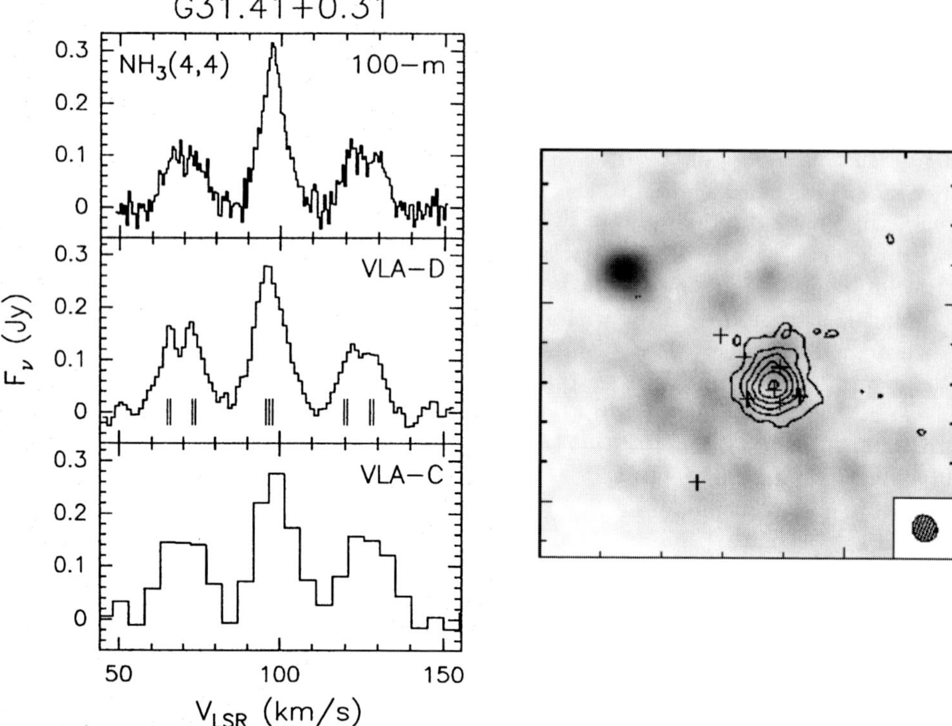

Figure 2. *Left:* The ammonia spectra of the G31.41+0.31 hot molecular core. The fact that we see the (4,4) transition indicates the gas is *hot*. The fact that we see the hyperfine satellite components indicates that the gas is *dense*. The fact that the gas is dense even at the high angular resolution of the VLA C-configuration indicates that the gas is *compact*. *Right:* An image of the G31.41+0.31 field. The greyscale indicates ionized gas (an UC H II region is seen to the northeast) while the contours show the ammonia (4,4) emission of the hot molecular core. Crosses show water masers. From Cesaroni *et al.* 1998.

Hatchell *et al.* (1998), Viti & Williams (1999), and van Dishoeck & Blake (1998, and references therein) to mention just a few.

In fact, chemical studies of these regions have now reached the point where they can serve as diagnostics of the star formation process. For example, studies by Viti *et al.* (2001) and Hatchell & Viti (2002) show that the NS/CS abundance ratio can be used to test for the passage of C-shocks within HMC, which in turn can be used to place restrictions on the lifetime of this phase. Similarly, CS and N$_2$H$^+$ observations of Zinchenko *et al.* (2005) coupled with models of Lintott *et al.* (2005) suggest an accelerated collapse rate within HMC, providing dynamically important information about the star formation process. Further information about such studies can be found in this volume, in the chapters by S. Doty and S. Viti.

4. HMCs and HMSCs: Another Pass of the Baton

Just as it was worthwhile to pass from UC H II regions to HMC as the earliest tracer of massive star formation, it is natural to seek an even early stage than HMC. If we ask "what comes before a hot molecular core?" presumably the answer is a *warm* molecular core. And presumably before a warm core, one has a *cold* core. The core presumably is cold because there is no internal energy source, so one might also refer to these cold cores

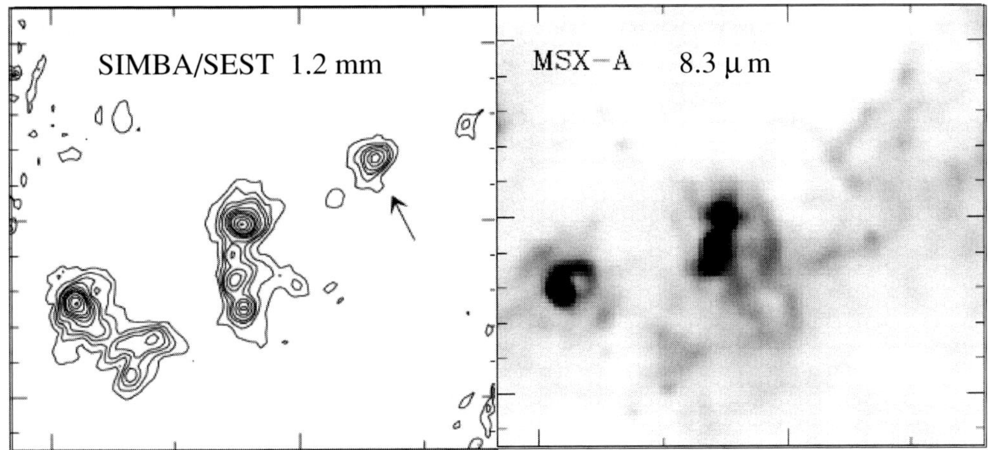

Figure 3. *Left:* The 1.2 mm continuum emission from IRAS 13080−6229. *Right:* The MSX A-band (8.3 μm) emission from the same sky area. Apart from the diffuse PAH emission seen in the MSX image, the most striking difference is appearance of the northwestern source in the SEST image (indicated by the arrow). To be seen at 1.2 mm and absent in all MSX and IRAS bands indicates a very cold source. Images from Garay *et al.* (2004).

as high-mass starless cores (HMSC). Not only would this take us closer to the beginning stages, which are least understood, but hopefully it would provide a means of distinguishing between different mechanisms of star formation (for example, disk accretion versus coalescence) and—something of a holy grail—it might show us if and how the physical conditions within the cloud core determine the initial mass function of the stellar cluster that ultimately will form.

Perhaps the most cogent prediction of this earlier phase was made by Evans *et al.* (2002) at the Boulder Hot Star Workshop. That meeting was infamous for generating a frightful array of abbreviations, and the reference in question was no exception, introducing Pre-Protostellar Cores (PPCs) and Pre-Proto-cluster Cores (PPclCs). A casual perusal of this section should convince the reader that the nomenclature has, if anything, degenerated even further. Evans *et al.* (2002) developed a model flux density distribution for these massive (4600 M$_\odot$), cold (10 K) putative PPclCs, which, not surprisingly for so cold an object, peaks in the sub-mm. More crucial, perhaps, is that in their model the flux density falls by 2–3 orders of magnitude going into the IRAS bands (100 to 12 μm).

So the stage was set for the search and discovery of these objects. All-sky infrared surveys, combined with pointed millimeter or sub-millimeter observations might detect these objects: something this cold should appear in the radio, but not in the infrared. A number of groups picked up the gauntlet, and they have succeeded in finding a substantial number of candidate sources.

Garay *et al.* (2004) report the detection of four molecular cores in the 1.2 mm continuum, which do not have IRAS or MSX counterparts. They derive sizes of 0.2–0.3 pc, densities around 2×10^5 cm^{-3}, and total masses from 400 to 2,000 M$_\odot$. The lack of emission at the IRAS and MSX (4 to 26 μm) bands implies a 17 K upper limit to the dust temperature within the cores. They confirm the presence of dense molecular gas with CS(2–1) observations (Figure 4) which shows highly supersonic linewidths of ΔV $\sim$4 km s^{-1}. If these linewidths arise from turbulence, rather than bulk motions in the gas, it may have important consequences for these massive cold cores. In particular, the contribution of turbulent pressure would allow hydrostatic equilibrium to occur at much

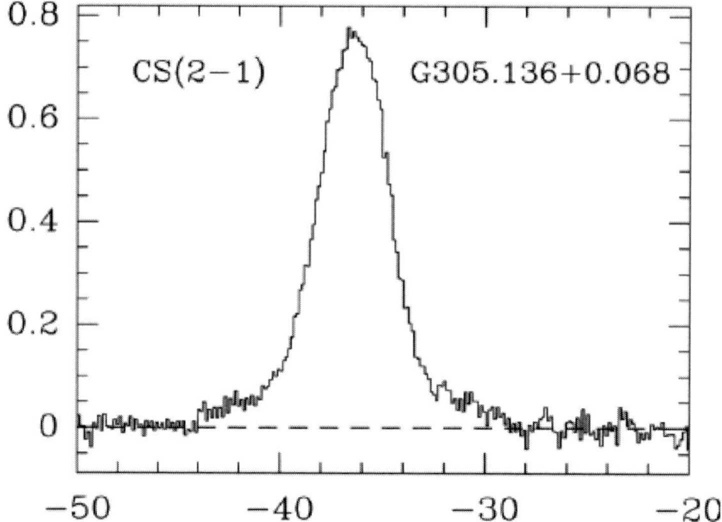

Figure 4. The CS(2–1) spectrum obtained from the IRAS 13080−6229 field (also identified as G305.136+0.069). The vertical axis is antenna temperature in Kelvins; the horizontal axis is LSR velocity in kilometers per second. Spectrum from Garay *et al.* (2004).

higher densities than in low-mass starless cores, which could have important implications for their chemistry.

Faúndez *et al.* (2004) report many additional sources from the same Garay *et al.* (2004) study, with typical temperatures of 32 K. These sources may be more representative of warm cores, but they report ten sources with dust temperatures of 25 K or less which may prove to be cold cores. An additional candidate was reported by Forbrich *et al.* (2004). Egan *et al.* (1998), Molinari *et al.* (2002), Williams *et al.* (2004), and Ao *et al.* (2004) have probably observed cold and warm cores as well, although additional observations are needed to confirm this.

Several large-scale surveys of high-mass starless cores are underway. The first is SCAMPS (SCubA Massive Precluster Survey) being carried out by M. Thompson, F. Wyrowski, and collaborators. This survey, which combines JCMT observations with data from MSX and IRAS, has identified about 25 high-mass starless cores. Preliminary results indicate that core temperatures are <35 K, densities are of order 10^5–10^6 cm^{-3}, and masses are from 1,000–4,000 M$_\odot$ (Thompson *et al.* 2006).

Klein *et al.* (2005) have made an extensive, multi-line survey based on a sample of 47 FIR-bright objects. Using a combination of near- and mid-IR data, together with radio observations, they have identified 12 pre-protocluster core candidates, with densities of order 10^4 cm^{-3} and masses of order 10^3 M$_\odot$. Their full sample includes objects in a range of evolutionary stages, and they develop a tentative sequence for a 5-stage star formation process, ranging from an initially static massive cloud core up to the dispersal of the parental cloud by the newly formed cluster. Their initial, single-dish results are being followed-up by high angular resolution interferometric observations (see the discussion following the bibliography).

Sridharan *et al.* (2005) have examined a sample of 69 candidate high-mass protostellar objects (HMPO), comparing the 1.2 mm continuum with 8.3 μm MSX data, and have identified 56 candidate high-mass starless cores (HMSC). Ammonia observations of these sources suggest that there are distinct populations: the starless cores candidates are

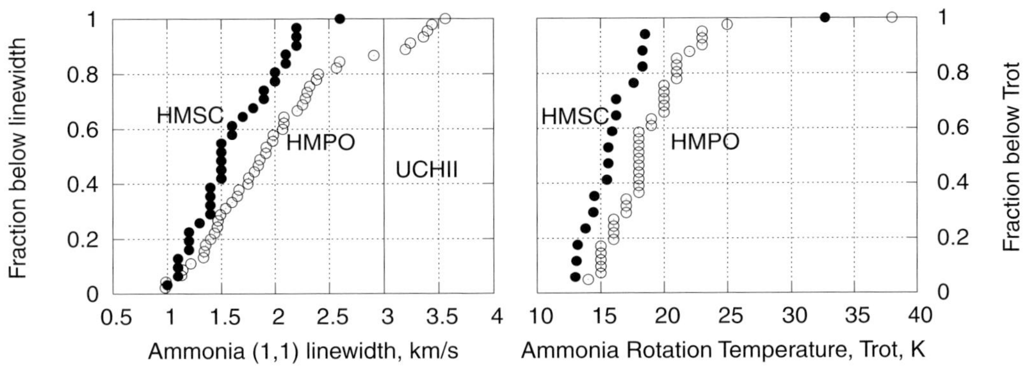

Figure 5. Cumulative probability plots (taken from Sridharan *et al.* 2005) show that the sample of high-mass starless core candidates is more quiescent (left) and cooler (right) than the sample of high-mass protostellar object candidates.

systematically more quiescent and colder than the protostellar objects (see Figure 5). Presumably (but not yet confirmed) the HMPOs are warmer and more turbulent because they already have a central heating source, and are more indicative of warm cores, rather than cold cores.

The above studies are all biased in the sense that they searched for cold cores in the vicinity of known or suspected massive star formation regions. Naturally it is desirable to have an un-biased search, and this is precisely the goal of the CHaMP survey (Census of High- and Medium-mass Protostars) presented at this conference by Barnes *et al.* (2006). Using the NANTEN telescope, they have identified $\sim$140 molecular cores from a 60 deg^2 area of the galactic plane. Follow-up observations in a variety of molecular tracers are currently underway with the Mopra and ATCA telescopes. The CHaMP survey should identify a complete population of cloud cores, which will provide constraints on timescales for the various evolutionary states, among other important results.

5. Where Do We Go from Here?

Perhaps one of the more urgent tasks awaiting us is to confirm that in fact these are *starless* cores. It is conceivable, for example, that low-mass stars are merrily forming away within these cores and are simply unable to heat or otherwise disturb the core sufficiently for us to have noticed. Sensitive far- and mid-IR observations are needed to confirm this.

Assuming this is confirmed, then the studies mentioned above will present a substantial sample of cold cores. But given this sample, how will we study it? What molecular tracers will we use? The workhorse molecule of HMC, ammonia, will be relatively useless for such cores: only the (1,1) inversion line is expected to be excited (see Figure 6). An alternative might be formaldehyde, which does have a significant number of transitions in the < 20 K range. At molecular hydrogen densities below about 10^5 cm^{-3} it is super-cooled and seen in absorption (Townes & Cheung 1969). At densities above 10^5 cm^{-3}, however, the collisional process that populates the lower levels is quenched, and the lines can appear in emission (Garrison *et al.* 1975). Whether formaldehyde will actually be useful may depend on other factors, however. For example, Young *et al.* (2004) have shown that in some low-mass star forming regions there is substantial depletion of formaldehyde. Such depletion, or some other as yet unrecognized chemical process, could severely restrict the usefulness of otherwise valuable tracers such as formaldehyde.

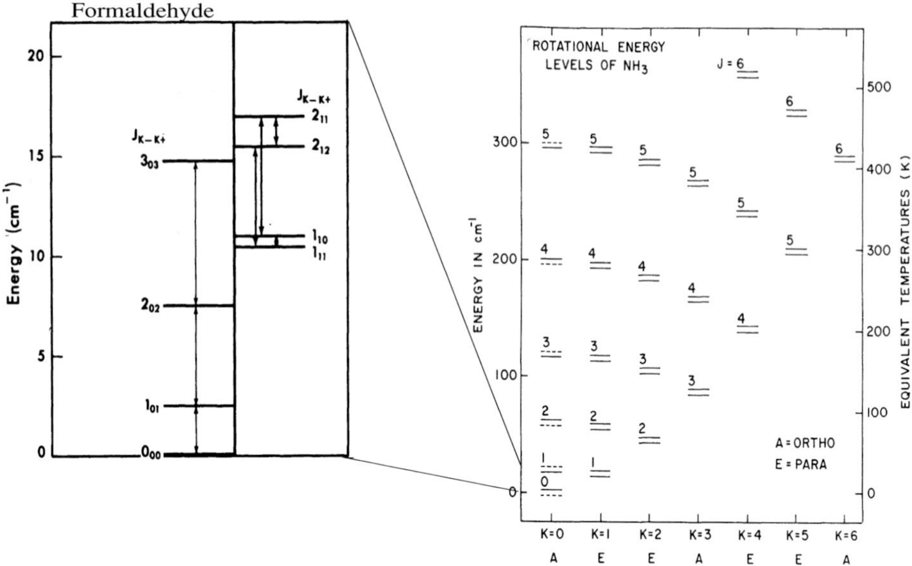

Figure 6. Level diagrams for formaldehyde and ammonia (from Townes & Cheung 1969 and Ho & Townes 1983, respectively). The ammonia (1,1) transition is at 23 K, the (2,2) at 65 K, and the (3,3) at 125 K; of these, only the (1,1) transition at 23 K will be useful to probe cold cores. Formaldehyde, however, has numerous transitions at even lower excitation levels.

One of the first tasks awaiting us is the identification and verification of useful cold core tracers. Once identified, we can begin a similar process of characterizing cold cores as was done previously for hot cores. And, ultimately, we can use the knowledge gained to further our understanding of the massive star formation process.

Acknowledgements

I am grateful to Th. Henning and P. Barnes for helpful comments, and to the many conference attendees with whom I had helpful, enlightening, and very enjoyable conversations.

References

Ao, Y., Yang, J., & Sunada, K. 2004 *A. J.* 128, 1716

Bally, J. & Zinnecker, H. 2005 *A. J.* 129, 2281

Barnes, P. J., Yonekura, Y., Miller, A., Agars, L., Mühlegger, M., Wong, T., Ladd, E. F., Mizuno, N., & Fukui, Y. 2006 *IAU 231*, (poster)

Bonnell, I. A., Bate, M. R., & Zinnecker, H. 1998 *MNRAS* 298, 93

Churchwell, E., Walmsley, C. M., & Cesaroni, R. 1990 *A&A Suppl.* 83, 119

Cesaroni, R., Hofner, P., Walmsley, C. M., & Churchwell, E. 1998 *A&A* 331, 709

Cesaroni, R., Walmsley, C. M., Koempe, C., & Churchwell, E. 1991 *A&A* 252, 278

Cesaroni, R., Walmsley, C. M., & Churchwell, E. 1992 *A&A* 256, 618

Charnley, S. B. 1997 *Ap. J.* 481, 396

Deharveng, L., Zavagno, A., & Caplan, J. 2005 *A&A* 433, 565

Doty, S. & Neufeld, D. 1997 *Ap. J.* 489, 122

Egan, M. P., Shipman, R. F., Price, S. D., Carey, S. J., Clark, F. O., & Cohen, M. 1998 *Ap. J.* 494, L199

Evans, N. J., Shirley, Y., Mueller, K. E., & Knez, C. 2002 *ASP Conf. Ser.* 267, 17

Faúndez, S., Bronfman, L., Garay, G., Chini, R., Nyman, L.-Å, & May, J. 2004, *A&A* 426, 97

Forbrich, J., Scheyer, K., Posselt, B., Klein, R. & Henning, Th. 2004, *Ap. J.* 602, 843

Franco, J., García-Segura, G., & Kurtz, S. 2006 *Ap. J.* submitted

Garay, G., Faúndez, S., Mardones, D., Bronfman, L., Chini, R. & Nyman, L.-Å 2004, *Ap. J.* 610, 313

Garrison, B. J., Lester, W. A., Miller, W. H., & Green, S. 1975, *Ap. J.* 200, L175

Hatchell, J., Millar, T. J., & Rodgers, S. D. 1998 *A&A* 332, 695

Hatchell, J. & Viti, S. 2002, *A&A* 381, L33

Ho, P. T. P. & Townes, C. H. 1983, *Ann. Rev. Astron. Astroph.* 21, 239

Kim, K.-T. & Koo, B.-C. 2001, *Ap. J.* 549, 979

Klein, R., Posselt, B., Schreyer, K., Frobich, J., & Henning, Th. 2005, *Ap. J.S* 161, in press

Königl, A., & Pudritz, R. 2000 *Protostars & Planets IV* Eds. V. Mannings, A. Boss, & S. Russell (Tucson: Univ. of Arizona Press), p. 759

Lintott, C. J., Viti, S., Rawlings, J. M. C., Williams, D. A., Hartquist, T. W., Caselli, P., Zinchenko, I., & Myers, P. 2005, *Ap. J.* 620, 795

Millar, T. J., MacDonald, G. H., & Gibb, A. G. 1997 *A&A* 325, 1163

Molinari, S., Testi, L., Rodríguez, L. F., & Zhang, Q. 2002 *Ap. J.* 570, 758

Shu, F., Adams, F., & Lizano, S. 1987 *Ann. Rev. A&A* 25, 23

Sridharan, T. K., Beuther, H., Saito, M., Wyrowski, F. & Schilke, P. 2005, *Ap. J.* submitted, and astro-ph/0508421

Thompson, M., Pestalozzi, M., Gibb, A., Hatchell, J., & Wyrowski, F. 2006 *IAU Symposium 227*, in press

Townes, C. H. & Cheung, A. C. 1969, *Ap. J.* 157, L103

van Dishoeck, E. & Blake, G. 1998 *ARAA* 36, 317

Vázquez-Semadeni, E., Ostriker, E., Passot, T., Gammie, C., & Stone, J. 2000 *Protostars & Planets IV* Eds. V. Mannings, A. Boss, & S. Russell (Tucson: Univ. of Arizona Press), p. 3

Viti, S., Caselli, P., Hartquist, T. W., & Williams, D. A. 2001, *A&A* 370, 1017

Viti, S. & Williams, D. A. 1999, *MNRAS* 305, 755

Williams, S. J., Fuller, G. A., & Sridharan, T. K. 2004 *A&A* 417, 115

Young, K. E., Lee, J.-E., Evans, N. J., Goldsmith, P. F., & Doty, S. D. 2004, *Ap. J.* 614, 252

Zinchenko, I. et al. 2005, *Ap. J.*, submitted

Discussion

STRELNITSKI:Is it possible that the "highly supersonic" line width you mentioned ($\Delta v \simeq 4$ km s^{-1}) is due to rotation of the core rather than to turbulence?

KURTZ: The vast majority of these sources have not yet been observed with sufficient angular resolution to address that issue. Molecular line profiles (single-dish, single-pointing) do show broad wings in some cases, suggesting bulk motion of the gas. This issue must be addressed by future observations; see the following comment by Thomas Henning.

HENNING: In our survey (Klein *et al.* 2005) we found a number of good candidates for high-mass protostellar cores (millimeter detection, no MIR counterparts). We found that at least one of these cores shows molecules that trace outflow activity.

KURTZ: Thanks for that comment, Thomas.

Photo: T. Snow

Astrochemistry: Recent Successes and Current Challenges
Proceedings IAU Symposium No. 231, 2005
D.C. Lis, G.A. Blake & E. Herbst, eds.

Chemistry as a Probe of the Structure and Processes in Massive YSO Envelopes

Steven D. Doty

Department of Physics and Astronomy, Denison University, Granville, OH 43023, USA
email: doty@denison.edu

Abstract. The distribution and composition of the dust and gas surrounding Young Stellar Objects (YSOs) is of continuing interest. Fortunately, rapid advances in observational capabilities have led to data of high spatial and spectral resolution, as well as the opportunity to observe in previously unavailable windows using satellites. Such high-quality data have motivated enhancements in theoretical models. Key to these models is the chemical evolution of the gas. Since the chemical evolution depends upon temperature, density, and time, the state and history of the source is encoded in the spatial distribution of the chemical abundances.

It is possible, using both parametric and detailed physical-chemical modeling, to constrain many source properties, and identify potential reactions of further laboratory interest. Using specific examples, I discuss some successes toward constraining the source properties, as well as challenges posed by current problems. Finally, I discuss the potential effect of infall dynamics and recent laboratory measurements of temperature programmed desorption of ices from grains on inferring source properties.

Keywords. astrochemistry — ISM: abundances — ISM: molecules — stars: formation

1. Introduction

Star formation is of significant and continuing interest. Not only does it give us insight into the dominant form of luminous matter in the universe, but it also helps provide connections between an understanding of the general interstellar medium on the large scale, and planet formation on the small scale. The scenario for formation of low-mass stars is relatively well understood in at least general terms (see, *e.g.*, Shu, 1997). However, the same is not true for massive star formation. While the timescales are presumably shorter than for low-mass star formation, the triggering, evolution, and structure of high-mass star forming regions is unclear.

In this paper I discuss the construction of detailed, self-consistent models of the envelopes surrounding massive young stellar objects (YSOs). These models are compared with observations to constrain the source structure. By way of example, these comparisons are used to point out recent successes and future challenges in understanding massive YSO envelopes.

2. The Challenge

From an astronomical perspective, the challenges are great. Can we understand the chemical and physical structures, the evolution, and origin of massive YSOs? To attempt to answer some of these questions requires detailed knowledge of the history and evolution of the region. Fortunately, the chemical structure of the envelope (in principal) encodes the evolution of the temperature and density of each parcel of gas. Since both the chemical evolution and molecular excitation/radiative transfer are temperature and density dependent, it may be possible to infer the source structure and evolution from

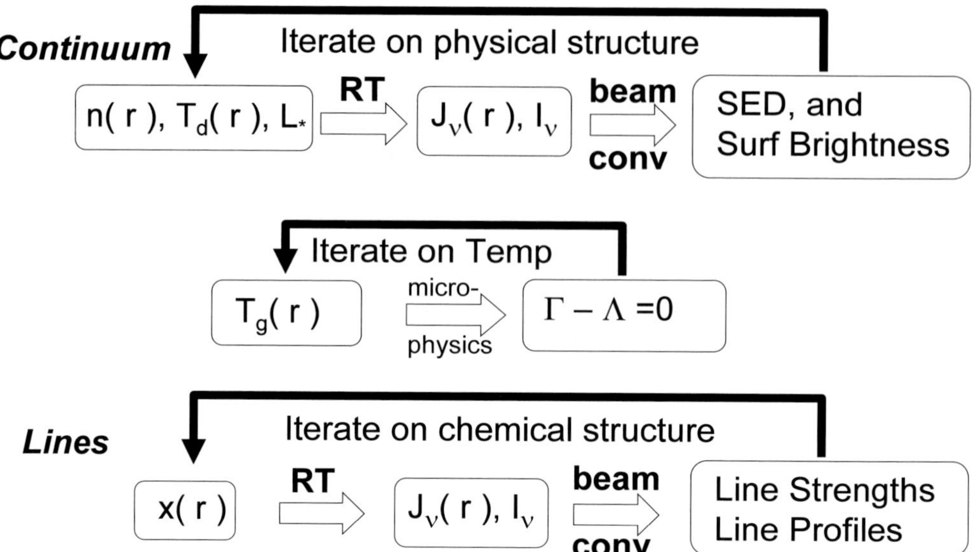

Figure 1. Schematic of comprehensive modeling. The inputs and unobservable internal source properties are on the left, the processes in the middle, and the outputs for comparison with observations are on the right.

multi-wavelength molecular line observations spanning a wide range of densities and temperatures.

From a modeling perspective, the challenges are also great. The temperature ranges from roughly ~ 10 K in the cool exterior to > 300 K in the warm interior. The densities vary from $< 10^4$ cm^{-3} in the exterior to $> 10^8$ cm^{-3} in the interior. Grains may hold, process, and liberate molecules from their mantles. Radiation from the ultraviolet to the X-ray may be present to process the gas. In short, the chemistry must respond to a wide range of physical conditions.

These two situations – together with recent (*e.g.*, SWAS, ODIN, ISO, JCMT, Spitzer, SMA) and upcoming (*e.g.*, SOFIA, ALMA, Herschel) observatories—suggest the need for detailed, comprehensive models of massive YSOs. The goal is to consider the combined effects of physical, thermal, and chemical structure as well as radiative transfer. A schematic of the modeling process is shown in Figure 1.

In this figure, the inputs/unobservable internal properties are on the left, and the outputs for comparison with observations are on the right. This is an important point. In particular, significant physics can be simulated between chemical evolution and detection of a signal at the telescope. Given the wide range of processing that can occur, it is preferable to compare simulations and direct observables such as line strengths and profiles, rather than intermediate quantities such as fractional abundances and column densities which require uncertain assumptions to infer directly from observations. In our modeling, we attempt to compare exclusively to line strengths and profiles. In cases where the column densities or abundances are presented, they have almost always been confirmed by radiative transfer models and simulated observations.

As can be seen, the physical structure is normally constrained by the dust continuum radiation. Likewise, the thermal and chemical structures are constrained by molecular line observations. Together this yields a comprehensive picture of the envelope.

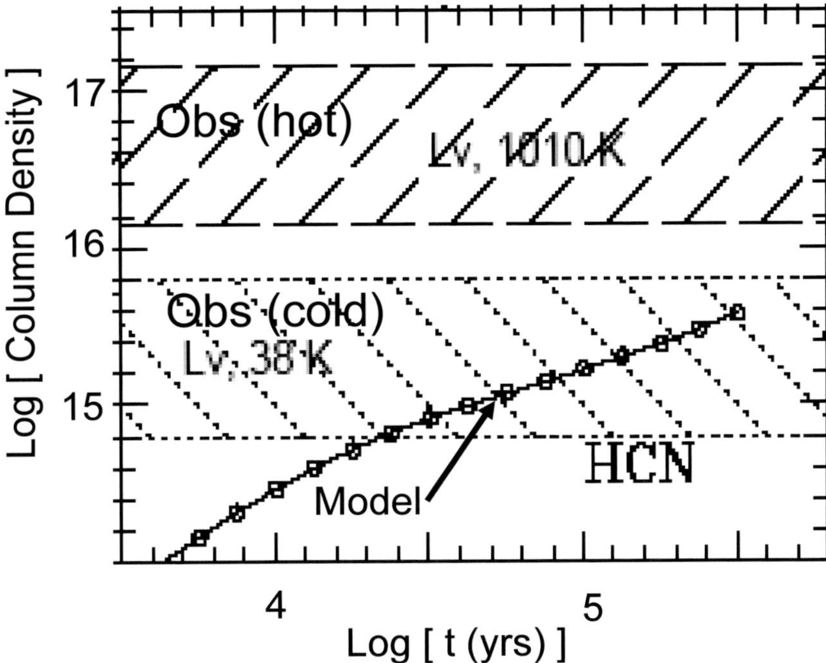

Figure 2. Comparison of predicted (line + symbols) and observed (shaded regions) column densities of HCN toward AFGL 2591. Notice that the cold HCN can be reproduced, but that the model cannot reproduce the hot HCN.

3. Discussion

3.1. *AFGL 2591*

AFGL 2591 is a massive YSO toward Cygnus X. It lies at a distance of approximately 1 kpc, though this is uncertain. AFGL 2591 has the advantage that it is a relatively isolated region of massive star formation, with far less influence from surrounding sources than other massive YSOs. While there exist broad wings in CO lines, the volume is not shock dominated. Most importantly, it has been well observed, with 29 molecular species detected. As a result, it is a good region in which to test our ability to understand the comprehensive chemistry and physics necessary to constrain the structures and processes in massive YSOs. In the following subsections, I discuss three sets of diagnostic species: HCN, sulphur-bearing species, and water.

3.2. *HCN*

HCN is observed toward AFGL 2591 with two apparent components: a cold ($T \sim 38$ K) less dense component, and a warm ($T \sim 1000$ K) and more dense component. The comparison of models with observations is shown in Figure 2. The modeling can explain the cold HCN abundance, but underproduces the hot HCN.

Radiative transfer modeling suggests an inner (warm) abundance of $x(\text{HCN}) >$ few $\times$ 10^{-6} for $r < 4 \times 10^{15}$ cm ($T > 400$ K). Normal hot gas-phase chemistry based upon the UMIST (Millar *et al.*, 1997) reaction network cannot produce this level of HCN, even for very high temperatures (~ 800 K) and long times ($\sim 3 \times 10^5$ yrs).

However, Staeuber *et al.* (2005) recently considered the potential for X-ray driven chemistry. In this scenario, X-rays produce fast secondary electrons which drive an

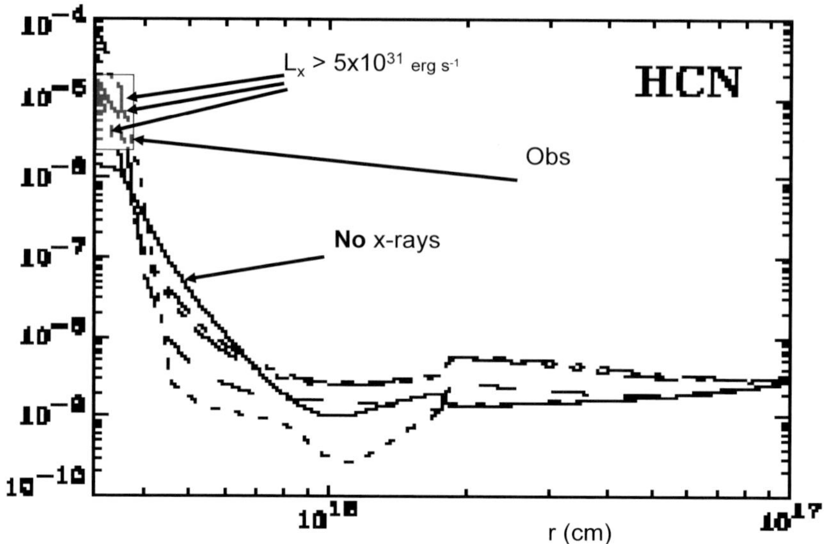

Figure 3. Comparison of model (lines) and observed (shaded region) HCN abundance as a function of position in the envelope. Notice that models with X-rays (broken lines) can enhance the HCN abundance to the observed level, while normal gas-phase chemistry (solid line) cannot.

efficient ion-molecule chemistry deep into the envelope at short times. They found that the inclusion of X-rays meaningfully improved the fit of models to observations. The case of HCN is shown in Figure 3.

In this figure, the broken lines signify different amounts of X-ray luminosity. The solid line is for a model without X-rays. Notice that including X-rays at a level $L_X > 5 \times 10^{31}$ erg s^{-1} produces sufficient HCN to reproduce the observations.

Combined, these results suggest that our current molecular databases have an incomplete description of HCN chemistry. In particular, other pathways (if HCN is driven by "normal" gas-phase chemistry) or processes (such as X-rays) must be considered in these interior regions.

3.3. *Sulphur Species*

Sulphur chemistry is generally regarded to be somewhat uncertain. As a result, we consider only those species which contain a significant fraction of the sulphur independent of the specific rates adopted. Such a sensitivity analysis with respect to reaction rates and initial state of the sulphur suggests that CS contains at least 25% of the sulphur in the cold ($T < 100$ K) gas phase. Above 100 K, much of the sulphur is shuttled into SO_2.

Radiative transfer modeling of CS and SO_2 lines suggest that $x(CS) \sim 3 \times 10^{-9} - 10^{-8}$ throughout the envelope. Similar modeling requires $x(SO_2) \sim 10^{-6}$ for $T > 100$ K. These abundances correspond to a jump of over 2 order of magnitude near $T \sim 100$ K in the dominant forms of sulphur.

This suggests that the cold sulphur is tied up in the ice mantles. Once $T > 100$ K, the mantles evaporate, liberating the sulphur into the gas phase. What is unclear, however, is the specific initial form of the cold sulphur (*e.g.*, H_2S, OCS, etc.) This is an open question, and will probably require more detailed and sophisticated gas-grain chemistry to answer.

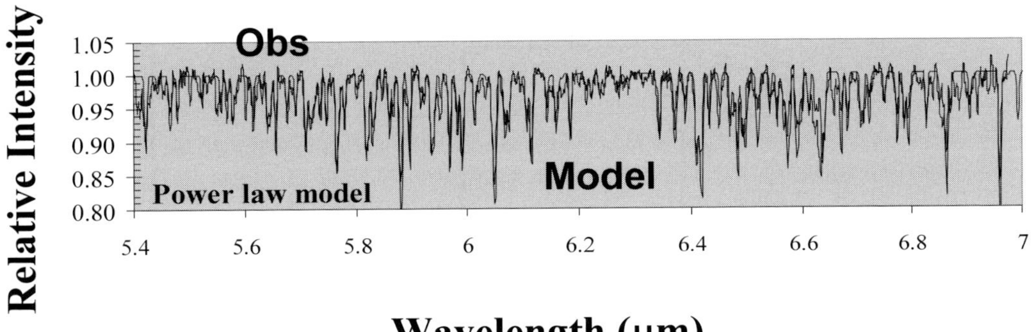

Wavelength (μm)

Figure 4. Comparison of observed (thick line) and model (thin line) ro-vibrational spectrum of water toward AFGL 2591. The rms deviation is less than a few percent.

3.4. *Water*

Water is well known as an important astrophysical molecule, due to both its role in thermal balance and as an important oxygen reservoir. It also can be a significant test of structure due to its wide range of excitable energy levels as well as its solid/gas phase-change at $T \sim 100$ K at astrophysical pressures.

Utilizing the model above together with detailed radiative transfer and parameterization of the water abundance throughout the envelope, Boonman *et al.* (2003) fit multi-wavelength observations of water toward AFGL 2591.

The upper limits on a range of ISO-LWS lines (and in particular the 108 μm $2_{21} - 1_{10}$ transition) requires that there be little cold gas-phase water to absorb against the warm central source. The strength of the SWAS 557 GHz line limits the cold water abundance, as well as the effective size of the warm inner region where water evaporates from grain surfaces. Finally, the 6 μm ro-vibrational absorption lines constrain the relative warm/cold water abundance. The fit for the 6 μm ro-vibrational lines is given in Figure 4.

The resulting best-fit water abundances are shown in Figure 5. As can be seen, the observational data constrain the water abundance in three ways: (1) a warm region having $x(\mathrm{H_2O})_{\mathrm{warm}} \sim 2 \times 10^{-4}$, (2) a cold region having $x(\mathrm{H_2O})_{\mathrm{cold}} < 10^{-8}$, and (3) a jump-like transition at $T \sim 100$ K. These results are consistent with recent laboratory data Fraser *et al.* (2001), showing that water ice evaporates from grain mantles near $T \sim 100$ K.

3.5. *Age Dating*

The comparison of model results and observations can be extended over the 29 species detected toward AFGL 2591. Since the physical and chemical models link the evolution of all species, it is possible to attempt to simultaneously age-date the envelope since protostellar turn-on. In Figure 6 the quality of fit over all species is shown as a function of time.

Note the best fit for $t \sim 1 - 3 \times 10^4$ yrs. The limits on the early and late times are shown in the figure. Note that both the early and late times are constrained by multiple species. This provides some confidence of the approximate age dating in the face of uncertainties in the reaction rates for any single species.

3.6. *Infall & Realistic Mantle Evaporation*

Last, Doty *et al.* (2005) consider the potential effects of stellar evolution and mantle evaporation. The protostellar evolution is taken from the recent model of McKee & Tan

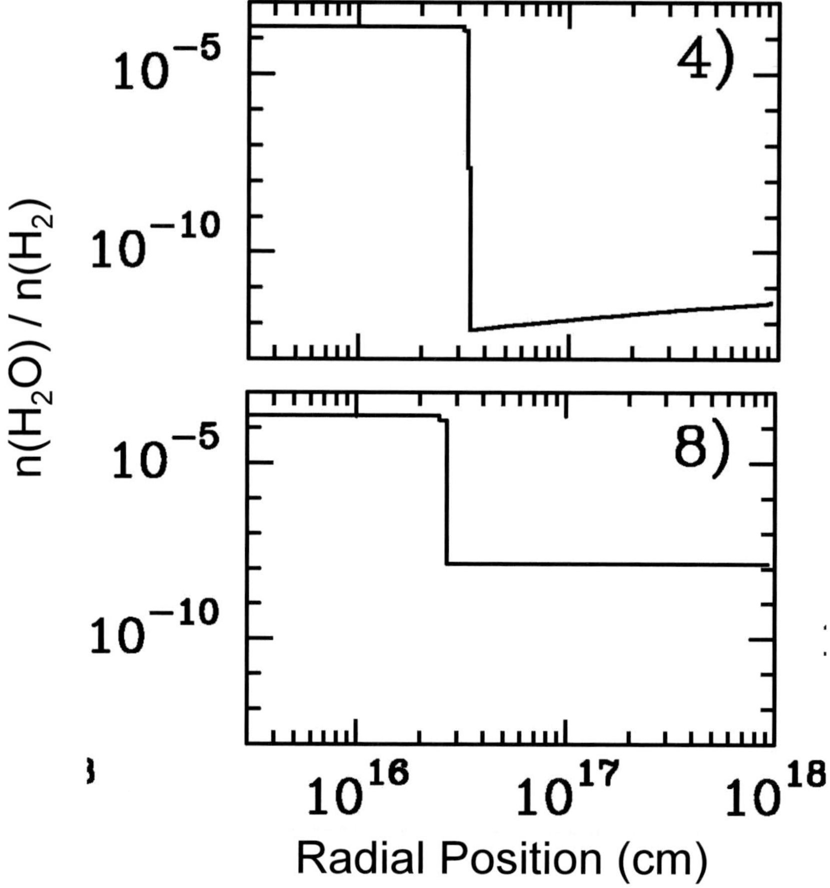

Figure 5. Best-fit water abundance profiles for AFGL 2591, based upon multi-wavelength observations of water lines.

(2003), which allows a determination of the stellar luminosity (and thus temperature distribution throughout the envelope) as a function of time. The stellar evolution is coupled with material infall. Likewise, we have removed the simple water abundance parameterization, and replaced it with laboratory-based Temperature Programmed Desorption studies of water by Fraser *et al.* (2001). As a result, we follow the gas and dust as it infalls into the warming envelope, and treat the mantle evaporation in detail.

The resulting gas-phase water abundance profiles are shown in Figure 7. Notice that the step function distribution that was inferred parametrically from the water line observations is naturally reproduced. As the protostar heats the envelope, a 'warming front' moves outward. The grain mantles evaporate nearly instantaneously on the timescale of the infall and protostellar evolution, leading to a near Heaviside-Lorentz abundance profile. While this is quite encouraging for water, it would be essential to be able to perform similar analyses for other grain mantle components.

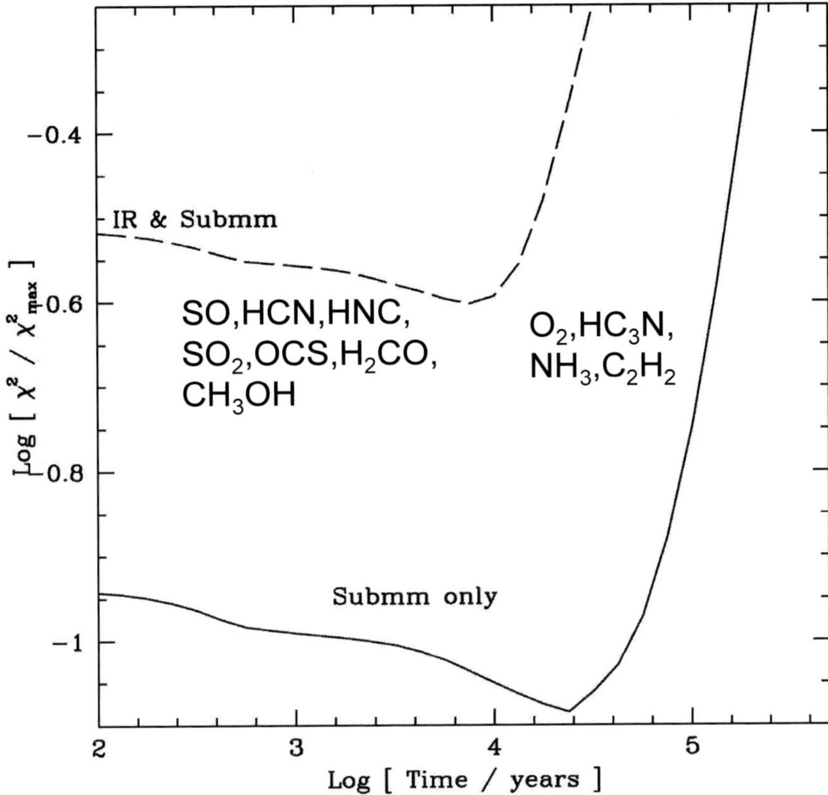

Figure 6. Quality of fit between models and observations for all 29 species observed toward AFGL 2591 as a function of time. The species best constraining the upper and lower times are noted.

4. Conclusions

Detailed, coupled, physical-chemical models of the envelopes of evolved stars have been constructed, and compared to observations with good success. We find that it is possible to simultaneously match many of the observed abundances in the envelope, and thus constrain the physical, thermal, and chemical structure, as well as (potentially) the source age. However, a number of open questions remain. In particular, is the high-temperature hydrocarbon chemistry incomplete? Are X-rays important? What is the low-temperature reservoir of solid-phase sulphur? To what extent can a better understanding of realistic grain surface chemistry aid the modeling process? Each of these questions will play key roles in any attempt to extend coupled, detailed models to the warmer interiors, earlier times, and/or more complicated regions.

Acknowledgements

This research was made possible through the contributions of many collaborators, including: Ewine van Dishoeck, Annemieke Boonman, Floris van der Tak, Pascal Staeuber, Arnold Benz, Ted Bergin, and many others. Support for this work was provided by the Research Corporation.

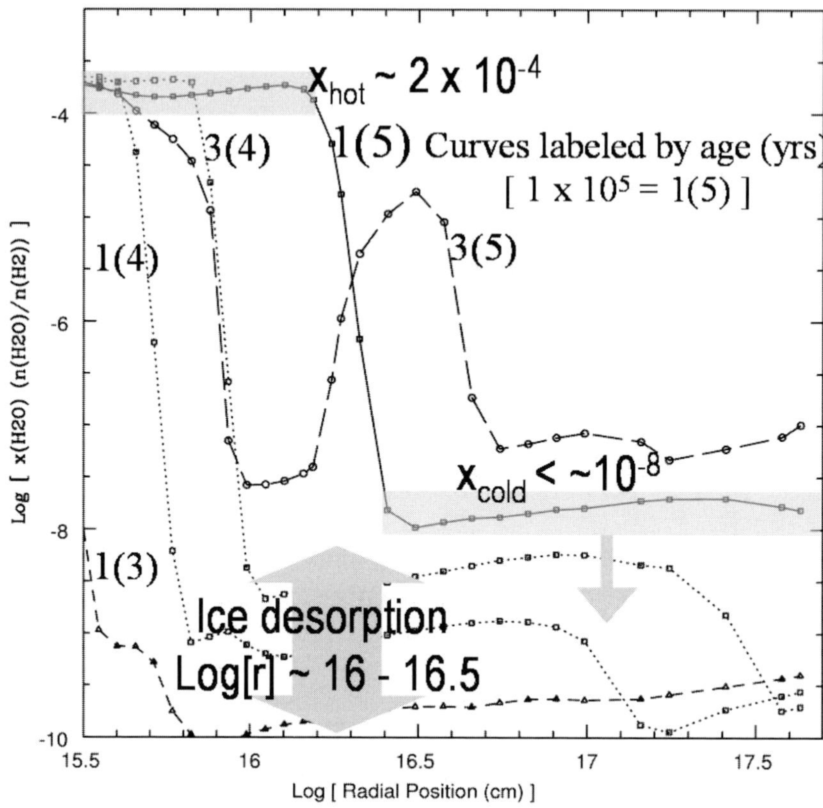

Figure 7. Water abundance profiles where the effects of protostellar evolution, infall, and exact water desorption model are included. Note that the step-function distribution inferred from the water line observations is naturally reproduced.

References

Boonman, A.M.S., Doty, S.D., van Dishoeck, E.F., Bergin, E.A., Melnick, G.J., Wright, C.M., & Stark, R. 2003, *A&A* 406, 937

Doty, S.D., van Dishoeck, E.F., & Tan, J.C. 2005, in preparation

Fraser, H.J., Collings, M.P., McCoustra, M.R.S., & Williams, D.A. 2001, *MNRAS* 327, L165

McKee, C.F., & Tan, J.C. 2003, *Ap. J.* 585, 850

Millar, T.J., Farquhar, P., & Willacy, K. 1997, *A&AS* 121, 193

Shu, F. 1997, in *Molecules in Astrophysics: Probes and Processes*, ed. E.F. van Dishoeck (Dordrecht: Kluwer), p. 19

Discussion

BAUCHTER: In an attempt to understand the erosion of graphite files of fusion reactor walls, Thomas Zedro and co-workers from the group of Prof. Kippers at Bayreuth University, and the Max-Planck Institute of Plasma Physics in Garching, conducted surface science experiments which shows carbon hydride formation at hydrogenated carbonaceous surfaces. In their TPD (Temperature Programmed Description) experiments they observed that light carbon hydrides are desorbed from edge related defect sites on a HOPG graphite surface (HOPG: highly oriented pyrolitically grown graphite) at temperatures above 700 K on a laboratory time scale. Did you include the possibility of carbon hydride formation on carbonaceous grains in your model?

DOTY: The experiment you describe is interesting. In our current model, we do not include carbon hydride formation on such surfaces. We also do not include surface erosion yet, though this may be of some importance in such high radiation field environments, especially if X-rays do play a role.

WALMSLEY: Is it possible that there exists a disk and that this may help explain the warm HCN?

DOTY: We do not include disks, so such a thing *may* be possible. However, to be consistent with linewidths, I would imagine it would have to be very face-on, and I am not sure this is consistent with the outflow geometry.

HUNTER: This object, AFGL 2591, was observed a dozen years ago in a VLA survey of massive protostellar water masers. There is a compact H II region that is many arc seconds in diameter. So, clearly, X-rays are present. How does this fit into your models?

DOTY: Unfortunately, to my knowledge X-rays have not yet been directly measured toward massive YSOs. The radio sources observed by Campbell and Trinidad *et al.* seem to lie within an ~ 5 arsecond region. This corresponds to 5000 AU at 1 kpc. In our model, the central envelope is relatively transparent to X-rays, with enough column to be X-ray opaque occurring in the 3000–10,000 AU region. Therefore, so long as any X-ray source(s) is/are within this transparent center, our results should be reasonably valid.

Photo: E. Herbst

Astrochemistry: Recent Successes and Current Challenges
Proceedings IAU Symposium No. 231, 2005
D.C. Lis, G.A. Blake & E. Herbst, eds.

© 2006 International Astronomical Union
doi:10.1017/S1743921306007058

Tracking the Early Stages of Massive Star Formation

Serena Viti

Department of Physics and Astronomy, University College London, Gower Street,
WC1E 6BT, UK
email: sv@star.ucl.ac.uk

Abstract. Until recently hot cores were the first indirect manifestation of a young massive star, but recent observational successes are now providing us with the first candidates for the very early phases of massive star formation, probably the precursors of hot cores. The characteristic hot core chemistry is believed to arise from the evaporation of the icy mantles that accumulate on the dust grains during the collapse the leads to the formation of the massive star. The duration over which the grains are warmed is determined by the time taken for a pre-stellar core to evolve to the Main Sequence. Hence, hot cores and their precursors contain an integrated record of the physics and chemistry occurring during the collapse that led to the star. In this article I will review recent advances in the chemical modelling of hot cores and their precursors in light of the recent TPD experiments on a variety of ices and I will briefly discuss the implications of such studies for the interpretation of high mass protostellar objects.

Keywords. astrochemistry — ISM: clouds — ISM: molecules — stars: formation — stars: pre-main sequence

1. Introduction

Unlike the case of low-mass stars, the process that leads to the formation of a massive star is not well understood. This is partly due to the fact that observationally it is still rather difficult to classify the early stages of massive star formation as we routinely do for low-mass stars; *e.g.*, there is not yet a definite equivalent of a class 0 or a class I for massive stars, although recent observational successes may be providing us the first candidates for such classification (*e.g.*, Garay *et al.* 2004; Faundez *et al.* 2004; Sridharan *et al.* 2005; see also chapter by Kurtz in this volume). The process that leads to the formation of a high mass star, whether it is by accretion (free-fall or accelerated collapse, *e.g.*, Lintott *et al.* 2005), or coalescence and accretion (Bonnell *et al.* 1998), must be extremely fast, $\sim 10^4$–10^5 yr, and should depend on the final mass of the star (*e.g.*, Bernasconi & Maeder, 1996). Thus, the contraction of a core from a low to a very high density ($\sim 10^7$ cm^{-3}) occurs very quickly and the hydrogen starts burning while the new born star is still well embedded in the parent cloud, which is probably still collapsing. The star will reach ZAMS (Zero Age Main Sequence, which is when the star reaches its minimum radius, its maximum mass and its hottest effective temperature) while still embedded and the dust surrounding the star is still very abundant. Hence a star may be quite evolved but have still enough dust to have the same spectral energy distribution of a younger object, making the task of detecting the early stages of high-mass star formation a hard one. In addition, it is well known that massive stars do not form in isolation but rather in so called OB associations, large aggregates of hundreds to several thousand young stars.

Nevertheless, there are indeed indirect ways of tracking and studying the early stages of a massive star, when still embedded. Until a couple of decades ago, massive star formation was traced by the detection of compact and ultracompact H II regions (UCHII), regions of dense gas ionized by the newly formed star. However, UCHII only trace the late stages of the formation of a massive star. Some two decades ago molecular observations in the submillimeter revealed the presence of the so called 'hot cores', remnants of the collapsing cloud in the process of formation of the massive star. These cores, in principle, may enable us to sample the pre-stellar material and infer the conditions in the natal cloud prior to the collapse. Hot cores are small ($\sim 10^{-2}$–10^{-1} pc), dense ($\geqslant 10^7$ H$_2$ cm^{-3}), relatively warm ($\geqslant 10^2$ K), optically thick ($A_V \geqslant 10^2$ mag.), and transient ($\leqslant 10^5$ yr) cores. They show a richer chemistry than that found in quiescent lower density ($\sim 10^3$–10^5 cm^{-3}) and cool (~ 10 K) molecular clouds. For example, in hot cores the abundance of small saturated molecules (*e.g.*, H$_2$O, NH$_3$, H$_2$S, CH$_3$OH) and large organic species (CH$_3$CN, CH$_2$CHCN, CH$_3$CH$_2$CN, C$_2$H$_5$OH) can be enhanced by over two orders of magnitude. Such high abundances are believed to arise from the evaporation, induced by the nearby (proto)star, of ice mantles frozen-out on to the dust grains during the collapse of the parent cloud. Since the freeze-out timescale is short compared to the chemical timescale, material freezing-out includes reactive species that may undergo further processing on the ice, in particular hydrogenation (although at high densities oxidation may also occur). So, for example, atomic N can be converted to NH$_3$, oxygen to water and so on. Large organics may then arise from a more drastic processing of the molecular ice. In summary, there are three main processes that affect and characterize the chemistry of hot molecular cores: depletion (or freeze-out on to grains); surface reactions; and thermal evaporation. A detailed interpretation of the observed molecular abundances require chemical models that take into account all three of these processes.

Since their discovery, hot cores have been studied intensively, both observationally and theoretically (see *e.g.*, reviews by Walmsley & Schilke 1993; Millar 1993; Kurtz *et al.* 2000). Models of hot cores in which a collapse phase including chemistry and freeze–out is followed by a static warm phase in which evaporation and gas-phase chemistry occurs have had considerable success in understanding their chemical richness (see for example Brown *et al.* 1988; Caselli *et al.* 1993; Charnley 1997; Hatchell *et al.* 1998). Although these studies do differ in the emphasis given to the chemistry in pre– and post–stellar phases, a common assumption has been that the evaporation of the ices occurs effectively instantaneously. The justification for such assumption goes as follow: if the evaporation is driven by the stellar radiation field, then the grain temperature must rise rapidly from around 10 K (in the molecular cloud) to around 200 K (in the hot core). However, such assumption may not be always appropriate as I will show in the next two sections.

2. Thermal Desorption of Ices

The instantaneous thermal evaporation approximation has always been adopted because contraction of a massive core occurs so quickly that hydrogen burning starts while the new–born star is still well embedded in the parent cloud which is probably still collapsing (Hanson, 1998). However, the contraction time (t_c), i.e the time after which hydrogen starts burning and the star becomes a ZAMS, is not well known for hot stars and certainly its dependence on the initial mass of the star has not yet been defined. In fact, it is unlikely to be less than 20,000 years even for the most massive stars. In such a time different species will then evaporate at different times.

Viti & Williams (1999) explored the idea that the duration in which the grains are warmed from very low (~ 10 K) temperatures to the temperature at which H$_2$O ice desorbs and above, is determined by the time taken for a pre-stellar core to evolve towards the Main Sequence and found that the chemistry of the hot cores must reflect this temperature-driven evolution; the evolutionary chemistry can therefore provide, in principle, a direct measure of the stellar turn-on time. Thus, the formation of hot cores contains a record of the collapse process as well as the ignition history of the star. Their models, in the absence of reliable data, made a gross simplification in the description of desorption from mixed ices; however, from their study it was already clear that the first $\sim$60,000 years of star formation differ significantly among models with different contraction times and that some key species show strong selective effects, in particular ions, sulfur-bearing species and some large molecules such as CH_3OH and CH_3CN.

Moreover, we know that shocks in the form of molecular outflows appear at an early stage of high-mass protostellar evolution (*e.g.*, Cesaroni *et al.* 1999). Hence it is possible that a slow shock could propagate through a hot core before or during its warming up phase, removing the grain mantles. Since the order of onset of hydrogen burning, outflows, and reaching the ZAMS in high-mass star formation is unknown, a core could be shocked before or during the thermal heating stage. Viti *et al.* (2001) have investigated whether it is possible to infer if and when a shock passed through a hot core by the use of chemical models. They find that, as in previous work (*e.g.*, Hatchell *et al.* 1998), the fractional abundances of various sulfur-bearing species is quite sensitive to the environmental influences. This was followed up by Hatchell & Viti (2002) who looked for variations in the NS/CS ratio in a sample of hot core sources. However, although the NS/CS value found excluded the 'standard' scenario (where everything evaporates at once) and indicated that hot cores are shocked at an early stage and that they are short-lived, this ratio was the same throughout the sample. This was a consequence of the fact that their sample was biased by their common association with UCHII regions, usually indicating an already evolved stage of high-mass star formation. What the Hatchell & Viti (2002) study clearly indicated, however, is that the selective chemical effects that should occur during the temperature increase of the dust grains will be relevant during the *formation* of a hot core rather than at a late stage when the hot core is already formed. In simple terms we can view the hot cores as the third stage of massive star formation with massive cold cores and high-mass protostellar objects (HMPOs) as the first two phases (but see chapter by Kurtz in this volume for a more thorough description of the different phases).

3. Chemical Characterization of HMPOs

From the mid-1990s HMPOs have been routinely discovered in systematic surveys (Molinari *et al.* 2002 and references therein; Sridharan *et al.* 2002; Beuther *et al.* 2002). Whether or not all HMPOs are in fact hot cores precursors, it is now clear that HMPOs and hot cores contain the history of the early phases of high-mass star formation. It is also probable that HMPOs are cooler than hot cores, hence that modelling such phases *must* take into consideration the time dependent increase of the temperature. Then, the time variation in the abundance ratio of evaporated molecular species can be used as a 'chemical clock' which may constrain the contraction time of hot stars.

The initial studies on time dependent evaporation already emphasized that appropriate temperature programmed desorption (TPD) experiments on mixed ices under near-interstellar conditions are essential. The recent observational searches for cold and warm massive cores highlight such needs even more. TPD measurements on a variety of ices have now been made (Fraser *et al.* 2001; Collings *et al.* 2003; Collings *et al.* 2004), and showed that the assumptions made in the models are indeed wrong (see Collings & McCoustra in this volume). In addition, the properties of some species for which experiments are not yet available have been considered (Collings *et al.* 2004). In summary, it was found that it is possible to categorize all species and that, depending on what

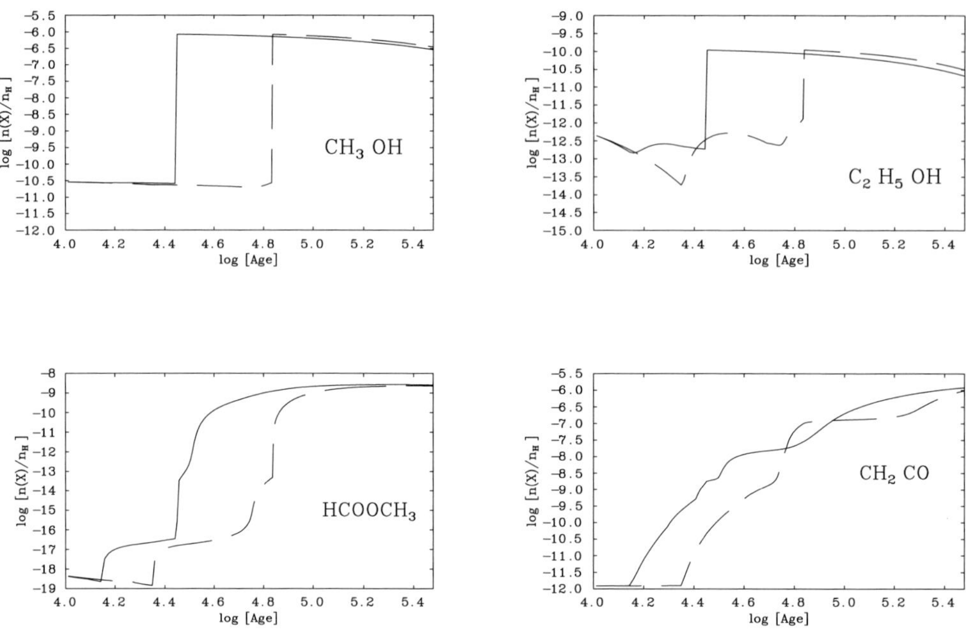

Figure 1. Fractional abundances of selected large molecules as a function of time for a 25 $M_\odot$ (solid line) and a 10 $M_\odot$ (dotted line) star.

category a species belongs to, desorption occurs in as many as four temperature bands. The proportion of each species that evaporates at each band depends on the total amount present on the grains at the time the temperature starts increasing from 10 K. These recent experimental results have now been incorporated into chemical models of high-mass star formation (Viti *et al.* 2004) and it was found that distinct chemical events occur at specific grain temperatures and these differ depending on the mass of the star. The details of such models can be found in Viti *et al.* (2004), but here it is worth noting some of the differences that appropriate TPD measurements can make to the abundances of some species. Figure 1 shows the time evolution of the fractional abundance of four molecular species for a 25 $M_\odot$ (solid line) and a 10 $M_\odot$ (dotted line). From the figure we note that large species are in fact good indicators of old cores as these strongly bound species are abundant only in the gas at late times. This implies that in HMPOs we should not find the same chemical richness of organic species that is routinely found in hot cores. However, the species that are most affected by the inclusion of the new evaporation treatment are, not surprisingly, sulfur-bearing species. In fact while larger species such as methanol and CH_2CO maintain the same trends from a 5 $M_\odot$ to a 25 $M_\odot$ star, the behavior of sulfur bearing species strongly depend on the heating rate and hence on the mass of the star. This can be of great relevance to the problem of sulfur depletion, as we shall see in the next section.

3.1. *The Sulfur Problem*

It is often assumed that simplest species, once frozen out, react with hydrogen atoms to form saturated compounds. However, recently, doubts have been raised regarding the

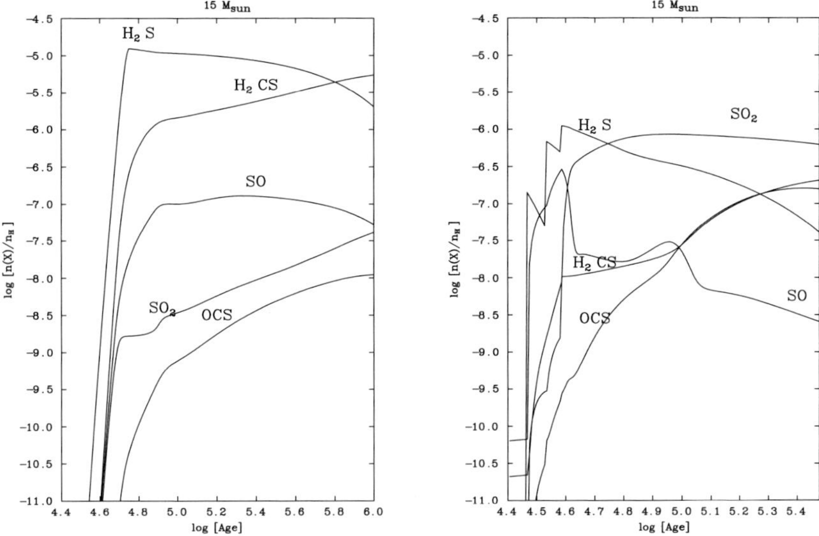

Figure 2. Time evolution of selected sulfur bearing species for the Vitti *et al.* (2004) models (*right panel*) and for the Vitti & Williams (1999) models (*left panel*).

form that sulfur takes once it depletes onto the grains. Attempts at detecting H_2S on ices with the Infrared Space Observatory have been unsuccessful (van Dishoeck & Blake 1998). In fact, the only sulfur bearing species detected in the solid state has been OCS (Palumbo *et al.* 1997), with an abundance of $\sim 10^{-7}$. The possibility that OCS could be the main reservoir of sulfur on the grains seems to have gained support recently by observations of the envelopes of massive young stars (van der Tak *et al.* 2003). Other species have been proposed as the main reservoir of sulfur on grains. For example, iron sulfide (FeS) grains have recently been discovered in protoplanetary disks and proposed as the reservoir of sulfur (Keller *et al.* 2002). However, there is no doubt that H_2S is abundant in hot cores (*e.g.*, Hatchell *et al.* 1998) and pure gas-phase models fail to reproduce it in the abundance observed; hence there seems to be little alternative but for some sulfur to hydrogenate and form H_2S once depleted. In fact, according to the Viti *et al.* (2004) models, by taking into consideration the selective chemical nature of time dependent evaporation, sulfur bearing species vary by substantial amount during the evolution of massive stars. Figure 2 shows the time evolution of selected sulfur bearing species for the Viti *et al.* (2004) models (right panel) and for the Viti & Williams (1999) models (left panel). Both plots show that sulfur bearing species vary in abundances with time as a consequence of the time dependent nature of the mantle evaporation but, the Viti *et al.* (2004) model, which incorporates the latest TPD experiments from Collings *et al.* (2004), shows much more variation in the gas-phase abundances of sulfur-bearing species. Although, to start with, all the main source of sulfur on the grains is H_2S, its evaporation at different evolutionary stages affects all other sulfur bearing species: In the 25 $M_\odot$ star model, for example, the abundance of H_2S itself varies by about one order of magnitude with time (see Figure 1 in Viti *et al.* 2004). This difference becomes larger as the mass of the star decreases, as can be seen from Figure 2. This fast and changing behavior has several consequences on other sulfur bearing species including OCS, which becomes abundant, even more than H_2S, for older cores, indicating that the

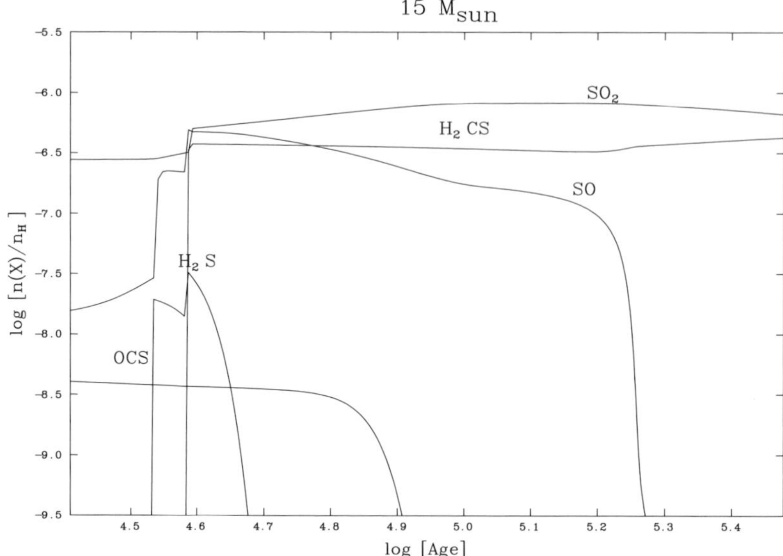

Figure 3. Time evolution of selected sulfur bearing species for a model where surface reactions during the collapse as well as the warm up phase are included (Viti, Garrod, & Herbst, in preparation).

high abundances of OCS observed by, for example, van der Tak (2003) may not lead necessarily to the conclusion that OCS is evaporated *directly* from the icy mantles.

The highly-reactive nature of sulfur bearing species during the HMPOs evolution means that although it is true that ratio of sulfur bearing species can be used as a time 'tracer', one must pay particular attention to the chemical treatment of desorption. In fact, it is possible that, for some species, the simple assumption of hydrogenation as the only type of surface reaction occurring during the formation of a hot core is not valid. Chemical models where more complex surface reactions are included have been presented before: Caselli *et al.* (1993), for example, follow the chemistry which consists of both gas-phase and grain-surface processes during the isothermal collapse that leads to the star, and purely gas-phase processes after the surrounding region becomes hot. They show that a better understanding of grain chemical processes is necessary in order to account even for the most abundant simple species. In fact, it is also likely that surface reactions continue to take place during the initial increase of the temperature, at the same time as thermal desorption. This is not usually taken into account in chemical models of massive star formation, but it turns out that it may be important, especially for sulfur-bearing species. Figure 3 shows an example of a new preliminary model where more complex surface reactions were also included during the warm up phase. The models used for Figures 2 and 3 only differ in their treatment of surface reactions: the models in Figure 2 only include surface reactions during the collapse phase and they consist only of fast hydrogenation (*e.g.*, any oxygen atom that sticks to the surface hydrogenate to water), while, for the model of Figure 3, a more thorough treatment of surface reactions, using the Ohio surface reaction rate file (Ruffle *et al.* 2000) was used and surface reactions continue into the warm up phase. A comparison of the two figures clearly shows that there is still plenty to learn about the role of surface chemistry!

The few examples given above clearly indicate that the uncertainties on the initial conditions during collapse such as the freeze out efficiency, cosmic ray induced desorption, as well as surface reactions make the problem harder: ratios of selected species as 'chemical clocks' can therefore be useful only if one first can calibrate such clocks, as also shown by Wakelam *et al.* (2004). In summary, the recent advances discussed above clearly indicate that chemical differentiation can be used as evolutionary tracer only if (i) a large sample of massive cores at different evolutionary ages is available and (ii) several molecular species are used at once. In the next section I will briefly mention one example of work in progress that aims at calibrating chemical clocks.

3.2. *The CH_3OH/CH_3CN Ratio in a Sample of HMPOs*

Theoretical studies (Viti & Williams 1999; Viti *et al.* 2004) showed that the time variation in the abundance ratio of selected evaporated molecular species could provide us with a 'chemical clock' which could constrain the contraction time of hot stars. In particular, CH_3OH/CH_3CN and NS/CS were found to be particularly sensitive to the pre-hot-core phase and to whether a shock chemistry occurs, respectively. An extended observational survey of CH_3OH/CH_3CN in a sample of HMPOs has now been carried out (Beardsmore *et al.* in preparation). The aim of this survey was quite simple: measure and compare the CH_3OH/CH_3CN ratio in a sample of probable HMPOs believed to be at different evolutionary stages. These two species were chosen because their detection would immediately confirm their identity as high-density warm cores, as both species are not easily produced in the gas-phase chemistry of dark clouds. Moreover, TPD experiments have shown that CH_3CN comes off the dust grains earlier than CH_3OH and can therefore be observed individually during that time. Observations of G29.96–0.02 (Pratap *et al.* 1999) and G9.62 + 0.19 (Hofner *et al.* 2001) show that the CH_3CN in the hot core is displaced from the CH_3OH peak which in both cases appears to arise from an unrelated source; this could be due to time-dependent evaporation. The sample of Beardsmore *et al.* was selected from a larger sample of 73 northern methanol masers (Walsh *et al.* 2003) and included 9 UCHII-associated and 9 unassociated cores, all with similar high mass and luminosity and dust temperature in the range 30–300K, *i.e.* sufficiently hot for the selective desorption of grain-mantle molecules to be occurring. Two molecular clumps from the list of Molinari *et al.* (1996), the so-called 'high' sources, believed to be at an even younger evolutionary stage than the cores not associated with UCHII, were also selected.

Preliminary results indeed show that the CH_3OH/CH_3CN ratio varies substantially among objects (see Figure 4) and that the cooler (youngest?) objects (such as IRAS 18089–1732 and IRAS 18236–1205) also have the lowest CH_3OH/CH_3CN ratio. However, it is worth noting that we are still uncertain about the routes of formation of CH_3CN, which probably occurs on grains. Hence models tend to underestimate its abundance so, although at a first sight, this result may be consistent with a young age when methanol has yet to evaporate completely, our inability to account for the CH_3CN abundance observed in evolved hot cores makes this ratio an unreliable calibrator on its own. Nevertheless Figure 4 indicates that, unlike with the Hatchell & Viti (2002) study, the ratio of CH_3OH/CH_3CN does vary by over one order of magnitudes among the selected objects.

Similar efforts have been carried out by Araya *et al.* (2005) who conducted a survey in CH_3CN for hot and dense molecular cores towards a sample of 17 southern sources with the aim of identifying very young sites of massive star formation. They use CH_3CN as an tracer of warm massive cores and they find that at least 4 of their sources are new candidates for hot molecular cores (although probably at a quite evolved stage).

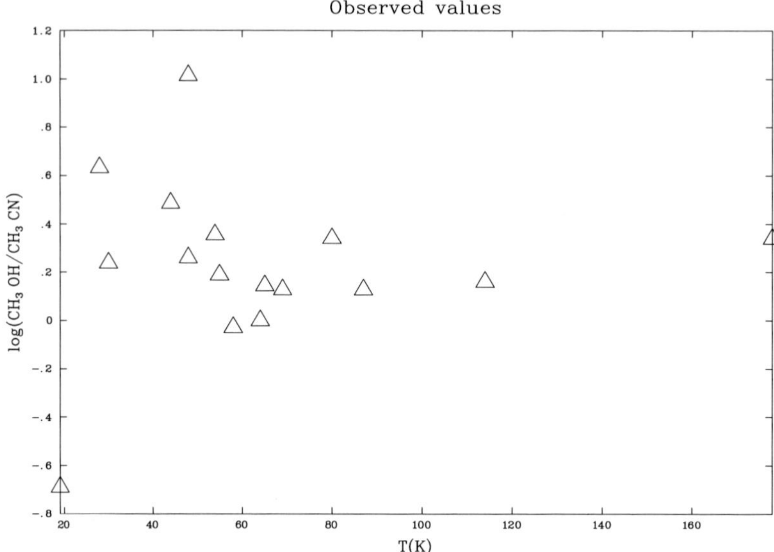

Figure 4. Log ratio of fractional abundances of methanol and CH$_3$CN for a sample of HMPOs as a function of their estimated temperature.

4. Conclusions

Over the last few years, studies of high-mass star formation have advanced enormously, and we are now at the stage where the cold sites of massive star formation may have been detected (see chapter by Kurtz in this volume). At the same time chemical models have evolved to a stage where surface reactions and thermal desorption need to be properly treated and help from experiments must be sought. The complexity of massive star formation is such that in order to use chemical models as an evolutionary tool to track high-mass protostellar objects they *must* include a proper treatment of evaporation of the icy mantle. The icy mantle also needs to be better characterized as the initial conditions (initial density, collapse, depletion, etc.) significantly impact its composition during and at the end of the collapse phase. These uncertainties are of particular importance for S-bearing species, which are often proposed as evolutionary tracers or 'chemical clocks'.

Acknowledgements

SV would like to thank PPARC for the financial support of an individual advanced fellowship.

References

Araya, E., Hofner, P., Kurtz, S., Bronfman, L., & DeDeo, S. 2005, *Ap. J. Suppl.* 157, 279
Bernasconi, P.A. & Maeder, A. 1996, *A&A* 307, 829
Beuther, H., Schilke, P., Menten, K.M., Motte, F., Sridharan, T.K., & Wyrowski, F. 2002, *Ap. J.* 566, 945
Bonnell, I., Bate, M.R., & Zinnecker, H. 1998, *MNRAS* 298, 93
Brown, P.D., Charnley, S.B., & Millar, T.J. 1988, *MNRAS* 231, 409
Caselli, P., Hasegawa, T.I., & Herbst, E. 1993, *Ap. J.* 408, 548
Cesaroni, R., Felli, M., Jenness, T., *et al.* 1999, *A&A* 345, 949
Charnley, S.B. 1997, *Ap. J.* 481, 396
Collings, M.P., Dever, J.W., Fraser, H.J., & McCoustra, M.R.S. 2003, *Ap. J.* 583, 1058

Collings, M.P., Anderson, M.A., Chen, R., Dever, J.W., Viti, S., Williams, D.A., & McCoustra, M.R.S. 2004, *MNRAS* 354, 1133

Faundez, S., Bronfman, L., Garay, G., Chini, R., Nyman, L.-A., & May, J. 2004, *A&A* 426, 97

Fraser, H.J., Collings, M.P., McCoustra, M.R.S., & Williams, D.A. 2001, *MNRAS* 327, 116

Garay, G., Faundez, S., Mardones, D., Bronfman, L., Chini, R., & Nyman, L.-A. 2004, *Ap. J.* 610, 313

Hanson, M.M. 1998, *Bolder-Munich II: Properties of Hot, Luminous Stars* ed. I. Howarth (ASP Conf. Ser. 131), 1

Hatchell, J. Thompson, M.A., Millar, T.J., & Mcdonald, G.H. 1998, *A&A* 338, 713

Hatchell, J. & Viti, S. 2002, *A&A* 381, L33

Hofner, P., Wiesemeyer, H., & Henning, T. 2001, *Ap. J.* 549, 425

Keller, L.P., Hony, S., Bradley, J.P., Molster, F.J., Waters, L.B.F.M. *et al.* 2002, *Nature* 417, 148

Kurtz, S., Cesaroni, R., Churchwell, E., Hofner, P., & Walmsley, C.M. 2000, in *Protostars and Planets IV*, eds. V. Mannings, A.P. Boss, & S.S. Russell, p. 299

Lintott, C.J., Viti, S., Rawlings, J.M.C., Williams, D.A., Hartquist, T.W., Caselli, P., Zinchenko, I., & Myers, P. 2005 *Ap. J.* 620, 795

Millar, T.J. 1993, in *Dust and Chemistry in Astronomy*, eds. T. Millar & D. Williams (Istitute of Physics, Bristol), p. 249

Molinari, S., Brand, J., Cesaroni, R., & Palla, F. 1996, *A&A* 308, 573

Molinari, S., Testi, L., Rodriguez, L.F., & Zahang, Q. 2002, *Ap. J.* 570, 758

Palumbo, M.E., Geballe, T.R., & Tielens, A.G.G.M. 1997, *Ap. J.* 479, 839

Pratap, P., Megeath, S.T., & Bergin, E.A. 1999, *Ap. J.* 517, 799

Ruffle, D.P. & Herbst, E. 2000, *MNRAS* 319, 837

Sridharan, T. K., Beuther, H., Schilke, P., Menten, K.M. & Wyrowski, F. 2002, *Ap. J.* 566, 931

Sridharan, T.K., Beuther, H., Saito, M., Wyrowski, F., & Schilke, P. 2005, *Ap. J.*, submitted, astro-ph/0508421

van der Tak, F., Boonman, A.M.S., Braakman, R., & van Dishoeck, E.F. 2003, *A&A* 412, 133

van Dishoeck, E. & Blake, J. 1998, *ARA&A* 36, 317

Viti, S. & Williams, D.A. 1999, *MNRAS*, 305, 755

Viti, S., Caselli, P., Hartquist, T.H., & Williams, D.A. 2001, *A&A* 370, 1017

Viti, S., Collings, M.P., Dever, J.W., McCoustra, M.R.S., & Williams, D.A. 2004, *MNRAS* 354, 1141

Wakelam, V., Caselli, P., Ceccarelli, C., Herbst, E., & Castets, A. 2004 *A&A* 422, 159

Walmsley, C.M. & Schilke, P. 1993, in *Dust and Chemistry in Astronomy*, eds. T. Millar and D. Williams (Istitute of Physics, Bristol), p. 37

Walsh, A.J., Macdonald, G.H., Alvey, N.D.S., Burton, M.G., & Lee, J.-K. 2003, *A&A* 410, 597

Discussion

GEBALLE: You did not explicitly mention photolysis, was it included in your modelling of grain warm-up? Is it also important in grain chemistry?

VITI: No, photolysis is not included in these models. Although it may indeed be important, I think that in the case of hot cores and their precursors thermal desorption will dominate.

Photo: E. Herbst

Astrochemistry: Recent Successes and Current Challenges
Proceedings IAU Symposium No. 231, 2005
D.C. Lis, G.A. Blake & E. Herbst, eds.

© 2006 International Astronomical Union
doi:10.1017/S174392130600706X

What Do We Know and What Do We Need to Know?

T. J. Millar†

Jodrell Bank Centre for Astrophysics, School of Physics and Astronomy,
University of Manchester, Sackville Street Building, Manchester M60 1QD, UK
email: Tom.Millar@manchester.ac.uk

Abstract. I review recent progress in determining the rate coefficients appropriate to modelling interstellar chemistry, give some information on appropriate databases from which rate coefficients can be obtained, and point to the importance of the gas-grain interaction in determining molecular abundances. Although many of the fundamental gas-phase reactions have been studied in the laboratory, the failure of the models to explain the observations of water and methanol in cold clouds indicates that grains may have an important role, both in acting as a surface for freeze-out and in the synthesis of complex molecules. The major challenge in astrochemistry is to develop a more quantitative model for the role of grains and, in some cases, to incorporate a better, probably more complex, physical model for interstellar clouds.

Keywords. astrochemistry — molecular data — ISM: molecules

1. Introduction

The synthesis of molecules in space is driven at low temperatures by ionisation – from ultraviolet photons in regions of low extinction, the so-called 'diffuse clouds' – and from cosmic ray particles in regions of high extinction, the so-called 'dense' or 'dark clouds'. In the latter objects, cosmic ray ionisation of molecular hydrogen produces the H_2^+ ion which reacts 'instantaneously' with H_2 to produce the ion H_3^+. As the proton affinity (PA) of H_2 is very low – only N, O_2, F, and the noble gases have lower PAs – this ion undergoes proton transfer reactions with almost all interstellar species, in particular with atoms, initiating chains of reactions, most importantly with H_2, which produce many of the molecules we detect in these clouds. Cosmic rays also ionise helium to produce He^+ which has an ionisation potential (IP) of some 24 eV. Thus collisions of He^+ with neutral molecules are very destructive and tend to break bonds and fragment the molecules. Thus, He^+ plays a destructive role in interstellar chemistry; at the same time the radicals and ions it produces are themselves able to take part in synthetic reactions which produce more complex species. The reaction between He^+ and H_2 is exothermic but, fortunately for astrochemistry, does not occur. If it did occur at the collisional rate, H_2 would be destroyed much faster than it is and few molecules would be formed.

The presence of ionisation in cold clouds leads to a number of classes of reaction which are important in molecular clouds. Ion-neutral reactions generally proceed via a complex which is lower in energy than the reactants so that such reactions tend to occur at the collisional rate if exothermic. For neutrals with no or only a small permanent electric dipole moment, the rate coefficient is given by the well known Langevin expression

---

††Current address: School of Mathematics and Physics, Queen's University Belfast, Belfast BT7 1NN, Northern Ireland; email: Tom.Millar@qub.ac.uk

which is independent of temperature. This leads to the happy situation that theory or experiments at room temperature provide rate coefficients which can be used at low (10 K) temperatures. For neutrals which have a significant dipole moment a correction to the Langevin formula is required. Several forms are possible but the Average-Dipole-Orientation (ADO) method provides a reasonably accurate value for the rate coefficient, many of which behave as $T^{-0.5}$ below 300 K. The invention of the Selected Ion Flow Tube (SIFT) in the early 1980s by David Smith and Nigel Adams revolutionised the laboratory study of ion-neutral reactions so that today well over 1000 reactions relevant to astrochemistry have been measured in the laboratory, with a significant number of these measured at temperatures down to 80 K and lower (Anicich 2004).

Neutral-neutral reactions can also occur at low temperatures. In general such reactions proceed through an intermediate complex which has energy larger than that of the reactants so that, even if exothermic, they tend to possess activation energy barriers. The weaker interaction between neutrals also tends to give rate coefficients with a $T^{0.5}$ behaviour. In recent years, however, there has been a great advance in the laboratory study of neutral reactions at low temperatures, particularly with the CRESU technique, as described by Sims in this volume. It has long been known that radical-radical reactions often proceed without activation barriers; the recent experiments show that they often possess rate coefficients with a T^{-n} dependence, and that radical-molecule reactions can also be fast at low temperature.

The ionisation caused by UV photons and cosmic ray particles produces electrons as well as positively charged molecular ions. Collisions between these species have large rate coefficients and are usually destructive in nature, with several product channels. In general one usually assumes that the major channel involves the loss of a H atom to produce an observed species, in line with the expectation that the channel involving the least rearrangement of orbitals might be preferred. Until recently only a few such reactions have been studied at the low energies appropriate to interstellar clouds but the use of storage rings as a means for producing low energy interactions between cooled species and to measure the product distribution has thrown up several surprises (see Geppert this volume).

One final class of reaction important in interstellar clouds is that of radiative association in which two species, most commonly an ion and a neutral, stabilise in collision through the emission of a photon. Such reactions are difficult to study in the laboratory where higher densities than in interstellar clouds means that the association complex is stabilised by collisions rather than radiation. Thus, most of the information we have on radiative association rate coefficients depends on three-body measurements coupled with a theoretical calculation to provide the appropriate radiative rate coefficient.

In order to identify the important reactions in a molecular cloud we consider the time-scales in which a molecular ion M^+ is destroyed by reaction with H_2, electrons and other neutral species, X. Using typical reaction rate coefficients these time-scales are $10^9/n(H_2)$, $10^6/n(e)$, and $10^9/n(X)$, respectively. Since $n(e) \sim 10^{-4}n$ and $n(X) \sim 10^{-6}n$ in diffuse clouds, loss of M^+ is dominated by reaction with H_2 as long as the fractional abundance of H_2 is greater than about 0.1. Dissociative recombination is only important if the rate coefficient of the H_2 reaction is less than 10^{-10} cm^3 s^{-1}. To a first approximation, the chemistry of diffuse clouds is 'simple'; one needs to consider only reactions of molecular ions with H_2 and electrons. The inclusion of radiative recombination with atomic ions and photodissociation of neutral molecules 'completes' the chemistry. In dense clouds, the fractional ionisation is less by a factor of 10^3–10^4, so that dissociative recombination and reactions with X are important only if the ion does not react or reacts very slowly with

H_2. Again, to a first approximation, the chemistry of dense clouds is 'simple'. Finally, we note that the since $n(H_2) \sim 10\text{--}10^6$ cm^{-3} in interstellar clouds, ions can be assumed to react 'instantaneously' with H_2 when they are able to do so.

In the following, I will briefly describe the important processes in diffuse and dense interstellar clouds and indicate areas in which we have good fundamental data. As we shall see there are several instances in which, despite having good gas-phase data, the observations are not understood and perhaps point to the importance of the gas-grain interaction and, in some cases, to the naive physical models adopted.

2. Diffuse Clouds

Diffuse interstellar clouds are characterised by having a low extinction which enables both visible and, more importantly, UV photons to penetrate them completely. The photons also provide a significant source of heating, through photoelectric emission and photoionisation, so that diffuse cloud temperatures are typically in the range 50–100 K. The gas phase association of H atoms is a very slow process so that the H_2 molecule is formed on interstellar dust particles and subsequently ejected vibrationally and rotationally excited by its formation energy into the gas-phase. Although the photodissociation of H_2 is a line process – so that the molecule self-shields effectively – the resulting fraction of H_2 is often small, so that the cloud is predominantly atomic in nature. In these objects the flux of UV photons is such that cosmic ray ionisation is always much less important than photoionisation of atoms such as C, S, Si and Mg and it is this process which enables the ionisation fraction to be large, $\sim 10^{-4}$. Nevertheless, cosmic ray ionisation of atomic hydrogen is important since it cannot be ionised, in common with O and N, by the interstellar UV field.

Although the formation of the simplest interstellar molecule, H_2, is on grain surfaces, the relatively large abundance of the HD, together with the fact that its photodissociation rate is 100–1000 times faster than that of H_2, since it does not self-shield efficiently, indicates that this species must be formed 100–1000 times faster than H_2, that is in the gas phase rather than on grains. The initiating reaction in the formation of HD is the charge exchange:

$$H^+ + D \longrightarrow D^+ + H$$

followed by

$$D^+ + H_2 \longrightarrow H^+ + HD.$$

Observations of HD in diffuse clouds allow one to estimate the rate at which H atoms are ionised by cosmic rays and lead to values in the range $5\text{--}50 \times 10^{-17}$ ionisations per H atom per second. The ionisation rate can also be estimated from observations of OH in warm diffuse clouds. In this case the chemistry is initiated by a charge transfer reaction which is slightly endothermic, by an energy equivalent to 240 K:

$$H^+ + O \longrightarrow O^+ + H$$
$$O^+ + H_2 \longrightarrow OH^+ + H$$
$$OH^+ + H_2 \longrightarrow H_2O^+ + H$$
$$H_2O^+ + H_2 \longrightarrow H_3O^+ + H.$$

The H_3O^+ produced by this fast chain of reactions then undergoes dissociative recombination with electrons to produce O, OH and H_2O, with the water being photodissociated rapidly to form OH. By comparing the loss rate of H^+ with O atoms to that of loss via radiative recombination with electrons, one concludes that in clouds with temperatures greater than about 50 K, every H^+ ion produced leads to the formation of an OH radical. Assuming that the chemistry is in steady state, a good assumption in diffuse clouds, one derives the abundance of OH as:

$$n(OH) = \frac{\zeta n(H)}{\beta},$$

where ζ is the cosmic ray ionisation rate of H atoms and β is the photodissociation rate of OH, which has been calculated quantum mechanically. Since all of the above reactions involving H_2 have been measured in the laboratory, as has the dissociative recombination of H_3O^+, observation of OH and H leads to an estimate for ζ which agrees with the values derived from observations of HD.

In general, the conclusion is that the observations of diffuse clouds are well described by gas-phase chemistry. Surface chemistry is important only for H_2 and possibly for NH (Wagenblast *et al.* 1993) although, to date, no models have yet incorporated the latest experimental results on the dissociative recombination of N_2H^+ which produces NH twice as efficiently as N_2 (Geppert *et al.* 2004); previous models had assumed that NH was not formed by this means.

3. Dense Clouds

In dense clouds, the cosmic ray ionisation of H_2 drives the chemistry through proton transfer reactions involving H_3^+ with atoms such as C, O, and S. Again, taking the O-H chemistry as an example, the initiating reaction is:

$$H_3^+ + O \longrightarrow OH^+ + H_2$$

followed by reactions with H_2 and the dissociative recombination of H_3O^+ with electrons as described in § 2. In dense clouds, there are no externally-produced photons, except in surface layers, and destruction of H_2O occurs through reactions with abundant ions such as He^+, C^+, H_3^+, and HCO^+. All of these reactions have been studied experimentally; all are fast, proceeding at the collisional rate. OH is also destroyed by these same ions, although in this case no experimental information is available. Nonetheless, we believe that theoretical estimates of the appropriate rate coefficients are probably accurate to within 30%. A more important loss for the radical OH is through fast reactions with other neutral atoms and radicals, in particular with the abundant atoms O, N, C, S, and Si. These reactions, which produce oxides, are thought to be fast. Indeed this is found experimentally for the reaction of OH with O, measured at temperatures down to 160 K, with N, measured down to 100 K, and S, measured at 300 K. The reaction between O and OH is thought to be the main production route to O_2 which itself, as a radical, is destroyed most efficiently in reaction with C atoms to form CO, measured to be fast down to 15 K, and also by ion reactions, the most important of which have all been measured. One concludes from this that the abundances of O, OH, and H_2O should be calculated, even from such a simple chemical model, to a fair degree of accuracy. In fact, the opposite is the case; dense clouds are severely under-abundant, particularly in O_2 and H_2O, compared to model predictions (see Liseau, this volume). This might be because the gas-grain interaction affects abundances through the process of freeze-out or that a more complex description of the physical model is required.

4. New Experimental Rate Coefficients

Recent experimental advances are covered elsewhere in this volume by Sims, Geppert & McCall; here I highlight a few results and their influence on chemical models.

4.1. *Dissociative Recombination*

The use of storage rings to study dissociative recombination (DR) has led to several new, and sometimes very surprising, results, particularly in regard to product channels. In a very beautiful experiment, McCall *et al.* (2004) combined a supersonic expansion jet source with the CRYRING to measure the DR rate coefficient, α, of rotationally cold H_3^+ down to very low energies and found $\alpha = 6.7 \; 10^{-8}(T/300)^{-0.52}$ cm^3 s^{-1}. This experiment finally ends the debate on the magnitude of this rate coefficient, a debate which has raged for over 20 years. DR is an inefficient loss of H_3^+ in dense clouds, where proton transfer reactions with abundant neutrals such as CO dominate, but is important in the diffuse gas, including those lines of sight in which the infrared absorption lines of H_3^+ have been detected and leads to even higher cosmic ray ionisation rates than derived from the OH and HD observations mentioned in § 2 (see McCall, this volume).

The CRYRING has been extensively used by Geppert and collaborators to study the DR of many important astrophysical ions including HCO^+, N_2H^+, $HOCO^+$, HC_3NH^+, and $CH_3OH_2^+$, among others. Some of the results have been surprising. Although HCO^+ recombines to form CO, only one-third of the recombinations of N_2H^+ form N_2. This new route to NH formation in dense clouds increases its calculated abundance by a factor of three. More importantly, DR now provides a mechanism by which N_2 and N_2H^+ can be depleted in high-density regions, even if N_2 has a low binding energy on interstellar grains since the NH formed in DR can freeze-out more effectively on grains than N_2. One observational consequence is that ND should possess the same deuteration fractionation as N_2D^+ and should be detectable close to 492 GHz. In many cases, product channels consisting of three or more neutrals are produced; for $CH_3OH_2^+$ only 5% of the product is methanol, compared to the usual assumptions in model calculations that methanol is produced in either 50% or 100% of recombinations. Figure 1 shows the influence of the small branching ratio on the calculated methanol abundance in a dense cloud model. The methanol abundance is almost directly related to the reduced branching ratio, in fact the reduction is slightly greater than the factor of 10 in the branching ratios due to less efficient recycling of protonated methanol. This figure shows that the observed methanol abundance in dense clouds, typically 10^{-9}–10^{-8}, is difficult to reproduce. We return to this point in § 4.3.

4.2. *Neutral-Neutral Reactions*

A significant number of neutral-neutral reactions have been measured at low temperature, particularly in the CRESU experiment. Among the important astrophysical species measured have been reactions of CH, CN, and C_2H. Smith, Herbst & Chang (2004) have analysed the low-temperature behaviour of some 300 neutral-neutral reactions included in astrochemical models while the NIST Chemical Kinetics Database (URL: `www.kinetics.nist.gov/index.php`) contains rate coefficients, based on both theory and experiment, of close to 12,000 neutral reactions, although the data is only complete until the end of 2000. More recent results from the CRESU include reactions of C_2 and C_4H (Sims, this volume). Although total rate coefficients are measured, it is much more difficult to get information on product channels.

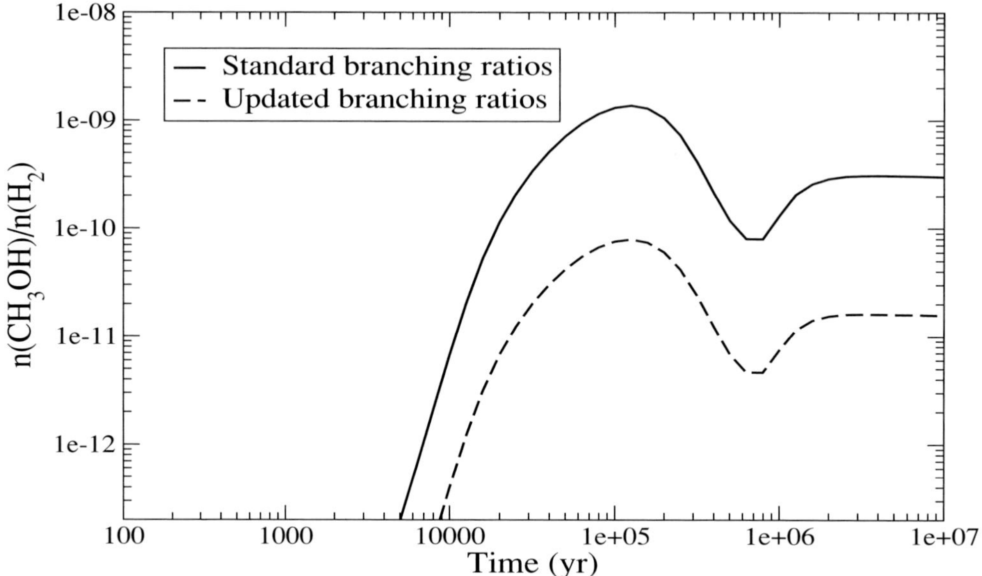

Figure 1. Time-dependent abundance of methanol using a standard branching ratio of 50% and the laboratory-determined value of 5% (Geppert *et al.* 2005).

4.3. *Radiative Association*

Radiative association reactions such as:

$$CH_3^+ + H_2 \longrightarrow CH_5^+ + h\nu$$

and

$$CH_3^+ + H_2O \longrightarrow CH_3OH_2^+ + h\nu$$

are important in building complex species from simpler species. As mentioned in §1, there are very few relevant experimental studies at pressures appropriate to interstellar conditions and rate coefficients are usually evaluated through a combination of theory and experiment. Important new results have been presented for the two systems above. Luca *et al.* (2002) studied the $CH_3^+ + H_2O$ system and derived a rate coefficient of 2×10^{-12} cm^3 s^{-1}, independent of temperature at low temperatures, compared to the standard value of $5.5 \times 10^{-12}(T/300)^{-1.7}$ cm^3 s^{-1}, based on a theoretical interpretation of a three-body association (Herbst 1985). Such theoretical analysis relies on appropriate experimental information. Recently Bacchus-Montabonel, Talbi & Persico (2000) have used the laboratory data of Gerlich and collaborators to reduce the usually adopted value of the rate coefficient of the $CH_3^+ + H_2$ radiative association from 1.3×10^{-14} $(T/300)^{-1}$ cm^3 s^{-1} (Smith 1989) to $2.4 \times 10^{-15}(T/300)^{-0.55}e^{-30/T}$ cm^3 s^{-1}, a difference of over 1000 at 10 K. Inclusion of this rate coefficient shows that the calculated fractional abundance of methanol never rises above 10^{-10} and in steady-state is only a few times 10^{-12}, failing to reproduce the observed abundance in dense clouds by a factor of 1000. This failure of gas-phase chemistry perhaps indicates a role for the grain surface formation of methanol in such sources. The critical issue then is how solid-state methanol is returned to the gas phase at 10 K.

4.4. *Photodissociation*

Photodissociation processes are particularly important in diffuse clouds and also in photo dominated regions (PDRs) and in the external envelopes of late-type stars. Here, the data situation is mixed. For small species there is extensive information on cross-sections available from theory or experiment, although in many cases there is no cross-comparison able to be made. Many stable molecules have measured UV absorption but the situation for radicals is poor. There are several excellent articles describing the estimation of photorates for interstellar and cometary molecules (van Dishoeck 1988; Roberge *et al.* 1991; Huebner *et al.* 1992) and web sites are available, through the efforts of J. Crovisier and W. Huebner (URLs: `http://www.usr.obspm.fr/~crovisier/basemole/` and `http://amop.space.swri.edu/`, respectively). However, when one looks in detail the situation is not always clear. As an example consider the unshielded photodissociation rate of methanol. Van Dishoeck (1988) quotes one set of products, H_2CO and H_2, with a rate coefficient of 1.4×10^{-9} s^{-1}. Roberge *et al.* (1991), on the other hand, have two photoionisation channels and one neutral channel, to CH_3 and H_2O, with a rate of 1.2×10^{-9} s^{-1}. Huebner *et al.* (1992) gives 5 channels, 4 of which are applicable to photons in the interstellar medium. The dominant channel in this compilation is H_2CO and H_2, with the rate coefficient for the CH_3 plus H_2O channel being some 12 times smaller. In fact, it turns out that Huebner *et al.* (1992) *assumed* that 5% of the product channel below threshold gives CH_3 and H_2O. Furthermore, although Roberge *et al.* (1991) quote an experiment to justify the dominance of their preferred channel, the experiment was such that it only monitored the OH production rate and, in fact, suggested that the yield to this channel was low (Nee, Suto & Lee 1985); seemingly, this was mis-interpreted by Roberge *et al.* (1991).

5. Databases

There are a number of databases which deal with fundamental data required for astrochemical modelling. In addition to the sites mentioned in § 4.2 and § 4.4, V. Anicich has provided a compilation of 'all' ion-neutral reactions (JPL Publication 03-19) measured in the period 1936-2003, containing references to some 2300 articles. The data are not evaluated but all individual measurements are listed. Some values for specific rate coefficients are labelled as 'Evaluated' but the values are, in fact, simple averages of all measurements.

With the exception of the NIST Chemical Kinetics Database, all other databases generally reflect the work of one particular individual. As a result they are at the mercy of changing scientific interests but, more usually, of retirement. Thus the Anicich and Huebner databases will no longer be updated, while the NIST database has not been updated since 2000 and Crovisier's since 2002. The lack of appropriate funding support for these databases is certainly an issue of concern.

Two groups have made available specific reaction sets appropriate to interstellar chemistry to the community. Both contain around 4000 reactions among about 400 species. E. Herbst's group at Ohio State University have 3 models available which differ slightly in details (URL: `http://www.physics.ohio-state.edu/~eric/index.html`). T.J. Millar's group at UMIST make similar reaction scts available: RATE99 (URL: `http://www.rate99.co.uk`) and RATE05 (URL: `http://www.udfa.net`). There are some differences in philosophy between the OSU and UMIST databases. The OSU database adopts ion-dipole enhanced rate coefficients where appropriate, and many unmeasured C atom reactions to produce hydrocarbon chains, balancing these with unmeasured

O atom reactions which destroy hydrocarbons. The UMIST database relies more heavily on measured rate coefficients – over one-third of all reactions in RATE05 have been measured. In addition the UMIST databases include high-temperature neutral-neutral reactions so that shock chemistry can be modelled.

RATE04 has been improved over its predecessors in a number of ways. An error in the implementation of cosmic-ray photon-induced reactions was noted by S. Doty. Both the UMIST and OSU implementations incorporate probabilities which are a factor of two too large. This is because the ionisation rate is proportional to $n(H_2)$ while the extinction due to grains, implicit in the determination of the probability, is proportional to the total density, that is, $2n(H_2)$. A critical study of the rate coefficients of neutral-neutral reactions at low temperature (Smith, Herbst & Chang 2004) was included, although not all their individual rate coefficients were included, in some cases the NIST data were preferred. New dissociative recombination rates and branching ratios were included, as discussed in § 4.1, and photorates were up-dated using results from the Huebner database and improved literature searches. In addition to the basic ratefile, the UMIST database also contains tables of heats of formation, electric dipole moments, a set of deuterium fractionation reactions, and codes for the calculation of interstellar cloud chemical models.

6. Dust Grains

The discussion on the failure of gas-phase astrochemical models to reproduce the observed methanol abundance in § 4.1 and 4.3 indicated one possible solution – the formation of methanol in the ice mantles from the hydrogenation of CO (see Watanabe, this volume). One of the key parameters in surface chemistry is binding energy, which determines not only mobility on the surface but also residence time, desorption, and thus any resulting gas-phase chemistry. In this regard the recent work by Collings *et al.* (2004) which shows that a molecule can have a number of binding energies depending on its binding site, has important chemical consequences, which have been discussed by Viti *et al.* (2004) and Nomura & Millar (2004). In particular, the spatial distribution of molecules in hot molecular cores shows much more small-scale structure than models in which molecules thermally evaporate at one particular temperature. The determination of accurate binding energies is a critical aspect of surface chemistry since time-scales depend exponentially on them. Although there have been recent efforts at studying surface hydrogenation, there is little quantitative information available on the possibility that radical-radical reactions are effective in cold ices. However, the detections of the highly unsaturated formic acid, HCOOH, in interstellar ices and methyl formate, $HCOOCH_3$, in comets, indicate that radical reactions might be important. If so, we are a long way from any quantitative modelling of surface chemistry.

7. Conclusions

The rate coefficients of most ion-neutral reactions are well-determined from 25 years of experimental study which have also measured products, and associated theoretical calculations. In particular, reactions of almost all interstellar ions with H_2 have been measured. Of the 2900 or so ion-neutral reactions in the RATE05 database, over 1100 have been measured. However, some important classes of reaction are poorly studied, particularly those involving atoms such as C, N, and O, and several important thermoneutral reactions need to be investigated at very low temperatures. Although there are many critical radiative association reactions in interstellar chemistry, there are very few measurements of the two-body processes at the low temperatures of interstellar clouds, and those systems which have been measured have rate coefficients orders of magnitude less than those inferred from three-body association experiments, indicating that previous theoretical estimates of radiative association may be greatly in error.

There has been a great deal of new experimental data on low-temperature neutral-neutral reactions although products are rarely determined. Progress in this field is difficult, both theoretically since interaction energies are small, and experimentally since each system needs a specific detection system to monitor the disappearance of one of the reactants or the identification of products. Smith, Herbst & Chang (2004) have made substantial progress in suggesting rate coefficients at low temperature and have identified several key reactions, mostly involving OH, for further study.

The use of storage rings to determine the rate coefficients and branching ratios of dissociative recombination reactions has produced many surprises, some of which have important implications for the efficiency of molecule formation through ion-molecule chemistry. The importance of the gas-grain interaction has been fully recognised and will present a major challenge to a quantitative understanding of molecular synthesis in cold interstellar clouds over the next 5–10 years.

Acknowledgements

This research has been supported by a grant from PPARC.

References

Anicich, V.G. 2004, *JPL Publication 03-19*

Bacchus-Montabonel, M.C., Talbi, D., & Persico, M. 2000, *J. Phys. B* 33, 955

Collings, M.P., Anderson, M.A., Chen, R., Dever, J.W., Viti, S., Wuilliams, D.A., & McCoustra, M.R.S. 2004, *MNRAS* 354, 1133

Geppert, W.D. *et al.* 2004 *Ap. J.* 608, 459

Geppert, W.D. *et al.* 2005, *this volume*

Herbst, E. 1985, *A&A* 153, 151

Huebner, W.F., Keady, J.J., & Lyon, S.P. 1992, *Ap. Sp. Sci.* 195, 1

Luca, A., Voulot, D., & Gerlich, D. 2002, *WDS02 Proceedings of Contributed Papers, Part II*, 294

McCall, B.J. *et al.* 2004, *Phys. Rev. A* 70, 052716

Nee, J.B., Suto, L.C., & Lee, L.C. 1985, *Chem. Phys.* 98, 147

Nomura, H., & Millar, T.J. 2004, *A&A* 414, 409

Smith, I.W.M. 1989, *Ap. J.* 3437, 282

Smith, I.W.M., Herbst, E., & Chang, Q 2004, *MNRAS* 350, 323

Roberge, W.G., Jones, D., Lepp, S., & Dalgarno, A. 1991, *Ap. J. Suppl.* 77, 287

van Dishoeck, E.F. 1988, in *Rate Coefficients in Astrochemistry*, eds. T.J. Millar & D.A. Williams (Dordrecht: Kluwer), p. 49

Viti, S., Collings, M.P., Dever, J.W., McCoustra, M.R.S., & Williams,, D.A. 2004, *MNRAS* 354, 1141

Wagenblast, R.M., Williams, D.A., Nejad, L.A.M., & Millar, T.J. 1993, *MNRAS* 260, 420

Discussion

GUÉLIN: The reaction rates measured in the laboratory show that some interstellar molecules (*e.g.*, methanol) cannot form in the gas phase and *must* result from grain surface reactions. Can we say equally that some molecules (*e.g.*, HD) *cannot* be formed on grains and must result from gas-phase reactions and, if yes, what fractions of the known astrophysical molecules is concerned?

Millar: First, I note that the laboratory measurement for the $CH_3^+ + H_2O$ radiative association reaction has not been published in a refereed journal some three years after publication in a conference proceeding, so the status of this one reaction is still uncertain. Certainly the abundance of HD in *diffuse* clouds indicates that it must form in the gas phase. I would also argue that all observed molecular ions – all of which can be formed by proton transfer reactions of H_3^+ with common neutral molecules – are the result of gas phase reactions. This gives a lower limit of 10% or so. I also believe that the vast majority of the molecules observed in TMC-1 are produced in the gas phase, although in some sources, some of these same molecules can be formed most efficiently on grains.

Tornow: Charnley has proposed a number of proton-transfer reactions relevant for hot cores. Do you have included them into your new data-base RATE 05?

Millar: We have been careful to use the Anicich database on ion-neutral reactions to include all proton-transfer reactions appropriate to our species set. However, our database has been set up specifically for diffuse and dense interstellar clouds. We do not include species such as methyl ethyl either; methyl formate, ethanol and dimetheyl either are included.

van Dishoeck: Does the new UMIST database give explicitly the rate coefficients of $C(J) + X$, $O(J) + X$ etc., *i.e.* as a function of the atomic fine structure level?

Millar: No, the database does not give such rate coefficients. Few, if any, such reactions have been studied experimentally.

Kroes: CH_3OH is now assumed to form through grains surface reactions in dense clouds. It is experienced as problematic that it is then released from the grains. Why is that so? Could not the energy released in the formation be used to desorb the CH_3OH?

Millar: That, of course, is a possibility although it depends on how exactly the excess energy is partitioned. If the process works in dense clouds, it presumably also happens to the CH_3OH molecules formed in 'pre-hot-core' ices. One major issue to understand is why the CH_3OH abundance in hot cores – which is also due to the hydrogenation of CO in an ice – is at least 100 times larger than that in dense clouds.

Herbst: Comment: We have just included desorption via the exothermicity of chemical reaction in our gas-grain code using simple statistical theory.

Astrochemistry: Recent Successes and Current Challenges
Proceedings IAU Symposium No. 231, 2005
D.C. Lis, G.A. Blake & E. Herbst, eds.

© 2006 International Astronomical Union
doi:10.1017/S1743921306007071

Terahertz Rotational Spectroscopy

T. F. Giesen[1], S. Brünken[2], M. Caris[1], P. Neubauer-Guenther[1], U. Fuchs[1], G. W. Fuchs[3] and F. Lewen[1]

[1]I. Physikalisches Institut, Universität zu Köln, Zülpicher Str. 77, 50937 Cologne, Germany

[2]Harvard-Smithsonian Center for Astrophysics,
60 Garden Street, Cambridge, MA 02138, USA

[3]Raymond and Beverly Sackler Laboratory for Astrophysics, Leiden Observatory,
Leiden University, Postbus 9513, 2300 RA, Leiden, The Netherlands

Abstract. The detection of interstellar molecules relies on the precise knowledge of spectral line positions from laboratory measurements. Technical developments of recent years have led to an extension of the accessible spectral range towards shorter wavelengths. New telescopes like SOFIA, the HIFI instrument aboard the Herschel satellite, and ALMA will be used for astrophysical observations in the terahertz region. The Cologne group has developed precise spectrometers to study molecules of astrophysical importance under laboratory conditions and to obtain characteristic spectra for their possible detection in space. We present recent results on light hydrides, carbon-chain molecules and more complex species.

Keywords. ISM: molecules — methods: laboratory

1. Introduction

Radioastronomy has been quite successful over the past decades in detecting new molecules in space. So far, more than 130 species have been clearly identified by means of characteristic spectral lines, and the number of new detections grows on average by four per year. While most of the detections were made at centimeter- and millimeter-wavelengths, new receiver technologies are opening up the spectral range towards shorter wavelengths and leading into the sub-millimeter wavelength region, also known as the terahertz domain. A number of new telescopes, such as the Atacama Pathfinder Experiment, APEX, the Stratospheric Observatory for Infrared Astronomy, SOFIA, and the HIFI instrument aboard the Herschel satellite will explore the terahertz domain and will certainly lead to the detection of new interstellar molecular species. This identification of new molecules relies on precise laboratory spectra. It is, thus, a challenging task for laboratory groups world wide to provide precise spectral data of molecules which might be of astrophysical importance.

Over the last two decades, the Cologne group has developed precise spectrometers to cover the spectral range from 50 GHz to 2.7 THz. In this paper we emphasize particular investigations of molecules that have characteristic spectra in the terahertz region, and which can be assigned to three important classes:

(1) pure rotational spectra of light hydrides,
(2) low ro-vibrational bending modes of carbon-chain molecules,
(3) spectra of large molecules with internal motions.

We will discuss these three aspects of terahertz spectroscopy by giving examples of molecules, which have recently been studied at the Cologne laboratories, and which are of particular interest for astronomy. In § 2 we present our recent results on the methylene radical, CH_2, a light hydride having pure rotational transitions from the lowest energy

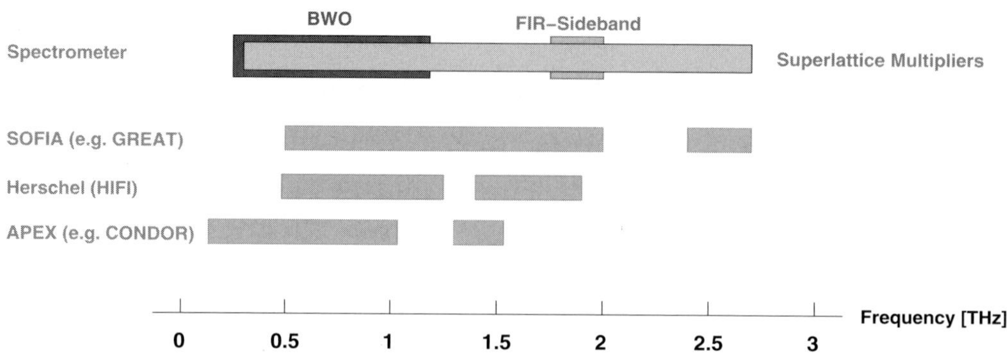

Figure 1. Frequency coverage of the Cologne spectrometers and the observatories SOFIA, Herschel and APEX.

levels at 2–3 THz. Ro-vibrational bending mode transitions of many molecules occur predominantly in the mid-infrared region, whereas carbon chains possess unusually low frequency bending vibrations. For example, the ν_2 mode of C_3 has been found at 2 THz and high-level *ab initio* calculations predict the lowest vibrational modes of longer pure carbon chains, such as C_4, C_5, . . . ,C_{10} to lie in the range of 1–5 THz. In § 3 we discuss the spectroscopy of carbon-chain molecules and present recent results on C_3H. § 4 deals with the spectral complexity of larger molecules and recent results on ethyl methyl ether, $C_2H_5OCH_3$. § 5 contains some general remarks on future spectroscopic investigations in the THz region and on the necessity of a data base for molecular spectroscopy.

The Cologne THz-spectrometers use backward wave oscillators (BWOs) as radiation and fundamental frequency sources to cover the frequency range from 50 GHz to 1.1 THz. Two different approaches are followed to extend the accessible frequency range beyond this point. First, the BWO output radiation is mixed with the radiation of a fixed frequency far infrared laser to generate sideband radiation between 1.75–2.01 THz. Secondly, to cover the gap between 1–2 THz and, furthermore, to access frequencies up to 2.7 THz, efficient superlattice multiplier devices (Scheuerer *et al.* 2003), pumped by the radiation of BWOs at 100–300 GHz are used. Figure 1 shows the spectral coverage of our spectrometers and of some present and future telescope receivers.

High spectral accuracy, $\Delta f/f = 10^{-11}$, is obtained by phase-locking the BWO to a Rubidium reference, and spectra can be recorded with Doppler- and partially sub-Doppler limited resolution. Detailed descriptions of the spectrometers can be found elsewhere (Winnewisser *et al.* 1994). The spectrometers have been combined with several devices for the production of unstable and highly-reactive molecules. These molecular sources include discharge cells, radio plasma techniques, and supersonic jets.

2. Light Hydrides – The Methylene Radical, CH_2

Large rotational constants are characteristic for light hydrides, defined here as molecules of the form XH_n (n = 1, 2). Consequently, even their ground state rotational transitions appear at comparatively high frequencies. At Cologne, the rotational spectra of a multitude of light hydride species have been measured during the last years by employing the Cologne terahertz spectrometers. Among these are the water molecule, H_2O, (Brünken *et al.* 2005), the NH radical (Lewen *et al.* 2004), the amidogen radical NH_2 (Gendriesch *et al.* 2001), and the phosphidogen radical PH_2 (Margulès *et al.* 2002).

The methylene radical, CH_2, is an important molecule for both spectroscopists and astrochemists. It is expected to be abundant in both dense and diffuse interstellar clouds and to be presumably produced quite early in the sequence of ion-molecule reactions, which makes it a test molecule for chemical models (Lee *et al.* 1996). Due to its lightness, the lowest levels are already high in energy and are widely spaced in energy. Thus transitions which probe the lowest rotational states occur in the terahertz spectral region. In its electronic $^3B_1$ ground state it has a bent, symmetric structure with an H-C-H equilibrium bond angle of 133.9308(21) degrees (Jensen & Bunker 1988) and a low barrier to linearity of less than 2000 cm^{-1}. A very shallow potential energy curve causes its floppiness, which in turn leads to strong centrifugal distortion effects. The approach of a standard Watson-type A-reduced Hamiltonian fails in the case of light and floppy molecules and leads to large uncertainties for the prediction of transition frequencies, especially at higher rotational quantum numbers.

CH_2 is an asymmetric top molecule with a permanent dipole moment of 0.57 Debye (Bunker & Langhoff 1983) pointing along the b molecular axis, which is the bisecting line of the H-C-H angle. Although CH_2 is a simple triatomic molecule, its spectrum is fairly complicated due to fine and hyperfine interactions. The electron spin in a $^3B_1$ ground state (S $=1$) causes a splitting of each rotational energy level into three fine structure components. Due to the presence of two protons, CH_2 has both an *ortho* (total nuclear spin, I $=1$) and a *para* (I $=0$) form. In the case of *ortho*-CH_2, the nuclear spin causes a further splitting of each fine structure rotational level into three hyperfine levels.

The first laboratory measurements on CH_2 were performed by Herzberg & Shoosmith (1959), which led to a clear identification of the $^3B_1$ ground state and later to the determination of its structure. In the beginning of the 1980s, Sears *et al.* used laser magnetic resonance spectroscopy to record the first ro-vibrational spectra in the mid-infrared (Sears *et al.* 1981, 1982) and pure rotational spectra of the vibrational ground state in the far infrared region (Sears *et al.* 1982a). These data were used by Lovas *et al.* (1983) to perform the first measurements on the $4_{04} - 3_{13}$ pure rotational transition at $\sim$70 GHz with microwave accuracy. Hollis *et al.* (1989) used the 70 GHz transition to start several attempts for an interstellar detection of CH_2 towards the Orion nebula. The tentative detection was later confirmed by the same authors using a receiver of improved sensitivity (Hollis *et al.* 1995). Ozeki & Saito (1995) measured two rotational transitions of *para*-CH_2, which correspond to excitation temperatures as high as 135 K and 310 K respectively. A search for CH_2 in the cold interstellar medium requires precise data of transitions, where the lowest energy levels are involved. Because of its floppiness these data can not be obtained to high accuracy from predictions based on a standard Watson Hamilton approach. To overcome these constraints, we started extensive laboratory studies on CH_2 in 2003 with the clear intention of extending the sparse data set towards terahertz frequencies and supporting a decisive astrophysical search for CH_2 in cold interstellar clouds.

We produced the methylene radical from ketene, CH_2CO, which was disintegrated in a DC-discharge cell cooled to a temperature of roughly 170 K to populate the lower-energy levels of CH_2. A 4 kHz Zeeman modulation was employed for phase sensitive signal detection, making use of the permanent magnetic moment of methylene. We used the Cologne THz spectrometer for measuring the hyperfine components of the $N_{Ka,Kc} = 1_{1,1} - 2_{0,2}$ transition of *ortho*-CH_2 centered at 943 GHz (Michael *et al.* 2003) and recorded two energetically low-lying transitions near 2 THz (Brünken *et al.* 2004) by operating the **Co**logne **S**ideband **S**pectrometer for **T**erahertz **A**pplications (COSSTA; Lewen *et al.* 1997). One of those transitions, the $N_{KaKc} = 1_{1,0} - 1_{0,1}$ at 1.915 THz stems from the lowest *para* energy level and is best suited for a detection of *para*-CH_2 in a cold interstellar environment. To fit the data to an effective Hamiltonian, we used a new approach introduced by Pickett *et al.* (2005), which is based on an expansion of the angular momentum operators in terms of an Euler series. All our data and all available precise data from the literature were included in the fit to give the most reliable set of molecular constants available (Brünken *et al.* 2005a). Based on this analysis, transition frequencies from the lowest energy levels of CH_2 were calculated and used by Polehampton *et al.* (2005) to search for CH_2 in the cold interstellar medium. For this purpose, the authors analyzed data obtained with the Long Wavelength Spectrometer (LWS; Clegg *et al.* 1996) aboard the Infrared Space Observatory (ISO). Several fine structure components of three rotational transitions of CH_2 were detected in cold molecular clouds in the line of sight towards the galactic center source Sagittarius B2 (Sgr B2). These transitions involve the three lowest rotational energy levels of methylene. The detections allowed the determination of the CH_2 column density and the estimation of the *ortho/para* ratio in the observed regions. Furthermore, the simultaneous observation of CH and CH_2 absorption lines gave insight into the underlying chemistry by comparison of the measured abundance ratio to that derived from different theoretical models. These astrophysical investigations were triggered by and based upon state-of-the-art laboratory measurements and their subsequent analysis.

3. Low Bending Modes of Carbon-Chain Molecules

Carbon is the fourth most abundant element in space. A large fraction of the more than 130 different interstellar molecules contain at least one carbon atom. Due to its unique ability to form linear chains, rings and branched, three dimensional structures, carbon is the most important element of a complex interstellar chemistry. It is conspicuous that many interstellar detected molecules are highly unsaturated linear carbon chains. Chemical models also predict *pure* carbon chains to be quite abundant in interstellar clouds and in the environment of late type carbon stars; nevertheless, only C_3 and C_5 have been undoubtedly detected so far. In the 19th century, Huggins (1881) discovered a spectral feature in the optical spectrum of comet b 1881 which was later assigned to C_3 by Douglas (1951). Based on ro-vibrational transitions in the infrared Hinkle *et al.* (1988) detected C_3 in the shell of the carbon star IRC+10216. Recent detections by rotationally resolved spectra in the visible range were published by Maier *et al.* (2001), who found C_3 in three diffuse interstellar clouds towards reddened stars and by Adamkovics *et al.* (2003) in translucent sight lines. The only secure detection of C_5 was reported by Bernath *et al.* (1989) towards IRC+10216 based on infrared spectra.

The main reason for the almost complete absence of pure carbon chains from the list of interstellar molecules is the missing of a permanent dipole moment, which would allow their detection by pure rotational transitions in the millimeter-wave region. However, pure carbon chains are predicted to have low-lying bending vibrations in the terahertz region. Precise knowledge of these transition frequencies would enable astronomers to search for longer linear carbon chains in cold interstellar clouds. *Ab initio* calculations on the structure and the vibrational frequencies were performed by e.g. Martin & Taylor (1996) and Botschwina (2005) by means of coupled cluster calculations. The bending mode vibrational dipole moment is on the order of 0.2–0.3 Debye for most of the C_n chains with $n = 3$–14.

C_3 is the only pure carbon-chain molecule for which measurements of the lowest bending mode transitions have been performed, using tunable far-infrared laser sideband spectroscopy at 2 THz (Schmuttenmaer *et al.* 1990; Gendriesch *et al.* 2003). The measurements by Schmuttenmaer *et al.* led to the detection of C_3 in the source Sgr B2,

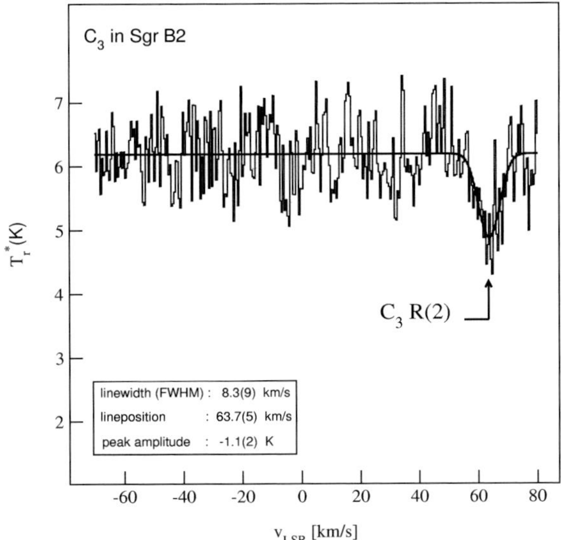

Figure 2. C$_3$ absorption line toward Sgr B2 using the KAO. The peak was assigned to the R(2) transition of the ν_2 bending mode.

which was the first detection of a pure carbon molecule in an interstellar cloud at all (Van Orden *et al.* 1994 and Giesen *et al.* 2001; see Fig. 2). This one-line detection was later confirmed by Cernicharo *et al.* (2000), using the long wavelength Fabry-Perot spectrometer (LWS) aboard the ISO satellite to detect C$_3$ in Sgr B2 and IRC+10216. The terahertz region will be discovered by a new generation of telescopes, such as SOFIA and the HIFI instrument aboard the Herschel satellite. Due to the lack of accurate laboratory data, it is inevitable to perform high-resolution measurements on the bending modes in this spectral region. For this purpose we have combined our supersonic jet molecular sources with a new Terahertz-Spectrometer. Figure 3 shows the experimental setup. The **Su**personic **Jet S**pectrometer for **T**erahertz **A**pplications (SuJeSTA) is an instrument that allows high-resolution mm- and submm-wave spectroscopy of unstable, highly-reactive molecules, such as radicals and ions at low rotational temperatures of a few Kelvin. Highly-reactive molecules are produced in either an electrical discharge, placed in the throat of the source, or by laser ablation of a graphite target. Both techniques are well established and have been described in various papers (see e.g. Linnartz *et al.* 2000; Giesen *et al.* 1994).

A variety of molecules can be produced by an appropriate mixture of gases entering the discharge source. The jet is probed by the radiation of tunable BWOs operating between 300 and 1200 GHz and very recently in combination with superlattice devices up to 2.7 THz. The spectrometer can be operated in two modes, a frequency-stabilized mode for highly-precise measurements over a small range of up to 500 MHz (typically 10–20 MHz), or in a free-running fast-scanning mode, which covers a large frequency range of several GHz but with less accuracy. The fast-scanning mode is used for searching the absorption lines of less known, or even completely unknown species, whereas the frequency-stabilized mode is used for highly-precise measurements (10 kHz accuracy) on selected, identified lines and for long integration times of weak signals. The technique of frequency stabilization of BWOs is well established at the Cologne

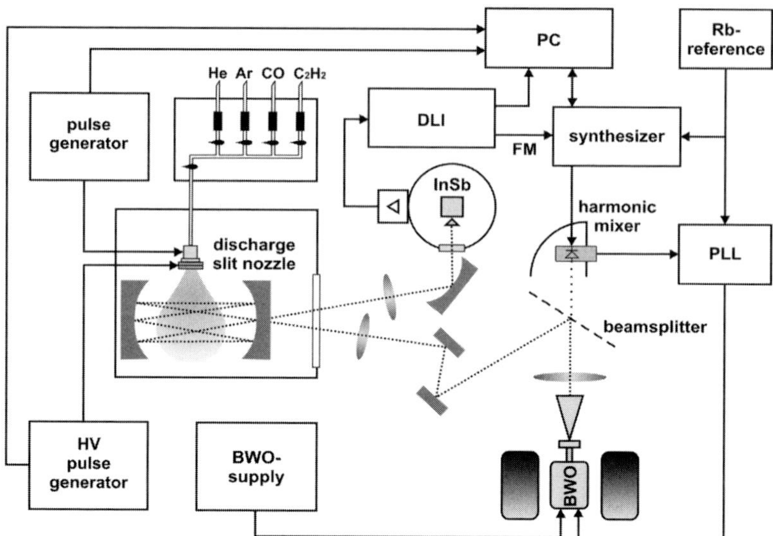

Figure 3. Experimental setup of the Cologne Supersonic Jet Spectrometer for Terahertz Applications (SuJeSTA). Backward Wave Oscillators (BWOs) are used as radiation sources. The signal is detected by a liquid helium cooled InSb-hot electron bolometer. A pulsed discharge slit nozzle produces C_3H radicals in a supersonic expansion of a C_2H_2/CO gas mixture diluted in He or Ar at 1–6 bar backing pressure. The probing microwave beam intersects the molecular jet six times in a multi-pass optical array, which leads to an overall absorption pathlength of approximately 30 cm.

laboratories (Winnewiser *et al.* 1994) and is used routinely in all of our measurements to achieve most reliable spectroscopic data for both laboratory and astrophysical applications.

Among a number of carbon-containing molecules detected in space, C_3H is considered a key molecule of carbon-chain growth in dark clouds. There are two different isomers of C_3H known, the linear (l-C_3H) and the cyclic (c-C_3H) forms, both of which have been identified in space by means of radioastronomy (Turner *et al.* 2000). The *ab initio* geometry for l-C_3H in the ground state has been calculated to be slightly bent with a rather small barrier to linearity of 100–200 cm^{-1}. The linear C_3H radical was detected by Thaddeus *et al.* (1985) towards IRC+10216 and TMC-1; laboratory spectra in the ground vibrational state were observed and analyzed by Gottlieb *et al.* (1986). The first information on rotational transitions of vibrationally excited states was obtained by Yamamoto *et al.* (1990). Although they deduced the existence of an extremely low lying $^2\Sigma$ bending state from their measurements, they were unable to directly measure transitions from the two $^2\Pi$ electronic ground state levels into the lowest $^2\Sigma$ bending state.

We have measured high-resolution spectra of the $^2\Pi$ vibrational ground state of C_3H and rotationally resolved transitions of the $^2\Pi - ^2\Sigma$ bending vibration. We used SuJeSTA to perform measurements in the frequency range between 370 and 600 GHz to extend the data set of Yamamoto *et al.* (1990) towards higher frequencies. A Hamiltonian for $^2\Pi$ molecules in a Hund's case (a) basis set has been used for a least squares fit analysis of all new data, including the measurements up to 350 GHz published by Yamamoto *et al.* (1990). We obtained improved molecular parameters and were able to determine the exact position of the low-lying $^2\Sigma$ band for the first time.

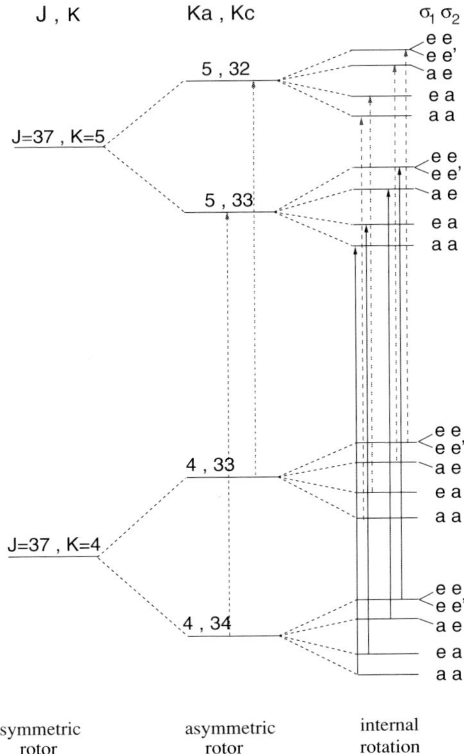

Figure 4. Energy level diagram for an asymmetric top molecule with internal rotation for the energy levels $J = 37$ with $K_a = 4, 5$. Asymmetric splitting, as well as splitting due to internal rotation, is depicted.

4. Complex Molecules – The Case of Ethyl Methyl Ether

During the process of high-mass star formation in hot-core regions of interstellar clouds temperatures can rise to $100 - 300$ K and subsequently lead to evaporation of icy mantles from interstellar dust grains (Millar *et al.* 1998). The mantles are thought to contain alcohols such as methanol (CH_3OH) and ethanol (C_2H_5OH), which have been produced via surface reactions (Charnley *et al.* 1995). In the gas phase, these alcohols can react with protonated alcohols to form ethers. Dimethyl ether (CH_3OCH_3) is the simplest representative of this class and has been found in hot cores (Ikeda *et al.* 2001). It is thus very likely to find larger ethers, such as ethyl methyl ether ($CH_3CH_2OCH_3$), in hot-core regions. Charnley *et al.* (2001) observed a single line in W51 e1/e2 and Orion KL at 160.1 GHz and another at 79.6 GHz in Sgr B2(N), which they assigned to trans-ethyl methyl ether. Since warm interstellar regions possess a rich interstellar chemistry, they have fairly dense spectra, and consequently these single line detections remained tentative. An unambiguous confirmation was highly desirable. We therefore started a laboratory investigation of ethyl methyl ether, using the FASSST spectrometer at Columbus, Ohio and the Cologne Terahertz spectrometer to record precise spectra from 55 GHz to 350 GHz. Ethyl methyl ether is an asymmetric top molecule with a strong *b*-type and a weaker *a*-type spectrum. Figure 4 shows the energy levels of the $J_K = 37_4$ and $J_K = 37_5$ rotational states. Asymmetry splitting lifts the degeneracy of the rotational levels, and the energy level diagram is further complicated by the internal rotation of the two methyl groups, which causes each level to split into five sub-levels.

From the analysis of more than 1000 individual lines including the lower-frequency transitions up to 40 GHz, which had been published earlier by Hayashi & Kuwada (1975), precise molecular parameter were derived to give reliable line frequency predictions up to 400 GHz (Fuchs *et al.* 2003). In 2002 and 2003, we used the IRAM 30-m telescope at Pico Veleta, Spain, and the SEST 15m telescope at La Silla, Chile, to search for ethyl methyl ether in several hot-core regions (Fuchs *et al.* 2005). The spectra contain lines from other organic molecules; most prominent are methanol lines, which overlap with many line positions of ethyl methyl ether. For all sources, we were able to find a set of ethyl methyl ether transition frequencies that are not blended by any other constituent. We clearly detected ethyl methyl ether towards W51 e2 and estimated a column density of 2×10^{14} cm^{-2} at $T_{rot} = 70$ K. Although these measurements confirm the tentative detection towards W51 e2 published by Charnley *et al.* (2001), we cannot confirm their tentative detection of ethyl methyl ether in Orion KL. In W51 e2 we also determined the relative abundance of ethyl methyl ether with respect to ethanol to be 0.13, which indicates a high abundance of ethers.

5. Conclusions

Ground-based observations in the terahertz region are hampered or completely blocked by the earth atmosphere. With the advent of new telescopes such as SOFIA and the Herschel/HIFI instrument, this important spectral window will be opened up for astronomical observations. We are thus expecting a wealth of line detections that carry information on well-known but also on completely unknown interstellar molecules. These new measurements will provide a deeper insight into physical conditions of astronomical sources and their molecular inventory. Accomplishing laboratory measurements in the terahertz region will assist the identification of spectral features and will play a key role in the detection of new molecules in space. The development of highly-precise terahertz spectrometers is thus of essential importance. New multiplier techniques based on super-lattice devices will expand the spectral range to several THz. In addition, monochromatic radiation sources such as quantum cascade lasers, which are commercially available for the mid-infrared region, will reach to the far-infrared region in the near future.

The Cologne spectrometers are well suited for the investigation of highly-reactive molecules, radicals and molecular ions in the terahertz region. Here we have presented three categories of molecules, which are of astrophysical importance. Light hydrides are thought to occur early in the formation of interstellar molecules and are appropriate candidates to test chemical models of ion-molecule and grain-gas phase reactions. Especially for di-hydrogenated species in cold environments the *ortho/para* ratio can deviate from the one expected from spin statistics and gives clues about the molecular formation.

Carbon-chain molecules are abundant in cold interstellar clouds and have been found in the shells of late type carbon stars. It is a challenging task to find *pure* carbon chains in theses objects. Although they have been predicted from chemical networks, they have so far eluded astronomical detections owing to the lack of laboratory data. We will use the Cologne spectrometers to study the terahertz spectra of pure carbon chains in the laboratory and will give precise data for their detection in space. Not only pure carbon chains, but all centro-symmetric chain molecules such as HC_nH, OC_nO, and NC_nN, ($n = 3, 4, \ldots$) have no permanent electrical dipole and thus no pure rotational spectrum. However, they have their lowest bending modes in the terahertz region, as has been predicted from *ab initio* calculations (Botschwina 2005).

Interstellar carbon chemistry bears a variety of complex molecules, which usually have quite complicated spectra spread over a wide frequency range. Laboratory data are rather sparse, but highly demanded for a more complete inventory of complex molecules in space. The Cologne spectrometers are well suited to record spectra over a wide frequency range from 50 GHz to 2.7 THz and will be used to investigate molecules such as dimethyl ether, ethyl formate, and ethyl cyanide at high frequencies.

Laboratory spectroscopists and astrophysicists use different languages when they talk about molecules. While spectroscopists use parameters and constants to describe molecules, astrophysicists prefer frequencies and line positions. The Cologne spectroscopy group has therefore created the **C**ologne **D**ata Base for **M**olecular **S**pectroscopy (CDMS), as a link between laboratory spectrocopy and astronomy. The CDMS, which contains line catalogs of more than 300 species, uses the same format as the JPL catalog. Free access is granted via internet `http://www.cdms.de`.

Acknowledgements

The present investigations were supported by the Deutsche Forschungsgemeinschaft under the contracts DFG-SFB 494 and LEA *HiRes.*

References

Adamkovics, M., Blake, G.A., & McCall, B.J. 2003, *Ap. J.* 595, 235

Bernath, P.F., Hinkle, K.W., & Keady, J.J. 1989, *Science* 244, 562

Botschwina, P. 2005, *Mol. Phys.* 103, 1441 and references therein

Brünken, S., Michael, E.A. Lewen, F., Giesen, T., Ozeki, H., Winnewisser, G., Jensen, P., & Herbst, E. 2004, *Can. J. Chem.* 82, 676

Brünken, S. 2005, Ph.D. thesis, *High Resolution Terahertz Spetctroscopy on Small Molecules of Astrophysical Importance* (Göttingen: Cuvillier Verlag)

Brünken, S., Müller, H.S.P., Lewen, F., & Giesen, T.F. 2005a, *J. Chem. Phys.* 123, 164315

Bunker, P.R. & Langhoff, S.R. 1983, *J. Mol. Spect.* 102, 204

Cernicharo, J., Goicoechea, J.R., & Caux, E. 2000, *Ap. J.* 534, L199

Charnley, S.B., Kress, M.E., Tielens, A.G.G.M., & Millar, T.J. 1995, *Ap. J.* 448, 232

Charnley, S.B., Ehrenfreund, P., & Kuan, Y.-J. 2001, *Spectrochim. Acta A* 57, 685

Clegg, P.E., Ade, P.A.R., Armand, C. *et al.* 1996, *A&A* 315, L38

Douglas, A.E. 1951, *Ap. J.* 94, 466

Fuchs, G.W., Fuchs, U., Giesen, T.F., & Wyrowski, F. 2005, *A&A* 444, 521

Fuchs, U., Winnewisser, G., Groner, P., De Lucia, F. C., & Herbst, E. 2003, *Ap. J. Suppl.* 144, 277

Gendriesch, R., Lewen, F., Winnewisser, G., & Müller, H.S.P. 2001, *J. Mol. Struct.* 599, 293

Gendriesch, R., Pehl, K., Giesen, T., Winnewisser, G., & Lewen, F. 2003, *Z. Naturforsch.* 58a, 129

Giesen, T.F., Van Orden, A., Hwang, H.J., Fellers, R.S., Provencal, R.A., & Saykally, R.J. 1994, *Science* 265, 756

Giesen, T.F., Van Orden, A.O., Cruzan, J.D., Provencal, R.A., Saykally, R.J., Gendriesch, R., Lewen, F., & Winnewisser, G. 2001, *Ap. J.* 551, L181

Gottlieb, C.A., Gottlieb, E.W., Thaddeus, P., & Vrtilek, J.M. 1986, *Ap. J.* 303, 446

Hayashi, M. & Kuwada, K. 1975, *J. Mol. Struc.* 28, 147

Herzberg, G. & Shoosmith, J. 1959, *Nature* 183, 1801

Hinkle, K.W., Keady, J.J., & Bernath, P.F. 1988, *Science* 241, 1319

Hollis, J.M., Jewell, P.R., & Lovas, F.J. 1989, *Ap. J.* 346, 794

Hollis, J.M., Jewell, P.R., & Lovas, F.J. 1995, *Ap. J.* 438, 259

Huggins, W. 1881, *Proc. R. Soc. London* 33, 1

Ikeda, M., Ohishi, M., Nummelin, A., Dickens, J.E., Bergman, P., Hjalmarson, A., & Irvine, W.M. 2001, *Ap. J.* 560, 792

Jensen, P. & Bunker, P.R. 1988, *J. Chem. Phys.* 89, 1327

Lee, H.-H., Bettens, R.P.A., & Herbst, E. 1996, *A&A* 119, 111

Lewen, F., Michael, E., Gendriesch, R., Stutzki, J., & Winnewisser, G. 1997, *J. Mol. Spect.* 183, 207

Lewen, F., Brünken, S., Winnewisser, G., Simeckova, M., & Urban, S. 2004, *J. Mol. Spect.* 226, 113

Linnartz, H., Vaizert, O., Motylewski, T., & Maier, J.P. 2000, *J. Chem. Phys.* 112, 9777

Lovas, F.J., Suenram, R.D., & Evenson, K.M. 1983, *Ap. J.* 267, L131

Maier, J.P., Lakin, N.M., Walker, G.A.H., & Bohlenender, D.A. 2001, *Ap. J.* 553, 267

Margulès, L., Herbst, E., Ahrens, V., Lewen, F., Winnewisser, G., & Müller, H.S.P. 2002, *J. Mol. Spect.* 211, 211

Martin, J.M.L. & Taylor, P.R. 1996, *J. Phys. Chem.* 100, 6047

Michael, E.A., Lewen, F., Winnewisser, G., Ozeki, H., Habara, H., & Herbst, E. 2003, *Ap. J.* 596, 1356

Millar, T.J. & Hatchell, J. 1998, *Faraday Discuss.* 109, 15

Ozeki, H. & Saito, S. 1995, *Ap. J.* 451, L97

Pickett, H.M., Pearson, J.C., & Miller, C.E. 2005, *J. Mol. Spect.* 233, 174

Polehampton, E., Menten, K., Brünken, S., Winnewisser, G., & Baluteau, J.-P. 2005, *A&A* 431, 203

Scheuerer, R., Haeussler, M., Renk, K.F., Schomburg, E., Koschurinov, Y.I., Pavelev, D.G., Maleev, N., Ustinov, V., & Zhukov, A. 2003, *Appl. Phys. Lett.* 82(17), 2826

Schmuttenmaer, C.A., Cohen, R.C., Pugliano, N., Heath, J.R., Cooksy, A.L., Busarow, K.L., & Saykally, R.J. 1990, *Science* 249, 897

Sears, T.J., Bunker, P.R., & McKellar, A.R.W. 1981, *J. Chem. Phys.* 75, 4731

Sears, T.J., Bunker, P.R., & McKellar, A.R.W., 1982, *J. Chem. Phys.* 77, 5363

Sears, T.J., Bunker, P.R., McKellar, A.R.W., Evenson, K.M., Jennings, D.A., & Brown, J.M. 1982a, *J. Chem. Phys.* 77, 5348

Thaddeus, P., Gottlieb, C.A., Hjalmarson, A., Johansson, L.E.B., Irvine, W.M., Friberg, P., & Linke, R.A. 1985, *Ap. J.* 294, L49

Turner, B.E., Herbst, E., & Terzieva, R. 2000, *Ap. J. Suppl.* 126, 427

Van Orden, A., Cruzan, J.D., Provencal, R.A., Giesen, T.F., Saykally, R.J., Boreiko, R.T. Boreiko, & Betz, A.L. 1994, *Proceedings of the 1994 Kuiper Airborne Astronomy Symposium, NASA-Ames Research Center*

Yamamoto, S., Saito, S., Suzuki, H., Deguchi, S., Kaifu, N., Ishikawa, S., & Ohishi, M. 1990, *Ap. J.* 348, 363

Winnewisser, G., Krupnov, A.F., Tretyakov, M.Y., Liedtke, M., Lewen, F., Saleck, A.H., Schieder, R., Shkaev, A.P., & Volokhov, S.V. 1994, *J. Mol. Spec.* 165, 294

Discussion

GUÉLIN: Did you study the bending modes of C_4H in the same way as you studied those of C_3H? Rotational transitions within vibrationally excited states of C_4H are observered in IRC+10216 and the lines are surprisingly strong. It would be interesting to learn how difficult it is to excite these states.

GIESEN: C_3H was the first molecule we studied with the new SuJeSTA instrument. C_4H indeed would be a very interesting molecule to study in the lowest bending mode, although the predicted frequency of 226 cm^{-1} or 7 THz is too high for our spectrometers.

HERBST: Can you help identify ethyl methyl ether in space by its internal rotation splittings?

GIESEN: In W51 e2 and Orion KL the internal dynamics of the source cause a line broadening that is larger than the internal rotation splittings of ethyl methyl ether. But in more quiescent sources, or at higher spatial resolution, the splittings would be partially resolved and could be used as characteristic patterns to identify ethyl methyl ether lines.

Astrochemistry: Recent Successes and Current Challenges
Proceedings IAU Symposium No. 231, 2005
D.C. Lis, G.A. Blake & E. Herbst, eds.

© 2006 International Astronomical Union
doi:10.1017/S1743921306007083

Experimental Investigation of Neutral-Neutral Reactions and Energy Transfer at Low Temperatures

Ian R. Sims

Laboratoire PALMS – UMR CNRS-Université 6627,
Equipe 'Astrochimie Expérimentale', Campus de Beaulieu, Bât. 11C,
Université de Rennes 1, 35042 RENNES Cedex, France

Abstract. Over the last few years, we have applied the CRESU (Reaction Kinetics in Uniform Supersonic Flow) technique to the study of neutral-neutral reactions and energy transfer processes in the gas phase. This has enabled the rates of a wide range of reactions between electrically neutral species to be measured down to temperatures as low as $\sim$10 K. These results have generated significant interest amongst theoretical chemists, and especially amongst astrochemists. Our measurements of low-temperature rate coefficients have had a significant impact on the models used to simulate the chemistry of dense interstellar clouds. In this article the motivations for the experimental study of reaction kinetics at low temperatures are described, and a detailed description of the CRESU technique is given, followed by a brief overview of the results obtained in Rennes and Birmingham. Four recent low-temperature kinetic studies are then described: the reaction O + OH and the interstellar oxygen problem, reactions of the C_2 radical, energy transfer in collisions of $C(^3P_J)$ with He and H_2, and preliminary results on the reaction kinetics of the C_4H radical.

Keywords. astrochemistry — ISM: molecules — methods: experimental — molecular processes

---

1. Neutral-Neutral Reactions and Energy Transfer at Low Temperatures

The temperature dependence of rate coefficients $k(T)$ for many reactions between electrically neutral species are found to obey, to a good approximation, the Arrhenius expression:

$$k(T) = A \exp(-E_a/RT)$$

with E_a, the activation energy, which is closely related to the barrier to reaction along the minimum energy path. This expression, first proposed by Arrhenius in 1889, has enjoyed such widespread success that 'Arrhenius plots' (of $\ln k(T)$ against $1/T$) have become the standard way of reporting kinetic data as a function of temperature.

The validity of the Arrhenius equation, however, is based on the idea of activation energy, and relies on the presence of an energetic barrier to reaction. Rates of barrierless reactions are not expected to display Arrhenius behaviour, and are good candidates for study at low temperatures. These include: ion-molecule reactions, radical-radical reactions, and even many radical-molecule reactions.

An example of such a barrierless reaction is that between the CN radical and the ethyne molecule C_2H_2, which was found by us to display a rapid rate constant that increases with decrease in temperature (Sims *et al.* 1993). Subsequent quantum chemical calculations by Woon & Herbst (1997) have shown that the potential energy surface for this reaction has the form shown in Figure 1.

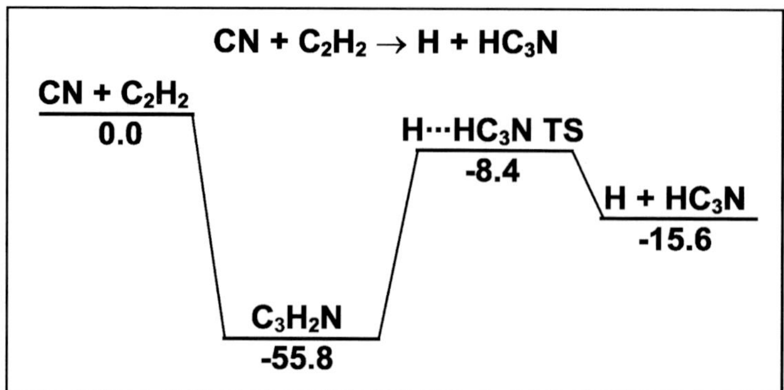

Figure 1. Minimum-energy potential pathway for the reaction of CN radicals with ethyne, as calculated by Woon & Herbst (1997). The computed relative energies (kcal mol^{-1}) listed for the stationary points (reactants, complex [energy minimum], transition state [TS; energy saddle point], and products) are given without zero-point energy corrections.

Prediction of rate coefficients at low temperatures for use in astrochemical models can be problematical. In the first place, extrapolation of rate data gathered at higher temperatures is not recommended as the form of temperature dependence may be complex. This is illustrated in Figure 2, which shows a sequence of Arrhenius plots for the reaction of CN radicals with ethane, C_2H_6.

While *ab initio* quantum chemical calculations have progressed rapidly over the past decade, they cannot yet predict arbitrary neutral-neutral rate constants over wide ranges of temperature with the necessary degree of accuracy.

It is therefore apparent that accurate measurements of neutral-neutral rate constants are necessary down to temperatures as low as 10 K prevailing in dense interstellar clouds (ISCs). Furthermore, low temperature measurements of rate coefficients for collisional energy transfer – rotational excitation and de-excitation – are also sorely needed. These constitute some of the principal motivations for the experimental work described in this brief review.

2. Techniques for Measuring Rate Coefficients at Low Temperatures: The CRESU Technique

The CRESU (Cinétique de Réaction en Ecoulement Supersonique Uniforme, or Reaction Kinetics in Uniform Supersonic Flow) apparatus was devised by Rowe and co-workers (Rowe *et al.* 1984; Rowe & Marquette 1987) initially for the study of ion-molecule reactions at extremely low temperatures. The strength of the technique lies in the nature of the isentropic expansion of gas through a Laval nozzle, which produces a flow of gas that is uniform in temperature, density and velocity, and which endures for several tens of centimetres and some hundreds of microseconds downstream of the nozzle exit. Frequent collisions occur during the controlled expansion within the nozzle and in the subsequent uniform region downstream where the gas density ($10^{16} - 10^{17}$ molecule cm^{-3}) is relatively high. Consequently, thermal equilibrium is maintained at all times, in contrast to the situation in free jet/molecular beam environments. The expansion is slow enough to maintain thermal equilibrium, but rapid enough that condensation is avoided (strongly supersaturated conditions prevail in the subsequent flow). A uniform, 'collimated' flow results at the exit of the nozzle.

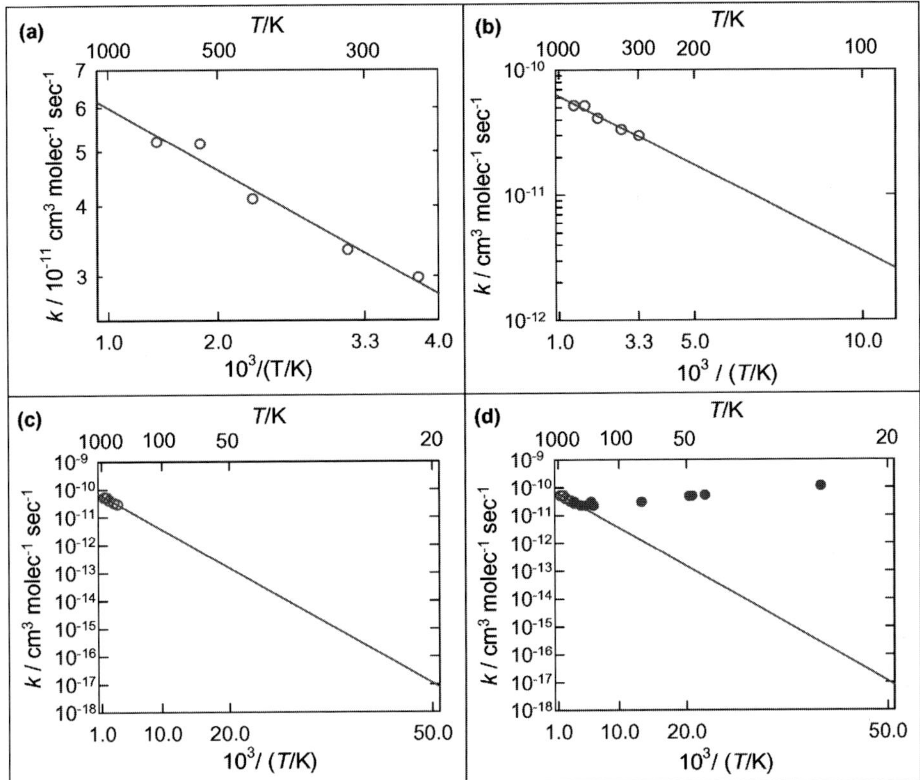

Figure 2. Rate constants measured for the reaction of CN radicals with ethane, C_2H_6, presented in the form of Arrhenius plots (Sims *et al.* 1993): (*a*) measurements made in heated or cooled cells (open circles) which fit well to an Arrhenius expression (solid line); (*b*) data extrapolated to 100 K (*c*) unjustifiably long extrapolation needed to reach 20 K, where the predicted rate constant is vanishingly small; (*d*) rate constants measured using supersonic flow techniques down to 25 K (solid circles), demonstrating the danger of extrapolating to low temperature the behaviour of rate constants at higher temperatures!

The CRESU technique provides an excellent environment in which to perform experiments on collisional processes at extremely low temperatures, not only between ionic and neutral species, but also between exclusively neutral species. This realisation, and the subsequent adaptation of the CRESU technique for the study of neutral-neutral reactions by the incorporation of the PLP-LIF technique, has enabled kinetic measurements on neutral-neutral reactions down to temperatures as low as 13 K (Bocherel *et al.* 1996; Sims *et al.* 1994).

A schematic diagram of the CRESU apparatus as used for the study of neutral-neutral reactions is shown in Figure 3. Photolysis and probe laser beams co-propagate through the throat of the Laval nozzle and out along the axis of the supersonic flow. The pulsed photolysis laser (or, sometimes, in the case of energy transfer studies, the excitation laser) creates a population of radicals by photodissociation of a precursor species introduced at low concentration into the gas flow. These radicals then collide with a co-reagent species also present in the flow, but in large excess with respect to the radical, whose concentration then decays according to (pseudo-)first-order kinetics, and is usually probed at set time delays by a second tunable laser which excites fluorescence (LIF) from the

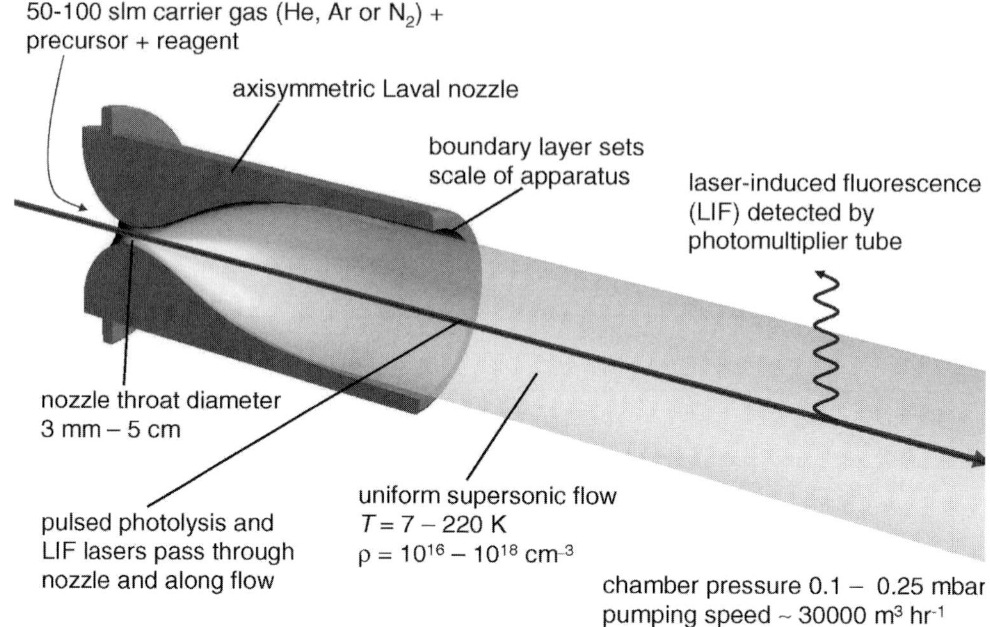

Figure 3. Schematic diagram of the CRESU apparatus depicting the Laval nozzle and supersonic flow, combined with pulsed laser photolysis – laser induced fluorescence (LIF) detection permitting the study of neutral-neutral reaction kinetics at extremely low temperatures.

radical species. The resultant exponential decay is fitted to yield a pseudo-first-order rate constant, and the experiment repeated at various co-reagent concentrations to yield a second-order plot, the gradient of which yields the rate constant of interest at the temperature of the supersonic flow. Full details of the technique may be found in Sims *et al.* (1994) and James *et al.* (1998). An example of the signal and results obtained for the reaction (vibrational relaxation) of CH $(v = 1)$ with CO may be seen in Figure 4.

The Rennes CRESU facilities currently comprise:

- One chamber dedicated to ion-molecule and electron attachment processes;
- Two chambers (Rennes/ex-Birmingham) dedicated to neutral chemistry (with room temperature or cooled reservoirs);
- One chamber with heated reservoir dedicated to very condensable species (PAHs); and
- One chamber dedicated to IR spectroscopy of aggregates.

3. A Brief Survey of Past Results

A surprising range of reactions studied in the CRESU apparatus have been found to remain rapid down to extremely low temperatures, including:

Radical-radical metathesis reactions, e.g. CN + O$_2$ (Sims *et al.* 1994), C(^{3}P) + O$_2$ (Chastaing *et al.* 2000);

Radical-unsaturated molecule reactions, e.g. CN + C$_2$H$_2$ (Sims *et al.* 1993), C$_2$H + C$_2$H$_2$ (Chastaing *et al.* 1998), C + C$_2$H$_2$, C$_2$H$_2$, C$_3$H$_4$ (Chastaing *et al.* 2001);

Radical-saturated molecule reactions, e.g. CH + CH$_4$ (Canosa *et al.* 1997), CN + C$_2$H$_6$ (Sims *et al.* 1993).

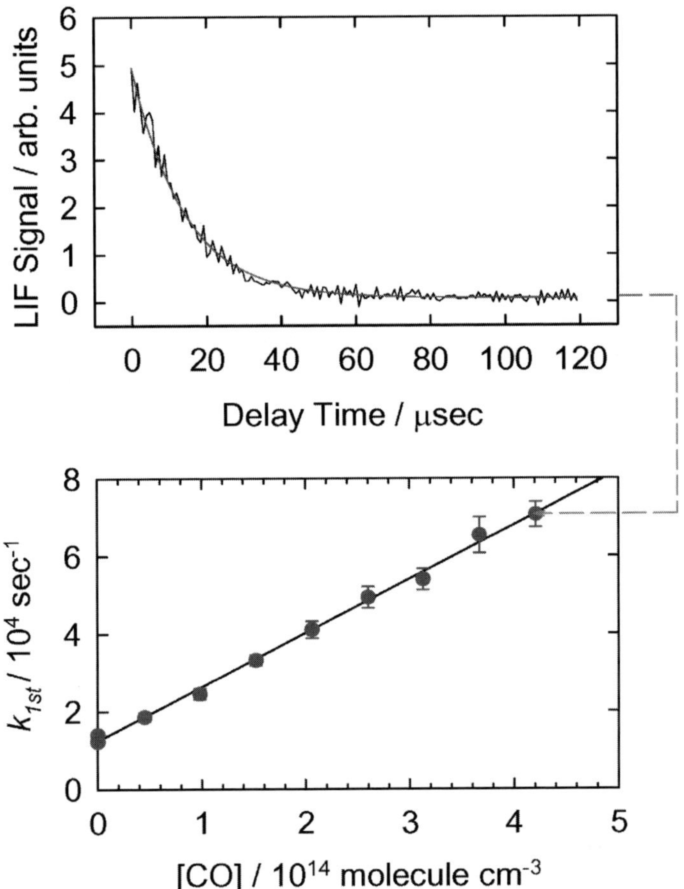

Figure 4. *Upper panel:* First-order decay of LIF signal from CH $(v = 1)$ in the presence of 4.2×10^{14} molecule cm^{-3} of CO at 44 K in Ar, fitted to a single exponential decay. *Lower panel:* First-order decay constants for CH $(v = 1)$ at 44 K in Ar plotted against the concentration of CO.

A number of reviews are available (Sims 1997; Sims & Smith 1995; Smith & Rowe 2000). These results have generated significant interest amongst theoretical chemists, and especially amongst scientists working in the area of astrochemistry. These measurements of low-temperature rate coefficients have had a significant impact on the models used to simulate the chemistry of dense ISCs, as has been recognised by leading astrochemical modellers such as Millar (Herbst *et al.* 1994) and Herbst (Herbst 2001; Terzieva & Herbst 1998).

4. Some Recent Studies

4.1. *The O + OH Reaction*

The reaction of ground state oxygen atoms with ground state OH radicals is of major interest from theoretical, experimental and astrochemical points of view. It is thought to be the main source of interstellar dioxygen:

$$O(^3P) + OH(X^2\Pi) \rightarrow O_2(^3\Sigma) + H(^1S)$$

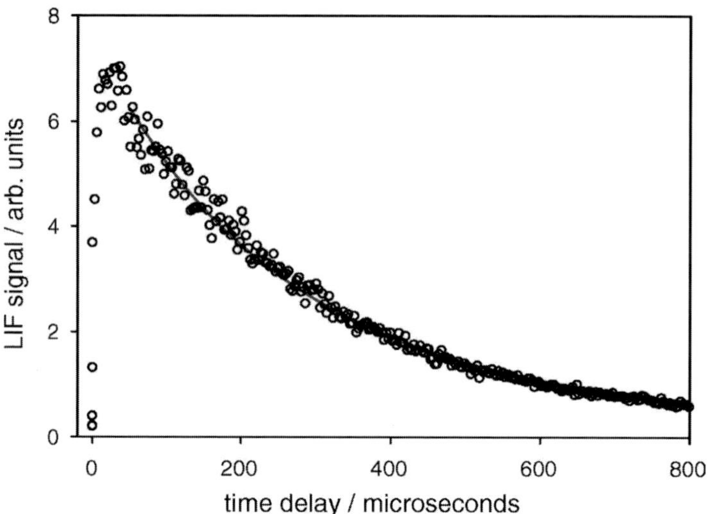

Figure 5. Decay of OH LIF signal in the presence of O(^{3}P) atoms at 83 K, fit to a single exponential decay.

However, successive attempts to detect interstellar O_2 have pushed the upper limit for O_2 abundance well below model predictions. A number of explanations have been advanced for this discrepancy, but clearly one important factor may be that the rate constant for the O + OH reaction currently employed at low ($T < 50$ K) temperatures is overestimated, as it is derived by extrapolation of measurements made at 160 K and above.

The O + OH reaction has long been a challenging system both from a theoretical and an experimental point of view. The number of possible potential energy surfaces in this reaction between two open shell species renders theoretical calculations especially difficult. The usual techniques for the experimental measurement of gas-phase bimolecular reactions rely on the ability to achieve pseudo-first order conditions, requiring one of the reagents (usually the stable one) to be placed in a known excess of concentration, while the concentration of the other (usually radical) reagent is monitored in a relative but time resolved fashion by, e.g., laser induced fluorescence. For a reaction between two unstable species this obviously presents significant difficulties: the need to produce a significant concentration of an unstable species; and the requirement to measure or otherwise know this concentration. For reactions between atomic and molecular radicals, such as O + OH, this has been achieved using a microwave discharge to produce the atomic radical in excess, and gas-phase titration to measure its concentration, coupled with, for example, pulsed photolysis and LIF on the molecular radical to measure the kinetics. For this reaction, such techniques have permitted the measurement of the rate constant down to temperatures as low as 160 K using cryogenic cooling (Smith & Stewart 1994). However, adaptation of these techniques for use in the CRESU apparatus presents a number of major problems: firstly, as the mass flow rates are much higher than in a typical flow tube, higher power discharges are needed; secondly, the gas pressures before and after the Laval nozzle expansion are fixed by aerodynamical considerations, resulting in conditions which are non-ideal for the production and maintenance of high concentrations of atomic radicals; and finally, the usual gas titration techniques are not possible, or at least difficult in the supersonic flow. Solutions to these problems are currently being sought in Rennes.

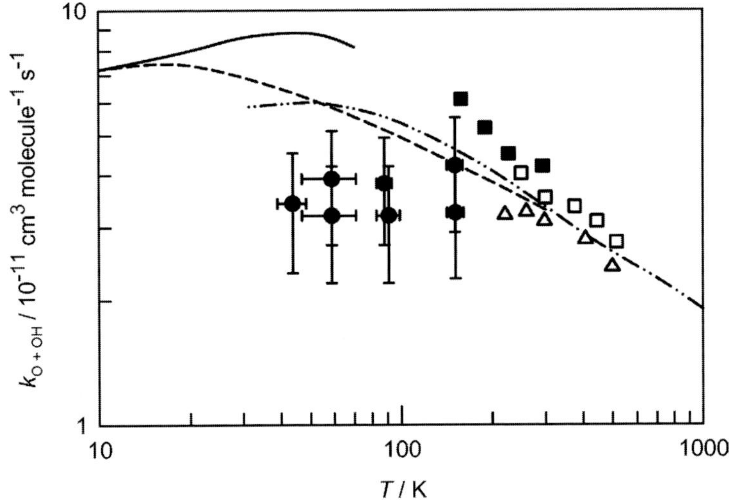

Figure 6. Comparison of temperature-dependent experimental and theoretical data for the reaction between OH radicals and O(^{3}P) atoms. The present corrected values of k are shown as closed filled circles. The corrected values of the rate constants are plotted at a temperature at the midpoint of the temperature range resulting from heating of the supersonic flow by laser absorption, with the horizontal lines through the points indicating the range of estimated temperatures. The vertical lines through the corrected data represent the range of the estimated uncertainties. The data of Howard & Smith (1980, 1981) and Smith & Stewart (1994) are represented by open and filled squares, respectively, and those of Lewis & Watson (1980) by open triangles. The lines represent theoretical results: dashed – the recommended rate expression of Davidsson & Stenholm (1990); solid – the quantum SACM calculations of Troe & Ushakov (2001); dashed-dotted – the classical trajectory calculations of Troe & Ushakov (2001).

In a significant breakthrough in Birmingham, a new co-photolysis method has been employed enabling the measurement of this rate constant down to around 40 K. We have employed 157 nm radiation from an F_2 excimer laser to produce oxygen atoms from the in-situ photolysis of O_2 within the cold supersonic flow, while at the same time generating a much smaller concentration of OH radicals from the photolysis of H_2O. The oxygen atoms are initially formed in both the ground (^{3}P) and excited (^{1}D) states, but the use of N_2 buffer gas ensures rapid relaxation to the ground state. The OH concentration was followed in the usual way by pulsed LIF detection. The oxygen atom concentration was determined by careful measurements of the photolysis laser fluence and O_2 concentrations used, coupled with a knowledge of the absorption cross section at 157 nm. An example of a typical decay trace obtained at 83 K is shown in Figure 5. The results obtained (after correction for thermal expansion induced by the photolysis laser) are shown in Figure 6. These results have been accepted for publication in a special issue of the Journal of Physical Chemistry A in honour of Jürgen Troe (Carty *et al.* 2005).

4.2. *Reaction Kinetics of the C_2 Radical*

The C_2 molecule is found throughout the Universe. It has been observed in widely differing environments ranging from cold regions such as interstellar clouds (Cecchi-Pestellini & Dalgarno 2002), circumstellar envelopes (Bakker *et al.* 1997) and comets (A'Hearn *et al.* 1995) to hot media such as hydrocarbon flames (Baronavski & McDonald 1977), plasmas (Rennick *et al.* 2004) or stellar atmospheres (Brault *et al.* 1982). Understanding its chemistry and more specifically the temperature dependence of rate constants for reactions with other reagents is therefore of major interest. To date, most of the available

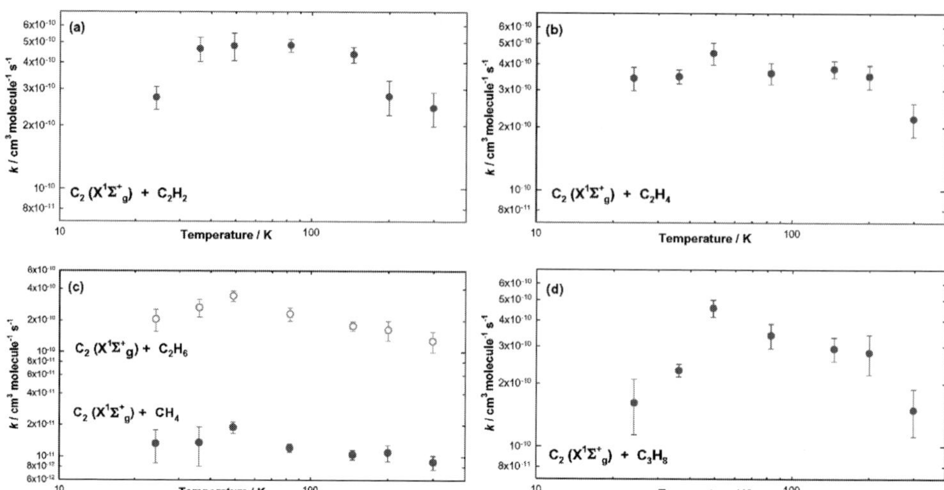

Figure 7. Rate constants for reaction of $^1C_2$ with various hydrocarbons displayed on a log-log scale as a function of temperature: (a) $^1C_2 + C_2H_2$; (b) $^1C_2 + C_2H_4$; (c) open circles, $^1C_2 + C_2H_6$; closed circles, $^1C_2 + CH_4$; (d) $^1C_2 + C_3H_8$.

experimental data, however, have been obtained at room temperature and essentially relate to the reactivity of C_2 with hydrocarbons. While studies have also been carried out at higher temperatures, no measurements have been performed below 300 K in the gas phase.

The reaction kinetics of ground state $X^1\Sigma_g^+$ C_2 (hereafter $^1C_2$) are somewhat complicated by the presence of a low-lying triplet state $a^3\Pi_u$, (hereafter $^3C_2$) at only 610 cm^{-1} (*i.e.*, $\sim$880 K) above the ground state. In our experiments, both $^1C_2$ and $^3C_2$ are thought to be formed by the 193 nm laser photolysis of C_2Cl_4, and we have exploited this fact to study down to temperatures as low as 24 K the kinetics of both species with O_2 and NO (to appear in a special issue of the Journal of Physical Chemistry A in honour of Jürgen Troe; Páramo *et al.* 2005a) as well as the hydrocarbons CH_4, C_2H_2, C_2H_4, C_2H_6 and C_3H_8 (Páramo *et al.* 2005b). $^1C_2$ and $^3C_2$ were detected via LIF in the Mulliken $(D^1\Sigma_u^+ \leftarrow X^1\Sigma_g^+)$ and Swan $(d^3\Pi_g \leftarrow a^3\Pi_u)$ bands respectively. Further details are to be found in Páramo *et al.* (2005a,b).

$^1C_2$ was found not to react with O_2, while $^3C_2$ was rapidly quenched by O_2, a fact which was exploited in subsequent measurements where O_2 was added to the supersonic flow in order to quench (or equilibrate with $^1C_2$) the initially formed $^3C_2$. A selection of results for the reactions of $^1C_2$ with hydrocarbons is shown in Figure 7.

As can be seen from Figure 7, $^1C_2$ is an extremely reactive species not only at room temperature, but also down to extremely low temperatures. In the case of the reaction $^1C_2 + C_2H_4$, the rate constant is found to be rapid and virtually temperature independent to 24 K. Balucani *et al.* (2001) in a crossed molecular beam study found C_4H_3 + H products at 15 and 29 kJ mol^{-1} collision energy (1200 and 2300 K, respectively). Their *ab initio* calculations predict a barrierless potential energy surface, in good agreement with the high value of the rate constant at low temperatures. For $^1C_2 + C_2H_2$, similar behavior is observed, and Kaiser *et al.* (2003) found C_4H + H products at 24.1 kJ mol^{-1} collision energy (2000 K). Remarkably, $^1C_2$ also displays rapid rate constants with the saturated hydrocarbons CH_4, C_2H_6 and C_3H_8. While the rate constant for reaction of $^1C_2$ with methane is an order of magnitude less than for the other hydrocarbons, it appears to show little or no temperature dependence down to the lowest temperatures, thus indicating a significant role for the $^1C_2$ species in methane-rich environments such as the atmosphere of Titan.

4.3. *Collisional Energy Transfer between Spin Orbit States of Atomic Carbon*

$C(^3P_J)$ is relatively abundant in interstellar space, and its intramultiplet transitions play an important role in cooling of interstellar environments. Energy transfer between these fine structure states is therefore of interest, particularly via collisions with He and H_2, two of the most abundant interstellar collision partners.

$C(^3P_J)$ was generated by the 193 nm photolysis of C_3O_2, and detected by resonant vacuum-ultraviolet laser-induced fluorescence (VUV-LIF); that is, with both excitation and observation of the fluorescence on the $(2s^22p3s\ ^3P_J - 2s^22p^2\ ^3P_J)$ transition in atomic carbon. Pulses of VUV laser radiation were generated using four-wave mixing in Xe: dye laser radiation was tuned to 255.94 nm in order to excite Xe in a two-photon transition to its $5p^56p[2\frac{1}{2}, 2]$ state and a second pulsed dye laser tuned through the wavelength range between 560 and 565 nm to generate VUV radiation between 165.9 and 165.4 nm.

Following previous kinetic measurements made using direct VUV-LIF probing of $C(^3P_J)$ in the CRESU, we were able to take advantage of the observation that the atomic carbon formed in the photolysis of C_3O_2 (and CBr_4 for subsequent experiments) had a warm nascent distribution over spin orbit states to make measurements of the relaxation of each of these states in collisions with He at 15 K. Fitting these relaxation data provided unique data on state-to-state rate coefficients for energy transfer between spin orbit states in $C(^3P_J)$–He collisions at 15 K (Le Picard *et al.* 2002).

While data on collisional energy transfer with He is of both fundamental and astrochemical interest, it is evident that the greatest astrochemical interest is in H_2 as collision partner, as this is by far the majority species (by a factor of 10^4 over CO) in dense ISCs. However, producing a pure H_2 flow in the CRESU presents considerable difficulties. The combination of a large rotational energy gap spacing in molecular hydrogen, coupled with the restriction to $\Delta J = \pm 2$ arising from nuclear spin statistics, results in very slow rates of H_2–H_2 rotational energy transfer (RET). Normally in the CRESU expansion, RET occurs at a rate which maintains thermal equilibration during the passage through the Laval nozzle. In H_2 this would not be the case. This would furthermore result in a ratio of specific heat capacities ($\gamma = C_p/C_v$, a key quantity in aerodynamics) which would vary during the expansion, destroying the uniform flow.

In a major breakthrough, a working H_2 Laval nozzle expansion was achieved by precooling the reservoir and nozzle with liquid nitrogen, allowing the H_2 to relax into the two lowest rotational states before the expansion, at which point the gas behaves aerodynamically as a monatomic gas with a value of γ of 5/3, no further rotational relaxation being possible. A temperature profile of this flow obtained by impact pressure measurements is shown in Figure 8. Using this nozzle, we have been able to make preliminary measurements on spin orbit energy transfer in $C(^3P_J)$–H_2 collisions at 7 K, as well as (using He as carrier gas) extending our earlier $C(^3P_J)$–He measurements down to 7 K. The results for $C(^3P_J)$–H_2 collisions at 7 K are displayed in Figure 9, along with a comparison to the calculations of Schröder *et al.* (1991).

While agreement between experiment and theory is excellent for the rate constant for transfer from $J = 1$ to $J = 0$, the other rate constants measured are in strong disagreement with the theoretical results. A similar pattern is observed for $C(^3P_J)$–He collisions, and these systems are the subject of further detailed experimental investigation to confirm or otherwise this potentially very significant disagreement with theory at low temperatures.

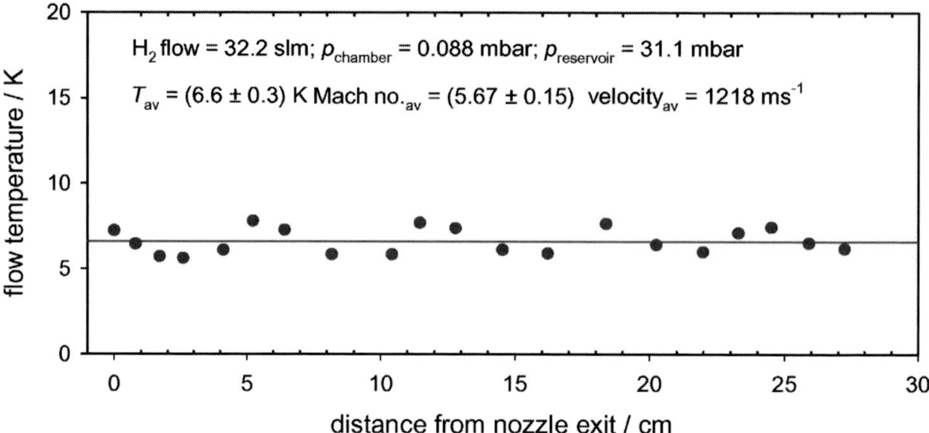

Figure 8. Temperature profile of new H_2 Laval nozzle flow.

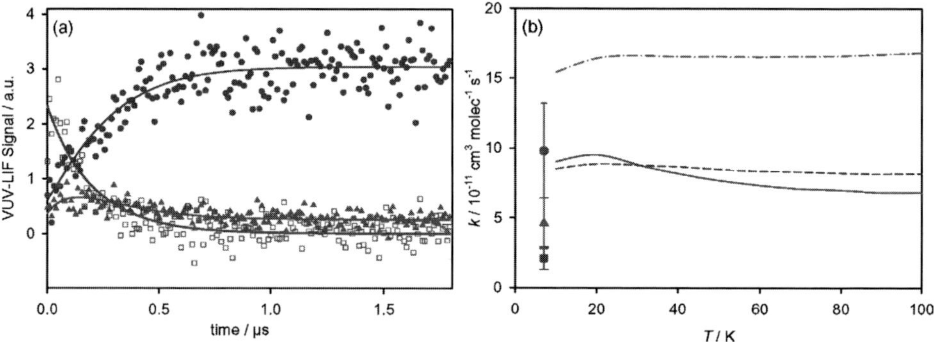

Figure 9. (*a*) VUV-LIF signal from $C(^3P_J)$ as a function of time delay after photolysis in a 7 K H_2 buffer. The closed circles show relaxation of the $J = 0$ state, the filled triangles relaxation of the $J = 1$ state, and the open squares relaxation of the $J = 2$ state. The solid lines indicate a simultaneous fit to these data, yielding the experimental points shown to the right; (*b*) experimental intramultiplet state-to-state rate constants for $C(^3P_J)$ in collision with normal H_2 at 7 K (solid symbols), along with the calculations (lines) of Schröder *et al.* (1991). The circle and solid line refer to transfer from $J = 1$ to $J = 0$, the square and dashed line to transfer from $J = 2$ to $J = 0$, and the triangle and dashed-dotted line to transfer from $J = 2$ to $J = 1$.

4.4. Reaction Kinetics of the C_4H Radical

Work is currently underway to elucidate the reaction kinetics of the C_4H radical, which has been detected in both interstellar clouds (Olano *et al.* 1988) and circumstellar envelopes (Dayal & Bieging 1993), and is implicated in the synthesis of complex hydrocarbons in the chemistry of planets such as Titan (Zwier & Allen 1996). To our knowledge there have been no kinetic studies to date of C_4H reactions at any temperature. In our experiments, C_4H is formed by 193 nm excimer laser photolysis of diacetylene, C_4H_2, and detected by laser induced fluorescence around 415 nm (Hoshina *et al.* 1998). Preliminary results have been obtained at 52 K for reaction of C_4H with C_2H_2, C_2H_6, C_3H_8 which all show fast rate constants in the range 1.54×10^{-10} cm³ molecule⁻¹ s⁻¹. The reaction of C_4H with CH_4 at 52 K is too slow for us to measure ($k < 10^{12}$ cm³ molecule⁻¹ s⁻¹). We propose that likely products of the reactions of the C_4H radical with unsaturated hydrocarbons are long carbon chains, e.g.,

$$C_4H + C_2H_2 \rightarrow HCCCCCCH + H$$

Further measurements are underway at different temperatures and with a range of co-reagents.

Acknowledgements

I would like to express my thanks to all my co-workers, in particular Ian W.M. Smith and Bertrand Rowe, who have been strongly involved in the development and exploitation of the CRESU technique. For the recent work described in this review I am grateful to my co-workers Coralie Berteloite, André Canosa, David Carty, Andrew Goddard, Sven Köhler, Alejandra Páramo, Sébastien Le Picard, Daniel Travers and Ian W.M. Smith. I thank the numerous agencies that have funded the work presented here including the UK EPSRC, the French national programmes PCMI and PNP, the European Commission, the Region of Brittany and Rennes Metropole. I am also very grateful to the European Commission for the award of a Marie Curie Chair for part of the period during which this work was performed under contract MEXC-CT-2004-006734.

References

A'Hearn, M.F., Millis, R.L., Schleicher, D.G., Osip, D.J., & Birch, P.V. 1995, *Icarus* 118, 223

Bakker, E.J., van Dishoeck, E.F., Waters, L., & Schoenmaker, T. 1997, *A&A* 323, 469

Balucani, N., Mebel, A.M., Lee, Y.T., & Kaiser, R.I. 2001, *J. Phys. Chem. A* 105, 9813

Baronavski, A.P. & McDonald, J.R. 1977, *J. Chem. Phys.* 66, 3300

Bocherel, P., Herbert, L.B., Rowe, B.R., Sims, I.R., Smith, I.W.M., & Travers, D. 1996, *J. Phys. Chem.* 100, 3063

Brault, J.W., Delbouille, L., Grevesse, N., Roland, G., Sauval, A.J., & Testerman, L. 1982, *A&A* 108, 201

Canosa, A., Sims, I.R., Travers, D., Smith, I.W.M., & Rowe, B.R. 1997, *A&A* 323, 644

Carty, D., Goddard, A., Khler, S., Sims, I.R., & Smith, I.W.M. 2005, *J. Phys. Chem. A*, in press

Cecchi-Pestellini, C. & Dalgarno, A. 2002, *MNRAS* 331, L31

Chastaing, D., James, P.L., Sims, I.R., & Smith, I.W.M. 1998, *Faraday Discuss.* 165

Chastaing, D., Le Picard, S.D., & Sims, I.R. 2000, *J. Chem. Phys.* 112, 8466

Chastaing, D., Le Picard, S.D., Sims, I.R., & Smith, I.W.M. 2001, *A&A* 365, 241

Davidsson, J. & Stenholm, L.G. 1990, *A&A* 230, 504.

Dayal, A. & Bieging, J.H. 1993, Astrophys. J. 407, L37

Herbst, E. 2001, *Chem. Soc. Rev.* 30, 168

Herbst, E., Lee, H.H., Howe, D.A., & Millar, T.J. 1994, *MNRAS* 268, 335

Hoshina, K., Kohguchi, H., Ohshima, Y., & Endo, Y. 1998, *J. Chem. Phys.* 108, 3465

Howard, M.J. & Smith, I.W.M. 1980, *Chem. Phys. Lett.* 69, 40

Howard, M.J. & Smith, I.W.M. 1981, *J. Chem. Soc.—Faraday Trans. II* 77, 997

James, P.L., Sims, I.R., Smith, I.W.M., Alexander, M.H., & Yang, M.B. 1998, *J. Chem. Phys.* 109, 3882

Kaiser, R.I., Balucani, N., Charkin, D.O., & Mebel, A.M. 2003, *Chem. Phys. Lett.* 382, 112

Le Picard, S.D., Honvault, P., Bussery-Honvault, B., Canosa, A., Laube, S., Launay, J.M., Rowe, B., Chastaing, D., & Sims, I.R. 2002, *J. Chem. Phys.* 117, 10109

Lewis, R.S. & Watson, R.T. 1980, J. Phys. Chem. 84, 3495

Olano, C.A., Walmsley, C.M., & Wilson, T.L. 1988, *A&A* 196, 194

Páramo, A., Canosa, A., Le Picard, S.D., & Sims, I.R. 2005a, *J. Phys. Chem. A*, in press

Páramo, A., Canosa, A., Le Picard, S.D., & Sims, I.R. 2005b, in preparation

Rennick, C.J., Smith, J.A., Ashfold, M.N.R., & Orr-Ewing, A.J. 2004, *Chem. Phys. Lett.* 383, 518

Rowe, B.R., Dupeyrat, G., Marquette, J.B., & Gaucherel, P. 1984, *J. Chem. Phys.* 80, 4915

Rowe, B.R. & Marquette, J.B. 1987, *Int. J. Mass Spectrom. Ion Proc.* 80, 239

Schröder, K., Staemmler, V., Smith, M.D., Flower, D.R., & Jaquet, R. 1991, *J. Phys. B—At. Mol. Opt. Phys.* 24, 2487

Sims, I.R. 1997, *Research in Chemical Kinetics*, eds. R.G. Compton, & G. Hancock (Oxford: Blackwell Science), vol. 4, pp. 121

Sims, I.R., Queffelec, J.L., Defrance, A., Rebrionrowe, C., Travers, D., Bocherel, P., Rowe, B.R., & Smith, I.W.M. 1994, *J. Chem. Phys.* 100, 4229

Sims, I.R., Queffelec, J.L., Travers, D., Rowe, B.R., Herbert, L.B., Karthauser, J., & Smith, I.W.M. 1993, *Chem. Phys. Lett.* 211, 461

Sims, I.R. & Smith, I.W.M. 1995, *Annu. Rev. Phys. Chem.* 46, 109

Smith, I.W.M. & Rowe, B.R. 2000, *Acc. Chem. Res.* 33, 261

Smith, I.W.M. & Stewart, D.W.A. 1994, *J. Chem. Soc. Faraday Trans.* 90, 3221

Terzieva, R. & Herbst, E. 1998, *Ap. J.* 501, 207

Troe, J. & Ushakov, V.G. 2001, *J. Chem. Phys.* 115, 3621

Woon, D.E. & Herbst, E. 1997, *Ap. J.* 477, 204

Zwier, T.S. & Allen, M. 1996, *Icarus* 123, 578

Discussion

DALGARNO: Comment: I have calculations that would enable you to determine whether or not your measurements are correct (Krems *et al.* 2002; *Phys. Rev. A* 66, 030702).

SIMS: We would be pleased to check your calculations against our experimental data. However, as the paper referred to only contains calculations for $C(^3P_J) + {}^3He$ at temperatures below 1K, further calculations would be necessary to enable the comparison, and these we would welcome.

BLACK: In the old combustion literature there was a hint that $C_2 + H_2 \rightarrow C_2H + H$ goes at a rate coefficient sufficient to be astrophysically interesting. Can this be measured in your experiment?

CANOSA: The $C_2 + H_2$ reaction was studied by Pitts et al. (1982; *Chem. Phys.* 68, 417) within 300 K and 500 K. They found a significant rate at high temperatures (7.9×10^{-12} cm^3 s^{-1} molecule^{-1}), but the rate falls to 1.38×10^{-12} cm^3 s^{-1} molecule^{-1} at room temperature. Such a behavior is not favorable to get a measurable rate coefficient at lower temperatures using our technique. Therefore we did not consider this system.

Astrochemistry: Recent Successes and Current Challenges
Proceedings IAU Symposium No. 231, 2005
D.C. Lis, G.A. Blake & E. Herbst, eds.

© 2006 International Astronomical Union
doi:10.1017/S1743921306007095

Collisional Excitation Rates in the ISM

M. L. Dubernet[1], P. Valiron[3], F. Daniel[1], A. Grosjean[2],
F. Lique[1], N. Feautrier[1], A. Spielfiedel[1], A. Faure[3],
M. Wernli[3], L. Wiesenfeld[3], and C. Rist[3]

[1]LERMA, UMR CNRS 8112, Paris Observatory, 5 Place Janssen, 92195 Meudon, France
email: marie-lise.dubernet@obspm.fr

[2]LAOB, Observatoire de Besançon, UMR CNRS 6091, Université de Franche-Comté,
41 bis avenue de l'Observatoire, BP 1615, 25010 Besançon Cedex, France

[3]Laboratoire d'Astrophysique de Grenoble, UMR 5571 CNRS-UJF, OSUG,
Université Joseph Fourier, BP 53, F-38041 Grenoble Cedex 9, France

Abstract. The paper focuses on collisional excitation rates of molecules by He and H_2 relevant
to the interstellar medium. It discusses currently available data, presents very recent work and
outlines new work being carried out by various teams.

Keywords. ISM: molecules — molecular data — molecular processes

1. Introduction

Over the next few years, ground-based and space-based missions will open up the
universe to high spatial and spectral resolution studies at infrared and submillimeter
wavelengths. This will allow us to study, in much greater detail, the composition and the
origin and evolution of molecules in space. These new missions can be expected to lead
to the detection of many thousands of new spectral features. Identification, analysis and
interpretation of these features in terms of the physical and chemical characteristics of
the astronomical sources will require detailed astronomical tools supported by laboratory
measurements and theoretical studies of chemical reactions and of collisional excitation
rates on species of astrophysical relevance.

The present paper focuses on the latest theoretical studies on collisional excitation
rate coefficients for molecules colliding with H_2 and He; these processes are important as
they contribute to molecular excitation in competition with radiative processes. In cold
environnements such as the interstellar medium, collisions with H_2 are most important
because of its high abundance, and collisions with He are often considered as a model for
collisions with H_2.

All the published rate coefficients used in astrophysical applications can be found on
personal websites, such as those of S. Green (`http://www.giss.nasa.gov/data/mcrates/`)
and D. Flower (`http://ccp7.dur.ac.uk/`) and in recent compilations , such as the Leiden
Atomic and Molecular Database LAMDA (`http://www.strw.leidenuniv.nl/~moldata/`)
by Schröier *et al.* (2005) and the BASECOL database (`http://www.obspm.fr/basecol`)
by Dubernet *et al.* (2005). Those recent compilations have different approaches. The
LAMDA database contains ready-to-use rate coefficients of the rotational excitation of
molecules by H_2 with extrapolations of rate coefficients in temperature and in transition, and connections of those rate coefficients to spectroscopic data from the JPL and
CDMS catalogues. The BASECOL database contains all published rate coefficients and
gives a complete information on the origin of the data with all relevant references. All

the information given in the present paper on previous calculations is collected from the BASECOL database. The BASECOL database is a continuously evolving service that will offer standardized tools for manipulating fundamental rate coefficients and spectroscopic data and that will be included in the future theoretical services of the Virtual Observatory.

2. Methodology

Within the Born Oppenheimer approximation, the determination of these collisional rate coefficients requires two steps: the calculation of the interaction between nuclei and electrons for fixed nuclei, leading to so-called potential energy surfaces (PESs) and the scattering calculations. In the context of observations in the mm and submm range, the collisions concern regions with relatively low temperatures (5 K < T < 1000 K), there is no electronic excitation, nor any reactive channels, and the collisions generally involve a limited number of PESs describing a van der Waals system : a single PES for Σ states and two surfaces for Π states.

2.1. *Potential Energy Surfaces*

Recent progress in *ab-initio* quantum chemistry permits very accurate determinations of the PESs of small non-reactive systems. As the complexity of the calculations increases very steeply with the target accuracy, significant savings can be obtained using hierarchies of basis sets, correlation methods and meshes. By combining advanced treatments of both the electronic correlation and nuclear motion, a typical cm^{-1} accuracy can be achieved for intra-molecular PESs, as demonstrated by Polyansky *et al.* (2003) and Rajamaki *et al.* (2004) for vibrational investigations of H_2O and NH_3.

For inter-molecular PESs involving non-reactive molecules the coupled cluster theory with singles, doubles and perturbative triples, CCSD(T), is generally assumed to be very accurate. However, the effects resulting from theoretical levels beyond CCSD(T) are poorly known for such systems. From a comparison to experimental virial coefficients Jankowski & Szalewicz (2005) estimated they might amount to several cm^{-1} in the vicinity of the CO–H_2 minimum. Basis set superposition errors are sizeable even for the largest basis set sizes, and should be removed by counterpoise. Various techniques can be used to approximate the infinite basis set limit at the CCSD(T) level of theory. A popular approach is to associate bond functions to large augmented correlation basis sets in order to improve the description of the wavefunction between the interacting molecules. This approach may be complemented with basis set extrapolation, as in Jankowski & Szalewicz (2005) and references therein.

An alternative method is to describe properly the electron-electron correlation cusp that is responsible for the slow convergence of all conventional approaches with respect to the basis set size. Such explicitly correlated methods offer a direct way of reaching the basis set limit values within a single calculation, i.e. without extrapolation. In particular the so-called CCSD(T)-R12 approach is computationally practical and proven to be highly accurate, in particular using adequate R12-suited basis sets. Appropriate references can be found in Faure *et al.* (2005a). Such CCSD(T)-R12 calculations can also be used to monitor the accuracy of conventional PESs, as was done for He and H_2 interacting with CO, NH_3, HF and HC_3N in Wernli *et al.* (2005), Valiron (unpublished) and Wernli *et al.* (2006).

2.2. *Scattering Calculations*

Quantum close-coupling calculations are feasible in the low energy range and lead to nearly exact results. Calculations can be extended to higher energies using approximate quantum or semi classical approaches. The fast and popular Infinite Order Sudden (IOS) quantum approximation usually converges well whenever the relevant Δj energies are small in comparison to the available collisional energy. When the energy is augmented, this condition is fulfilled for the lower-j levels, but not for the highest-j levels. For vibrational quenching or excitation rates, the corresponding VCC-IOS approximation may also prove questionable for molecule-molecule pairs involving large rotational constants, as pointed out by Krems *et al.* (2001) and Faure *et al.* (2005b). In that respect the extension of present scattering calculations to the obtaining of inelastic rates for warm gas or shocked regions may be problematic, if accurate results are sought.

3. New Results

3.1. $H_2O + H_2$

The available excitation rate coefficients of H_2O by H_2 are those calculated by Phillips *et al.* (1995) and Phillips *et al.* (1996), using the close-coupling and the coupled states methods with the potential energy surface (PES) calculated by Phillips *et al.* (1994). Those authors provided data between 20 K and 140 K for a number of ortho/para H_2O – ortho/para H_2 pure rotational transitions. These calculations were extended to 5 K by Dubernet *et al.* (2002) and Grosjean *et al.* (2003), using pure close-coupling calculations over the same PES (Phillips *et al.* 1994). The results presented in Dubernet *et al.* (2002) were significantly different from those presented earlier by Phillips *et al.* (1996) at 20 K. In particular, the excitation rate coefficient for the 1_{01}–1_{10} transition of water was more than 50% larger than in Phillips *et al.* (1996). This was a consequence of an insufficient fine energy grid used by Phillips *et al.* (1996) in integrating the cross-sections over a Maxwellian distribution of kinetic energies.

A new 9D PES of the H_2–H_2O system has recently been calculated by Faure *et al.* (2005a) using a three-step procedure: (i) a 5D rigid-rotor PES reference was computed at the CCSD(T) level of theory; (ii) this reference surface was then calibrated to few cm^{-1} accuracy using 812 "high cost" CCSD(T)-R12 calculations; (iii) the R12-corrected rigid rotor surface was then extended to 9D using a new set of CCSD(T) calculations. The resulting 9D PES contains all relevant information to describe the interaction of all H_2O–H_2 isotopomers in zero-point or excited bending states. As a first application, high temperature ($1500 < T < 4000$ K) rate constants for the relaxation of the ν_2 bending mode of H_2O were estimated from quasiclassical trajectory calculations by Faure *et al.* (2005a) and the role of rotation in the vibrational relaxation of water is emphasized by Faure *et al.* (2005b). Their results are larger by an order of magnitude than those currently used in the astrophysical literature and might lead to a thorough reinterpretation of vibrationally excited water emission spectra from space.

A 5D PES was also obtained by averaging the 9D PES over the ground vibrational states of H_2O and H_2. Using this averaged 5D PES, new scattering calculations were carried out for the rotational excitation of $H_2O(000)$ by $H_2(v=0)$ using the same procedure as in Dubernet *et al.* (2002) and Grosjean *et al.* (2003) with a larger basis set as reported in Daniel *et al.* (2006). The rates involving ortho-H_2 are changed by 10 to 20% only, while low energy para-H_2 rates may differ by a factor of 2 or more.

These changes reflect the improvements in the PES, where surprisingly the largest contribution is attributed to the vibrational averaging. However Valiron *et al.* (2006a) showed that the vibrationally averaged PES can be accurately approximated by rigid rotor calculations using properly vibrationally averaged geometries for both interacting molecules, as already postulated by Mas & Szalewicz (1996) and Jeziorska *et al.* (2000). This simplification may facilitate accurate PES calculations whenever inelastic vibrational effects are not to be considered. An application is underway for H_2CO–H_2 by Valiron (2006b).

3.2. $N_2H^+ + He$

Up until now, the only available rate coefficients for N_2H^+ excited by He were pure rotational excitation rate coefficients among rotational levels $j < 6$ and for temperature up to 40 K. There were calculated by Green (1975), using an electron gas model for the potential energy surface. Using an adiabatic decoupling approximation a two-dimensional potential energy surface corrected for the influence of the ν_1 (mainly NH-local mode) stretching vibration was calculated by Meuwly & Bemish (1997) at the CCSD(T)/aug-cc-pVQZ level. Its reliability has been assessed by comparing energies of bound states and rotational constants with experimental data (Meuwly *et al.* 1996). The percentage differences between earlier rotational results by Green (1975) and new calculations by Daniel *et al.* (2005) are larger for transitions with large Δj and vary in the range from a few percent to 100%. In overall the new rates are larger for all transitions and the differences decrease with increasing temperature.

In astrophysical applications only the splittings due to couplings with the nuclear spins of the two nitrogen atoms are resolved and there are 9 hyperfine components for a given rotational level $j > 0$. The cross sections between hyperfine levels are obtained using a recoupling technique, first introduced by Corey & McCourt (1983) for the case of a single electronic spin and extended in Daniel *et al.* (2004) to the case of 2 nuclear spins. The dynamical problem reduces to spin-less equations, which is solved using close-coupling (CC) methods. It is found that the only well-defined propensity rule among hyperfine rate coefficients is $\Delta F = \Delta F_1 = \Delta j$ and that calculations are required in order to obtain the relative intensities of the 2 spin hyperfine rate coefficients at temperature below 50 K. Daniel *et al.* (2005) note that the usual simple approaches such as Infinite Order Sudden scaling and proportionality of the hyperfine rate coefficients to the final degeneracy of the hyperfine levels are inadequate. The authors provide Boltzmann averages of the so-called opacity factors; these allow calculation of pure rotational rate coefficients, rate coefficients among hyperfine levels as well as among magnetic sublevels of hyperfine levels for the 7 first rotational levels and for temperature less than 50 K.

These new results have been used in astrophysical applications (Daniel *et al.* 2005; Daniel *et al.* 2006a) and scattering calculations are currently extended to reach a temperature of 300 K. The $N_2H^+ + H_2$ system is currently under investigation by M. Meuwly and M.L. Dubernet.

3.3. $SO + He$

The previous available rate coefficients were provided for the SO–H_2 system by Green (1994) at temperatures ranging from 50 K to 350 K among the 70 lowest fine-structure energy levels corresponding to a maximal rotational level $N_{max} = 23$. The author used a CS–H_2 PES, which had been adapted (Green 1975) from an electron gas model (Gordon *et al.* 1972) of CS–He, treating the CS molecule as a rigid rotor. The long range part of the CS–He PES had been modified to account for the electrostatic interaction between CS and H_2. The rate coefficients among fine-structure levels were obtained with an IOS scaling relationship similar to the one used by Neufeld & Green (1994) for HCl + He.

New collisional excitation cross sections of the fine-structure levels of SO colliding with He have recently been calculated by Lique *et al.* (2005a) at low energies using a full close-coupling treatment and a new ab initio potential energy surface calculated at the CCSD(T) level with an aug-cc-pVQZ basis set complemented by bond functions. Rate coefficients were obtained for temperatures ranging from 5 K to 50 K. The results displayed the expected propensity rules $\Delta J = \Delta N$. The use of recoupling techniques from spin-free cross sections was investigated at low energy. This approximation is not valid for excitation between the first levels where the fine structure splitting is large compared to the rotational splitting, but it should provide a reasonable estimate of rates for high-J levels at large temperatures. Some preliminary tests by Lique *et al.* (2005b) show that the use of the new rate coefficients compared to the use of rate coefficients previously calculated by Green (1994) can significantly change the diagnostics of the SO abundance in dark clouds. Calculations for temperatures up to 300 K are under way.

3.4. $HC_3N + He/H_2$

The currently available rate coefficients are those of Green & Chapman (1978); they are calculated among the 21 lowest rotational levels for T = 10, 20, 40, 80 K using quasiclassical calculations combined with the Infinite Order scaling relationship. Their PES is computed with an electron gas model PES (Gordon *et al.* 1972), treating the HC_3N molecule as a linear rigid rotor, with bonds fixed at experimental values.

Two new PESs have been calculated at a CCSD(T) level for the HC_3N–He and HC_3N–H_2 systems by Wernli *et al.* (2006). The new HC_3N–He agrees pretty well with the PES obtained by Akin-Ojo *et al.* (2003). Steric hindrance problems involving the HC_3N rod limit the convergence of the angular expansion of the PES, as anticipated by Green & Chapman (1978). However for low energy collisions Wernli *et al.* (2006) showed it is feasible to regularize the PES by smoothing out the repulsive walls and to achieve a perfectly converged angular expansion. Corresponding close-coupling calculations led to surprising results due to the rod-like features of the PES. Firstly quantum interferences strongly defavour odd Δj transitions and favour even Δj ones. This propensity rule is likely to favour the $J = 1$ population for H_2 densities in the 10^3–10^4 cm^{-3} range. Secondly, despite the very large HC_3N dipole moment, the para-H_2 and ortho-H_2 rates are nearly identical. While the even Δj propensity rule could not be found in the quasiclassical calculations by Green & Chapman (1978), the new rates remain within the same order of magnitude despite the very crude electron gas model PES. This is not too surprising as the rod-like features of the PES dominate the scattering.

3.5. $CO + H_2$

Rotational rate coefficients among the lowest 29 levels of CO with para-H_2 ($j = 0$) and the lowest 20 levels of CO with ortho-H_2 ($j = 1$) were provided using the close-coupling method by Flower (2001) for temperatures ranging from 5 K to 400 K. Those calculations were of better quality and superseded the results of Mengel *et al.* (2001). It should be noted that both Flower and Mengel used the PES of Jankowski & Szalewicz (1998), but Mengel used a scaled version of the PES, while Flower used it without modifications. Quantum mechanical studies of this system had previously been undertaken by Green & Thaddeus (1976), who used a scaled CO–He interaction PES, and by Flower & Launay (1985) and Schinke *et al.* (1985), who used a different ab initio CO–H_2 PES.

The Jankowski & Szalewicz (1998) PES has recently been improved by Jankowski & Szalewicz (2005) and new close-coupling calculations were carried out for the rotational excitation of CO by H_2 (see Wernli *et al.* 2005). The authors confirmed the quality of the PES by independent CCSD(T)-R12 calculations, and found that a 10% difference in the PES brings about a 30% to 50% difference in the final rate coefficients at low temperature compared to the earlier results by Flower (2001). They also reported fair agreement above a temperature of 70 K.

3.6. $NH_3 + He$

Numerous studies have been carried out on the rotational excitation of NH_3. The available set of data concerns the rotational excitation of NH_3 by He, obtained by Green (1981) using coupled states calculations and a PES of Davis *et al.* (1979).

A new set has been calculated by Machin & Roueff (2005) using rigid-body close-coupling calculations (no inversion) with a PES by Hodges & Wheatley (2001). They report de-excitation rate coefficients for transitions up to $j = 4$ for para-NH_3, and up to $j = 7$ for ortho-NH_3 for temperatures between 5 K and 300 K. Large differences are obtained with the previous results of Green (1981), mainly attributed to the PES changes. Comparisons with experiments fare well when the rate coefficients are summed over the 2 final parity components. It is believed that the discrepancies with experimental results come from the neglect of the inversion motion.

Interaction energies were checked against CCSD(T)-R12 calculations by Valiron (unpublished) for several interacting geometries and intermolecular separations. The PES by Hodges & Wheatley (2001) is slightly too repulsive. Agreement is better along the attractive orientations, with a 2 cm^{-1} offset for the absolute minimum, and poorer for the less attractive orientations with a typical 5 cm^{-1} offset at 7 bohrs, to be compared to the $\sim$35 cm^{-1} absolute minimum. As the scattering is expected to be most sensitive to the minima and to the positioning of the repulsive walls this surface should provide a fair accuracy for astrophysical applications. PES and rate calculations involving the dominant collision partner H_2 are planned within the Molecular Universe European network.

3.7. $HF + He$

Various studies have been carried out on ro-vibrational excitation of HF by He. The latest study involves the calculation of a new PES by Stoecklin *et al.* (2003) at a CCSD(T) level with an aug-cc-pVQZ basis set complemented by bond functions. Interaction energies for both the primary and secondary minima were checked against CCSD(T)-R12 calculations by Valiron (unpublished) and agreed to within 0.3 cm^{-1}. This extremely accurate PES by Stoecklin *et al.* (2003) was subsequently used for close-coupling calculations of the rotational excitation of HF by Reese *et al.* (2005). The authors provide rotational de-excitation rate coefficients for levels up to $j = 9$ and for temperatures up to 300 K. It should be noted that Moszynski *et al.* (1994) and Moszynski *et al.* (1996) had provided a SAPT PES that agreed well with a semi-empirical PES of Lovejoy & Nesbitt (1990) obtained from near-infrared spectroscopy.

The predicted rates are small and lead to a high critical density, in agreement with the modeling of the pioneering HF$_{2-1}$ observations in absorption towards Sgr-B2 by Neufeld *et al.* (1997). However due to the large dipole moment of HF the collisional rates with ortho-H_2 might be substantially higher.

4. Conclusions

Accurate predictions of collisional excitation rates are now feasible for a large set of small-molecule interactions with He and H_2.

The accuracy of the potential energy surfaces (PES) can be monitored at a given level of theory using basis set extrapolation techniques or preferably by explicitly correlated approaches, and few cm^{-1} accuracy is now attainable for small non-reactive molecules.

Ideally, the PES should be averaged over the relevant vibrational states of both interacting molecules. However it is sufficient to perform rigid rotor calculations using vibrationally averaged geometries. Recent investigations using new PESs for CO, H_2O, NH_3, N_2H^+, SO and HC_3N permitted the estimation of the influence of PES errors on the final inelastic rates. Except at very low temperature or for the smallest rates the error amplification is moderate, generally much below an order of magnitude. Consequently accurate PESs with a few cm^{-1} accuracy should lead to 5–10% accuracy in the rates if appropriately converged and energy-averaged close-coupling calculations can be performed. However that level of PES accuracy is highly computer-time consuming. It requires an optimized choice of the grid geometries for the ab initio calculations and performance of an analytic fit of the PES leading to some additional inaccuracy.

Moreover the best level of accuracy is out of reach of current approximate approaches for scattering. Accordingly, the main source of uncertainties may come in the future from the scattering calculations instead of the PES, especially for higher energies or inelastic ro-vibrational rates for which brute force close-coupling calculations are not feasible, especially as mixing between electronic states may occur.

On the other hand, astronomers are welcome to express their needs in terms of accuracy, as much cheaper PESs would be sufficient to obtain moderate rate accuracies.

Acknowledgements

The authors thank E. Roueff, L. Machin, and T. Stoecklin for providing information on their results. New results presented here illustrate the French ongoing developments supported by the CNRS program "Physique et Chimie du Milieu Interstellaire" and by the "Molecular Universe" European network (2005–2008). Most calculations benefited from substantial time allocations on the French supercomputers IDRIS and CINES.

References

Akin-Ojo, O., Bukowski, R., & Szalewicz, K. 2003, *J. Chem. Phys.* 119, 8379

BASECOL database, http://www.obspm.fr/basecol

Corey, G. C. & McCourt, F. R. 1983, *J. Phys. Chem.* 87, 2723

Daniel, F., Dubernet, M.-L., & Meuwly, M. 2004, *J. Chem. Phys.* 121, 4540

Daniel, F., Dubernet, M.-L., Meuwly, M., Cernicharo, J., & Pagani, L. 2005, *MNRAS*, in press

Daniel, F., Cernicharo, J., & Dubernet, M.-L. 2005, *Semaine de l'Astrophysique Française*, meeting held in Strasbourg, France, Eds.: F. Casoli, T. Contini, J. M. Hameury and L. Pagani. EdP-Sciences, Conference Series, in press

Daniel, F., Dubernet, M.-L., Grosjean, A., Faure, A., Valiron, P., Wernli, M., Wiesenfeld, L., Rist, C., Noga, J., & Tennyson, J. 2006, *A&A* to be submitted

Daniel, F., Cernicharo, J., & Dubernet, M.-L. 2006, *Ap. J.* to be submitted

Davis, S. L., Boggs, J. E., & Mehrotra, S. C. 1979, *J. Chem. Phys.* 71, 1418

Dubernet, M.-L. & Grosjean, A. 2002, *A&A* 390, 793

Dubernet, M.-L., Grosjean, A., Flower, D., Roueff, E., Daniel, F., Moreau, N., & Debray, B. 2005 *J. Plasma Fusion Research Series*, volume 7

Faure, A., Valiron, P., Wernli, M., Wiesenfeld, L., Rist, C., Noga, J., & Tennyson, J. 2005, *J. Chem. Phys.*, 122, 221102

Faure, A., Wiesenfeld, L., Wernli, M., & Valiron, P. 2005b, *J. Chem. Phys.* 123, 104309

Flower, D. R. 2001, *J. of Phys. B* 34, 2731

Flower, D. R. & Launay, J.-M. 1985, *MNRAS* 214, 271

Gordon, R. G. & Kim, Y. S. 1972, *J. Chem. Phys.* 56, 3122

Green, S. 1975, *Ap. J.* 201, 366

Green, S. 1981, *NASA Technical Memorandum 83869*

Green, S. 1994, *Ap. J.* 434, 188

Green, S. & Chapman, S. 1978, *Ap. J. Suppl.* 37, 169

Green, S. & Thaddeus, P. 1976, *Ap. J.* 205, 766

Grosjean, A., Dubernet, M.-L., & Ceccarelli C. 2003, *A&A*, 408, 1197

Hodges, M. P. & Wheatley, R. J. 2001, *J. Chem. Phys.* 114, 8836

Jankowski, P. & Szalewicz, K. 1998, *J. Chem. Phys.* 108, 3554

Jankowski, P. & Szalewicz, K. 2005, *J. Chem. Phys.* 123, 104301

Jeziorska, M., Jankowski, P., Szalewicz, K., & Jeziorski, B. 2000, *J. Chem. Phys.* 113 2957

Krems, R. V., Marković, N., Buchachenko, A. A., & Nordholm, S. 2001, *J. Chem. Phys.* 114, 1249

Lique, F., Spielfiedel, A., Dubernet, M.-L., & Feautrier, N. 2005, *J. Chem. Phys.*, in press

Lique, F., Spielfiedel, A., Dubernet, M.-L., & Feautrier, N. 2005, *Semaine de l'Astrophysique Française*, meeting held in Strasbourg, France, Eds.: F. Casoli, T. Contini, J.M. Hameury and L. Pagani. EdP-Sciences, Conference Series, in press

Lovejoy, C. M. & Nesbitt, D. J. 1990, *J. Chem. Phys.* 93, 5387

Machin, L. & Roueff, E. 2005, *J. of Phys. B* 38, 1519

Mas, E. M. & Szalewicz, K. 1996, *J. Chem. Phys.* 104, 7606

Mengel, M., de Lucia, F. C., & Herbst, E. 2001, *Can. J. of Phys.* 79, 589

Meuwly, M. & Bemish, R. J. 1997, *J. Chem. Phys.* 106, 8672

Meuwly, M., Nizkorodov, S. A., Maier, J. P., & Bieske, E. J. 1996, *J. Chem. Phys.* 104, 3876

Moszynski, R., de Weerd, F., Groenenboom, G. C., & van der Avoird, A. 1996, *Chem. Phys. Lett.* 263, 107

Moszynski, R., Wormer, P. E. S., Jeziorski, B., & van der Avoird, A. 1994, *J. Chem. Phys.* 101, 2811

Neufeld, D. A. & Green, S. 1994, *Ap. J.* 432, 158

Neufeld, D. A., Zmuidzinas, J., Schilke, P., & Phillips, T. G. 1997, *A&A* 488, L141

Phillips, T. R., Maluendes, S., & Green, S. 1995, *J. Chem. Phys.* 102, 6024

Phillips, T. R., Maluendes, S., & Green, S. 1996, *Ap. J. Suppl.* 107, 467

Phillips, T. R., Maluendes, S., McLean, A. D., & Green, S. 1994, *J. Chem. Phys.* 101, 5824

Polyansky, O. L., Csaszar, A. G., Shirin, S. V., Zobov, N. F., Barletta, P., Tennyson, J., Schwenke, D. W., & Knowles, P. J. 2003, *Science* 299, 539

Rajamaki, T., Kallay, M., Noga, J., Valiron, P., & Halonen, L. 2004, *Mol. Phys.* 102, 2297

Reese, C., Stoecklin, T., Voronin, A., & Rayez, J. C. 2005, *A&A* 430, 1139

Schinke, R., Engel, V., Buck, U., Meyer, H., & Diercksen, G. H. F. 1985, *Ap. J.* 299, 939

Schöier, F. L., van der Tak, F. F. S., van Dishoeck, E. F., & Black, J. H. 2005, *A&A* 432, 369

Stoecklin, T., Voronin, A., & Rayez, J. C. 2003, *Chem. Phys.* 294, 117

Valiron, P., Wernli, M., Faure, A., & Rist, C. 2006, *J. Chem. Phys.*, in preparation

Valiron, P. 2006b, *J. Chem. Phys.*, in preparation

Valiron, P., unpublished

Wernli, M., Valiron, P., Faure, A., Wiesenfeld, L., Jankowski, P., & Szalewicz, K. 2005, *A&A*, in press

Wernli, M., Valiron, P., Faure, A., & Wiesenfeld, L. 2006, *A&A*, in preparation

Discussion

TOWNES: Are there any experimental results, which directly test your calculations of cross-sections, and if so how well do they agree?

DUBERNET: No there are, unfortunately, no direct experimental tests as yet.

Proceedings Title IAU Symposium
Proceedings IAU Symposium No. 231, 2005
A.C. Editor, B.D. Editor & C.E. Editor, eds.

© 2006 International Astronomical Union
doi:10.1017/S1743921306007101

Dissociative Recombination of $CD_3OD_2^+$

W. D. Geppert[1], **F. Hellberg**[1], **F. Österdahl**[2], **J. Semaniak**[3],
T. J. Millar[4], **H. Roberts**[4], **R. D. Thomas**[1],
M. Hamberg[1], **M. af Ugglas**[1], **A. Ehlerding**[1],
V. Zhaunerchyk[1], **M. Kaminska**[3], **and M. Larsson**[1]

[1]Molecular Physics Division, Department of Physics, Stockholm University, AlbaNova, SE
10691, Stockholm, Sweden, e-mail: wgeppert@physto.se

[2]Institute of Physics, Royal Institute of Technology, AlbaNova, SE 10691, Stockholm, Sweden

[3]Świętokrzyska Akademy, ul. Świętokrzyska 15, Pl-245406 Kielce, Poland

[4]University of Manchester, Manchester M13 9PL, United Kingdom

Abstract. The branching ratios of the different reaction pathways and the overall rate of the
dissociative recombination of $CD_3OD_2^+$ were measured at the CRYRING storage ring located
at the Manne Siegbahn Laboratory in Stockholm, Sweden. A preliminary analysis of the data
yielded that formation of methanol accounts for only 6 ± 2 % of the total reaction rate. Largely,
dissociative recombination of $CD_3OD_2^+$ involves fragmentation of the C–O bond, the major
process being the three-body break-up forming CD_3, OD and D (branching ratio 0.59). A non-
negligible formation of interstellar methanol by the previously proposed mechanism is therefore
very unlikely.

Keywords. ISM: molecules — methods: laboratory — molecular processes

1. Introduction

Methanol is one of the most important and interesting interstellar compounds. It is well
known that masers of methanol are associated with the early stages of formation of high-
mass stars. Observations of methanol densities are also used to determine the temperature
and density of a molecular cloud simultaneously (Leurini *et al.* 2004). Furthermore, the
compound can be used as an evolutionary indicator during the embedded phase of massive
star formation (van der Tak, van Dishoeck & Caselli 2000).

The formation of interstellar methanol has puzzled scientists for a long time. Since
grain-surface processes were not thought to be able to produce the observed methanol
densities (Allen & Robinson 1977), gas-phase reactions have been regarded as the major
source of this molecule in space. Up to now, it had been widely accepted that the first step
of interstellar methanol formation is a radiative association of CH_3^+ and H_2O followed
by dissociative recombination (DR) leading to CH_3OH and H:

$$CH_3^+ + H_2O \rightarrow CH_3OH_2^+, \tag{1.1}$$

$$CH_3OH_2^+ + e^- \rightarrow CH_3OH + H. \tag{1.2}$$

Generally, radiative association reactions tend to have very low thermal rates. However,
in the case of formation of a large stable molecular ion, the radiative lifetime of the intermediately formed excited complex can exceed its dissociative lifetime and, consequently,
a stable ion can be formed. This has, e. g. been observed in the radiative association of
C_3H^+ and H_2 (Sorgenfrei & Gerlich 1994). For larger systems it has then been assumed

that rates of radiative associations are collisional and can exceed 10^{-9} cm^3 s^{-1} (Herbst & Dunbar 1991). To explain the observed interstellar abundances of methanol it has been postulated that the rate for radiative association should lie in the range between 8×10^{-12} cm^3 s^{-1} and 8×10^{-8} cm^3 s^{-1} (Gottlieb *et al.* 1979). Indeed, a radiative association reaction of $\sim 2 \times 10^{-10}$ cm^3 s^{-1} at 300 K has been obtained in a SIFT experiment (Smith & Adams 1977).

But recently, it has been shown in an ion trap experiments that the rate of the radiative recombination of CH_3^+ and H_2O amounts to maximally 2×10^{-12} cm^3 s^{-1} and therefore is at least a factor of 10 too low to explain the interstellar abundances of methanol if one applies the common astrochemical models used (Luca, Voulot & Gerlich 2002; Herbst 1985). However, since CH_3^+ has never been observed, its abundances in interstellar clouds and other astronomical environments might differ considerably from model predictions. Therefore a substantial contribution of the proposed mechanism to the formation of interstellar methanol cannot yet be ruled out. Nevertheless, this only holds under the assumption that the dissociative recombination (DR) of $CH_3OH_2^+$ mostly or exclusively leads to methanol. Exact knowledge of the branching ratio of $CH_3OH_2^+$ is therefore necessary to be able to finally disregard the gas-phase formation mechanism according to equations (1.1) and (1.2).

This contribution presents the determination of the rate and the branching ratios of the DR of $CD_3OD_2^+$ at relative kinetic energies encountered in dark interstellar clouds.

2. Experimental

The DR experiments have been performed at the heavy-ion storage ring CRYRING at the Manne Siegbahn Laboratory, Stockholm University. Since the experimental procedure has been described in detail elsewhere (Neau *et al.* 2000), it is summarized only briefly here. For technical reasons related to the data analysis (better mass resolution on the surface barrier detector), $CD_3OD_2^+$ was used in the present experiment. The $CD_3OD_2^+$ ions were produced in a discharge ion source from a mixture of methanol vapor and deuterium (in excess). After extraction of the ions from the source at 40 keV energy, they were mass selected, injected into the ring and accelerated to 2.53 MeV translational energy. The stored ion beam was merged with a mono-energetic electron beam in an electron cooler, the length of the interaction region being 0.85 m. During the first 3 s after acceleration, the electron and ion beams were kept at the same average velocity to allow heat transfer from the ion to the electron beam in order to reduce the translational temperature of the ions which results in an increase of their phase-space density. Furthermore, such a storage time enables radiative vibrational cooling of the ions, which might partially have been formed in a rovibrationally excited state.

Neutral products generated by DR reactions in the electron cooler leave the ring tangentially and were detected by an energy-sensitive silicon surface barrier detector (SBD) with a diameter of 17 mm mounted at a distance of 4.30 m from the center of the interaction region. A background signal due to neutral products emerging from collisions of the ions with residual gas was also present; this was measured with the relative translational energy between the ions and electrons tuned to 1 eV, where the DR cross section is very low and the measured neutral fragments are therefore almost exclusively produced by rest gas collisions. This background was subsequently subtracted from the total SBD signal.

3. Results

3.1. *Absolute Cross Section and Thermal Reaction Rate*

During cross section measurements, the relative translational energy between the ions and the electrons was continuously varied between 1 and 0 eV. This was achieved by changing the cathode voltage of the electron cooler over 1 s from a high value corresponding to a centre-of-mass energy of 1 eV, the electrons being faster than the ions, down to a low value also corresponding to 1 eV but where the electrons were slower than the ions. Thus, a voltage corresponding to a centre-of-mass energy of 0 eV is reached during the scan. Before the measurement was started, 3 s of cooling, with the electrons tuned to 0 eV collision energy, was carried out. The signal from the SBD was monitored by a single-channel analyser, selecting signals arising from all fragments reaching the detector simultaneously and thereafter recorded by a multichannel scaler, yielding the number of counts vs. storage time and, therefore, at a given relative translational collision energy.

The experimental DR rate coefficient in the electron cooler is expressed by the formula:

$$\langle v_{cm}\,\sigma \rangle = \left(\frac{dN}{dt} \right) = \frac{v_i v_e e^2 r_e^2 \pi}{I_e I_i l} \tag{3.1}$$

where dN/dt is the number of counts per unit time, v_i and v_e are the electron and ion velocities, respectively, r_e is the radius of the electron beam, l the length of the interaction region, and I_e and I_i are the electron and ion current, respectively.

Simultaneously with the measurement of the dissociative recombination count rate, the ion current was monitored using a Bergoz beam charge monitor, with continuous averaging, and an AC integrating current transformer combination with a neutral particle detector installed at the end of one of the straight sections of the ring (Paal *et al.* 2004). Continuous measurement of neutral products due to the ion-rest gas collisions also allowed the decay of the ion beam to be monitored during each injection. The following corrections to the measured data had to be performed. Firstly, the voltage of the electron cooler cathode (and therefore v_e) had to be corrected for space charge effects. Secondly, the measured rate coefficient $\langle v_{cm}\,\sigma \rangle$ had to be adjusted because of the toroidal effect (Lampert *et al.* 1996). The toroidal effect stems from the zones at both ends of the interaction region where the electron beam is bent into or out of the ion beam. In these regions, the transversal electron velocity is higher than in the merged interaction region leading to larger collision energies. Thirdly, the electron beam has (in contrast to the ion beam) a non-neglectible energy spread, and the measured reaction rate $\langle \sigma v_{cm} \rangle$ has to be deconvoluted according to the formula:

$$\langle v_{cm}\,\sigma \rangle = \int_{-\infty}^{\infty} v_e f(v_e)\sigma(v_e)d^3 v_e \tag{3.2}$$

where $f(v_e)$ is the velocity distribution. From the resulting rate constants, the (comparatively) very small contribution to the data due to charge transfer processes with the rest gas had to be subtracted. Since the cross section of the DR is very low at 1 eV collision energy, and the rest gas collisions are independent of the electron velocity, the signals measured at this energy were assumed to be due solely to charge transfer. The cross-sectional data are best fitted by the expression $\sigma = 9.6 \pm 1.9 \times 10^{-15}\,E^{-0.96\pm0.02}$ cm^2. The thermal reaction rate constant can be deduced from the cross sections by applying the formula

$$kT = \frac{8\pi m_e}{(2\pi m_e kT)^{3/2}} \int_0^{\infty} E\sigma(E)e^{-\frac{E}{kT}}\,dE \tag{3.3}$$

where m_e is the mass of the electron. The temperature dependence of the rate coefficient can then be fitted to the expression $9.1 \pm 1.8 \times 10^{-7}(T/300)^{-0.63}$ cm^3 s^{-1}.

3.2. *Branching Ratios*

In the DR of $CD_3OD_2^+$ the following exoergic reaction channels exist:

$$CD_3OD_2^+ + e^- \rightarrow CD_3 + D_2O \qquad \Delta H = -6.87 eV \qquad (3.4a)$$

$$CD_3OD_2^+ + e^- \rightarrow CD_3 + OD + D \qquad \Delta H = -1.70 eV \qquad (3.4b)$$

$$CD_3OD_2^+ + e^- \rightarrow CD_2 + D + D_2O \qquad \Delta H = -2.12 eV \qquad (3.4c)$$

$$CD_3OD_2^+ + e^- \rightarrow CD + D_2 + D_2O \qquad \Delta H = -2.32 eV \qquad (3.4d)$$

$$CD_3OD_2^+ + e^- \rightarrow CD_3OD + D \qquad \Delta H = -5.70 eV \qquad (3.4e)$$

$$CD_3OD_2^+ + e^- \rightarrow CD_3O + D_2 \qquad \Delta H = -5.70 eV \qquad (3.4f)$$

$$CD_3OD_2^+ + e^- \rightarrow CD_3O + 2D \qquad \Delta H = -1.18 eV \qquad (3.4g)$$

$$CD_3OD_2^+ + e^- \rightarrow CD_2O + D + D_2 \qquad \Delta H = -4.81 eV \qquad (3.4h)$$

$$CD_3OD_2^+ + e^- \rightarrow CD_2O + 3D \qquad \Delta H = -0.29 eV \qquad (3.4i)$$

$$CD_3OD_2^+ + e^- \rightarrow CD_4 + O + D \qquad \Delta H = -1.81 eV \qquad (3.4j)$$

$$CD_3OD_2^+ + e^- \rightarrow CD_4 + OD \qquad \Delta H = -6.24 eV \qquad (3.4k)$$

$$CD_3OD_2^+ + e^- \rightarrow CD_2 + OD + D_2 \qquad \Delta H = -1.47 eV \qquad (3.4l)$$

$$CD_3OD_2^+ + e^- \rightarrow CD_3 + O + D_2 \qquad \Delta H = -1.78 eV \qquad (3.4m)$$

$$CD_3OD_2^+ + e^- \rightarrow CDO + 2D_2 \qquad \Delta H = -5.42 eV \qquad (3.4n)$$

$$CD_3OD_2^+ + e^- \rightarrow CDO + D_2 + 2D \qquad \Delta H = -0.94 eV \qquad (3.4o)$$

$$CD_3OD_2^+ + e^- \rightarrow CO + 2D_2 + D \qquad \Delta H = -4.76 eV \qquad (3.4p)$$

$$CD_3OD_2^+ + e^- \rightarrow CO + D_2 + 3D \qquad \Delta H = -0.28 eV \qquad (3.4q)$$

With the use of a storage ring, branching ratios can be determined even for DR reactions possessing a multitude of product pathways such as the case for $CD_3OD_2^+$. The fragments produced by a DR event reach the detector within a very short time interval compared with the integration time of the detection system. Owing to that fact, the pulse height of the SBD signal is proportional to the kinetic energy carried by the products of the reaction and therefore their total mass. To measure the branching ratios of the DR channels, a metal grid with a transmission $T = 0.297 \pm 0.015$ is inserted in front of the detector (Neau *et al.* 2000). Particles stopped by the grid do not reach the detector, and DR events where one of the fragments has been stopped result in a signal whose amplitude is proportional to the kinetic energy of the detected fragment. The registered DR spectrum therefore splits into a series of peaks with different energies, the intensities of which can be expressed in terms of the branching ratios and the probabilities of the particles passing the grid. For example, the intensity of the CD_3 peak emerging from reaction (3.4a) is proportional to $T(1-T)\alpha$, with α being the branching ratio of reaction (3.4a) and $T(1-T)$ the probability of only the CD_3 fragment passing the grid, respectively. The energy spectrum of the $CD_3OD_2^+$ DR reaction is shown in Figure 1. The different signal peaks are reasonably resolved and were therefore could be fitted to double gaussian functions to allow the peaks to be slightly asymmetric. As can easily be seen in Figure 1, the final fitting curve is in very good agreement with the data.

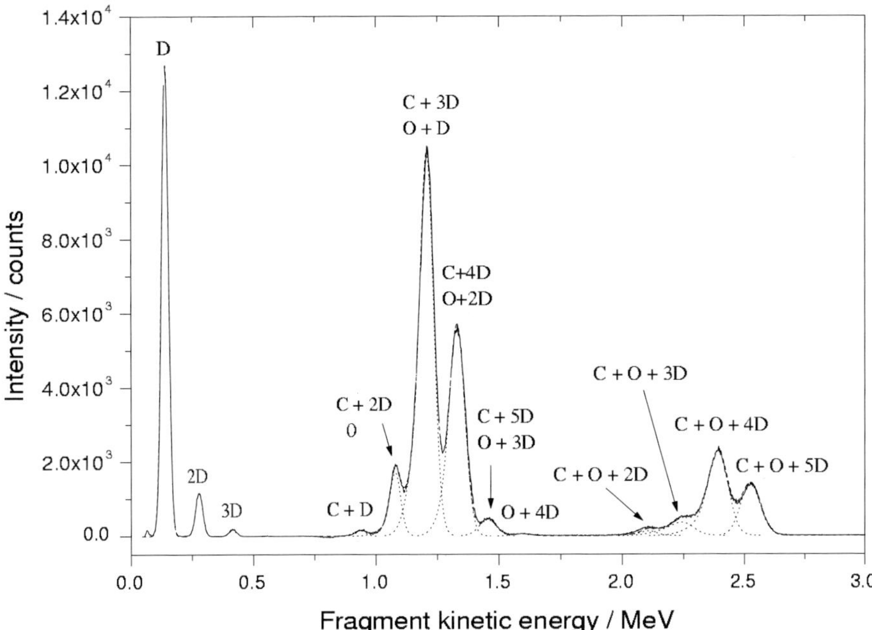

Figure 1. Fragment energy spectrum of protonated methanol with the grid in place. The solid line shows the data, the dotted lines the double-Gaussian fitting curves, and the dashed line the total fit.

Concerning the evaluation of the data, two problems arise. Firstly, some of the different channels produce products with the same mass distribution, leading to an identical signal distribution in the fragment energy spectrum spectrum. This holds for channels $(3.4a)$ and $(3.4k)$, $(3.4c)$ and $(3.4j)$ as well as $(3.4l)$ and $(3.4m$; therefore, there exist three pairs of indistinguishable channels. However, since formation of CD_4 would involve a considerable rearrangement of the intermediately formed neutral, it can be argued that channels $(3.4j)$ and $(3.4k)$ are unlikely to play a major role. We will see later that the branching ratio of channel $(3.4l)$ and $(3.4m)$ is 0. Furthermore, since there is no signal corresponding to the mass of CO and DCO, channels $(3.4n)$, $(3.4o)$, $(3.4p)$ and $(3.4q)$ can be disregarded.

Secondly, owing to the high-energy release of reactions $(3.4e)$ and $(3.4f)$ and the comparatively low mass of the D and D_2 fragments, some of the deuterium atoms and molecules produced by these pathways might receive a large transversal velocity relative to the propagation direction, and therefore miss the detector. In the case of reaction $(3.4e)$ this reduces the contribution of reaction $(3.4e)$ to the C + 5H + O energy (mass) peak to $T^2(1 - L)\varepsilon$, where L is the D-atom loss factor and ε denotes the branching ratio of channel $(3.4e)$. Conversely, the C + 4H + O energy channel is augmented to $T(1 - T)(1 - L)\varepsilon + T\varepsilon L$. If one assumes that the whole reaction enthalpy is converted to kinetic energy, one would obtain a D loss of 67% for reaction $(3.4e)$ and a D_2 loss of 13% for reaction $(3.4f)$. Using the transmission probabilities together with the necessary corrections, a matrix is formulated for the relative intensities of the different energy (mass) peaks. To evaluate the loss of the deuterium atoms, an energy spectrum, shown in Figure 2, was measured with the grid removed. Since for this case $T = 1$ and the sum of all branching ratios amounts to 1, it can easily be worked out that the ratio between the

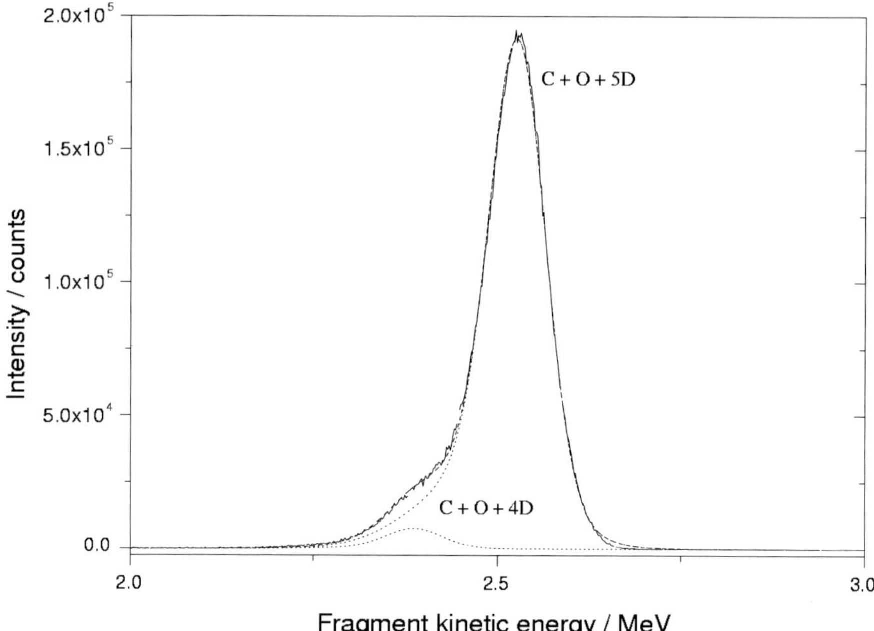

Figure 2. Fragment energy spectrum of protonated methanol with the grid removed. The solid line shows the data, the dotted lines the double Gaussian fitting curves and the dashed line the total fit.

Table 1. Branching ratios of the dissociative recombination of $CD_3OD_2^+$.

Reaction channel	Products	Branching ratio
$a\ (+k)$	$CD_3 + D_2O$	0.11
b	$CD_3 + OD + D$	0.59
$c(+j)$	$CD_2 + D + D_2O$	0.17
d	$CD + D_2 + D_2O$	0.01
e	$CD_3OD + D$	0.06
f	$CD_3O + D_2$	0.05
g	$CD_3O + 2D$	0.00
h	$CD_2O + D + D_2$	0.02

intensities of the $C+O+4D$ peak $(I_{(C+O+4D)})$ and the $C+O+5D$ $(I_{(C+O+5D)})$ peak will be:

$$\frac{I_{(C+O+4D)}}{I_{(C+O+5D)}} = \frac{\varepsilon L}{1 - \varepsilon L} \tag{3.5}$$

No contribution of the C+O+3D mass was detected, therefore $L\zeta$ equals 0. By iteratively solving the matrix with a certain loss and calculating L according to (3.5), the loss of D-atoms, L, was determined to be 0.56 and the branching ratios listed in Table 1 were obtained.

3.3. *Model Calculations*

The new data on the rate and the branching ratios of dissociative recombination were used as an input for a model calculation of a dark cloud resembling TMC-1 using the

UMIST code (Markwick, Millar & Charnley 2000). Using the previous data, the model predicted a peak abundance relative to H_2 of 1×10^{-9}, which is in good agreement with the observations of Friberg *et al.*, which yielded a methanol abundance in the low digits of 10^{-9} for TMC-1. (Friberg *et al.* 1988). With the present rate and branching rations of the dissociative recombination of protonated methanol, the peak abundance is lowered to 8×10^{-11}, while the one for steady-state sinks from 3×10^{-10} to 2×10^{-11}. If also the new rate for the radiative association of CH_3^+ and H_2O is included, the peak abundance is further lowered to 7×10^{-13} and the steady state to 1×10^{-13}, which is clearly below the observed values in dark clouds and probably below the detection limits for existing telescopes.

4. Discussion

The most striking feature of the DR of $CD_3OD_2^+$ is the low branching ratio of the "expected" channel leading to CD_3OD and D and the predominance of pathways involving 3-body break-ups. This is contrary to the theory formulated by Bates (1986), according to which the channel involving the least rearrangement of orbitals should be favored. However, recent research on DR branching ratios showed that these processes very often do not follow this prediction and three-body break-ups are actually quite common (see e.g. Larsson & Thomas 2001; McCall *et al.* 2004). It is interesting to compare the present findings to the ones obtained for the somewhat related H_3O^+ and D_3O^+ ions. With these species, pathways leading to three products also dominate with a total branching ratio of 0.61 and 0.60, respectively and the channels leading to H_2O and D_2O (branching ratios 0.25) are of minor importance (Jensen *et al.* 2000). These findings are similar to our results for $CD_3OD_2^+$, where the branching ratios of all three-body pathways add up to 0.78.

From the model calculations it ultimately becomes clear that the proposed gas-phase process (radiative association of CH_3^+ and H_2O with subsequent DR) cannot account for the production of interstellar methanol. Therefore, repetitive H-atom addition to CO on interstellar CO-containing ices seems the only plausible mechanism for methanol formation in the interstellar medium. This is corroborated by the fact that recent observations found a correlation between the abundances of methanol and CO in star-forming regions (Bisschop, Jørgensen & van Dishoeck 2005). Also, comparison of modelled and and observed deuterium fractionation in methanol points to a grain-surface process (Parise *et al.* 2004).

However, two problems are associated with this assumption: (a) Is the addition of H on solid CO efficient enough in forming methanol at the low temperatures that are encountered in dark interstellar clouds? (b) Can methanol be released from grain surfaces in sufficient quantities to explain the observed abundances not only in hot cores but also in dark clouds?

Very recently, it has been established experimentally, that the formation of methanol on pure CO and mixed CO/H_2O ices proceeds efficiently at 10 K (Watanabe *et al.* 2004) and the obtained yields of methanol are consistent with observations. Hydrogenation of the precursor methoxy or hydroxymethyl radicals is highly exoergic (-435.1 and -393.1 $kJ\,mol^{-1}$, respectively (Lias *et al.* 1988)) and might lead to methanol in a highly excited states and the resulting methanol molecule might use a part of this energy to recoil from the surface and end up in the gas phase. The question to what extent this is possible still lacks a definitive answer.

Acknowledgements

Financial support from the Swedish Space Board and the Swedish Science Council is gratefully acknowledged.

References

Allen, M. & Robinson, G.W. 1977, *Ap. J.* 212, 396

Bates D.R. 1986, *Ap. J. Lett.* 306, 45

Bisschop, S.E., Jørgensen, J.K., & van Dishoeck, E.F. 2005, *IAU Symposium No. 231*, poster

Friberg, P., Madden, S.C., Hjalmarson, Å., & Irvine, W.M. 1988, *A&A* 195, 281

Gottlieb, C.A., Ball, J.A., Gottlieb, E.W., & Dickinson, D.F. 1979, *Ap. J.* 227, 422

Herbst, E. 1985, *Ap. J.* 291, 226

Herbst, E. & Dunbar, R.C. 1991, *MNRAS* 253, 341

Jensen, M.J., Bilodeau, R.C., Safvan, C.P., Seiersen, K., Andersen, L.H., Pedersen, H.B., & Heber, O. 2000, *Ap. J.* 543, 764

Lampert, A., Wolf, A., Habs, D., Kenntner, J., Kilgus, G., Schwalm, D., Pindzola, M., & Badnell, N. 1996, *Phys. Rev. A.* 53, 1413

Larsson, M. & Thomas, R.D. 2001, *Phys. Chem. Chem. Phys.* 3, 4471

Lias, S.G., Bartmess, J.E., Liebman, J.E., Holmes, J.L., Levin, R.D., & Mallard, W.G. 1988, *J. Phys. Chem. Ref. Data* 17

Leurini, S., Schilke, P., Menten, K.M., Flower, D.R., Pottage, J.T., & Xu, L.-H. 1985, *A&A* 422, 573

Luca, A., Voulot, D., & Gerlich, D. 2002, *WDS02 Proceedings of Contributed Papers, PART II* 294

Markwick, A., Millar, T.J., & Charnley, S.B. 2000, *Ap. J.* 535, 256

McCall, B.J., Huneycutt, A.J., Saykally, R.J., Djuric, N., Dunn, G.H., Semaniak, J., Novotny, O., Al-Khalili, A., Ehlerding, A., Hellberg, F., Kalhori, S., Neau, A., Thomas, R.D., Paal, A., Österdahl, F., & Larsson, M. 2004, *Phys. Rev. A* 70, 052716

Neau, A., Al-Khalili, A., Rosén, S., Le Padellec, A., Derkatch, A.M., Shi, W., Vikor, L., Larsson, M., Nagard, M.B., Andersson, K., Danared, H., & af Ugglas, M. 2000, *J. Chem. Phys.* 113, 762

Paal, A., Simonsson, A., Källberg, A., Dietrich, & Mohos, I. 2004, in *Proceedings of EPAC 2004 Lucerne*, 2743

Parise, B., Castets, A., Herbst, E., Caux, E., Ceccarelli, C., Mukhopadhyay, I., & Tielens, A.G.G.M 2004, *A&A* 416, 159

Smith, D. & Adams, N.G. 1977, *Ap. J.* 217, 741

Sorgenfrei, A. & Gerlich, D. 1994, in *Physical Chemistry of Molecules and Grains in Space*, ed. I. Nenner (AIP Press, New York), 505

van der Tak, F.F.S., van Dishoeck, E.F., & Caselli, P. 2000, *A&A* 361, 327

Watanabe, N., Nagaoka, A., Shiraki, T., & Kouchi, A. 2004, *Ap. J.* 616, 638

Astrochemistry: Recent Successes and Current Challenges
Proceedings IAU Symposium No. 231, 2005
D.C. Lis, G.A. Blake & E. Herbst, eds.
© 2006 International Astronomical Union
doi:10.1017/S1743921306007113

Deuterium Fractionation and Ion-Molecule Reactions at Low Temperatures

Stephan Schlemmer[1,2], Oskar Asvany[2], Edouard Hugo[2], and Dieter Gerlich[3]

[1]I. Physikalisches Institut, Universität zu Köln, Zülpicher Strasse 77, 50937 Köln, Germany
email: schlemmer@ph1.uni-koeln.de

[2]Sterrewacht Leiden, P.O. Box 9513, 2300 RA Leiden, The Netherlands

[3]Fakultät für Naturwissenschaften, Technische Universität Chemnitz,
09107 Chemnitz, Germany

Abstract. Understanding deuterium fractionation is currently one of the greatest challenges in astrochemistry. In this contribution deuteration experiments of the series CH_n^+, n = 2–5, in a low temperature 22-pole ion trap are used to systematically test a simple chemical rule predicting which molecular ion undergoes deuterium exchange in collisions with HD. CH_4^+ turns out to be a problem case, where prediction fails. The method of laser induced reaction (LIR) is used to determine the population ratio of the lowest ortho-to-para states of H_2D^+ relaxed in collisions with H_2. Preliminary results indicate that the ortho-to-para ratio of H_2D^+ is substantially reduced in para-H_2. This points at the important role of nuclear spin in deuterium fractionation, in particular at the destruction of ortho-H_2D^+ in collisions with ortho-H_2. More systematic LIR experiments are needed for a chemical model of deuterium fractionation including state-to-state modifications of the species involved.

Keywords. astrochemistry — ISM: molecules — molecular processes

1. Introduction

About 25 deuterated molecules have been found to date in the interstellar medium, Millar (2003). Much attention has been drawn to recent detections of doubly and even triply deuterated species in cold interstellar clouds. Triply deuterated ammonia (Lis *et al.* 2002) and methanol (Parise *et al.* 2004) show an isotopic enhancement (fractionation) which is up to twelve orders of magnitude larger than the simple predictions based on the cosmic D/H ratio of 1.5×10^{-5}. The recent detection of D_2H^+ (Vastel *et al.* 2004), stimulated new chemical modelling calculations for very cold dense clouds which find that even D_3^+ can be the dominant molecular species (Walmsley *et al.* 2004; Flower *et al.* 2004).

The main source of deuterium in molecular clouds is HD. Three primary molecular ions, H_3^+, CH_3^+ and $C_2H_2^+$, are known to transfer the deuterium in collisions with HD. These reactions are exothermic by a few hundred Kelvin due to differences in zero-point vibrational energies. Since typical cloud temperatures are much smaller than these exoergicities, isotopic enhancement of the singly deuterated ions can easily reach factors up to 10^4. Reactions in subsequent collisions of these ions lead to the deuteration of other molecules like HCO^+.

Due to the Arrhenius-type behavior of the back reactions (hydrogenation) the fractionation strongly depends on the temperature of the particular environment. Also the fractional electron abundance plays an important role for isotopic fractionation as dissociative recombination competes with deuteration. For the same reason the presence of atoms and molecules which react with the primary ions also reduce the isotopic enhancement. These species include almost all C, N and O bearing molecules. In contrast, freeze out of molecules on grains in very cold environments removes these species from the gas phase. As a consequence isotopic enrichment is found to be increased. Many of these dependencies are included in current chemical models for fractionation.

Another level of complication arises from collision systems with identical nuclei. This is for example the case in all three key initial deuteration reactions. Besides conservation of energy, momentum, angular momentum and parity, conservation of the total nuclear spin imposes further restrictions on the outcome of such reactions (Quack (1977)). Because of this, even more importantly, inelastic collisions with the most dominant H_2 molecule can influence the D/H ratio significantly. Thorough descriptions of appropriate theoretical methods to include nuclear spin effects as well as experiments to investigate the role of nuclear spin have been given by Gerlich *et al.* (2002a), Gerlich & Schlemmer (2002b), and Gerlich (2004).

Current laboratory challenges on understanding deuterium fractionation include (i) the search for more primary deuteration reactions, (ii) testing of theoretical predictions for possible deuteration reactions, and (iii) the role of nuclear spin conservation on the efficiency to replace H vs. D. More systematic studies on possible deuteration reactions of small hydrocarbons, CH_n^+, $(n = 3 - 5)$, Asvany *et al.* (2004a), and of $C_3H_n^+$, $(n = 1 - 3)$ have been carried out recently, Savic & Gerlich (2005), Savic *et al.* (2005). In this contribution the series $CH_n^+ + HD$ will be used to compare the results to theoretical predictions worked out by Henchman *et al.* (1988) and modified by Maluendes *et al.* (1992).

In order to detemine the role of nuclear spin conservation, state-to-state reactive and inelastic rate coefficients are needed. Experimentally this is very difficult. However, as described below, a method has been developed in recent years by Schlemmer *et al.* (1999), Schlemmer *et al.* (2002), Mikosch *et al.* (2004), Schlemmer *et al.* (2005), Asvany *et al.* (2005a), and Asvany *et al.* (2005b) to infer ro-vibrational spectra of a cloud of molecular ions in a trap experiment. Based on a few assumptions the rotational population of the ion cloud can be derived. First results on H_2D^+ using the free electron laser FELIX are presented.

2. Experimental

To measure the rate coefficients of low-temperature ion-molecule reactions, a 22-pole ion trap has been used. This apparatus has been developed and thoroughly described by (Gerlich & Horning 1992) and recent applications to isotopic fractionation are well documented by Gerlich & Schlemmer (2002b), Asvany *et al.* (2004b), Savic & Gerlich (2005), and Savic *et al.* (2005).

In brief, ions are produced in a storage ion source by electron bombardment, mass filtered in a quadrupole analyzer and then injected into the 22-pole ion trap, which is held at cryogenic temperatures: T > 10 K. On entrance, the ion temperature is adapted to the trap temperature by a short helium pulse. During the storage time of several seconds the ions can react with a neutral gas, here mainly HD, of known constant but adjustable number density. After the trapping period, the trap content is extracted, mass analyzed and the parent and product ions are counted with high efficiency. From this procedure, the rate coefficients of the involved processes are inferred by repeating the mentioned cycle of *trapping/reaction/detection* for all possible reaction products and different trapping times.

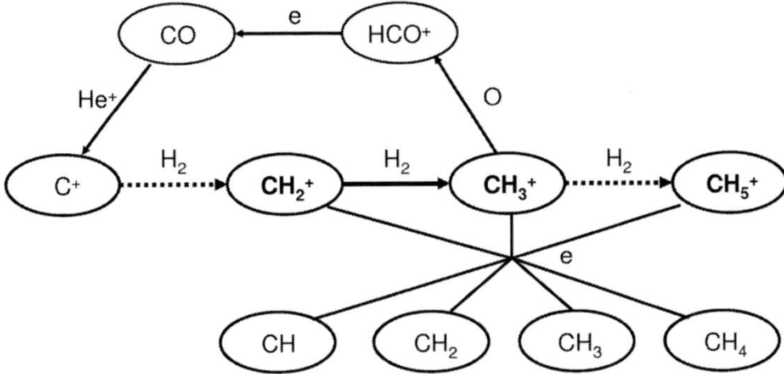

Figure 1. Formation routes for small hydrocarbon molecules in interstellar clouds as described by Black & Dalgarno (1977).

An extension of this technique is the method of laser induced reactions (LIR), in which an endothermic ion-molecule reaction (e.g. $H_2D^+ + H_2 \rightarrow H_3^+ + HD$, as presented below) is enhanced by resonantly exciting the trapped parent ions with a laser. This offers not only the possibility to record a spectrum by counting the product ions as a function of the laser wavelength, but LIR also yields information on state-selected rate coefficients, lifetimes of the excited states and the population of the pumped states (Schlemmer *et al.* 2002, 2005).

3. Results and Discussion

3.1. *Kinetic Studies*

Figure 1 shows the network of small hydrocarbon ions as proposed by Black & Dalgarno (1977) for diffuse interstellar cloud conditions. Similar schemes have been found also for cold dense clouds. In this scheme carbon is cycled between the small hydrocarbon ions, CH_n^+, $n = 0-3$, and CO. Thus these ions play an important role in molecular formation in interstellar clouds. Starting with C^+, hydrogen is either added via radiative association (adding H_2, dashed line) or in an abstraction reaction (adding H, solid line). Growth stops at CH_5^+, since CH_n^+ with $n > 5$, does not form a strongly bound molecule. Instead, dissociative recombination of this molecule is a major source for interstellar methane.

Based on such a scheme, it is not obvious why only CH_3^+ should be a source for deuteration. Therefore more systematic studies of the deuteration of the CH_n^+, $n = 2 - 5$, ions have been conducted in recent years (Asvany *et al.* 2004a). The results of the kinetic measurements at a trap temperature of 15 K are summarized in Table 1. Various outcomes are possible in the collision of $CH_n^+ + HD$. CH_2^+ undergoes hydrogen or deuterium abstraction to form CH_3^+ or CH_2D^+. This process is so fast, that any deuteration, forming CHD^+, can hardly be detected, especially since the abstraction reaction product, CH_3^+, has the same mass as the deuteration product, CHD^+. Comparison of the measured total rate coefficient for the abstraction reaction to the limiting collisional rate coefficient, k_L, reveals that this process is happening practically in every collision. In fact the experimental value is about 20% larger than the predicted value. One possible reason for this discrepancy lies in the long-range attraction of HD.

Since abstraction is also very fast for CH_2^+ in collisions with H_2, the less frequent collisions with HD hardly yield a substantial fraction of CH_2D^+. Because of this reason

Table 1. Rate coefficients, branching ratios and reaction efficiencies (k/k_L) for the deuteration reactions of $CH_n^+ + HD$, $n = 2..5$, at $T = 15$ K.

Reaction		Branching ratio	k $[10^{-9}cm^3s^{-1}]$	k/k_L
$CH_2^+ + HD$	$\rightarrow CH_2D^+ + H$	0.8	1.6	1.21
	$\rightarrow CH_3^+ + D$	0.2		
$CH_3^+ + HD$ (+He)	$\rightarrow CH_2D^+ + H_2$		1.65	1.25
$CH_3^+ + HD$ (+n-H_2)	$\rightarrow CH_2D^+ + H_2$		1.5	1.14
$CH_3^+ + HD$ (+p-H_2)	$\rightarrow CH_2D^+ + H_2$		0.4	0.30
$CH_2D^+ + HD$ (+He)	$\rightarrow CD_2H^+ + H_2$		1.59	1.21
$CD_2H^+ + HD$ (+He)	$\rightarrow CD_3^+ + H_2$		1.5	1.15
$CH_4^+ + HD$	$\rightarrow CH_3D^+ + H_2$	<0.002	0.45	0.34
	$\rightarrow CH_5^+ + D$	0.68		
	$\rightarrow CH_4D^+ + H$	0.32		
$CH_5^+ + HD$	$\rightarrow CH_4D^+ + H_2$		$< 5 \cdot 10^{-9}$	0

CH_2^+ does not play a role in isotopic fractionation. Instead, only those ions are relevant for isotopic fractionation in collisions with HD which react very inefficiently with H_2. This is the case for CH_3^+, which forms CH_5^+ only very slowly via radiative association and for CH_5^+, which is effectively the end of the reaction chain. Therefore at least these two ions are reasonable candidates to contribute to isotopic fractionation. CH_4^+ is not shown in the network, see Figure 1, and will be considered separately.

For CH_3^+ sequential deuteration has been measured and rate coefficients for all individual steps have been determined at a trap temperature of 15 K, see Table 1. As in the case of CH_2^+ the rate coefficients are significantly larger than the capture rate coefficient. Moreover, each step of deuteration happens slightly faster. Besides these peculiarities, it turns out that deuteration of CH_3^+ is happening practically in every collision. It is quite surprising, that this process can be so fast, because simple statistics would predict a 40% chance to find the D-atom in HD and a 60% chance to find it in CH_3^+ after the collision. The deviation from this rather simple counting argument shows that the potential energy surface of the collision system has to be taken into account to understand the large deuteration efficiency.

As described in detail in Asvany *et al.* (2004a), the results for collisions with pure HD have been compared to experiments, where the dominant collision partner is H_2. In the case of pure HD, CH_3^+ has been cooled in the trap to the ambient temperature via collisions with He. The value derived from this experiment is very similar to the value when CH_3^+ is cooled in n-H_2, which contains 1/4 p-H_2 and 3/4 o-H_2, as well as the terrestrial abundance of $[HD]/[H_2] = 3 \times 10^{-4}$. However, when conducting the experiment in p-H_2 with the natural abundance of HD (Gerlich & Schlemmer 2002b), the deuteration reaction is found to proceed about four times slower when compared with the other experiments, as shown in Table 1. The presence of the various collision partners changes the relaxation of the ion cloud. This result is linked to the influence that nuclear spin conservation imposes on the rotational state distribution when most collisions are with p-H_2 rather than with o-H_2 or HD. This finding will be discussed below in more detail.

As shown in Table 1, the collision of CH_4^+ with HD leads primarily to hydrogen abstraction, a process that has been described in a previous publication (Asvany *et al.* 2004b). The total rate coefficient for abstraction at 15 K, however, amounts to about 30% of the collision limit. Interestingly, the experimental data show no indication of deuteration in this case. Although the detection of deuterated species has been limited by some spurious isotopomers coinciding with the mass of the products, an upper limit for the rate coefficient of deuteration could be derived. According to this result, shown in Table 1, more than 1000 collisions are necessary to exchange a hydrogen for a deuterium in CH_4^+ in collisions with HD. Another interesting result concerns the branching ratio in the abstraction reaction. The less exoergic product CH_5^+ is formed twice as often as compared to the energetically more favorable CH_4D^+ product. Moreover, the abstraction reaction shows a negative temperature dependence, meaning, that abstraction becomes more favorable at low temperatures. These results indicate that the dynamics of the CH_4^+ + HD collision system does not follow a simple downhill reaction without barrier.

Finally the collision system CH_5^+ + HD has been studied. As for CH_4^+, no indication for the exchange of a deuterium for a hydrogen atom has been found. In fact, product detection in this case is more sensitive than for the CH_4^+ measurements. Therefore the upper limit of deuteration has been determined to be much lower: $k \leqslant 5 \times 10^{-18}\,\mathrm{cm^3\,s^{-1}}$; see Table 1. As pointed out before and shown in Figure 1, CH_5^+ is the end of the hydrogenation route in interstellar clouds. The four valence electrons form a closed shell molecule. Therefore abstraction of D or H does not occur in collisions with HD. Since radiative association is a very slow process for this system, $k_r = 2 \times 10^{-16}\,\mathrm{cm^3\,s^{-1}}$, dissociative recombination dominates under interstellar conditions, leading to the various neutral small hydrocarbons depicted in Figure 1.

In summary, CH_n^+ exhibits fast deuteration in collisions with HD only for $n = 3$. All other molecular ions show a negligible chance for the exchange of a deuterium atom. Therefore no additional primary deuteration reaction has been found in these experiments. However, due to the systematic studies presented, the results can be compared to predictions or at least simple chemical rules. Such rules have been first formulated by Henchman *et al.* (1988). Here reactions are characterized distinguishing reactants with filled and unfilled valence shells. According to this work, no barriers inhibit deuterium exchange for exoergic reactions if at least one of the reactants has an unfilled valence shell. In contrast, barriers can inhibit the exchange if both reactants have filled valence shells. This simple rule is illustrated in Figure 2. Here the minimum energy path for an exoergic deuterium exchange, $AH^+ + HD \rightarrow AD^+ + H_2$, is depicted. At the low temperatures of interstellar clouds the two reactants form an entrance complex $AH^+ \cdot \cdot HD$, where the two partners remain as distinct molecules. The H-D exchange needs a closer approach of the two reactants and occurs via a transition state. In case of closed shell species, this transition state is substantially higher in energy than the entrance channel. Therefore the exchange reaction is hindered. However, in case of one open shell reactant the complex formed is much stronger bound and the transition state is expected to be below the entrance channel, such that the exchange happens within the collision complex. Due to the exoergicity of the reaction the collision complex dissociates preferentially into AD^+ and H_2 based on statistical arguments counting the number of energetically accessible states. This simple classification is able to explain why H_3^+, CH_3^+ and $C_2H_2^+$ do undergo H-D exchange in collisions with HD, while HCO^+, N_2H^+ and NH_4^+, for example, do not.

The fact that the "filled valence shell" ions form only weakly bound complexes is related to the fact that electronic reorganization is hindered since empty AH^+ orbitals are energetically inaccessible due to a large HOMO-LUMO ("highest occupied molecular orbital", "lowest unoccupied molecular orbital") gap. Therefore the terms "filled" and "unfilled valence shells" have been replaced by Maluendes *et al.* (1992) for the chemically more precise terms of large and small HOMO-LUMO gap, respectively.

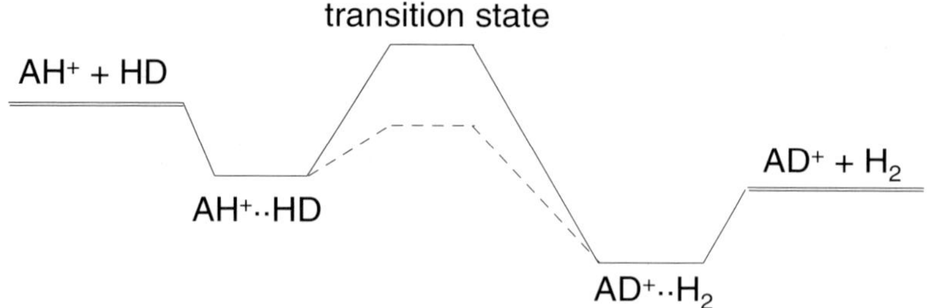

Figure 2. Energetics of an exoergic deuterium exchange reaction with barrier (solid) without a barrier (dashed).

HD has a very large HOMO-LUMO gap, therefore only the small hydrocarbon ions have to be considered in the present case. In the CH_n^+, n = 2–5, series, only CH_5^+ has a fully occupied valence shell with a very large HOMO-LUMO gap. The experimental result that more than 10^8 collisions are necessary for H-D exchange is in very good agreement with the prediction. For CH_2^+ the abstraction reaction is so fast, that deuterium exchange could not be detected. Therefore no comparison with the simple prediction can be made. H-D exchange is very fast for CH_3^+, which is in perfect agreement with the prediction. In fact, the CH_4D^+ complex which is formed is a strongly bound molecule in its ground state. Therefore the collision complex, the transition state and the complex in the exit channel are more or less the same entity. Apparently, the various H-atoms and the D-atom swap places in this complex very efficiently without substantial energetic hindrance. This process is known as *scrambling*. Thus the picture drawn in Figure 2 describes this case very well. Even the fact that deuteration is happening in almost 100% of the cases, rather than only in three out of five cases, as predicted by the statistical arguments given above has been explained by Maluendes *et al.* (1992). Upon forming the entrance channel complex the system can reach the transition state with a certain probability, x. In all other cases, probability $1 - x$, it re-dissociates back into the reactants. From the transition state the complex can fall back into the entrance channel again or react. From the entrance channel it can again form the transition state etc. In effect the chance of forming the deuterated product reaches almost unity.

In contrast, as an open valence shell ion CH_4^+ should undergo rapid H-D exchange, but does not, as shown above. Only little is known about the electronic structure of this molecule (Jacox 1994). Therefore the HOMO-LUMO gap cannot be quantified. This hampers a more quantitative comparison. Despite these limitations, as will be discussed in the following, deuteration is hindered even without a substantial barrier. The exoergic abstraction reaction in a CH_4^+ + HD collision proceeds faster by at least one order of magnitude at 10 K compared to 300 K. Such a behavior is associated with an extended lifetime of the collision complex in the entrance channel. This longer lifetime enhances substantially the chance to overcome the transition state and reaching the products. Even a small barrier can be surpassed via tunneling. Nevertheless, if the reaction is hindered by a barrier, higher temperatures will increase the reaction probability as well. Thus a reaction over a barrier (barrier height E_a) will follow the strongly positive temperature dependence of an Arrhenius type behavior, $\exp(-E_a/kT)$, in the high-temperature limit. In summary, the temperature dependence can be negative (collision complex) for low temperatures but will be positive (barrier) for high temperatures. For the CH_4^+ + H_2 system the negative temperature dependence prevails up to more than 1000 K based on the trap and selected ion flow drift tube (SIFDT) experiments (Asvany *et al.* 2004b; Federer *et al.* 1985). Such a behavior would imply a very high barrier, which makes it difficult for the reaction to occur at low temperatures at all. This is in contradiction to the fact that abstraction proceeds in about 1/3 of the cases. In fact, abstraction of CH_4^+ + H_2 occurs about 100 times faster than in the isoelectronic NH_3^+ + H_2 system, which shows a turnover to a positive temperature dependence already at 150 K. Therefore it is concluded that a barrier is not hindering the abstraction reaction of CH_4^+ + H_2. Rather, the transition state seems to form a dynamical bottleneck that is hard to surpass. Note, that for the abstraction an atomic species has to be liberated to form CH_5^+ + D or CH_4D^+ + H, indicating a very intimate chemical interaction along this path. Apparently, this path can be followed without a substantial barrier. The products of a hypothetical reverse reaction, e.g., CH_4D^+ + H $\rightarrow$ CH_3D^+ + H_2 are energetically accessible when starting from a low temperature CH_4^+ + HD collision. Therefore deuterium exchange is obviously not hindered by a barrier. Nevertheless it does not occur. This finding is in sharp contrast to the simple model made by Henchman *et al.* (1988), where the hinderance of the H-D exchange is associated with a barrier in the transition state.

3.2. *Laser Induced Reactions*

In the series of small hydrocarbon ions only CH_3^+ undergoes rapid H-D exchange. It has been found in the kinetics studies above that the speed of this exchange depends on the nature of inelastic collision partners, p-H_2, o-H_2 or He, see Table 1. It is the aim of this section to (i) understand the possible reasons for this, as well as the consequences on isotopic fractionation and (ii) to describe an experiment, which allows a derivation of rotational populations of a low temperature ion cloud.

Inelastic collisions with H_2 are much more frequent in space than collisions with HD. Thus CH_3^+ relaxes to the lowest rotational states accessible. Due to the Pauli exclusion principle antisymmetric/symmetric rotational states of this Fermion are linked to the symmetric/antisymmetric nuclear spin state. The population of these states therefore depends on the H_2 temperature and the ortho-to-para ratio of H_2. As a consequence, the two nuclear spin configurations can be treated as separate molecules when dealing with inelastic collisions, which do not change the nuclear spin state of CH_3^+. In contrast, there is an interchange between the two configurations when nuclear spin changing collisions are permitted. Still in the latter case restrictions are imposed by the conservation of the total nuclear spin in those collisions. The fast deuteration of CH_3^+ shows that scrambling of the protons (and the deuteron) is a common process in the CH_4D^+ collision complex. This complex decomposes into different nuclear spin configurations, thus allowing for nuclear spin changes. It has been worked out by Bates (1991) for the CH_5^+ system that conservation of the total nuclear spin is one restriction for the possible outcome of such collisions. As a consequence, relaxation of CH_3^+ in H_2 leads to different rotational states of CH_3^+, depending on the ortho-to-para ratio of the H_2 collision partner. Apparently this must be the reason why deuteration is slower when CH_3^+ is relaxed in p-H_2 rather than in He or n-H_2.

Determination of the rotational population of such an ion cloud would yield direct insight into the role of the nuclear spin configurations. Such an experiment has been described recently by Mikosch *et al.* (2004). In this work a 22-pole trap experiment has been devised to determine the rotational population of an H_3^+ ensemble at $T = 55$ K by means of action spectroscopy. For this purpose H_3^+ is excited with more than two quanta of vibration in the ν_2-mode using a diode laser. Chemical probing in collisions with Ar is used to detect the excitation. The number of ArH^+ products is a direct measure for the number of H_3^+ found in the rotational state prior to excitation. Three rotational lines have been detected. From the relative intensities of the absorption lines the rotational temperature and from the Doppler width the kinetic temperatures have been derived. As a result H_3^+ is found to be thermalized.

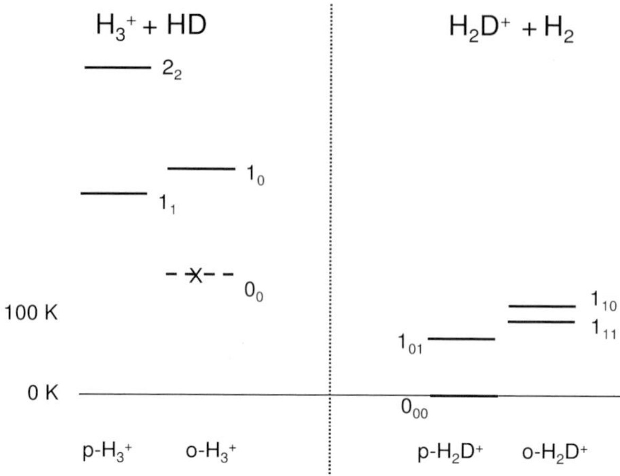

Figure 3. State specific energetics of the H_3^+ + HD $\leftrightarrow$ H_2D^+ + H_2 reaction. Only $J = 0$ states of HD and H_2 are considered. For collisions with, $J = 1$, o-H_2, 175 K extra energy has to be added.

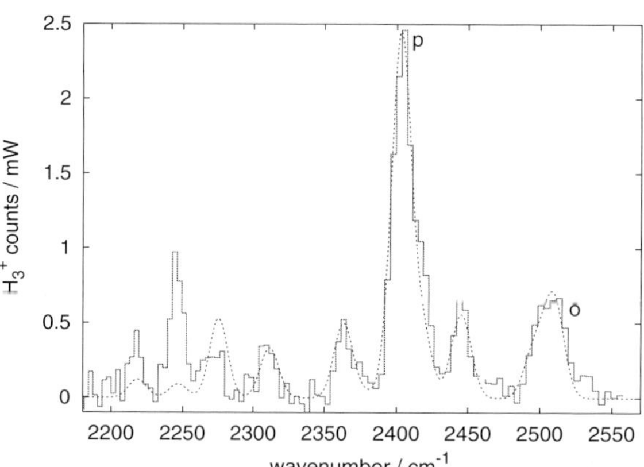

Figure 4. LIR spectrum of H_2D^+ in collisions with n-H_2. The simulation (dotted line) represents contributions from the ν_2 and ν_3 vibrational modes. Transitions of the lowest ortho (o) and para (p) state used for the analysis are marked. Deviations between simulation and spectrum hint at saturation effects which are discussed in the text.

It has been pointed out by Gerlich *et al.* (2002a) that the relative population of the lowest ortho state of H_2D^+ plays a key role for the efficiency of the reverse reaction of deuteration of H_3^+, see Figure 3. In effect the back reaction becomes exoergic in collisions of o-H_2D^+ with o-H_2 since it adds some 175 K of rotational energy. Therefore it is most interesting to determine the rotational distribution of H_2D^+. This has been achieved in the present work by ro-vibrational excitation of H_2D^+ using the free electron laser FELIX, in the range of the fundamental ν_2 and ν_3 vibrational modes. In this way an equivalent of some 3500 K of energy is added, thus enhancing the back reaction with n-H_2 substantially. By detecting the number of H_3^+ products as a function of wavelength, the spectrum of H_2D^+ shown in Figure 4 has been recorded. From the intensities of these lines relative rotational populations of the vibrational ground state H_2D^+ could be derived. For the analysis, recently calculated Einstein A coefficients of the corresponding transitions were used (Ramanlal & Tennyson 2004), from which the corresponding Einstein B coefficients were derived. In addition, it has been assumed that all excited ro-vibrational states undergo hydrogenation with the same efficiency and that the rate of fluorescence, A, is much smaller than the collisional rate at $[H_2] = 7 \times 10^{10}\,\mathrm{cm}^{-3}$. The nominal temperature of the trap was $T = 17$ K. According to the preliminary analysis, the rotational temperature lies between 20 and 30 K, and the population ratio of the lowest ortho and para state of H_2D^+ is about $\sim 3/4$. Preliminary experiments using p-H_2 show much smaller values. These results are in agreement with the general trend that p-H_2 leads to a more efficient relaxation than n-H_2. This is an important result hinting at the crucial role of nuclear spin in low temperature collisions. However, higher populations of the lowest ortho state of H_2D^+ would be expected in both cases based on the previous study of Gerlich *et al.* (2002a).

Before final conclusions can be drawn, several more systematic studies are necessary. The validity of the assumptions made for the analysis have to be checked in future experiments, e.g., by determining the dependence of the signal on the H_2 number density, thus comparing fluorescence, dominating at *low* densities, with collisions, dominating at *high* densities. Moreover, it has to be taken into account that FELIX is a pulsed laser system with micro-pulses of several picosecond duration only. Therefore saturation effects influence the experimental outcome. In order to overcome such a limitation, experiments with continuous laser excitation are underway. Also a narrower bandwidth excitation allows for the determination of the kinetic temperature of the ion cloud.

4. Conclusions

Deuteration of CH_4^+ in collisions with HD shows that the simple classification of "unfilled" and "filled valence shells" even in the more sophisticated version of Maluendes *et al.* (1992) cannot be used as a general guidance to distinguish reactants, which do or do not undergo H-D exchange. In contrast, this example demonstrates that not even the knowledge of particular points along the minimum energy path is sufficient. Due to the dynamical restrictions occurring along the reaction path dynamical calculations are necessary to predict the fate of the reactants.

Scrambling of H and D is an important way for efficient H-D exchange as seen for $CH_3^+ + HD$ because atoms can swap places without the need of breaking and forming the very strong bonds of the reactants. This process is linked to the chemical equivalence of the atoms involved. Conservation of total nuclear spin in collisions with identical nuclei has to be considered. Therefore state-to-state rate coefficients are needed. One experimental route towards this aim is the determination of rotational state populations. Only several hundred H_2D^+ are needed to detect a ro-vibrational LIR-spectrum. It is possible to determine the rotational population from such spectra based on several assumptions. A first example spectrum has been analyzed. However, more systematic LIR studies are necessary. Once reliable rotational state populations have been derived at low temperatures, more extensive modelling of the H_3^+ and CH_3^+ systems can be used to derive state-to-state rate coefficients for inelastic and reactive collisions. With the advent of SOFIA and the Herschel satellite several observations detecting transitions between the lowest rotational states of H_2D^+ and D_2H^+ are planned. It will be interesting to see how experiment, astrophysical modelling and observation compare.

Acknowledgements

This work has been financially supported by the Deutsche Forschungsgemeinschaft (DFG) via the Forschergruppe FOR 388 "Laboratory Astrophysics" and by the Nederlandse Organisatie voor Wetenschappelijk Onderzoek (NWO) via grant 614.000.415. We gratefully acknowledge support by the Stichting voor Fundamenteel Onderzoek der Materie (FOM) in providing the required beam time on FELIX and highly appreciate the skilful assistance by the FELIX staff (René van Buuren, Lex van der Meer, Britta

Redlich). We thank Wim van der Zande and Danny Ramsamoedj for supplying parahydrogen. Furthermore the authors are thankful for the excellent support of the electronic and mechanical workshops of Leiden University (Ewie de Kuyper, Koos Benning, René Overgauw, Arno van Amersfoort).

References

Asvany, O., Schlemmer, S., & Gerlich. D. 2004a, *Ap. J.* 617, 685

Asvany, O., Savic, I., Schlemmer, S., & Gerlich., D. 2004b, *Chem. Phys.* 298, 97

Asvany, O., Giesen, T., Redlich, B., & Schlemmer, S. 2005a, *Phys. Rev. Lett.* 94, id.073001

Asvany, O., Kumar, P., Redlich, B., Hegemann, I., Schlemmer, S., & Marx, D. 2005b, *Science* 309, 1219

Bates, D.R. 1991, *Ap. J.* 375, 833

Black, J.H., & Dalgarno, A. 1977, *Ap. J. Suppl.* 34, 405

Federer, W., Villinger, H., Tosi, P., Bassi, D., Fergusson, E., & Lindinger, W. 1985, in *Molecular Astrophysics-State of the art and future directions*, eds. G.H.F. Diercksen *et al.* (Reidel, Boston), p. 649

Flower, D.R., Pineau des Forêts, G., & Walmsley., C.M. 2004, *A&A* 427, 887

Gerlich D. & Horning, S. 1992, in *Ion-Molecule Reactions*, ed. Z. Herman, *Chemical Reviews* 92, 1509

Gerlich, D., Herbst, E., & Roueff, E. 2002a, *Planetary and Space Science* 50, 1275

Gerlich, D. & Schlemmer, S. 2002b, *Planetary and Space Science* 50, 1287

Gerlich, D. 2004, in *XIV Symposium in Atomic, Cluster and Surface Physics*, ed. P. Casavechia (La Thuile, Italy), 1

Henchman, M., Paulson, J.F., Smith, D., Adams, N.G., & Lindinger, W. 1988, in *Rate coefficients in astrochemistry*, eds. T.J. Millar and D.A. Williams, p. 201

Jacox, M.E. 1994, *J. Phys. Chem. Ref. Data* 3

Lis, D.C., Roueff, E., Guerin, M., Phillips, T.G., Coudert, L.H., van der Tak, F.F.S., & Schilke, P. 2002, *Ap. J.* 571, L55

Maluendes, S.A., McLean, A.D., & Herbst, E. 1992, *Ap. J.* 397, 477

Mikosch, J., Kreckel, H., Wester, R., Plašil, R., Glosík, J., Gerlich, D., Schwalm, D., & Wolf, A. 2004, *J. Chem. Phys.* 121, 11030

Millar, T.J. 2003, *Space Science Reviews* 106, 73

Quack, M. 1977, *Mol. Phys.* 34, 477

Parise, S., Castets, A., Herbst, E., Caux, E., Ceccarelli, C., Mukhopadhyay, I., & Tielens, A.G.G.M. 2004, *A&A* 416, 159

Ramanlal, J. & Tennyson, J. 2004, *MNRAS* 354, 161

Savic, I. & Gerlich, D. 2005, *Phys. Chem. Chem. Phys.* 7, 1026

Savic, I., Schlemmer, S., & Gerlich, D. 2005, *Ap. J.* 621, 1163

Schlemmer, S., Kuhn, T., Lescop, E., & Gerlich, D. 1999, *Int. J. Mass Spectrom.* 185, 589

Schlemmer, S., Lescop, E., v. Richthofen, J., & Gerlich, D. 2002, *J. Chem. Phys.* 117, 2068

Schlemmer, S., Asvany, O., & Giesen, T. 2005, *Phys. Chem. Chem. Phys.* 7, 1592

Vastel, C., Phillips, T.G., & Yoshida, H. 2004, *Ap. J.* 606, L127

Walmsley, C.M., Flower, D.R., & Pineau des Forêts. G. 2004, *A&A* 418, 1035

Discussion

KROES: CH_4^+ +HD produces mostly CH_5^+ through a dynamic mechanism (H farther away from centre of mass of HD). For which other systems do you expect this mechanism to be important as well?

SCHLEMMER: The effect of producing more hydrogenated than deuterated species has been predicted by David Clary's group for the collision of O^+ +HD. However, no experiments have been carried out so far. Based on the arguments given in their publication one could expect this effect in many systems exhibiting hydrogen abstraction.

Astrochemistry: Recent Successes and Current Challenges
Proceedings IAU Symposium No. 231, 2005
D.C. Lis, G.A. Blake & E. Herbst, eds.

© 2006 International Astronomical Union
doi:10.1017/S1743921306007125

Shock Models Revisited

Malcolm Walmsley[1], Guillaume Pineau des Forêts[2], and David Flower[3]

[1]INAF-Osservatorio Astrofisico di Arcetri, Largo E.Fermi 5, I-50125 Italy
email: walmsley@arcetri.astro.it
[2]IAS, Université de Paris–Sud, 92405 Orsay Cedex, France
[3]Physics Department, The University, Durham DH1 3LE, UK

Abstract. We review progress in the area of the modelling of shocks in molecular clouds. In particular, we consider what has been learnt about shock structure evolution in situations where the steady-state assumption is no longer valid. We discuss the interpretation of the observed water abundance from SWAS, Odin, and ISO. We also consider the erosion of grains in shocks as well as the effect of the presence of grains on shock structure.

1. Introduction

Molecular clouds exhibit supersonic linewidths and hence one might expect that their structure and chemical make-up is influenced by shocks. Shock waves cause compression and heating of the pre-shock material. Both effects have consequences for cloud structure as well as for the observed molecular abundance distribution in molecular clouds as shown in numerous articles over the past three decades (e.g., Hollenbach & McKee 1979; McKee & Hollenbach 1980; Draine *et al.* 1983; Neufeld & Dalgarno 1989; Hollenbach 1997). In fact, the very origin of molecular clouds is likely due to shock compression (Bergin *et al.* 2004) and so shock chemistry may have considerable relevance for star formation and galactic evolution.

However, in this brief review, we will confine ourselves to a few topics which have attracted attention during the past 5 years due to either observational or theoretical developments. One of these is the consequences for shock structure and chemistry of departures from steady-state. We examine these in § 2 and consider the observational predictions of evolution in shock structure. In § 3, we consider what we have learnt from recent satellite observations of water emission of shocks and in § 4, the behavior of dust grains in shocks (as well as their effect on shock structure). Finally, in § 5, we discuss future directions.

2. Non Steady Shocks

The timescale on which matter flows through a C-shock is relatively long (dependent on the ionization degree of the pre-shock medium) and can be of the same order (a few thousand years) as estimated ages for outflows from YSOs. Thus, it is possible (see Chièze *et al.* 1998; Flower *et al.* 2003a; Lesaffre *et al.* 2004a; Lesaffre *et al.* 2004b) that the shock structures in outflow sources are evolving giving rise to a "mixed" situation where a J-type (or Jump) discontinuity is embedded in a C-type (or Continuous) flow. To complicate matters still further, the details of the chemistry and in particular the balance between atomic and molecular hydrogen influences the structure. Last, but not least, grains can considerably influence shock structure (see discussion in § 4).

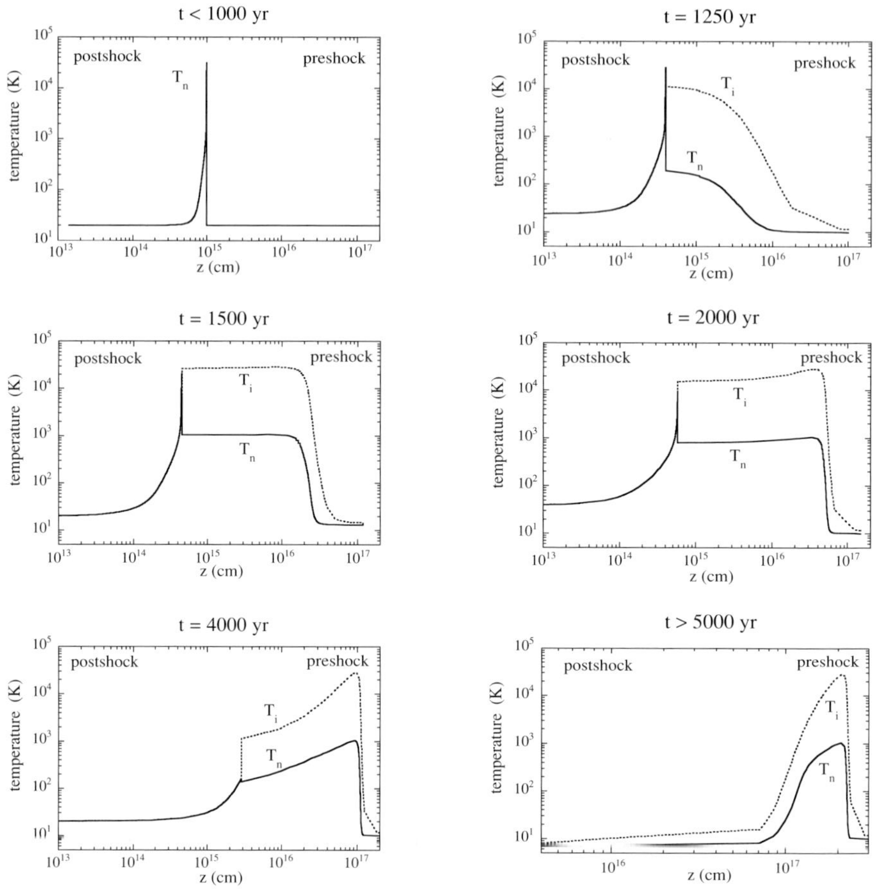

Figure 1. Computed temperature structure of C-J shock as function of time for a 25 km s^{-1} shock incident on a medium of density 10^4 cm^{-3} and magnetic field 100 μG. At early times (*top left*), the temperature shows a J-type discontinuity with rapid cooling in the post-shock gas whereas beyond 1000 years, the shock precursor heats both ions (dotted curve) and neutrals (full curve). Beyond 5000 years (*bottom right*), the shock has become stationary C-type. From Pineau des Forêts & Flower (2000).

An example showing the evolution of shock structure with time is given in Figure 1. One sees that with time, a discontinuous J-type structure evolves into a C-shock. How in detail this happens depends on both ionization degree and magnetic field among other things. One concludes that many observed outflows could be in the latter phase of this evolution. Are there observational discriminants which mark a C-J shock of this type? In fact, ISO observations of the rotational H$_2$ lines towards some outflows do need a mixture of C and J to explain them. Figure 2 from the work of Flower *et al.* 2003b (see also Cabrit *et al.* 2004) is an example of this. One sees that whereas a C-shock fits the low-excitation lines well and a J-shock fits the high-excitation lines, one needs a C-J model to fit both.

We note, however, that these models are one dimensional and that a proper treatment of a jet-outflow system will presumably involve a variety of shocks of differing obliquities and velocities. Distinguishing between different models will require consideration of line

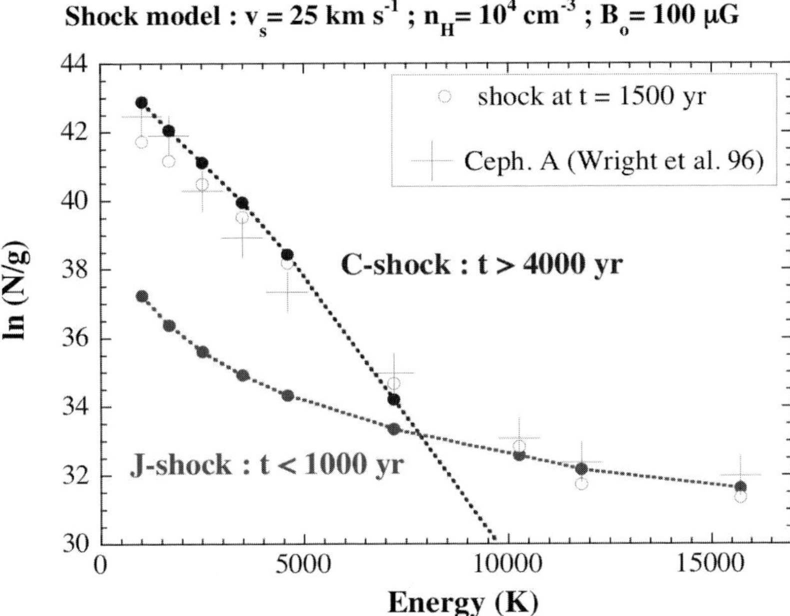

Figure 2. Log of level column density against level excitation of H_2 rotational transitions for a C-J shock model with pre-shock H number density 10^4 cm^{-3} and magnetic field 100 μG. The empty circles show model results for a "shock age" of 1500 years while the crosses show observations from Wright *et al.* (1996). The dotted lines connecting filled circles show model expectations both for a C-shock (fitting the low-excitation data) and a J-shock (fitting high-excitation lines).

profiles at well as fine scale spatial information. It will also require consideration of species such as water whose abundance varies dramatically in different contexts.

3. The Water Conundrum

It has been clear for many years that interstellar water plays a central role in the oxygen balance of molecular clouds. This is demonstrated by the fact that in *solid* form, it is often extremely abundant (see e.g., van Dishoeck 2004). Moreover, it has been known for a long time that in post-shock gas, above about 250 K, all available oxygen gets converted into water and that for shock velocities above 25 km s^{-1}, ice mantles become eroded (Draine *et al.* 1983; Flower and Pineau des Forêts 1994). Thus much shocked gas should be water rich and the searches for gas-phase water with ISO, SWAS, and Odin were based upon this assumption. The observations however only partially confirmed the theoretical predictions (though see Harwit *et al.* 1998, Wright *et al.* 2000) and in general, the water abundance was lower than expected (e.g., Snell *et al.* 2005; Liseau *et al.* 1996; Neufeld *et al.* 2000; Benedettini *et al.* 2003; González–Alfonso *et al.* 2002; Olofsson *et al.* 2003). We briefly consider in this section measurements of the water abundance in shocked regions and their interpretation.

One of the problems interpreting the water observations is that observed intensities are very sensitive to the distribution of density and temperature of the observed structures which are typically not resolved by the beams of, e.g., ISO and SWAS. This can easily cause order of magnitude uncertainty in the inferred water abundances, although one can improve matters using multi-line studies. It is also true that using CO as a surrogate for the H_2 column density can be very misleading due to its differing excitation characteristics. Nevertheless, there are cases such as the interaction of the supernova remnant IC443 with surrounding molecular gas (Snell *et al.* 2005) where the water under-abundance is presently a challenge to the modellers. Here, one is faced with the problem not only of fitting the water data but also OH and atomic oxygen (Burton *et al.* 1990).

There are a variety of possible solutions that have been proposed to resolve the "water conundrum" but no single solution appears to pass all observational tests. In unshocked dense gas, it is possible that a large fraction of oxygen is in the form of water ice and hence is unavailable to form gas-phase water (or OH, O_2 etc). This can also be important in low-velocity shocked gas (below $25 \, \mathrm{km \, s^{-1}}$) where sputtering is not sufficiently efficient to desorb the outflows. But the velocities detected in regions such as the IC443 clumps are much higher than this suggesting that sputtering of ice mantles is likely. It is also the case that the most suitable venue for producing hot gas-phase water in in C-shocks of, say, 20–$40 \, \mathrm{km \, s^{-1}}$. These however may be suppressed if the ionization degree is much higher than that obtained assuming cosmic rays to be the main ionizing agents. Could that be true? It is not clear but certainly direct estimates of ionization degree are difficult and the question of the influence of shock produced UV is important (e.g., Sternberg & Dalgarno 1989; Viti *et al.* 2003).

4. Grains and Their Consequences

The presence of grains influences shock structure and the destruction of grains in shocks has observational consequences. The first question has had considerable study over the past few years (Ciolek & Roberge 2002; Ciolek *et al.* 2004; Flower & Pineau des Forêts 2003) but it seems likely that the last word has not been said. Results are very sensitive to the grain mass distribution and thus, to the extent to which grains have coagulated and acquired ice mantles. Figure 3 gives a recent estimate of the boundary between C and J-shocks. We see that at low densities, the critical speed for H_2 dissociation (Le Bourlot *et al.* 2002) is the dominant effect whereas at high pre-shock densities, the effective magnetosonic speed of the grains is most important (Cabrit *et al.* 2004).

Linked to this question is that of grain survival in shocks (see review of Jones 2000). One of the interests here is that some grains survive at shock velocities well above $100 \, \mathrm{km \, s^{-1}}$. On the other hand, some erosion does occur in shocks with velocities of order 50–$80 \, \mathrm{km \, s^{-1}}$ as evidenced by the iron abundance measured along jets from young stars (30% in the gas phase; Nisini *et al.* 2002; Nisini *et al.* 2005). Moreover, there is also evidence for dust emission (Smith *et al.* 2005) from some illuminated jets and there appear to be good prospects for comparing these results with theory. In the jets, one is presumably dealing with J-type shocks but there is also evidence for dust erosion in C-type shocks. This mainly comes from the measurements of mm-wavelength SiO lines (e.g., Jiménez–Serra *et al.* 2005) where one observes greatly enhanced SiO abundances (factor 1000) in the outflow from some young Class 0 objects. One of the peculiarities here is that one does not observe evidence for erosion of refractory elements other than Si (suggesting perhaps that silicon oxide in some form is a minor component of the dust which can be eroded in C-shocks). FeO , for example (Walmsley *et al.* 2002), has been searched for in many outflows but only found in the mysterious absorbing layer towards SgrB2.

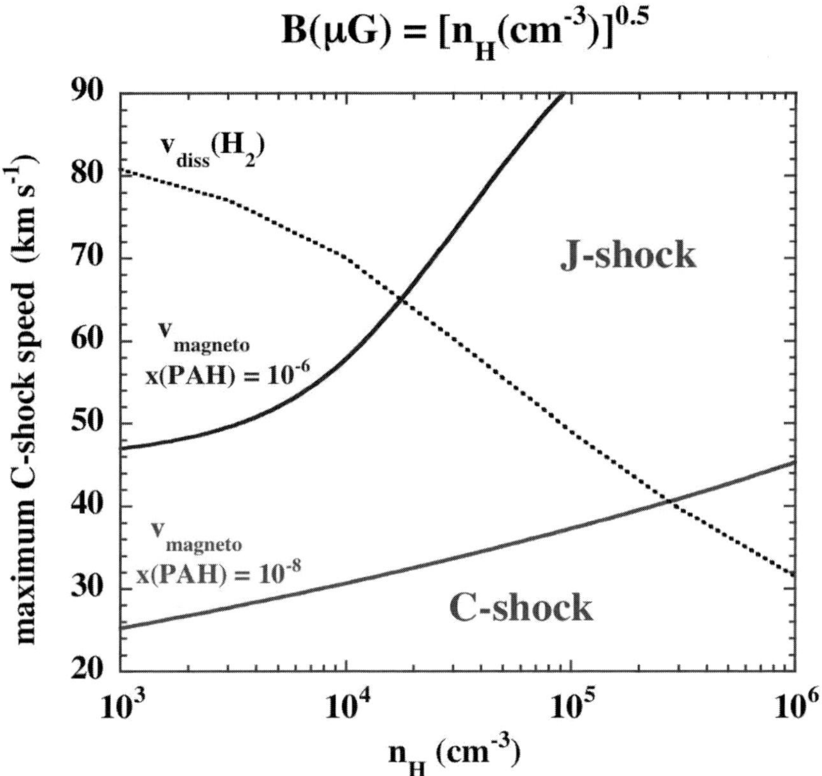

Figure 3. Critical velocity for the transition between a C-shock and a J-shock as a function of pre-shock H number density. The full lines show the magnetosonic speed taking grain coupling into account for two assumed abundances for PAHs. The dotted line shows the critical shock velocity for H_2 dissociation from Le Bourlot *et al.* (2002).

5. Future Prospects

Clearly, the new generation of instruments is going to give new impetus to this field and some of the topics discussed here already demonstrate that. A tenth of an arc second is 50 AU at 500 parsec (the distance of Orion). Studies at this resolution would allow one to probe the details of C shocks and thereby test current models with much more confidence. It is also worth noting that the increasing sophistication of MHD models may allow the insertion of chemistry and grain characteristics in a self-consistent fashion into shock models.

References

Benedettini, M., Viti, S., Giannini, T., Nisini, B., Goldsmith, P.F., & Saraceno, P. 2003, in *Chemistry as a Diagnostic of Star Formation*, eds. C.L. Curry & M. Fich (NRC Research Press), p. 248

Bergin, E.A., Hartmann, L.W., Raymond, J.C., & Ballesteros-Paredes J. 2004, *Ap. J.* 612, 921

Burton, M.G., Hollenbach, D.J., Haas, M.R., & Erickson, E.F. 1990, *Ap. J.* 355, 197

Cabrit, S., Flower, D.R., Pineau des Forêts, G., Le Bourlot, J., & Ceccarelli, C. 2004, *Ap. Sp. Science* 292, 501

Chièze, J.-P., Pineau des Forêts, G., & Flower, D.R. 1998, *MNRAS*, 295, 672

Ciolek, G.E. & Roberge, W.G. 2002, *Ap. J.* 567, 947

Ciolek, G.E., Roberge, W.G., & Mouschovias, T.Ch. 2004, *Ap. J.* 610, 781

Draine, B.T., Roberge, W.G., & Dalgarno, A. 1983, *Ap. J.* 264, 485

Flower, D.R. & Pineau des Forêts, G. 1994, *MNRAS* 268, 724

Flower, D.R., Le Bourlot, J., Pineau des Forêts, G., & Cabrit, S. 2003a, *Ap. Sp. Science* 287, 183

Flower, D.R., Le Bourlot, J., Pineau des Forêts, G., & Cabrit, S. 2003b, *MNRAS* 341, 70

Flower, D.R. & Pineau des Forêts, G. 2003, *MNRAS* 343, 390

González–Alfonso, E., Wright, C.M., Cernicharo, J., Rosenthal, D., Boonman, A.M.S., & van Dishoeck, E.F. 2002, *A&A* 386, 1074

Harwit, M., Neufeld, D.A., Melnick, G.J., & Kaufman, M.J. 1998, *Ap. J.* 497, L105

Hollenbach, D. & McKee, C.F. 1979, *Ap. J. Suppl.* 41, 555

Hollenbach, D. 1997, *Proceedings of IAU Symposium 182*, ed. B. Reipurth

Jiménez–Serra, I., Martin–Pintado, J., Rodríguez–Franco, A., & Martín, S. 2005, *Ap. J.* 627, L121

Jones, A.P. 2000, *J. Geophys. Res.* 105, 10, 257

Le Bourlot, J. *et al.* 2002, *MNRAS* 332, 985

Lesaffre, P., Chièze, J.-P, Cabrit, S., & Pineau des Forêts, G. 2004, *A&A* 427, 147

Lesaffre, P., Chièze, J.-P, Cabrit, S., & Pineau des Forêts, G. 2004, *A&A* 427, 157

Liseau, R., Ceccarelli, C., Larsson, B., Nisini, B., White, G.J. *et al.* 1996, *A&A* 315, L181

McKee, C.F. & Hollenbach, D.J. 1980, *ARAA* 18, 219

Neufeld, D.A. & Dalgarno, A. 1989, *Ap. J.* 340, 869

Neufeld, D.A., *et al.* 2000, *Ap. J.* 539, L107

Nisini, B., Caratti o Garatti, A., Giannini, T., & Lorenzetti, D. 2002, *A&A* 393, 1035

Nisini, B., Bacciotti, F., Giannini, T., Massi, F., Eislöffel, J., Podio, L., & Ray, T.P. 2005, *A&A*, in press

Olofsson, A.O.H., *et al.* 2003, *A&A* 402, L47

Pineau des Forêts, G. & Flower, D.R. 2000, in *Molecular Hydrogen in Space*, eds. F. Combes & G. Pineau des Forêts (CUP), p. 117

Smith, N., Bally, J., Shuping, R.Y., Morris, M., & Kassis, M. 2005, *A. J.*, in press

Snell, R.L., Hollenbach, D., Howe, J.E., Neufeld, D.A., Kaufman, M.J., Melnick, G.J., Bergin, E.A., & Wang Z. 2005, *Ap. J.* 620, 758

Sternberg, A. & Dalgarno, A. 1989, *Ap. J.* 338, 197

van Dishoeck, E.F. 2004, *ARAA* 42, 119

Viti, S., Girart, J.M., Garrod, R., Williams, D.A., & Estalella R. 2003, *A&A* 399, 187

Walmsley, C.M., Bachiller, R., Pineau des Forêts, G., & Schilke, P. 2002, *Ap. J.* 566, L109

Wright, C.M., Drapatz, S., Timmermann, R., van der Werf, P.P., Katterloher R., & de Graauw, T. 1996, *A&A* 315, L101

Wright, C.M., van Dishoeck, E.F., Black, J.H., Feuchtgruber, H., Cernicharo, J., González Alfonso, E., & de Graauw, Th. 2000, *A&A* 358, 689

Astrochemistry: Recent Successes and Current Challenges
Proceedings IAU Symposium No. 231, 2005
D.C. Lis, G.A. Blake & E. Herbst, eds.

© 2006 International Astronomical Union
doi:10.1017/S1743921306007137

PDRs and XDRs: Theory and Observations

Amiel Sternberg

School of Physics & Astronomy, Tel Aviv University, Ramat Aviv, 69978, ISRAEL
amiel@wise.tau.ac.il

Abstract. I present an overview of theory and observations of FUV-photon and X-ray dominated regions (PDRs and XDRs), with a discussion of some recent results.

Keywords. PDRs — XDRs

---

1. Introduction

In this talk I present an overview of interstellar "photon-dominated regions" or "photodissociation regions" (PDRs), and "X-ray dominated regions" (XDRs), with a discussion of both theory and observations. The reader is invited to read this document while clicking through the power-point presentation.†

PDRs are most broadly defined as neutral components of the interstellar medium (ISM) where far-ultraviolet (FUV; 6–13.6 eV) radiation from external sources, usually massive hot OB stars and clusters, dominates the gas heating, and controls the chemical structure. In PDRs, the FUV photon penetration depth is limited by dust absorption and scattering, and the gas is heated primarily by photoelectric emission of electrons from grain surfaces. The chemistry is driven by FUV ionization of species with ionization thresholds longward of the hydrogen Lyman limit. In contrast (slide 2) XDRs refer to neutral hydrogen clouds in which keV X-rays dominate the gas heating and chemistry. In XDRs the X-ray penetration is limited by photoionization of the heavy elements ("metals") rather than by dust particles. For this reason the photon penetration lengths in XDRs are much larger than in PDRs. In XDRs the gas is heated by direct photoionization of H and H_2. As I will explain, this leads to more efficient gas heating in XDRs compared to PDRs.

PDRs include the classical "warm neutral medium" (WNM) and "cold neutral medium" (CNM), optically thin diffuse clouds, thicker translucent clouds with moderate visual extinctions ($A_V \lesssim 5$) and intermediate gas densities ($n_H \lesssim 10^3$ cm^{-3}), and optically thick interfaces in dense ($n_H \gtrsim 10^3$ cm^{-3}) molecular clouds with high ($A_V \sim 10$) visual extinctions (slide 3). Dense PDRs exposed to intense FUV fields in star-forming regions are of particular interest as they are sources of luminous millimeter, submillimeter, and far-infrared line emissions, such as O I 63 μm and C II 157.7 μm fine-structure lines, and rotational and vibrational lines of a variety of molecules including the most abundant CO and H_2 molecules. XDRs may be prominent near intense X-ray sources such as accreting black holes in active galactic nuclei.

The study of PDRs, and subsequently XDRs, dates back to the late 70s and early 80s when advances in detector technology opened up the submillimeter and far-infrared spectral windows. In this talk I will comment on some recent results and puzzles. Comprehensive reviews have been presented by Hollenbach & Tielens (1997, 1999).

---

† `http://asilomar.caltech.edu/sternberg.ppt`

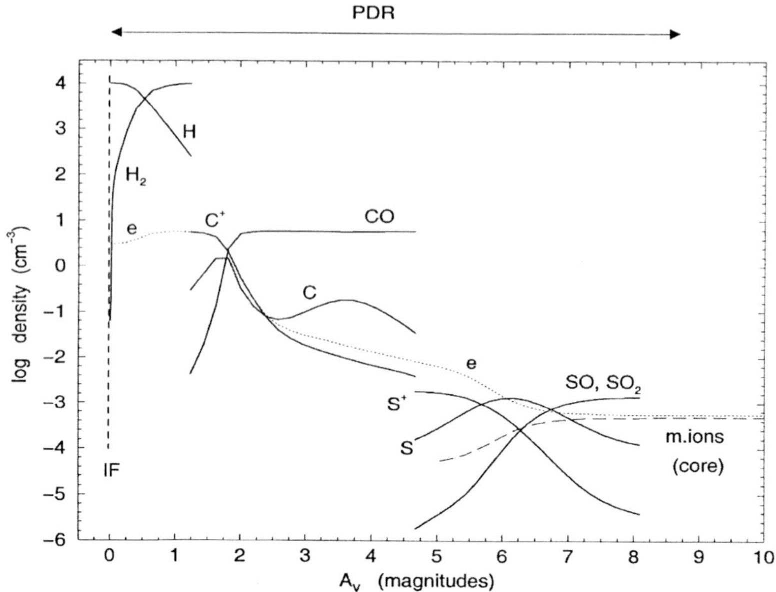

Figure 1. Schematic structure of a PDR. FUV radiation is incident from the left.

2. Physical Processes

The starburst galaxy M82 (3.3 Mpc) provides a nice example of a global PDR (slide 4). Optical imaging shows a central dusty starburst with large-scale $H\alpha$ filaments extending out of the galaxy in a galactic superwind driven by the starburst. Infrared, submillimeter, and millimeter-wave spectroscopy of this source reveals a thermal dust continuum, and a wealth of PDR emission lines. Of particular importance are the observed fine-structure line strengths, with line-to-continuum ratios of 0.1–1%. This is typical of many Galactic and extragalactic PDRs.

The PDR line strengths may be understood by considering theoretical models. In the simplest picture (Fig. 1; slide 5) a dense PDR may be considered as a plane-parallel slab of gas that absorbs FUV radiation incident from one side. Any UV radiation capable of ionizing hydrogen is assumed to be absorbed in a thin ionization front adjacent to the PDR. The penetrating FUV radiation dissociates molecules, and ionizes atoms, as it penetrates the cloud. The photorates are attenuated with increasing cloud depth, or visual extinction, as the FUV radiation is absorbed by the dust. The energy absorbed by the dust is reradiated as thermal infrared. Hence atoms such as oxygen (IP 13.62 eV) remain neutral in the PDR, but atoms such as carbon (IP 11.26 eV) are singly ionized (in the outer parts). PDRs consist of several distinct zones and transition layers, from outer photodissociated gas to inner fully molecular gas (Jansen *et al.* 1995; Sternberg & Dalgarno 1995). The H/H$_2$ transition occurs closest to the cloud surface where the combined effects of dust opacity and H$_2$ absorption-line "self-shielding" reduce the photodissociation rate. The H/H$_2$ transition point depends on the efficiency with which H$_2$ forms on the grain surfaces. The transition from C$^+$ to C to CO occurs deeper in the cloud, and a similar S$^+$ to S to SO, SO$_2$, CS transition zone occurs at still larger A_V. The sizes of the various transition layers are controlled by several parameters, including most importantly the cloud density (or pressure), and the intensity of the illuminating radiation field. Additional controlling parameters are the grain FUV scattering properties, the cloud geometry and clumpiness, and the elemental gas-phase abundances.

The FUV intensity is a crucial paramter. Near hot OB stars in star-forming regions the FUV fields can be orders of magnitude more intense than the mean value in the ISM. The FUV intensity is usually given in terms of a scaling factor (G_0) relative to the empirically based mean value in the ISM in the solar neighborhood (Habing 1968; Draine 1978; slide 6). Recent *ab initio* computations of the interstellar FUV field (Parravano, Hollenbach & McKee 2003) show that dust opacity limits the contributing OB associations out to distances of 500 pc from typical points within the Galactic disk. The mean FUV field fluctuates in response to the birth and death of OB associations. Intense fields near individual clusters can also vary on shorter timescales (Myr) as the massive stars evolve (Sternberg, Hoffmann & Pauldrach 2003; slide 7).

The primary gas heating mechanim in PDRs is photoelectric emission from dust grains, first considered by Spitzer (1948) and investigated in many subsequent studies (Draine 1978; Bakes & Tielens 1994; Weingartner & Draine 2001). In this process energetic electrons are ejected from the grains at a rate that depends on the FUV intensity and the grain charge, set by the balance between photoemission and recombination (slide 8). A second important process, occurring in dense ($n \gtrsim 5 \times 10^4$ cm^{-3}) gas, is the collisional deexcitation of FUV pumped H_2 (Sternberg & Dalgarno 1989).

The gas is cooled by neutral impact excitations of atomic fine-structure, and molecular rotational energy levels, followed by radiative decays. The gas temperature is then set by the balance between heating and cooling. Predicted, and observed, gas temperatures range from a few 100 K up to 1000 K in the outer cloud layers, down to several 10 K at large depths where the FUV is fully attenuated and where other process such as cosmic-ray heating dominates.

Fine-structure cooling plays a particularly important role in PDRs. Indeed the first PDR models (Tielens & Hollenbach 1985) were constructed in order to account for the observed strengths of fine-structure emission lines in PDRs. A famous example is the $^2P_{1/2}$–$^2P_{1/2}$ 157.7 μm line of C^+. Long before it was first detected in Orion (Russell *et al.* 1980), the C II line was recognized theoretically as a likely major cooling transition for the neutral hydrogen component of the ISM (Spitzer 1948; Dalgarno and McCray 1972). C II cooling plays a crucial role in the theory of the two-phase CNM/WNM structure of the H I gas in the ISM (Field, Goldsmith, & Habing 1969; Wolfire, McKee, & Hollenbach 2003), dominating the energy losses of the cool phase. However, the early theoretical considerations focussed primarily on the relatively low density gas ($n_H \lesssim 50$ cm^{-3}), where the excited level is subthermally populated and the C II emissivities are low (Fig. 2; slide 9). For fixed density, the cooling efficiency decreases sharply for temperatures T below the transition energy of 92 K, and levels off at higher T, where the Boltzmann factor in the collisional excitation rate coefficient approaches unity. At fixed gas temperature the efficiency increases linearly with density up to the critical hydrogen density of 10^4 cm^{-3}, above which collisional deexcitations quench the radiative emissions. For conditions in the classical WNM ($n_H \sim 0.4$ cm^{-3}, $T \approx 8000$ K) and CNM ($n_H \approx 60$ cm^{-3}, $T \approx 50$ K), the C II emission efficiencies are well below the maximum possible. However, for densities greater than 10^2 and up to at least 10^4 cm^{-3}, and for T above $\sim 10^2$ K, the cooling efficiencies are significantly enhanced. This range of elevated density and temperature corresponds to conditions characteristic of dense PDRs.

Thus, in PDRs, most of the incident FUV energy is absorbed by dust particles and reradiated as quasi-thermal IR, but a small fraction of the FUV heats the gas via the photoelectric ejection of electrons from the grain surfaces. The typical "work-function", for interstellar grains is $W = 6$ eV, and the expected photoelectric yield, or the fraction of electrons that are actually dislodged off the grains, is typically $Y = 0.1$. Therefore, for a mean incident FUV photon energy of 10 eV, a fraction ($[10 - 6]/10) \times 0.1 = 0.04$ of

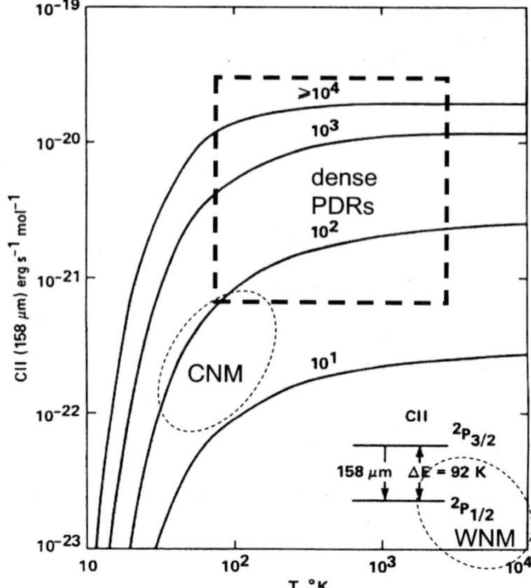

Figure 2. C II line cooling per particle as a function of the gas temperature and hydrogen density, for neutral impact excitation (adapted from Tielens & Hollenbach 1985).

the original photon energy is available to heat the gas (slide 10). This fraction, or "heating efficiency", decreases if the grains become positively charged, as they do when the photoelectron emission rates become large relative to the grain-electron recombination rates. This argument predicts cooling emission lines with intensities of order 1% of the thermal IR dust continua, consistent with observations. Photoelectric heating is probably dominated by the smallest (< 15 Å) grains or PAHs for typical grain size distributions (Bakes & Tielens 1994; Weingartner & Draine 2001).

The fine-structure line intensties are, therefore, probes of the incident FUV fluxes. For fixed density the line intensities saturate with increasing G_0 as the heating efficieny decreases (Tielens & Hollenbach 1985; Sternberg & Dalgarno 1989). Line ratios, such as [O I] 63.2 μm / [C II] 157.7 μm, are sensitive to the density, because O I is quenched at higher density than is C II (*e.g.*, Kaufmann *et al.* 1999; slide 11).

3. PDR Chemistry, CN/HCN

The $C^+/C/CO$ transition layer is one of several locations in PDRs where specific types of molecules are formed (Sternberg & Dalgarno 1995). The depth at which the $C^+/C/CO$ transition occurs depends on the C^+ recombination efficiency. Because radiative recombination is slow, processes such as mutual neutralization of C^+ ions with negative PAH^- ions (Lepp & Dalgarno 1988; Bakes & Tielens 1998), and "grain assisted" recombination (Weingartner & Draine 2001) are important (slides 13 and 14).

High abundances of molecular radicals are expected in the outer parts of PDRs where the $C^+/C/CO$ transition occurs. One nice example of this is the behavior of CN/HCN in PDRs (Boger & Sternberg 2005). Fuente *et al.* (1993, 2003) presented molecular maps of the edge on PDR in the reflection nebula NGC 7023. In this object, the illuminating star is a Herbig B3Ve star, and $G_0 = 2.4 \times 10^3$ at the PDR interface. The cloud density is about 10^4 cm^{-3}. Fuente *et al.* found that the CN radical peaks closer to the illuminating

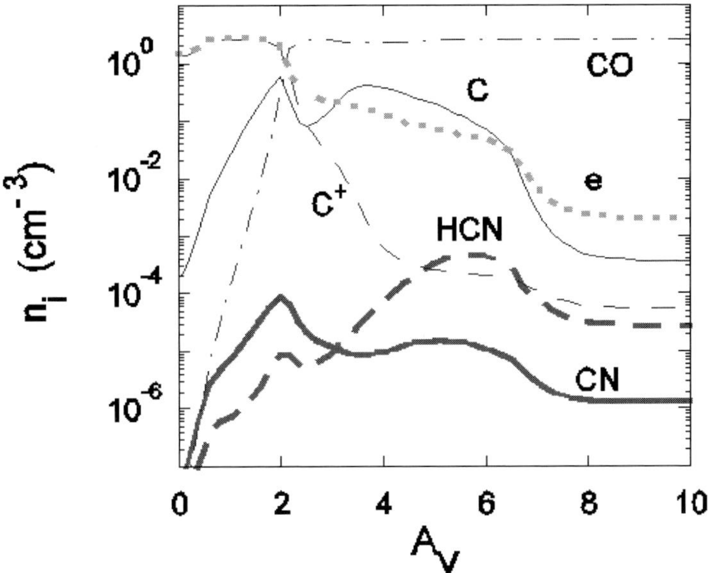

Figure 3. CN and HCN densities as functions of cloud depth, for $G_0 = 10^3$ and $n_{\rm H} = 10^4$ cm^{-3} (adapted from Boger & Sternberg 2005).

star than does the saturated molecule HCN, and that the CN/HCN abundance ratio decreases with increasing cloud depth (slide 15).

The explanation for the CN peak is as follows (Boger & Sternberg 2005; slides 17 and 18). Because the hydrogen becomes fully molecular within the C$^+$ layer, CN is rapidly produced by a gas-phase sequence initiated by radiative association

$$C^+ + H_2 \rightarrow CH_2^+ + \nu. \tag{3.1}$$

This is then followed by rapid abstraction, and dissociative recombination

$$CH_2^+ + H_2 \rightarrow CH_3^+ + H \tag{3.2}$$

$$CH_3^+ + e \rightarrow CH + H_2 \tag{3.3}$$

$$CH_3^+ + e \rightarrow CH_2 + H, \tag{3.4}$$

leading to the radicals CH and CH$_2$. These then react with abundant N atoms

$$CH + N \rightarrow CN + H \tag{3.5}$$

$$CH_2 + N \rightarrow HCN + H. \tag{3.6}$$

The CN and HCN molecules are removed by photodissociation

$$CN + \nu \rightarrow C + N \tag{3.7}$$

$$HCN + \nu \rightarrow CN + H. \tag{3.8}$$

A CN density peak occurs at cloud depths where the C$^+$ density is still large, but where the CN photodissociation rate is attenuated as much as possible, *i.e.*, at the inner edge of the C$^+$ zone. Photodissociation keeps the CN/HCN ratio high at this location. Because HCN is more vulnerable to photodissociation than is CN, as the FUV radiation is attenuated, the HCN density grows depth while the CN density saturates. The CN/HCN

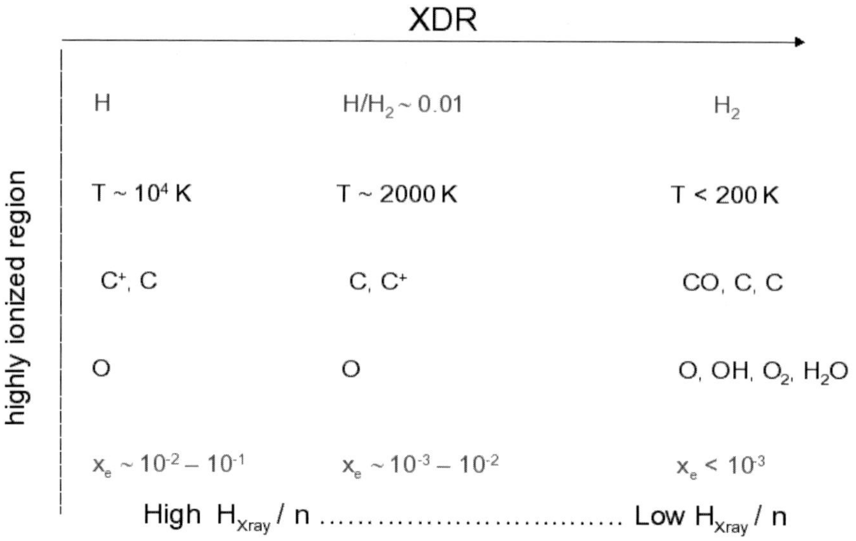

Figure 4. XDR structure (adapted from Maloney *et al.* 1996).

ratio therefore decreases with depth, and this abundance ratio is a probe of the local FUV field strength. This behavior is displayed in Figure 3.

Additional chemical diagnostics of PDRs are discussed in Sternberg & Dalgarno (1995).

4. XDRs

In X-ray dominated regions (XDRs), penetrating (keV) X-rays heat the gas and drive the chemistry. As already mentioned, there are two important differences compared to PDRs. First, because the X-rays are absorbed in photoionization events, as opposed to the FUV dust absorption that occurs in PDRs, the X-ray intensity falls off much more slowly in XDRs. Second, the radiative heating efficiency, that is, the fraction of X-ray energy converted to heat and then rereadiated in cooling emission lines, is larger in XDRs than in PDRs.

As discussed by Maloney *et al.* (1996) a key parameter in XDRs is the ratio, H_X/n_H, of the local X-ray energy deposition rate H_X to hydrogen gas density n_H (slide 19). For large H_X/n_H, the gas remains primarily atomic, with large (0.1–0.01) ionization fractions, and at high ($\sim 10^4$ K) temperature. The H/H$_2$ ratio, ionization fraction, and gas temperature, decrease as H_X/n_H becomes small. Molecules form with high efficiency at intermediate values of H_X/n_H where the gas temperature is maintained at a level of a few hundred K.

It is of interest to consider the X-ray energy deposition in the limit of a fully molecular gas (slide 20). In this case, energetic electrons are produced by photoionization

$$H_2 + Xray \rightarrow H_2^+ + e^*. \tag{4.1}$$

The energetic electrons, e*, lose energy via Coulomb scattering with free thermal electrons, by electronic excitations of the H$_2$ molecules (and trace H atoms), and by collisional ionizations of the H$_2$. For small fractional ionizations Couloumb scattering is a minor channel. The excitations are followed by emissions of UV photons which are absorbed by, and heat, the dust. The induced UV photons also mediate the chemistry by photodissociating molecules. Importantly, the electron impact ionizations

$$e^* + H_2 \rightarrow H_2^+ + e + e^* \tag{4.2}$$

are immediatley followed by proton transfer

$$H_2^+ + H_2 \rightarrow H_3^+ + H, \tag{4.3}$$

and dissociative recombination

$$H_3^+ + e \rightarrow H_2 + H. \tag{4.4}$$

Most of the ionization energy of 15.4 eV is converted into kinetic energy of the dissociating fragments, *i.e.*, heats the gas (Glassgold & Langer 1973). However, the energy lost to each energetic electron per ionization event is on average 37.7 eV (Dalgarno *et al.* 1999). Therefore a heating efficiency of about 15.4/37.7=0.4 is expected in XDRs, significantly larger than in PDRs. This effect is seen in computations of the fine-structure emission line intensities in XDRs, with large resulting line-to-continuum ratios (slide 21).

Dale *et al.* (2004) suggest that the enhanced [O I] 63 μm emission they observed in a sample of Seyfert galaxies arises in XDRs.

5. Galaxies

5.1. *PDRs*

Another nice example of a PDR is the starburst in the "Antennae" pair of colliding galaxies NGC 4038/4039 (slides 22 and 23). Optical HST imaging (Whitmore & Schweizer 1995) shows that the system contains hundreds of compact clusters of young OB stars that are interacting with surrounding dense and dusty molecular clouds. Near-IR Keck spectroscopy of the embedded mid-IR star cluster detected in ISO observations reveals a spectrum containing a wealth of vibrational H_2 emission lines (Gilbert *et al.* 2000; slide 23). The luminosity in just the 1–0 S(1) line alone is 9600 $L_\odot$. The relative intensities of lines in the $v = 1$ and $v = 2$ bands indicate that the H_2 excitation mechanism is FUV-pumping (slide 24). In this process FUV photons are absorbed in the Lyman and Werner bands, followed by rapid radiative decays to excited vibrational levels of the ground electronic state. These then decay in a cascade of quadrupole transitions giving rise to a rich infrared spectrum (Black & Dalgarno 1976; Black & van Dishoeck 1987; Sternberg 1988; Sternberg & Dalgarno 1989; Draine & Bertoldi 1996; Shaw *et al.* 2005). The ortho-to-para ratio in excited states is 1.6±0.07. This is nicely consistent with what is expected for FUV-pumping in optically thick absorption lines, for a ground vibrational state with an equilibrium rotational ortho-to-para ratio of 3 indicative of warm gas (Sternberg & Neufeld 1999).

Far-infrared imaging spectroscopy of the Antennae shows that the C II line emission peaks at the location of the "interaction zone" located between the individual galaxy nuclei (Nikola *et al.* 1998; slide 26). A global FUV intensity of $G_0 \approx 500$ is inferred for the interaction zone. The FUV radiation is interacting with a total gas mass of at least a few 10^7 M$_\odot$. The total luminosity in the C II line is 3.7×10^8 L$_\odot$, corresponding to about 1% of the far-IR (IRAS) luminosity.

The C II to FIR luminosity ratio in the Antennae pair is consistent with expectations from PDR theory. This contrasts with recent observations of more luminous systems. For example, Maiolino *et al.* (2005) recently detected the [C II] line in the high redshift, $z = 6.42$, quasar J1148+5251 (slide 26). The observations were carried out at the IRAM

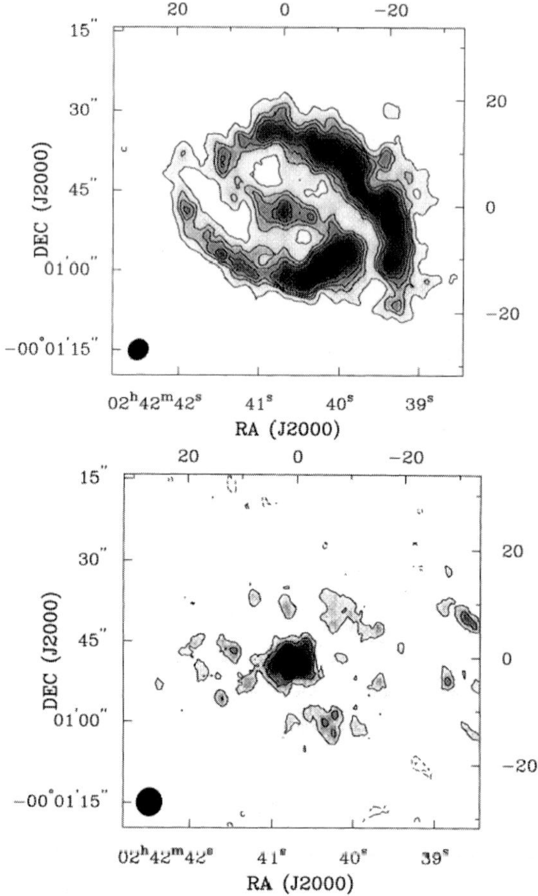

Figure 5. CO (top) and HCN (bottom) in NGC 1068 (from Helfer & Blitz [1995]).

30-meter telescope. They find $L_{\mathrm{FIR}} = 2.2 \times 10^{13}$ $L_{\odot}$, and $L_{\mathrm{CII}} = 4.4 \times 10^{9}$ $L_{\odot}$, and infer a total star-formation rate of 3000 $M_{\odot}$ yr^{-1}. Interestingly, the small FIR to C II ratio of 2×10^{-4} is comparable to what is observed in ultraluminous infrared galaxies (Luhman *et al.* 2003; slide 27). Possible explanations for the C II "line deficit" in these objects include (a) large ratios of FUV flux to gas density in the ULIRG (and quasar) PDRs leading to inefficient photoelectric gas heating, (b) an (unlikely) underabundance of OB stars relative to cooler stars, (c) line "self absorption" in optically thick clouds, and (d) a significant absorption of FUV photons in dusty H II regions with large ionization parameters. This last possibility is favored by Luhman *et al.* (2003), although it remains unclear whether this idea is compatible with the nebular emission line spectra from photoionized gas in these objects.

5.2. *XDRs*

For some examples of extragalactic XDRs, I first turn to the the active galaxy NGC 4258 (7.3 Mpc; slide 28). This object is famous for the prominent 22 GHz H_2O water masers that have been resolved via VLBI in a rotating circumnuclear disk (*e.g.*, Herrnstein *et al.* 2005; slide 29). The Keplerian orbital motions imply that the central supermassive black hole mass is 3×10^{7} $M_{\odot}$. Neufeld *et al.* (1994) suggested that the water masers are produced in XDRs formed in molecular material exposed to the X-rays emitted by the

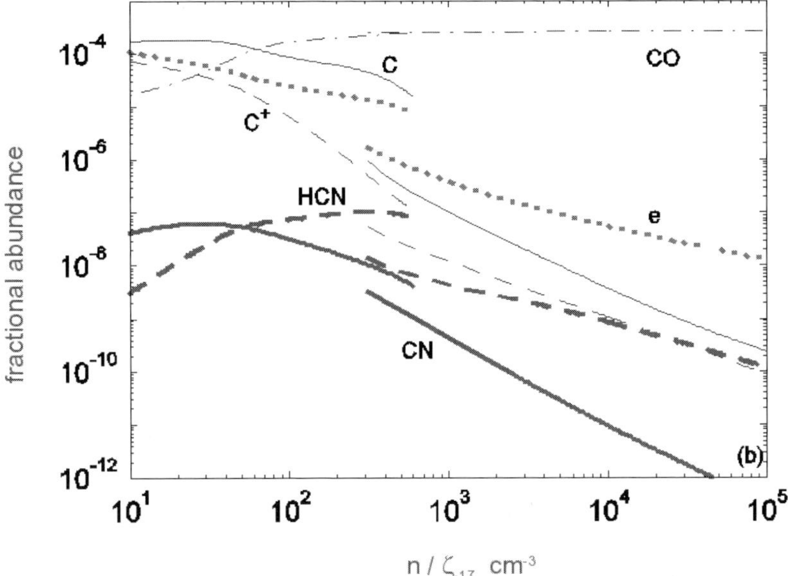

n/ζ_{-17} cm⁻³

Figure 6. CO (top) and HCN (bottom) in NGC 1068 (from Helfer & Blitz 1995).

accreting black hole (slide 30). In this model water is formed efficiently via the direct neutral-neutral reactions

$$O + H_2 \to OH + H \qquad (5.1)$$

$$OH + H_2 \to H_2O + H \qquad (5.2)$$

in the X-ray heated gas (slide 31). For temperatures above 300 K, the above sequence drives a large fraction of the available oxygen into water. In cooler gas, H_2O is formed less efficiently via ion-molecule reactions (*e.g.*, Herbst & Klemperer 1973; Prasad & Huntress 1980). However, while the kinematic maser disk in NGC 4258 is spectacular, the interpretation in terms of XDRs remains uncertain because the H_2O abundances cannot be reliably inferred from the nonthermally amplified maser emissions.

The Seyfert 2 galaxy NGC 1068 (15.5 Mpc; slide 32) may be a better source for a quantative study of XDR chemistry. Millimeter-wave interferometry shows that in NGC 1068 molecular emissions are produced in a kpc scale star-forming ring, and in an inner 100 pc scale region surrounding the active nucleus (slide 33). A variety of molecules have been detected in the nuclear region, including CO, HCN, CN, CS, HCO⁺, HOC⁺, and SiO (Tacconi *et al.* 1994; Sternberg *et al.* 1994; Helfer & Blitz 1995; Usero *et al.* 2004). The mm-wave maps show clear differences between the circumnuclear ring, and the nucleus. In particular the HCN/CO intensity ratio is small in the ring, and strikingly large in the nucleus (Fig. 5; slide 34). Part of this difference can be attributed to a higher gas density in the nucleus which then favors emission from molecules with large dipole moments such as HCN. However, a radiative transfer analysis (Sternberg, Genzel & Tacconi 1994) shows that the large HCN/CO ratio in the nucleus is due in part to an enhanced HCN/CO *abundance* ratio. Usero *et al.* (2004) infer an HCN/CO abundance ratio of $\sim 10^{-3}$ in the nucleus, an extraordinarily large value compared to Galactic molecular clouds.

Lepp & Dalgarno (1996) suggested that the large HCN/CO abundance ratio in the nucleus of NGC 1068 might be due to an enhanced ionization rate due to X-rays from the accreting black hole. They also suggested that CN would persist at high ionization rates.

Figure 6 (slide 35) shows recent computations by Boger & Sternberg (2005) that support these ideas. Figure 6 shows the computed abundances (relative to H_2) for several species including CN, HCN, and CO, as functions of the parameter n_H/ζ_{-17} (cm^{-3}), where ζ_{-17} is the H_2 ionization rate in units of 10^{-17} s^{-1}. The abundances are for a pure gas-phase chemistry in steady state, for Galactic ISM abundances of the heavy elements carbon, nitrogen, and oxygen. At high density or low ionization rate, the gas is in a low-ionization phase (LIP). In the LIP the CN and HCN abundances, and the HCN/CO and CN/HCN density ratios, are small. At low density or high ionization rate the gas is in a high-ionization phase (HIP). In the HIP, the CN and HCN abundances become large. Furthermore, HCN/CO approaches $\sim 10^{-3}$ in the HIP, similar to what is observed in the nucleus of NGC 1068. Furthermore, the CN/HCN ratio increases with the ionization rate. This is nicely consistent with the recent observations of Usero et $al.$ (2004) who find CN/HCN$\gtrsim 1$ in the nucleus of 1068.

We have seen that large CN/HCN ratios are expected in both PDRs and in XDRs, though for somewhat different reasons. This raises the general question of whether there are any unique chemical diagnostics of XDRs that may be used to distinguish them from PDRs, beyond the enhanced ionization rates or large line-to-continuum ratios expected in XDRs. One idea is to make use of the fact that in XDRs multiply charge atomic ions are expected to coexist together with the neutral molecular gas. The multiply charged species are produced by inner-shell X-ray ionization and Auger decay, processes that do not occur in FUV irradiated gas in PDRs. Ions such as C^{2+} (Dalgarno 1976; Langer 1978) and S^{2+} (Yan & Dalgarno 1996) have been put forth as possible candidates, provided they are not rapidly neutralized by charge transfer or gas-grain reactions. Fine structure line transtions of S^{2+} at 33.5 and 18.7 μm would then be potentially useful probes. If these ions react rapidly with H_2 they may lead to enhanced CH^+ and SH^+ in XDRs (see also Sternberg, Yan & Dalgarno 1996).

6. Conclusion

The study of PDRs and XDRs will remain important for many years to come, especially with the recent advent of the Spitzer Space Telescope, and other upcoming facilities including SOFIA, ALMA, and Herschel (slide 37).

Acknowledgements

I thank the organizers for an excellent meeting, and for inviting me to speak. My research in molecular astrophysics is supported by the Israel Science Foundation.

References

Bakes, E.L.O. & Tielens, A.G.G.M. 1994, *Ap. J.* 427, 822
Bakes, E.L.O. & Tielens, A.G.G.M. 1998, *Ap. J.* 499, 258
Black, J.H. & Dalgarno, A. 1976, *Ap. J.* 203, 132
Black, J.H. & van Dishoeck, E.F. 1987, *Ap. J.* 322, 412
Boger, G.I. & Sternberg, A. 2005, *Ap. J.* 632, 302
Crawford, M.K., Genzel, R., Townes, C.H., & Watson, D.M. 1985, *Ap. J.* 291, 755
Dale, D.A., Helou, G., Brauher, J.R., Cutri, R.M., Malhotra, S., & Beichman, C.A. 2004, *Ap. J.* 604, 565
Dalgarno, A. & McCray, R.A. 1972, *ARAA* 10, 375
Dalgarno, A., Yan, M., & Weihong, L. 1999, *Ap. J. Suppl.* 125, 237
Draine, B.T. 1978, *Ap. J. Suppl.* 36, 395
Draine, B. & Bertoldi, F. 1996, *Ap. J.* 468, 269

Field, G.B., Goldsmith, D.W., & Habing, H.J. 1969, *Ap. J.* 155, L149

Fuente, A., Martin-Pintado, J., Cernicharo, J., & Bachiller, R. 1993, 276, 473

Fuente, A., Rodriguez-Franco, A., Garcia-Burillo, S., Martin-Pintado, J., & Black, J.H. 2003, *A&A* 406, 899

Genzel, R., Watson, D.M., Townes, C.H., Dinerstein, H.L., Hollenbach, D., Lesterm D.F., Werner, M., & Storey, J.W.V. 1984, *Ap. J.* 276, 551

Gilbert, A.M. *et al.* 2000, *Ap. J.* 533, L57

Glassgold, A.E. & Langer, W.D. 1973, *Ap. J.* 186, 859

Goldshmidt, O. & Sternberg, A. 1995, 439, 256

Habing, H.J. 1969, Bull. Astr. Inst. Netherlands, 19, 421

Helfer, T.T. & Blitz, L. 1995, *Ap. J.* 450, 90

Herbst, E. & Klemperer, 1973, *Ap. J.* 185, 505

Herrnstein, J.R., Moran, J.M., Greenhill, L.J., & Trotter, A.S. 2005, *Ap. J.* 629, 719

Hollenbach, D. & Tielens, A.G.G.M. 1997, *ARAA* 35, 179

Hollenbach, D. & Tielens, A.G.G.M. 1999, Rev. Mod. Phys., 71, 173

Jansen, D.J., Spaans, M., Hogerheijde, M.R., & van Dishoeck, E.F. 1995, *A&A* 303, 541

Kaufman, M.J., Wolfire, M.G., Hollenbach, D.J., & Luhman, M.L. 1999, *Ap. J.* 527, 795

Lepp, S. & Dalgarno, A. 1988, *Ap. J.* 335, 769

Lepp, S. & Dalgarno, A. 1996, *A&A* 306, L21

Lugten, J.B., Watson, D.M., Crawford, M.K., & Genzel, R. 1986, *Ap. J.* 311, L51

Luhman, M.L., Satyapal, S., Fischer, J., Wolfire, M.G., Sturm, E., Dudley, C.C., Lutz, D., & Genzel, R. 2003, *Ap. J.* 594, 758

Maloney, P.R., Hollenbach, D.J., & Tielens, A.G.G.M. 1996, *Ap. J.* 466, 561

Maiolino, R. *et al.* 2005, *A&A* 440, L51

McKee, C.F. 1989, *Ap. J.* 345, 782

Melnick, G., Gull, G.E., & Harwit, M. 1979, *Ap. J.* 227, L29

Nikola, T., Genzel, R., Herrmann, F., Madden, S.C., Poglitsch, A., Geis, N., Townes, C.H., & Stacey, G.J. 1998, *Ap. J.* 504, 749

Neufeld, D.A., Maloney, P.R., & Conger, S. 1994, *Ap. J.* 436, L127

Parravano, A., Hollenbach, D.J., & McKee, C.F. 2003, *Ap. J.* 584, 79

Prasad, S.S. & Huntress, W.T. 1980, *Ap. J. Suppl.* 43, 1

Russell, R.W., Melnick, G., Gull, G.E., & Harwit, M. 1980, *Ap. J.* 240, L99

Shaw, G., Ferland, G.J., Abel, N.P., Stancil, P.C., & van Hoof, P.A.M. 2005, *Ap. J.* 624, 794

Spitzer, L.Jr., 1948, *Ap. J.* 107, 6

Sternberg, A. 1988, *Ap. J.* 332, 400

Sternberg, A. & Dalgarno, A. 1989, *Ap. J.* 338, 197

Sternberg, A., Genzel, R., & Tacconi, L. 1994, *Ap. J.* 436, L131

Sternberg, A. & Neufeld, D. 1999, *Ap. J.* 516, 371

Sternberg, A. & Dalgarno, A. 1995, *Ap. J. Suppl.* 99, 565

Sternberg, A., Yan, M., & Dalgarno, A. 1996 in *Molecules in Astrophysics: Probes & Processes*, ed. E.F. van Dishoeck (Dordrecht: Kluwer), IAU Symp., 178, p. 141

Sternberg, A., Hoffmann, T., & Pauldrach A.W.A. 2003, *Ap. J.* 599, 1333

Storey, J.W.V., Watson, D.M,. & Townes, C.H., 1979, *Ap. J.* 233, 109

Tacconi, L.J., Genzel, R., Blietz, M., Cameron, M., Harris, A.I., & Madden, S. 1994, 426, L77

Tielens, A.G.G.M. & Hollenbach, D. 1985, 291, 722

Usero, A., Garca-Burillo, S., Fuente, A., Martn-Pintado, J., & Rodrguez-Fernndez, N.J. 2004, *A&A* 419, 897

Weingartner, J.C. & Draine, B.T. 2001, *Ap. J. Suppl.* 134, 263

Whitmore, B.C. & Schweizer, F. 2005, *A. J.* 109, 960

Wolfire, M.G., McKee, C.F., Hollenbach, D., & Tielens, A.G.G.M. 2003, 587, 278

Photo: E. Herbst

Astrochemistry: Recent Successes and Current Challenges
Proceedings IAU Symposium No. 231, 2005
D.C. Lis, G.A. Blake & E. Herbst, eds.

© 2006 International Astronomical Union
doi:10.1017/S1743921306007149

Carbon Chemistry in Photodissociation Regions

M. Gerin[1], E. Roueff[2], J. Le Bourlot[2], J. Pety[3], J. R. Goicoechea[1], D. Teyssier[4], C. Joblin[5], A. Abergel[6], and D. Fossé[1]

[1]LERMA, Observatoire de Paris and ENS, CNRS UMR8112,
24 Rue Lhomond, 75231 Paris cedex 05, France
email: gerin@lra.ens.fr, fosse@lra.ens.fr, javier@lra.ens.fr

[2]LUTH, Observatoire de Paris, CNRS UMR8102,
place J. Janssen, 92195 Meudon cedex, France
email: evelyne.roueff@obspm.fr, jacques.lebourlot@obspm.fr

[3]J. Pety, IRAM, 300 rue de la piscine, 38406 Grenoble cedex, France
email: pety@iram.fr

[4]European Space Astronomy Centre, Villafranca del Castillo, Madrid 28080, Spain
email: dteyssier@sciops.esa.int

[5]CESR, CNRS and Univ. Paul Sabatier, 9 Avenue du colonel Roche, 31028 Toulouse, France
email: joblin@cesr.fr

[6]Institut d'Astrophysique Spatiale, CNRS et Univ. Paris-Sud,
Bât. 121, 91405 Orsay Cedex, France
email: alain.abergel@ias.u-psud.fr

Abstract. We present recent results on the carbon chemistry in photodissociation regions. We show that carbon chains and rings (CCH, c-C_3H_2 and C_4H) are tightly spatially correlated with each other, and with the mid-infrared emission due to PAHs (7 and 15 μm), mapped by ISOCAM. Neither the spatial distribution, nor the abundances of these species can be fit by state-of-the-art PDR models, which calls for another production mechanism. We discuss model predictions for carbon clusters and simple hydrocarbons. We show how selected abundance ratios can be used as a diagnostic of the physical conditions. We stress the need for more theoretical and laboratory work on fundamental processes relevant for the interstellar medium, which should be taken into account in the astrochemical models, but whose rates are not known accurately enough.

Keywords. astrochemistry — ISM: clouds — ISM: evolution — ISM: molecules — ISM: structure

1. Introduction

As carbon is one of the most abundant elements in the interstellar medium, which can bind to other elements in many ways, most of the interstellar and circumstellar molecules contain at least one carbon atom. Table 1 lists the detected interstellar and circumstellar species through September 2005. Carbon species account for 75% of the 144 detected molecules, and include the most complex and heaviest molecules detected.

Carbon chemistry is also important for understanding the structure and evolution of the interstellar medium (ISM). Carbon monoxide is the most abundant molecule after molecular hydrogen and a widely used tracer of molecular regions, as H_2 lines are usually weak and not easily accessible. Carbon chemistry affects the gas cooling processes as C^+ and C fine-structure line emission, together with CO and H_2O rotational line emission,

153

Table 1. Interstellar and Circumstellar Molecules

Molecular Hydrogen and Related Ions						
H_2	HD	H_3^+	H_2D^+	D_2H^+		
Carbon Chains and Rings						
CH	CH^+	C_2	CH_2	CCH	C_3	CH_3
C_2H_2	$l\text{-}C_3H$	$c\text{-}C_3H$	CH_4	C_4?	$c\text{-}C_3H_2$	$l\text{-}C_3H_2$
C_4H	**C_2H_4**	C_5	C_5H	$l\text{-}H_2C_4$	**HC_4H**	CH_3CCH
C_6H	C_6H_2	**HC_6H**	**C_7H**	CH_3C_4H	C_8H	**C_6H_6**
Species with O, or H and C						
OH	H_2O	O_2?	CO	CO^+	HCO^+	HOC^+
HCO	C_2O	CO_2	H_3O^+	$HOCO^+$	H_2CO	C_3O
HCOOH	CH_2CO	H_2COH^+	CH_3OH	CH_2CHO	HC_2CHO	C_5O
CH_3CHO	$c\text{-}C_2H_4O$	CH_3OCHO	CH_2OHCHO			
CH_3COOH	CH_2CHOH	$(CH_3)_2O$	CH_3CH_2OH	CH_2CHCHO	$CH_3CH_2\text{-}$	$HOCH_2\text{-}$
$(CH_3)_2CO$	$CO(CH_2OH)_2$				-CHO	$\text{-}CH_2OH$
Species with N, or H and C						
N_2	NH	CN	NH_2	HCN	HNC	N_2H^+
NH_3	$HCNH^+$	H_2CN	HCCN	C_3N	CH_2CN	CH_2NH
HC_3N	HC_2NC	NH_2CN	C_3NH	**HC_4N**	CH_3CN	CH_3NC
HC_3NH^+	C_5N	CH_3NH_2	C_2H_3CN	HC_5N	CH_3C_3N	C_2H_5CN
HC_7N	CH_3C_5N?	HC_9N	$HC_{11}N$			
Species with N, O or H and C						
NO	HNO	N_2O	HNCO	NH_2CHO		
Species with S, Si and Other Elements						
SH	CS	SO	**SO^+**	NS	SiH	**SiC**
SiN	SiO	SiS	HCl	**NaCl**	**AlCl**	**KCl**
HF	**AlF**	**CP**	PN	H_2S	C_2S	SO_2
OCS	HCS^+	$c\text{-}SiC_2$	**SiCN**	**NaCN**	**MgCN**	**MgNC**
H_2CS	HNCS	C_3S	**c-SiC_3**	**SiH_4**	**SiC_4**	CH_3SH
C_5S	FeO	**AlNC**	**SiNC**	CF^+		

Note: Species detected in **circumstellar envelopes** only are written in **boldface**. Species detected in external galaxies are underlined.

and neutral oxygen fine structure line emission are the main gas coolants over a wide range of gas densities and temperatures (see, e.g., Kaufman *et al.* 1999; Wolfire *et al.* 2003). Finally, carbonaceous grains and PAHs contain a large fraction of the total carbon. These small solid particles and macro-molecules control the gas heating rate through the photoelectric effect, since the photoelectric efficiency is largest for small particles. Although grains are thought to be formed in circumstellar envelopes, their properties evolve as they travel through the interstellar medium. Various observational pieces of evidence for grain processing have been found (see, e.g., Jones 2005; Dartois *et al.* 2005). In addition to evolution of the grain material itself, dust grains can either grow, or fragment, in the interstellar medium, depending on their surroundings. Sputtering and fragmentation of carbonaceous grains is likely to release fresh carbon back into the ISM.

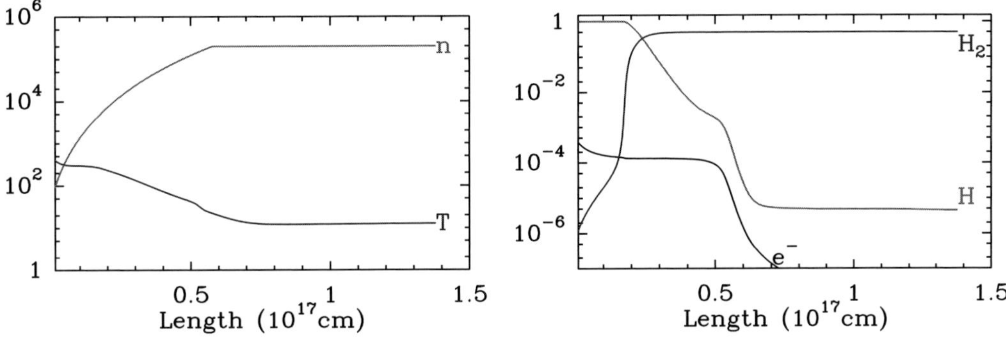

Figure 1. *Left:* Gas density and temperature for the Horsehead nebula model as a function of distance from the illuminated edge in cm. *Right:* Abundances of neutral (H) and molecular (H$_2$) hydrogen, and electron abundance (e$^-$). The FUV radiation field is 100 in Draine's unit.

Photodissociation regions (PDRs) are defined as the warm interfaces, where the gas is molecular but its heating is dominated by FUV photons ($<$13.6 eV). Such an environment can be found in reflection nebulae—where a massive star illuminates a nearby molecular cloud—but also at the surface of protostellar disks, in circumstellar envelopes and in (proto)planetary nebulae. Because of the enhanced heating, and the efficient cooling due to the moderate gas density, PDRs produce intense continuum and line radiation over a wide range of wavelengths. The intense cooling lines from PDRs are easily detected in external galaxies, and the most distant galaxies become accessible with state-of-the-art instruments (e.g., the detection of the [C II] 158 μm line by Maiolino *et al.* 2005). Therefore PDRs are attractive sources, where the physical and chemical processes relevant for the ISM evolution can be studied in detail, and state-of-the-art models compared with accurate observations.

2. Carbon Chemistry in the Horsehead Nebula

2.1. *Observations*

Unsaturated carbon chains and rings are rather abundant molecules in the ISM, and are even more abundant in carbon rich circumstellar envelopes around evolved stars, although these species are extremely reactive in the laboratory. In circumstellar envelopes (most notably the carbon rich star IRC+10216) and in dark clouds, the abundance ratios of these species follow a well-defined pattern with a slow decrease of the abundance as the molecule size increases (Guélin *et al.* 1977; Teyssier *et al.* 2004). Until recently, no such systematic studies had been performed for PDRs. Using the IRAM-30m telescope Fuente *et al.* (2003) observed reactive ions in PDRs and measured high abundances of HOC$^+$ and CO$^+$. They also observed c-C$_3$H$_2$ lines in the same sources and concluded that the observed abundance is larger than PDR model predictions. Teyssier *et al.* (2004) studied the distribution and abundance of selected carbon chains and rings (CCH, the cyclic and linear isomers of C$_3$H and C$_3$H$_2$, and C$_4$H) in three PDRs (ρ Ophiuchi, IC 63 and the Horsehead nebula). They concluded that the abundance and spatial distribution of these species are not well fitted by PDR models. However, these observations reached a moderate spatial resolution ($\sim$20$''$), which did not allow the authors to resolve the steep abundance gradients predicted by the PDR models.

The Horsehead nebula was selected for a follow-up study at higher spatial resolution, using the IRAM Plateau de Bure interferometer. The Horsehead nebula is particularly

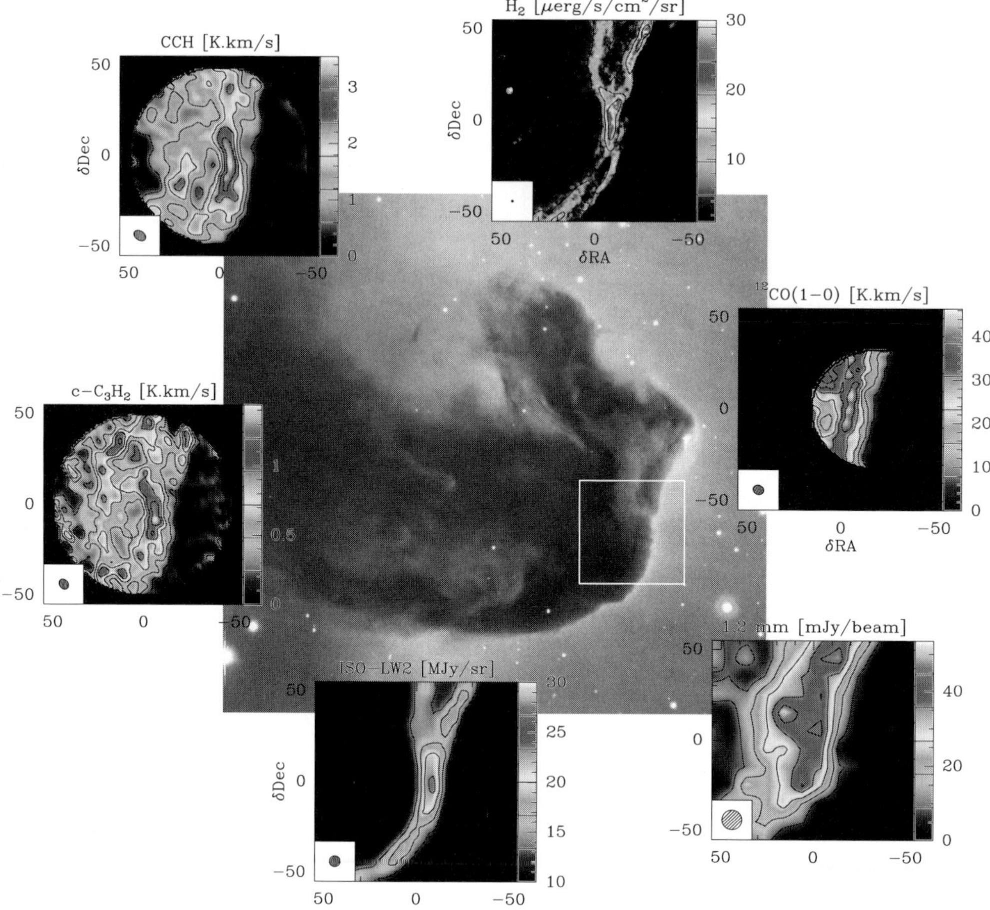

Figure 2. Summary of observations of the Horsehead nebula in LDN1630 from Pety *et al.* (2005) and Habart *et al.* (2005). The background image is taken from the ESO archive. The white square indicates the observed field of view of the IRAM Plateau de Bure interferometer. The CCH (*top left*), c-C$_3$H$_2$ (*left*), and ^{12}CO(1–0) (*right*) maps have been obtained with this facility. The PdBI data have been combined with short spacings from the IRAM-30m telescope to recover all spatial frequencies. The H$_2$ v=1–0 S(1) image (*top*) is from ESO-NTT, the 1.2 mm image (*bottom right*) from MAMBO on the IRAM-30m telescope, and the ISO-LW2 image (*bottom*) has been taken with ISOCAM (Abergel *et al.* 2003). Note the similarity of the CCH and c-C$_3$H$_2$ maps with the ISO-LW2 image, while the CO(1–0) and dust maps look different.

well suited as a template PDR, as it is viewed nearly edge-on (Habart *et al.* 2005 for a thorough discussion on the viewing geometry) and is nearby. At a distance of 400 pc, $1''$ corresponds to a linear size of 6.0×10^{15} cm, or 0.002 pc. The Horsehead nebula is illuminated by the O9.5 star σ Ori, $\sim$3.5 pc away, resulting in a far UV radiation field $G_0 \sim 60$ times the mean interstellar radiation field, as determined by Draine (1978). As shown by Habart *et al.* (2005) and Pety *et al.* (2005), the gas density profile is well constrained by the infrared maps of the H$_2$ rovibrational emission, $v = 1$–0 S(1) at 2.1 μm, and of the PAH emission near 7 μm. The best-fit density profile rises steeply (in $\sim$10$''$) from $n_H \sim 10^4$ cm^{-3} to $n_H \sim 10^5$ cm^{-3}, the resulting model being almost isobaric (constant pressure). Figure 1 presents the adopted model.

A summary of the maps is shown in Figure 2. As discussed extensively by Pety *et al.* (2005), the spatial distribution of hydrocarbons is very similar to the PAH emission,

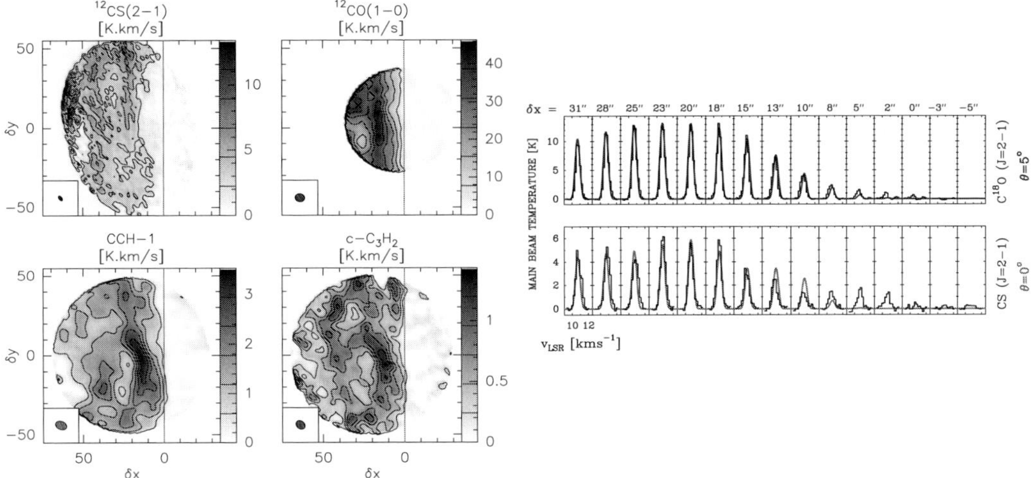

Figure 3. *Left:* PdBI Maps of the CS(2-1), CO(1–0), CCH(1–0) and c-$C_3H_2(2_{1,2} - 1_{0,1})$ lines. The maps have been rotated by 14° in to ease the comparison with 1-D models. *Right:* Comparison of observed and modelled $C^{18}O$ and CS spectra (from Goicoechea *et al.* 2005).

and different from CO and other species. Both the spatial distribution and the molecule abundance could not be fitted by PDR models, even by trying different chemical networks or small modifications of the source geometry and density structure. Because the molecule destruction is dominated by photodissociation in the external layers where the carbon chains are most abundant, detecting species with high abundance implies that their formation is efficient. Pety *et al.* (2005) reached similar conclusions as Fuente *et al.* (2003) and Rizzo *et al.* (2003, 2005), namely that the PAHs responsible for the mid-infrared emission features play a role in the formation of small carbon chains and rings. The most likely explanation is that, when these particles are released from larger structures (Rapacioli *et al.* 2005; Jones 2005), a fraction of these PAHs fragment and feed the gas with small carbon clusters and molecules.

2.2. *Carbon Molecules and Grains*

Indeed, much observational evidence has now been presented on the evolution of carbonaceous material in the interstellar medium. For instance, Rapacioli *et al.* (2005) have isolated three characteristic spectra in the reflection nebula NGC 7023, which neatly trace the evolution of PAHs as a function of the distance to the illuminating star. Vijh *et al.* (2005) discuss the spatial distribution of the recently discovered blue luminescence and also conclude that the carriers are liberated in situ, from larger carbonaceous structures.

While PDR models take into account the role of PAHs in the ionization balance, the influence of PAHs on the carbon chemistry has not been studied yet, because of the lack of basic information. Le Page *et al.* (2001, 2003) discuss the processes governing the size distribution, ionization balance and hydrogen coverage of PAHs in the ISM. The destruction of PAHs is limited to photodissociation because they do not include possible reactions with species other than hydrogen. Data on reaction rates between oxygen, carbon or nitrogen and PAHs are critically needed to assess possible growth paths of carbon species in the ISM.

2.3. *Further Comparison with Observations*

While the above analysis is based on comparisons of molecular abundances, it is now possible to perform a more accurate comparison between models and observations,

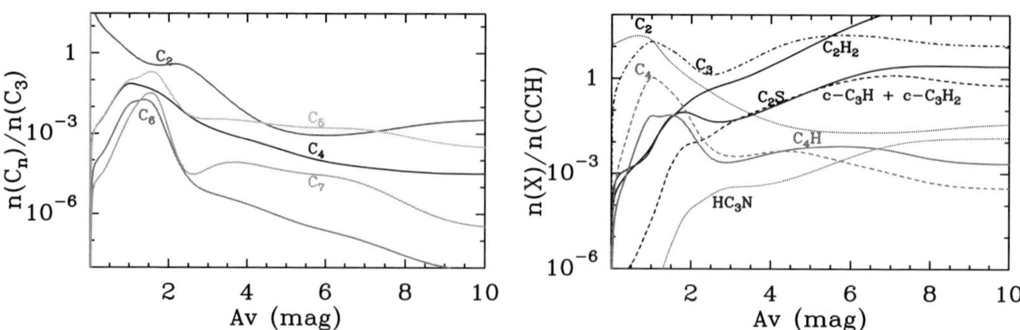

Figure 4. Prediction of the abundance ratio of carbon clusters (*right*) and other hydrocarbons (*left*) compared to C_3 and CCH respectively, for physical conditions of the Horsehead nebula.

using the full power of the high spectral resolution provided by millimeter observations. Figure 3 shows how well observed spectra can be fitted by combined PDR and Monte-Carlo radiative transfer models. A very small inclination along the line of sight (6°) can be identified when performing such detailed modelling work.

2.4. *Carbon Budget*

As shown in Fig. 4, C_2 is predicted to be abundant at low A_V, while C_3 and heavier carbon clusters are predicted to be most abundant at moderate extinctions, $A_V \sim 1$–3 mag. Saturated species, such as acetylene, C_2H_2, participate in the carbon budget at higher A_V's. Though the carbon chemistry is still not well understood, these conclusions are qualitatively in agreement with the observational data. Carbon clusters (C_2, C_3 and maybe even C_4) have been widely observed in the ISM (Oka *et al.* 2003; Cernicharo *et al.* 2002), with a tight correlation between C_2 and C_3 abundances. The low abundance of both HC_3N and CH_3CCH in the Horsehead nebula PDR (Teyssier *et al.* 2004) is also consistent with model predictions.

3. Carbon Chemistry as a Tracer of Gas Properties

Chemical models show that interstellar molecular abundances are sensitive to the gas physical conditions, such as the density, temperature, but also the electron abundance. This important parameter for the gas physics cannot be measured directly, as the main carriers of the charges cannot be observed. However, a few molecular abundance ratios have been shown to scale with the electron abundance. In dense and shielded regions, the degree of deuterium fractionation of molecular ions (HCO^+, N_2H^+, etc.) can be used to measure the electron abundance (Guélin *et al.* 1982; Caselli *et al.* 1999). Fossé *et al.* (2001) have showed that the $[c\text{-}C_3H_2]/[l\text{-}C_3H_2]$ abundance ratio is sensitive to the electron abundance in the dark cloud TMC-1. Teyssier *et al.* (2005) have presented observational evidence for a similar behaviour in the Horsehead PDR. Figure 5 presents model predictions of abundance ratios as a function of the electron density. While [HCN]/[HNC] is not sensitive to the electron abundance, the ratio of C_3H_2 and C_3H isomers can indeed be used to estimate the fractional ionization. As suggested by Sternberg (2005, this volume), [CN]/[HCN] can also be used for probing the electron abundance. [CN]/[HCN] gradients have been observed in the Orion Bar and NGC 7023 by A. Fuente and co-workers (e.g., Fuente *et al.* 1993), though the spatial resolution provided by a single dish telescope was

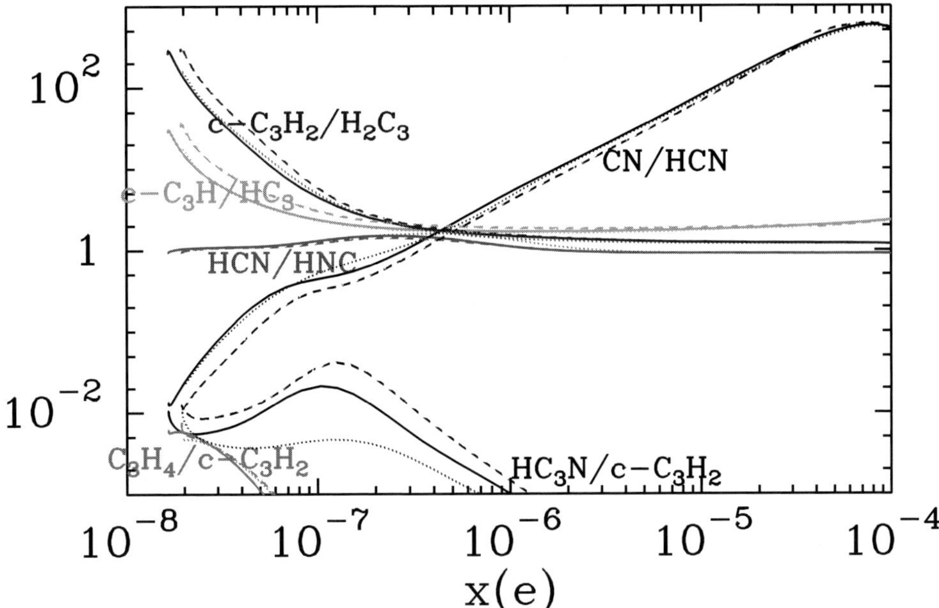

Figure 5. Molecular abundance ratios as a function of electron density for conditions suited to the Horsehead nebula. The radiation field is $G_0 = 100$. The full line corresponds to a varying gas density, with a complete thermal balance. The dashed line corresponds to a model using the same density profile, but with a minimum temperature set to 30 K, as used by Goicoechea *et al.* (2005) to fit the CS and $C^{18}O$ data. The dotted line corresponds to a model with a constant gas density of $n_H = 10^5$ cm^{-3}. High electron abundances are found in the warm outer layers, while the low values correspond to the dense, well shielded regions (see Fig. 1).

not quite sufficient. Higher spatial resolution maps, which can now be obtained with millimeter interferometers, can be used to constrain the models accurately.

4. Conclusions

With the current generation of instruments, and even more with the upcoming generation, excellent data can be obtained, which will allow us to probe interstellar chemistry accurately. State-of-the-art chemical models can now be combined with sophisticated radiative transfer codes to obtain accurate predictions of the line and continuum emission, which can be compared with the observed data. This method requires a simple source geometry (an inaccurate geometry can lead to important differences in line predictions and invalid conclusions). PDRs will remain template sources where physical processes, which operate on a wide range of physical conditions, can be studied with the greatest detail. Carbon chains and rings, and carbon clusters, are promising probes of the interstellar chemistry as they can be used for studying the ionization balance, the relationship between dust grains, macromolecules and gas phase species, as well as the formation routes of complex molecules.

However, PDR models still need improvement. In the coming years, advances are expected on the following aspects:

• Many parameters which are used in current models suffer from large uncertainties. Chemical reaction rates, such as those governing the sulfur chemistry for instance, are not known to the required accuracy over the temperature range relevant for interstellar chemistry.

- Although PDR models solve self-consistently for the gas chemistry and thermal balance, the predicted temperatures are too low at moderate extinction. This old problem has important consequences, as the predicted intensities of millimeter rotational lines are strongly dependent on the assumed gas kinetic temperature. Hence molecular abundances deduced from those lines can be affected by an error in the gas temperature determination.

- Observations of both gas-phase and solid-phase species point to the necessity of including a tight coupling of the dust and gas phase chemistry in the models. This requires, however, the knowledge of reaction rates for reactions involving abundant ions and atoms and PAHs, for instance. Laboratory studies are underway, as shown by posters presented at this conference (Betts *et al.* 2005; Joblin *et al.* 2005).

- Most PDR models solve the gas chemical abundances at steady state, in order to be able to correctly treat the FUV radiative transfer and H_2 photodissociation and ro-vibrational excitation. However, this hypothesis is not completely valid, as PDR are usually bordered by an H II region (associated with the illuminating star), which could be in expansion. Further developments could include a combined treatment of the ionized, neutral and molecular gas, as presented by R. Ferland and co-workers for the CLOUDY model (Ferland *et al.* 2005). Work aiming at understanding the temporal evolution of a PDR (see e.g; Morata & Herbst 2005) should also be pursued. Finally dynamical models, such as those presented by Hosokawa & Inutsuka (2005), provide a complementary view of the evolution of the gas density structure.

Acknowledgements

We thank the organizers for a very nice conference in beautiful Asilomar. We would like to acknowledge funding by the CNRS/INSU Programme PCMI and CNFA.

References

Abergel, A., Teyssier, D., Bernard, J.P., *et al.* 2003, *A&A* 410, 577

Betts, N., Bierbaum, V.M., & Snow, T.P. 2005, *IAUS 231*, poster

Caselli, P., Walmsley, C.M., Tafalla, M., Dore, L., & Myers, P.C. 1999, *Ap. J.* 523, L165

Cernicharo, J., Goicoechea, J.R., & Benilan, Y. 2002, *Ap. J.* 580, L157

Dartois, E., Muñoz-Caro, G.M., Deboffle, D., Montagnac, G., & D'Hendecourt, L. 2005, *A&A* 432, 895

Draine, B.T. 1978, *Ap. J. Suppl.* 36, 595

Ferland, G.J., Abel, N., Elwert, T., Porter, R., & Shaw, G. 2005, *IAUS 231*, poster

Fossé, D., Cernicharo, J., Gerin, M., & Cox, P. 2001, *Ap. J.* 552, 168

Fuente, A., Martin-Pintado, J., Cernicharo, J., & Bachiller, R. 1993, *A&A* 276, 473

Fuente, A., Rodriguez-Franco, A., Garcia-Burillo, S., Martin-Pintado, J., & Black, J.H. 2003 *A&A*, 406, 899

Goicoechea, J.R., Pety, J., Gerin, M., Roueff, E., Teyssier, D., Abergel, A., Habart, E., & Joblin, C. 2005, *IAUS 231*, poster

Guélin, M., Langer, W.D., & Wilson, R.W. 1982, *A&A* 107, 107

Guélin, M., Cernicharo, J., Travers, M.J. *et al.* 1997 *A&A* 317, L1

Habart, E., Abergel, A., Walmsley, C.M., Teyssier, D., & Pety, J. 2005, *A&A* 437, 177

Hosokawa, T. & Inutsuka, S. 2005, *Ap. J.* 623, 917

Joblin, C., Abergel, A., Bernard, J.-P., *et al.* 2005, *IAUS 231*, poster

Jones, A. 2005, *IAUS 231*, poster

Kaufman, M.J., Wolfire, M.G., Hollenbach, D.J., & Luhman, M.L. 1999, *Ap. J.* 527, 795

Le Page, V., Snow, T.P., & Bierbaum, V.M. 2001, *Ap. J. Suppl.* 132, 233

Le Page, V., Snow, T.P., & Bierbaum, V.M. 2003, *Ap. J.* 584, 316

Maiolino, R., Cox, P., Caselli, P., *et al.* 2005, *A&A*, in press

Morata, O. & Herbst, E. 2005, *IAUS 231*, poster

Oka, T., Thorburn, J.A., McCall, B.J. *et al.* 2003, *Ap. J.* 582, 823

Pety, J., Teyssier, D., Fossé, D., Gerin, M., Roueff, E., Abergel, A., Habart, E., & Cernicharo, J. 2005, *A&A* 435, 885

Rapacioli, M., Joblin, C., & Boissel, P. 2005, *A&A* 429, 193

Rizzo, J.R., Fuente, A., Rodriguez-Franco, A., & Garca-Burillo, S. 2003, *Ap. J.* 597, L153

Rizzo, J.R., Fuente, A., & Garcia-Burillo, S. 2005, *Ap. J.*, in press

Sternberg, A. 2005, this volume

Teyssier, D., Fossé, D., Gerin, M., Pety, J., Abergel, A., & Roueff, E. 2004, *A&A* 417, 135

Teyssier, D., Hily-Blant, P., Gerin, M., Cernicharo, J., Roueff, E., & Pety, J. 2005, in *The dusty and molecular universe: a prelude to Herschel and ALMA*, ed. A. Wilson (ESA SP-577), p. 423

Vijh, U.P., Witt, A.N., & Gordon, K.D. 2005, *Ap. J.*, in press

Wolfire, M.G., McKee, C.F., Hollenbach, D., & Tielens, A.G.G.M. 2003, *Ap. J.* 587, 278

Discussion

PAPADOPOULOS: It has been 20 years since the first PDR models were published. Since then the "sandwich" distribution of $C^+/C\,\textsc{i}/CO$ at the edges, $A_v \sim 1 - 3$, of clouds has failed to show up anywhere in the sky. So why should we believe/trust any other predicted distribution of species coming out of such models?

GERIN: First, the model fares well for other species: CO, $C^{18}O$, CS for instance. An accurate fit needs to consider the cloud geometry: the predictions for an infinite edge-on slab need to be tailored for a finite slab of gas, including two surfaces along the line of sight which contribute some of the emission for C and ^{12}CO. A careful understanding of the source geometry and structure is mandatory before assessing model success and failures. Of course, the question of the atomic carbon abundance and spatial distribution is probably not totally understood yet.

GUÉLIN: What about desorption from dust grains in PDRs?

GERIN: Desorption from dust grains could be considered, but it is most relevant for species which are abundant in the mantles (H_2O, CH_3OH, H_2CO, etc.), and not so much for the hydrocarbons, which we study.

Photo: E. van Dishoeck

Astrochemistry: Recent Successes and Current Challenges
Proceedings IAU Symposium No. 231, 2005
D.C. Lis, G.A. Blake & E. Herbst, eds.
© 2006 International Astronomical Union
doi:10.1017/S1743921306007150

First Astronomical Detection of the CF^+ Ion

**D. A. Neufeld[1], P. Schilke[2], K. M. Menten[2], M. G. Wolfire[3],
J. H. Black[4], F. Schuller[2], H. Müller[2,5], S. Thorwirth[2], R. Güsten[2],
and S. Philipp[2]**

[1]Department of Physics and Astronomy, Johns Hopkins University, Baltimore, USA

[2]Max-Planck-Institut für Radioastronomie, Bonn, Germany

[3]Astronomy Department, University of Maryland, College Park, USA

[4]Onsala Space Observatory, Onsala, Sweden

[5]Universität zu Köln, Köln, Germany

Abstract. We report the first astronomical detection of the CF^+ (fluoromethylidynium) ion, obtained by recent observations of its $J = 1 - 0$ (102.6 GHz), $J = 2 - 1$ (205.2 GHz), and $J = 3 - 2$ (307.7 GHz) pure rotational emissions toward the Orion Bar. Our search for CF^+— carried out using the IRAM 30m and APEX 12m telescopes—was motivated by recent theoretical models that predict CF^+ abundances of a few $\times\ 10^{-10}$ in UV-irradiated molecular regions where C^+ is present. The measurements confirm the predictions. They provide support for our current theories of interstellar fluorine chemistry, which suggest that hydrogen fluoride should be ubiquitous in interstellar gas clouds.

Keywords. astrochemistry — ISM: clouds — ISM: molecules — molecular processes — radio lines: ISM

The fluoromethylidynium ion, CF^+, sounds like a rather exotic species, but in some ways it is very familiar, being isoelectronic with carbon monoxide, the most widely observed interstellar molecule. Indeed, adding a proton and two neutrons to the oxygen nucleus in CO would yield CF^+. The rotational constant is smaller by about 10%, and the dipole moment larger by a factor of 10, but there are still 14 electrons in a $^1\Sigma^+$ ground state. Furthermore, the $J = 1 - 0$, $J = 2 - 1$ and $J = 3 - 2$ rotational transitions still lie respectively in the 3, 1.5, and 1 mm bands accessible to ground-based observatories.

Figure 1 shows the $J = 1 - 0$, $J = 2 - 1$, and $J = 3 - 2$ spectra obtained toward one of several positions where CF^+ has been detected in the Orion Bar ($\alpha = 05^{\rm h}\,35^{\rm m}\,22.80^{\rm s}$, $\delta = -05^{\rm d}\,25'\,01.0''$), a well-studied photodissociation region with an edge-on geometry that is favorable for the detection of molecules of low abundance. The lower two transitions were observed at the IRAM 30 m telescope, while the submillimeter $J = 3 - 2$ line was detected at the new APEX 12 m telescope, located at an altitude of 5100 m on the Chajnantor plateau in the Atacama desert.

When we construct a rotational diagram, we find—rather fortunately—that the $J = 1$, $J = 2$, and $J = 3$ states that we have observed are almost certainly the three most highly-populated rotational states, and that for any reasonable extrapolation to higher J, the total CF^+ column density lies close to $1.7 \times 10^{12}\ \mathrm{cm}^{-2}$, averaged over our beam.

Our discovery of CF^+ was not obtained by serendipity, nor in a line survey, but rather through a targeted search that was prompted by a recent theoretical study undertaken by Neufeld, Wolfire & Schilke (2005). Motivated by the detection of HF toward the Sgr B2 cloud (Neufeld *et al.* 1997), we considered the chemistry of interstellar fluorine-bearing molecules and reached three key conclusions:

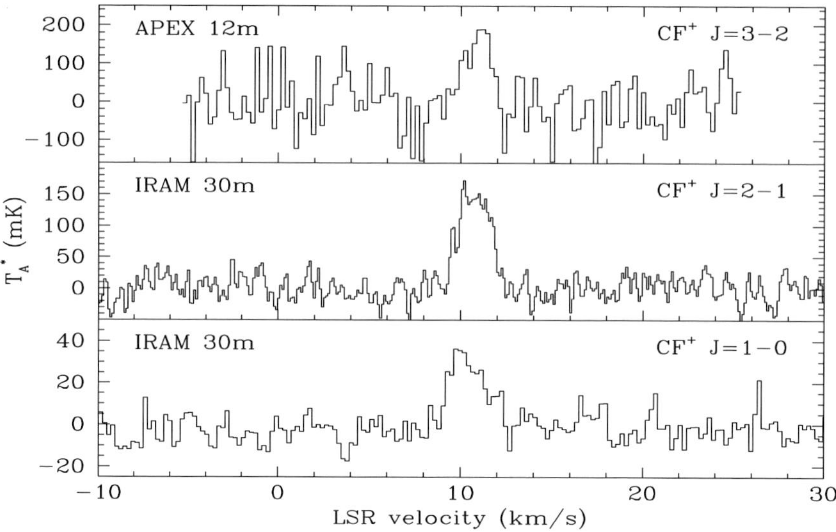

Figure 1. CF$^+$ $J = 1 - 0$, $J = 2 - 1$, and $J = 3 - 2$ spectra obtained toward the Orion Bar.

(1) Hydrogen fluoride forms rapidly by the reaction of fluorine atoms with H$_2$. This is a simple consequence of thermochemistry: HF has the highest dissociation energy of any neutral diatomic hydride, and is the only such molecule to be more strongly bound than molecular hydrogen. Fluorine is therefore the only element in the periodic table to have a neutral atom that reacts exothermically with H$_2$. As Zhu *et al.* (2002) have shown, the reaction is expected to be fairly rapid even in cold molecular clouds.

(2) Hydrogen fluoride is the dominant reservoir of fluorine nuclei (solar abundance: $n_F/n_H \sim 3 \times 10^{-8}$) in the gas phase, even near cloud surfaces. Beneath the surface of a UV-illuminated cloud, HF forms precisely where hydrogen becomes molecular, and long before carbon gets incorporated into CO.

(3) Hydrogen fluoride is destroyed primarily by photodissociation and by reaction with C$^+$ to form CF$^+$. There is a substantial region where C$^+$ and HF overlap, and it is in this region where the CF$^+$ abundance is largest, accounting for a few percent of the total fluorine nuclei. The total predicted CF$^+$ column density is $\sim 10^{12}\,\mathrm{cm}^{-2}$ over a wide range of physical conditions.

To summarize, we have obtained the first astronomical detection of the CF$^+$ ion, toward the Orion Bar. This observation supports a theoretical model that predicts large abundances of HF close to cloud surfaces where the C$^+$ abundance is high, and it suggests that Herschel and SOFIA will detect widespread absorption by the $J = 1 - 0$ transition of HF in diffuse clouds along lines-of-sight to far-infrared continuum sources.

Acknowledgements

IRAM is supported by INSU/CNRS (France), MPG (Germany), and IGN (Spain). APEX is a joint project of the Max Planck Institute for Radio Astronomy, the European Southern Observatory, and the Swedish National Facility for Radio Astronomy.

References

Neufeld, D. A., Zmuidzinas, J., Schilke, P., & Phillips, T. G. 1997, *Ap. J.* 488, L141
Neufeld, D. A., Wolfire, M. G., & Schilke, P. 2005, *Ap. J.* 628, 260
Zhu, C., Krems, R., Dalgarno, A., & Balakrishnan, N. 2002, *Ap. J.* 577, 795

Astrochemistry: Recent Successes and Current Challenges
Proceedings IAU Symposium No. 231, 2005
D.C. Lis, G.A. Blake & E. Herbst, eds.

© 2006 International Astronomical Union
doi:10.1017/S1743921306007162

Optical and Infrared Observations of Diffuse Clouds

Benjamin J. McCall

Departments of Chemistry and Astronomy, University of Illinois at Urbana-Champaign,
Urbana, IL 61801, USA
email: bjmccall@uiuc.edu

Abstract. In the past several years, great progress has been made on the spectroscopy of polyatomic molecules in diffuse interstellar clouds. In this talk, I will review recent developments involving H_3^+, C_3, and the Diffuse Interstellar Bands (DIBs).

The simplest polyatomic molecular ion, H_3^+, has long been recognized as the cornerstone of ion-neutral chemistry in dense molecular clouds (Herbst & Klemperer 1973; Watson 1973). However, in diffuse clouds (where electrons are abundantly produced from photoionization of atomic carbon) the H_3^+ number density was expected to be considerably lower than in dense clouds, owing to the efficiency of electron recombination. It was, therefore, a surprise when a large column density of H_3^+ was detected (McCall *et al.* 1998) in the diffuse line of sight towards Cygnus OB2 12, and subsequently in a sample of heavily reddened diffuse sightlines (McCall *et al.* 2002). Recently, we have detected H_3^+ even in the classical diffuse cloud sightline towards ζ Persei; together with a new measurement of the electron recombination rate coefficient, this result suggests that the cosmic-ray ionization rate is much higher in diffuse clouds than in dense clouds (McCall *et al.* 2003a)!

In 2001, interstellar C_3 was first detected by J. P. Maier and colleagues (Maier *et al.* 2001) in three diffuse cloud sightlines. This was quickly followed up by another detection (Roueff *et al.* 2002) and a survey conducted at low-resolution (Oka *et al.* 2003). This was followed by a high-resolution survey (Ádámkovics, Blake, & McCall 2003) that yielded rotationally resolved spectra of C_3 in 10 sightlines. Much like C_2, C_3 has no permanent dipole moment, and therefore its rotational distribution serves as a sensitive diagnostic of both temperature and density.

The existence of larger polyatomic molecules in diffuse clouds is clear from the presence of the DIBs, which have remained an enigma since their discovery some eight decades ago. A recent survey of the DIBs at the Apache Point Observatory has resulted in a uniform sample of DIB spectra with much higher signal-to-noise ratios than previously available. Some early results from the analysis of the survey data include the identification of a set of DIBs that are related to C_2 (Thorburn *et al.* 2003) as well as some well-correlated pairs of DIBs. I will discuss our recent results.

---

1. Introduction

The first molecules to be detected in the interstellar medium (CH, CH^+, and CN) were found using optical spectroscopy in what we now refer to as diffuse clouds. Polyatomic molecules, such as NH_3, H_2O, and H_2CO, were then detected by radioastronomers in dense molecular clouds, and much of the work in the young field of astrochemistry has been focused on these clouds and the star-forming regions within them.

However, in recent years many polyatomic molecules have been detected in *diffuse* clouds, both at optical and infrared wavelengths (as discussed in this contribution) and at radio wavelengths (as discussed by Liszt, this volume). This is somewhat surprising, considering that diffuse clouds have lower number densities, higher radiation fields, and

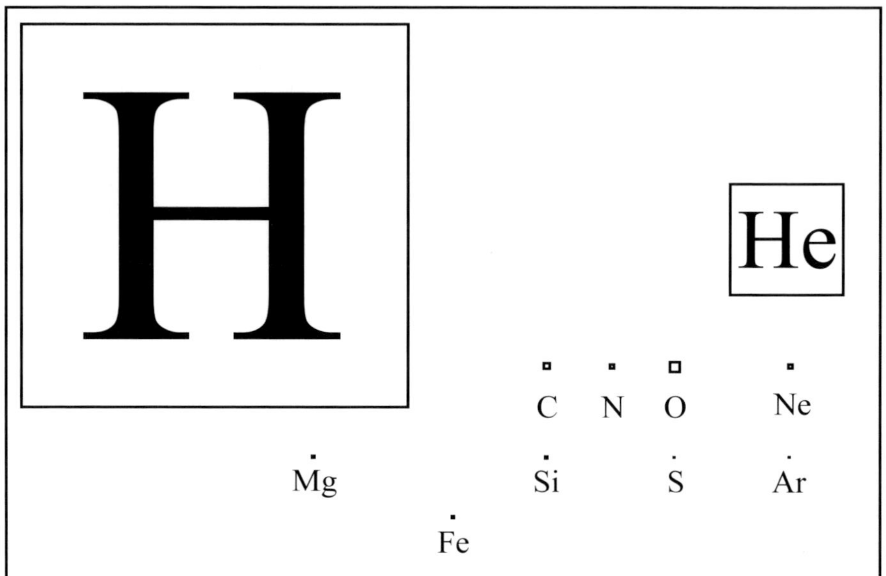

Figure 1. The Astronomer's Periodic Table, in which the area of each element's box is proportional to its cosmic abundance.

higher electron fractions than dense clouds, and these conditions do not seem particularly kind to an active ion-molecule chemistry. For additional background on the physical and chemical conditions in diffuse clouds, the reader is referred to an upcoming *Annual Review* article (Snow & McCall 2006).

2. H_3^+ in Diffuse Clouds

As Figure 1 illustrates, about 90% of all nuclei in the universe are hydrogen nuclei, and consequently hydrogenic species are very important in astrochemistry. In fact, most of the chemistry in the interstellar medium begins when H_2 is ionized by cosmic-ray impact to form H_2^+, which then quickly reacts with another H_2 to form H_3^+. H_3^+ is the simplest polyatomic molecule, and can be thought of as molecular hydrogen with an extra proton attached, in an equilateral triangle configuration. Once H_3^+ is formed, it can donate its "extra" proton to just about any atom or molecule, thus initiating a chain of ion-neutral reactions that leads to the production of many of the molecules observed by radioastronomers in dense clouds. This scheme of interstellar chemistry was first proposed independently by Herbst & Klemperer (1973) and Watson (1973).

As the formation rate of H_3^+ is limited by the cosmic-ray ionization of H_2, this rate can be expressed as $\zeta n(H_2)$, where ζ is the cosmic-ray ionization rate (typically assumed to be $\sim 3 \times 10^{-17}$ s^{-1}). In dense molecular clouds, the destruction of H_3^+ is dominated by chemical reaction with CO to form HCO^+, with a rate of $kn(H_3^+)n(CO)$, where k is the rate constant for this reaction (measured in the laboratory to be $\sim 2 \times 10^{-9}$ cm^3 s^{-1}). In steady state, the rates of formation and destruction are equal, and one can solve for the number density of H_3^+:

$$n(H_3^+) = \frac{\zeta}{k} \frac{n(H_2)}{n(CO)} \sim 10^{-4} \text{ cm}^{-3} \quad \text{[dense clouds]} \tag{2.1}$$

This quantity is essentially constant through a dense cloud, because ζ/k is a constant and because the ratio $n(\mathrm{H}_2)/n(\mathrm{CO}) \sim 6700$ is approximately constant (since nearly all H is in the form of H_2 and nearly all C is in the form of CO). The constancy of $n(\mathrm{H}_3^+)$ makes H_3^+ a particularly powerful probe of dense clouds: if one observes the column density $N(\mathrm{H}_3^+)$ spectroscopically, one can easily infer the column length $L \sim N/n$ and the average hydrogen number density $\langle n(\mathrm{H}_2)\rangle \sim N(\mathrm{H}_2)/L$.

In diffuse clouds, the formation of H_3^+ is by the same mechanism, but the destruction is dominated by dissociative recombination with the abundant electrons generated by photoionization of atomic carbon. The rate expression for the destruction is then $k_e n(\mathrm{H}_3^+)n(e)$, where k_e is the rate constant for the dissociative recombination. In steady state, the number density of H_3^+ is then:

$$n(\mathrm{H}_3^+) = \frac{\zeta}{k_e}\frac{n(\mathrm{H}_2)}{n(e)} \sim 4 \times 10^{-7} \ \mathrm{cm}^{-3} \quad \text{[diffuse clouds]}. \tag{2.2}$$

The quantity k_e has been measured many times in the laboratory over the past few decades, but different experiments have given different results, leaving the value of this key quantity somewhat controversial. It has been suggested that a possible reason for the variations between experiments is varying degrees of internal (rotational and vibrational) excitation of the H_3^+ ions used in the various measurements. Recently, storage ring techniques (which ensure complete vibrational relaxation of the H_3^+) have been combined with rotationally cold ion sources to measure k_e under conditions which most closely approximate the interstellar case. Using a supersonic expansion ion source at CRYRING, McCall *et al.* (2004) measured $k_e \sim 1.3 \times 10^{-6} \ T^{-0.5} \ \mathrm{cm}^3\,\mathrm{s}^{-1}$, and this result has been confirmed by Kreckel *et al.* (2005) using a cryogenic ion trap at TSR. Here we have adopted a temperature of ~ 60 K as used by Le Petit, Roueff, & Herbst (2004) for the diffuse cloud sightline towards ζ Persei, and find $k_e \sim 1.7 \times 10^{-7} \ \mathrm{cm}^3\,\mathrm{s}^{-1}$. The ratio $n(\mathrm{H}_2)/n(e) \sim 2400$ is approximately constant in diffuse clouds, where roughly half of the hydrogen is in molecular form and nearly all of the carbon is ionized to produce electrons.

Once again, we find a constant number density of H_3^+ (just as in the case of dense clouds), but here the value is considerably lower. Given this low number density, one might expect not to observe H_3^+ in diffuse clouds, but in fact it has been seen in many diffuse cloud sightlines (McCall *et al.* 1998, 2002, 2003a; Geballe *et al.* 1999) with column densities of $\sim 10^{14} \ \mathrm{cm}^{-2}$, comparable to those in dense clouds. If we take the case of ζ Persei as an example, the column density $N(\mathrm{H}_3^+) = 8 \times 10^{13} \ \mathrm{cm}^{-2}$, so $L = N/n \sim 2 \times 10^{20} \ \mathrm{cm} \sim 67$ pc. Given the known total column density of hydrogen $N_{\mathrm{H}} \equiv N(\mathrm{H}) + 2N(\mathrm{H}_2) \sim 1.6 \times 10^{21} \ \mathrm{cm}^{-2}$, we would infer an average number density $\langle n_{\mathrm{H}}\rangle \sim 8 \ \mathrm{cm}^{-3}$. However, this is considerably lower than the density inferred from observations of other molecules, which give densities around 150–$500 \ \mathrm{cm}^{-3}$.

This represents quite a discrepancy, even by astronomical standards! Essentially, the predicted value of $n(\mathrm{H}_3^+)$ yields a path length that is too long (or equivalently a density that is too low) by a factor of ~ 20–60. But equation (2.2) above is quite simple and really contains only three parameters: the cosmic-ray ionization rate ζ, the dissociative recombination rate constant k_e, and the electron fraction $n(e)/n(\mathrm{H}_2)$. In the case of ζ Persei, the electron fraction can be directly inferred from spectroscopic observations of the column densities of H_2 and C^+, and is consistent with complete ionization of carbon as suggested above. This leaves only ζ and k_e as possible suspects for the "overabundant" H_3^+. It is, of course, conceivable that both storage ring measurements of k_e are incorrect and that the "real" value of k_e is a factor of 20–60 lower than adopted here. But it is somewhat difficult to accept that both experiments could be wrong, especially since their value of k_e has found recent support from theoretical calculations (Kokoouline & Greene 2003).

It seems more likely that the cosmic-ray ionization rate ζ is a factor of $\sim$20–60 times higher than expected, or roughly $\sim$1.2 $\times$ 10^{-15} s^{-1} rather than $\sim$3 $\times$ 10^{-17} s^{-1}. Such a high ionization rate was originally envisaged by Spitzer & Tomasko (1968), but was later abandoned when observations suggested the value $\sim$3 $\times$ 10^{-17} s^{-1} (see the discussion by Dalgarno). The concept of a high value of ζ in *diffuse* clouds has recently regained some support, however (Liszt 2003; Le Petit, Roueff, & Herbst 2004; Padoan & Scalo 2005). It now seems that spectroscopic observations of H_3^+ in diffuse clouds may serve as a valuable probe of the cosmic-ray ionization rate, which is otherwise difficult to constrain observationally.

3. Carbon Chains in Diffuse Clouds

The shortest carbon chain, C_2, was discovered in diffuse clouds by Souza & Lutz (1977) using the *A-X* 1–0 band near 1.015 μm in the sightline towards Cygnus OB2 12. Following this initial detection, many spectra of C_2 in diffuse clouds have been obtained, primarily using the 2–0 band near 8750 Å, which is more easily accessible with CCD-based spectrographs. As described by van Dishoeck & Black (1982), the observed rotational excitation of C_2 can be used as a probe of both the density and the kinetic temperature of the absorbing gas. An online "calculator" for extracting density and temperature from observed C_2 spectra is available at dibdata.org.

It was not until 2001 that the next longest carbon chain, C_3, was detected in diffuse clouds, by Maier *et al.* (2001) using the *A-X* 0–0 band near 4051 Å. These authors detected C_3 towards ζ Oph, ζ Per, and 20 Aql, with rotational resolution. Shortly thereafter, Roueff *et al.* (2002) detected C_3 towards HD 210121 and developed a full excitation model (like in the case of C_2) to infer the temperature and number density. Unfortunately, the details of this model have yet to be published, and so it is not available to analyze the many spectra of C_3 that have been subsequently obtained. In 2003, Oka *et al.* (2003) published a number of C_3 spectra resulting from the Apache Point Observatory diffuse interstellar band survey (described below). Although these spectra had lower resolution and were therefore not rotationally resolved, they demonstrated the widespread occurrence of C_3 and showed a rough correlation between the C_2 and C_3 column densities.

To follow up on the C_3 results of the Oka survey, Máté Ádámkovics, Geoff Blake, and I began a campaign to obtain higher resolution spectra of Oka's sightlines using the Lick and Keck observatories. From this campaign we obtained rotationally resolved spectra of C_3 in 10 sightlines (Ádámkovics, Blake, & McCall 2003). In the course of this survey, we noted that the wavelength of the $R(0)$ line did not seem to match the laboratory values reported by Gausset *et al.* (1965); a subsequent experiment using cavity ringdown spectroscopy on a supersonically expanding plasma (which produces rotationally cold C_3) confirmed that the original laboratory value was in error (McCall *et al.* 2003b).

Having been the first to detect C_3, Maier and co-workers turned their attention to the next longest carbon chains C_4 and C_5, whose laboratory spectra in the visible region were already known. They obtained very high signal-to-noise ratio spectra towards ζ Oph, but were not able to detect either molecule (Maier, Walker, & Bohlender 2004). Although ζ Oph is very bright (V = 2.56), and therefore one can obtain high S/N in a short time, it does not have particularly high C_2 and C_3 column densities. A sightline such as HD 204827, which has very high column densities of both C_2 and C_3, seemed a better bet, although it is much fainter (V = 7.94). Recently, Máté Ádámkovics, Geoff Blake, and I spent two nights at Keck to obtain a very high S/N spectra of HD 204827—sadly, our effort did not yield a detection of either C_4 or C_5 (Ádámkovics, Blake, & McCall 2006). Our limit on the C_4 column density is roughly a third of the C_3 column density, and that on the C_5 column density is roughly one-sixteenth of C_3.

The lack of observed C_4 and C_5 is somewhat surprising, and must be telling us something about the chemistry of these clouds. To our knowledge, the only chemical model of diffuse clouds that includes C_4 and C_5 is that of Roueff *et al.* (2002). This model targets the sightline towards HD 210121, and suggests that at visual extinctions greater than about 1.3, the C_4 column density *exceeds* that of C_3. At $A_V \sim 1.8$, C_5 overtakes C_3. If we naïvely extrapolate the figure in their paper to $A_V = 2.6$ (for HD 204827), we find the C_4 should have twice the column density of C_3! Clearly there is a conflict between the observations and the model, and we encourage chemical modellers to work on this very interesting sightline towards HD 204827.

4. The Diffuse Interstellar Bands

The diffuse interstellar bands (DIBs) are an enigmatic set of at least 200 absorption lines observed in the visible spectra of nearly all stars whose light is filtered through interstellar gas and dust. The first of the DIBs, $\lambda 5780$ and $\lambda 5797$ were observed by Mary Lea Heger at Lick Observatory in 1919. The DIBs seem, for the most part, to be associated with diffuse atomic clouds; their intensity is roughly correlated with the visual extinction, but at high visual extinctions their intensity tends to level off. The DIBs have been reviewed by Herbig (1995) and are also discussed by Snow & McCall (2006).

For a long time, the rough correlation with the visual extinction led researchers to favor solid state carriers for the DIBs. However, several characteristics of the DIBs argue strongly against dust as the carrier. These characteristics include the constancy of the central wavelengths, the lack of any associated emission features, and the fine structure that has been observed at high resolution in some DIBs (e.g., Sarre *et al.* 1995). Today, it is generally accepted that the DIBs are caused by large polyatomic molecules in the gas phase. Most likely the DIB carriers are carbon-based, and constitute a vast reservoir of organic material in the interstellar medium. The problem of the DIBs has been called the greatest unsolved mystery in all of spectroscopy. The eventual identification of the molecules causing the DIBs will more than double the known inventory of interstellar molecules, and will provide a powerful multidimensional probe of the chemical and physical conditions in interstellar clouds.

In an effort to compile a high-quality set of DIB spectra that can be used both for studying the DIBs themselves as well as for comparison with laboratory spectra of candidate carriers, a large-scale survey has recently been conducted at the Apache Point Observatory in New Mexico. This survey has utilized the ARCES (Astrophysical Research Consortium Echelle Spectrograph), which offers a resolving power $\lambda/\Delta\lambda \sim 37,500$ and complete spectral coverage from 3600–10,200 Å. Our main campaign involved observations on 119 nights from January 1999 through January 2003, and yielded spectra with signal-to-noise ratios (at 5780 Å) greater than 500 for 160 stars (115 of which are reddened). The measurements and analysis of the DIBs in these spectra are still very much underway, but I'd like to discuss some of our early results.

As mentioned above, most of the DIBs appear to be associated with diffuse atomic clouds. For example, Herbig (1993) found a better correlation between the strength of the $\lambda 5780$ band with the column density of atomic hydrogen than with that of molecular hydrogen. Preliminary results from the APO survey are shown in Figure 2 and confirm Herbig's findings with much greater clarity (York *et al.* 2006). The fact that most DIBs

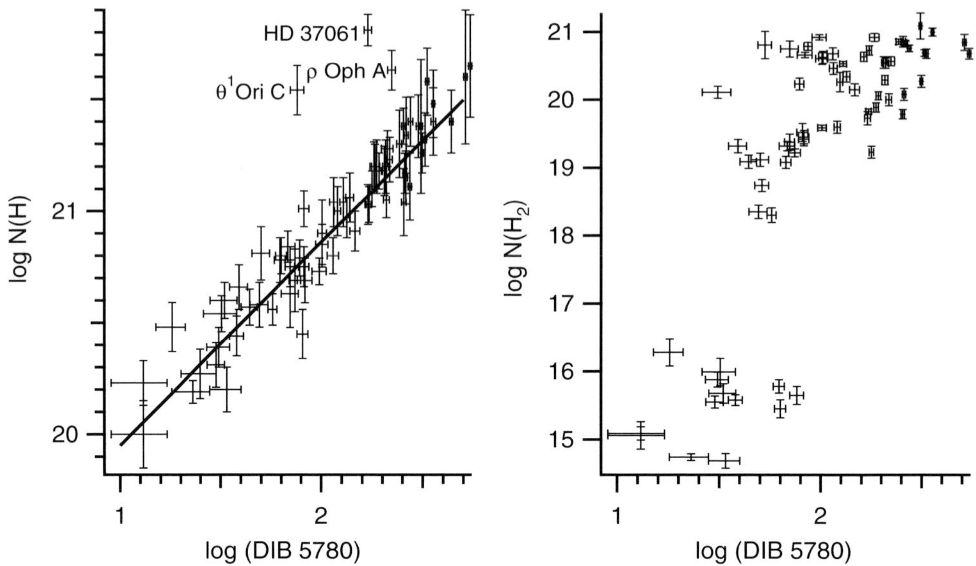

Figure 2. The good correlation (*left*) between $\lambda5780$ and the column density of H I, compared with the bad correlation (*right*) between $\lambda5780$ and H_2. For H I and $\lambda5780$, the correlation coefficient is $r = 0.94$, excluding the three outliers (which have unusual radiation fields).

are found in gas where the hydrogen molecular fraction is low represents an important chemical constraint on the types of molecules responsible for the DIBs.

However, not all DIBs seem to follow this behavior. The APO survey has yielded the identification of a set of DIBs (now called the "C_2 DIBs") whose intensities show some correlation with the column density of C_2 (Thorburn *et al.* 2003). Since C_2 is known to exist in the denser regions of diffuse clouds, which are probably more molecular, these C_2 DIBs seem to form a distinct class from the classical DIBs like $\lambda5780$, which correlate with atomic hydrogen. Further evidence for this separation of the DIBs into two classes has been provided by observations of the peculiar line of sight towards HD 62542, which appears to have been stripped of its outer, more diffuse layers by a shock. Snow *et al.* (2002) failed to detect DIBs in this sightline, but a more recent higher S/N study by Ádámkovics, Blake, & McCall (2005) detected weak DIBs. Most interestingly, the C_2 DIBs were clearly observed, while many of the classical DIBs could not be detected at all. This strongly suggests that the C_2 DIBs are abundant in the "cores" of diffuse clouds (that is, in diffuse molecular clouds, using the classification scheme of Snow & McCall 2006) whereas the classical DIBs are abundant in the "envelopes" of diffuse clouds (that is, in diffuse atomic clouds).

As part of our APO survey, we have been searching for pairs of DIBs whose intensities are well-correlated in different sightlines. The motivation for this search is that most molecules show more than one vibronic band in their electronic spectra, and that (in the absence of low-lying splittings in the electronic ground state) the relative intensities of these vibronic transitions are governed only by the Franck-Condon factors, not by environmental conditions. Therefore, a set of DIBs that are "perfectly" correlated could be expected to represent the electronic spectrum of a single molecular carrier, and the spacings among those DIBs could give insight into the nature of that carrier. To date, we have only searched for correlations among the strongest DIBs, and have found relatively

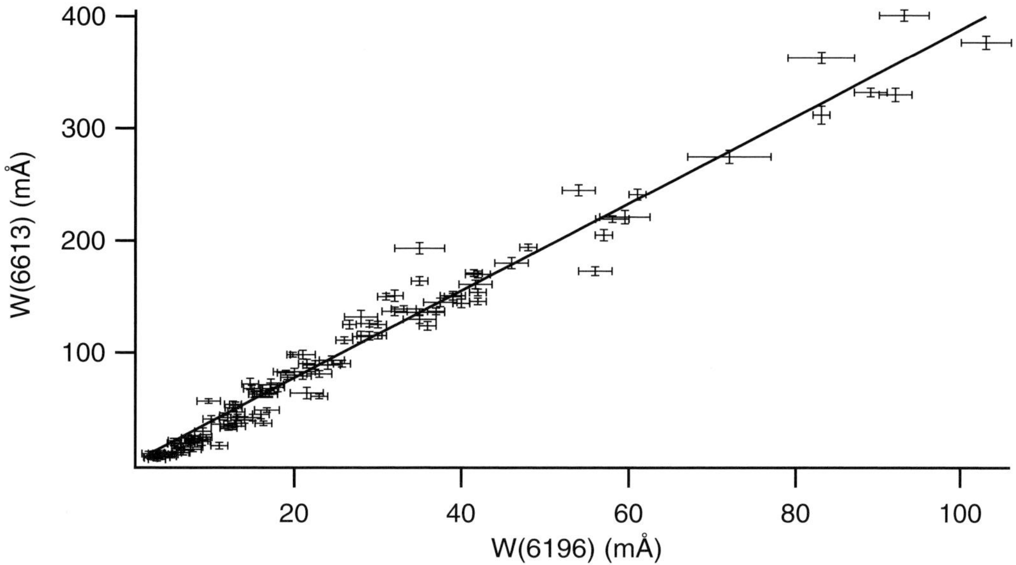

Figure 3. The remarkably good correlation between $\lambda 6196$ and $\lambda 6613$. The correlation coefficient is $r = 0.986$.

few correlations which might be considered "perfect." The best correlation we have found so far is that between $\lambda 6196$ and $\lambda 6613$, as shown in Figure 3. The correlation coefficient for this pair of DIBs is $r = 0.986$ in our sample, which is not unity but is pretty close. It appears that the deviation from a perfect straight line is larger than would be expected from our estimated uncertainties, which suggests either that we have underestimated our errors, or that these two DIBs are not due to the same molecule but instead to two very closely related molecules. However, we consider the quality of this correlation to be quite striking, especially by astronomical standards.

5. Conclusions

Just ten years ago, the conventional wisdom was that diffuse interstellar clouds were relatively uninteresting chemically, consisting mainly of atoms and a few diatomic molecules. In the past decade, infrared and optical spectroscopy have demonstrated the ubiquity of H_3^+ and C_3 in diffuse clouds, and observations at radio wavelengths have detected many other polyatomics (Liszt, this volume). It is becoming increasingly clear that diffuse clouds are chemically rich, and that our understanding of their chemistry is still very immature. Perhaps our greatest reminder of this fact is the existence of the diffuse interstellar bands and the fact that the identities of their carriers still elude us over 85 years since their first observation. We can expect the next decade to be an incredibly exciting one for chemical studies of diffuse clouds, in terms of both observations and modelling. With increasing sensitivity of both optical and infrared spectroscopy, many new molecules may be detected. With increasing sophistication of chemical models of diffuse clouds, we may finally understand some of the puzzling aspects of diffuse cloud chemistry. And, with a high-quality database of DIB spectra, coupled with ever more sophisticated laboratory experiments, we may even identify some of the carriers of the DIBs!

Acknowledgments

It is a pleasure to acknowledge the contributions of my collaborators. The H_3^+ observations were conducted with T. R. Geballe and T. Oka. The storage-ring measurements of the H_3^+ dissociative recombination rate involved a large team headed up by M. Larsson which included A. J. Huneycutt, R. J. Saykally, N. Djuric, G. H. Dunn, J. Semaniak, O. Novotny, A. Paal, F. Österdahl, A. Al-Khalili, A. Ehlerding, F. Hellberg, S. Kalhori, A. Neau, and R. Thomas. The observations of C_3, C_4, and C_5 were conducted with M. Ádámkovics and G. A. Blake. The APO DIB survey has included contributions from S. D. Friedman, L. M. Hobbs, T. Oka, B. L. Rachford, T. P. Snow, P. Sonnentrucker, J. A. Thorburn, D. E. Welty, and D. G. York.

I also wish acknowledge financial support from the NASA Laboratory Astrophysics program, a National Science Foundation CAREER award, a Dreyfus New Faculty award, an American Chemical Society Petroleum Research Fund grant, and the University of Illinois.

References

Ádámkovics, M., Blake, G.A., & McCall, B.J. 2003, *Ap. J.* 595, 235

Ádámkovics, M., Blake, G.A., & McCall, B.J. 2005, *Ap. J.* 625, 857

Ádámkovics, M., Blake, G.A., & McCall, B.J. 2006, in preparation

Gausset, L., Herzberg, G., Lagervist, A., & Rosen, B. 1965, *Ap. J.* 142, 45

Geballe, T.R., McCall, B.J., Hinkle, H.K., & Oka, T. 1999, *Ap. J.* 510, 251

Herbig, G.H. 1993, *Ap. J.* 407, 142

Herbig, G.H. 1995, *ARA&A* 33, 19

Herbst, E. & Klemperer, W. 1973, *Ap. J.* 185, 505

Kokoouline, V. & Greene, C.H. 2003, *Phys. Rev. A*, 68, 012703

Kreckel, H. *et al.* 2005, *Phys. Rev. Lett.* 95, 263201

Le Petit, F., Roueff, E., & Herbst, E. 2004, *A&A* 417, 993

Liszt, H. 2003, *A&A* 398, 621

Maier, J.P, Lakin, N.M, Walker, G.A.H., & Bohlender, D.A. 2001, *Ap. J.* 553, 267

Maier, J.P, Walker, G.A.H., & Bohlender, D.A. 2004, *Ap. J.* 602, 286

McCall, B.J., Geballe, T.R., Hinkle, K.H., & Oka, T. 1998, *Science* 279, 1910

McCall, B.J. *et al.* 2002, *Ap. J.* 567, 391

McCall, B.J. *et al.* 2003a, *Nature* 422, 500

McCall, B.J., Casaes, R.N, Ádámkovics, M., & Saykally, R.J. 2003b, *Chem. Phys. Lett.* 374, 583

McCall, B.J. *et al.* 2004, *Phys. Rev. A* 70, 052716

Oka, T. *et al.* 2003, *Ap. J.* 582, 823

Padoan, P. & Scalo, J. 2005, *Ap. J.* 624, L97

Roueff, E., Felenbok, P., Black, J.H., & Gry, C. 2002, *A&A* 384, 629

Sarre, P.J. *et al.* 1995, *MNRAS* 277, L41

Snow, T.P. *et al.* 2002, *Ap. J.* 573, 670

Snow, T.P. & McCall, B.J. 2006, *ARA&A*, in preparation

Souza, S.P. & Lutz, B.L. 1977, *Ap. J.* 216, L49

Spitzer, L. & Tomasko, M.G. 1968, *Ap. J.* 152, 971

Thorburn, J. *et al.* 2003, *Ap. J.* 584, 339

van Dishoeck, E.F. & Black, J.H. 1982, *Ap. J.* 258, 533

Watson, W.D. 1973, *Ap. J.* 183, L17

York, D.G. 2006, in preparation

Discussion

KREŁOWSKI: (1) Despite the very good correlation between the 6196 and 6614 DIBs, (found by Moutou *et al.* 1999, *A&A* 351, 680) their profiles do not change in unison (Galazutdinov *et al.* 2002, *A&A* 384, 215). (2) In the case of C_5, nature proves once again its complexity. While the 5109 Å feature, predicted by the laboratory experiments is apparently absent in spectra of reddened stars, the second feature, 4975 Å, has a counterpart, most likely a weak DIB (Galazutdinov *et al.* 2002, *A&A* 395, 223). (3) The "C_2 DIBs" reported in your lecture correlate in our sample to a various degree with the C_2 column density. However, the strength ratios of these DIBs are variable. What do they have in common?

McCALL: As discussed in detail by Thorburn *et al.* 2003 (*Ap. J.* 584, 339), the "C_2 DIBs" are a set of DIBs whose equivalent widths (normalized to $\lambda 6196$) are well correlated with $N(C_2)/E(B-V)$. We do not suggest that they belong to the same molecule, only that (unlike other DIBs) they are correlated with C_2.

GALAZUTDINOV: While relating the abundances of C_2 and C_3 molecules, have you applied the data from the highest resolution study of C_3 published until now (Galazutdinov *et al.* 2002, *A&A* 395, 969)?

McCALL: In our comparison of C_2 and C_3 column densities, we have utilized the equivalent widths for the three sources (HD 148184, 149757, and 152236), for which you reported C_3 detections. However, we had to recompute the column densities as the values in your paper were incorrect (too low) by nearly an order of magnitude.

SARRE: Concerning the "C_2" diffuse bands, have you investigated if there is any link between the C_2 rotational temperatures and the diffuse band strengths?

McCALL: No, we have not looked into that, but it is a very interesting suggestion!

ROUEFF: Extrapolation of C_4 and C_5 modeled column densities is *not* acceptable. Can you make precise the errorbars in the determination of the column density of H_3^+?

McCALL: I hope that my unacceptable extrapolation will serve as encouragement for new chemical models of this sightline! I do not have the calculated error bars off the top of my head, but we are preparing a manuscript with such observational details. I would estimate that the H_3^+ column density towards ζ Per is good to at least a factor of two.

Photo: E. Herbst

Astrochemistry: Recent Successes and Current Challenges
Proceedings IAU Symposium No. 231, 2005
D.C. Lis, G.A. Blake & E. Herbst, eds.

© 2006 International Astronomical Union
doi:10.1017/S1743921306007174

Ultraviolet Observations of Interstellar Molecules

Theodore P. Snow†

University of Colorado, Boulder, CO, USA
email: tsnow@casa.colorado.edu

Abstract. The benefits of placing ultraviolet (UV) spectrographs in space were first pointed out by Spitzer (1946) and Spitzer & Zabriskie (1959), with much of the emphasis on absorption-line measurements of interstellar atoms and molecules. The first UV observations of the diffuse interstellar medium took place in the early 1970s, using sounding rockets, which were followed by a series of orbital instruments starting in the later 1970s and extending almost to the present time. Many of these observations have provided information on the chemistry of the diffuse ISM, and will be emphasized here. The most abundant molecule in space, H_2, is not readily observed in other wavelength bands and was first detected by a UV sounding rocket experiment (Carruthers 1970). Thereafter a series of orbital instruments, starting with the *Copernicus* satellite in 1972 and culminating with the launch of *FUSE* in 1999, carried out broad surveys of molecular hydrogen in a variety of ISM environments (*e.g.*, Spitzer *et al.* 1973; Snow *et al.* 2000; Shull *et al.* 2000; see also the review by Shull & Beckwith 1982). The UV observations of electronic transitions of H_2, including lines arising from excited rotational levels of the ground state, have provided a wealth of information on not only the chemistry of the diffuse ISM, but also the composition and physical conditions. The second most abundant molecule in the diffuse ISM, CO, also has numerous electronic transitions in the UV, and has been widely observed by many different instruments, most usefully the *Hubble Space Telescope* (*HST*) (*e.g.*, Lambert *et al.* 1994; Sheffer *et al.* 2002). Valuable information on diffuse ISM chemistry, isotopic composition, and physical conditions has been provided by these observations. A few additional molecules, all of them diatomics such as OH and N_2, have been detected through UV absorption-line observations, and many others have been sought unsuccessfully. The Cosmic Origins Spectrograph, expected to be installed aboard the HST in 2007, will have enhanced UV sensitivity and will provide information on molecular abundances in denser clouds than previously studied. One of the high priorities will be the search for electronic transitions of complex organic molecules such as PAHs.

Keywords. molecular data — molecular processes — techniques: spectroscopic — ISM: abundances — ISM: atoms — ISM: molecules — ultraviolet: ISM

---

1. Introduction

For reasons described in detail below, ultraviolet studies of molecules in the interstellar medium (ISM) are restricted to diffuse clouds. Therefore UV measurements do not apply to the dense molecular clouds where star formation takes place. Nevertheless, because the chemistry in diffuse clouds is simpler yet still applicable to dense clouds, it is very useful to apply UV spectroscopy to the study of molecules in interstellar clouds. This paper therefore provides a brief review of what has been learned from ultraviolet observations of interstellar molecules in diffuse clouds.

In the diffuse interstellar medium, collisions are rare and most atoms, ions, and molecules are in their ground electronic state. For molecules, only very low-lying rotational states of the ground vibrational level are likely to be populated, except in regions where

† Present address: CASA—University of Colorado, 389 UCB, Boulder, CO 80309-0389

there is a local excitation source such as a strong radiation field. For the majority of these species transitions from the ground electronic state to higher levels require sufficient energy to fall in the ultraviolet portion of the spectrum. Therefore, perhaps a bit ironically, observations of absorption lines formed by cold interstellar atoms and molecules require the high-energy photons of ultraviolet radiation.

Despite the richness of infrared and mm-wave radio observations of interstellar molecules, there are a few species that are best observed through their electronic transitions in the UV. These include the most abundant interstellar molecule of all, H_2, as well as its isotopologue, HD. In addition the second most abundant molecule in space, CO, can be observed in diffuse clouds through its UV electronic transitions, while IR and radio methods are superior for denser clouds. Many other species have electronic transitions in the UV, including both OH and H_2O as well as various large organic species suspected of being common in the diffuse ISM but not yet identified.

Ultraviolet radiation does not penetrate dense interstellar material, so as noted above, UV observations of interstellar molecules are restricted to diffuse clouds. Here we refer to diffuse atomic clouds and diffuse molecular clouds, which are defined on the basis of their molecular hydrogen content. Clouds containing little or no H_2 are identified as diffuse atomic clouds, while those containing a significant molecular fraction are diffuse molecular clouds. The preceding paper by B. J. McCall contains further comments on this distinction between cloud types, as does the review by Snow & McCall (2006).

Despite being limited to diffuse atomic and molecular clouds, UV observations of interstellar molecules have made significant contributions to our understanding of astrochemistry. Important experiments and missions in the past include the sounding rockets used to make the initial discoveries of molecular hydrogen (Carruthers 1970) and the UV transitions of CO (Smith & Stecher 1971); the *Copernicus* observatory, which operated in the 1970s and carried out pioneering studies of interstellar atomic and molecular abundances (particularly H_2); the *International Ultraviolet Explorer (IUE)*, which was especially useful in measuring dust extinction curves; both the GHRS and STIS instruments on the *Hubble Space Telescope (HST)*; and the *Far Ultraviolet Spectroscope Explorer (FUSE)*, which at this writing was the only UV spectroscopic instrument still operating.

No new UV missions are being planned, so the future for UV astronomy is grim, though we can hope that the Cosmic Origins Spectrograph (Green 2000), already built and calibrated and ready to go, will eventually be deployed on the *HST*.

2. Technical Challenges to UV Spectroscopy

The primary challenge facing observers of UV interstellar spectra is brought about by the Earth's atmosphere, which refuses to allow UV photons to enter from space. Thus UV telescopes must be launched into space, an expensive prospect.

A second challenge, more readily surmounted, is to design and build telescope optics that will be reflective at UV wavelengths. Most materials become absorptive in the UV, more so with decreasing wavelength. But great success has been achieved with aluminum mirrors that are overcoated with substances such as magnesium fluoride (MgF_2) or lithium fluoride (LiF), which transmit UV radiation while protecting the aluminum from oxidation, the chief cause of reflectivity loss. An alternative far-UV reflective surface is silicon carbide (SiC), which has been used successfully as well, for example in the *FUSE* telescope optics.

An additional obstacle to UV observations of interstellar lines, one which cannot be avoided, is that the interstellar medium itself blocks out UV radiation, in proportion to distance or column density. The most drastic effect is caused by interstellar dust, which absorbs and scatters photons with increasing efficiency toward short UV wavelengths. The resulting extinction curve rises from the infrared through to the far-UV, reaching levels below 1200 Å as high as 10 or 15 times the color excess E_{B-V}, which translates to UV attenuation factors of 10,000 to 100,000 in comparison with unreddened stars. Unfortunately, the diffuse clouds with the highest molecular abundances are also the ones with the steepest and most obstructive UV dust extinction curves.

Another difficulty in observing interstellar molecules at UV wavelengths is caused, ironically, by interstellar molecules. In reddened lines of sight H_2 bands wipe out large segments of the spectrum below 1110 Å, preventing observations of interstellar lines due to other species. The saturated (often damped) lines arising from the $J=0$ and $J=1$ rotational levels of the ground electronic and vibrational states of H_2 are strong enough to obliterate several Å at the location of each of the many Lyman and Werner bands that are spaced regularly from 1108 Å all the way through to the Lyman limit at 912 Å.

Finally, yet another complication for observations of UV interstellar lines is that the hot stars that are used as background sources themselves have complex spectra. Whereas O and B stars have smooth continua at visible wavelengths, in the UV their spectra are crowded with lines, adding to the difficulty of detecting and analyzing interstellar features. The problem is ameliorated by selecting stars with high rotational velocities, whose photospheric lines are much broader than most interstellar lines—but even so, the continuum against which the interstellar lines are measured is rarely smooth and level.

Despite the many difficulties, UV spectroscopy has proven to be an extremely valuable tool for studies of the diffuse ISM, including observations of molecular species.

3. Aside: What We Can Learn about the Atomic Gas from UV Spectroscopy

In order to place molecular observations into context, for example in order to test observations against chemical models, it is necessary to have a general picture of what we can learn from UV observations of atomic lines formed in the same diffuse clouds where the observed molecules reside. Several useful reviews of UV atomic line observations have been published, including those by Jenkins & Spitzer (1975), Savage & Sembach (1996); and Snow (1997).

As noted in the Introduction, many species, both atomic and molecular, have their ground state absorption transitions in the ultraviolet. Most of the common elements, from hydrogen through the iron peak, can be observed in the UV, often in their dominant ionization state. In the diffuse ISM the ambient interstellar radiation field, which approximates a black body having a temperature of around 10,000 K, ensures that most species whose ionization potentials are below the 13.6 eV Lyman limit of hydrogen are singly ionized. Atomic hydrogen permeating the diffuse ISM obliterates all photons with energies immediately above 13.6 eV (*i.e.*, wavelengths just below 912 Å). Thus species such as nitrogen and oxygen, whose ionization potentials are near or above 13.6 eV, remain mostly in the neutral atomic form while most other common elements through the iron peak are found in the singly ionized state (with the notable exception of helium, which is simply unobservable in diffuse clouds except for the very local ISM where extreme UV spectra can be obtained).

Observations of the UV absorption lines of common elements lead to derivations of their relative abundances, and in turn to estimates of elemental depletions; *i.e.*, the factors by which the elements in the diffuse ISM are underabundant relative to a reference standard, which is often taken to be the solar composition. Most elements are depleted to some extent (relative to hydrogen), ranging from less than a factor of two for volatiles such as C, N, and O to factors from 10^2 to 10^4 for some of the refractory elements. What is missing from the gas is assumed to be present in dust grains, so depletion measures are an indirect means of determining the composition of the dust.

Of greater relevance for understanding the chemistry of diffuse clouds is what the atomic line observations can tell us about the physical conditions in diffuse clouds, because factors such as the volume density of atoms, the electron density, and the radiation field intensity, all of which can be derived from data on atomic lines, are important in assessing the chemical environment. The volume density of atoms (or molecules) can be estimated from the analysis of fine-structure collisional excitation in neutral atoms such as atomic carbon (C I) and ionized carbon, nitrogen, and silicon (C II, N II, and Si II) (see Bahcall & Wolf 1968 or Jenkins & Shaya 1981 for details). The electron density can be deduced from the ionization ratios of elements where both the neutral and cationic forms are observed, such as carbon, magnesium, iron, and others (see Morton 1975 for a summary of this analysis for the line of sight toward ζ Oph). The radiation field intensity is best derived from the analysis of rotational excitation in molecules such as H_2 and C_2 whose excitation is dominated by UV pumping (this will be described in a later section).

Finally, high-resolution spectra of atomic interstellar line profiles can reveal the velocity distribution along a line of sight, thereby constraining the interpretation of observed molecular lines. Most such observations to date have been made at visible wavelengths where spectrographs are available with resolving powers ($\lambda/\Delta\lambda$) as high as 10^6. Application of velocity structure information to lower-resolution UV spectra requires line profile modeling, but has proven very useful in constraining and interpreting interstellar abundances (ideally we would have an ultraviolet spectrograph with ultra-high resolving power, but as noted above, the prospects for this any time soon are poor).

4. Interstellar Molecules Detected through UV Observations

A number of molecules have been detected through UV absorption lines (Table 1). By far the two most abundant species, not only for UV measurements but in dense clouds as well, are molecular hydrogen and carbon monoxide. But a few other species have been detected as well, and are important for constraining models of diffuse cloud chemistry.

Table 1. Molecules detected through UV observations

Species	log column density (for ζ Oph)	Abundance relative to H	Reference
H_2	20.62	0.31	Morton 1975
H_2 ($v = 3$)	11.81	4.8×10^{-10}	Federman *et al.* 1995
HD	14.80	6.8×10^{-7}	Lacour *et al.* 1995
CO	15.40	1.9×10^{-6}	Lambert *et al.* 1994
^{13}CO	13.17	1.1×10^{-8}	Lambert *et al.* 1994
OH	13.70	3.7×10^{-8}	Crutcher & Watson 1976
HCl	11.43	2.0×10^{-10}	Federman *et al.* 1995
N_2[†]	13.66	1.6×10^{-7}	Knauth *et al.* 2004

[†] The values for N_2 is not for ζ Oph but instead for the star HD 124314, which has a total visual extinction of about 1.5 mag as opposed to about 1.0 mag for ζ Oph.

4.1. *Ultraviolet Observations of H_2 and CO*

4.1.1. *Molecular Hydrogen*

H_2 is a homonuclear diatomic species having no dipole moment and hence no allowed rotational or vibrational transitions. Therefore, with the exception of forbidden quadrupole emission seen near 2.2 μm in the near infrared in radiatively excited nebulae, H_2 is virtually unobservable in the diffuse ISM except through its allowed electronic transitions in the ultraviolet. However, H_2 was always expected to be the most abundant molecule in space, leading to the development of spectrographs specifically aimed at observing this species. The initial detection of interstellar H_2 was achieved by Carruthers (1970) with a small rocket-borne spectrograph, and the first extensive surveys of molecular hydrogen in the ISM were carried out by the *Copernicus* observatory (Rogerson *et al.* 1973; Spitzer *et al.* 1973; Spitzer, Cochran, & Hirshfeld 1974; Savage *et al.* 1977; Jenkins & Spitzer 1975; Shull & Beckwith 1982). More recently interstellar H_2 absorption has been recorded by the IMAPS and ORFEUS Space Shuttle payloads (Jenkins *et al.* 1996; Hurwitz *et al.* 1998) and extensively by the *Far-Ultraviolet Spectroscopic Explorer* (*FUSE*) mission (Moos *et al.* 2000). The *FUSE* observations in particular have extended observations of molecular hydrogen to more reddened galactic lines of sight than previously explored; have provided sensitive surveys of H_2 throughout the Milky Way's disk and halo; and have provided the first surveys of interstellar H_2 in the Magellanic Clouds.

The essential results of the UV surveys of interstellar H_2 are: (a) molecular hydrogen is *EVERYWHERE* in space, including lines of sight with very low reddening; (b) in diffuse clouds the fraction of hydrogen atoms in the form of H_2 ranges from very small values (0.01 or less) to as large as 0.8; (c) there is a sharp transition from very low to very significant molecular fraction that occurs at a total hydrogen column density of about 5×10^{19} cm^{-2} (corresponding to $E_{B-V} \approx 0.08$ mag); (d) the relative populations of H_2 molecules in rotationally excited states of the ground electronic state typically fit two different excitation temperatures; (e) vibrationally excited H_2 is rarely seen in absorption in diffuse clouds; and (f) H_2 in the Magellanic Clouds shows a reduced molecular fraction per reddening as compared with the Galaxy. Each of these points is expanded below.

On the ubiquity of interstellar H_2: Early data from *Copernicus* observations indicated that even for lines of sight with interstellar dust color excess $E_{B-V} \leqslant 0.1$ molecular hydrogen absorption could be detected. Now *FUSE* observations, notably the diffuse sightline surveys of Shull *et al.* (2001; 2002) have confirmed that H_2 is nearly always detectable, with a range of column densities from about 10^{14} cm^{-2} to over 10^{20} cm^{-2}. Few or no lines of sight have been found which do not contain detectable molecular hydrogen absorption.

The molecular fraction: Molecular hydrogen represents only a small fraction of all hydrogen atoms in lightly-reddened lines of sight (below about $E_{B-V} = 0.1$ or so). For higher reddenings the molecular fraction rises, to a maximum value (to date) of roughly 0.8. No sightlines have been found with 100 percent of the hydrogen in molecular form, though it is possible that relatively dense clouds in some cases may approach this value (the difficulty is that every line of sight contains at least some diffuse material where much of the hydrogen is in atomic form; we cannot isolate specific dense clouds to ascertain their molecular fractions individually). One current issue is whether translucent clouds (as defined by van Dishoeck & Black 1989, in which the hydrogen should be fully molecular while there is still sufficient penetration by UV photons to affect the chemistry) actually exist. While to date *FUSE* observations have found no molecular fraction of unity, a few sightlines have been identified which may contain translucent clouds where all the hydrogen is molecular—but as yet this has been difficult to establish due to the mix of clouds and environments along every line of sight.

The transition at $N_H \approx 5 \times 10^{19}$ cm^{-2}: There is a very significant break in the molecular fraction when total column densities exceed about 10^{19} cm^{-2}, as shown first by Spitzer *et al.* (1973) and Savage *et al.* (1977) on the basis of *Copernicus* data and then verified by the various *FUSE* surveys. The explanation for this transition from low to high molecular fraction is straightforward: When the column density of H_2 reaches about 10^{19} cm^{-2}, the strong electronic transitions in the far-UV Lyman and Werner bands become saturated, so that the molecules in the outer portion of a cloud shield those in the interior from the UV photons that would otherwise destroy them.

Rotational excitation of H_2 in diffuse clouds: Typically most of the observed H_2 in diffuse clouds is in the first two rotational states ($J = 0,1$) of the ground vibrational level of the ground electronic state. But observations reveal in addition substantial populations in higher rotational states, up to $J = 6$ or so in diffuse clouds (Spitzer *et al.* 1973; Spitzer & Cochran 1973; Spitzer, Cochran, & Hirshfeld 1974). In clouds sufficiently dense for the strong lines of H_2 to be self-shielded, radiative processes do not affect the $J = 0$ and $J = 1$ levels, so that local chemical reactions govern their relative populations. As a result the ratio $N(J = 1)/N(J = 0)$ is a local thermometer, indicating the kinetic temperature of the gas. In diffuse clouds kinetic temperatures range from 50 to 150 K. The higher-J lines are believed to be populated by radiative pumping, as first pointed out by Black & Dalgarno (1973) and then explored further by Jura (1975a,b), who developed useful models for deriving both the radiation field intensity and the local density for the optically thin and thick cases. In this process the high-J levels are populated due to UV absorption resulting in excitation to an excited electronic state, from which the molecules either dissociate (a small fraction of the time) or cascade back down to the ground electronic state through a series of vibrational and rotational levels. Thus in equilibrium, in the presence of a strong UV radiation field, there is a constant population of H_2 molecules in high-J rotational states.

Vibrationally excited H_2 in diffuse clouds: In rare cases absorption lines arising from excited vibrational states are seen, as in ζ Oph (Federman *et al.* 1995), HD 37903 (Meyer *et al.* 2001), and HD 34078 (Boisse *et al.* 2005). The vibrational states show up only in H ii regions close to a hot exciting star—and are accompanied by much higher rotational states (up to $J = 11$ in the case of HD 34078) than normally seen in diffuse clouds. Also, in H ii regions and photon-dominated regions (PDRs), forbidden quadrupole emission from vibrationally and rotationally excited H_2 is commonly seen near 2.2 μm in the infrared, as is ultraviolet fluorescence emission in some cases (*e.g.*, IC 63, a small reflection nebula excited by the star γ Cas; Witt *et al.* 1989; Hurwitz 1998).

Molecular hydrogen in the Magellanic Clouds: Using *FUSE* data, Tumlinson *et al.* (2002) amassed a survey of some 70 lines of sight toward hot stars in both Magellanic Clouds, finding that the molecular fraction in both is significantly lower on average than in the Galaxy. Tumlinson *et al.* attribute this to two effects: an enhanced UV radiation field in the Clouds, which results in a higher photodissociation rate; and a reduced H_2 formation rate due to the lower dust-to-gas ratio in the metal-deficient Clouds, which reduces the grain surface area available for molecular hydrogen formation. It will be interesting, in the distant future when more sensitive far-UV spectroscopy is possible, to see how the hydrogen molecular fraction depends on metallicities in other galaxies. Tumlinson *et al.* found gas kinetic temperatures based on the $N(J = 1)/N(J = 0)$ ratio to be similar to those found for diffuse sightlines in the Galaxy.

4.1.2. *UV Observations of HD*

Detections of the deuterated form of molecular hydrogen were made by *Copernicus* (Spitzer *et al.* 1973; Morton 1975), and extended with *FUSE* observations (Ferlet *et al.* 2000; Lacour *et al.* 2005). The ratio of $N(\mathrm{HD})/2N(\mathrm{H_2})$ is typically around 10^{-6}, about an order of magnitude lower than the atomic D/H ratio derived from *Copernicus* and *FUSE* observations. While the atomic D/H ratio is an important cosmological indicator related to early big bang nucleosynthesis, it is not clear that the ratio of HD to $\mathrm{H_2}$ provides useful information constraining D/H. The main problem is that $\mathrm{H_2}$ is self-shielded in diffuse molecular clouds to a much greater extent than is HD, so that HD is selectively destroyed by photodissociation. Lacour *et al.* (2005) argue that in sufficiently dense diffuse clouds, where the molecular fraction approaches one, then the ratio of HD to $\mathrm{H_2}$ would be a valid tracer of the atomic D/H ratio. But to date no such cloud has been observed.

4.1.3. *UV Observations of Carbon Monoxide*

Unlike $\mathrm{H_2}$, CO has allowed transitions between rotational and vibrational states as well as electronic transitions. Thus CO is widely observed through its mm-wave emission, which is collisionally excited in dense clouds; mm-wave absorption where there is a continuum radio source in the background; and infrared emission and absorption. In most cases these rotational and vibrational transitions are observed in dense clouds, but there is the possibility of obtaining such data for the same diffuse clouds that are observed in ultraviolet absorption.

Because CO has allowed mm-wave transitions and $\mathrm{H_2}$ does not, CO has been widely used as a surrogate for molecular hydrogen in estimating the masses of interstellar molecular clouds. But for the most part the correlation between CO and $\mathrm{H_2}$ is derived indirectly through intermediate assumed correlations between $\mathrm{H_2}$, CO, and dust extinction, a method that is fraught with uncertainties. So there is a premium on obtaining direct measurements of both CO and $\mathrm{H_2}$ in the same lines of sight. An opportunity to do this is offered by diffuse cloud lines of sight where ultraviolet observations of both CO and $\mathrm{H_2}$ are possible. There are complications in the interpretation, though, because these are likely to be clouds with sufficient column density for $\mathrm{H_2}$ to be self shielding, while CO will not be, at least not to the same extent. Therefore the $\mathrm{CO/H_2}$ ratio will be artificially low and a correction for $\mathrm{H_2}$ self-shielding must be applied.

The first UV detection of interstellar CO was achieved with a rocket-borne spectrometer flown by Smith & Stecher (1971), who detected CO toward ζ Oph at a level of about 10^{-5} of the total hydrogen column density. A few *Copernicus*-based detections followed (Jenkins *et al.* 1973; Morton 1975; Snow 1975), but given the modest spectral resolving power of that instrument, it was difficult to disentangle the rotational line structure in order to derive unambiguous column densities.

Much greater success has been achieved with the high-resolution UV spectrographs such as the GHRS and the STIS aboard the *Hubble Space Telescope*. With spectral resolving power of about 100,000, these instruments provided UV spectra of CO bands in which the rotational lines are well resolved (*e.g.*, Lambert *et al.* 1994). While the allowed electronic transitions of CO are typically saturated, leading to ambiguities in deriving column densities, there are a number of inter-system bands with smaller f-values, and from these it is possible to obtain very accurate CO column densities—though these became useful only after f-values had been determined through a series of laboratory and empirical astronomical studies (*e.g.*, Sheffer *et al.* 2002; Eidelsberg & Rostas 2004). Sheffer *et al.*, taking advantage of a little-used very small entrance slit in the STIS that provides a resolving power over 200,000, found a $\mathrm{CO/H_2}$ ratio toward X Per of 5×10^{-6}.

This can be considered a lower limit to the CO/H_2 ratio that might be found in dense clouds, due to the selective self-shielding effect mentioned above.

Through UV spectra of CO it is possible to measure isotopic variations such as ^{13}CO, $C^{17}O$, and $C^{18}O$, which are indicators of interest for galactic nucleosynthesis studies. While the isotopically substituted species usually have sufficient wavelength shifts to be readily distinguished from the more common form $^{12}C^{16}O$, there are complications, most notably the fact that the isopotologues are usually less saturated and less self-shielding than CO. In general the $^{12}CO/^{13}CO$ ratio is consistent with a lower galactic ratio of $^{12}C/^{13}C$ than the terrestrial ratio of about 90. The isotopic ratios $C^{17}O/C^{16}O$ and $C^{18}O/C^{16}O$ can also be derived from UV carbon monoxide observations, yielding far higher dominance for $C^{16}O$ than is observed on Earth or in atomic interstellar gas (Sheffer *et al.* 2002). Sheffer *et al.* explain this as due to the favored photodestruction of $C^{17}O$ and $C^{18}O$ relative to $C^{16}O$, which is self-shielded in moderately dense clouds.

4.2. *Other Molecules Detected through UV Spectroscopy*

Many molecules have electronic transitions in the ultraviolet, but only a few have been detected in interstellar clouds—largely due to the opacity of the dust, as noted earlier in this review. Here we describe results on the small number of species that have been detected. For a list of upper limits for diffuse clouds (including species whose most favorable transitions lie at wavelengths longward of the ultraviolet) see Snow & McCall (2006). Here we confine ourselves to UV detections only.

4.2.1. *UV Studies of Interstellar OH*

The first molecule detected through radio observations, the OH radical, has electronic transitions at 1222 Å and 3078 Å. The far-UV line was detected by Snow (1976) using the *Copernicus* satellite, while the near-UV line was first detected from the ground, with great effort due to atmospheric ozone interference, in the same year by Crutcher and Watson (1976) and later by Chaffee and Lutz (1977) and Felenbok and Roueff (1996). The near-UV OH features are more easily observed from space, though the transition lies near the long-wavelength cut-off for instruments such as the *IUE* and the *HST*, and was not covered at all by *Copernicus* or *FUSE*. While OH λ3078 feature was readily detectable with the GHRS and the STIS, only a few observations have been reported (*e.g.*, Snow *et al.* 1994). Microwave absorption due to OH in diffuse sightlines was observed by several authors, as summarized by Crutcher (1979) and Liszt & Lucas (1996). Based on both the optical and the radio observations, the column densities of OH, at the level of 10^{-8} of the total hydrogen, are generally consistent with gas-phase chemistry models for diffuse clouds (*e.g.*, Black & Dalgarno 1977; van Dishoeck & Black 1986).

4.2.2. *UV Observations of Diatomic Carbon*

It is noteworthy that diatomic carbon (C_2) has a prominent band near 2313 Å, which is potentially a useful tool for measuring and analyzing this molecule. But this transition has been little exploited, largely because there are several useful transitions of C_2 in the far red, which can be observed from the ground. The first attempt to detect C_2 through its UV band (Snow 1978) was marginally successful, but at about the same time the utility of using the red bands became obvious (*e.g.*, Chaffee & Lutz 1978). A recent GHRS/*HST* observation of the UV band toward X Persei has produced a beautiful spectrum (Welty *et al.*, in preparation) but in general the far-red bands are used instead for this molecule. Hence we do not include C_2 in Table 1.

4.2.3. *UV Observations of N_2 in Diffuse Clouds*

Another diatomic not expected to be abundant according to gas-phase models is N_2, which has electronic transitions only in the far-UV. An early search based on *Copernicus* data proved negative (Lutz *et al.* 1979), but recently Knauth *et al.* (2004) have claimed a detection toward HD 124314 and have reported an additional detection toward 20 Aquilae (Knauth 2005; private communication), based on *FUSE* spectra in both cases. Because of blending with interstellar H_2 bands as well as a prominent stellar photospheric line, and with additional concern about the possibility of telluric N_2 contamination, it is very difficult to conclude unambiguously that interstellar N_2 has been detected. But, as discussed by Snow (2004), Knauth *et al.* have addressed these issues and make a convincing case for detection. The inferred column density of N_2 in the sightlines where the detections are claimed is higher than expected from gas-phase models for diffuse clouds, but lower than expected for dense molecular clouds, suggesting that dust grain surface reactions may be responsible. In view of the observational complexities, clearly it is important to search for N_2 in additional lines of sight, and Knauth and colleagues are doing so, again using *FUSE*.

4.2.4. *Hydrogen Chloride in Diffuse Clouds*

Jura (1974), followed by Dalgarno *et al.* (1974), realized that ionized chlorine can undergo rapid reactions with H_2 to form HCl^+, which then leads to H_2Cl^+ and HCl. Jura predicted that HCl might be detectable through an electronic transition at 1291 Å, but this line was not found in a subsequent search using *Copernicus* data (Jura & York 1978). The idea lay dormant for some time, until Federman *et al.* (1995), using HST/*HST* spectra, detected HCl (along with vibrationally excited H_2) absorption toward ζ Oph. The detected column density was consistent with the predictions of models.

4.3. *Molecules on the Edge of Detection?*

Chemical models call for significant abundances of a few species that have not yet been detected, but for which ultraviolet observations may offer the best hope. These include water, hydrogen fluoride, and complex organic species such as those widely thought to be responsible for the diffuse interstellar bands.

4.3.1. *Water in Diffuse Interstellar Clouds*

Models of diffuse cloud chemistry (*e.g.*, van Dishoeck & Black 1986) predict an abundance of H_2O of around 10^{-8} of the total hydrogen abundance. Water has a moderately strong UV transition near 1240 Å, and a few searches have been conducted, without success so far. Snow (1975) and then Smith & Snow (1978) used *Copernicus* spectra to search for H_2O, and more recently, using STIS/*HST* data, Spaans *et al.* (1998) conducted an intensive search for the 1240 Å transition of H_2O toward HD 154368, placing a limit on the column density of 9×10^{-12} cm^{-2}, or about 10^{-8} times the total hydrogen column density, which is more stringent than the *Copernicus* limit for ζ Oph.

The observational limit for H_2O is very close to the value predicted by the models, so we can expect that either H_2O will be detected by slightly more sensitive searches, or that the models are in error. Unfortunately, due to the current hiatus in UV astronomy, it is not clear when such a search will be conducted.

4.3.2. *Interstellar Hydrogen Fluoride*

Fluorine is among the most reactive of all the common elements (*e.g.*, atomic fluorine is the only neutral atom that undergoes an exothermic reaction with H_2), and it has been shown that most of the fluorine in dense clouds—and likely in diffuse clouds as well—should be in the form of HF (Neufeld *et al.* 2005). Given the cosmic abundance of fluorine, this suggests that HF should be readily detectable, even in diffuse clouds.

There are some far-UV transitions of HF, the strongest lying near 951 Å. An intensive search for this feature has been conducted (Sonnentrucker & Neufeld, private communication), but interference by H_2 bands makes detection difficult. Currently the upper limit on HF is close to, but not in conflict with, the model predictions. Additional searches of the *FUSE* archive, for stars with varying H_2 columns, is probably worth the effort.

4.3.3. *Interstellar Organic Molecules in Diffuse Clouds*

By now it is apparent that there is a large population of complex molecules permeating the diffuse interstellar medium, which manifest themselves by virtue of their infrared vibrational emission in excited regions and by the optical diffuse interstellar absorption bands (for a recent review of the evidence for large organics in diffuse clouds, see Snow & McCall 2006).

Among the leading candidates for these large organic molecules in space are polycyclic aromatic hydrocarbons (PAHs), which consist of interlocking carbon hexagons with external attached hydrogen atoms (for a comprehensive overview, see Allamandola *et al.* 1992). Estimates of the total abundance of these molecules, which have an enormous variety of specific forms, are in the vicinity of 10^{-7} times the total hydrogen column density. Thus these species, most of which have electronic transition at ultraviolet wavelengths, might be detectable through direct absorption measurements.

Searches for UV absorption by PAHs are hampered, as usual, by the dust opacity in the UV, and by the complexity of stellar photospheric spectra, which is especially problematic when trying to detect bands that might be broader then typical atomic interstellar lines. The most sensitive search to date, based on STIS/*HST* spectra, revealed no detections (Clayton *et al.* 2003)—but plans are afoot to do a better job, when and if the Cosmic Origins Spectrograph (COS) is installed aboard the *HST* (for comments on this, see §5).

5. Future Prospects

Unfortunately, the future prospects for improved UV observations of molecules are very dim at this time. The two high-resolution spectrographs on the *HST* (i.e, the GHRS and the STIS) have been removed or have died; the *FUSE* observatory is limping due to pointing problems and probably will not last much longer; and the only planned new UV instrument, the Cosmic Origins Spectrograph (COS; Green 2000), is awaiting a launch that may never occur. Worse, there is no UV instrument even in the planning stages, except for small rocket payloads that can provide only limited data.

So it is difficult to discuss future prospects for UV observations of interstellar molecules. The best hope for the near future is that the Space Shuttle will be returned to service and that the long-awaited *Hubble* Servicing Mission 4 will be flown. Currently the future of the Shuttle program is uncertain, pending a successful and accident-free launch in the spring of 2006. If this goes as planned, then we can realistically hope for deployment of the COS by the end of 2007.

If and when the COS is installed and operating aboard the *HST*, there will be an unprecedented opportunity to conduct sensitive searches for the electronic transitions of large organic molecules thought to be responsible for the infrared emission features and possibly also the unidentified optical diffuse interstellar bands. The COS may also help in the search for water in diffuse clouds, though with only marginal spectral resolution. Because molecular observations are a high priority within the COS Science Team, we can be confident that such observations will be pursued, when and if the COS is deployed.

Acknowledgements

We gratefully acknowledge NASA grants NAG5-11487 and NG04GL34G to the University of Colorado, as well as NASA contract NAG5-12279, for support of research used in the preparation of this review. Joshua Destree was indispensable in helping prepare this manuscript for publication.

References

Allamandola, L.J., Tielens, A.G.G.M., & Barker, J.R. 1992, *Ap. J. Suppl.* 71, 733

Black, J.H. & Dalgarno, A. 1973, *Ap. J.* 184, L101

Black, J.H. & Dalgarno, A. 1977, *Ap. J. Suppl.* 34, 405

Boissé, P., Le Petit, F., Rollinde, E. Roueff, E., Pineau des Forêts, G., Andersson, B.-G., Gry, C., & Felenbok, P. 2005, *A&A* 429, 509

Carruthers, G.R. 1970, *Ap. J.* 161, L81

Chaffee, F.H. & Lutz, B.L. 1977, *Ap. J.* 213, 394

Chaffee, F.H. & Lutz, B.L. 1978, *Ap. J.* 221, L91

Clayton, G.C., Gordon, K.D., Salama, F., Allamandola, L., Martin, P.G., Snow T.P., Whittet, D.C.B., Witt, A.N., & Wolff, M.J. 2004, *Ap. J.* 592, 947

Crutcher, R.M. 1979, *Ap. J.* 31, L151

Crutcher, R.M. & Watson, W.D. 1976, *Ap. J.* 203, L123

Dalgarno, A., de Jong, T., Oppenheimer, M., & Black, J.H. 1974, *Ap. J.* 192, L37

Eidelsberg, M. & Rostas, F. 2003, *Ap. J. Suppl.* 145, 89

Federman, S.R., Cardell, Jason A., van Dishoeck, E.F., Lambert, D.L., & Black, J.H. 1995, *Ap. J.* 445, 325

Felenbok, P. & Roueff, E. 1996, *Ap. J.* 465, L57

Ferlet, R. *et al.* 2000, *Ap. J.* 583, L69

Green, J.C. 2000, *Proc. Soc. Photo-Opt. Inst.* 4013, 352

Hurwitz, M. 1998, *Ap. J.* 500, L67

Hurwitz, M. *et al.* 1998, *Ap. J.* 500, L1

Jenkins, E.B., Reale, M.A., Zucchino, P.M., & Sofia, U.J. 1996, *Ap&SS* 239, 315

Jenkins, E.B. & Shaya, E.J. 1981, *Ap. J.* 231, 55

Jenkins, E.B. & Spitzer, L. 1975, *ARA&A* 13, 133

Jura, M. 1974, *Ap. J.* 190, 33

Jura, M. 1975a, *Ap. J.* 197, 575

Jura, M. 1975b, *Ap. J.* 197, 581

Jura, M. & York, D.G. 1978, *Ap. J.* 219, 861

Knauth, D.C., Andersson, B.-G.; McCandliss, S.R., & Moos, H.W. 2004, *Nature* 429, 636

Lacour, S.; André, M.K., Sonnentrucker, P., Le Petit, F., Welty, D.E., Desert, J.-M., Ferlet, R.; Roueff, E., & York, D.G. 2005, *A&A* 430, 967

Lambert, D.L., Sheffer, Y., Gilliland, R.L., & Federman, S.R. 1994, *Ap. J.* 420, 756

Liszt, H. & Lucas, R. 1996, *A&A* 314, 917

Meyer, D.M., Lauroesch, J.T., Sofia, U.J., Draine, B.T., & Bertoldi, F. 2001, *Ap. J.* 553, L59

Moos, H.W. *et al.* 2000, *Ap. J.* 583, L1

Morton, D.C. 1975, *Ap. J.* 197, 85

Neufeld, D.A., Wolfire, M.G., & Schilke, P. 2005, *Ap. J.* 628, 260

Rachford, B. 2002, *Ap. J.* 577, 221

Rogerson, J.B., Spitzer, L., Drake, J.F., Dressler, K., Jenkins, E.B., Morton, D.C., & York, D.G. 1973, *Ap. J.* 181, L97

Savage, B.D. & Sembach, K.R. 1996, *ARA&A* 34, 279

Savage, B.D., Drake, J.F., Budich, W., & Bohlin, R.C. 1977, *Ap. J.* 216, 291

Sheffer, Y., Federman, S.R., & Lambert, D.L. 2002, *Ap. J.* 572, L95

Shull, J.M. & Beckwith, S. 1982, *ARA&A* 20, 163

Shull, J.M. *et al.* 2000, *Ap. J.* 538, L73

Smith, W.H. & Snow, T.P. 1978, *Ap. J.* 228, 435

Smith, A.M. & Stecher, T.P. 1971, *Ap. J.* 164, L43

Snow, T.P. 1975, *Ap. J.* 201, L21

Snow, T.P. 1976, *Ap. J.* 204, L127

Snow, T.P. 1978, *Ap. J.* 220, L93

Snow, T.P. 1997, in *Formation & Evolution of Solids in Space*, eds. J.M. Greenberg and A. Li (Kluwer), p. 1

Snow, T.P. 2004, *Nature* 429, 615

Snow, T.P. *et al.* 2000, *Ap. J.* 538, L69

Snow, T.P. & McCall, B.J. 2006, *ARA&A*, in press

Snow, T.P., Hansen, M.M., Black, J.H., van Dishoek, E.F., Crutcher, R.M., & Lutz, B.L. 1998, *Ap. J.* 504, 55

Spaans, M., Neufeld, D., Lepp, S., Melnick, G.J., & Stauffer, J. 1998, *Ap. J.* 503, 780

Spitzer, L. 1946, *Astr. Quart.* 7, 131

Spitzer, L. & Cochran, W.D. 1973, *Ap. J.* 191, L127

Spitzer, L., Cochran, W.D., & Hirshfeld, A. 1974, *Ap. J. Suppl.* 28, 373

Spitzer, L., Drake, J.F., Jenkins, E.B., Morton, D.C., Rogerson, J.B., & York, D.G. 1973, *Ap. J.* 181, L116

Spitzer, L. & Zabriskie, F.R. 1959, *PASP* 71, 412

Tumlinson, J. *et al.* 2002, *Ap. J.* 566, 857

van Dishoeck, E.F. & Black, J.H. 1986, *Ap. J. Suppl.* 62, 109

van Dishoeck, E.F. & Black, J.H. 1989, *Ap. J.* 340, 273

Witt, A.N., Stecher, T.P., Boroson, T.A., & Bohlin, R.C. 1989, *Ap. J.* 336, L21

Astrochemistry: Recent Successes and Current Challenges
Proceedings IAU Symposium No. 231, 2005
D.C. Lis, G.A. Blake & E. Herbst, eds.

© 2006 International Astronomical Union
doi:10.1017/S1743921306007186

Millimeter-wave Observations of Polyatomic Molecules in Diffuse Clouds

Harvey Liszt[1], Robert Lucas[2], and Jerome Pety[2,3]

[1]National Radio Astronomy Observatory, 520 Edgemont Road Charlottesville, VA 22903, USA
[2]Institut de Radioastronomie Millimétrique,
300 rue de la Piscine F38406 Saint Martin d'Hères, France
[3]Observatoire de Paris, 61 av. de l'Observatoire, F75014 Paris, France

Abstract. Millimeter-wave (mm-wave) absorption profiles toward extragalactic sources consistently find just diffuse (occasionally perhaps translucent) neutral gas—low/moderate density and extinction—along even some very long, dark lines of sight. CO, often heavily fractionated and mimicking the appearance of dark gas in emission, occasionally absent in emission even when present in absorption, is not the dominant form of carbon in these regions (presumably it is C^+) yet the abundances of many other molecules resemble those seen in TMC-1. Some species (OH, HCO^+, C_2H and C_3H_2) turn on with high abundances just when H_2 does; others (HCN, HNC, CN) require slightly higher $N(H_2)$ and yet others (CS and other sulfur-bearing species, NH_3 and H_2CO) even higher $N(H_2)$. The systematics and implications of these recent discoveries are discussed here.

Keywords. astrochemistry — ISM: clouds — ISM: molecules — ISM: structure

1. Introduction

Within a few years of the discovery of CO $J = 1$–0 rotational emission from Orion (Wilson *et al.* 1970), a wealth of mm-wave observations quickly established the existence of complex polyatomic chemistries in dense interstellar clouds, largely driven by cosmic-ray ionization of H_2 to form H_3^+ (Herbst & Klemperer 1973). The rest is, if not exactly history, at least the subject of this well-attended meeting on interstellar chemistry. Today, observations of molecular rotational emission are synonymous with the idea of "molecular gas" or "molecular clouds", and such emission traces the star-forming gas reservoirs both nearby and in the most distantly-observable systems in the Universe.

However, the existence of molecules in the ISM was first noted nearly 70 years ago in optical spectra of regions which are not at all heavily extinguished or very dense: accumulating gradually over this long interval, the list of molecules observed in such "diffuse gas" or "diffuse clouds" is both fairly extensive and in some regards very difficult to understand. Recently, the pace of observational discovery has quickened considerably owing to the application of mm-wave absorption techniques to difuse cloud studies, which is the subject of this contribution.

2. Precedents for RF Observations of Molecules in Diffuse Clouds

It was the observation of $\lambda 18$ cm OH emission from dark clouds (Heiles 1969) which as much as anything else first convincingly showed the prevalence of H_2 in such regions, where the H-nuclei (directly observed only in H I) seemed otherwise untraceable. Thereafter, surveys of OH emission and absorption in directions marked by known H I absorption certainly and often knowingly sampled molecular gas in diffuse clouds (Dickey *et al.* 1981), but provided little opportunity for follow-up. When CO and CH emission

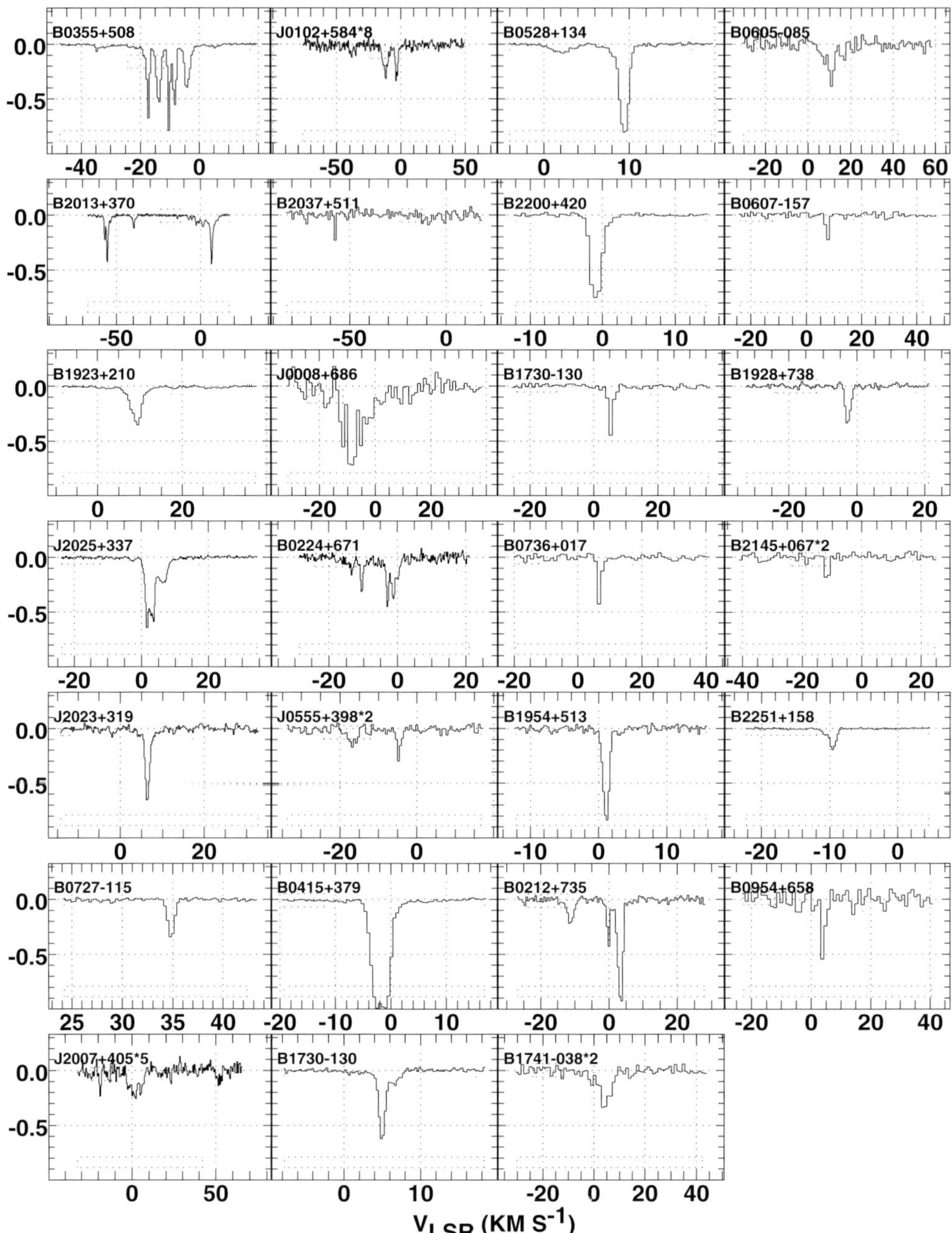

Figure 1. Rogue's gallery of all HCO$^+$ profiles.

were observed in diffuse clouds with high resolution, as toward ζ Oph (Liszt 1979), the observed line profiles were narrower than optical spectroscopists had in general been willing to consider, but such studies also cannot be said to have had much influence because the abundaces of CO and CH were more straightforwardly determined in optical/UV absorption spectra.

With hindsight it becomes obvious that radio-frequency observations of molecules in diffuse gas were often not recognized as such when first performed—for instance, in H_2CO absorption surveys over the galactic plane (see the remarks in Liszt & Lucas 1995) and in spectra of NH_3 and H_2CO in the work of Nash (1990)—largely owing to disbelief that such species would be detectable there (see Federman & Allen 1991). But by the end of the 1980's, with PAHs becoming an increasing topic of discussion, Cox *et al.* (1988) certainly and tellingly stated the matter quite clearly in showing the ubiquity of the hydrocarbon C_3H_2 in the diffuse ISM.

Marscher *et al.* (1991) subsequently showed that sensitive mm-wave absorption profiles could be taken with small arrays (3×10 m). Specifically, using OVRO, they produced a profile of ^{12}CO $J = 1$–0 absorption toward BL Lac, an (approximately) 1 Jy mm-continuum source which was known to suffer an optical extinction about equal to that of ζ Oph (0.32 mag) and which had long been used as a target for H I and OH absorption studies. The line was, unfortunately, quite optically thick, complicating their efforts to infer the physical structure of the intervening medium, but it sufficed to demonstrate the practicality of taking such spectra.

We subsequently were in the process of observing ^{13}CO absorption against BL Lac and other compact extragalactic mm-wave sources toward which CO had been seen in emission surveys (Liszt & Wilson 1993; Liszt 1994), when two fortuitous events intervened: we became aware of a serendipitous detection of HCO^+ absorption toward BL Lac during an unrelated instrumental phase calibration for another project, and BL Lac flared (actually not an uncommon event among blazars). We asked for time to observe HCO^+ and other species, and found them surprisingly handily toward BL Lac and elsewhere (Lucas & Liszt 1993, 1994). What is discussed here follows from these two happy coincidences now back a baker's dozen of years.

3. HCO^+ and Implications for Diffuse Cloud Chemistry

We surveyed HCO^+ absorption toward an approximately flux-limited sample of a few dozen strong background sources (this is complicated when all the sources are variable) and reported the results in Lucas & Liszt (1996). A slightly updated version of those results is shown in the first two figures. Figure 1 shows the actual absorption profiles we have accumulated, in order (left to right and top to bottom) of increasing galactic latitude ($|b|$). The profiles are summarized in two ways in Figure 2, where the profile integrals, normalized to column density assuming excitation at the temperature of the cosmic background, are plotted twice. At left, we illustrate our old result that HCO^+ absorption was found in essentially all directions at galactic latitudes below about $15°$, but much less commonly above; although HCO^+ absorption is somewhat more frequent than CO emission, the same effect had been seen in CO surveys and it is attributable to the local galactic geometry. Basically, because of the "local bubble" we look out as from the bed of a wide but shallow crater; to look out in unobscured fashion it is necessary also to look up (or down) some $12°$–$15°$. The local neutral gas is dominated not by a plane parallel vertical stratification but by the excavatory efforts of the bubble.

At right in Figure 2 the same HCO^+ integrals are plotted against the reddening taken from the work of Schlegel *et al.* (1998). From this, four particular physical inferences can be drawn: (1) The absorption becomes detectable rather abruptly at $E_{B-V} > 0.08$ mag, which is the famous turn-on of high H_2-fractions in the diffuse ISM according to Bohlin *et al.* (1978); see also Lucas & Liszt (2000b). (2) At higher reddening, $N(HCO^+)$ increases *linearly* with E_{B-V} suggesting that HCO^+ turns on, at full-strength, about when H_2 does. (3) The HCO^+ abundance is very large, nearly as high as in TMC-1,

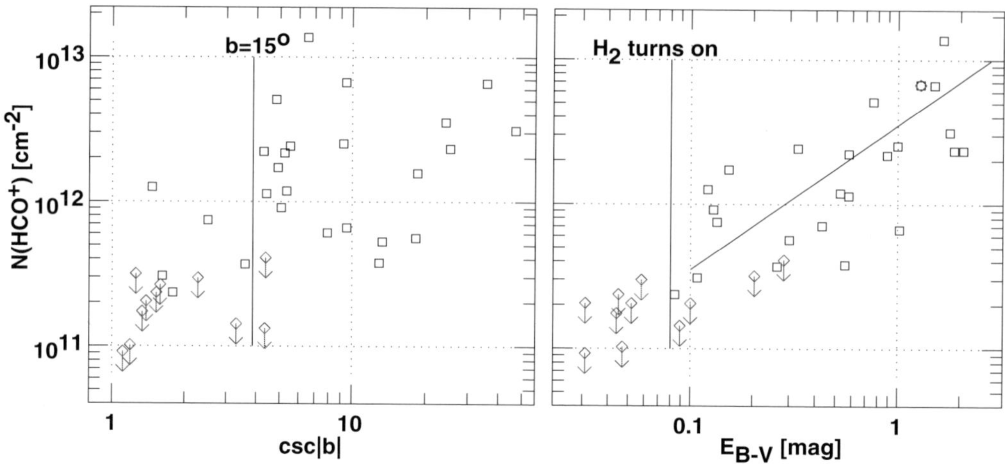

Figure 2. Total HCO$^+$ column density *vs.* csc($|b|$) and $E_{\text{B}-\text{V}}$.

and (4) much smaller abundances are predicted by conventional gas-phase chemistry of diffuse clouds; for (3) and (4) see just below.

Using the usual gas-reddening ratio, whereby $E_{\text{B}-\text{V}} = 1$ mag corresponds to a column density $N(\text{H}) = 5.8 \times 10^{21}$ cm^{-2} (Bohlin *et al.* 1978), we have that $N(\text{HCO}^+)/E_{\text{B}-\text{V}} \approx 3 \times 10^{12}$ mag^{-1} or $N(\text{HCO}^+)/N(\text{H}) = 5 \times 10^{-10}$. Then, if the typical fraction of H-nuclei in H$_2$ is 0.2 in the diffuse ISM, as per the *Copernicus* observations, it follows that $X(\text{HCO}^+) = N(\text{HCO}^+)/N(\text{H}_2) = 5 \times 10^{-9}$; the usual value quoted for TMC-1 is $X(\text{HCO}^+) = 10^{-8}$ (Ohishi *et al.* 1992). In fact in our work we typically use smaller values, $X(\text{HCO}^+) = 2 - 3 \times 10^{-9}$, but these also are far too large to be easily understood.

In diffuse gas, the dominant form of gas-phase carbon is C$^+$, not CO, somewhat turning dense-gas CO chemistry on its head (Glassgold & Langer 1976; Black & Dalgarno 1977). That is, conventional gas-phase chemistry forms CO by first producing CO$^+$ via C$^+$ + OH $\rightarrow$ CO$^+$ + H, after which CO forms either by CO$^+$ + H $\rightarrow$ CO + H, or by CO$^+$ + H$_2$ $\rightarrow$ HCO$^+$ + H, HCO$^+$ + e $\rightarrow$ CO + H. All these reactions are relatively fast, and radiative recombination CO$^+$ + e $\rightarrow$ C + O is unimportant with the e/H fractions expected for diffuse gas (Liszt 2003). So CO is the dominant product of the C$^+$– OH interaction no matter the H$_2$-fraction, but, if the region is substantially molecular, most CO$^+$ will proceed to HCO$^+$. Then, because n(e) $\approx$ n(C$^+$) (*ibid.*), it follows pretty directly that the OH/HCO$^+$ ratio will be largely independent of the density and H$_2$-fraction *etc.* and somewhat dependent on temperature owing to the functional dependence of the HCO$^+$ recombination rate coefficient. With values $X(\text{OH})/X(\text{HCO}^+) \approx 1000$ for typical conditions, it follows that $X(\text{HCO}^+) = 10^{-10}$ if $X(\text{OH}) = 10^{-7}$ as observed along a few sightlines (Felenbok & Roueff 1996; Liszt & Lucas 2002).

So the discrepancy between observation and theory is quite substantial, if not as extreme as for CH$^+$. The column densities of C$^+$, OH, and CO are all observable in optical/UV absorption, so the profound failure of this scheme to explain either $X(\text{HCO}^+)$ or $X(\text{CO})$ would ordinarily lead to its dismissal. However, a fixed ratio $N(\text{HCO}^+)/N(\text{OH})$ is actually observed, and no generally accepted alternatives for explaining either this ratio or $X(\text{HCO}^+)$ and $X(\text{CO})$ have appeared. The high abundances of H$_3{}^+$ discussed by McCall at this meeting do not by themselves account for the creation of HCO$^+$ from CO, by large factors.

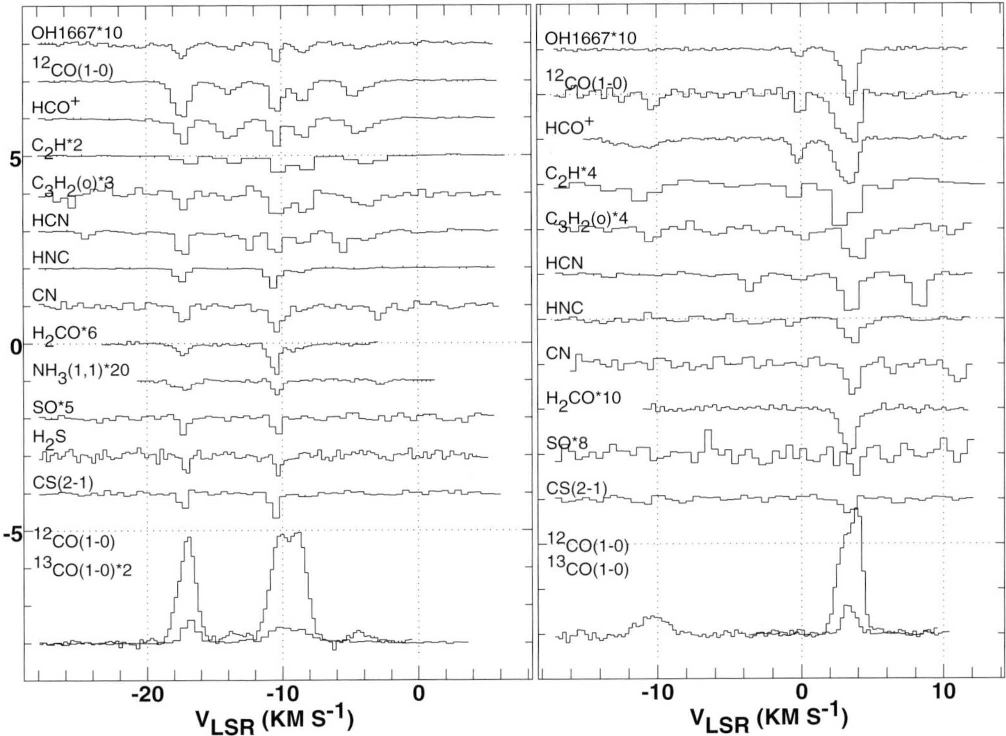

Figure 3. *Left:* Spectra toward B0355 + 508 at $b = -1.6°$. *Right:* B0212 + 735 at $b = 12.0°$.

Although not understood, the observation of a high abundance of HCO^+ is by itself sufficient to eliminate yet another failure of chemical models, to explain the abundance of CO itself (Van Dishoeck & Black 1988). If an abundance $X(HCO^+) = 2 - 3 \times 10^{-9}$ can be maintained, the usual radiative recombination to form CO is sufficient to account for the observed $X(CO)$ when the formation of H_2 and CO is calculated self-consistently, including their self and mutual shielding (see Liszt & Lucas 2000). But the solution of the CO problem is actually somewhat more robust than even this discussion indicated, because $CO^+ + H \rightarrow CO + H^+$ may form CO *without* first forming HCO^+. Molecular chemistry in only partially-molecular regions is really more different than at first might meet the eye.

4. Chemical Complexity and Systematics in Diffuse Clouds

Figure 3 shows an exhaustive summary of our molecular profiles taken toward two sources with rich spectra. The most commonly occurring species—OH, HCO^+(Liszt & Lucas 1996; Lucas & Liszt 1996), CO (Liszt & Lucas 1998), C_2H and C_3H_2 (Lucas & Liszt 2000a)—appear at top, followed by the CN-family (Liszt & Lucas 2001), which is slightly more common than those further below, namely, H_2CO, NH_3 (see Lucas & Liszt 1995; Nash 1990; and a forthcoming paper of ours) and the sulfur-bearing species (Lucas & Liszt 2002).

The long line of sight toward B0355 at $b = -1.6°$ is itself dark ($E_{B-V} = 1.5$ mag), but the features seen along it are most likely diffuse; there are about a half-dozen of these with very nearly the same $N(HCO^+) \approx 1 - 1.5 \times 10^{12}$ cm^{-2} and $N(OH) \approx 3 - 5 \times 10^{13}$ cm^{-2}, just as toward ζ Oph and other classic optical diffuse-cloud lines of sight (see Fig. 4).

The absorption features generally appear in CO emission (bottom) albeit rather weakly in some. Of the three absorption features which emit more strongly in ^{12}CO, two are chemically much richer than the other.

At right, toward B0212 at $b = 12°$ ($E_{\mathrm{B-V}} = 0.76$ mag) the strongest absorption component has higher column densities than any feature seen toward B0355, while the absorption feature at 0-velocity lacks a CO emission counterpart at quite low levels. This is an example of the fact that CO and some polyatomics form at much lower pressure than is needed to excite even CO to detectable levels in emission. A search for HCO$^+$ emission revealed only a single very weak (0.02 K) line toward each source (Lucas & Liszt 1996).

HCO$^+$ emission at such low levels is consistent with rather low density when due to electron excitation in a gas where the carbon is largely in the form of C$^+$. That is, for a typical thermal pressure of $p/k = 2000$–3000 cm^{-3} K, at temperatures 30–70 K, $n(\mathrm{H}) \approx$ 30–100, with $n(e)/n(\mathrm{H}) \approx 2 \times 10^{-4}$ (Liszt & Lucas 1994; Lucas & Liszt 1996). Consistent with this, ^{13}CO is often observed to be heavily fractionated into CO in our work, yielding relative abundance ratios $N(^{12}\mathrm{CO})/N(^{13}\mathrm{CO})$ as small as 12–15 (Liszt & Lucas 1998) in gas where the isotope ratio is actually 60–65 (Lucas & Liszt 1998). In one case the fractionation is so strong that HCN appears to be starved for ^{13}C (*ibid.*) Observations of CO toward hot stars seem to show the opposite effect of selective photodissociation of ^{13}CO (see the contribution by Snow in this volume).

5. Implications for Interpretation of CO Emission

For weak excitation (say, a few K above the CMB) a feature with $N(\mathrm{CO}) \approx 10^{15}$ cm^{-2} will have an optical depth of unity in the $J = 1$–0 line over a span of 1 km s^{-1} (Liszt & Lucas 1998); the optical depth of the $J = 1$–0 line of CO toward ζ Oph should be about 2, according to optical spectroscopy. The point is that even diffuse clouds with relatively small amounts of carbon in the form of CO may have quite optically thick lines, and, depending on internal conditions, they may emit either fairly strongly or not at all. Examples of both sorts of behaviour are plentiful in our data. In both cases shown in Figure 3, the CO column densities of even the stronger features are small compared to the amount of free gas-phase carbon in a gas column having $A_{\mathrm{V}} = 1$ mag, *i.e.* $N(\mathrm{CO}) \approx 10^{16}$ cm^{-2} even toward B0212 (Liszt & Lucas 1998). Yet, they emit at levels (3 K) typically associated with dark gas. What is more, the ^{13}CO/^{12}CO ratios seen in emission are often "fake", resulting from fractionation but giving the impression of arising from very optically thick CO.

6. Chemical Systematics

Figure 4 shows that the chemistry is well-organized, however mystifying. At upper left we have copied a diagram from our initial survey work of OH and HCO$^+$ (Liszt & Lucas 1996; Lucas & Liszt 1996) illustrating the very tight empirical relationship between the integrated optical depths of OH and HCO$^+$. To quote $N(\mathrm{OH})$ it is necessary to ascertain the excitation temperature, which we did based on a comparison of OH emission and absorption results; the same small excitation temperatures are seen optically (Felenbok & Roueff 1996) and in fact have been a puzzle for nearly 25 years. The OH should be thermalized in collisions with electrons in diffuse gas as noted by Dickey *et al.* (1981). Parenthetically, we note that the same discrepancy appears in H$_2$CO, which appears in anomalous absorption in some directions in our work (Liszt & Lucas 1995): the λ 6 cm transition, too, should be thermalized.

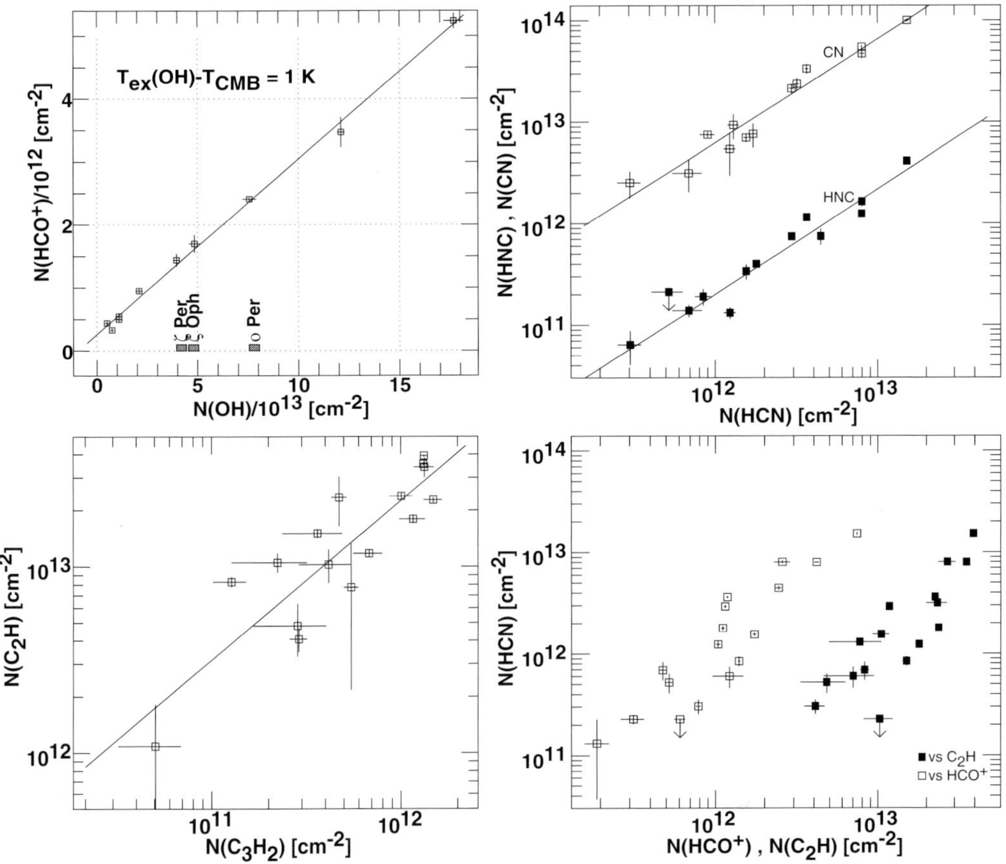

Figure 4. Relationships between chemical species and families.

In any case, the relationship expected between OH and HCO$^+$ is manifest, although with $N(\mathrm{OH})/N(\mathrm{HCO}^+) \approx 30-50$, rather than 1000 or more. Also noted in that diagram are the OH column densities along three of the four directions where it is seen optically; clearly our clouds span the same regime. Later versions of this diagram had a bit more scatter. The OH/HCO$^+$ abundance ratio toward TMC-1, far off to the right at $N(\mathrm{OH}) = 2.5 \times 10^{15}$ cm^{-2} (Ohishi *et al.* 1992), is only 50% higher than given by the extrapolation of the fit shown here.

Other panels in Figure 4 show that there are similarly strong linear relationships in other chemical families as well (CN, HCN, HNC; C$_2$H and C$_3$H$_2$) and, generally strong non-linearities otherwise. No species beside OH is linear with HCO$^+$, and none beside C$_3$H$_2$ with C$_2$H. CS is approximately linear with HCN and NH$_3$ is linear with H$_2$CO. The steep slope of HCN's variation with HCO$^+$ around $N(\mathrm{HCO}^+) = 1-3 \times 10^{12}$ cm^{-2} in Figure 4 can be directly related to the turn-on of CN with H$_2$ in optical data compiled by Federman (*e.g.*, Federman *et al.* 1994). The large HCN/HNC ratios seen in our work are typical of warm gas, *i.e.* not especially like TMC-1 and other dark clouds, but, except in this regard, the abundance ratios are all too often astoundingly like those quoted for TMC-1.

7. Conclusions

We began these studies by trying to detect ^{13}CO absorption in classic "molecular" gas, for the purpose of understanding CO line formation. Instead, we stumbled upon the existence of an unexpectedly rich chemistry in diffuse clouds, repeating the experience of optical observers 60 years earlier. Although our directions were at first chosen blindly (in a flux-limited sample) and although many of our sightlines at low latitude are long and very dark, we have been unable to find any dense or dark gas. But perhaps we found the next best thing. The abundances we derived for HCO^+, C_2H and C_3H_2 generally, or (for example) for HCN, CS, NH_3 *etc.* in the richer regions, are surprisingly (and mystifyingly) like those seen toward TMC-1 (the HNC/HCN and HOC^+/HCO^+ ratios are typical of warm gas). And although the diffuse gas is largely silent in emission, even in some cases in CO, its CO emission may also be surprisingly strong and, due to the peculiar chemistry of carbon isotope fractionation, also somewhat misleadingly dark-cloud like.

Although clearly following some well-worn paths, we sincerely hope that the chemical problems elucidated in our work are not as persistent as, say, that for CH^+. In summarizing the conference, Alex Dalgarno noted that the appearance of a dark-cloud-like set of chemical abundances in diffuse gas might be a clue that the chemistry was okay (his word) and the solution is to be found in gas dynamics (shock-dynamical models have been unsuccessful for CH^+, of course). Or perhaps he said demonics (as in Maxwell's); we were sitting somewhat off to the side. There are a number of strong hints that the diffuse neutral ISM is not the simple undifferentiated woofly thing of idealized models. Linewidths are somewhat supersonic; some studies show very small scale structure (10–100 AU) in optical and RF absorption; there is a ubiquitous small fraction of material at very high pressure in C I excitation studies; the excitation of most OH and some HCO^+ seems to be oblivious to the high electron fractions in diffuse gas.

Diffuse clouds, usually noted for and defined by their seeming transparency, nonetheless manage to have a rich interior life to which we are not entirely privy, and it is these presently hidden aspects, not our idealizations, that are most important to understanding what is happening in them.

Acknowledgements

The National Radio Astronomy Observatory is operated by AUI, Inc. under a co-operative agreement with the US National Science Foundation. IRAM is operated by CNRS (France), the MPG (Germany) and the IGN (Spain). We owe the staff at IRAM (Grenoble) and the Plateau de Bure our thanks for their assistance in taking the data.

References

Black, J. H. & Dalgarno, A. 1977, *Ap. J., Suppl.* 34, 405
Bohlin, R. C., Savage, B. D., & Drake, J. F. 1978, *Ap. J.* 224, 132
Cox, P., Güsten, R., & Henkel, C. 1988, *A&A* 206, 108
Dickey, J. M., Crovisier, J., & Kazes, I. 1981, *A&A* 98, 271
Federman, S. R. & Allen, M. 1991, *Ap. J.* 375, 157
Federman, S. R., Strom, C. J., Lambert, D. L., Cardelli, J. A., Smith, V. V., & Joseph, C. L. 1994, *Ap. J.* 424, 772
Felenbok, P. & Roueff, E. 1996, *Ap. J.* 465, L57
Glassgold, A. E. & Langer, W. D. 1976, *Ap. J.* 206, 85
Heiles, C. 1969, *Ap. J.* 157, 123
Herbst, E. & Klemperer, W. 1973, *Ap. J.* 185, 505
Liszt, H. 1979, *Ap. J.* 233, L147
Liszt, H. 1994, *Ap. J.* 429, 638

Liszt, H. 2003, *A&A* 398, 621

Liszt, H. & Lucas, R. 1994, *Ap. J.* 431, L131

Liszt, H. & Lucas, R. 1995, *A&A* 299, 847

Liszt, H. & Lucas, R. 1996, *A&A* 314, 917

Liszt, H. & Lucas, R. 1998, *A&A* 339, 561

Liszt, H. & Lucas, R. 2000, *A&A* 355, 333

Liszt, H. & Lucas, R. 2001, *A&A* 370, 576

Liszt, H. & Lucas, R. 2002, *A&A* 391, 693

Liszt, H. & Wilson, R. W. 1993, *Ap. J.* 403, 663

Lucas, R. & Liszt, H. 1993, *A&A* 276, L33

Lucas, R. & Liszt, H. 1994, *A&A* 282, L5

Lucas, R. & Liszt, H. 1996, *A&A* 307, 237

Lucas, R. & Liszt, H. 1998, *A&A* 337, 246

Lucas, R. & Liszt, H. 2000a, *A&A* 358, 1069

Lucas, R. & Liszt, H. 2000b, *A&A* 355, 327

Lucas, R. & Liszt, H. 2002, *A&A* 384, 1054

Marscher, A. P., Bania, T. M., & Wang, Z. 1991, *Ap. J.* 371, L77

Nash, A. G. 1990, *Ap. J., Suppl.* 72, 303

Ohishi, M., Irvine, W., & Kaifu, N. 1992, in Astrochemistry of cosmic phenomena: proceedings of the 150th Symposium of the International Astronomical Union, held at Campos do Jordao, Sao Paulo, Brazil, August 5–9, 1991. Dordrecht: Kluwer, ed. P. D. Singh, 171–172

Schlegel, D. J., Finkbeiner, D. P., & Davis, M. 1998, *Ap. J.* 500, 525

Van Dishoeck, E. F. & Black, J. H. 1988, *Ap. J.* 334, 771

Wilson, R. W., Jefferts, K. B., & Penzias, A. A. 1970, *Ap. J.* 161, L43

Photo: E. van Dishoeck

Astrochemistry: Recent Successes and Current Challenges
Proceedings IAU Symposium No. 231, 2005
D.C. Lis, G.A. Blake & E. Herbst, eds.

© 2006 International Astronomical Union
doi:10.1017/S1743921306007198

Modelling Diffuse Interstellar Environments

Evelyne M. Roueff[1] and Franck Le Petit[1,2]

[1]LUTH and UMR 8102 du CNRS, Observatoire de Paris,
Place J. Janssen, F-92190 Meudon, France
email: evelyne.roueff@obspm.fr

[2]Onsala Space Observatory, S-439 92 Onsala, Sweden
email: franck.lepetit@obspm.fr

Abstract. Diffuse clouds are defined as low-density clouds permeated by the ultraviolet interstellar radiation field. We discuss the ensemble of atomic, ionic and molecular observations in the context of our Photon Dominated Region model. The comparison with all observational constraints is limited and implies that the diffuse environment is more complex than originally thought.

Keywords. ISM: abundances — ISM: molecules — molecular processes

1. Introduction

A number of observational challenges for chemical models of diffuse clouds is addressed in the advertisement of the present symposium. These include the explanation of high column densities of molecular ions such as H_3^+, CH^+, HCO^+ and polyatomic molecules C_3, CCH and c-C_3H_2. In addition, one may also emphasize the discovery of molecular nitrogen by Knauth *et al.* (2004), HOC^+ by Liszt, Lucas & Black (2004) and the excitation of H_2, which has received recent attention with the observations of the Far Ultraviolet Spectrograph Explorer (FUSE). Diffuse clouds are subject to ultraviolet radiation coming from the surrounding stars, which drives the atomic to molecular ratio via photophysical and photochemical processes. A. Sternberg in this volume has summarized the various physical and chemical aspects involved in Photon Dominated Regions (PDRs), which encompass diffuse clouds environments. We recall in § 2 the main properties of our numerical code and define the physical parameters for a "standard" homogeneous diffuse cloud model. We discuss in § 3 some aspects of diffuse clouds and compare model results with recent observations in § 4. We present our conclusions and perspectives in § 5.

2. The "Meudon" PDR Code

Our studies of diffuse clouds will be discussed in the frame of the "Meudon" PDR code which has been developed over more than 10 years (Abgrall *et al.* 1992; Le Bourlot *et al.* 1993; Le Petit, Roueff & Herbst (2004); Le Petit *et al.* 2005) and which is downloadable on the website `http://aristote.obspm.fr/MIS`. The main distinctive features are:

• UV radiative transfer is solved by considering both dust absorption and discrete transitions of H_2, HD, CO and its isotopes. Explicit calculations of photodissociation probabilities are performed for these particular species. The impinging radiation field can be introduced for 1 or 2 sides of a plane-parallel slab of gas. The presence of a close bright star on one side of the cloud can be introduced.

• Chemical balance is solved with versatile chemical networks.

• Thermal balance is solved by considering the different heating and cooling processes. The photoelectric effect is calculated by including the extinction properties of the grains (silicates and carbon particles) and the actual computed value of the radiation field. The size distribution of dust particles is assumed to follow the Mathis, Rumpl & Nordsieck (MRN) law (Mathis, Rumpl & Nordsieck 1977). The cooling processes are determined via the emission of submillimeter and millimeter radiation of the main coolants following collisional excitation by H, He, H_2...

• The main shortcoming of such a model is the stationary approximation. However, the (photo)chemical time scales associated with diffuse regions are short, and such an assumption is reasonable.

The major input parameters and subsequent results are:

Parameters	*Results*
Equation of state: density, pressure, temperature profile	Abundance of the various species at each point in cm^{-3}
Incident radiation field to the left and right sides	Excitation state of C^+, C,O,...,H_2, HD, CO, CS, HCO^+ ...
Elemental abundances in the gas phase	Individual heating and cooling efficiencies in erg cm^{-3} s^{-1}
Extinction curve, size distribution of grains	Temperature profile of gas and grains (per size)
Cosmic ray ionisation rate, turbulent velocity ...	Integrated quantities (column densities, emitting intensities of the "cloud")
Atomic and molecular parameters (Inelastic collision and chemical reaction rate coefficients ...)	Chemical analysis at each point

We define a standard cloud model with a density, $n_H = 100$ cm^{-3}, and a cosmic ionisation rate, $\zeta = 5 \times 10^{-17}$ s^{-1}, which is irradiated by the interstellar standard radiation field (ISRF) expressed in Draine's units† (the scaling factor $\chi = 1$). The size of the cloud is defined by the total visual extinction value $A_V^{tot} = 1$ corresponding to a colour excess E_{B-V} of about 0.3, representative of diffuse cloud conditions. The chemical network contains 120 species linked by 1900 chemical reactions, which are mainly taken from the most recent Ohio State University chemistry (*cf.* http://www.physics.ohio-state.edu/~eric/research.html).

3. Discussion

In the following, we discuss some common approximations performed in diffuse cloud calculations and test the role of different parameters.

3.1. *1-Sided versus 2-Sided Models*

For computing time reasons and numerical convenience, diffuse clouds are often represented as semi infinite slabs illuminated on one side. Then, column densities are obtained by integrating abundances up to $A_V^{tot}/2$ and multiplying the results by two. This assumption is justified when photochemical effects are negligible at this visual extinction. We test this hypothesis and compare in Table 1 the results obtained for three different representations of an interstellar cloud of total visual magnitude of 1, immersed in the standard ultraviolet radiation field.

We have checked that the ionization fraction and the temperature profile are very similar in the range of visual magnitude between 0 and 0.5. The derived column densities

† *Cf.* Draine 1978

Table 1. Column densities (cm^{-2}) obtained for three different models.

Molecule	A	B	C
H	2.2(20)	2.3(20)	1.4(20)
H_2	8.2(20)	8.1(20)	8.6(20)
C	1.3(15)	1.9(15)	8.(15)
CO	4.4(13)	5.3(13)	1.3(14)
CH	4.1(13)	5.2(13)	7.5(13)
C_2	7.8(12)	1.7(13)	6.1(13)
C_3	9.4(11)	3.2(12)	1.6(13)
OH	1.7(13)	1.7(13)	1.0(13)

Notes: **Model A**: the effect of the radiation field on the two sides of the cloud ($A_V^{tot} = 1$) is computed. **Model B**: the radiation field is assumed to come from one side; the integration is performed until $A_V = 0.5$ and the column densities are computed by multiplying the results by 2. **Model C**: the radiation field is assumed to come from one side; the integration is performed until $A_V = 1$. Values in parenthesis refer to powers of ten.

of molecular hydrogen are marginally different in the 3 cases, since self-shielding effects are very efficient at low values of A_V. The results displayed above reflect the sensitivity of the various species to photoprocesses such as CO, which is often used as a tracer of H_2. The results also show, as known from previous studies, that OH is mainly formed in the external part of the cloud so that models A and B lead to similar values of the corresponding column density. Substantial differences are however obtained for carbon chain molecules such as C_2 and C_3 and it is therefore advisable to consider explicitly the two-sided illumination for these species.

3.2. *Role of the Shielding of H_2*

Photodissociation of H_2 occurs via discrete absorption in the Lyman and Werner band systems followed by emission in the continuum of the electronic ground state. This leads to very efficient self-shielding of H_2, which may be evaluated by following the approximation of Federman, Glassgold & Kwan (1979), known as the FGK approximation.

Table 2. Column densities (cm^{-2}) calculated with the FGK approximation and with the "exact" solution of the ultraviolet radiative transfer.

Model	FGK self-shielding of H_2	"Exact" full radiative transfer
f	0.88	0.92
column density		
C	1.3(15)	2.1(15)
CH	4.1(13)	4.8(13)
CN	2.0(11)	2.2(11)
CO	4.4(13)	7.0(13)
OH	1.7(13)	1.4(13)
CS	1.0(11)	1.2(11)
C_2	7.8(12)	9.4(12)
C_3	9.4(11)	1.2(12)

Note: Values in parenthesis refer to powers of ten.

However, the numerous electronic bands of H_2 may reduce the intensity of the radiation field considerably as they overlap between themselves and with other ultraviolet transitions such as those of CO, C, ... This effect is seldom taken into account as it requires a full treatment of the radiative transfer. We test its relevance on the molecular fraction $f = \frac{2N(\mathrm{H_2})}{N(\mathrm{H})+2N(\mathrm{H_2})}$ and on the column densities of the various species displayed in Table 2. The major effect is on CO (and C) for which we perform the explicit calculation of photodissociation probability. Other molecules whose photodissociation rates are computed from analytic expressions are indirectly affected. Figure 1 displays the total absorption of the radiation field in the 912–1100 Å window computed in the two approaches.

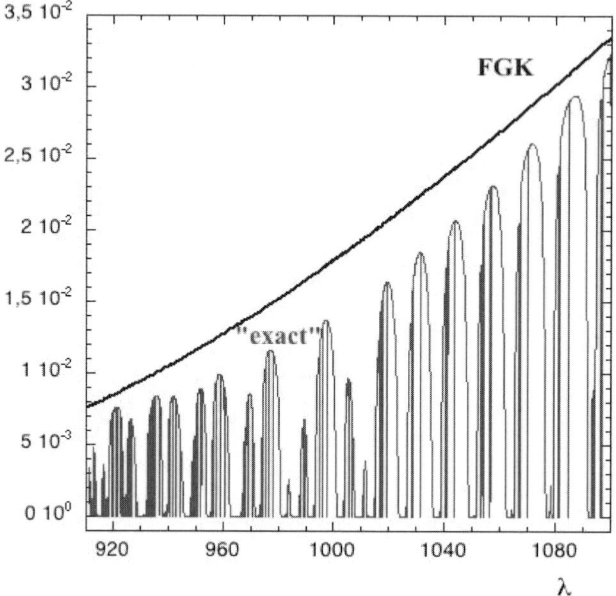

Figure 1. Total absorption factor of the radiation through the cloud

3.3. *Role of the Equation of State*

Constant density models are often considered, but isobaric conditions are more appropriate, as derived from the analysis of the populations of the atomic carbon fine structure levels (Jenkins 2002). The fractal structure of interstellar clouds is also a subject of discussion. Table 3 shows results for three models corresponding to different equations of states. The corresponding densities as a function of the visual extinction are displayed in Figure 2. Models B and C involve higher densities than model A. Then larger values of molecular column densities are obtained. Carbon chains are the most sensitive to density fluctuations.

3.4. *Chemical Issues*

3.4.1. *One Among Many Chemical Questions*

Atomic and molecular properties may also be considered as parameters. Not only are many reaction rate coefficients not available for interstellar temperature conditions, but radicals, reactive species found in the ISM are difficult to study in the laboratory. Then, a sensitivity analysis for the chemistry is an important issue (*cf.* Roueff, Le Bourlot & Pineau des Forêts 1996; Vasyunin *et al.* 2004). We address one single question amongst many others: this concerns the reaction between H and CH. This reaction was supposed to have an activation energy of ~ 2200 K (Aannestad 1973). However, recent theoretical

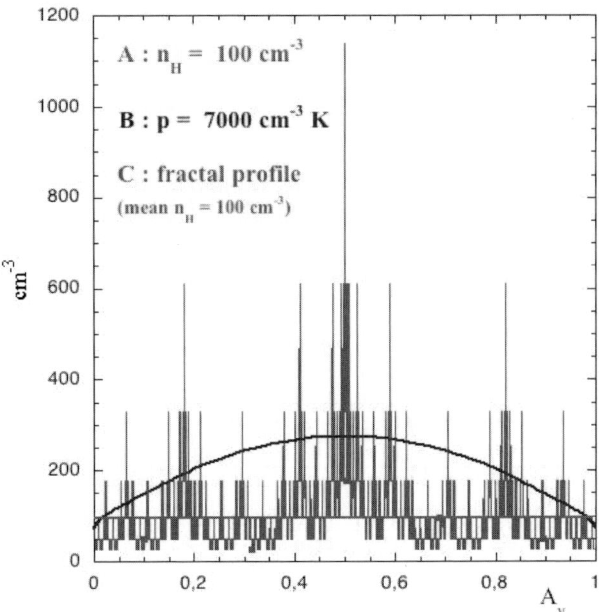

Figure 2. Density profiles in models A, B and C.

Table 3. Model A: $n_{\mathrm{H}} = 100$ cm^{-3}. **Model B**: $p = 7 \times 10^3$ cm^{-3} K. **Model C**: fractal density profile with a mean value of $n_{\mathrm{H}} = 100$ cm^{-3}.

Models	**A**	**B**	**C**
size (pc)	6.1	3.2	9.7
mean value of T (K)	80	57	81
f	0.88	0.93	0.82
column density (cm^{-2})			
CO	4.4(13)	6.3(13)	3.6(13)
CH	4.1(13)	8.5(13)	5.9(13)
C	1.3(15)	6.3(15)	1.8(15)
C_2	7.8(12)	4.8(13)	2.1(13)
C_3	9.4(11)	1.2(13)	5.1(12)
H_3^+	0.8(13)	0.4(13)	1.0(13)

Note: Values in parenthesis refer to powers of ten.

calculations by van Harrevelt, van Hemert & Schatz (2002) find that this reaction should be rapid, and a discrepancy by a factor of 7 is found with the most recent experiment at 300 K (Becker *et al.* 1989). The question remains open. In diffuse cloud conditions, there is a competition between CH photodissociation, which is known accurately from ab-initio calculations by van Dishoeck (1987) and the reaction with H, which may occur in the absence of any activation barrier.

3.4.2. *Empirical Introduction of Non-Thermal Effects*

The modelling of CH$^+$ has been a longstanding problem for diffuse cloud models since this ion is destroyed by both atomic and molecular hydrogen. In addition, the reaction of C$^+$ with H$_2$ is endothermic by about 0.36 eV. Then, the computed abundance is smaller than the observations by more than 2 orders of magnitude in all quiescent interstellar

Table 4. Effect on the "standard model" of the empirical introduction of non-thermal effects for ion-molecule reactions.

Models	standard	$v_{in} = 2\ \mathrm{km\,s^{-1}}$	$v_{in} = 3\ \mathrm{km\,s^{-1}}$	$v_{in} = 4\ \mathrm{km\,s^{-1}}$
column density $(\mathrm{cm^{-2}})$				
CH	4.1(13)	5.2(13)	9.6(13)	1.3(15)
CN	2.0(11)	1.9(11)	5.9(12)	2.1(13)
CO	4.4(13)	4.6(13)	2.5(14)	7.5 (14)
OH	1.7(13)	3.6(13)	8.0(13)	1.5(14)
C_2	7.8(12)	2.9(12)	3.8(13)	3.6(13)
H_3^+	8.3(12)	8.5(12)	5.4(12)	1.4(12)
CH^+	1.6(11)	1.9(11)	3.2(13)	2.4(14)
HCO^+	7.0(10)	6.0(10)	2.8(11)	1.1(12)

Note: Values in parenthesis refer to powers of ten.

models. To allow this reaction to proceed, various hypothesis on energy inputs have been proposed, among them: reaction with vibrationally excited H_2 (Garrod *et al.* 2003), MHD shocks (Pineau des Forêts *et al.* 1986), intermittent turbulent energy (Falgarone, Pineau des Forêts & Roueff 1995; Joulain *et al.* 1998; Falgarone *et al.*, this conference). A simple formulation to include non thermal effects was given by Federman *et al.* (1996), where an effective temperature, defined for each ion-neutral reaction, as $\frac{3}{2}kT_{eff} = \frac{3}{2}kT_{kin} + \frac{1}{2}\mu v_{in}^2$, is used in chemical reactions rates. In this expression, v_{in} is the relative ion-neutral drift speed and μ the reduced mass of the system. In Table 4 we compare models for different values of the ion-neutral drift velocity.

We note that this modus operandi allows us indeed to increase significantly the amount of predicted CH^+ and to a lesser extent CO, OH and HCO^+. The values obtained for an ion-neutral drift velocity of 3 $\mathrm{km\,s^{-1}}$ are close to the observed order of magnitudes of $N(CH^+)$. However, this procedure is empirical and does not account for the actual physical processes at work, contrary to what has been done by Joulain *et al.* (1998).

3.5. *Role of the Cosmic Ionization Rate*

The recent detection of H_3^+ towards the diffuse ζ Per line of sight by McCall *et al.* (2003) has been explained by an enhancement of the cosmic ionization rate, ζ, by 2 orders

Table 5. Role of the cosmic ionization rate ζ on the "standard model" conditions compared to the observed values towards ζ Per.

$\zeta\ (\mathrm{s^{-1}})$	5×10^{-17} (standard)	1×10^{-16}	5×10^{-16}	1×10^{-15}	5×10^{-15}	ζ Per
f	0.88	0.87	0.76	0.66	0.36	0.53 - 0.66
T (K)	80	71	69	67	68	45 - 75
column density $(\mathrm{cm^{-2}})$						
CH	4.1(13)	3.4(13)	1.2(13)	5.8(12)	9.0(11)	1.9(13) - 2(13)
CN	2.0(11)	2.2(11)	2.4(11)	2.2(11)	4.4(10)	2.7(12) - 3.3(12)
CO	4.4(13)	6.4(13)	1.8(14)	1.4 (14)	2.8(13)	5.4(14)
OH	1.7(13)	2.5(13)	6.0(13)	4.8(13)	1.6(13)	4.0(13)
C_2	7.8(12)	6.3(12)	1.3(12)	4.0(11)	1.5(10)	1.6(13)
C_3	0.9(12)	0.7(12)	0.6(12)	7.0(9)	5.3(6)	1.0(12)
H_3^+	8.3(12)	1.4(13)	3.0(13)	3.0(13)	1.8(13)	8.(13)
CH^+	1.6(11)	1.7(11)	1.6(11)	1.4(11)	6.7(10)	3.5 (12)
HCO^+	7.0(10)	8.610)	1.1(11)	5.2(10)	3.2(9)	

Note: Values in parenthesis refer to powers of ten.

of magnitude on the basis of its formation and destruction processes. We have tested this assumption with our chemical network and display in Table 5 the column densities for different values of ζ. We see that the H_3^+ column density increases with the cosmic ionization rate until the latter reaches a value of 5×10^{-16} s^{-1}. We find however that for higher values of the cosmic ionization rate, the molecular fraction becomes too low to account for the observed H_3^+ and that the column density of all molecular species decrease dramatically, in contradiction to the finding of McCall *et al.* (2003).

4. Comparison with Observations

4.1. *Excitation of H_2*

We discuss now the problem of the rotational excitation of H_2. Figure 3 displays the H_2 excitation diagram for different lines of sight, as well as for models with different values of the scaling factor, χ_{left}, of the radiation field impinging the left side of the "cloud". The lines of sight include Copernicus data (ζ Oph and ζ Per), high galactic clouds in the Chameleon (HD 96675, HD 102065 and HD 108927), which have been studied with FUSE (Gry *et al.* 2002), HD 37903 which corresponds to a cloud illuminated by the bright θ Ori star (Meyer *et al.* 2001) and HD 34078 (Boissé *et al.* 2005). We note that the highest rotational levels are not well reproduced by any homogeneous model, even with radiation fields 10 times the standard values. Moreover, a high value of χ leads to a substantial increase of the temperature and the ratio between $J = 0$ and 1 rotational levels is no more satisfactory. Other processes involving external input of energy can be at work and lead to excitation of molecular hydrogen as well as formation of CH$^+$. MHD shock models (*cf.* Heck *et al.* 1993) and turbulent chemical models of dissipative diffuse structures (Falgarone *et al.* 2005) are promising in this respect. However, the coupling of these macroscopic mechanisms with the detailed UV radiative transfer driving the atomic to molecular transition has not yet been fully introduced.

4.2. *The ζ Per Line of Sight*

The discovery of H_3^+ by McCall *et al.* (2003) towards ζ Per has led Le Petit, Roueff & Herbst (2004) to consider *all* observational constraints available in a chemical model, using the Meudon PDR code. It is clear from this study that a single density component is not able to reproduce the observations. The combination of a diffuse ($n_H = 100$ cm^{-3} with $A_V = 0.69$ corresponding to 4.2 pc) and dense slab ($n_H = 2 \times 10^4$ cm^{-3} with $A_V = 0.013$ corresponding to 76 AU) model is required to fit both atomic and molecular column densities in a reasonable way. A cosmic ray ionization rate of 2.5×10^{-16} s^{-1} is derived. The column density of H_3^+ is reproduced within a factor of 3. Similar conclusions had been formerly obtained by Cecchi-Pestellini & Dalgarno (2002) in the analysis of C_2 towards diffuse interstellar clouds. Introducing density fluctuations, length-scales and filling factors could be important for the interpretation of diffuse cloud chemistry. It should be noted that the excitation of rotational levels of H_2 above $J = 2$ and the observed column density of CH$^+$ are not reproduced in such a 2-phase model. An additional energetic phase is required for these additional observational constraints.

5. Conclusions

We have studied the dependence of a steady-state diffuse cloud model on various physical and astrophysical parameters. Much care has been paid to detailed microphysical processes. Carbon chains appear to be very sensitive to various parameters such as density profiles and geometry (one-sided or two-sided irradiation). The influence of cosmic ray

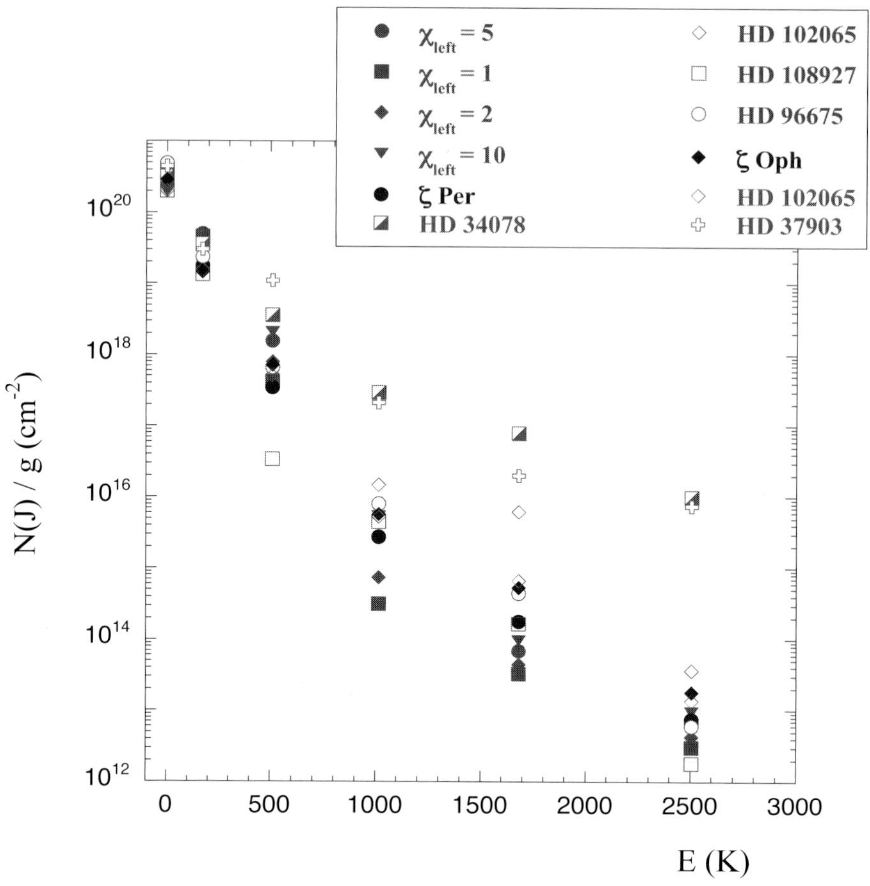

Figure 3. Rotational diagram of H_2.

ionization is decisive on the chemistry of H_3^+. However, too large a value for the cosmic ionization rate has a destructive impact on the molecular composition (via the secondary UV photons in particular) and the modelled column density of this molecular ion remains a factor 2 to 3 below the observed value.

Present steady-state models appear unsatisfactory to explain the large variety of observations in diffuse environments. However, such preliminary detailed studies are mandatory before introducing specific processes dealing with the astrophysical environment (time dependence, density fluctuations, MHD shocks, turbulence, ...). In spite of many observations and models, the chemistry of diffuse regions remains elusive.

Acknowledgements

We would like to acknowledge the many discussions we had with J. Le Bourlot and the support from the French Program "Physico-Chimie du Milieu Interstellaire" (PCMI).

References

Aannestad, P. 1973, *Ap. J. Suppl.* 25, 223

Abgrall, H., Le Bourlot, J., Pineau des Forêts G., Roueff, E., Flower, D.R., & Heck, L. 1992, *A&A* 253, 525

Becker, K.H., Engelhardt, B., Wiesen, P., Bayes, K.D. 1989, *CPL* 154, 342

Boissé, P., Le Petit, F., Rollinde, E., Roueff, E., Pineau des Forêts, G., Andersson, B.G., Gry, C., & Felenbok, P. 2005, *A&A* 429, 509

Cecchi-Pestillini, C., & Dalgarno, A. 2002 *MNRAS* 334, L31

Draine, B.T. 1978, *Ap. J. Suppl.* 36, 595

Falgarone, E., Pineau des Forêts, G., & Roueff, E. 1995, *A&A* 300, 870

Falgarone, E., Verstraete, L., Pineau des Forêts, G., & Hily-Blant, P. 2005, *A&A* 433, 997

Falgarone, E., Hily-Blant, P., Pineau des Forêts, G., & Schilke, P. 2005, *IAUS231*, poster I.21

Federman, S.R., Glassgold, A.E., & Kwan J. 1979, *Ap. J.* 227, 466

Federman, S.R., Rawlings, J.M.C., Taylor, S.D., & Williams D.A. 1996, *MNRAS* 279, L41

Garrod, R.T., Rawlings, J.M.C., Williams, D.A. 2003, *Ap&SS* 286, 487

Gry, C., Boulanger, F., Nehmé, C., Pineau des Forêts, G., Habart, E., & Falgarone, E. 2002, *A&A* 391, 675

Heck, E.L., Flower, D.R., Le Bourlot, J., Pineau des Forêts, G., & Roueff, E. 1993, *MNRAS* 258, 233

Jenkins, E.B. 2002, *Ap. J.* 580, 938

Joulain, K., Falgarone, E., Pineau des Forêts, & Flower, D.R. 1998, *A&A* 340, 241

Knauth, D.C., Andersson, B.-G., McCandiss S.R., & Warren Moos, H. 2004, *Nature* 429, 636

Le Bourlot, J., Pineau des Forêts G., Roueff, E., & Flower, D.R. 1993, *A&A* 267, 233

Le Petit, F., Roueff, E., & Herbst, E. 2004, *A&A* 417, 993

Le Petit, F., Nehmé, C., Hily-Blant P., Le Bourlot, J., Roueff, E., & Le Bourlot, E. 2005, *A&A*, to be submitted

Liszt, H., Lucas, R., & Black, J.H. 2004, *A&A* 428, 117

Mathis, J. S., Rumpl, W., & Nordsieck, K.H. 1977, *Ap. J.*, 217, 425

McCall, B.J. *et al.* 2003, *Nature* 422, 500

Meyer, D.M., Lauroesch, J.T., Sofia, U.J., Draine, B.T. & Bertoldi, F. 2001, *Ap. J.* 553, L59

Pineau des Forêts, G., Flower, D.R., Hartquist, T.W. , & Dalgarno, A. 1986, *MNRAS* 230, 801

Roueff, E., Le Bourlot, J., & Pineau des Forêts G. 1996, in *Dissociative Recombination: Theory, Experiment and Applications*, rds. D. Zajfman, J.B.A. Mitchell, D. Schwalm, & B.R. Rowe (World Scientific, Singapore), 11

van Harrevelt, R., van Hemert, M.C., & Schatz, G.C. 2002 , *JCP* 116, 6002

van Dishoeck, E. F. 1987, *JCP*, 86, 196

Vasyunin, A.I., Sobolev, A.M., Wiebe, D.S. and Semenov D.A. 2004, *Astr. Letters*, 30, 566

Discussion

VAN DISHOECK: You have not said much about HD, whose abundance is also sensitive to the cosmic ray ionization rate. Can you comment?

ROUEFF: A promising way to determine HD column densities was indeed via observations in the far UV in absorption (Copernicus, FUSE). However, the accurate determination of HD column density is hampered by saturation effects because the corresponding transitions lie in the flat portion of the equivalent width. The column density of HD is then determined to within a factor 5–10 only and does not constrain model results.

BENSCH: The 2-phase model for ζ Per has a much higher CO column density in the diffuse component than in the dense component, and the diffuse models shown before for $A_V = 1$. Why is the CO column density here (for $X = 2$, $A_V < 1$) much higher than shown before for $X = 1$, $A_V = 1$?

ROUEFF: The CO is indeed produced in the diffuse component: the higher cosmic ionization rate aids the formation of CO as OH is substantially increased and the $C^+ + OH$ reaction is the main channel to CO production.

NEUFELD: What happens if you throw up your hands, say that there is something we simply don't understand about H_3^+, and put in the observed abundance into the models *by hand* (i.e. treat the H_3^+ abundance as a fixed parameter rather than a quantity obtained by solution of the rate equations)? What problems would this solve in diffuse cloud chemistry (and what problems might it create)?

ROUEFF: This is an interesting suggestion and would probably help to build polyatomic species—more detailed consequences deserve further study.

McCALL: I have two concerns with the dense component included in your model of zeta Persei. First, the observed rotational distribution of C_2 and C_3 seems to be inconsistent with such a high density as 20,000 cm^{-3}. Second, if the size of the densest cloudlet on the sky were comparable to its supposed length (76 AU), it seems quite lucky that we happen to be looking through it. If we wait for a couple of years, it will likely no longer be along the line of sight, due to proper motion. Do you have any comment on these concerns?

ROUEFF: Your question refers to the results reported in the paper by Le Petit, Roueff & Herbst (2004). Concerning your first concern, we have run your code available at the URL http://dib.uiuc.edu/c2/ computing the excitation of C_2 to check the relevance of a dense component, and we found that such a solution was indeed compatible with the observations except for $J = 6$. However, a similar problem arises for a lower density component such as one with density 100 cm^{-3}. The observed $J = 6$ column density may be over-estimated. Indeed, only one transition, R(6), has been found whereas the two others, P(6) and Q(6), have not been detected.

   Your argument about possible small scale structure is perfectly reasonable. However, there is mounting observational evidence that interstellar matter can have a clumpy or filamentary structure on the scale of 10s to a few 1000s of AU and Cordiner, Fossey and Sarre (this conference) report such occurrences from diffuse band strength studies. Such an hypothesis is thus in accordance with present observational knowledge.

Astrochemistry: Recent Successes and Current Challenges
Proceedings IAU Symposium No. 231, 2005
D.C. Lis, G.A. Blake & E. Herbst, eds.

© 2006 International Astronomical Union
doi:10.1017/S1743921306007204

What Constitutes Spectroscopic Proof for the Detection of Large "Hot Core" Molecules?

Lucy M. Ziurys and Aldo J. Apponi

Departments of Chemistry and Astronomy, University of Arizona, 933 N. Cherry Avenue,
Tucson, AZ 85721 USA

Abstract. The presence of large organic species in interstellar gas has important implications
for the origin of life and pre-biotic chemistry. Accurate identifications of such molecules, however,
are problematic in molecular clouds. There are many reasons for such difficulties. One is that the
spectral density is very high in objects where the chemistry is sufficiently complex to produce
such species—at least 10 lines per 100 MHz in Sgr B2(N), for example, at 3 mm, at a sensitivity
of 10 mK peak-to-peak. Hence, the possibility of chance coincidences is large. Another reason
is the presence of many large organic molecules at relatively high temperatures; these large
asymmetric tops have vast numbers of favorable transitions under these conditions, including
those originating from low-lying vibrational states. Confusion and blending of transitions of one
large molecule with those of another add to the risks of an inaccurate identification.

A case in point is that of glycolaldehyde, CH_2OHCHO. In order to confirm the identification
of this molecule in Sgr B2(N), we searched for its most favorable transitions in the 2 and 3 mm
windows, spanning the energy range of $\sim$10–100 K—a total of 43 individual transitions. Of all
these lines, only seven were not heavily blended or contaminated by other molecules, i.e. so-called
"clean" features. Emission, however, was visibly detected at 34 of the other transitions, but one
transition was clearly absent. The "missing" line corresponds to a weak transition originating in
the $K_a = 3$ ladder, and its absence is consistent with the other detected features. Based on the
clean features only, glycolaldehyde has a $V_{LSR} = 61.7 \pm 1.5$ km s^{-1} and $\Delta V_{1/2} = 7.8 \pm 1.8$ km s^{-1}
with intensities consistently in the range $T_R^* \sim 20-70$ mK. Given these data, the identification of
glycolaldehyde is 99.9% secure. A rotational diagram from this data set yields a column density
of $N_{tot} \sim 6 \times 10^{13}$ cm^{-2} for CH_2OHCHO—roughly a factor of 27 less than that of H_2CO. These
data illustrate the problems and subtleties in identifying large organic molecules in space.

Keywords. astrobiology — astrochemistry — ISM: abundances — ISM: molecules — molecular
processes — radio lines: ISM

---

1. Introduction

The large molecules found in dense interstellar clouds are typically simple organic
species containing common functional groups. Ketones, aldehydes, acids, esters, and
amides, for example, are well represented in interstellar molecules, as demonstrated by
the presence of compounds such as dimethyl ether (CH_3OCH_3), formic acid (HCOOH),
methyl formate ($HCOOCH_3$), acetaldehyde (CH_3CHO), and formamide (NH_2CHO). In
fact, interstellar space is the one common place that organic molecules are routinely
found other than in living organisms on Earth.

The presence of organic molecules in these two vastly different environments suggests
that there may be a connection. In fact, it is possible that interstellar chemistry actually
supplied the initial compounds that led to terrestrial life. It is now thought that the
evolution of life on Earth did not always result in the best biochemical solutions (e.g.
Benner *et al.* 2004). There are many other chemical possibilities for living systems. For

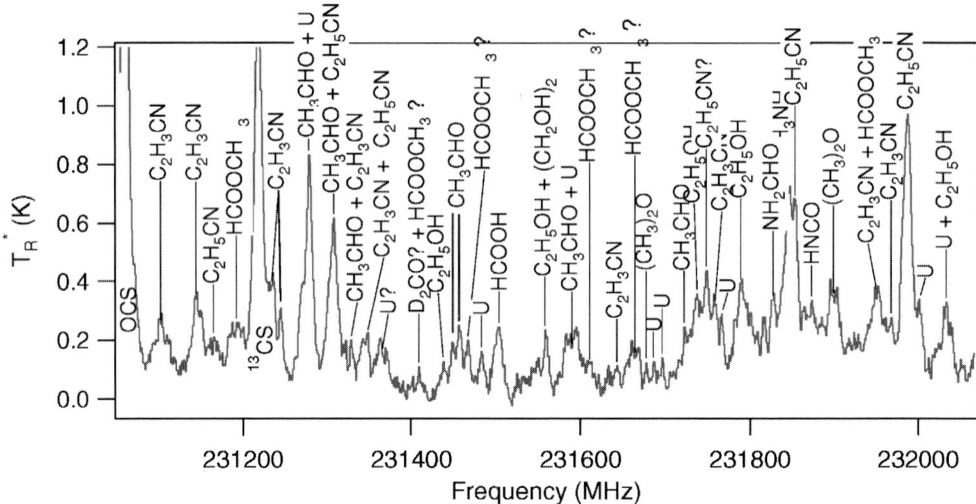

Figure 1. Single-sideband spectrum of Sgr B2(N) taken with the ARO SMT 10-meter telescope near 231.6 GHz. With an integration time of 6 hours, the spectrum is near the confusion limit with no baseline regions at the theoretical 3 σ noise level ($\sim$10 mK).

example, it is not clear why ribose and deoxyribose are used in DNA and RNA. Why not glycol, a tetrose or even a hexose? Work by Albert Eschenmoser (1999) has suggested that other sugars are viable substitutes for ribose in the DNA structure. Why are only twenty standard amino acids used in living proteins? Studies have shown that there are no biochemical reasons to exclude other unnatural amino acids in proteins (e.g. Hohsaka & Masahiko 2002). However, life may have chosen one set of chemical possibilities over others not because they were the best solution to the problem, but because of synthetic contingency, namely, that what evolved is a result of what was available as starting materials. Could interstellar chemistry have provided these contingencies via molecule formation in dense clouds? Such compounds may have been transported to planet surfaces via comets and meteorites, and hence became the substances that led to living systems.

To examine such a hypothesis, it is clearly necessary to establish what chemical compounds are present in dense molecular clouds and in what quantities. Only then can the pieces be put together to examine the possible contribution of interstellar molecules to early biochemistry. However, accurate identification of such large organic species is not trivial. In this paper, we discuss such difficulties and formulate criteria for secure detections of large molecules in dense interstellar clouds.

2. The Millimeter-Wave Spectrum of Sgr B2(N)

The spectrum of Sgr B2(N) shown in Figure 1 is a good illustration of a confusion limited spectrum. This 6-hour integration on Sgr B2(N) using the ARO SMT 10-meter telescope has a theoretical 3 σ noise level of about 10 mK. No significant baseline regions are found in this spectrum at this level, indicating that the entire bandpass is made up of partially overlapping emission features. Only the strongest of these features are easily identifiable; i.e., those belonging to vinyl cyanide, carbon monosulfide, ethyl cyanide, and carbonyl sulfide. The intensity of these lines typically are greater than 500 mK. Lines below that level start to become contaminated with other features and show up as blends of two or more transitions. These data illustrate the difficulty in unambiguously assigning the lines near the confusion limit.

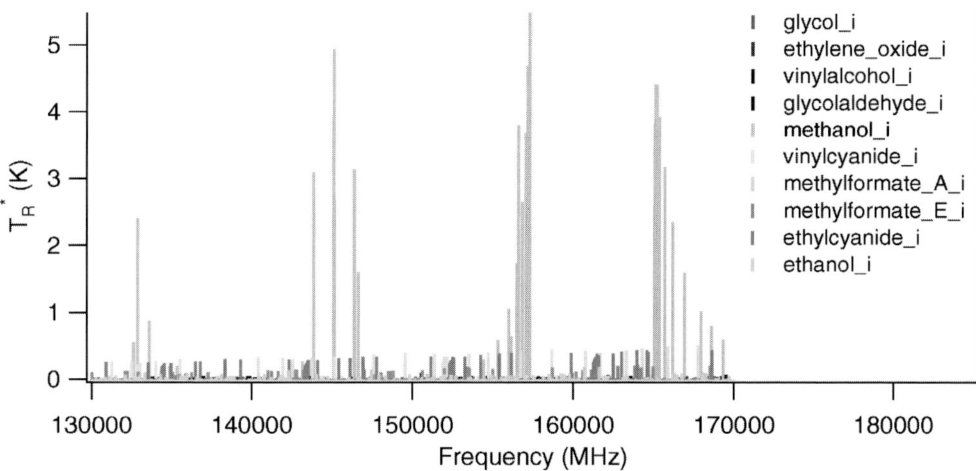

Figure 2. Simulated 2 mm spectrum of Sgr B2(N) using the column density and rotational temperatures given in the literature for 11 interstellar molecules of varying abundance. The combination of only these few molecules quite effectively shows that source confusion starts to become very significant in Sgr B2(N) at about the 500 mK level. See the on-line presentation for the color version of this figure.

The lines that make up the confusion-limited spectra of molecular clouds in general can be attributed to relatively large asymmetric top organic molecules. These types of species possess intrinsically complex spectra owing to their three unique moments of inertia and their system of complex energy levels that contribute to large rotational partition functions. The result is a dense population of rotational transitions that peak in the millimeter region of the spectrum.

Figures 2 and 3 illustrate the three levels of complexity that can be expected while observing astronomical sources containing large organic molecules. Figure 2 is a model of the 2 mm spectrum of Sgr B2(N), only considering the 11 molecules listed in the legend. At first glance, the spectrum is dominated by strong emission lines of the most abundant species present (in this case methanol). These lines are very strong (several Kelvin) and are sparsely distributed throughout the spectrum. It is noteworthy that the density of lines this strong is small (only a few per 5000 km s^{-1}); hence, any contamination of weak features on these lines will be insignificant. These lines are easily identifiable.

The next level of complexity found in Sgr B2(N) is at about the 300 to 500 mK level. Here a pattern of lines of increasing density is found, resulting from molecules that are of lesser abundance, as well as weaker features from molecules that are of high abundance. The emission features of methyl formate, ethyl cyanide and vinyl cyanide were selected for this illustration. These lines are about the same size as many of the weak methanol lines, which adds a level of complexity. The transitions are more densely spaced because these molecules are generally heavier and more asymmetric. The line density at this level of sensitivity is about an order of magnitude higher.

The last level of complexity is at about the 40 mK level, where there is nearly a continuum of closely space lines (see Figure 3). The weak features of the molecules of high and intermediate abundance plus the strongest features of yet lesser abundant species all contribute at this level. The molecules selected to illustrate this case are ethanol, ethylene glycol, ethylene oxide, vinyl alcohol and glycolaldehyde. Even with these few additional species, the problem in assigning spectra becomes very apparent. The line density increases yet again by about an order of magnitude to about 5 lines per

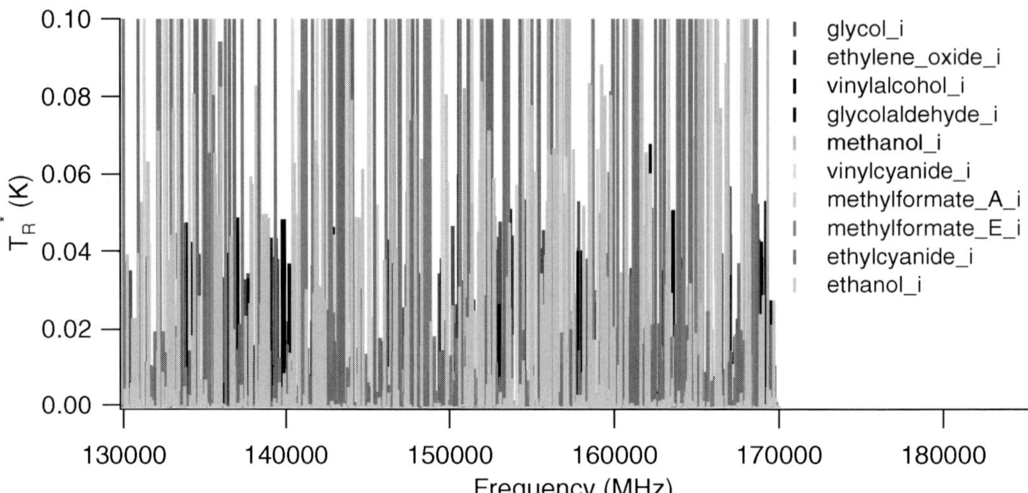

Figure 3. Same as figure 2, but the intensity scale is magnified to illustrate the increased complexity that occurs at about the 40 mK level in Sgr B2(N).

$100 \ \mathrm{km \, s^{-1}}$ in Sgr B2(N). Contamination at this level is inevitable and clean transitions will be few in number.

3. Approaching This Complex Problem

The statistical probability of finding a particular spectral feature of a molecule at a particular frequency is a function of: (i) the physics associated with the transition under investigation (N_{tot}, T_{rot}, S_{ij}, E_u, etc.) and (ii) the number of abundant contaminating species with favorable transitions that significantly overlap the same spectral region. Ideally, we would like the first of these two factors to dominate, and we have all become accustomed to assuming that this is true. However, as the sensitivity of an observation is increased, the second factor becomes significant. Hence, only through the observation of a sufficient number of clean lines (lines free from significant contamination) can a molecule be unambiguously identified near the confusion limit.

To date, spectral surveys at millimeter wavelengths have not been sensitive enough to identify new molecules or even some weak known molecules. For instance, Nummelin *et al.* (2000) surveyed the 1.3 mm atmospheric window down to about 100 mK peak to peak and Friedel et al. (2004) sparsely covered the 3 mm window at modest sensitivity compared to that required for identifying new molecules. Neither of these surveys of Sgr B2(N) was able to detect glycolaldehyde or ethylene glycol, nor were they able to help confirm or dismiss other recently claimed molecular identifications. To identify new asymmetric organic molecules with low abundance, it is clear that higher sensitivity is needed. Moreover, owing to high contamination levels, a vast number of transitions need to be covered to ensure that clean lines are observed over the most probable energy ranges.

The approach that we have adopted to help solve this problem is to conduct deep, confusion-limited, spectral line surveys that cover a large frequency range (65–270 GHz). This large dataset will provide the means to locate clean transitions of new molecules over a large energy range. The overall method can be summarized as follows: (i) obtain an accurate set of rest frequencies, preferably by direct laboratory measurement, (ii) locate clean transitions that match in line shape and width, (iii) analyze the intensity of those lines to ensure that they represent a physically connected system, (iv) locate a sufficient number of lines that sample an appropriate energy range, (v) identify and model intensities of all known species, particularly weak lines that may overlap to make strong lines, (vi) attempt to separate out contaminating features to strengthen the case, and (vii) account for beam-dilution where appropriate (Snyder *et al.* 2005; Halfen *et al.* 2005).

4. Case Studies of Glycolaldhyde and Dihydroxyacetone

Two case studies illustrate the power of this technique for identification of new interstellar molecules. These studies involve the molecules glycolaldehyde and dihydroxyacetone. Glycolaldehyde was first reported by Hollis *et al.* (2000) to be present in Sgr B2(N) using the Kitt Peak 12-meter telescope at millimeter wavelengths. These observations were followed by measurements at the NRAO GBT telescope in the K-band region (Hollis *et al.* 2004). Dihydroxyacetone was first reported by Widicus Weaver & Blake (2005) to be present in Sgr B2(N) using the CSO 10-meter telescope.

4.1. *Confirmation of Glycolaldehyde*

In Hollis *et al.* (2000), several lines were reported in the 3 mm window; however, most of these are apparently contaminated with other lines because they exhibit unusually broad line widths of 20 to 30 $\mathrm{km\,s^{-1}}$. Recently, Halfen *et al.* (2005) have searched for an additional 43 transitions of this molecule. In this work, we discovered that only about 20% of the transitions were free from contamination, but we were able to observed 7 clean features that agree in line width and line center (V_{LSR}), and their intensities are consistent for a gas with a rotational temperature of 34 K.

Figure 4 is an example of some of the 40 spectra that we have measured for glycolaldehyde. These spectra represent the complete dataset. The degree of contamination ranges from total obscuration (102.5 GHz) to partial blending (103.4 GHz). The spectrum at 103.7 GHz shows an example of a clean feature, which exhibits the canonical line width and line center of glycolaldehyde in Sgr B2(N). Still, some ambiguities remain. For example, in the spectrum at 95.1 GHz, vibrationally excited lines of vinyl cyanide are seen throughout the spectrum (Nummelin & Bergman 1999; Halfen *et al.* 2005). The relative intensities of these lines closely match the observed data. However, there does seem to be some excess emission near the center frequency, which can be attributed to glycolaldehyde although it is difficult to quantify how much. Contamination such as this is seen throughout the dataset and highlights the importance of a complete accounting of all the emission from known constituents.

To help prove that the lines observed for glycolaldehyde are all physically connected, a rotational diagram was used (see Figure 5). As shown in the figure, both the clean lines and the partially blended lines all fit quite closely to the same physical model.

4.2. *A Rigorous Attempt to Confirm Interstellar Dihydroxyacetone*

Widicus Weaver & Blake (2005) reported the detection of 9 transitions of dihydroxyacetone (DHA) in Sgr B2(N) at 1.3 mm that closely matched rest frequencies measured by the same group (Widicus *et al.* 2004). From their observations, they derived a column density of 5×10^{15} $\mathrm{cm^{-2}}$ and a rotational temperature of 220 K. Figure 6 shows a single sideband spectrum of DHA taken at the ARO SMT 10-meter telescope that reproduces Figure 1 shown in Widicus Weaver & Blake (2005). The 220 K rotational temperature does not appear to reproduce the relative intensities of the observed lines as well as had been reported. The figure also shows that if the temperature is lowered, the predicted

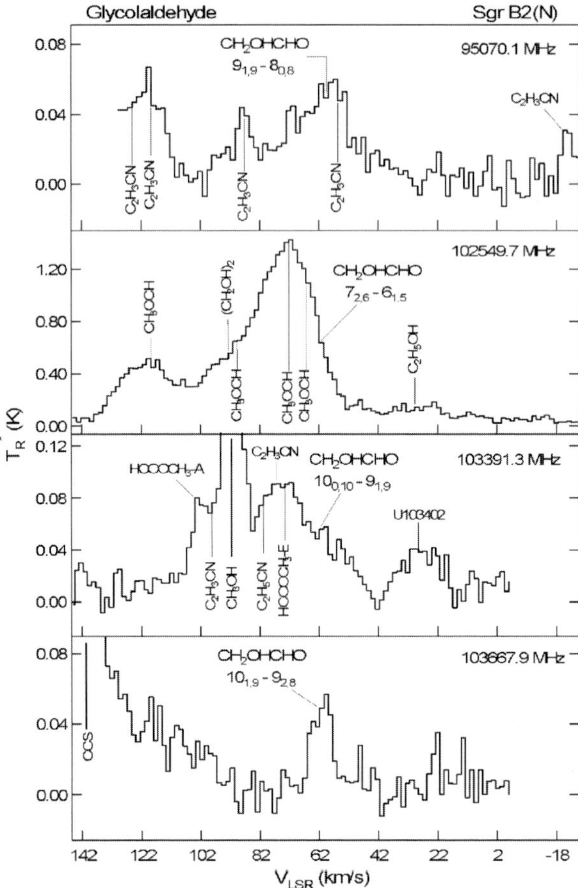

Figure 4. Single-sideband spectra of Sgr B2(N) taken with the ARO KP 12-meter telescope (Halfen *et al.* 2005).

relative intensities change dramatically. Also, the reported column density is about a factor of 6 higher than that of the common interstellar molecule H_2CO towards the same region (Apponi *et al.*, in preparation), clearly raising some questions about the validity of their detection. The intense lines observed at 1.3 mm (400 mK) and the high rotational temperature proposed indicate that DHA should possess a very intense spectrum from 100 GHz through 300 GHz. Figure 7 shows the simulated spectrum of DHA from 100 GHz to 245 GHz. The blue trace (see on-line presentation for color version) indicates the positions of the transitions, while the red trace is the modeled spectrum assuming 10 $km\,s^{-1}$ line width and 10 mK rms noise. It is easy to see that there should be no problem detecting this molecule at lower frequencies.

In addition, a closer examination of the rotational diagram of the original 9 lines recorded at 1.3 mm further shows that their intensities are not reproduced well (see Figure 8). On the left is the rotational diagram reproduced from the results of Widicus Weaver & Blake (2005). At first glance, the data all seem reasonable, except that not all of the data points fall on the fitted line. The right panel of the figure shows the optical depth for each transition that would be required to bring the respective point onto the fitted line. These data show that the optical depths vary from -1 to 1.5 and suggest that these lines are not all physically connected.

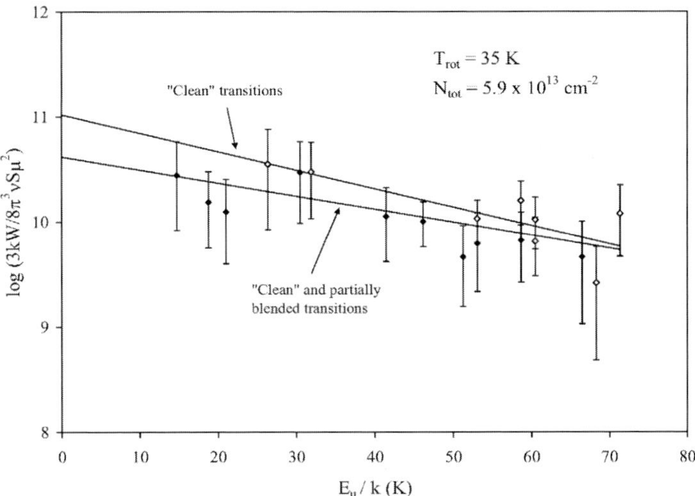

Figure 5. Rotational energy diagram for clean and partially blended lines of glycolaldehyde in Sgr B2(N).

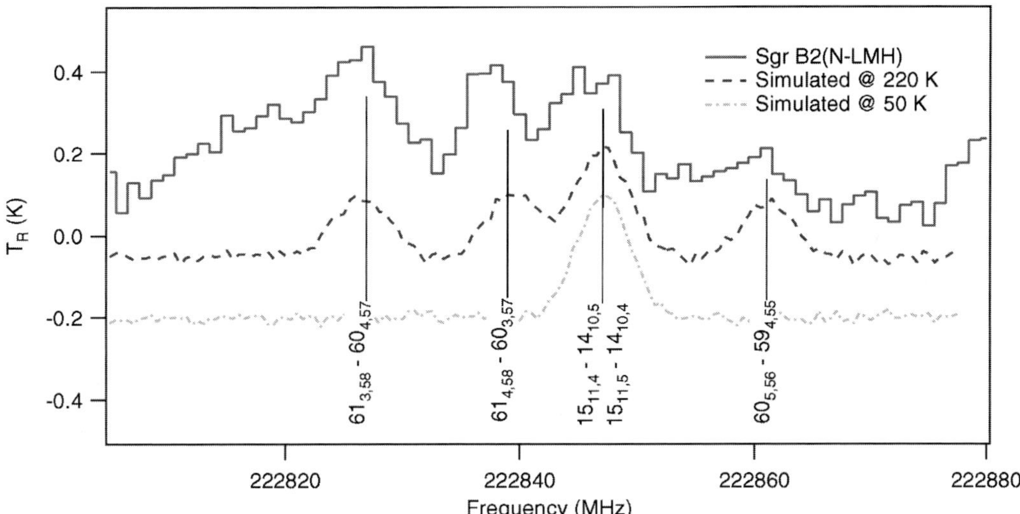

Figure 6. Single sideband spectrum taken at the ARO SMT 10-meter telescope covering the same spectral features as shown in Figure 1 of Widicus Weaver & Blake (2005). Overlaid on the spectrum are simulated spectra at 220 K and 50 K showing that these data are not well reproduced at either temperature even though they do show a close match in frequency.

The Ziurys group, in collaboration with Hollis and coworkers, also conducted searches for DHA in the same source in the K-band at the NRAO GBT and at 2 and 3 mm at the ARO 12-meter telescope at nearly the same time as the CSO measurements. Of 40 separate frequencies covering 61 favorable rotational transitions of DHA, only 2 of them show reasonable intensity correlation with the original work reported by Widicus Weaver & Blake (2005). Fourteen of the regions were contaminated with other molecules and 20

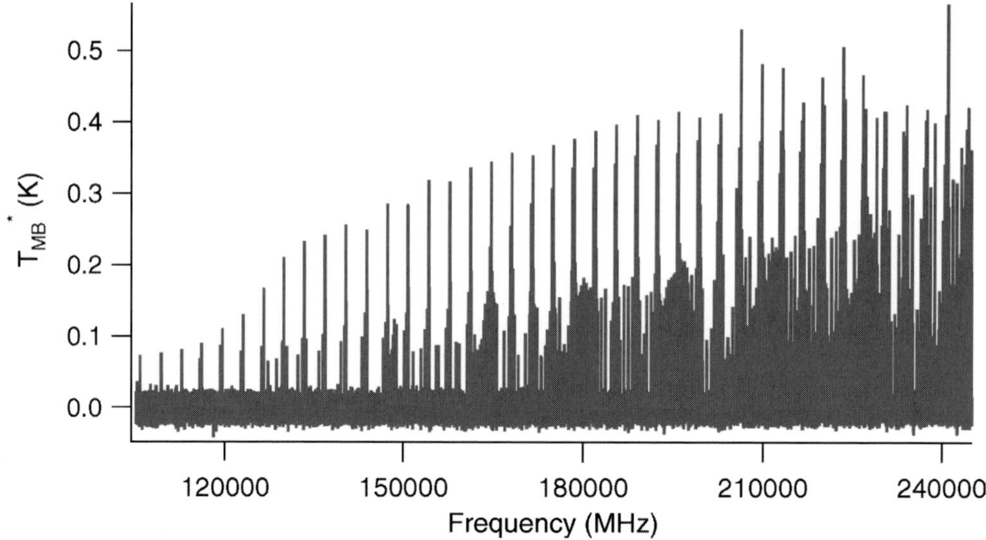

Figure 7. Simulated spectrum of DHA in Sgr B2(N) using the physical parameters listed in Widicus Weaver & Blake (2005). For a color version of this figure see the on-line presentation.

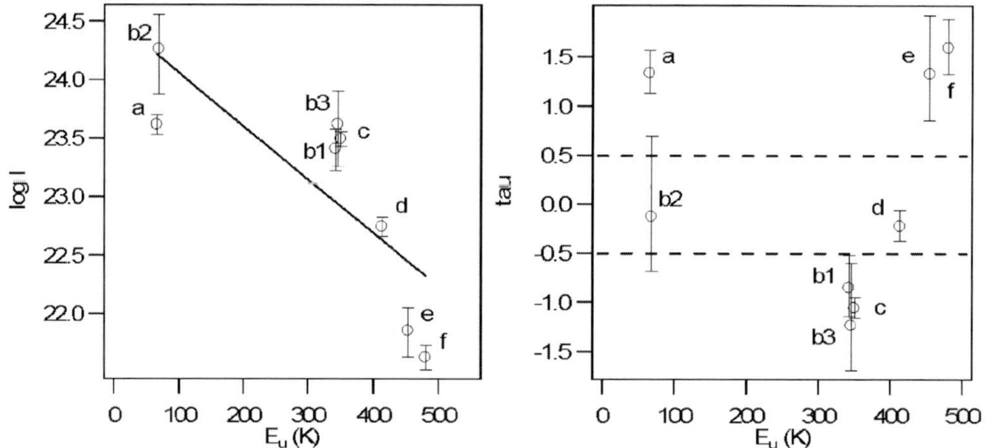

Figure 8. Rotational diagram analysis of the 9 lines taken from Widicus Weaver & Blake (2005) (*left*) and the respective opacities that would be required to correct each data point back onto the fitted line (*right*).

of them showed no significant emission. Figure 9 shows a few of these spectra. With these new results, we have constrained the temperature of DHA to below 50 K and have reduced the upper limit of the column density to 5×10^{13} cm^{-2}. It is doubtful from these new data that DHA is abundant enough in Sgr B2(N) for a definitive detection.

5. Conclusion

To obtain reliable spectroscopic evidence for the presence of large molecules, a careful and consistent dataset needs to be assembled. Although such data collection is time consuming, it is clearly necessary for a reliable molecular identification. Previous

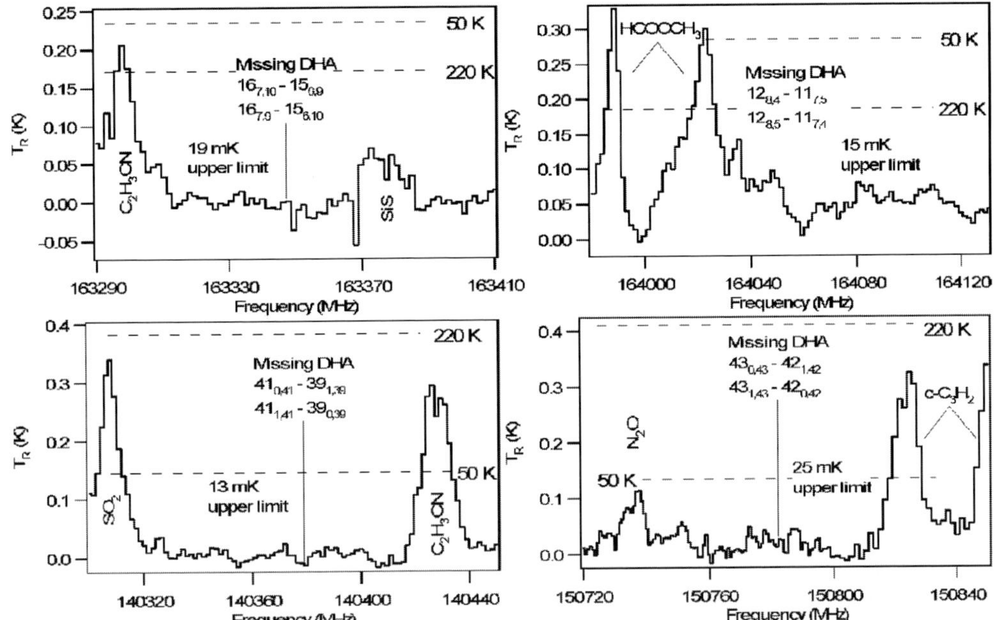

Figure 9. Spectra covering 4 favorable regions that should have significant DHA emission. The spectra were taken with the ARO 12-meter telescope in single sideband mode (Apponi *et al.* in preparation). The dashed lines indicate the predicted intensity of DHA emission based on T_{rot} = 220 K and T_{rot} = 50 K as labeled.

identifications of large species based on a limited number of transitions need to be closely re-examined. An accurate picture of interstellar organic chemistry will never be obtained otherwise.

Acknowledgements

This material is based upon the work supported by the National Aeronautics and Space Administration through the NASA Astrobiology Institute under Cooperative Agreement No. CAN-02-055-02 issued through the office of Space Science.

References

Apponi, A. J., Halfen, D. T., Ziurys, L. M., Hollis, J. M., Lovas, F. J, & Remijan, A., in preparation.
Benner, S. A., Ricardo, A., & Carriga, M. A., 2004, *Curr. Opin. Chem. Biol.* 8, 672
Eschenmoser, A., 1999, *Science* 284, 2118
Friedel, D. N., Snyder, L. E., Turner, B. E., & Remijan, A., 2004, *Ap. J.* 600, 234
Halfen, D. T., Apponi, A. J., Woolf, N., Polt, R., & Ziurys, L. M. 2005, *Ap. J.*, accepted
Hohsaka, T. & Masahiko, S. M., 2002, *Curr. Opin. Chem. Biol.* 6, 809
Hollis, J. M., Lovas, F. J., & Jewell, P. R. 2000, *Ap. J.* 540, L107
Hollis, J. M., Jewell, P. R., Lovas, F. J., & Remijan, A. 2004, *Ap. J.* 613, L45
Nummelin, A. & Bergman, P., 1999, *A&A* 341, L59
Nummelin, A. et al., 2000, *Ap. J. Suppl* 128, 213
Snyder, L. E. et al. 2005, *Ap. J.* 619, 914
Widicus, S. L., Braakman, R., Kent, D. R., & Blake, G. A., 2004, *J. Mol. Spectrosc.* 224, 101
Widicus Weaver, S. L., & Blake, G. A., 2005, *Ap. J.* 624, 33

Photo: E. Herbst

Astrochemistry: Recent Successes and Current Challenges
Proceedings IAU Symposium No. 231, 2005
D.C. Lis, G.A. Blake & E. Herbst, eds.

Interferometric Observations of Complex Organic Molecules

Sheng-Yuan Liu

Academia Sinica, Institute of Astronomy and Astrophysics,
Taipei, Taiwan
email: syliu@asiaa.sinica.edu.tw

Abstract. Rich molecular complexity has been shown to exist in star forming regions primarily through millimeter to submillimeter single-dish spectral line surveys or searches over the years. With a few case studies, we demonstrate that interferometric observations are a powerful tool for advancing our knowledge of the chemical processes related to the formation of complex organic molecules in these regions. In particular, interferometers are sensitive to compact structures, such as those of the so called "hot molecular core" sources where complex organics are almost exclusively found. The high angular resolutions achieved by interferometers allow us to better spatially discriminate chemical differentiation and, possibly, evolutionary effects.

Keywords. astrochemistry — ISM: abundances — ISM: individual (Orion KL, Sgr B2, G9.62 + 0.13) — line: identification

1. Introduction

To date more than 130 molecular species have been detected in the interstellar and circumstellar environment. Observations have shown that while simpler and long chain carbon molecules are often found to be abundant in diffuse/dark clouds, complex and more saturated organic species, such as vinyl cyanide (C_2H_3CN), ethyl cyanide (C_2H_5CN), dimethyl ether (($CH_3)_2O$), methyl formate ($HCOOCH_3$), etc. are found predominantly or even exclusively in the so-called "hot molecular cores," or HMCs. HMCs are dense ($n_{H_2} > 10^6$ cm^{-3}) and warm (T > 100 K) molecular condensations often associated with molecular masers, H II regions, or embedded near-infrared sources. These signatures suggest HMCs are active massive star-forming regions, possibly at the earliest phase of star formation (Kurtz *et al.* 2000 and this volume).

The formation mechanisms of complex organic species are of great interest. This is particularly true for those molecules of potential biological importance. The understanding of such molecular synthesis may give profound insight into the pervasiveness of organic chemistry in the Universe or even how life started on Earth. Based on models with extensive reaction networks, quiescent gas-phase reactions have quite successfully accounted for the observed molecular composition of diffuse/dark clouds. On the other hand, it appears very unlikely that such processes can explain the formation of complex organic species in the large abundances found toward HMCs. Surface processing and gas-grain interactions were often restored to explain the enhanced abundances of complex organics (*e.g.*, Miao *et al.* 1995; Mehringer & Snyder 1996). Detailed quantitative models for grain-surface molecular formation, however, remain difficult to develop due to the lack of our understanding of grain surfaces, molecular mobilities, desorption processes, etc.

To better understand surface reaction processes and their relative importance with respect to gas-phase reactions for molecular formation, two types of comparison may help in establishing guidelines (Turner & Apponi 2001; see also Hollis, this volume). The first comparison is to utilize backbone chemistry: comparison among species with a common backbone or sharing similar structural elements but different degrees of saturation may provide clues to the efficiency of hydrogenation on grain surfaces. Comparison of families with different backbones, on the other hand, may shed light on the mobility of heavy atoms or small functional groups on grain surfaces. Second, complex molecules can form isomeric families, in which the same number and type of atoms can form more than one molecular species. Relative abundances within an isomeric family may contain information on the rearrangement processes on grains and on the relative desorption efficiencies of the various members.

Past molecular line surveys, mainly carried out by single-dish telescopes toward limited lines of sight, have established the paradigm for investigating relative abundances of various molecular species in star-forming regions, including those sharing common backbones or that are members of the same isomeric family ($e.g.$, Blake et $al.$ 1987; Turner 1991; Schilke et $al.$ 1997). Nevertheless, the angular resolutions achieved by these studies are often sufficiently coarse that different species, though they trace different cloud components, are found within the same beam and thus cannot be easily distinguished spatially or even kinematically. The derived relative abundances between various species are therefore averages over different cloud components.

To remedy the above difficulties, observations at higher angular resolution, which can be naturally achieved by interferometers, are highly desired. This is especially important for studying HMCs, given that these sources are often located at substantial distances, typically several kiloparsecs away. In addition to the their critical imaging capabilities, interferometers also act as spatial filters to remove extended structures, hence better matching and isolating the emission from sources that are compact in nature, like HMCs.

We highlight here two case studies of complex organics carried with interferometric observations, mainly toward HMCs. The first is the imaging of HCOOH and CH_3COOH; the second involves the aperture synthesis imaging of the C_2H_4O isomeric triplet. We also present a new observational example of chemical differentiation between complex nitrogen- and oxygen-bearing species in star-forming cores.

2. The Imaging of HCOOH and CH_3COOH

Among the complex organic species, formic acid (HCOOH) and the related species acetic acid (CH_3COOH) are two of great interest. HCOOH is the simplest organic acid; while acetic acid and HCOOH share common structural elements with glycine (NH_2CH_2COOH), the simplest and biologically important amino acid. In the laboratory a bimolecular synthesis of NH_2CH_2COOH can occur when acetic acid combines with NH_2^+. CH_3COOH is also important for astrochemical studies because it contains the elusive C–C–O backbone; interstellar molecules with this structure appear to have less favorable formation routes than their counterparts with the C–O–C backbone (Millar et $al.$ 1988). For example, an isomer of acetic acid, methyl formate ($HCOOCH_3$), has the C–O–C backbone and is detected ubiquitously in a host of HMCs.

Interstellar HCOOH was first reported by Zuckerman, Ball, & Gottlieb (1971) and by Winnewisser & Churchwell (1975), both toward Sgr B2; and has been subsequently identified in single-dish telescope spectral line surveys since then ($e.g.$, Turner 1991; Ziurys, & McGonagle 1993), primarily toward two massive star forming regions—Sgr B2 and Orion KL. More recently, HCOOH was observed in comet Hale-Bopp (Crovisier & Bockelée-Morvan 1999) and tentatively identified in interstellar ice absorption spectra (Schuttee et $al.$ 1996, 1999). CH_3COOH, on the other hand, was not found in ISM until the first detection with an interferometric array (Mehringer et $al.$ 1997).

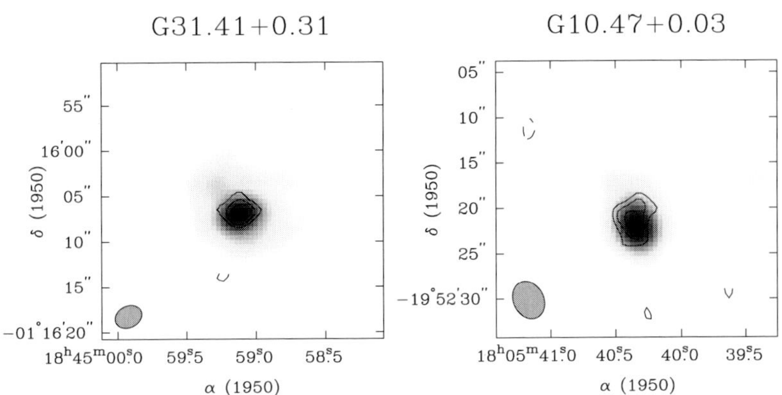

Figure 1. Contours in both panels represent HCOOH emission while grayscale represent the 3 mm continuum emission in both regions.

To investigate the distribution of HCOOH and CH₃COOH, survey observations with the BIMA Array were carried out. HCOOH was successfully imaged toward the massive star forming cores in Sgr B2 (N), Orion KL, and W51 (Liu, Mehringer, & Snyder 2001) while CH₃COOH was mapped in Sgr B2(N) and W51 (Remijan *et al.* 2002). Further observations carried out with the BIMA Array as well as the OVRO Millimeter Array mapped HCOOH emission in G34.3 + 0.2, G19.61 − 0.23, W75 N (Remijan *et al.* 2003, 2004) as well as G31.41 + 0.31 and G10.47 + 0.03 (Liu *et al.* 2005a; Fig. 1), respectively. BIMA observations also detected CH₃COOH toward G34.3 + 0.2 (Remijan *et al.* 2004).

In these observations, both HCOOH and CH₃COOH emission was found to be compact and co-spatial with the dust emission from dense cores. The derived HCOOH abundances of 10^{-10} to 10^{-8} (Liu, Mehringer, & Snyder 2001) were at least comparable to, or a few orders of magnitudes higher than those suggested by gas-phase reaction pathways for cold dark clouds such as s L134 N and L183 (Irvine *et al.* 1990; Turner, Terzieva, & Herbst 1999). The CH₃COOH abundances of $10^{-2} \sim 10^{-1}$ relative to that of HCOOCH₃, were also found to be much higher than the predictions of pure gas-phase reaction networks. The formation mechanism for HCOOH and CH₃COOH detected in these surveys was therefore attributed to grain-surface process. Several grain-surface reaction scenarios (Tielens & Hagen 1982; Allamandola, & Sandford 1990; Charnley 1995) in fact suggested HCOOH readily produced in icy grain mantles. Upon mantle evaporation, HCOOH will be released into the gas phase.

Recent BIMA observations toward Orion KL at higher angular resolution demonstrated that HCOOH shows a partial shell morphology (Liu *et al.* 2002, see Fig. 2). Its unique distribution suggests that HCOOH is located in a layer that delineates the interaction region between the outflow and the ambient quiescent gas. This is supported by HCOOH emission being at the southwest rim of numerous H₂O masers spots as well as a shell of SO₂ emission, both considered to be strongly enhanced due to passages of shocks (Kaufman & Neufeld 1996; Pineau *et al.* 1993). The HCOOH distribution is also noticeably different from those of other complex O-bearing molecules such as HCOOCH₃ and (CH₃)₂O (Fig. 3), suggesting the former is freshly evaporated from grain surfaces while the latter are likely synthesized from evaporated species via gas-phase routes. It was noted that the observed HCOOH gas-phase abundances are, however, significantly less than those implied on grain surfaces by IR absorption studies. UV or soft X-ray destruction of HCOOH in ices or immediately after mantle evaporation has been proposed to account for this (Ehrenfreund *et al.* 2001; Boechat-Roberty, Pilling, & Santos 2005).

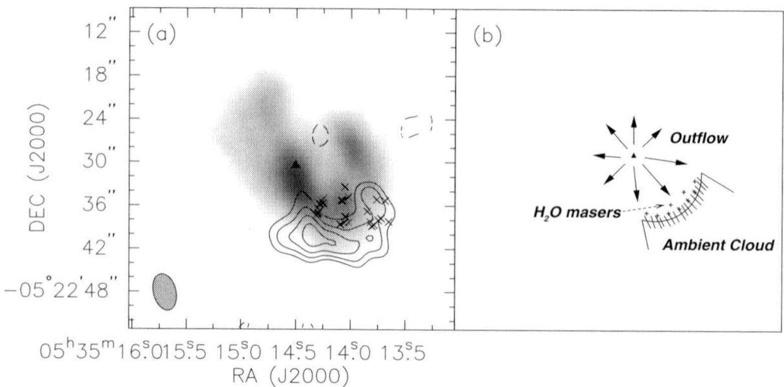

Figure 2. (a) Thick contours represent HCOOH emission. The SO_2 emission is shown in grayscale. The filled triangle denotes the radio continuum source I, while the crosses represent H_2O maser locations. (b) A diagram illustrating the cloud configuration (Liu 2002).

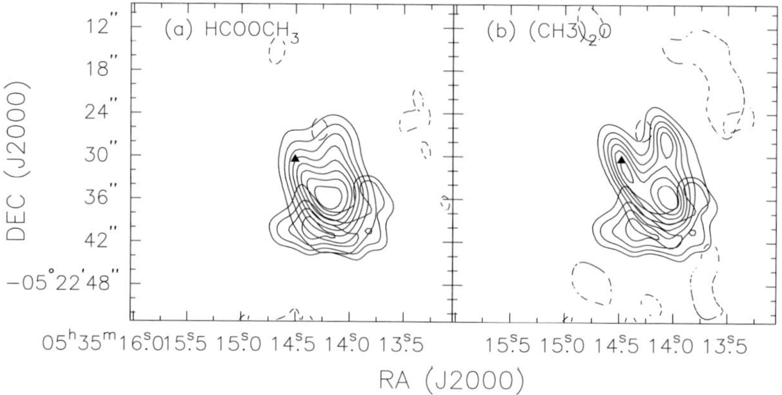

Figure 3. Thick contours in both panels represent HCOOH emission. (a) Thin contours represent $HCOOCH_3$ emission. (b) Thin contours represent $(CH_3)_2O$ emission. (Liu 2002).

3. The Imaging of the C_2H_4O Isomeric Triplet

Two important isomeric triplets contain species that are unable to be formed via quiescent gas-phase processes alone: one triplet is the $C_2H_4O_2$ family, including methyl formate ($HCOOCH_3$), acetic acid (CH_3COOH), and glycolaldehyde ($HOCH_2CHO$); the other is the C_2H_4O family which includes acetaldehyde (CH_3CHO), ethylene oxide (c-C_2H_4O), and vinyl alcohol (CH_2CHOH). While $HCOOCH_3$ and CH_3CHO were discovered in the 1970s, the remaining species (CH_3COOH, c-C_2H_4O, $HOCH_2CHO$, and CH_2CHOH) were identified only fairly recently, all toward Sgr B2 (Mehringer *et al.* 1997; Dickens *et al.* 1997; Hollis, Lovas, & Jewell 2000; Turner & Apponi 2001). For the $C_2H_4O_2$ family, observations of $HOCH_2CHO$ with the BIMA array showed, surprisingly, that it existed largely in an extended region toward Sgr B2(N). This led Hollis *et al.* (2001) to suggest the reactive nature of sugars may account for the $HOCH_2CHO$ deficiency in the Sgr B2(N) hot core (see also Hollis, this volume).

To directly probe the spatial distribution of the C_2H_4O isomeric triplet, observations of these species were carried out toward Sgr B2(N) and Orion KL using the BIMA array (Liu *et al.* 2005b). Sgr B2(N-LMH) is a uniquely rich hot core region where many complex species, including c-C_2H_4O and CH_2CHOH, were first identified. As evident in Figure 4, emission of c-C_2H_4O toward Sgr B2(N) is largely resolved out, indicating that

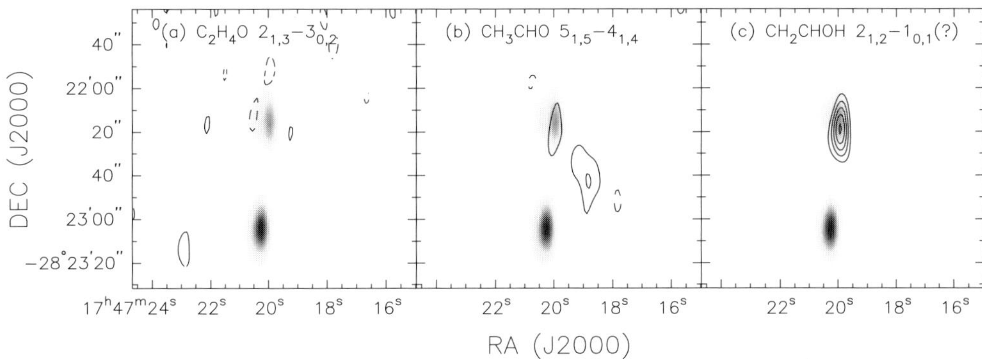

Figure 4. The grayscale in all three panels represents 3 mm continuum emission toward Sgr B2. (*a*) Contours represent c-C$_2$H$_4$O emission. (*b*) Contours represent CH$_3$CHO emission. (*c*) Contours represent possible CH$_2$CHOH emission.

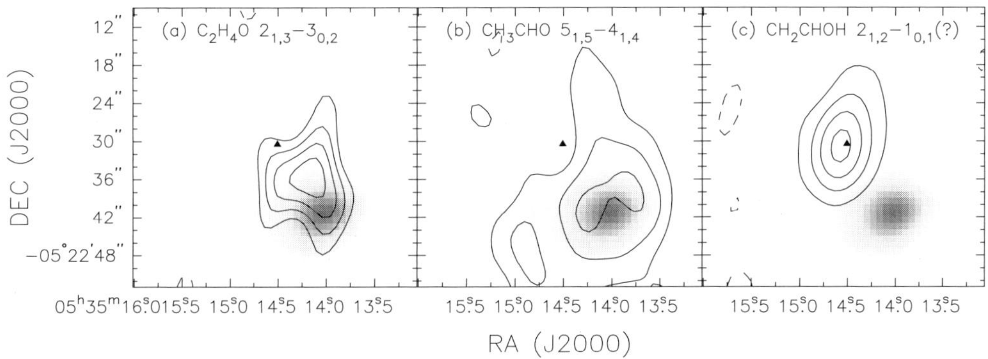

Figure 5. The grayscale in all three panels represents HCOOH emission toward Orion KL (Liu *et al.* 2001). (*a*) Contours represent c-C$_2$H$_4$O emission. (*b*) Contours represent CH$_3$CHO emission. (*c*) Contours represent possible CH$_2$CHOH emission.

its distribution is rather extended as was suggested by Dickens *et al.* (1997) based on their observed line-width *vs.* beam-size relation. CH$_3$CHO emission is found to arise from Sgr B2(N), extending toward the south-west. These features are in good agreement with the strongest part of the c-C$_2$H$_4$O $1_{1,1}$—$1_{1,0}$ emission imaged by Chengalur & Kanekar (2003) with the GMRT. Given that its relative abundance to H$_2$CO is considerably smaller than the ratio of 10^4 as expected from pure gas-phase models for the production of these molecules (Lee *et al.* 1996), Chengalur & Kanekar suggested that grain chemistry is likely to be important for the enhanced CH$_3$CHO; its wide-spread distribution is probably related to the presence of numerous shocks in Sgr B2. This is in accordance with the large number of hot expanding shells produced by massive stars in the galactic center region found by Martín-Pintado *et al.* (1999) and the the quasi-thermal 44 GHz methanol emission also likely arising from shocked gas at the boundaries of expanding ionized shells around the young stars, as suggested by Mehringer & Menten (1997).

Orion KL is the nearest archetypical hot core source, with the detection of c-C$_2$H$_4$O recently reported by Ikeda *et al.* (2001). Our interferometric observations with BIMA reveal that the c-C$_2$H$_4$O emission peaks are offset from the HCOOH emission feature, but rather coincident with the HCOOCH$_3$ and (CH$_3$)$_2$O emission peaks as shown in Figure 6 (similar to Fig. 3 but with different HCOOCH$_3$ and (CH$_3$)$_2$O transitions and at a lower angular resolution). In contrast, CH$_3$CHO emission is mostly located toward the HCOOH emission peak. The spatial coincidence of the CH$_3$CHO emission with the

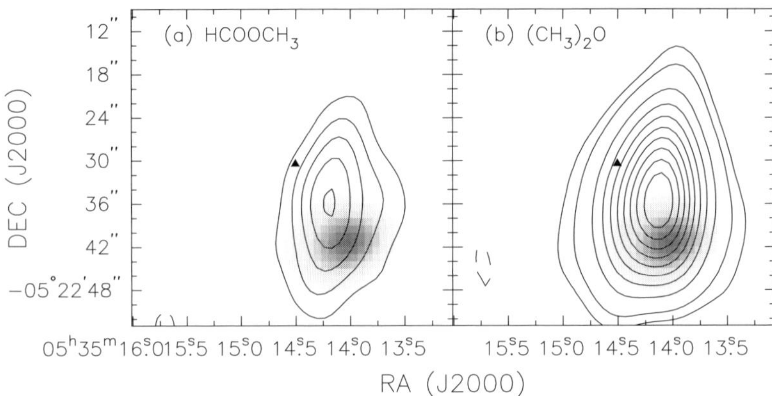

Figure 6. As in Figure 5, the grayscale in both panels depicts the HCOOH emission toward Orion KL mapped by Liu *et al.* (2001). (*a*) Contours represent HCOOCH$_3$ emission. (*b*) Contours represent (CH$_3$)$_2$O emission.

HCOOH emission and the c-C$_2$H$_4$O distribution with the HCOOCH$_3$ as well as (CH$_3$)$_2$O strongly suggests that CH$_3$CHO is directly released into the gas phase from mantle evaporation like HCOOH , while c-C$_2$H$_4$O is more likely a secondary species formed by warm gas phase reactions. Charnley (2004) suggested that pure reduction/hydrogenation reactions of CO primarily synthesize CH$_3$CHO in mantle environments while reducing-oxidizing reactions involving acetylene on grain surface are possible pathways for the formation of the C$_2$H$_4$O triplet members. If CH$_3$CHO is directly evaporated from gain mantles while c-C$_2$H$_4$O is a secondary species formed in the gas phase, as suggested by their spatial distribution, this evidence implies that pure reduction/hydrogenation is perhaps the dominate surface process for molecular synthesis on grain surfaces.

4. The Chemical Differentiation in HMCs

Past observations have often found chemical differentiation within/among star forming cores. One example is Orion KL. It has long been recognized that considerable differences in the chemical compositions exist between different core components, more specifically, the so called "hot core" and the "compact ridge" (Blake *et al.* 1987). The "hot core" is pronounced in complex and saturated N-bearing molecules, such as NH$_3$, C$_2$H$_5$CN, and CH$_3$CN; while the complex O-bearing molecules characterized by CH$_3$OH, HCOOCH$_3$, and (CH$_3$)$_2$O are most abundant in the nearby "compact ridge". High angular resolution observations toward the W3(OH)/W3(H$_2$O) cores also revealed similar chemical gradients (Wyrowski *et al.* 1999).

G9.62 + 0.19, located at a distance of 5.7 kpc, is a high-mass star forming region associated with a luminous far IR source IRAS 18032–2032 (3×10^5 L$_\odot$) that contains a cluster of H II regions, probably at different evolutionary stages. Multi-wavelength VLA observations identified nine radio continuum sources (denoted A–I) in this region (Garay *et al.* 1993; Testi *et al.* 2000). A and B have more extended ($\sim$15–30$''$) morphologies, indicating that they are probably more evolved. C–I are compact sources with angular sizes <5$''$ lying along a SE-NW ridge where a chain of H$_2$O maser spots were detected by Hofner & Churchwell (1996). As revealed in NH$_3$ (4,4), (5,5) and CH$_3$CN ($J = 6$–5), component F is a HMC and hence likely the youngest source in the region (Cesaroni *et al.* 1994; Hofner *et al.* 1996). Our observations with the OVRO Millimeter Array and the Submillimeter Array (SMA) show that, while CH$_3$CN emission is strongest toward

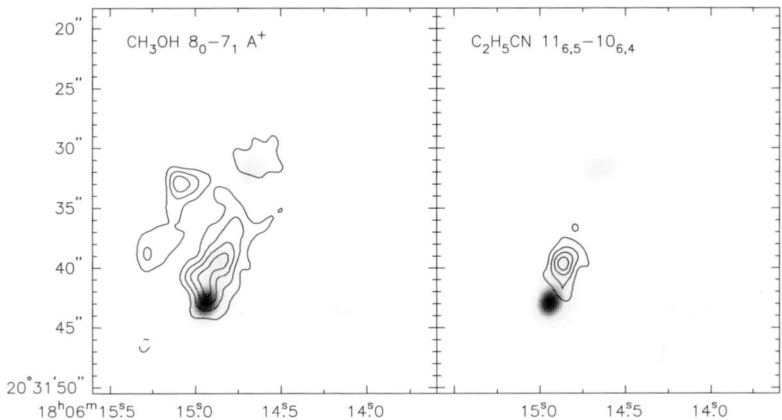

Figure 7. The grayscale in both panel represents 3 mm continuum emission from G9.62 + 0.19. (*a*) Contours represent integrated CH_3OH 8_0–7_1 A^+ line emission. (*b*) Contours represent integrated C_2H_5CN $11_{6,5}$–$10_{6,4}$ line emission.

the HMC-F source, $HCOOCH_3$ and $(CH_3)_2O$ are brightest in the H II region E (see Fig. 7 and 8). The E/F pair in G9.62 + 0.19 thus appears similar to the $W3(OH)/W3(H_2O)$ system, where a hot core is dominated by N-bearing species and a nearly H II region is pronounced mainly in emission from O-bearing species. Interestingly, such chemical differentiation is not exclusively found in massive star forming regions, such as those shown by SMA observations toward G9.62 + 0.19 (Su *et al.* 2005) and Orion KL (Beuther *et al.* 2005). Recent results from SMA also revealed chemical differentiation in the protobinary low-mass protostar IRAS 16293-2422 (Kuan *et al.* 2004).

Theoretic models for the origin of such chemical differentiation have been proposed by, for example, Casille, Hasegawa, & Herbst (1993), Rodgers & Charnley (2001), and recently by Rodgers & Charnley (2003). These models suggest that the thermal/evolutionary history of cloud cores may modify grain mantle composition or govern mantle desorption processes differently in various parts of the cloud, and hence result in distinct chemical signatures as a function of location. G9.62 + 0.19, being a region readily displaying chemical differentiation features among various cores/H II regions, warrants future high sensitivity/resolution observations for further verification of these proposed chemical models.

5. Conclusions

With the results from studies of HCOOH and CH_3COOH, as well as those of the C_2H_4O isomeric triplet, we demonstrate that interferometric observations which spatially resolve molecular emission are powerful tools for distinguishing their formation processes. We also show that G9.62 + 0.19 provides an example of chemical differentiation signatures that are possibly related to evolutionary effects.

Future interferometric arrays such as the Atacama Large Millimeter Array (ALMA) with wider bandwidths and more collecting area will be able to carry out similar but much more sensitive, multi-transition observations for diagnosing the chemical structures of molecular cloud cores. We stress that more laboratory measurements and theorctical calculations of molecular rest frequencies are strongly desired for identifying the numerous weak spectral features expected. More sophisticated but user friendly radiative transfer modeling and chemical schemes are also required for better interpreting the vast amount of observational data anticipated.

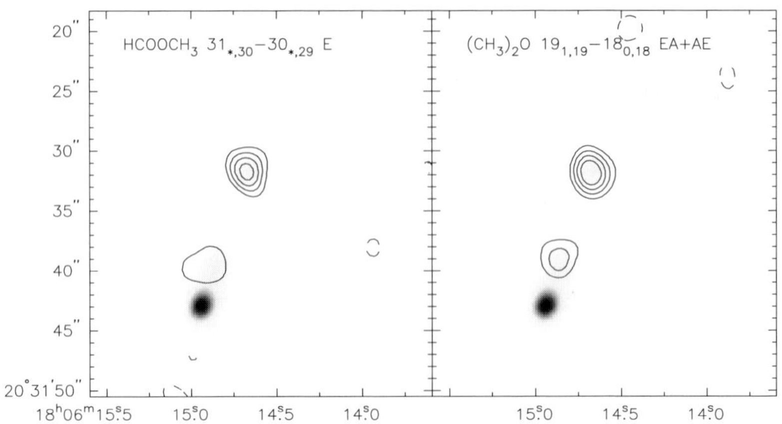

Figure 8. As in Figure 7, the grayscale in both panel represents 3 mm continuum emission from G9.62 + 0.19. (a) Contours represent integrated HCOOCH$_3$ $31_{*,30}$–$30_{*,29}$ line emission. (b) Contours represent integrated (CH$_3$)$_2$O $19_{1,19}$–$18_{0,18}$ EA + AE line emission.

Acknowledgements

S.-Y. Liu is grateful to the Scientific Organizing Committee for the invitation of make this presentation. Many thanks also go to L.E. Snyder, D. Mehringer, A. Remijan, G.A. Blake, Y.-J. Kuan, and Y.-N. Su for collaboration on the studies presented here.

References

Allamandola, L.J. & Sandford, S.A. 1990, *NASA Conf. Pub.* 3061, 113

Blake, G.A., Sutton, E.C., Masson, C.R., & Phillips, T.G. 1987, *Ap. J.* 315, 621

Beuther, H., *et al.* 2005, *Ap. J.*, in press

Boechat-Roberty, H.M., Pilling, S., & Santos, A.C.F. 2005, *A&A* 438, 915

Caselli, P., Hasegawa, T.I., & Herbst, E. 1993, *Ap. J.* 408, 548

Cesaroni, R., Churchwell, E., Hofner, P., Walmsley, C.M., & Kurtz, S. 1994, *A&A* 288, 903

Charnley, S.B. 1995, *Ap&SS*, 224, 251

Charnley, S.B. 2004, *ASR*, 33, 23

Chengalur, J.N. & Kanekar, N. 2003, *A&A* 403, L43

Crovisier, J. & Bockelée-Morvan, D. 1999, *Space Scien. Rev.*, 90, 19

Dickens, J.E., Irvine, W.M., Ohishi, M., Ikeda, M., Ishikawa, S., Nummelin, A., & Hjalmarson, Å. 1997, *Ap. J.* 489, 753

Ehrenfreund, P., D'Hendecourt, L., Charnley, S., & Ruiterkamp, R. 2001, *JGR*, 106, 33291

Garay, G., Rodriguez, L.F., Moran, J.M., & Churchwell, E. 1993, *Ap. J.* 418, 368

Hollis, J.M., Lovas, F.J., & Jewell, P.R. 2000, *Ap. J.* 540, L107

Hollis, J.M., Vogel, S.N., Snyder, L.E., Jewell, P.R., & Lovas, F.J. 2001, *Ap. J.* 554, L81

Hofner, P., Kurtz, S., Churchwell, E., Walmsley, C.M., & Cesaroni, R. 1996, *Ap. J.* 460, 359

Ikeda, M., Ohishi, M., Nummelin, A., Dickens, J.E., Bergman, p., Hjalmarson, Å., & Irvine, W.M. 2001, *Ap. J.* 560, 792

Irvine, W.M., Friberg, P., Kaifu, N., Matthews, H.E., Minh, Y.C., Ohishi,M., & Ishikawa, S. 1990, *A&A* 229, L9

Kaufman, M.J. & Neuleld, D.A. 1996, *Ap. J.* 456, L611

Kurtz, S., Cesaroni, R., Churchwell, E., Hofner, P., & Walmsley, C.M. 2000, *in Ptostars and Planets IV*, ed. V. Manning, A.P. Boss, & S. Russell (Tucson: Univ. Arizona Press), 299

Kuan, Y.-J., *et al.* 2004, *Ap. J.* 616, L27

Lee, H.-H., Bettens, R.P.A., & Herbest, E. 1996, *A&AS*, 119, 111

Liu, S.-Y., Mehringer, D.M., & Snyder, L.E. 2001, *Ap. J.* 552, 654

Liu, S.-Y., Girart, J.M., Remijan, A., & Snyder, L.E. 2002, *Ap. J.* 576, 255

Liu, S.-Y., *et al.* 2005a, in preparation

Liu, S.-Y., *et al.* 2005b, in preparation

Martín-Pintado, J., Gaume, R.A., Rodríguez-Fernández, N., de Vicente, P., & Wilson, T.L. 1999, *Ap. J.* 519, 667

Mehringer, D.M. & Menten, K.M. 1997, *Ap. J.* 474, 346

Mehringer, D.M. & Snyder, L.E. 1996, *Ap. J.* 471, 897

Mehringer, D.M., Snyder, L.E., Miao, Y., & Lovas, F.J. 1997, *Ap. J.* 480, L71

Miao, Y., Mehringer, D.M., Kuan, Y.-J., & Snyder, L.E. 1995, *Ap. J.* 445, L59

Millar, T.J., Olofsson, H., Hjalmarson, Å., & Brown, R.D. 1988, *A&A* 205, L5

Pineau des Forêts, G., Roueff, E., Schilke, P., & Flower, D.R. 1993, *MNRAS*, 262, 915

Remijan, A.J., Liu, S.-Y., Snyder, L.E., Mehringer, D.M., & Kuan, Y.J. 2002, *Ap. J.* 576, 264

Remijan, A., Shiao, Y.-S., Fridel, D.N., Meier, D.S., & Snyder, L.E. 2004, *Ap. J.* 617, 384

Remijan, A., Snyder, L.E., Friedel, D.N., Liu, S.-Y., & Shah, R.Y. 2003, *Ap. J.* 590, 314

Rodgers, S.D. & Charnley, S.B. 2001, *Ap. J.* 546, 324

Rodgers, S.D. & Charnley, S.B. 2003, *Ap. J.* 585, 355

Schilke, P. Groesbeck, T.D., Blake, G.A., & Phillips, T.G. 1997, *Ap. J.S*, 108, 301

Schutte, W.A., *et al.* 1996, *A&A* 315, L333

Schutte, W.A., *et al.* 1999, *A&A* 343, 966

Su, Y.-N., *et al.* 2005, in preparation

Testi, L., Hofner, P., Kurtz, S., & Rupen, M. 2000, *A&A* 359, L5

Tielens, A.G.G. & Hagen, W. 1982, *A&A* 114, 245

Turner, B.E. 1991, *Ap. J. Suppl.* 76, 617

Turner, B.E. & Apponi, A.J. 2001, *Ap. J.* 561, L207

Turner, B.E., Terzieva, R., & Herbst, E. 1999, *Ap. J.* 518, 699

Winnewisser & Churchwell 1975, *Ap. J.* 200, L33

Wyrowski, F. Schilke, P., Walmsley, C.M., & Menten, K.M. 1999, *Ap. J.* 514, L43

Ziurys, L.M. & McGonagle, D. 1993, *Ap. J. Suppl.* 89, 155

Zuckermann, B., Ball, J.A., & Gottlieb, A.C. 1971, *Ap. J.* 163, L41

Discussion

HERBST: How does interferometry help identify new complex molecules?

LIU: Interferometers provide both spectral as well as spatial information on line features. They therefore assist in new line identification only if all potential spectral features match both in frequency (velocity) and in their spatial distribution (existing in the same volume) with the correct corresponding velocity. Interferometers additionally filter out spectral line originated from extended structures which may otherwise blank or confuse weak target lines in single-dish telescope surveys.

VAN DISHOECK: How much do the column densities and—in particular—the column density/abundance *ratios* of complex molecules change when derived from interferometer data compared with single-dish observations (in your experience)?

LIU: Single-dish observations provide only abundances and (abundance) ratios averaged over the beam. Depending on the source structure, the molecular column densities can be modified (typically increased) by large factors when observed with interferometers. The relative abundance ratios between different species can also change drastically, on occassion by orders of magnitude. This is particularly true if there is a spatial differentiation between species, such as is the case for c-C_2H_4O/CH_3CHO toward Orion, as we demonstrate here.

Photo: E. Herbst

Astrochemistry: Recent Successes and Current Challenges
Proceedings IAU Symposium No. 231, 2005
D.C. Lis, G.A. Blake & E. Herbst, eds.

Complex Molecules and the GBT:
Is Isomerism the Key?

J. M. Hollis

NASA Goddard Space Flight Center, Greenbelt, MD 20771, USA
email: mhollis@milkyway.gsfc.nasa.gov

Abstract. Interstellar aldehydes have been called the "sugars of space" ever since the discoveries of formaldehyde (H_2CO) in 1969 and acetaldehyde (CH_3CHO) in 1973. At present, more than 135 interstellar molecular species have been identified. Excluding diatomic species, 30% of all interstellar molecules have isomeric counterparts. The newest instrument in the interstellar molecule search arsenal is the Green Bank Telescope (GBT), which is credited with the discovery of the large aldehydes propenal (CH_2CHCHO) and propanal (CH_3CH_2CHO). In addition, the GBT has been used to observe interstellar glycolaldehyde (CH_2OHCHO), the simplest possible aldehyde sugar, and interstellar ethylene glycol ($HOCH_2CH_2OH$), the sugar alcohol of glycolaldehyde. These new GBT observations that suggest a universal prebiotic chemistry are presented and discussed. While there is no consensus regarding how large complex interstellar molecules are formed, it may be that the first step in the polymerization of interstellar formaldehyde (H_2CO) with its isomer trans hydroxy methylene (t-HCOH) is responsible for the formation of interstellar glycolaldehyde. We discuss this polymerization mechanism that can result in the generation of more complex sugars. We assess the likelihood of interstellar trans hydroxy methylene and suggest a search strategy for it using the GBT.

Keywords. ISM: abundances — ISM: clouds — ISM: individual (Sagittarius B2(N-LMH)) — ISM: molecules — radio lines: ISM

---

1. Introduction

The giant molecular cloud complex Sgr B2 near the center of our Galaxy is undoubtedly the preeminent source for the study of large complex interstellar molecules. The Sgr B2 complex is a star-forming region containing compact hot molecular cores of arcsecond dimensions, molecular maser emitting regions, and ultracompact continuum sources surrounded by larger-scale continuum features as well as molecular material extended on the order of arcminutes. In addition, small-scale and large-scale shock phenomenon characterize the complex. In particular, the hot molecular core known as the Large Molecule Heimat (LMH) has for the last ten years been the first source searched to detect and identify new large interstellar molecules since many of the large organic species have been previously found confined to its $\sim 5''$ diameter. For example, Figure 1 is a VLA image of the ethyl cyanide (CH_3CH_2CN), a well-documented LMH molecule.

These results are consistent with conventional wisdom persisting for decades that the interesting complex molecules would be found and studied with high-resolution arrays able to couple to more compact molecular sources. In fact, this view was a driver in the design of future mm-wave arrays. In 2001 the picture radically changed for one large interstellar molecule. Having recently identified glycolaldehyde (CH_2OHCHO) with the NRAO 12-meter telescope (Hollis, Lovas, & Jewell 2000), an attempt was made to map one of the same mm-wave transitions with the BIMA Array (Hollis *et al.* 2001). The interferometer was unable to recover the flux measured by the single antenna, thereby indicating that glycolaldehyde toward the LMH has to be extended on a scale of at least

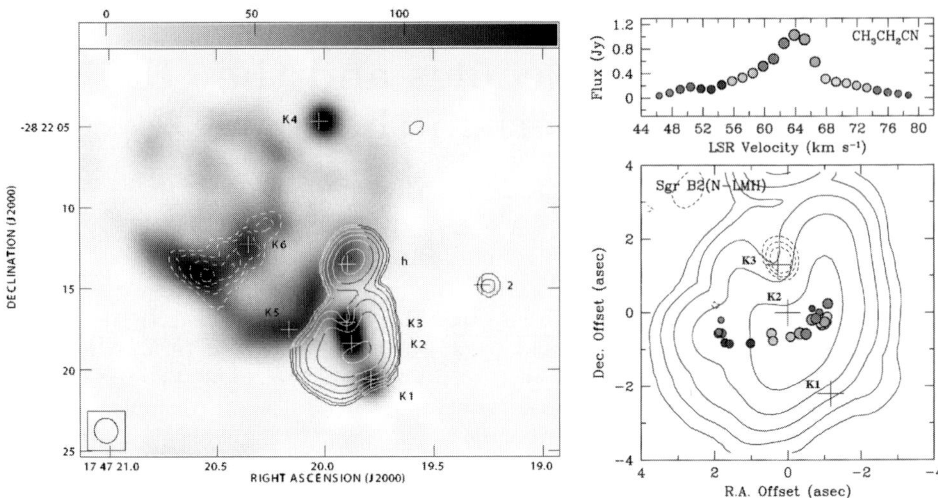

Figure 1. VLA images of the 5_{15}–4_{14} transition of ethyl cyanide (Hollis *et al.* 2003). *Left:* Contours of the integrated intensity Sgr B2(N-LMH) overlaid on a grey-scale image of 43.5 GHz continuum emission. *Right:* LSR velocity structure of the LMH. Centroid emission from each velocity channel shows a spatio-velocity gradient consistent with a rotating molecular disk seen approximately edge-on. This plot is color-coded in Hollis *et al.* (2003).

an arcminute. Subsequently, acetaldehyde (CH_3CHO) was imaged with the GMRT (see Figure 2), showing that another large aldehyde has a widespread spatial distribution of several arcminutes toward Sgr B2 (Chengalur & Kanekar 2003).

The Green Bank Telescope (GBT) may well be the ideal instrument for capitalizing on the spectral identification of both compact and spatially extended complex interstellar molecules toward Sgr B2. For example, the half-power beamwidth of the GBT is $740''/\nu(\mathrm{GHz})$. The GBT beam size will nearly match that of the LMH when observing in the vicinity 70–100 GHz, a spectral region characterized by a large number of rotational transitions of hot core molecules. Conversely, at lower frequencies <50 GHz, the GBT samples the colder gas of spatially extended molecules such as the low temperature transitions of the larger aldehydes. It is chiefly in this latter context that our experiences with the GBT as a cold molecular gas telescope will be discussed.

2. Recent Spectral Line Observations with the GBT

With a collecting area of 7854 m^2 and a mass of 7.6 million kg, the GBT is the largest and heaviest movable structure on land in the world. For spectral line searches at Ku- (12–15.4 GHz) and K-bands (18–22.5 GHz) discussed here, we define a "track" as 6 hrs, the amount of time that the GBT is able to continuously track Sgr B2 for elevations above 10°. The GBT was pointed at the Sgr B2(N-LMH) position which is coincident with the K2 ultracompact continuum source. The spectrometer is typically configured into four 200 MHz bands in two polarizations which yield a spectral resolution of 24.4 kHz. A scan consisted of observing OFF the source 60′ east for two minutes followed by observing ON the source for two minutes. These short scans were performed to minimize spectra baseline fluctuations. Figure 3 shows two raw spectral bandpasses representing the accumulation of scans for one track at Ku- and K-bands. In the presence of continuum emission, the instrumental slopes in a bandpass can be quite significant but can be reliably

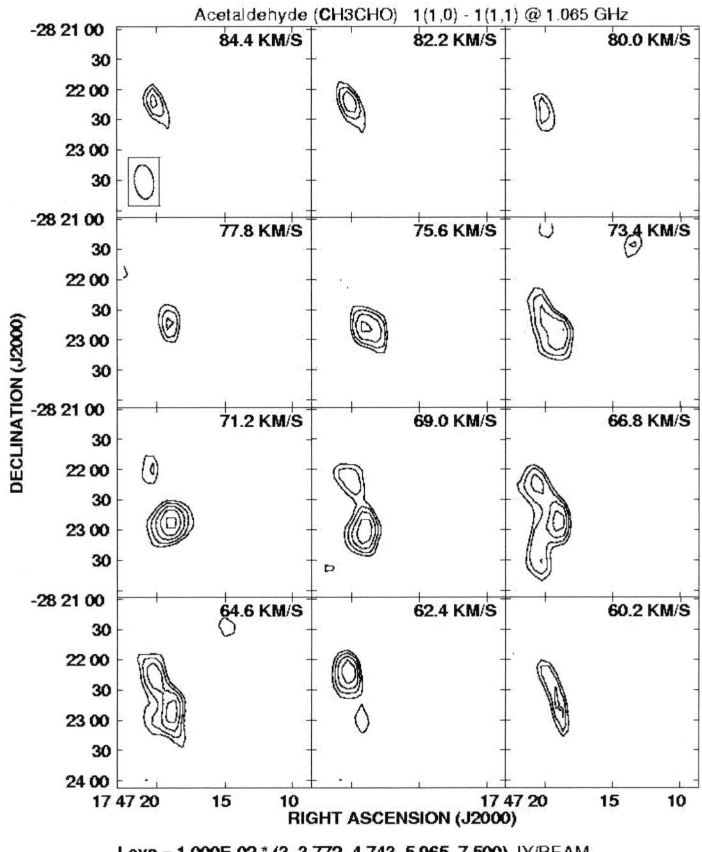

Figure 2. GMRT velocity channel contour images of the 1_{10}–1_{11} transition of acetaldehyde toward Sgr B2 (courtesy of J.N. Chengalur and N. Kanekar).

removed with a median filtering technique (G. Langston 2005, in preparation), taking care not to distort spectral line shapes. As can be seen, recombination lines are the dominant spectral features at the Ku- and K-bands and molecular spectral lines are few when compared to the mm-wave range.

2.1. *The Inverse P Cygni Profile of N-15 Ammonia*

The 6_5–6_5 transition of N-15 ammonia ($^{15}NH_3$) in Figure 4 is heretofore an unpublished GBT spectrum displaying a classic inverse P Cygni profile which is representative of other transitions of this molecule toward Sgr B2(N). The 6_5–6_5 transition of $^{15}NH_3$ has a lower energy level 450 K above ground, indicative of hot molecular gas. We interpret the inverse P Cygni spectra of $^{15}NH_3$ transitions as critical evidence for the collapse of the molecular cloud in proximity to the LMH. As Figure 4 shows, the absorption is clearly red-shifted relative to the LMH systemic velocity of $+64$ km s^{-1}, whereas the blue-shifted gas is seen entirely in emission. The spectra show that the absorbed gas on the near side of the continuum source and the emission gas on far side are both moving toward the continuum source. Moreover, the molecular ring has a characteristic LSR systemic velocity of $+64$ km s^{-1} as well, indicating that the collapse is probably towards the molecular ring (see Figure 1). Thus, the simplest interpretation of the $^{15}NH_3$

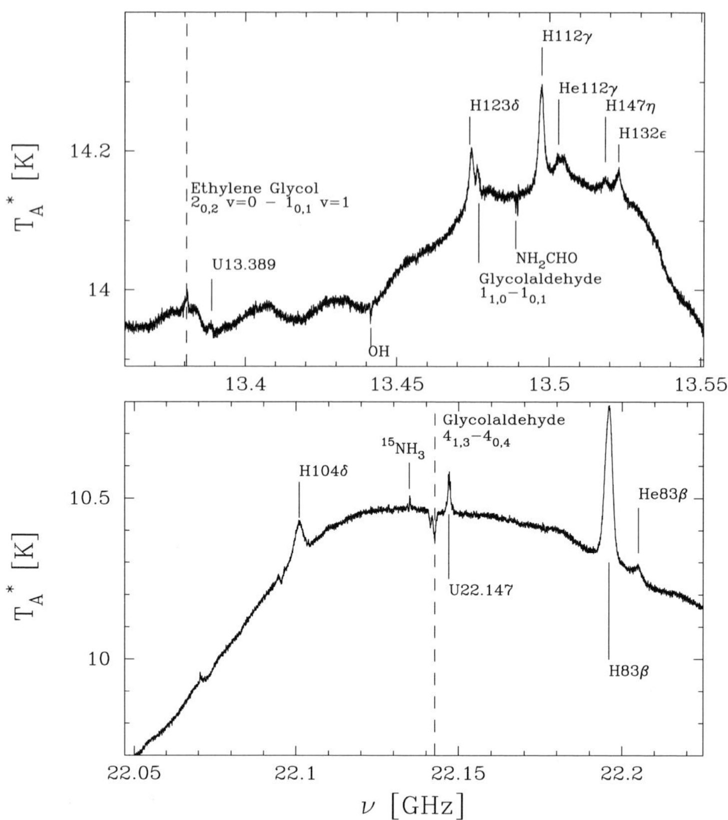

Figure 3. Typical raw GBT spectral bandpasses at Ku- and K-bands.

observations is that the ambient molecular cloud is in free-fall toward the molecular ring which contains a number of continuum sources. Observations of the normal isotope of NH_3 at high temperature have previously been interpreted as evidence for shock-heated gas toward Sgr B2(N) as would be expected for a collapsing molecular cloud (*e.g.*, Ceccarelli *et al.* 2002; Flower *et al.* 1995). On the other hand, our GBT observations of the normal isotopic form of ammonia did not show inverse P Cygni profiles, but, rather, were dominated by two simple emission features at LSR velocities of $+64$ and $+82$ $\mathrm{km\,s^{-1}}$. These strong components at $+82$ and $+64$ $\mathrm{km\,s^{-1}}$ are associated with star-forming cores that may have been triggered by collision between two molecular clouds with these systemic velocities as suggested by Mehringer & Menten (1997).

2.2. *Glycolaldehyde and Ethylene Glycol*

Glycolaldehyde (CH_2OHCHO), the simplest possible aldehyde sugar, was detected with the GBT toward Sgr B2(N) by means of the 1_{10}–1_{01} , 2_{11}–2_{02}, 3_{12}–3_{03}, and 4_{13}–4_{04} rotational transitions at 13.48, 15.18, 17.98, and 22.14 GHz, respectively (see Figure 5, and Hollis *et al.* 2004a). An analysis of these four rotational transitions yields a glycolaldehyde state temperature of $\sim$8 K. Previously reported emission line detections of glycolaldehyde with the NRAO 12-m telescope at mm-wavelengths (71 GHz to 103 GHz) are characterized by a state temperature of $\sim$50 K. By comparison, the GBT detections are surprisingly strong and seen in emission at 13.48 GHz, emission and absorption at 15.18 GHz, and absorption at 17.98 GHz and 22.14 GHz. The GBT detects glycolaldehyde absorption primarily at an LSR velocity of $+64$ $\mathrm{km\,s^{-1}}$ with a possible weaker component

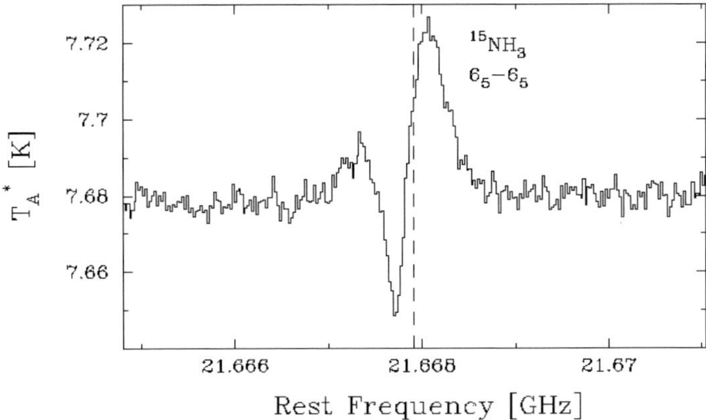

Figure 4. The inverse P Cygni profile of Nitrogen-15 ammonia observed with the GBT. The dashed line indicates an LSR velocity of $+64$ km s^{-1}.

at $+82$ km s^{-1}; any emission components detected are centered near $+73$ km s^{-1}. On the other hand, glycolaldehyde emission observed with the NRAO 12-m telescope (Hollis, Lovas, & Jewell 2000) is found to have an LSR velocity of $\sim +71$ km s^{-1} (see Hollis *et al.* 2002). Emission components with LSR velocities of $+71$ or $+73$ km s^{-1} suggest the emission comes from a cool extended cloud that surrounds the two star forming cores characterized by LSR velocities of $+64$ and $+82$ km s^{-1}. For the $+64$ km s^{-1} LSR velocity component, we believe that the hot ammonia is in proximity to the colder glycolaldehyde, suggesting that the $^{15}NH_3$ molecule characterizes the infall shock front while the glycolaldehyde represents much older, colder, post-shock gas.

Figure 6 shows two heretofore unpublished GBT spectra of ethylene glycol ($HOCH_2CH_2OH$), the reduced sugar alcohol of glycolaldehyde. Since both glycolaldehyde and ethylene glycol are easily seen with the GBT in a spectral region that is not dominated by molecular emission, it suggests that these two molecules are spatially and chemically related. This suggests that successive hydrogen addition could account for production of ethylene glycol from glycolaldehyde on the surface of interstellar grains:

$$CH_2OHCHO \xrightarrow{2H} HOCH_2CH_2OH \tag{2.1}$$

2.3. *Propynal, Propenal, and Propanal*

As a consequence that hydrogen addition appears important in creating successively larger interstellar molecules, we used the GBT to search for the molecular aldehyde hydrogen-addition sequence of propynal (HC_2CHO), propenal (CH_2CHCHO), and propanal (CH_3CH_2CHO) shown here:

$$HC_2CHO \xrightarrow{2H} CH_2CHCHO \xrightarrow{2H} CH_3CH_2CHO \tag{2.2}$$

Propynal had previously been identified as an interstellar molecule (Irvine *et al.* 1988; Turner 1991), and the GBT detected this species, as well as the new interstellar aldehydes propenal shown in Figure 7 and propanal shown in Figure 8. It is noteworthy that the line intensities of the propynal, propenal, and propanal sequence decrease. Each propanal spectrum in Figure 8 represents the accumulation of 2 full tracks. LSR velocity components for propenal and propanal are similar to those of glycolaldehyde.

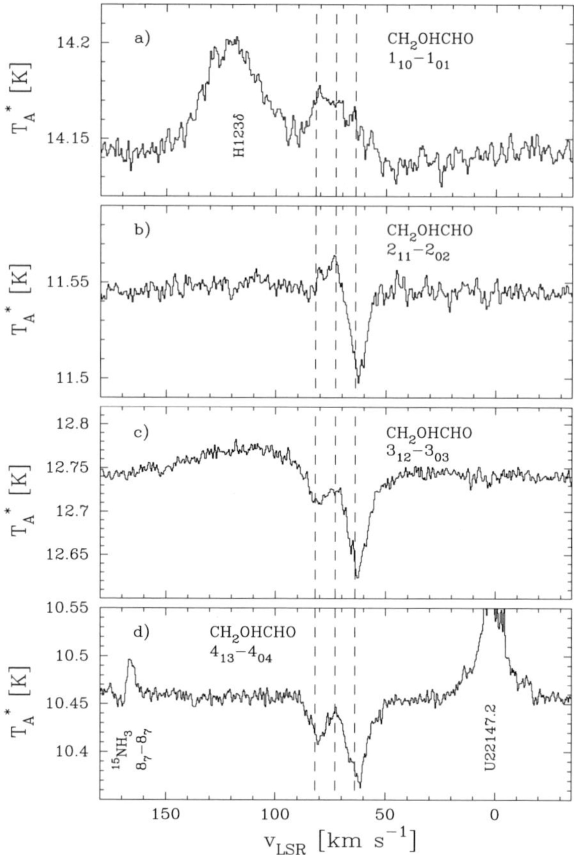

Figure 5. Glycoladehyde GBT spectra (Hollis *et al.* 2004a). Fiducial LSR velocities are +82, +73, and +64 km s^{-1} in this and following figures.

3. Implications of Complex Interstellar Molecules

Although the chemistry on Earth and in interstellar clouds is much different, the results can be very similar. For example, laboratory experiments in aqueous solution that start with mixtures of methane (CH_4), ammonia (NH_3), and molecular hydrogen (H_2) have been able to synthesize amino acids (Strecker synthesis). These experiments purport to simulate conditions of the early Earth and result in substantial quantities of hydrogen cyanide (HCN) and a number of aldehydes in the initial stage and the concentration of these intermediate products fall off as the experiment proceeds. Our GBT experiments confirm the presence of aldehydes that may be associated with a similar process in interstellar clouds. It could be that prebiotic chemistry—the formation of the molecular building blocks essential to life—occurs in an interstellar cloud long before the collapse to form a new solar system with planets. Since many of the interstellar molecules discovered to date are the same kinds detected in laboratory experiments specifically designed to synthesize prebiotic molecules, this suggests a universal prebiotic chemistry. Moreover, there appears to be a high degree of isomerism among interstellar molecules. Of all interstellar molecules that could have isomers (*i.e.*, molecules containing 3 or more atoms), 30% have isomeric counterparts. Additionally, it appears that isomerism may favor the more complex species. For example, of the nine interstellar molecules comprised of 8 atoms, acetic acid (CH_3COOH), methyl formate (CH_3OCHO), and glycolaldehyde

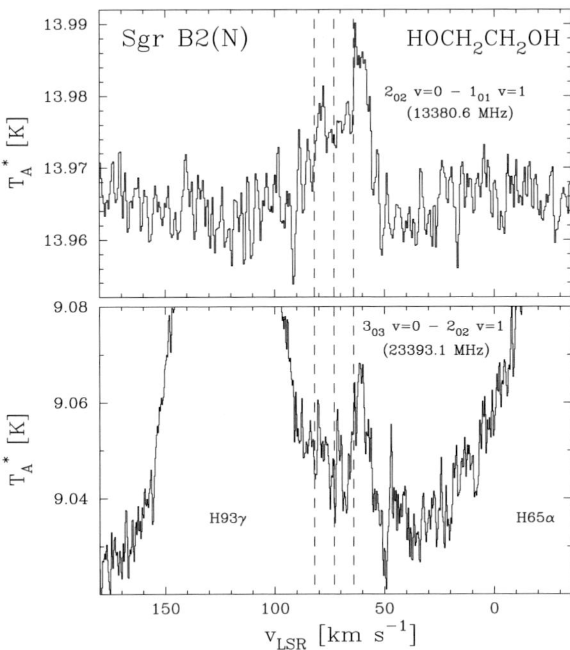

Figure 6. Ethylene glycol GBT spectra.

(CH$_2$OHCHO) are isomers, and hexapentaenylidene (H$_2$C$_6$) and triacetylene (HC$_6$H)
are isomers. Thus, the role of isomerism seems to be an important clue in the study of
complex interstellar molecules.

4. Formaldehyde Polymerization May Explain Sugars

The discoveries of interstellar formaldehyde (H$_2$CO) in 1969 (Snyder *et al.* 1969) and
interstellar glycolaldehyde(CH$_2$OHCHO) in 2000 (Hollis, Lovas, & Jewell 2000) suggest
that the polymerization of two formaldehyde molecules may produce glycolaldehyde.
However, the exact details as to how the polymerization occurs has not been amenable to
experimental verification. For example, it has been long thought that prebiotic synthesis
of sugars on the early Earth occurs via the so-called formose reaction which is an aqueous-
catalyzed polymerization of formaldehyde. The formose reaction itself is shrouded in
some mystery as to exactly how the bond-breaking/bond-making proceeds to produce a
molecule of glycolaldehyde from two formaldehyde molecules. One theoretical possibility
is that hydroxy methylene (HCOH—a formaldehyde isomer of unusual reactivity) is first
formed, which can more readily combine with formaldehyde to produce glycolaldehyde.
In such a scenario hydroxy methylene would be an intermediate product that would be
difficult if not impossible to observe in the laboratory. The first step in the polymerization
reaction is depicted in Figure 9.

The formation of glycolaldehyde in Figure 9 occurs because the highly reactive trans
hydroxy methylene divalent carbon adds to the formaldehyde carbon which causes the
formaldehyde oxygen to accept the trans hydroxy methylene hydrogen and the double
C=O bond characteristic of glycolaldehyde is created. This reaction is essentially instantaneous. Trans hydroxy methylene can produce 3-carbon glyceraldehyde by adding to
glycolaldehyde; similarly, more complex sugars can be synthesized. Strong experimental evidence for this polymerization mechanism was provided by Flanagan, Ahmed, &

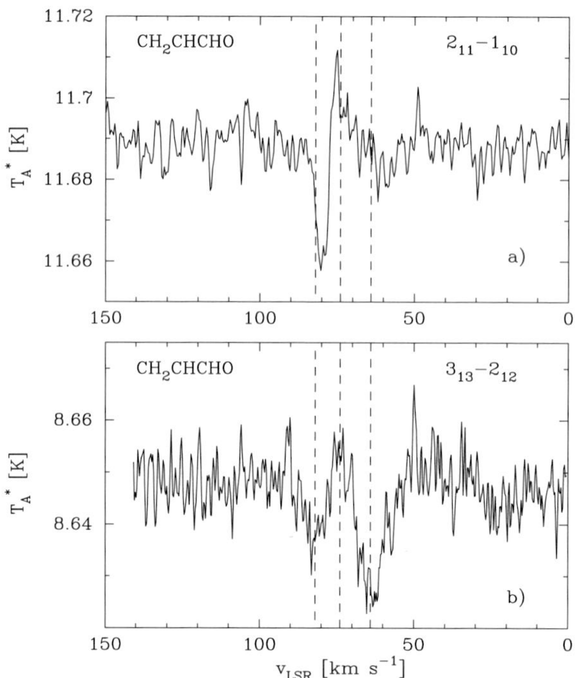

Figure 7. Propenal GBT spectra (Hollis *et al.* 2004b).

Shevlin (1992) in a laboratory study in which condensation of atomic carbon with water at 77 K generated a mixture of aldehyde sugars that included glycolaldehyde as a principal product. It is noteworthy that this experiment was conducted under conditions that are somewhat similar to those that exist in hot cores of interstellar clouds.

We next explore the likelihood that interstellar hydroxy methylene exists. For that we must consider the interstellar chemistry that produces formaldehyde itself. The production of interstellar formaldehyde is widely accepted to be produced primarily in the gas phase by the following reaction:

$$O + CH_3 \longrightarrow H_2CO + H \tag{4.1}$$

Succeeding reactions proceed by classic ion-molecule chemistry. Gas-phase formaldehyde is depleted to protonated formaldehyde, a known interstellar molecule, by the following reaction:

$$H_3^+ + H_2CO \longrightarrow H_2COH^+ + H_2 \tag{4.2}$$

This gas-phase protonated formaldehyde combines with an electron to again form formaldehyde and two conformers of hydroxy methylene (Hoffmann & Schaefer 1981):

$$e^- + H_2COH^+ \longrightarrow \text{trans } HCOH + H \tag{4.3a}$$

$$e^- + H_2COH^+ \longrightarrow \text{cis } HCOH + H \tag{4.3b}$$

$$e^- + H_2COH^+ \longrightarrow H_2CO + H \tag{4.3c}$$

The branching ratios in equations (4.3) are not known. The dissociative recombination of the interstellar molecule H_2COH^+ is considered very important in interstellar

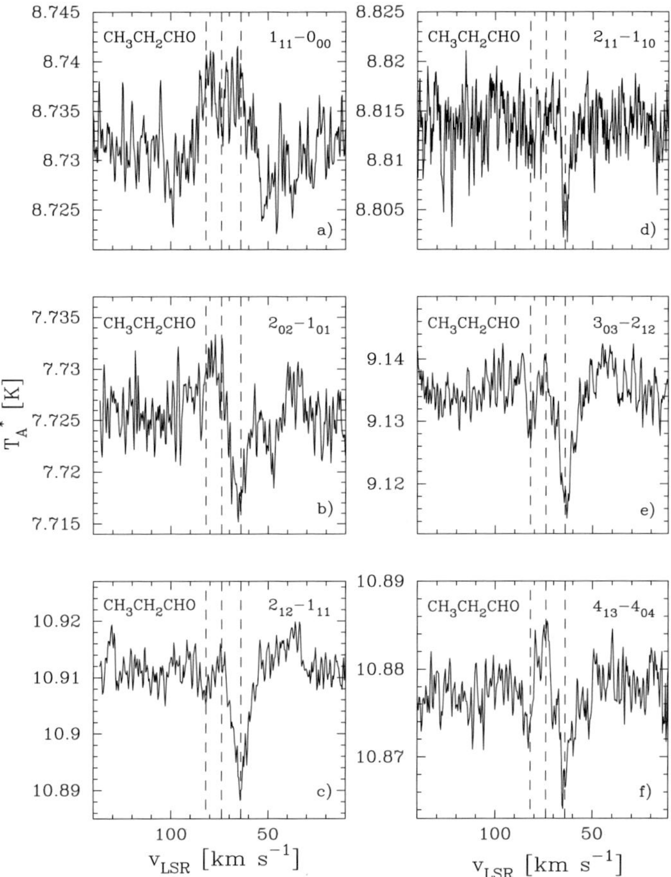

Figure 8. Propanal GBT spectra (Hollis *et al.* 2004b).

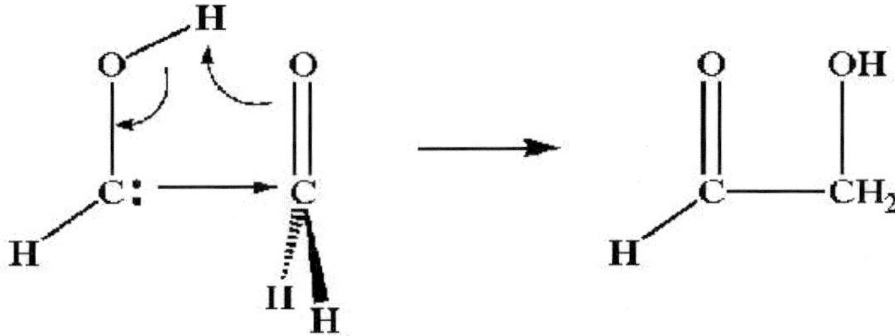

Figure 9. The first step in formaldehyde polymerization. Trans hydroxy methylene combining with formaldehyde to form glycolaldehyde.

clouds and hydroxy methylene (HCOH) should be a relatively stable and long- lived molecule under interstellar cloud conditions (Hoffmann & Schaefer 1981). Trans HCOH lies $\sim$5 kcal mole^{-1} (2516 K) below cis HCOH with a substantial isomerization barrier of $\sim$30 kcal mole^{-1} (15,098 K) to the formation of formaldehyde. Thus, it is evident that the likely form of interstellar hydroxy methylene would favor trans HCOH.

Ab initio quantum chemical models may be able to determine the molecular structure of hydroxy methylene with sufficient precision such that an accurate rotational spectrum can be calculated. An interstellar search could then be mounted for this important isomer of formaldehyde by searching for low energy transitions that occur in the K-band range. Such transitions would have the least frequency uncertainty and likely be detectable since low energy transitions of both formaldehyde and glycolaldehyde are readily detected by the GBT. If interstellar trans HCOH were found, it would provide strong evidence for the mechanism by which glycolaldehyde is formed and that the synthesis of more complex sugars similarly occurs.

Acknowledgements

It is a great pleasure to acknowledge the help of perennial collaborators Phil Jewell, Frank Lovas, and Anthony J. Remijan and research support from Harley Thronson of the NASA Science Mission Directorate.

References

Ceccarelli, C., *et al.* 2002, *A&A* 383, 603

Chengalur, J.N., & Kanekar, N. 2003, *A&A* 403, L43

Flanagan, G., Ahmed, S.N., & Shevlin, P.B. 1992, *J. Am. Chem. Soc.* 114, 3892

Flower, D.R., Pineau des Forêts, G., & Walmsley, M. 1995, *A&A* 294, 815

Hoffmann, M.R., & Schaefer, H.F. 1981, *Ap. J.* 249, 563

Hollis, J.M., Jewell, P.R., Lovas, F.J., & Remijan, A. 2004a, *Ap. J.* 613, L45

Hollis, J.M., Jewell, P.R., Lovas, F.J., Remijan, A., & Møllendal, H. 2004b, *Ap. J.* 610, L21

Hollis, J.M, Lovas, F.J., & Jewell, P.R. 2000, *Ap. J.* 540, L107

Hollis, J.M., Lovas, F.J., Jewell, P.R., & Coudert, L.H. 2002, *Ap. J.* 571, L59

Hollis, J.M., Pedelty, J.A., Boboltz, D.A., Liu, S.-Y., Snyder, L.E., Palmer, P. Lovas, F.J., & Jewell, P.R. 2003, *Ap. J.* 596, L235

Hollis, J.M., Vogel, S.N., Snyder, L.E., Jewell, P.R., & Lovas, F.J. 2001, *Ap. J.* 554, L81

Irvine, W.M., *et al.* 1988, *Ap. J.* 335, L89

Mehringer, D.M., & Menten, K.M. 1997, *Ap. J.* 374, 346

Snyder, L.E., Buhl, D., Zuckerman, B., & Palmer, P. 1969, *Phys. Rev.* 22, 679

Turner, B.E. 1991, *Ap. J. Suppl.* 76, 617

Discussion

HERBST: Could you discuss the current state of knowledge of the reaction t-HCOH + $H_2CO \rightarrow$ glycolaldehyde?

HOLLIS: The formation of glycolaldehyde occurs because the highly reactive t-HCOH divalent carbon adds to the H_2CO carbon. The reaction is fully discussed in Flanagan *et al.* (1992; *J. Am. Chem. Soc.* 114, 3892) along with experimental results that verify the reaction occurs in low temperature conditions.

IRVINE: How accurately can the lowest rotational transitions of t-HCOH be calculated *ab initio*?

HOLLIS: The lowest transition frequency is the difference between the rotational constants B and C. The uncertainty on $(B - C)$ is ~ 50 MHz.

BOTTA: Under what conditions did the polymerization reaction in the laboratory had been carried out (solution, temperature)?

HOLLIS: These are condensation experiments of atomic carbon and deuterated water at 77 K. The solid is later warmed and analyzed for sugar content.

Astrochemistry: Recent Successes and Current Challenges
Proceedings IAU Symposium No. 231, 2005
D.C. Lis, G.A. Blake & E. Herbst, eds.

© 2006 International Astronomical Union
doi:10.1017/S174392130600723X

Pathways to Molecular Complexity

S. B. Charnley and S. D. Rodgers

Space Science & Astrobiology Division, MS 245-3, NASA Ames Research Center,
Moffett Field, CA 94035, USA

Abstract. Pathways leading to the formation of complex organic molecules are described. Gas-phase processes that may build large carbon-chain species in cold molecular clouds are summarised. Catalytic reactions on grain surfaces can lead to a large variety of organic species, and model calculations of mantle formation by atom additions to multiply-bonded molecules are presented. The subsequent desorption of these mixed molecular ices can initiate a distinctive organic chemistry via ion-molecule pathways. The predictions of this theory are briefly compared with observations to show how possible organic formation pathways in the interstellar medium may be constrained.

Keywords. astrobiology — astrochemistry — ISM: chemistry — ISM: molecules

1. Introduction

Well over one hundred molecules have been discovered in the interstellar medium[†] (ISM), with the largest molecule detected to date containing 13 atoms ($HC_{11}N$; Bell *et al.* 1997). A large variety of different types of species are present, including alcohols, aldehydes, ethers, carboxylic acids, amines, nitriles, etc. Several of these organic molecules have also been observed in the gaseous comae of comets (e.g., Bockelée-Morvan *et al.* 2000), including the recent detection of ethylene glycol in comet Hale-Bopp (Crovisier *et al.* 2004). Laboratory analysis of meteoritic material has revealed a complex organic component (Botta, this volume), including numerous amino acids (Cronin & Chang 1993). The enhanced D/H ratios in both cometary and meteoritic molecules suggest an interstellar heritage for some solar system material, and raise the question of how much of the prebiotic organic material present on the young Earth can be traced to interstellar precursors (Ehrenfreund & Charnley 2000). The detection in the ISM of the smallest sugar and amino acid (Hollis *et al.* 2000; Kuan *et al.* 2003), indicates that significant complexity can arise in interstellar gas before planetary systems begin to form.

Several of the molecules observed in the ISM are isomers (e.g., ethylene oxide, acetaldehyde, vinyl alcohol); others correspond to more or less reduced forms of the same heavy-atom 'backbone' (e.g., hydrogen cyanide, methanimine, methylamine). A key challenge in astrochemistry is to determine the chemical mechanisms that lead to large molecules, whilst also explaining the relative abundances of different isomers and the differing degrees of hydrogenation. In this paper we discuss three pathways that have been proposed to account for the formation of complex molecules: (i) cold gas-phase processes occurring in dark clouds which lead to the growth of long carbon chains, (ii) atom addition reactions on the surfaces of interstellar dust grains, leading to more reduced species such as aldehydes and alcohols, and (iii) gas-phase processes in warm gas where icy grain mantles have evaporated, seeding the gas with large abundances of specific molecules, which proceed to react to form more complex species.

[†] See `http://www.molres.org/astrochymist/astrochymist_mole.html` for a current list of all detections.

237

2. Carbon Chains and Carbonaceous Dust

2.1. *Formation Mechanisms*

The rapid condensation of carbonaceous dust in the outflows of evolved stars is thought to be initiated by the formation of small polycyclic aromatic hydrocarbons (PAHs) driven by reactions of acetylene molecules (Frenklach & Feigelson 1989; Cherchneff *et al.* 1992). An important recent development has been the realization that nitrogen heterocycles (a.k.a. PANHs) may be abundant in space (see Hudgins, this volume; Charnley *et al.* 2005). Ricca *et al.* (2001) showed how additions of HCN as well as C_2H_2 can lead to PANHs rather than PAHs—Figure 1 illustrates this process for the formation of the two-ring PANHs quinoline and isoquinoline. Alternatively, grains may form via carbon-chain growth, followed by spontaneous ring closure (e.g., Tielens & Charnley 1997). The growth of small C-chains is driven by neutral-neutral reactions involving C_2H and C_2 (Cherchneff & Glassgold 1993; Millar *et al.* 2000). Cyanopolyyne formation is initiated by neutral-neutral reactions involving the CN radical (Fukuzawa *et al.* 1998). Similar processes are also able to form the observed abundances of carbon-chain molecules in dark interstellar clouds, following the removal of acetylene and methane from grain mantles (Markwick *et al.* 2000).

2.2. *Dust Destruction in PDRs and Comets*

Observations of hydrocarbons in photodissociation regions (PDRs) have revealed large abundances of many species, greatly elevated above the amounts predicted by traditional chemical models. For example, Teyssier *et al.* (2004) derive a C_6H abundance of 10^{-10} in the Horsehead nebula. These observations have been interpreted as being caused by UV-driven fragmentation of PAHs or small C-rich grains (Pety *et al.* 2005). Guélin *et al.* (1999) suggested that UV destruction of dust grains could be responsible for enhanced abundances of small C-chains in the outer regions of circumstellar envelopes.

Destruction of large organic macromolecules as a source of smaller species has also been proposed to occur in the comae of comets close to the Sun. Measurements by the Giotto probe as it passed through the coma of comet Halley showed that much of the H_2CO seen in the coma was not sublimating from the nucleus, but must have an extended coma source (e.g., Meier *et al.* 1993). This fact, together with regularly spaced peaks in the mass spectrometer data, led Huebner (1987) to suggest that the formaldehyde polymer polyoxymethylene (POM) was an important constituent of the refractory organic component of cometary dust. Laboratory experiments on the thermal

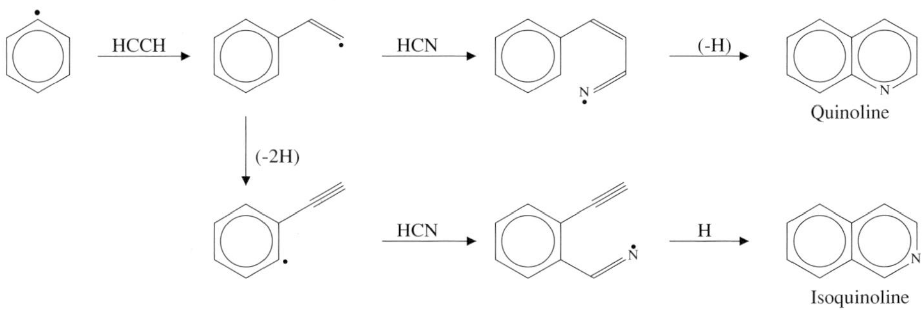

Figure 1. PANH formation via C_2H_2 and HCN addition to phenyl radicals (from Charnley *et al.* 2005). The position of the N in the final molecule depends on the details of the reaction mechanism (Ricca *et al.* 2001; Peeters *et al.* 2005).

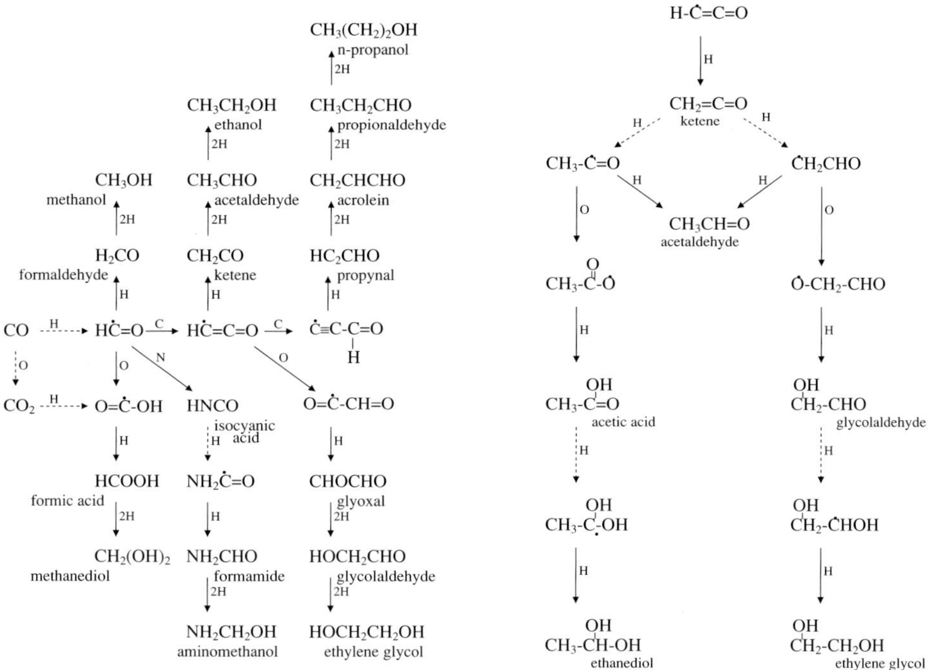

Figure 2. Grain-surface atom addition reactions starting from CO. The right hand side shows the alternative pathways that can arise if an O rather than an H atom adds to the C_2H_3O radical.

degradation of POM confirm that this scenario is feasible (e.g., Cottin *et al.* 2004), and POM is known to be produced when H_2CO-rich ices are warmed to 40–80 K (Schutte *et al.* 1993). Erosion of N-bearing macromolecules has been proposed to account for the excess HNC and CN production seen in comets at small heliocentric distances (Rodgers *et al.* 2003; Fray *et al.* 2005). This idea is corroborated by the large $C^{15}N/C^{14}N$ ratios detected in several comets—around twice that of the $HC^{15}N/HC^{14}N$ ratio—which have been ascribed to the destruction of ^{15}N-rich presolar material (Manfroid *et al.* 2004).

3. Grain-Surface Chemistry

3.1. *Atom Addition Reactions*

As species freeze out onto the surfaces of interstellar dust grains, they can combine with other species to form larger molecules. At the low temperatures in the ISM, only atoms are likely to be mobile on grains, and the dominant reaction mechanism will be atomic addition reactions. Laboratory experiments have demonstrated that H atoms can tunnel through barriers to react with stable species such as C_2H_2, CO and H_2CO (Hiraoka *et al.* 2000; Watanabe *et al.*, this volume), and it is probable that interstellar CO_2 ice forms via oxidation of CO (e.g., Roser *et al.* 2001). The type of chemistry occurring will depend on the relative accretion rates of different atoms: if the accretion rate of C atoms is high, unsaturated C-chains will result, whereas at late times after most heavy atoms have frozen out, the accretion will be dominated by H atoms and so we expect the upper layers of ice mantles to be dominated by reduced species such as H_2CO and CH_3OH.

An important constraint that will influence the type of molecules formed on the surface is that at each stage the most stable radical will be formed. Unfortunately, very little is known about the relative stability of many radical isomers, with the exception of the CH_2OH/CH_3O system (Jodkowski *et al.* 1999). An additional constraint is that H atoms will break double and triple bonds in sequence so that weaker bonds are broken first; $C\equiv O$, $C=C$, $C\equiv C$, $C=O$ (Charnley 2001a). Figure 2 shows an example of some of the surface pathways that are allowed under these constraints. Extensions to this scheme involving larger molecules or the addition of nitrogen atoms are also possible (e.g., Charnley 1997; Ehrenfreund *et al.* 2005). Finally, a further complication that can arise is that back reactions can occur, *i.e.*, rather than simply adding to a radical, an H atom may abstract another H to form H_2. These types of reactions may be particularly important in determining the deuterium fractionation in certain molecules, as D abstraction may be energetically inhibited (e.g., Tielens 1983; Nagaoka *et al.* 2005).

3.2. *Stochastic Simulation*

Although numerous models have been developed to model grain-surface chemistry, almost all suffer from two problems. The first is that these models typically solve the chemical evolution using the mass action kinetics (*i.e.*, systems of coupled ordinary differential equations) used to follow gas-phase chemistry. However, when the typical number of atoms on the surface of each individual dust grain is small, the use of mean abundances leads to gross errors in the model predictions (Charnley *et al.* 1997; Tielens & Charnley 1997). In this case, accurate solutions can only be obtained via stochastic simulation of the chemical master equation (Charnley 1998, 2001b). The second problem is that only species on the surface may take part in the chemical processing. Models which assume that all solid molecules are available to react will therefore give incorrect results, and are unable to follow the chemical differentiation that arises in sequential ice layers as the relative accretion rates of gas-phase species vary.

In order to accurately calculate the grain-surface chemistry it is essential that both the gas-phase and surface chemistry are followed simultaneously. We have developed a coupled gas-grain model that solves the master equation using the Monte Carlo algorithm of Gillespie (1976) (Rodgers & Charnley 2005). In addition to gas-phase and surface processes, we include accretion and desorption, and we also consider the covering and uncovering of surface species, assuming that there are 10^6 surface sites available. To resolve a single dust grain requires that a gas volume containing $\sim 10^{12}$ H_2 molecules is considered. Figure 3 shows the results for the surface abundances and bulk mantle fractions for a model in which surface species as large as propanol are considered. At early times, C atoms compete effectively with H atoms, and C-chain radicals are formed and subsequently hydogenated. At late times, H additions dominate the surface chemistry, and methanol becomes the primary product.

3.3. *Secondary Processing: Photolysis and Radiolysis*

Many experiments have been performed on the photolysis of interstellar ice analogs. UV irradiation can lead to the production of complex molecules such as HMT ($C_6H_{12}N_4$; Bernstein *et al.* 1995) and amino acids (Bernstein *et al.* 2002; Muñoz-Caro *et al.* 2002). However, all these experiments require methanol, yet photolysis of simple ices is unable to produce methanol. Hence, if UV processing is indeed important in interstellar ices, there must exist a previous phase of grain-surface chemistry to produce the CH_3OH required to initiate the formation of more complex species. Radiolysis of laboratory ices by energetic protons, simulating the processing of interstellar ice by cosmic rays, has also been investigated in many experiments (e.g., Moore & Hudson, this volume). Unlike photolysis, radiolysis is able to produce CH_3OH from mixed CO and H_2O ice (Hudson & Moore 1999). Another difference between photolysis and radiolysis is that the latter

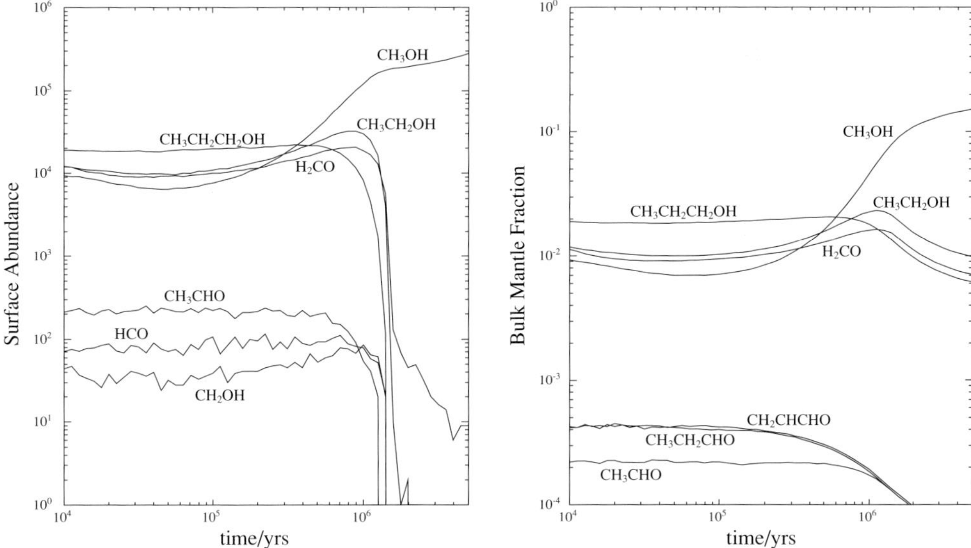

Figure 3. Solid-phase abundances calculated by our surface chemistry model (Rodgers & Charnley 2005). *Left:* surface populations. *Right:* fractional composition of the bulk ice.

appears to be much more effective in breaking the N≡N bond in molecular nitrogen, and so may be a source of nitriles (Moore & Hudson 2003). Although both photolysis and radiolysis are capable of producing complex molecules, quantitative models of the yield of specific molecules expected from these processes in interstellar ices are currently lacking.

4. Mantle-Seeded Gas: Hot Cores and Hot Corinos

Hot molecular cores are warm, dense regions of the ISM with large abundances of many complex molecules. They are thought to represent one of the earliest stages in massive star formation, where the growing protostar becomes sufficiently luminous to heat a significant mass of its natal cocoon, leading to the evaporation of interstellar ices (e.g., Kurtz *et al.* 2000). It is now known that similar regions surround low-mass protostars, albeit on a smaller scale, so-called 'hot corinos' (Kuan *et al.* 2004; Bottinelli *et al.* 2004). In this paper we are concerned principally with the chemical pathways to large molecules, which are similar in both hot cores and hot corinos, and hereafter we use the term hot core generically to refer to any region where interstellar ice mantles have recently evaporated.

Models of the chemistry in hot cores have demonstrated that reactions in the warm post-evaporation gas can lead to large abundances of second-generation species produced from the evaporated 'parent' species (e.g., Millar *et al.* 1991; Charnley *et al.* 1992, 1995). Small-scale chemical differentiation is often observed in hot cores (e.g., Kuan *et al.* 2004), and this is thought to be related to different dynamical and thermal histories of specific regions (Rodgers & Charnley 2001, 2003). A detailed description of the kinds of coupled dynamical-chemical models used to analyse protostellar envelopes appears in the chapter by Doty. Below, we describe the most important classes of chemical reactions leading to the production of large organics in hot cores.

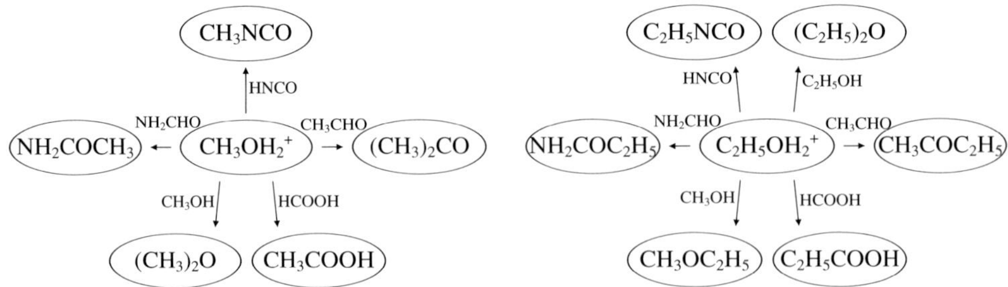

Figure 4. Alkyl cation transfer reactions from protonated methanol (*left*) and ethanol (*right*) can lead to a variety of large organic molecules.

4.1. *Reactions of Protonated Methanol and Ethanol*

Laboratory experiments involving protonated alcohols reacting with neutral coreactants have shown that, when the direct proton transfer channel is endothermic, transfer of the ionized alkyl group often proceeds rapidly (Karpas & Mautner 1989). Thus, protonated methanol and ethanol can respectively add CH_3^+ and $C_2H_5^+$ side groups to other species. Self-alkylation of alcohols can lead to large abundances of ethers in the gas phase, in particular dimethyl ether (Charnley *et al.* 1995). Similar reactions with other species can lead to a large variety of complex organic molecules. Considering reactions with several species which are thought to be formed on grain surfaces, Figure 4 shows some of the second-generation species that can be formed in the hot core.

Reactions of protonated alcohols with HCN have also been demonstrated to result in alkyl cation transfer (Mautner & Karpas 1986). In this case, protonated isocyanides are formed, *i.e.*,

$$CH_3OH_2^+ \; + \; HCN \; \rightarrow \; CH_3NCH^+ \; + \; H_2O \tag{4.1}$$

Recombination will most likely lead to the isocyanide rather than the cyanide. Therefore, observations of the isomer pairs CH_3CN and CH_3NC may be able to assess the importance of reaction (4.1) in hot cores.

4.2. *Interstellar Acids*

Alkyl cation transfer reactions from protonated methanol and ethanol to formic acid have been shown to lead to acetic acid and propanoic acid respectively (D. Bohme, private communication), e.g.,

$$CH_3OH_2^+ \; + \; HCOOH \; \rightarrow \; CH_3COOH_2^+ \; + \; H_2O \tag{4.2}$$

It is possible that similar reactions involving protonated aminoalcohols (which are readily formed via the putative surface chemistry scheme of the previous section; see Figure 2) and formic acid may lead to protonated amino acids, e.g.,

$$NH_2CH_2OH_2^+ \; + \; HCOOH \; \rightarrow \; NH_2CH_2COOH_2^+ \; + \; H_2O \tag{4.3}$$

(Charnley 2001a).

Alternatively, the experiments of Blagojevic *et al.* (2003) showed that protonated amino acids are formed via reactions of protonated hydroxylamine with carboxylic acids, *i.e.*,

$$NH_3OH^+ \; + \; CH_3COOH \; \rightarrow \; NH_2CH_2COOH_2^+ \; + \; H_2O \tag{4.4}$$

Although NH_2OH has not yet been detected in space, models of gas-grain processing suggest it should be abundant in interstellar ices (Charnley *et al.* 2001). In this scenario,

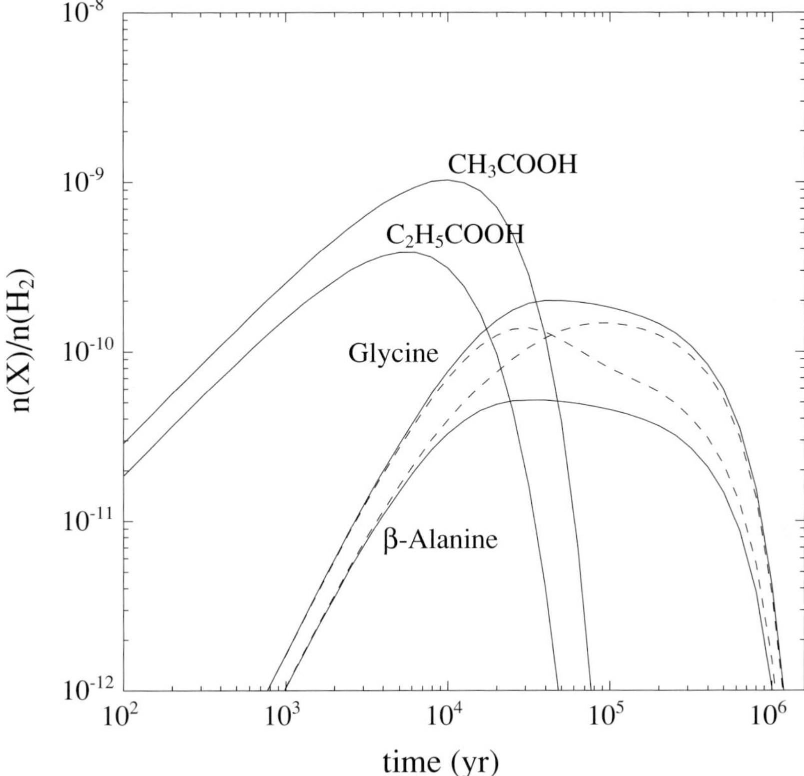

Figure 5. Abundances of organic acids produced via gas-phase chemistry following the evaporation of ices rich in CH_3OH, $HCOOH$, and NH_2OH (Charnley *et al.* 2005). Solid lines are for the standard model, and dashed lines show the effects of including a CH_3^+ transfer channel from protonated methanol to glycine.

evaporation of alcohols and HCOOH leads to the formation of carboxylic acids via reactions such as (4.2), and subsequent reactions with NH_3OH^+ lead to amino acids. To test this hypothesis, we ran our standard hot-core model (Rodgers & Charnley 2001) with these reactions included, and with initial abundances of CH_3OH, $HCOOH$, and NH_2OH of 10^{-7}. We also included an additional channel to β-alanine via alkyl cation transfer from $CH_3OH_2^+$ to glycine. Even though in this case the competing proton transfer channel is exothermic, it is possible that a few per cent of collisions result in CH_3^+ transfer (D. Bohme, private communication). The resulting abundances of the acids of interest are shown in Figure 5. The inclusion of the additional channel results in a β-alanine/glycine ratio greater than unity. Such a large ratio has been measured in the Orgueil meteorite, and led to the suggestion that this carbonaceous chondrite originated from a cometary, as opposed to an asteroidal, parent body (Ehrenfreund *et al.* 2001). If correct, a cometary origin would strengthen the connection with the type of organic chemistry proposed here.

4.3. *Methyl Formate, Acetic Acid and Glycolaldehyde*

Methyl Formate (CH_3OCHO), acetic acid (CH_3COOH) and glycolaldehyde ($HOCH_2CHO$) are isomers, and all three have been detected in hot cores (Remijan *et al.* 2003; Hollis *et al.* 2004). However, the abundances and spatial distributions of the molecules are different. The reason for this is that each of these species is produced by a different mechanism. Acetic acid can be synthesized post-evaporation in the gas phase, via reaction (4.2). Glycolaldehyde, on the other hand, can be formed on grain surfaces by simple atom addition reactions (see Fig. 2), and so will evaporate directly from the ice mantles. The origin of methyl formate is less certain. Blake *et al.* (1987) proposed that methyl cation transfer from $CH_3OH_2^+$ to H_2CO could lead to protonated methyl formate. However, both experiment and theory have demonstrated that this reaction leads to $CH_3OCH_2^+$ (Karpas & Mautner 1989; Charnley 1997; Horn *et al.* 2004). Grain-surface chemistry via atomic addition is also unable to form methyl formate. One possibility is that grain-surface chemistry is more complicated than described above, and that radicals such as HCO and CH_3O are able to encounter each other and combine directly. Another possibility is that these type of radical-radical reactions occur efficiently in the wake of cosmic ray particle passing through the ice, where it is expected that many radicals will be formed in close proximity via the hot secondary electrons generated by the cosmic ray.

5. Conclusions

We have discussed several pathways to molecular complexity in the ISM. In circumstellar envelopes and dark clouds, neutral-neutral reactions driven by small C-bearing radicals can generate long carbon chains and small aromatic molecules. Some of these species will coagulate to form C-rich dust particles, and the erosion of this dust by UV photons may be an important source of several species in PDRs and cometary comae. Grain-surface chemistry via atomic addition reactions can lead to the growth of larger molecules. When surface chemistry schemes are limited by realistic constraints, such as radical stability and the penetration of multiple bonds in order of activation energy, we find that these reactions tend to lead to aldehydes and alcohols. The degree of hydrogenation will depend on the competition between accretion of H atoms and other heavy atoms, and may be greater in the upper layers of interstellar ice mantles. When these ice species are evaporated in hot cores, subsequent gas-phase processing can synthesize a variety of new, even larger molecules. Ethers, carboxylic acids, and amino acids are all signatures of this type of gas-phase chemistry. Finally, there are some molecules that cannot be generated efficiently by any of these mechanisms, such as methyl formate and dihydroxyacetone. For these molecules, a possible origin is direct reactions of functional groups, driven by energetic processing of the ices.

Acknowledgements

We are grateful to Diethard Bohme and Simon Petrie for communication of their results prior to publication. This work was funded by NASA's Exobiology and Long-Term Space Astrophysics Programs through funds allocated by NASA Ames under Cooperative Agreement No. NCC2-1412 to the SETI Institute.

References

Bell, M.B. *et al.* 1997, *Ap. J.* 483, L61
Bernstein, M. *et al.* 2002, *Nature* 416, 401
Blagojevic, V., Petrie, S., & Bohme, D.K. 2003, *MNRAS* 339, L7
Blake, G.A., Sutton, E.C., Masson, C.R., & Phillips, T.G. 1987, *Ap. J.* 315, 621
Bockelée-Morvan, D. *et al.* 2000, *A&A* 353, 1101
Bottinelli, S. *et al.* 2004, *Ap. J.* 617, L69
Charnley, S.B. 1997, in *Astronomical and Biochemical Origins and the Search for Life in the Universe*, eds. C.B. Cosmovici *et al.* (Editrice Compositori), p. 89

Charnley, S.B. 1998, *Ap. J.* 509, L121

Charnley, S.B. 2001a, in *The Bridge Between the Big Bang and Biology*, ed. F. Giovanelli (Rome: Consiglio Nazionale delle Ricerche), p. 139

Charnley, S.B. 2001b, *Ap. J.* 562, L99

Charnley, S.B., Tielens, A.G.G.M., & Millar, T.J., 1992, *Ap. J.* 399, L71

Charnley, S.B., Kress, M.E., Tielens, A.G.G.M., & Millar, T.J., 1995, *Ap. J.* 448, 232

Charnley, S.B., Tielens, A.G.G.M., & Rodgers, S.D. 1997, *Ap. J.* 482, L203

Charnley, S.B., Rodgers, S.D., & Ehrenfreund, P. 2001, *A&A* 378, 1024

Charnley, S.B. *et al.* 2005, *Adv. Space Res.* 36, 137

Cherchneff, I. & Glassgold, A.E., 1993, *Ap. J.* 419, L41

Cherchneff, I., Barker, J.R., & Tielens A.G.G.M. 1992, *Ap. J.* 401, 269

Cottin, H., Bénilan, Y., Gazeau, M-C., & Raulin, F. 2004, *Icarus* 167, 397

Cronin J.R. & Chang S. 1993, in *The Chemistry of Life's Origins*, eds. J.M. Greenberg *et al.* (Kluwer), p. 209

Crovisier, J., Bockelée-Morvan, D., Biver, N., Colom, P., Despois, D., & Lis, D.C. 2004, *A&A* 418, L35

Ehrenfreund, P. & Charnley, S.B. 2000, *ARAA* 38, 429

Ehrenfreund, P., Botta, O., Glavin, D., Cooper, G., Kminek, G., & Bada, J. 2001, *Proc. Nat. Acad. Sci.* 98, 2138

Ehrenfreund, P., Charnley, S.B., & Botta O., 2005, in *Astrophysics of Life*, eds. M. Livio *et al.* (Cambridge University Press), p. 1

Fray, N., Bénilan, Y., Cottin, H., Gazeau, M-C., & Crovisier J. 2005, *Planet. Space Sci.* 53, 1243

Frenklach, M. & Feigelson, E.D. 1989, *Ap. J.* 341, 372

Fukuzawa, K., Osamura, Y., & Schaefer, H.F. 1998, *Ap. J.* 505, 278

Gillespie, D.T. 1976, *J. Comp. Phys.* 22, 403

Guélin, M., Neininger, N., Lucas, R., & Cernicharo, J. 1999, in *The Physics and Chemistry of the Interstellar Medium*, eds. V. Ossenkopf *et al.*, p. 326

Hiraoka, K., Takayama, T., Euchi, A., Handa, H., & Sata, T. 2000, *Ap. J.* 532, 1029

Hollis, J.M., Lovas, F.J., & Jewell, P.R. 2000, *Ap. J.* 540, L107

Hollis, J.M., Jewell, P.R., Lovas, F.J., & Remijan, A. 2004, *Ap. J.* 613, L45

Horn, A. *et al.* 2004, *Ap. J.* 611, 605

Hudson, R.L. & Moore, M.H. 1999, *Icarus* 140, 451

Huebner, W.F. 1987, *Science* 237, 628

Jodkowski, J.T. *et al.* 1999, *J. Phys. Chem. A*, 103, 3750

Karpas, Z. & Mautner, M. 1989, *J. Phys. Chem.* 93, 1859

Kuan, Y-J., Charnley, S.B., Huang, H-C., Tseng, W-L., & Kisiel, Z. 2003, *Ap. J.* 593, 848

Kuan, Y-J. *et al.* 2004, *Ap. J.* 616, L27

Kurtz, A., Cesaroni, R., Churchwell, E., Hofner, P., & Walmsley, C.M. 2000, in V. Mannings *et al.* (eds.), *Protostars & Planets IV*, (University of Arizona Press), p. 299

Manfroid, J. *et al.* 2004, *A&A* 432, L5

Markwick, A.J., Millar, T.J., & Charnley, S.B. 2000, *Ap. J.* 535, 256

Mautner, M. & Karpas, Z. 1986, *J. Chem. Phys.*, 90, 2206

Meier, R., Eberhardt, P., Krankowsky, D., & Hodges, R.R. 1993, *A&A* 277, 677

Millar, T.J., Herbst, E., & Charnley, S.B. 1991, *Ap. J.* 369, 147

Millar, T.J., Herbst, E., & Bettens, R.P.A. 2000, *MNRAS* 316, 195

Moore, M.H. & Hudson, R.L. 2003, *Icarus* 161, 486

Muñoz-Caro, G.M. *et al.* 2002, *Nature* 416, 403

Nagaoka, A., Watanabe, N., & Kouchi, A. 2005, *Ap. J.* 624, L29

Peeters, Z. *et al.* 2005, *A&A* 433, 583

Pety, J. *et al.* 2005, *A&A* 435, 885

Remijan, A. *et al.* 2003, *Ap. J.* 590, 314

Ricca, A., Bauschlicher, C.W., & Bakes, E.L.O. 2001, *Icarus* 154, 516

Rodgers, S.D., Butner H.M., Charnley, S.B., & Ehrenfreund, P. 2003, *Adv. Space Res.* 31, 2577

Rodgers, S.D. & Charnley, S.B. 2001, *Ap. J.* 546, 324

Rodgers, S.D. & Charnley, S.B. 2003, *Ap. J.* 585, 355

Rodgers, S.D. & Charnley, S.B. 2005, *MNRAS*, submitted
Roser, J.E., Vidali, G., Manicò, G., & Pirronello, V. 2001, *Ap. J.* 555, L61
Schutte, W.M., Allamandola, L.J., & Sandford, S.A. 1993, *Icarus* 104, 118
Teyssier, D., Fossé, D., Gerin, M., Pety, J., Abergel, A., & Roueff, E. 2004, *A&A* 417, 135
Tielens, A.G.G.M. 1983, *A&A* 119, 177
Tielens, A.G.G.M. & Charnley, S.B. 1997, *Origins Life Evol. Biosphere* 27, 23

Discussion

DOTY: The identification of pathways for complex molecules is very interesting. To make quantitative predictions for comparison with observations, however, the results presumably depend upon the structure of the grain surface and mantles. Could you comment on what you assume, and how changing these assumptions would affect the predicted abundances?

CHARNLEY: In these particular calculations a regular lattice of surface binding sites was assumed. Recent simulations by Cuppers, Chang and Herbst, reported at this meeting, consider the kinetics of H_2 formation on more realistic porous surfaces.

MILLAR: Do you know of any laboratory experiments, which indicate that the destruction of carbonaceous grains or PAH molecules which result in the efficient production of carbon chains.

CHARNLEY: I seem to recall a paper by Duley & Williams, where the discussion was based on the results of laboratory experiments. Carbon chains were reported to be produced from laser processing of solid carbonaceous material.

MÜLLER: Many of the species you have proposed may be present in the interstellar medium. The question is: will we ever be in the position to detect these species unambiguously? (E.g., because of issues such partition function, vapor presences, low lying vibrational rates.)

CHARNLEY: This is more a question for Aldo (Apponi).

Astrochemistry: Recent Successes and Current Challenges
Proceedings IAU Symposium No. 231, 2005
D.C. Lis, G.A. Blake & E. Herbst, eds.

© 2006 International Astronomical Union
doi:10.1017/S1743921306007241

Production of Complex Molecules in Astrophysical Ices

Marla H. Moore[1] and Reggie L. Hudson[2]

[1]Astrochemistry Laboratory, NASA/Goddard Space Flight Center,
Greenbelt, MD 20771, USA
email: Marla.H.Moore@nasa.gov

[2]Department of Chemistry, Eckerd College,
St. Petersburg, FL 33711, USA,
email: hudsonrl@eckerd.edu

Abstract. The inventory of interstellar and solar system molecules now numbers well over 100 species, including both ions and neutrals. Gas-phase formation routes to many of the observed organics are still uncertain, so that solid-phase syntheses are of interest. Low-temperature reactions are thought to occur within interstellar ices, on ice and grain surfaces in the interstellar medium (ISM), and on icy surfaces of solar system objects (e.g., Europa, Pluto). Ionizing radiation, such as cosmic rays, and far-UV photons are two possible initiators of such chemistry.

In the Cosmic Ice Lab at NASA's Goddard Space Flight Center, we can study both the photo- and radiation-induced chemistries of ices at 8–300 K. Our most-recent work has been motivated by the detections of ethylene glycol, $HOCH_2CH_2OH$, in the ISM (Hollis *et al.* 2002) and in comet Hale-Bopp (Crovisier *et al.* 2002). Ethylene glycol is currently the largest firmly-identified cometary molecule as well as one of the larger interstellar organics. It's formation and accompanying chemistry provide challenges and tests for current astrochemical thought.

Here we discuss laboratory experiments on ethylene glycol's solid-phase formation and destruction. Using infrared spectroscopy, we have identified low-temperature radiation-chemical pathways that lead from known interstellar ices, such as either CH_3OH or H_2O and CO, to ethylene glycol. We also have identified a role for ethylene glycol in the formation of glycolaldehyde, $HOCH_2C(O)H$, a simple sugar (Hollis, Lovas, & Jewell 2000). Related to these results are the 6- and 7-atom molecules formed by oxidation and reduction processes in irradiated H_2O + C_2H_2 ices. Analogous experiments have been conducted on two other triply-bonded molecules in H_2O-rich ice, HCN and N_2. An important focus of our work is the development of reaction schemes for the formation of complex molecules within astrochemically-relevant ices, and the use of such schemes to predict new molecules awaiting detection.

Keywords. astrobiology — astrochemistry — ISM: abundances — ISM: molecules — methods: laboratory — molecular processes

1. Introduction

The current inventory of interstellar (IS) and solar system molecules now numbers well over a hundred species including inorganics, organics, ions, and radicals. The most complex molecule has thirteen atoms. Of the stable organics, half can be considered nitrile and acetylene derivatives, with the other half being aldehydes, alcohols, ethers, acids, ketones, amides, and related molecules. Ethylene glycol, $HOCH_2CH_2OH$, is composed of ten atoms, and is the most complex organic identified in both comet Hale-Bopp (Crovisier *et al.* 2002) and IS environments (Hollis *et al.* 2000). Table 1 lists observed interstellar and circumstellar organic molecules, with cometary species indicated by underlining.

Although a variety of astrochemical problems can be approached successfully through gas-phase chemistry, efficient syntheses for many interstellar and cometary organics are

Table 1. Organic molecules observed in interstellar and circumstellar regions. A few related inorganic molecules are included.

Hydrocarbons and Simple Hydrides, Oxides, and Sulfides

CH_4	H_2O	NH_3	H_2S
C_2H_4	CO	N_2O	SO_2
C_2H_6	CO_2	N_2	OCS

Nitriles, Alkynes, and Related Molecules

CH_3CN	HCN	HNC	C_2H_2
$CH_3(CC)CN$	$H(CC)CN$	CH_3NC	$H(CC)_2H$
$CH_3(CC)_2CN$	$H(CC)_2CN$	$HCCNC$	$H(CC)_3H$
CH_2CHCN	$H(CC)_3CN$		$CH_3(CC)H$
CH_3CH_2CN	$H(CC)_4CN$	NH_2CN	$CH_3(CC)_2H$
	$H(CC)_5CN$		

Organics and Related Molecules

H_2CO	CH_3OH	$HCOOH$	CH_2CO
$HC(O)CH_3$	$CH_2CH(OH)$	CH_3COOH	
H_3C-CH_2-$C(O)H$	CH_3CH_2OH		$HNCO$
$H_2C{=}CH$-$C(O)H$	$HOCH_2CH_2OH$	$HC(O)OCH_3$	
$HCCH$-$C(O)H$			$HNCS$
$HOCH_2C(O)H$	c-C_2H_4O	$HC(O)NH_2$	H_2CS
	$(CH_3)_2O$	CH_2NH	CH_3SH
$(CH_3)_2CO$		CH_3NH_2	

still unknown, so that solid-phase reactions are of interest. Among the gas-phase interstellar observations in which solid-phase chemistry has been invoked are those involving formaldehyde, H_2CO (Federmann & Allen 1991); methanol, CH_3OH (e.g., Charnley *et al.* 1995; Tielens & Whittet 1997; Teixeira *et al.* 1998); acetic acid, CH_3COOH (Mehringer *et al.* 1997); ethylene oxide, c-C_2H_4O (Nummelin *et al.* 1998); methylenimene, CH_2NH (Dickens *et al.* 1997); ethyl cyanide, CH_3CH_2CN (Miao & Synder 1997); vinyl alcohol, $CH_2CH(OH)$ (Turner & Apponi, 2001); ethylene glycol, $HOCH_2CH_2OH$ (Hollis *et al.* 2002); glycolaldehyde, $HOCH_2C(O)H$ (Hollis *et al.* 2000); propenal, $CH_2CHC(O)H$; and propanal, $CH_3CH_2C(O)H$ (Hollis *et al.* 2004).

One way in which organics can be formed in space is through chemical reactions occurring within the icy mantles of interstellar grains. Reactions can be initiated by either particle radiation or high-energy photons, resulting in the synthesis of new molecular species which can then be ejected into the gas phase by various mechanisms (e.g., Bringa & Johnson 2004, and references therein). Possible evidence for energetic processing within IS ices is the OCN^- ion (Hudson *et al.* 2001), while evidence for energetic processing in solar system ices comes from identifications of H_2O_2, O_2, and O_3 on Galilean moons. In comets, as with many interstellar regions, it is the many detections of gas-phase species that imply an active ice chemistry. In all cases, the importance of solid-phase reactions will depend on parameters such as initial ice composition, radiation flux, exposure time, temperature, and competing reactions.

Table 2 gives estimates of radiation and photon fluxes and doses for ices in both IS and solar system environments. Galactic cosmic rays are estimated to be about 87% protons, 12% helium nuclei, and 1% heavier nuclei, with an overall isotropic flux assumed constant for the age of the solar system, 4.6×10^9 years. Since the IS electron cosmic-ray flux is 10–100 times smaller than the proton flux, for energies below 100 MeV the cosmic-ray processing of IS ices is dominated by protons whose greatest intensity lies in the low

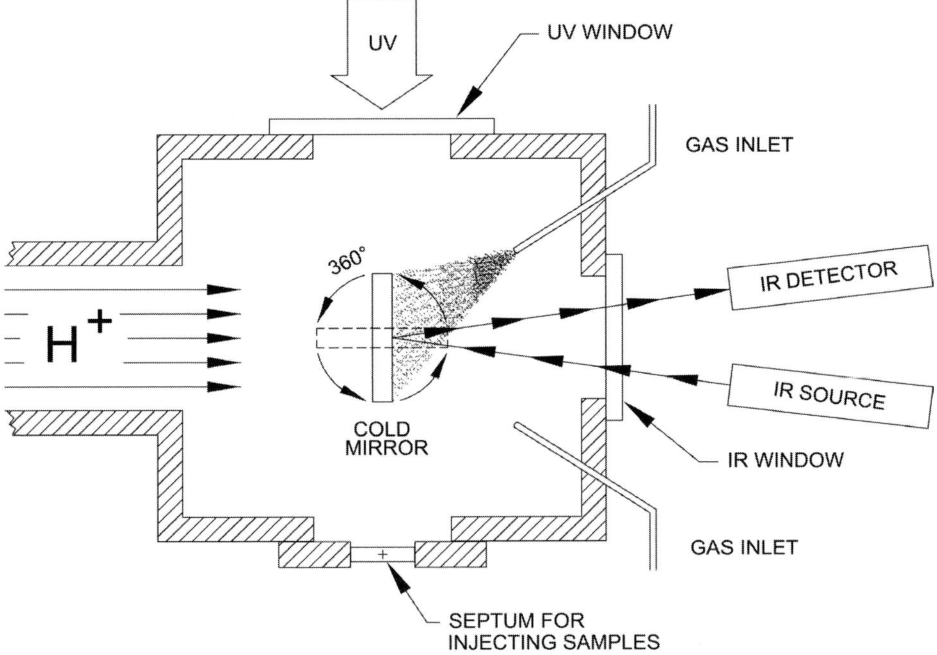

Figure 1. Schematic of laboratory set-up.

MeV range, with an energy spectrum decreasing with a power-law index of 2.5 at higher energies. The estimated energy dose from cosmic ray protons is $\sim$0.3 eV molecule^{-1} over 10^6 years, the approximate lifetime of a molecular cloud. Within this same time, $\sim$0.4 eV molecule^{-1} is deposited from UV photons that exist within dense cloud regions. Most of these IS photons are produced by hot O and B type stars, whose estimated abundance gives a calculated UV flux of 9.6×10^8 eV cm^{-2} s^{-1} between 91 and 180 nm.

The final products of photon- and radiation-induced chemical reactions within ices can be, and typically are, identical, but the processes by which they are made are quite different. The energy of a 1 MeV proton is deposited in an ice along a track containing numerous excitations and ionizations, with the concomitant production of thousands of secondary electrons causing most of the observed chemical changes. In contrast, each vacuum-UV photon is absorbed in a single excitation or ionization event. For comparison, in a 1 g cm^{-3} H$_2$O-ice the range of a 1 MeV proton is $\sim$20 μm, while there is a 37% drop in UV transmission in 0.15 μm, for a UV cross-section of 2×10^{-18} cm^2 (Okabe 1978).

In our lab we synthesize complex organics in cosmic-ice analogs, with reactions initiated by either high-energy protons ($\sim$1 MeV), to mimic cosmic rays, or vacuum-UV photons. We systematically vary reaction doses, temperatures, abundances of both minor and major (matrix) materials, and ice thickness. We also routinely employ isotopic labeling and isoelectronic systems to identify reaction products and test promising mechanisms. This approach allows us to understand and predict new interstellar molecules, and to identify processes and products that less comprehensive strategies would miss.

For this paper we describe the synthesis of complex organics from three triply-bonded molecules: carbon monoxide (C$\equiv$O), acetylene (HC$\equiv$CH), and hydrogen cyanide (HC$\equiv$N). These unsaturated, isoelectronic, linear molecules are all highly-reactive, and all are detected in both IS and solar system environments. These three species are the starting point for making about twenty different molecules, radicals, and ions where, for CO and C$_2$H$_2$, the primary syntheses are based on reactions of H and OH radicals; for HCN the

Table 2. Estimated UV photon and 1 MeV proton fluxes in different regions of space, and doses experienced by ices.

Region	Residence Time of Ices in Region (years)	UV[a]		Cosmic Ray Protons[a]	
		Flux 10 eV photon eV cm^{-2}s^{-1}	Energy Dose eV molecule^{-1}	Flux 1 MeV proton eV cm^{-2}s^{-1}	Energy Dose eV molecule^{-1}
IS Cold Dense Cloud (0.02 μm ice)	10^6 10^7	1.4×10^4	0.4 4	1×10^6	0.3 3
Oort Cloud and KBO Region	4.6×10^9	9.6×10^8	10^8 (top 0.015 μm layer)	Energy dependent flux ϕ(E)	150 (0.1 m) 5 (5 m)
Jovian Satellites e.g., Europa[b]	$<1\times10^8$	4×10^{13}	10^{13} (top 0.015 μm layer)	Magnetospheric ions 7.8×10^{13}	150 (1 cm, 10^4 years)
Typical Lab Experiment	4.6×10^{-4}	8.6×10^{14}	14	7×10^{10}	9

Notes: [a] Energy dose for UV and cosmic ray protons for the IS medium is from Moore *et al.* (2001); for the Oort cloud, KBO and Jovian satellite regions from Moore *et al.* (2003) and references therein. [b] Possible resurfacing lifetime from Stern and McKinnon (2000).

primary observed reaction path is oxidation of the cyano (-CN) functional group. Here we highlight radiation experiments in which we have identified the infrared (IR) signatures of complex species in H_2O-rich ices containing CO, C_2H_2, or HCN. Since actual cosmic ices are not simple binary systems, we have also had to consider the roles played by other molecules and reactions, such as the formation of ions through acid-base chemistry with NH_3. Finally, our results for these simple systems permit predictions of yet undetected, and more complex, organic molecules in comets and IS space.

2. Laboratory Methods

Details of our experimental set-up, ice preparation, IR spectral measurements, cryostat, and proton beam source have been published (Moore & Hudson 2000; Moore & Hudson 1998). Figure 1 is a schematic that represents an ice sample within our sample chamber. Briefly, ices were formed by condensation of gas-phase mixtures onto a pre-cooled aluminum mirror at 10 – 20 K, where mid-IR spectra (4000 – 400 cm^{-1}) of ices were taken by reflection from the mirror. Ices were 1 – 5 μm thick, as determined by a laser interference fringe system. Samples temperatures were controlled from 10 to 300 K with a heater on the cryostat tail section.

IR spectra were recorded before and after exposure of ice samples to a 0.8 MeV proton beam from a Van de Graaff accelerator. The use of proton irradiation to simulate cosmic-ray bombardment has been discussed earlier (e.g., Hudson & Moore, 2001; Moore *et al.*

1983). Since sample ice thicknesses were less than the range of the incident 0.8 MeV protons, the bombarding ions penetrated the ices and came to rest in the underlying metal substrate, not in the ices themselves. Radiation doses were determined by counting the proton fluence (p^+ cm^{-2}) at the ice, and converting to a common scale of eV (18-amu molecule)$^{-1}$. Proton stopping powers and ranges were calculated with Ziegler's SRIM program (Ziegler *et al.* 1985; `http://www.srim.org`).

3. Laboratory Studies of Unsaturated Carbon-Containing Species in H_2O-Rich Ices

3.1. *Study of CO in H_2O Ice*

To make a complex organic molecule, a laboratory chemist selects a simpler starting material and performs reactions such as those of addition or substitution of atoms or functional groups. In cometary and IS ices, carbon monoxide trails only H_2O in abundance, and so is a likely source of solid-phase organic chemistry. The H_2O-ice itself, subject to ionizing radiation, is a ready source of H atoms, OH radicals, H_2O^+, and e^-. These reactive species are available to participate in reactions such as H_2O_2 formation, protonation by H_2O^+ (or H_3O^+), OH radical addition, and reduction by e^-. The possibility that such chemistry might take place in astronomical ices is at least 10 years old (Tielens & Allamandola 1987; Mumma *et al.* 1996), and probably much older.

Reaction Scheme 1 summarizes the laboratory irradiations of H_2O + CO ices (Hudson & Moore 1999), where brackets indicate presumed but unobserved products. Radiation processing of H_2O + CO mixtures formed HCO, H_2CO, and CH_3OH through H-addition, and formic acid (HCOOH) in a branching reaction step. In addition, CO_2 was efficiently made by O-atom transfer from OH radicals to CO. Figure 2 shows the IR spectrum, in the 1900–1000 cm^{-1} region, of an H_2O + CO (5:1) ice after irradiation to 11 eV molecule^{-1} at 16 K. Bands of H_2CO, CH_3OH, and HCOOH in this region are identified by comparison with reference spectra of dilute mixtures of each of molecule in H_2O-ice. The HCO radical is at 1853 cm^{-1}. Similar experiments were performed at other H_2O/CO ratios (Hudson & Moore 1999) and in the presence of NH_3 (Hudson *et al.* 2001).

One of the most important findings from these experiments was the relatively high percentage of conversion of CO into other molecules. After about 22 eV molecule^{-1}, for an ice whose initial H_2O/CO ratio was 5, the conversion of CO into CO_2 was $\sim$20%, HCOOH $\sim$40%, H_2CO $\sim$7% and CH_3OH $\sim$9%. These percentages are a function of the initial CO abundance and radiation dose, and are calculated by knowing the integrated band strengths of each molecule and assuming an ice density (see Hudson & Moore 1999).

A comparison of our results with hydrogenation experiments of Watanabe *et al.* (2003, 2004) is shown in Figure 3. In their experiments, the source of H atoms was a cold atomic hydrogen beam simulating the emplacement of H-atoms onto IS ice grain mantles in cold dark clouds. Hydrogenation of H_2O + CO (4:1) ices resulted in production of both

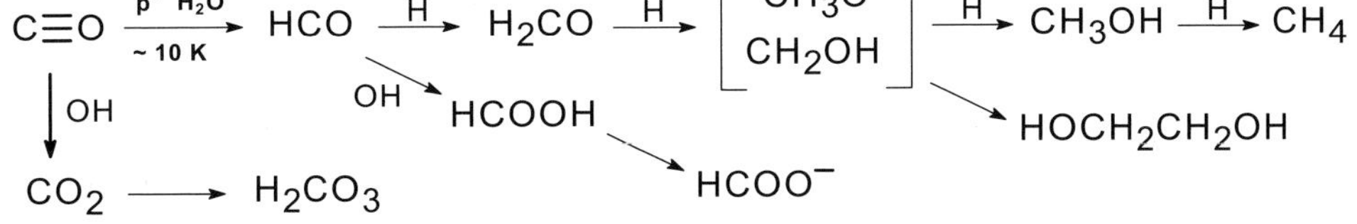

REACTION SCHEME 1

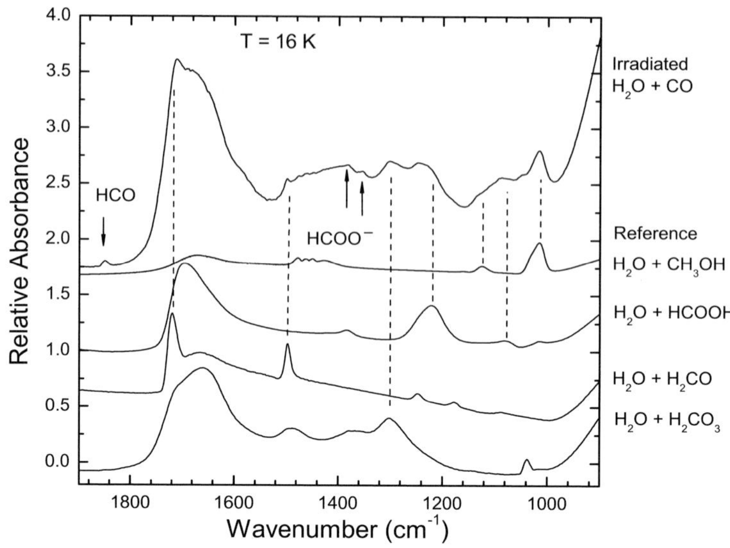

Figure 2. New species formed in an H_2O + CO (5:1) ice irradiated to 11 eV molecule^{-1} are identified by comparison with reference spectra of dilute mixtures of organics at ∼16 K. Spectra are offset for clarity.

H_2CO and CH_3OH. Watanabe *et al.* (2003, 2004) assumed a $10^4 - 10^6$ year exposure to an H-atom density of 1 cm^{-3} at 10 K. Ratios of column densities of CO/CH_3OH and H_2CO/CH_3OH are plotted in Figure 3. The left-hand data point is equivalent to 10^6 years, and the right-hand data point to 10^4 years (based on Watanabe *et al.* 2004, Figure 5). Our results are plotted to show data representing exposure of an H_2O + CO (5:1) ice to ∼1, 10, and 22 eV molecule^{-1} corresponding to cosmic-ray proton exposure for 3×10^6, 3×10^7, and 7×10^7 years in a dense 10 K cloud. Both types of experiments appear able to convert CO to H_2CO and CH_3OH at levels comparable to those detected in cometary comae and interstellar ices. Note, however, that the rate of conversion of CO by H atoms is still controversial (Hiraoka *et al.* 2002, 2005).

3.1.1. *CH₃OH as a Starting Point for Ethylene Glycol Formation*

Once formed, CH_3OH can undergo other reactions to make more complex molecules. Results from our experiments on H_2O + CH_3OH ice (3:1) showed that after irradiation at 16 K, CO, CO_2, HCO, H_2CO, HCOO⁻, and ethylene glycol ($HOCH_2CH_2OH$) are present (Hudson & Moore 2000a). Figure 4 shows the sample's IR spectra before and after irradiation. Absorptions at 1088 cm^{-1} (peak) and 1046 cm^{-1} (shoulder) are matched with a reference spectrum of ethylene glycol. This molecule's formation is explained by straightforward chemistry in which hydroxymethyl radicals (CH_2OH), produced by proton irradiation of CH_3OH, couple to make $HOCH_2CH_2OH$ (see Reaction Scheme 1).

3.1.2. *Relationship Between Ethylene Glycol and Glycolaldehyde*

Recently we reported the solid-phase synthesis of glycolaldehyde [$HOCH_2C(O)H$] in irradiated ices containing ethylene glycol, both with and without H_2O (Hudson *et al.* 2005). This reaction is consistent with other aldehyde–alcohol interconversions we have studied, such as for [H_2CO, CH_3OH], [$CH_3C(O)H$, CH_3CH_2OH] pairs (Hudson & Moore 2001). Figure 5 shows the spectral changes of ethylene glycol ice during proton irradiation.

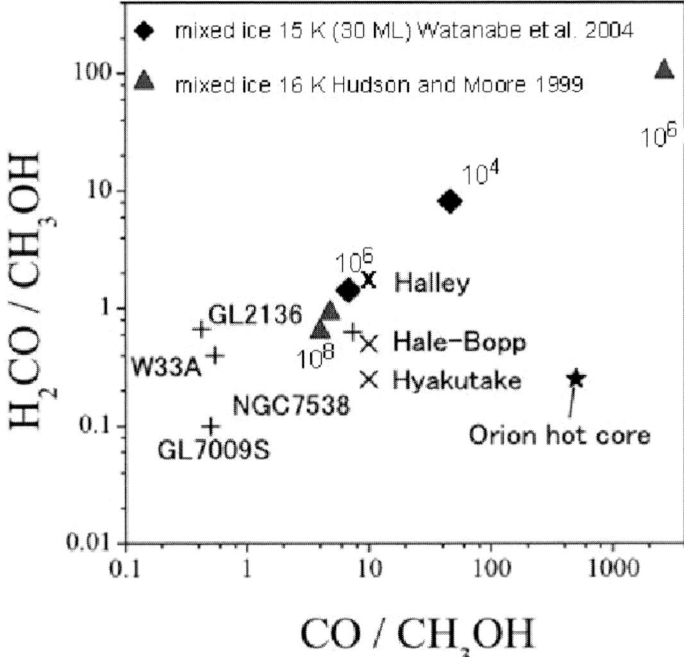

Figure 3. Abundance ratios of H_2CO and CO, relative to CH_3OH, are compared for hydrogenation results of Watanabe *et al.* (2004) on H_2O+CO (4:1) ice at 15 K, proton irradiation results of Hudson & Moore (1999) on $H_2O + CO$ (5:1) ice at $\sim$16 K, and observed ratios in interstellar ices, comets, and the Orion hot core. The graph is based on Watanabe *et al.* (2004, Fig. 5), and shows increasing processing from right to left. Powers of 10 indicate the ice's estimated exposure time (in years) to either the H fluence in molecular clouds (Watanabe's data) or the 1-MeV H^+ cosmic-ray fluence (Hudson's data).

Another route we examined for glycolaldehyde formation will apply when CH_3OH exists in a CO-rich (apolar) ice. Figure 6 shows the IR spectrum of CO + CH_3OH (100:1) after radiolysis. In this ice, rupture of a CH bond gives an H atom that reacts with the CO matrix to make HCO, with a subsequent radical-radical reaction between HCO and CH_2OH giving glycolaldehyde. Two weak features in Figure 6 indicate the presence of $HOCH_2C(O)H$, as shown by our reference spectrum. Other products include the HCO radical, H_2CO, and $CH_3C(O)H$.

3.2. *Study of C_2H_2 in H_2O Ices*

A second small, carbon-containing, unsaturated molecule we examined is acetylene (C_2H_2). Acetylene forms in the photospheres of AGB stars where $C/O \gg 1$, locking up most of the elemental carbon. Gas-phase acetylene and its derivatives are thought to be the building blocks of PAHs and soot (e.g., Tielens & Charnley 1997). When C_2H_2 is condensed into an icy grain mantle, however, H-atom addition reactions are possible, culminating in C_2H_6 (Charnley *et al.* 1992, 1995). In fact, it was the detection of nearly equal abundances of C_2H_6, C_2H_2, and CH_4 in Comets C/1996 B2 Hyakutake and C/1996 O1 Hale-Bopp that led Mumma *et al.* (1996) to point out that these cometary species did not originate in a thermochemically-equilibrated region of the solar nebula. Instead, the observed abundances were consistent with expected production of C_2H_6 in interstellar ices. Similar

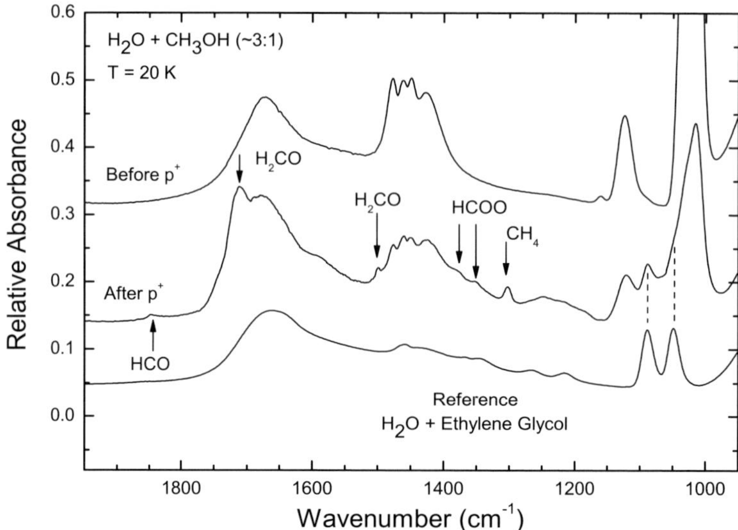

Figure 4. IR spectra of H$_2$O + CH$_3$OH ice before (upper) and after (middle) an irradiation to ~18 eV molecule^{-1} with 0.8 MeV protons. New products include H$_2$CO, CH$_4$, HCOO$^-$, and ethylene glycol. At the bottom is a spectrum of unirradiated H$_2$O + ethylene glycol for reference. Not shown are several other confirming bands for ethylene glycol between 500 and 1000 cm^{-1}.

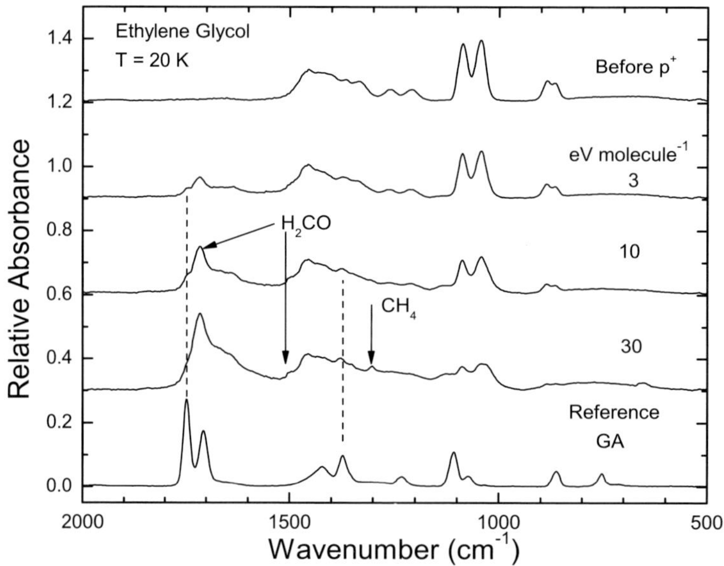

Figure 5. IR spectra of ethylene glycol before and after proton irradiation to a dose of ~3, 10, and 30 eV molecule^{-1} compared to a spectrum of glycolaldehyde (GA) at 20 K.

abundances for these organic volatiles were also found in Comet Lee, 153P/Ikeya-Zhang, and 9P/Tempel-1 (post-impact); see Mumma *et al.* (2005) and references therein.

We have shown experimentally that radiation processing of H$_2$O + C$_2$H$_2$ ices forms C$_2$H$_4$ and C$_2$H$_6$ through H addition (Moore & Hudson 1998) in the condensed phase.

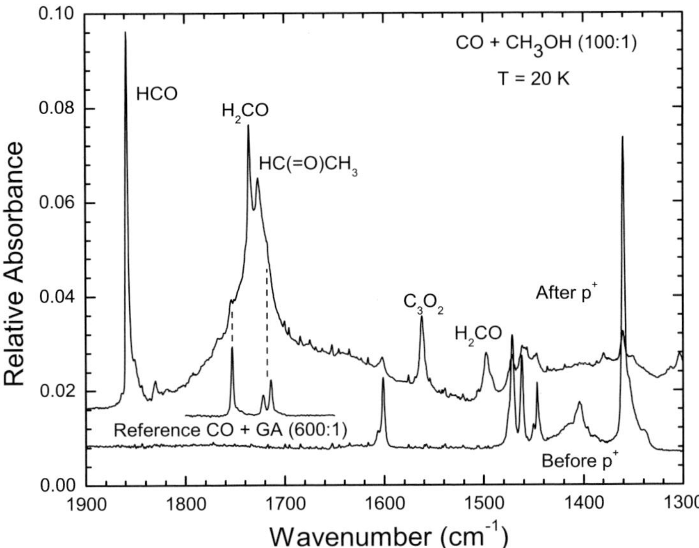

Figure 6. IR spectra of CO + CH$_3$OH (100:1) before (lower) and after (upper) irradiation with 0.8 MeV protons to a dose of $\sim$1 eV molecule^{-1}. A reference spectrum of CO + glycolaldehyde (GA) (600:1) is compared with weak features on the wings of the large 1730 cm^{-1} band.

$$HC\equiv CH \xrightarrow[\sim 10\ K]{p^+\ H_2O} \left[C_2H_3\right] \xrightarrow{H} C_2H_4 \xrightarrow{H} \left[C_2H_5\right] \xrightarrow{H} C_2H_6$$

$$\left[C_2H_3\right] \xrightarrow{OH} CH_2CH(OH) \longrightarrow CH_3C(O)H$$

$$\left[C_2H_5\right] \xrightarrow{OH} CH_3CH_2OH$$

REACTION SCHEME 2

CH$_3$C(O)H, C$_2$H$_6$, CH$_2$CH(OH), and C$_2$H$_5$OH form through the multi-step reactions seen in Reaction Scheme 2. In addition, both CO$_2$ and CO are detected along with H$_2$CO. Figure 7 shows IR spectra of H$_2$O + C$_2$H$_2$ (4:1) at 20 K before and after a dose of 17 eV molecule^{-1}. Additional C$_2$H$_6$ absorptions at 2880, 2939, and 2975 cm^{-1} were stronger than the weak 1464 cm^{-1} band shown in this region, and aided in the spectral assignments. The identifications of the isomers vinyl alcohol and acetaldehyde, CH$_2$CH(OH) and CH$_3$C(O)H respectively, are especially interesting (Hudson & Moore 2003) as both are known interstellar molecules. Our experiments support the idea that a similar chemistry produces these molecules in interstellar space.

The abundance ratios of CH$_4$, C$_2$H$_6$, and C$_2$H$_2$ in these experiments are sufficient to explain the cometary observations. Variables that did change the final abundances include the radiation dose, the dilution of C$_2$H$_2$ in the starting ice, and the initial presence of CH$_4$ in the ice. Since we did not observe C$_2$H$_2$ formation in irradiated H$_2$O + CH$_4$ mixtures, but did find CH$_4$ formation in H$_2$O + C$_2$H$_2$ ices, we conclude that relevant

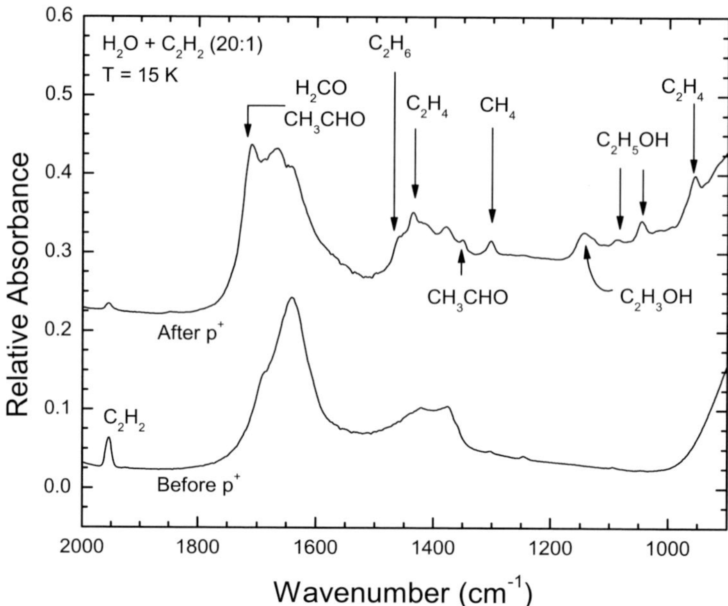

Wavenumber (cm^{-1})

Figure 7. IR spectra of H_2O + C_2H_2 before and after irradiation to a dose of 17 eV molecule^{-1}. New products are identified by comparison with appropriate reference spectra (see Moore & Hudson 1998).

$$HC \equiv C - C\overset{O}{\underset{H}{<}} \xrightarrow{2\,H} H_2C = CH - C\overset{O}{\underset{H}{<}} \xrightarrow{2\,H} H_3C - CH_2 - C\overset{O}{\underset{H}{<}} \xrightarrow{2\,H} H_3C - CH_2 - CH_2OH$$

REACTION SCHEME 3

hydrocarbon chemistry for these comets is founded on H addition to C_2H_2, as opposed to CH_4 reactions. This suggests that C_2H_2 is part of the natal cometary ice composition.

In extrapolating these results to a more complex molecule, propynal [HC≡CC(O)H], similar H-addition reactions are expected to form propenal and propanal. All three of these aldehydes have been observed in Sagittarius B2(N) (Hollis *et al.* 2004). The H-atom addition sequence connecting these molecules is expected to be that shown in Reaction Scheme 3. Although only the first three organics in Reaction Scheme 3 have been identified as interstellar, the conversion of aldehydes to alcohols is consistently found in our experiments. Therefore, we safely predict the conversion of propanal into *n*-propanol, and encourage observational searches for the latter.

3.3. *Study of HCN in H_2O Ices*

A third unsaturated, carbon-containing molecule we examined is HCN. This molecule is seen in the gas phase in the interstellar medium (e.g., Snyder & Buhl 1971; Boonman *et al.* 2001) and in cometary comae (e.g., Ip *et al.* 1990; Bockelée-Morvan *et al.* 1994; Magee-Sauer *et al.* 1999). It is also thought that HCN is a native nuclear ice component in comets. The idea that HCN might participate in the chemistry of interstellar ices was the motivation for our experiments on H_2O + HCN ices (Gerakines *et al.* 2004).

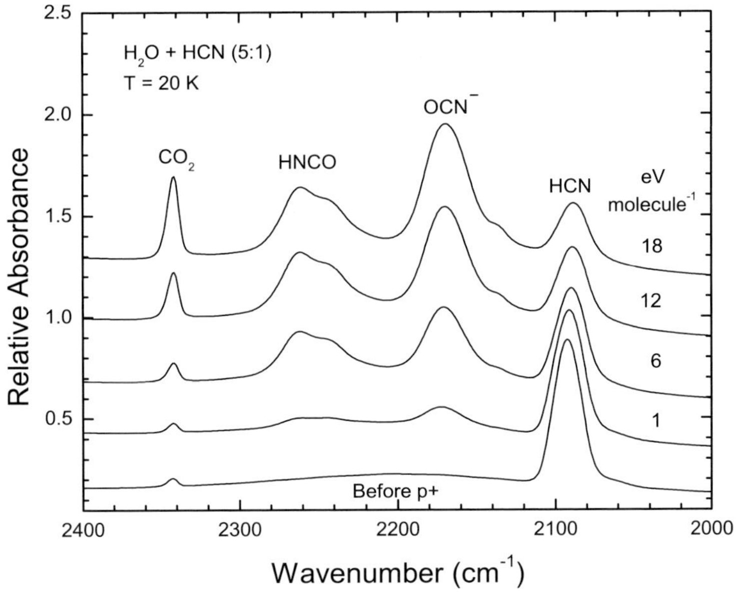

Figure 8. IR spectra of H_2O + HCN before and after irradiation to a dose of $\sim$1, 6, 12, and 18 eV molecule^{-1}. New products include HNCO, OCN$^-$, and CO_2.

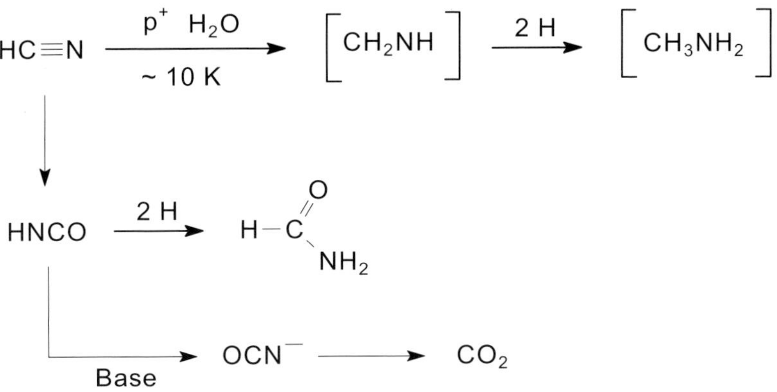

REACTION SCHEME 4

Solid-phase HCN reactions analogous to those with CO and C_2H_2 will result in methylenimine (CH_2NH) and methylamine (CH_3NH_2), both being IS gas-phase molecules. However, our experiments showed that oxidation products of HCN dominated the IR spectrum of H_2O + HCN ices after radiolysis. Figure 8 shows the 2400–2000 cm^{-1} region for H_2O + HCN (5:1) at 18 K before and after several radiation doses (Gerakines *et al.* 2004). Identified products are the cyanate anion (OCN$^-$), isocyanic acid (HNCO), and CO_2. Features in the 1800 – 1000 cm^{-1} region (not shown) were identified with the NH_4^+ cation at 1470 cm^{-1} and formamide ($HCONH_2$) at both 1686 and 1386 cm^{-1}. Steps leading to these products are shown in Reaction Scheme 4.

Although IS ices are dominated by H_2O, and neither HCN nor any organic nitrile has been identified in them, these ices often show an intense IR feature ascribed to OCN$^-$ (Hudson *et al.* 2001). This observation of OCN$^-$, as opposed to HCN and nitriles, can be understood from our experiments, since our work demonstrates the ease with which OCN^- forms in H_2O + HCN mixtures. Morover, HCN is not the only possible source of cyanate, as subsequent irradiations of many nitriles in H_2O-ice all showed OCN^- formation (Hudson & Moore 2004). This suggests that the cyanate ion may be a marker for energetic processing. Like CO_2, the cyanate ion is a stable, linear, 16-electron species. It has been seen in warming experiments to $\sim$200 K (Moore & Hudson 2003), and is predicted to be on the surfaces of outer solar system bodies.

HNCO, indicated in Figure 8, is detected only as a gas-phase species in both interstellar sources and comets (e.g., Gibb *et al.* 2000; Schutte & Khanna 2003 and references therein). Its absence in IS ices suggests that it reacts to form OCN^- and a counter-ion. Candidate bases include H_2O, formamide, and methylamine. More likely, HNCO combines with NH_3 to make OCN^- and NH_4^+. In fact, features in several interstellar sources are attributed to NH_4^+ (Schutte & Khanna 2003).

4. Conclusions

Our experiments demonstrate that several types of chemical reactions are involved in making complex molecules. In irradiated H_2O-rich ices containing CO or C_2H_2, H- and OH- additions produce many stable organics, some more oxidized than the reactants, and some more reduced. Radicals formed during irradiation can dimerize, thereby lengthening carbon chains. Examples include the dimerization of CH_3 to form C_2H_6 (Moore & Hudson 1998) and, more recently, the dimerization of CH_2OH to make $HOCH_2CH_2OH$ (Hudson *et al.* 2005). Current experiments involve studies of 2 CH_2CN → $(CH_2CN)_2$ in processed H_2O + CH_3CN. This is work in progress and relies on HPLC analysis of irradiated ice residues. It shows that identification of complex molecules are often at the limit of IR capabilities and require more-sensitive analytical techniques.

We have also found isomerization reactions help to explain the presence of isomeric pairs such as vinyl alcohol and acetaldehyde. Our work shows that vinyl alcohol is made in irradiated H_2O + C_2H_2, and then isomerizes to acetaldehyde. Among the nitriles, isomerizations include HCN → HNC and CH_3CN → CH_3NC (Hudson & Moore 2004), and formation of ketenimines such as CH_3CN → $H_2C{=}C{=}NH$. Extensions give a prediction of interstellar carbodiimide from cyanamide ($NH_2C{\equiv}N$ → $HN{=}C{=}NH$) and allene from methylacetylene ($CH_3C{\equiv}CH$ → $H_2C{=}C{=}CH_2$).

Interconversions of aldehydes and alcohols are examples of both H-atom addition and the reverse reaction (hydrogen elimination). Our work has included the following: [H_2CO, CH_3OH], [$CH_3C(O)H$, CH_3CH_2OH], and [$HOCH_2C(O)H$, $HOCH_2CH_2OH$]. This chemistry is expected to apply to more-complex pairs such as [$HC{\equiv}C{-}C(O)H$, $HC{\equiv}C{-}CH_2OH$], [$H_2C{=}CH{-}C(O)H$, $H_2C{=}CH{-}CH_2OH$], and [$H_3C{-}CH_2{-}C(O)H$, $H_3C{-}CH_2{-}CH_2OH$]. Therefore, our experiments suggest that IS aldehyde-alcohol pairs are likely if either member participates in condensed-phase radiation chemistry.

In all cases, formation paths compete with dissociation and elimination processes that lead to less-complex molecules. Acid-base reactions can also play a role, particularly regarding the formation of ions. Altogether, the reactions examined lead to about 50% of the known stable IS organics containing oxygen, and most of those found in comets. These observations show that discovering the identities of and reactions paths for complex interstellar molecules will require a firm understanding of simpler systems.

Acknowledgements

NASA funding through the Planetary Atmospheres and SARA programs, and the Goddard Center for Astrobiology, is acknowledged. R.L.H. acknowledges the support of

NASA for Grant NAG-5-1843. Claude Smith and Steve Brown are thanked for their work with the Van de Graaff accelerator at the Goddard Space Flight Center.

References

Allamandola, L.J., Sandford, S.A., & Valero, G.J. 1988, *Icarus* 76, 225

Bockelée-Morvan, D., Padman, R., Davies, J.K., & Crovisier, J. 1994, *Planet. Space Sci.* 42, 655

Boonman, A.M.S., Stark, R., van der Tak, F.F.S., van Dishoeck, E.F., van der Wal, P.B. Schäfer, F., de Lange, G., & Laauwen, W.M. 2001, *Ap. J.* 553, L63

Bringa, E.M. & Johnson, R.E. 2004, *Ap. J.* 603, 159

Charnley, S.B., Kress, M.E., Tielens, A.G.G.M., & Millar, T.J. 1995, *Ap. J.* 448, 232

Charnley, S.B., Tielens, A.G.G.M., & Millar, T.J. 1992, *Ap. J.* 399, L71

Crovisier, J., Bockelée-Morvan, D., Biver, N., Colom, P., Despois, D., & Lis, D.C. 2004, *A&A* 418, L35

Dickens, J.E., Irvine, W.M., & DeVries, C.H. 1997, *Ap. J.* 312, 307

Federman, S.R. & Allen, M. 1991, *Ap. J.* 375, 157

Gerakines, P.A., Moore, M.H., & Hudson, R.L. 2004, *Icarus* 170, 204

Gibb, E.L., Whittet, D.C.B., & Chair, J.E. 2001, *Ap. J.* 558, 702

Hiraoka, K., *et al.* 2005, *Ap. J.* 620, 542

Hiraoka, K., *et al.* 2002, *Ap. J.* 577, 262

Hollis, J.M., Lovas, F.J., & Jewell, P.R. 2000, *Ap. J.* 540, L107

Hollis, J.M., Lovas, F.J., Jewell, P.R., & Coudert, L.H. 2002, *Ap. J.* 571, L59

Hollis, J.M., Jewell, P.R., Lovas, F.J., Remijan, A., & Møllendal, H. 2004, *Ap. J.* 610, L21

Hudson, R.L., Moore, M.H., & Cook, A.M. 2005, *Adv. Space Res.* 36, 184

Hudson, R.L. & Moore, M.H. 2004, *Icarus* 172, 466

Hudson, R.L. & Moore, M.H. 2003, *Ap. J.* 586, L107

Hudson, R.L. Moore, M.H., & Gerakines, P.A. 2001, *Ap. J.* 550, 1140

Hudson, R.L. & Moore, M.H. 2001, *JGR-Planets* 106, 33275

Hudson, R.L. & Moore, M.H. 2000a, *Icarus* 145, 661

Hudson, R.L. & Moore, M.H. 2000b, *A&A* 357, 787

Hudson, R.L. & Moore, M.H. 1999, *Icarus* 140, 451

Ip, W.-H., *et al.* 1990, *Ann. Geophys.* 8, 319

Magee-Sauer, K., Mumma, M.J., DiSanti, M.A., Dello-Russo, N., & Rettig, T.W. 1999, *Icarus* 142, 498

Mehringer, D.M., Snyder, L.E., & Miao, Y. 1997, *Ap. J.* 480, L71

Miao, Y. & Snyder, L.E. 1997, *Ap. J.* 480, L67

Moore, M.H. & Hudson, R.L. 2003b, *Icarus* 161, 486

Moore, M.H., Hudson, R.L., & Ferrante, R.F. 2003a, *Earth, Moon & Planets* 92, 291

Moore, M.H. & Hudson, R.L. 2000, *Icarus* 145, 282

Moore, M.H., Hudson, R.L., & Gerakines, P.A. 2001, *Spectrochimica Acta* 57, 843

Moore, M.H. & Hudson, R.L. 1998, *Icarus* 135, 518

Mumma, M.J., *et al.* 2005, *Science* 310, 270

Mumma, M.J., DiSanti, M. A., Dello-Russo, N., Fomenkova, M., Magee-Sauer, K., Kaminski, C.D., & Xie, D. X. 1996, *Science* 272, 1310

Nummelin, A., Dickens, J.E., Bergman, P. Hjalmarson, Å., Irvine, W.M., Ikeda, M., & Ohishi, M. 1998, *A&A* 337, 275

Okabe, H. 1978, *Photochemistry of Small Molecules*, Wiley, New York

Schutte, W.A. & Khanna, R.K. 2003, *A&A* 398, 1049

Snyder, L.E. & Buhl, D. 1971, *Ap. J.* 163, L47

Teixcira, T.C., Emerson, J.P., & Palumbo, M.E. 1998, *A&A* 330, 711

Tielens, A.G.G.M. & Whittet, D.C.B. 1997, in *Molecules in Astrophysics: Probes and Processes*, ed., E. F. van Dishoeck (Kluwer Academic Pub.), 45

Tielens, A.G.G.M. & Charnley, S.G. 1997, in *Origins of Life and Evolution of the Biosphere* (Kluwer Academic Pub.), 23

Tielens, A.G.G.M. & Allamandola, L.J. 1987, in *Interstellar Processes*, eds. D. J. Hollenbach & H. A. Thronson (Reidel Pub., Dordrecht), 397.

Turner, B.E. & Apponi, A.J. 2001, *Ap. J.* 561, L207

Watanabe, N., Nagaoka, A., Shiraki, R., & Kouchi A. 2004, *Ap. J.* 616, L638

Watanabe, N., Shiraki, T., & Kouchi, A. 2003, *Ap. J.* 588, L121

Ziegler, J.P., Biersack, J.P., & Littmark, U. 1985, *The Stopping and Range of Ions in Solids* (Pergamon Press), New York, also see: `http://www.srim.org/`

Discussion

SCHLEMMER: The reaction schemes are written with neutral species. Protons induce electrons in the ice. Therefore the question is on the role of ions and excited species in the ice chemistry presented.

MOORE: In our proton-radiation experiments, the observed chemistry is dominated by reactions produced by secondary electrons. These electrons combine with a cation, producing free radicals such as H and OH that lead to the products we observe. In some cases, the electrons are sufficiently energetic to produce molecular excited states or bond dissociation. What we show in our paper are overall reaction sequences and not the mechanistic details.

BAURICHTER: How much UV light background do you have along with your ion beam coming out of your ion source?

MOORE: In our laboratory, the proton beam passes through a nickel metal foil before reaching our samples. This foil prevents any light from the proton beam tube from entering the ice chamber. The main purpose of the foil is to prevent samples from cryopumping contaminants from the vacuum system of the Van de Graaff accelerator.

HERBST: Would you consider modeling interstellar radiolysis chemistry based on your lab work? In the ISM, one must also worry about other processes such as a non-inert atmosphere?

MOORE: We do get rates of processing as a function of radiation dose. For example, we can measure how much CO is left in an $H_2O + CO$ ice after various doses, and we can measure column densities of new species produced. Since there are few alternatives to radiation processing for making many of the observed complex IS organics, data such as we present should be incorporated into models of IS chemistry. A knowledge of the expected cosmic-ray flux will be needed.

Astrochemistry: Recent Successes and Current Challenges
Proceedings IAU Symposium No. 231, 2005
D.C. Lis, G.A. Blake & E. Herbst, eds.

© 2006 International Astronomical Union
doi:10.1017/S1743921306007253

Chemistry in the Dense Molecular Gas of Starburst Galaxies and AGNs

Susanne Aalto

Department of Radio and Space Science, Onsala Space Observatory, S-439 92 Onsala, Sweden
email: susanne@oso.chalmers.se

Abstract. Understanding the molecular phase of the ISM in starburst and active galaxies is important for the modelling of the onset and evolution of their nuclear activity. Observations of high density gas tracers such as HCN, HNC, HC_3N, HCO^+ and CN are essential for probing physical and chemical conditions of the dense, star-forming gas. These tracers show great potential as indicators of the evolution of star formation as well as probes of X-ray illuminated molecular gas around an active galactic nucleus (AGN). In particular, towards the inner kpc of luminous and ultra luminous galaxies will molecular line ratios prove useful as diagnostic tools, since optical and even near infrared starburst tracers are difficult to apply in these highly obscured regions.

Keywords. astrochemistry — galaxies: active — galaxies: evolution — galaxies: ISM — galaxies: starburst — ISM: abundances — radio lines: galaxies — radio lines: ISM

1. Introduction

In order to understand the activities in the centers of luminous galaxies it is essential to also study the physical conditions of the dense gas component. The lower transitions of ^{12}CO and ^{13}CO typically trace gas in the density range $n(H_2) = 10^2 - 10^3$ cm^{-3} while the polar molecule HCN is commonly used as a tracer of dense molecular gas, i.e. gas at $n(H_2) \geqslant 10^4$ cm^{-3} (e.g., Solomon *et al.* 1992; Helfer & Blitz 1993; Paglione *et al.* 1995; Curran *et al.* 2000; Gao & Solomon 2004). Solomon *et al.* (1992) and Gao & Solomon (2004) find a tighter correlation between FIR and HCN luminosity than the one previously found between FIR and CO. They suggest that, in general, the IR luminosities originate from star formation rather than AGN activity in FIR luminous galaxies.

However, in order to understand both the nature and evolutionary stage of the processes involving the dense gas phase it is essential to both study the excitation of HCN as well as the relative intensities and excitation of other high density gas tracers. In this review I will briefly discuss CN, HNC, HC_3N and HCO^+ and their interpretation in terms of central activity. In general, it is found that galaxies with otherwise similar FIR/HCN, or CO/HCN, luminosity ratios may show vast differences in the properties and chemistry of their dense gas.

This paper is focussed around some aspects of chemistry of the dense gas in luminous ($L_{IR} > 10^{11}$ L$_\odot$, LIRGs) and ultraluminous ($L_{IR} > 10^{12}$ L$_\odot$, ULIRGs) galaxies. Significant advances have been made lately in the high-resolution study of chemistry in less luminous, nearby galaxies—for example of the starburst region in the nucleus of IC 342 (e.g., Meier and Turner 2005); NGC 253 (e.g., Martin *et al.*) and in the active nucleus of NGC 1068 (e.g., Usero *et al.* 2004).

2. Molecular Line Ratios as Diagnostic Tools

The starburst and AGN activities of many luminous IR galaxies are obscured by large columns of dust, aggravating – and sometimes prohibiting – the studies of the properties of the activity though optical, UV, and even NIR diagnostic methods. For example, Gonzáles-Alfonso et al. (2004) suggest an extremely high extinction of $A_V \approx 10^4$ towards the nuclei of the ultraluminous merger Arp 220. Spectral lines and continuum at longer wavelengths are therefore useful additional diagnostic tools to determine properties and evolution of the central activity of luminous galaxies, for example radio continuum and radio recombination lines (e.g., Yun et al. 2005).

The properties of the star-forming gas can also be studied via its chemistry and here we will briefly discuss a few line ratios for the dense gas, mainly HCO^+/HCN, HNC/HCN and CN/HCN. It is important to note that the regions studied are compact, and are often only marginally resolved even by mm-wave interferometers, such as the OVRO or Plateau de Bure arrays. Therefore, the properties of the gas traced by the measured line ratios are the *average* properties of the gas within the beam. When ALMA comes on line, this will change dramatically since the resolving power of ALMA for an ultraluminous galaxy like Arp 220 will allow us to study the ISM properties on a GMC scale—enabling us to probe the properties of the nuclear gas of Arp 220 in great detail.

Single dish and aperture synthesis studies of ^{12}CO and ^{13}CO have shown that the $^{12}CO/^{13}CO$ 1–0 and 2–1 line intensity ratios are efficient diagnostic tools for large scale ISM properties in galaxies. In particular the ratio between diffuse and self-gravitating gas – as well as the effects of temperature and density in the moderately dense ($n = 10^2 - 10^4$) phase of the molecular gas. Within galaxies there is a general trend of increasing $^{12}CO/^{13}CO$ 1–0 towards the central region where the gas is warmer and denser (e.g., Wall et al. 1993; Aalto et al. 1995; Paglione et al. 2001). Globally, there is a correlation between the $^{12}CO/^{13}CO$ 1–0 ratio and the FIR $f(60\mu m)/f(100\mu m)$ flux ratio (e.g., Young & Sanders 1986; Aalto et al. 1991b; Aalto 1995). The extreme ratios generally occur in ULIRGs with large dust temperatures. For various discussions of this important diagnostic line ratio see, for instance: Young & Sanders (1986); Aalto et al. (1991ab); Casoli et al. (1992); Aalto (1995); Hüttemeister et al. (2000); Paglione et al. (2001); Glenn & Hunter (2001); Tosali et al. (2002); Aalto (2005ab).

3. The HCO^+/HCN 1–0 Line Ratio

HCO^+ is suggested by Kohno et al. (2001, 2005) to be a tracer of the fraction of the dense gas which is involved in star formation. They compared the HCN/HCO^+ 1–0 line ratio in a selection of Seyfert and starburst galaxies and claim to find that the relative HCO^+ 1–0 luminosity is significantly higher in starbursts. According to Lepp & Dalgarno (1996) a deficiency of HCO^+ is expected near a hard X-ray source—while the HCN abundance may instead increase. Hence, in an AGN dominated nuclear ISM, one expects a brighter HCN line with respect to HCO^+ than in a softer starburst environment. *Lepp and Dalgarno propose using the HCN/HCO^+ ratio as an indicator of what scenario (AGN or starburst) that dominates the luminosity.* Kohno and his collaborators have demonstrated the potential observational usefulness of this line ratio. In Figure 1, a diagnostic diagram is presented showing the HCN/HCO^+ ratio vs. the HCN/CO ratio. Here, composite (i.e. harboring both an AGN and a nuclear starburst) galaxies are mixed in with starbursts, while the pure AGNs show brighter HCN emission relative to both the CO and HCO^+ luminosity. They have applied this to the dusty, edge-on IR-uminous

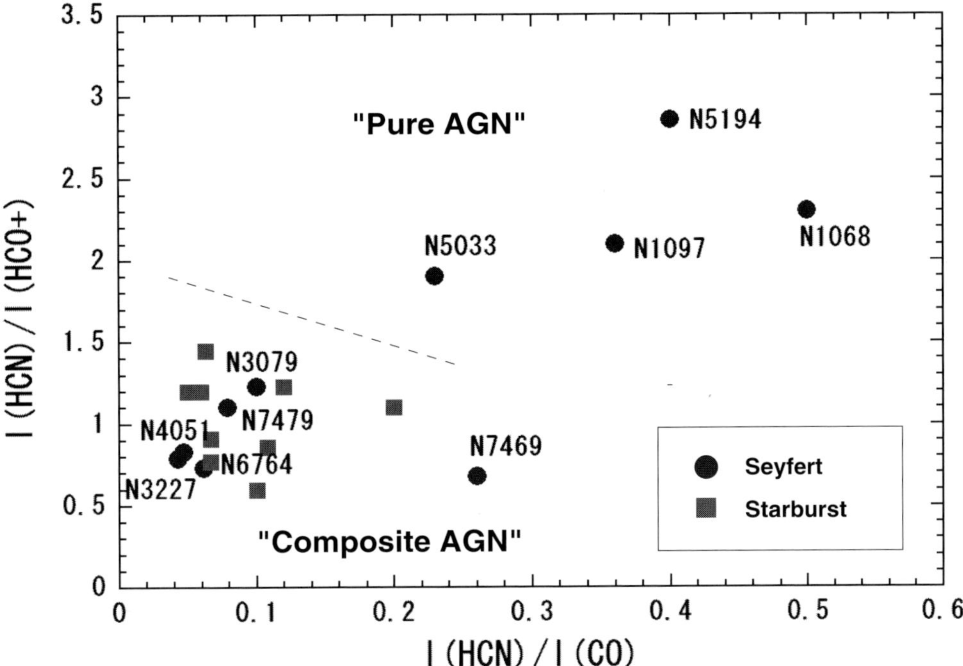

Figure 1. The HCN/HCO^{+} and HCN/CO line ratios as diagnostic tracers of AGN-Starburst activity (Kohno *et al.* 2001, 2005).

galaxy NGC 4418, and arrived at the conclusion that there is a significant contribution to the luminosity from an AGN since HCN(1–0)/HCO^{+}(1–0)=2.5.

This demonstrates again the importance of using tracers other than HCN to investigate what the dense gas is actually *doing* in the centers of external galaxies. The molecular ISM of the nearby Seyfert 2 galaxy NGC 1068 has been discussed in terms of X-ray dominated chemistry by Usero *et al.* (2004) and is another example where the HCN 1–0 emission outshines that from HCO^{+} in the nucleus.

4. The HNC/HCN Ratio

HNC, the isomer of HCN, traces gas of equally high density and in dense molecular cloud cores it may be a useful tracer of gas temperature since neutral-neutral chemical models predict that the [HCN]/[HNC] abundance ratio increases with increasing temperature. This is supported by the fact that the measured [HCN]/[HNC] abundance ratio is especially high in the vicinity of the hot core of Orion KL. Most of the temperature dependence is between 10 and 50 K, after which there is a considerable flattening (Schilke *et al.* 1992).

It is therefore surprising that the HCN/HNC J=1–0 intensity ratios are found to be low in luminous galaxies (e.g., Hüttemeister *et al.* 1995; Aalto *et al.* 2002, APHC02) and that the HCN/HNC line ratio appears to increase with galactic luminosity. APHC02 undertook a survey of HNC 1–0 line emission in a sample of 13 luminous IR galaxies ($L_{IR} > 10^{11}$ $L_{\odot}$) with previously measured HCN luminosities. They found that the HCN/HNC 1–0 ratios vary strongly within the sample—from 1 to $\gtrsim$ 6. From this we can learn that the actual *properties* of the dense gas vary significantly from galaxy to galaxy, even if their FIR luminosities, HCN luminosities (and global ^{12}CO/HCN 1–0 luminosity ratios) are similar. Galaxies where $I(\mathrm{HCN})=I(\mathrm{HNC})$ were all found to be luminous Seyfert galaxies. In general, it can be concluded that the *HNC emission is not a reliable tracer of cold (10 K) gas in the center of luminous IR galaxies*, the way it may be in clouds in the disk of the Milky Way. Standard interpretations based on molecular clouds in the disk of the Milky Way cannot be used for inner kiloparsec of luminous galaxies.

As an explanation for the abnormally bright HNC emission, APHC02 suggest that the chemistry is dominated by fast ion-neutral reactions in moderately dense ($n = 10^4 - 10^5\ \mathrm{cm}^{-3}$) PDR-like regions, instead of the neutral-neutral reactions that likely govern the hot dense cores of the Orion cloud. At lower densities, reactions with HCNH^+ (HCN and HNC reacts with H_3^+ to form HCNH^+) become more important. The ion abundance is higher and once HCN and HNC become protonated, HCNH^+ will recombine to produce either HCN or HNC with 50% probability. At higher densities, the ion abundance is likely lower and reactions like $\mathrm{HNC} + \mathrm{O} \rightarrow \mathrm{CO} + \mathrm{NH}$ become more prominent at high temperatures. This scenario is interesting, since the electron and ion abundance is likely higher in PDRs. Therefore, in a PDR chemistry, the connection between HNC abundance and kinetic temperature may also be weak since we there expect the HCNH^+ reactions to be important. It is easy to conceive of a chemistry dominated by fast ion-neutral reactions at the heart of an active starburst where there is strong interaction between the activity itself (in photon dominated regions, PDRs) and the surrounding ISM. It is interesting to note that for the nearby galaxy nucleus IC 342 Meier and Turner (2003) find no correlation between the HCN/HNC 1–0 intensity ratio and star forming activity.

Another possible scenario for more luminous galaxies is that, instead of being collisionally excited by H_2, HNC (and perhaps also HCN) is being excited by a number of processes. Both HCN and HNC may become excited via electron collisions (at $X(e) \approx 10^{-5}$) or be pumped by $14\,\mu\mathrm{m}$ (HCN) or $21.5\,\mu\mathrm{m}$ (HNC) continuum radiation through vibrational transitions in the degenerate bending mode. Furthermore, we note that some XDR models predict an overabundance of HNC compared to that of HCN (Meijerink & Spaans 2005), which should also be looked into as a possible explanation for bright HNC emission in Seyfert galaxies.

In order to investigate the underlying reasons behind the overluminous HNC emission in luminous galaxies, Aalto *et al.* have searched for HCN and HNC 3–2 emission in a sample of LIRG and ULIRG galaxies with the JCMT telescope. So far, the results show similar excitation for both HNC and HCN in most galaxies studied. Typically, the 3–2 line is fainter than the 1–0 line by at least a factor of 2, suggesting overall densities $\lesssim 5 \times 10^4\ \mathrm{cm}^{-3}$ in the HCN, HNC emitting gas.

4.1. *IR Pumping of HNC?*

An exception to the above finding of subthermally excited HNC is the ultraluminous galaxy Arp 220. For Arp 220, the HNC excitation is highly superthermal with a 3–2/1–0 ratio of $\gtrsim 2$ (Aalto *et al.* 2005c). Wiedner *et al.* find the corresponding HCN to be thermal with ratios close to 0.9. Furthermore, the HNC 3–2 line is significantly brighter than the HCN 3–2 line.

Both HCN and HNC have degenerate bending modes in the IR. For HNC this mode occurs at λ=21.5 $\mu\mathrm{m}$ with an energy level $h\nu/\mathrm{k}$=669 K and an A-coefficient of A_{IR}=5.2 s^{-1}. For HCN, the bending mode occurs at λ=14 $\mu\mathrm{m}$, energy level $h\nu/\mathrm{k}$=1029 K and A_{IR}= 1.7 s^{-1} (see, e.g., Aalto *et al.* 1994). Given that there is a sufficient HNC abundance, it is therefore easier to pump the bending state of HNC than HCN. A brief analysis show that the pumping starts to become effective when the IR background reaches an optically thick brightness temperature of $T_{\mathrm{B}} \approx 50$ K.

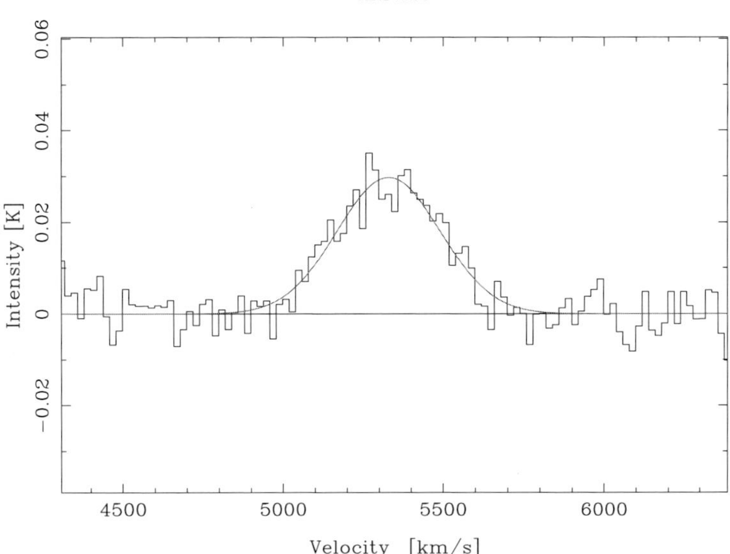

Figure 2. HNC 3–2 JCMT spectrum of Arp 220. Antenna temperature scale is in T_A^* (Aalto *et al.* 2005c).

4.2. *Pumping of HNC and HCN in Arp 220*

Soifer *et al.* (1999) discuss 24.5 μm brightness temperatures of Arp 220 in excess of 85 K. This result is model and source size dependent—the actual temperatures may in fact be higher. The HNC 3–2 line intensity is greater than that of HCN 3–2, which may indicate that the HNC line is affected by IR-pumping. In extremely cold ($\lesssim 10$ K) and dense conditions it is possible that X[HNC]>X[HCN]—conditions such as these may occur in the hearts of cold dark clouds (e.g Ziurys and Turner 1986). In the ULIRG ARP 220 however, the ISM conditions are very unlike those of cold dark clouds with an extreme starburst and average molecular gas temperatures exceeding 40 K (e.g Sakamoto *et al.* (1999) and an intense IR radiation field (Soifer *et al.* 1999). In contrast to HCN, CN is subthermally excited—indicating moderate densities around 5×10^4 cm^{-3}—while (if collisionally excited) the HCN line ratios suggest densities two orders of magnitude greater, $10^6 - 10^7$ cm^{-3}. How can we reconcile these two apparently conflicting results? Perhaps the CN emission is emerging from moderately dense PDRs at the surfaces of very dense clumps from which the HCN emission arises. Alternatively, the CN emission may reflect the density of the gas—while (at least a fraction of) the HCN emission may be pumped. Further excitation studies of HCN, HNC and CN and other high density tracer molecules are necessary to distinguish between these two scenarios.

5. CN and PDRs in Luminous Galaxies

The radical CN is another tracer of dense gas, with a somewhat lower (by $\sim 5\times$) critical density than HCN. The abundance of the CN radical becomes enhanced at the inner edge of a PDR (at an A_V of about 2 magnitudes). At larger depths into the cloud the CN abundance radically declines and the [HCN]/[CN] abundance ratio increases (e.g., Jansen *et al.* 1995). Observations of the CN emission toward the Orion A molecular clouds (Rodríguez-Franco *et al.* 1998) show that the morphology of the CN emission is dominated by the ionization fronts of H II regions. The authors conclude that this molecule is an excellent tracer of regions affected by UV radiation. Thus, the emission from the CN molecule should serve as a measure of the relative importance of gas in PDRs.

Aalto *et al.* (2002) also conducted a CN 1–0 (and 2–1) survey in the same galaxies they surveyed for HNC (see above). The goal was to probe the PDR phase of the molecular ISM of which CN should be a reliable tracer. Also in this study, the HCN/CN 1–0 intensity ratios show significant variation—ranging from 0.5 to $\gtrsim 6$. There is a surprising trend of *decreasing* CN luminosity with increasing FIR luminosity (but this trend needs to be confirmed with a larger sample). This is unexpected, since the luminosity of ULIRGs is suspected to be largely dominated by massive starbursts. It is also noteworthy that another classical PDR tracer is weak in ULIRGs: the [C II] 158 μm fine structure line is found to be unexpectedly faint compared to other, less FIR luminous, starburst galaxies like NGC 3690 (e.g., Luhman *et al.* 1998). Malhotra *et al.* (1997) report a decreasing trend in $F_{\rm [C II]}/F_{\rm FIR}$ with increasing $f(60)/f(100)$ μm flux ratio. Several possible explanations for the [C II] faintness have been brought forward (e.g., Malhotra *et al.* 1997; Luhman *et al.* 1998; van der Werf 2001). The PDRs may be quenched in the high pressure, high density environment in the deep potentials of the ULIRGs and the H II regions exist in forms of small-volume, ultracompact H II regions that are dust-bounded. The [C II] line may become saturated either in low density ($n \propto 10^2$ cm^{-3}) regions of very high UV fields ($G_0 \propto 10^3$) or in dense ($n \propto 10^5$ cm^{-3}) regions of more moderate UV fields ($G_0 = 5 - 10$). A soft UV field from an aging starburst is another possibility, and a higher dust-to-gas ratio would also decrease the expected $F_{\rm [C II]}/F_{\rm FIR}$ ratio.

There is not a one-to-one correspondence between CN and [C II] faint galaxies, but the general trend appears similar and should be further investigated. This emphasizes the need for care when using nearby, significantly less luminous, systems as prototypes for all starbursts. The molecular interstellar medium of ULIRGs seems to have different properties than that of more modest bursts and we still have to find out whether the cause is evolutionary or whether the bursts are intrinsically different in the ULIRGs. The presence of an AGN may also significantly affect the properties of the nuclear ISM.

5.1. *CN in Arp 299*

In the luminous merger Arp 299 there are three main regions of activity: the two galactic nuclei (IC 694 and NGC 3690) and the overlap region (where the galactic disks come together). Arp 299 has been studied in ^{12}CO, ^{13}CO 1–0, 2–1, HCN 1–0 at high resolution (Aalto *et al.* 1997; Aalto *et al.* 1999; Casoli *et al.* 1999). The dense gas, as traced by HCN, was found to be abundant in the three main regions of activity.

Aalto *et al.* imaged the CN 1–0 emission with the OVRO array with the goal of measuring the HCN/CN line ratio in the three main molecular emission regions. As is evident in Figure 3, *the CN 1–0 emission peaks towards the nucleus of IC 694 and is also bright in the overlap region, while no emission can be detected towards the center of NGC 3690.* A preliminary estimate of the HCN/CN ratios suggest that the HCN/CN ratio exceeds 5 in the NGC 3690 nucleus. This reveals that the properties of the molecular ISM in the two nuclei is quite different—something already indicated in the high resolution ^{13}CO data. The fact that bright CN emission is found in the extended emission of the overlap region shows that it is not only confined to starbursts in the inner regions of galaxies. Why is there no CN detection towards the nucleus of NGC 3690 despite bright HCN emission? From NIR and optical studies (Alonso Herrero *et al.* 2000) the starburst in NGC 3690 is intermediate (age $\sim$7.5 Myrs) in age between IC 694 (11 Myrs) and the overlap region (4–5 Myrs)—so there appears to be no obvious connection to the age of the starburst. *However, if the nucleus of NGC 3690 is deeply dust-enshrouded and*

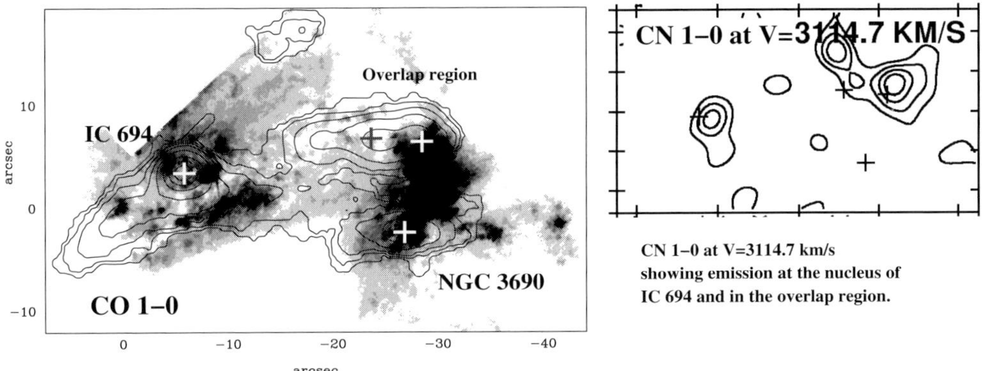

CN 1–0 at V=3114.7 km/s
showing emission at the nucleus of
IC 694 and in the overlap region.

Figure 3. The CN 1–0 emission (OVRO) towards the luminous merger Arp 299. The crosses mark the radio continuum peak positions of the IC 694 and NGC 3690 nuclei and in the overlap region.

compact it be unreachable by optical or near infrared diagnostic tools—then NGC 3690 may harbor a very young starburst or an AGN. In this context the CN-deficiency is the result of large masses of dust absorbing UV emission from the nuclear activity. From recent water maser observations, there is evidence of the presence of a deeply buried AGN in NGC 3690—and perhaps also in IC 694 (e.g., Della Ceca *et al.* 2002, Henkel *et al.* 2005). The CO emission is compact and unresolved in the OVRO beam towards the NGC 3690 nucleus, consistent with deeply buried activity (e.g., Aalto *et al.* (1997)).

Perhaps the cause behind the (apparently) faint CN emission in luminous galaxies is because the activity, and molecular distribution, becomes more nuclear with increasing luminosity. Compact, high-A_V regions are then expected to have bright HCN—but faint CN emission. The eastern nucleus of Arp 220 is another example of a CN-faint compact nucleus (see below).

6. HC$_3$N Tracing Young, Compact Starburst Activity?

From spectral line shapes one can deduce a difference in the HCN/CN 1–0 line ratio between the two nuclei of the ultraluminous merger Arp 220. Most of the CN emission appears to be emerging from the western nucleus while HCN 1–0 is coming from both nuclei (Aalto *et al.* 2002). The compact, deeply enshrouded eastern nucleus shows little or no CN emission, while instead it has a strong HC$_3$N signal (APHC02). Rodríguez-Franco *et al.* (1998) show that the emission from HC$_3$N is bright toward hot, dense cores, while the [HC$_3$N]/[CN] abundance ratio is only 10^{-3} in PDRs. Thus the eastern nucleus seems to be in an earlier evolutionary phase where star formation has just begun.

7. Concluding Remarks

Molecular line ratios are showing interesting, and sometimes surprising, results as tracers of the chemistry and physical conditions of the molecular gas in the inner regions of active galaxies. They are showing great potential in distinguishing between types of nuclear activity—as well as between evolutionary stages of a starburst. Together with other extinction-free tracers such as radio continuum, radio recombination lines and X-rays, the diagnostic value of the line ratios will be further enhanced. With the resolving power and sensitivity of the ALMA array, we will be able to study these regions in even more detail. This will also increase the complexity of the required chemical and radiative transport models hopefully intensifying contacts and collaborations between observers and theorists.

Acknowledgements

I am grateful to many people for discussions and advice. Most recently these include John Black, Henrik Olofsson, Peter Schilke, Martina Wiedner and Marco Spaans.

References

Aalto, S., Black, J.H., Booth, R.S., & Johansson, L.E.B. 1991a, *A&A* 247, 291

Aalto, S., Johansson, L.E.B., Booth, R.S., & Black, J.H. 1991b, *A&A* 249, 323

Aalto, S., Booth, R.S., Black, J.H., & Johansson, L.E.B. 1994, *A&A* 286, 365

Aalto, S., Booth, R.S., Black, J.H., & Johansson, L.E.B. 1995, *A&A* 300, 369

Aalto, S., Radford, S.J.E., Scoville, N.Z., & Sargent, A.I., 1997, *Ap. J.* 475, L107

Aalto, S., Radford, S.J.E., Scoville, N.Z., & Sargent, A.I. 1999, in *IAU Symposium 186, Galaxy Interactions at Low and High Redshift*, eds.J. E. Barnes & D.B. Sanders, 231

Aalto, S., Hüttemeister, S., & Walter, F. 2001, in *The Central Kiloparsec of Starbursts and AGN: The La Palma Connection*, eds. J.H. Knapen, J.E. Beckman, I. Shlosman, & T.J. Mahoney (ASP Conference Proceedings Vol. 249), p. 626

Aalto, S., Polatidis, A.G., Hüttemeister, S., & Curran, S.J. 2002, *A&A* 381, 783 (APHC02)

Aalto, S. 2005a, *Ap. S. S.* 295, 143

Aalto, S. 2005b, in *The Neutral ISM of Starburst Galaxies*, eds. S. Aalto, S. Hüttemeister and A. Pedlar, p.3

Aalto, S., Olsson, E., *et al.*, 2005c, submitted to *A&A*

Alonso-Herrero, A., Rieke, G., Rieke, M.J., & Scoville, N.Z. 2000, *Ap. J.* 532, 845

Casoli, F., Dupraz, C., & Combes, F. 1992, *A&A* 264, 55

Casoli, F., Willaime, M.-C., & Viallefond, F., & Gerin, M. 1999, *A&A* 346, 663

Curran, S.J., Aalto, S., & Booth, R.S. 2000, *A&AS* 141, 193

Della Ceca, R., Ballo, L., Tavecchio, F., *et al.* 2002, *Ap. J.* 581, L9

Gao, Y. & Solomon, P.M. 2004, *Ap. J.* 606, 271

Glenn, J. & Hunter, T.R. 2001, *Ap. J. S.* 135, 177

González-Alfonso, E., Smith, H.A., Fischer, J., & Cernicharo, J. 2004, *Ap. J.* 613, 247

Helfer, T.T. & Blitz L. 1993, *Ap. J.* 419, 86

Hüttemeister, S, Henkel, C., Mauersberger, R., *et al.* 1995, *A&A* 295, 571

Hüttemeister, S., Aalto, S., Das, M., & Wall, W.F. 2000, *A&A* 363, 93

Jansen, D.J., van Dishoeck, E.F., Black, J.H., Spaans, M., & Sosin, C. 1995, *A&A* 302, 223

Kohno, K., Matsushita, S., Vila-Vilaro, B., Okumura, S.K., Shibatsuka, T., Okiura, M., Ishizuki, S., & Kawabe, R. 2001, in *The Central Kiloparsec of Starbursts and AGN: The La Palma Connection*, eds. Knapen, Beckman, Shlosman, Mahoney (ASP Conference Series, Vol. 249), p. 672

Lepp, S. &, Dalgarno, A. 1996, *A&A* 306, L21

Luhman, M.L., Satyapal S., Fischer, J., *et al.* 1998, *Ap. J.* 504, L11

Malhotra, S., Helou G., Stacey G., *et al.* 1997, *Ap. J.* 491, L27

Meier, D.S., Turner, J.L., & Hurt, R.L. 2000, *Ap. J.* 531, 200

Meier, D.S. & Turner, J.L. 2005, *Ap. J.* 618, 259

Meijerink, R. & Spaans, M. 2005, *A&A* 436, 397

Paglione, T.A.D., Tosaki, T., & Jackson, J.M. 1995, *Ap. J.* 454, L117

Paglione, T.A.D., Wall, W.F., Young, J. S., Heyer, M.H., Richard, M., Goldstein, M., Kaufman, Z., Nantais, J., & Perry, G. 2001, *Ap. J.* 135, 183

Rodríguez-Franco, A., Martin-Pintado, J., & Fuente, A. 1998, *A&A* 329, 1097

Schilke, P., Walmsley, C.M., Pineau de Forêts, G., *et al.* 1992, *A&A* 256, 595

Scoville, N.Z., Sargent, A.I., Sanders, D.B., & Soifer, B.T. 1991, *Ap. J.* 366, L5

Soifer, B.T., Neugebauer, G., Matthews, K., *et al.* 1999, *Ap. J.* 513, 207

Solomon, P.M., Downes D., & Radford, S.J.E. 1992, *Ap. J.* 387, L55

Smith, J., *et al.* 1990, *Ap. J.* 362, 455

Tosaki, T., *et al.* 2002, *PASJ* 54, 209

Usero, A., *et al.* 2004, *A&A* 419, 897

van der Werf, P.P. 2001, in *Starburst galaxies—near and far*, proceedings of the Ringberg workshop, eds. D. Lutz, & L.J. Tacconi, in press

Wall, W.F., Jaffe, D.T., Bash, F.N., *et al.* 1993, *Ap. J.* 414, 98

Wiedner, M.C., *et al.* 2005, in *The Neutral ISM of Starburst Galaxies*, eds. S. Aalto, S. Hüttemeister, & A. Pedlar, p. 35

Wild, W., Harris, A.I., Eckart, A., *et al.* 1992, *A&A* 265, 447

Young, J.S. & Sanders, D.B. 1986, *Ap. J.* 302, 680

Yun, M.S., Scoville, N.Z., & Shukla, H. 2005, in *The Neutral ISM of Starburst Galaxies*, eds. S. Aalto, S. Hüttemeister & A. Pedlar, p. 27

Ziurys, L.M. & Turner, B.E. 1986, *Ap. J.* 302, L31

Photo: R. Paz

Astrochemistry: Recent Successes and Current Challenges
Proceedings IAU Symposium No. 231, 2005
D.C. Lis, G.A. Blake & E. Herbst, eds.

© 2006 International Astronomical Union
doi:10.1017/S1743921306007265

CO and [C I] in Nearby Galaxies: Probing Physical and Chemical Conditions

Christine D. Wilson[1,2]

[1]Department of Physics and Astronomy, McMaster University, Hamilton, Ontario L8S 4M1
Canada
email: wilson@physics.mcmaster.ca
[2]Harvard-Smithsonian Center for Astrophysics

Abstract. Observations of CO can provide constraints on the density and temperature of the dense star-forming gas in galaxies, while observations of [C I] trace photon-dominated regions and provide information on the amount of atomic carbon in molecular clouds. In this paper, I review CO and [C I] observations of nearby galaxies, with an emphasis on galaxies for which high-resolution observations of multiple transitions and isotopologues are available. I also briefly discuss upper limits on water emission from nearby starburst galaxies and on O_2 emission in the SMC obtained with the Odin satellite.

Keywords. galaxies — millimeter interferometry — molecular abundances

1. Introduction

Observations of gas-phase atoms and molecules in galaxies allow us, in principle, to study astrochemistry in physical environments that are quite different from those found in our own Galaxy, *e.g.*, lower abundances of heavy elements, stronger radiation fields, etc. However, one of the difficulties in observing molecular species in even the nearest galaxies is that the lines are quite weak, which limits us to observing only the most abundant species and only a few lines from each species. In addition, the relatively low spatial resolution that is obtained in observations of galaxies means that our data are probing a mixture of chemical and physical conditions, which makes the interpretation more difficult. Finally, the strongest molecular line, ^{12}CO, is generally optically thick and not in local thermal equilibrium across even its lowest rotational transitions, which requires more complicated radiative transfer models for its interpretation. All these factors produce the result that the very nearest galaxies are the easiest to study.

Aside from starburst galaxies (discussed, *e.g.*, in Aalto, this volume), the only galaxy which has been observed in a wide variety of molecular species is the Large Magellanic Cloud (LMC). Johansson *et al.* (1994) present observations of over 20 species and isotopologues towards the N159 complex of H II regions in the LMC. They are able to deduce a number of abundances from this data set as well as place constraints on the rotational temperature of several species. For example, they find a $[^{12}CO]/[^{13}CO]$ abundance ratio of 50 ± 20, very similar to that in the Milky Way (Langer & Penzias 1990), while the $[^{18}O]/[^{17}O]$ ratio of 2 ± 0.5 is a factor of two below the Galactic value.

Recent observations by Brouillet *et al.* (2005) illustrate the difficulties in observing species other than CO even in normal galaxies that are very nearby. They used the IRAM 30 m telescope to observe several regions in the nearby spiral galaxy M31 in the ground-state rotational transitions of HCN and HCO^+. Figure 1 in their paper shows their HCN and HCO^+ detections compared to the ^{12}CO J=1–0 spectra, which are roughly fifty times stronger than the HCN and HCO^+ lines in a given region. It is this extreme weakness of the lines from less abundant species that has limited detailed astrochemical work on nearby galaxies.

2. CO in Nearby Galaxies

There have been a large number of observations of CO emission from nearby galaxies in the last two decades and so I make no attempt to be complete in my discussion here. Good starting points for CO J=1–0 data are the FCRAO survey by Young et $al.$ (1995) and the BIMA SONG survey by Helfer et $al.$ (2003); Dumke et $al.$ (2001) discuss extended CO J=3–2 mapping of a number of nearby galaxies. In this section, I discuss CO observations of nearby galaxies with high intrinsic spatial resolution and observations of multiple transitions or isotopologues. I start with a somewhat older study of M33 by Wilson, Walker, & Thornley (1997), which gives a good illustration of the types of models and assumptions that are required. I then review some recent results from the Smithsonian Submillimeter Array (SMA) by Iono et $al.$ (2006) and Petitpas et $al.$ (2006) which illustrate the promise of submillimeter interferometry for this research.

2.1. *Giant Molecular Clouds in M33*

At a distance of 0.84 Mpc (Freedman, Wilson, & Madore 1991), M33 is close enough that large single dish telescopes can isolate the emission from individual giant molecular clouds in the disk. Wilson, Walker, & Thornley (1997) carried out a study of the ^{12}CO J=2–1 and J=3–2 and ^{13}CO J=2–1 emission from seven molecular clouds in M33 that had been previously identified in interferometric CO J=1–0 observations. The data revealed very uniform intensity ratios for ^{12}CO J=2–1/J=1–0 and ^{12}CO/^{13}CO J=2–1 emission, but showed significant variations in the ^{12}CO J=3–2/J=2–1 intensity ratio. Fitting all the data with large velocity gradient (LVG) models ($e.g.$, Scoville & Solomon 1974) gave typical values of 4×10^3 H$_2$ cm^{-3} for the molecular hydrogen volume density and 20 K for the kinetic temperature. The analysis presented in Wilson, Walker, & Thornley (1997) illustrates very clearly the effect that the adopted [^{12}CO]/[^{13}CO] abundance ratio has on the physical results of the models (Figure 1). The primary effect is to change the derived value for the ^{12}CO column density, $N(^{12}$CO$)$, in direct proportion to the adopted abundance ratio. Although it can generally be difficult to separate the effects of volume density and temperature in LVG models with only a few rotational transitions ($e.g.$, Petitpas & Wilson 2000), the apparent link between increased kinetic temperature and the presence of a (bright) H II region led Wilson, Walker, & Thornley (1997) to conclude that the variation in the ^{12}CO J=3–2/J=2–1 line ratio was primarily due to variations in the kinetic temperature of the gas.

2.2. *High-Resolution Observations of More Distant Galaxies*

For galaxies beyond the Local Group, the spatial resolution achievable with even large single dish telescopes declines to the point where we can no longer isolate individual molecular clouds or even small groups of clouds, but rather must focus on the properties of the interstellar medium on kiloparsec scales. Such observations can reveal interesting differences between spiral arm and interarm regions ($e.g.$, Walsh et al. 2002) as well as the dynamical properties of the molecular disk. However, to focus on the $\sim$ 100 pc scales where we may expect some degree of chemical and physical uniformity requires millimeter and submillimeter interferometry.

NGC 2903 is a barred spiral galaxy at a distance of $\sim$6 Mpc that has been mapped in the CO J=1–0 line as part of the BIMA SONG survey (Helfer et al. 2003). The CO J=1–0 observations show bright emission from the central region along with fainter

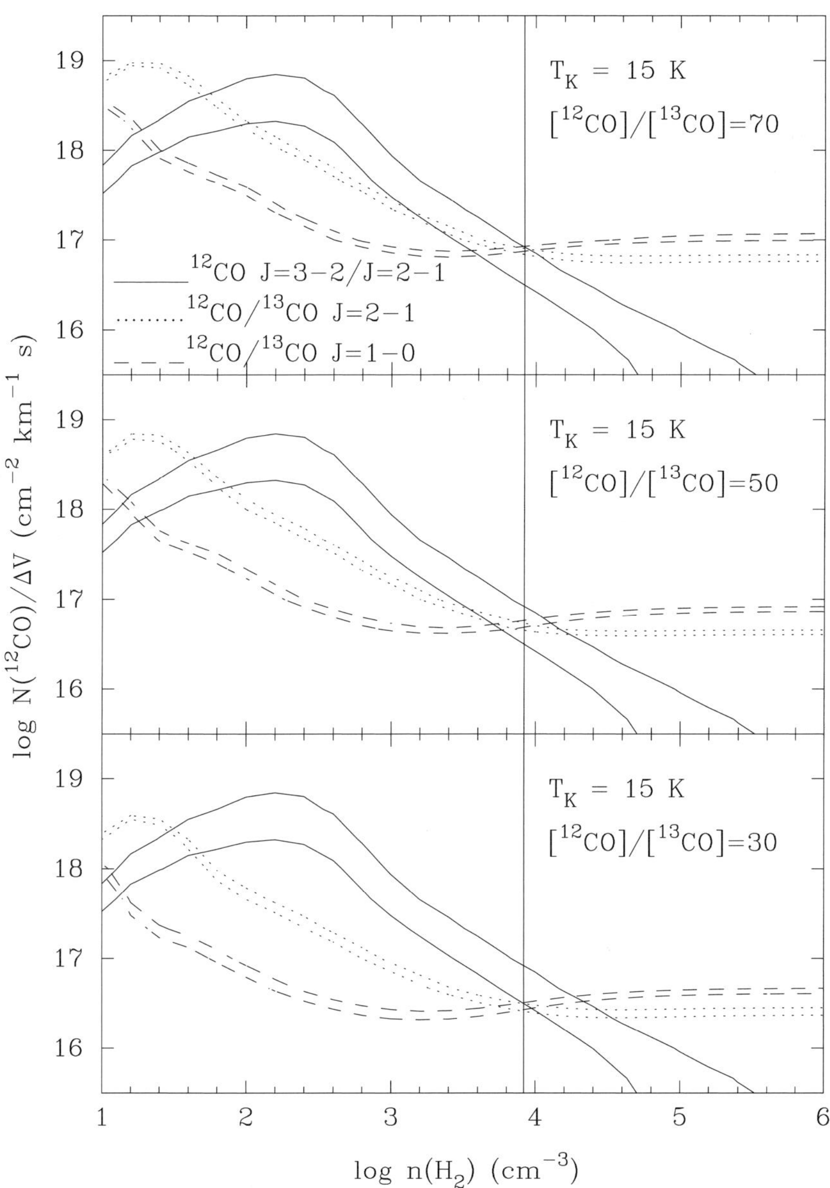

Figure 1. LVG models for molecular clouds in M33 for three different assumed values of the $[^{12}CO]/[^{13}CO]$ abundance ratio. The lines indicate the $\pm 1\sigma$ values for the observed line ratios of $^{12}CO/^{13}CO$ J=1–0 (dashed), $^{12}CO/^{13}CO$ J=2–1 (dotted), and ^{12}CO J=3–2/J=2–1 (solid). The vertical solid line indicates the common density solution for all three models. Figure from Wilson, Walker, & Thornley (1997).

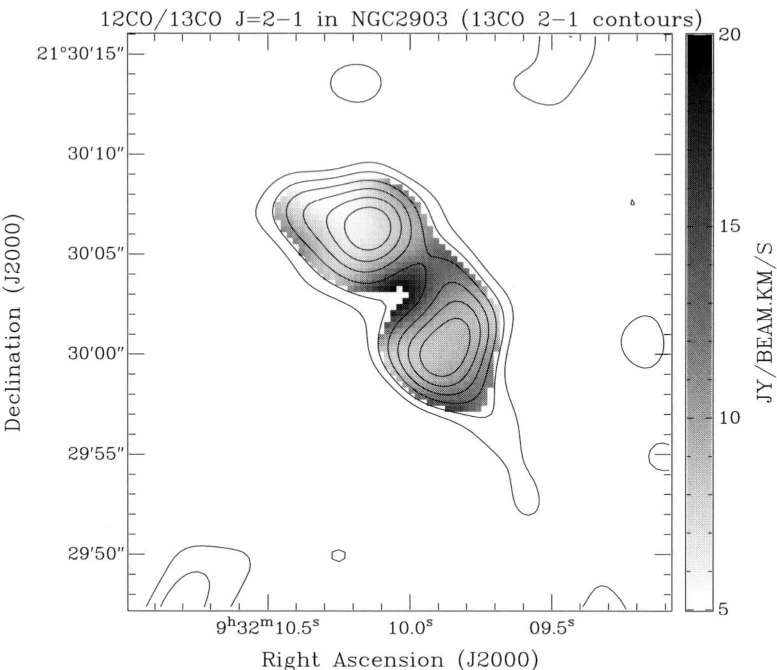

Figure 2. SMA map of the ^{12}CO/^{13}CO J=2–1 line intensity ratio (greyscale) in the central region of the barred spiral galaxy NGC 2903 with integrated intensity contours of the ^{13}CO J=2–1 emission overlaid. From Petitpas *et al.* (2006).

emission extending along the inner bar. Petitpas *et al.* (2006) have recently mapped the central region of NGC 2903 in the ^{12}CO and ^{13}CO J=2–1 lines with the SMA. The 2.5″ beam corresponds to a spatial resolution of 70 pc, comparable to the spatial resolution of the M33 study of Wilson, Walker, & Thornley (1997) discussed above. These data reveal a previously undetected double-peaked morphology in the central region; such morphologies are commonly seen in high-resolution CO J=1–0 observations of nuclear regions such as those of Kenney *et al.* (1992). What is new in the study of Petitpas *et al.* (2006) is the ability to trace the CO line ratios across the double-peaked morphology (Figure 2). The observations show lower ^{12}CO/^{13}CO integrated intensity ratios towards the two emission peaks; these lower line ratios most likely trace material with high optical depths and hence high ^{12}CO column densities. What is particularly interesting is the area with high line ratios located in the valley between the two emission peaks. Higher line ratios can indicate either a lower optical depth or a larger intrinsic $[^{12}$CO$]/[^{13}$CO$]$ abundance ratio. Since the ^{12}CO integrated intensity (not shown) is still strong where the line ratio is high, the low optical depth explanation would have to involve some rather unusual excitation conditions. On the other hand, rapid variations in the $[^{12}$CO$]/[^{13}$CO$]$ abundance ratio on the scale of a few hundred parsecs would also be unusual. It is clear there are interesting processes at work in the nucleus of this rather normal spiral galaxy.

Even more intriguing are recent results from SMA observations of the Antennae (NGC 4038/39) by Iono *et al.* (2006). At a distance of 19 Mpc, the Antennae is the closest example of a merger between two large spiral galaxies. Previous interferometric CO

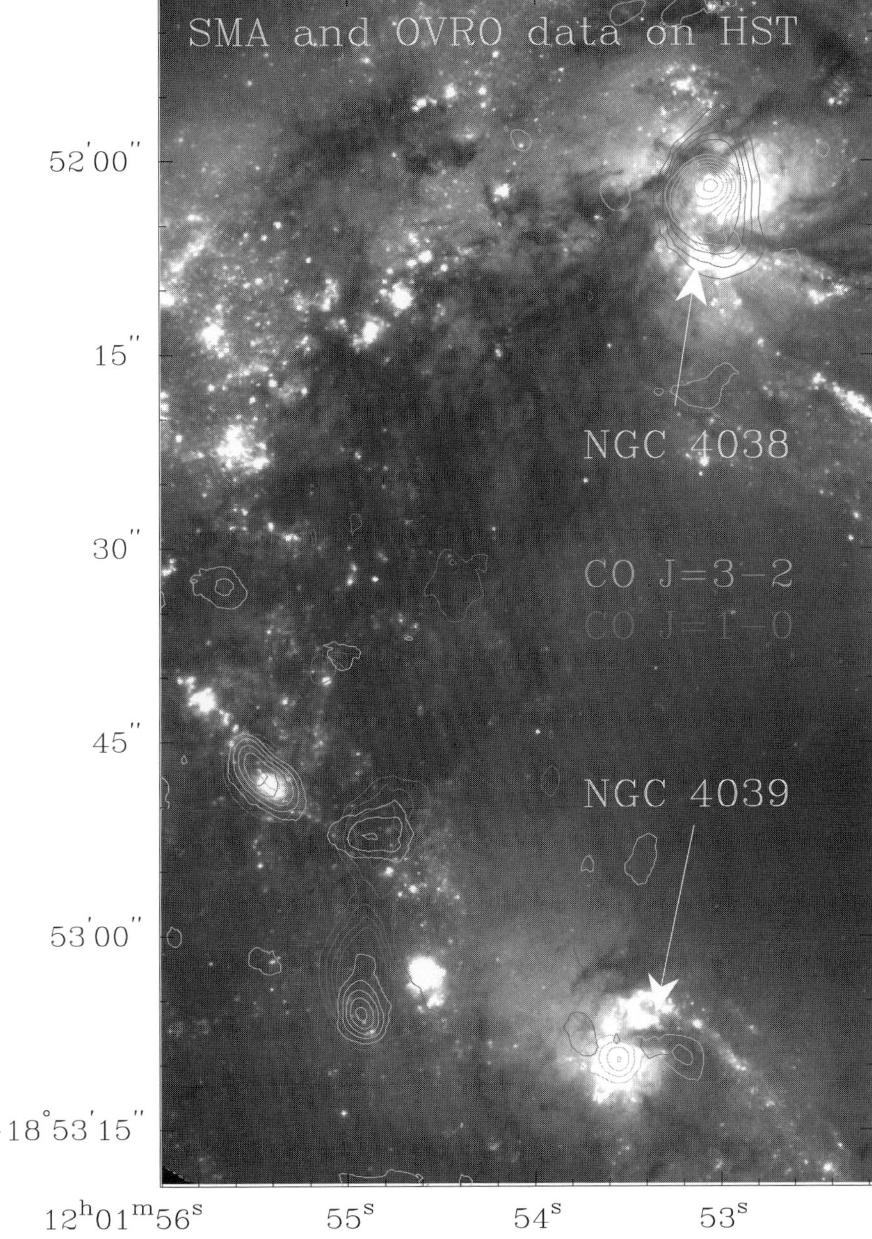

Figure 3. ^{12}CO $J=3$–2 and $J=1$–0 emission (from the SMA and OVRO, respectively) overlaid on the HST image from Whitmore *et al.* (1999) of the Antennae (NGC 4038/39). The two CO data sets have been filtered to have the same resolution and to sample the same range of spatial scales. From Iono *et al.* (2006).

$J=1$–0 observations identified a large number of super-massive molecular complexes with masses approaching 10^9 M$_\odot$, more than three orders of magnitude larger than the most massive giant molecular cloud in the Milky Way (Wilson *et al.* 2000, 2003). The unusually high masses of the cloud complexes in the Antennae combined with the shocks and high pressures expected in such a large collision have the potential to produce unusual physical and chemical conditions in the molecular gas. Iono *et al.* (2006) have observed the Antennae in the CO J=3–2 line (Figure 3) and have produced a high-quality map of the ^{12}CO J=3–2/J=1–0 line ratio by filtering the J=1–0 data to have exactly the same resolution and to probe the same range of spatial scales as the J=3–2 data. The resulting data cube reveals intriguing variations in this CO line ratio, both spatially *and* in velocity. These variations in line ratio may trace temperature or optical depth variations, while the velocity information helps to separate out physically distinct clouds located along the same line of sight.

3. [C I] in Nearby Galaxies

Gas-phase atomic carbon can be a significant fraction of the total carbon content in galaxies. The amount and distribution of atomic carbon can provide important constraints on the physical and chemical structure of molecular clouds. However, due to the difficulty in observing even the ground-state fine structure line of atomic carbon, there are relatively few published observations of atomic carbon in nearby galaxies. Thus, I have attempted to be complete in my review of published [C I] data in this section.

The first extragalactic observation of the $^3P_1 - {}^3P_0$ fine structure line of atomic carbon was made by Stark *et al.* (1997) of the LMC. Wilson (1997) observed four giant molecular clouds in M33 in [C I] emission and found that the [C]/[CO] abundance ratio ranged from 0.03 to 0.16 in the four clouds. The cooling by this single [C I] line accounted for 20–70% of the cooling of the cloud in the lowest three rotational transitions of CO (note that there is an error in the relative cooling rates calculated in Wilson 1997). Most Local Group galaxies have now been observed in the [C I] 492 GHz line [LMC: Kim, Walsh, & Xiao (2004); SMC: Bolatto *et al.* (2000a); M33: Taylor & Wilson (2000); M31: Israel, Tilanus, & Baas (1998); IC10: Bolatto *et al.* (2000b)], albeit in no more than a few positions.

Observations of individual, more distant galaxies in the [C I] line include Arp 220 (Gerin & Phillips 1998), M83 (Petitpas & Wilson 1998; older data are also presented in Israel & Baas 2001), NGC 6946 (Israel & Baas 2001), Maffei 2 and IC 342 (Israel & Baas 2003), and He2-10 and NGC 253 (Bayet *et al.* 2004). In addition, there have been two larger studies of nearby galaxies, one by Gerin & Phillips (2000) using the Caltech Submillimeter Observatory and one by Israel & Baas (2002) using the James Clerk Maxwell Telescope. The data in the Gerin & Phillips (2000) study is more modern than that of Israel & Baas (2002) and Gerin & Phillips also include a comprehensive discussion of published [C I] data from other authors and so I have chosen to focus my discussion on their results.

Gerin & Phillips (2000) find that the [C I]/^{12}CO J=1–0 intensity ratio is fairly constant, with an average value of 0.2 ± 0.2, in a wide range of environments (spirals, irregulars, interacting/merging galaxies; see their Figure 3), although there are some variations both within and between galaxies. When compared with the higher rotational transitions of ^{12}CO (J=2–1, J=3–2, and J=4–3), the [C I] emission is relatively weaker inside galactic nuclei but stronger in the disks, especially in regions of the disk that are not very active in star formation. They find that atomic carbon makes a significant contribution to the thermal budget, with cooling by C and CO generally of the same order of magnitude (although CO cooling is more important in the starburst galaxies). However, cooling by both C and CO is typically only 5% of the cooling due to ionized carbon and only 2×10^{-5} times the cooling due to the far-infrared continuum emission.

Gerin & Phillips (2000) were able to trace the spatial distribution of the various lines in the edge-on spiral galaxy NGC 891 and found that the [C I] emission traces the 1.3 mm dust continuum emission very well over the inner regions of the disk (see their Figure 5). Since these are the parts of the disk where molecular hydrogen is more abundant than

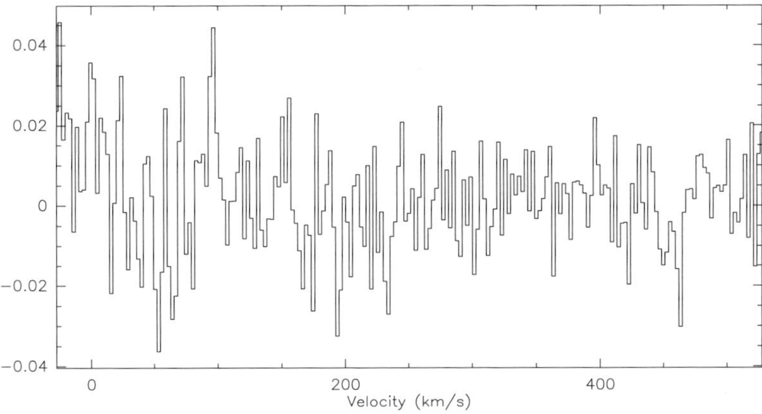

Figure 4. Deep H_2O integration toward the starburst galaxy NGC 253 made with the Odin satellite. The units on the y axis are K (T_A^*). Figure credit: the Odin team.

Table 1. Upper limits to H_2O in galaxies

Galaxy	$I(H_2O)$ ($K\ km\ s^{-1}$)	Predicted $I(H_2O)$
M82	< 1.8	0.4-1.7
NGC 253	< 1.0	0.7-2.8
Cen A	< 2.4	0.4-1.6
IC 342	< 0.5	0.4-1.5
NGC 4258	< 1.3	0.2-0.9

atomic hydrogen, they suggest that [C I] emission may be a good tracer of molecular gas in galaxies, perhaps even more reliable than CO.

4. Recent Results on H_2O and O_2 in Galaxies from Odin

We have used the Odin satellite (Hjalmarson *et al.* 2003) to search for emission in the 557 GHz H_2O line from five nearby galaxies (Table 1). Three of these galaxies are starburst galaxies (NGC 253, M82, IC 342) and two contain active nuclei (Cen A) and water masers (NGC 4258). We restricted our search to nearby galaxies (< 8 Mpc) for which strong emission from the central regions would not be too heavily diluted in the 2′ Odin beam. None of the galaxies has been detected in H_2O emission; a representative spectrum is shown in Figure 4.

We estimated 3σ upper limits to the H_2O integrated intensity using the observed CO $J=1$–0 line width; these are given in Table 1. The predicted H_2O intensities in Table 1 were calculated from an estimate of the CO $J=1$–0 intensity in the Odin beam, which was then scaled down by a factor of 12–50 (our estimate of the relative strength of the CO and H_2O emission in the Galactic star-forming region W3; see Wilson *et al.* 2003), to obtain a prediction of the H_2O line intensity. A comparison of the predicted values and the observed upper limits in Table 1 shows that the upper limits obtained with Odin may in some cases place interesting constraints on the structure of the dense interstellar medium in starburst galaxies, but more work is needed to interpret these results.

We have also used Odin to make a sensitive search for O_2 emission in the Small Magellanic Cloud (SMC) using the ground-state 118.75 GHz line (Wilson *et al.* 2005).

The SMC is an interesting galaxy in which to search for O_2 because its abundance of heavy elements is approximately $1/10^{th}$ that of the Milky Way. Astrochemical models by Frayer & Brown (1997) have suggested that the abundance of O_2 may be enhanced in lower metallicity gas. Our 3σ upper limit to the O_2 emission corresponds to an upper limit on the O_2/H_2 abundance ratio of $< 1.3 \times 10^{-6}$, a factor of 20 higher than the best O_2 abundance limit obtained for a Galactic source by Pagani *et al.* (2003). [However, it is important to note that O_2 has proved notoriously difficult to detect even in our own Galaxy (Goldsmith *et al.* 2002; Pagani *et al.* 2003). A possible detection of O_2 emission from ρ Oph is discussed by Liseau *et al.*, this volume.] By comparing our abundance limit with a variety of astrochemical models, we find that the low O_2 abundance observed in the SMC is most likely to be produced by the effects of photo-dissociation on molecular cloud structure in gas with a low abundance of heavy elements. Another possibility is freeze-out of molecules onto dust grains, but theoretical models involving this process in gas with a reduced abundance of heavy elements are not yet available.

5. Future Instruments and Prospects

The Atacama Large Millimeter Array (ALMA), now under construction in the Atacama desert of northern Chile, promises to be an incredible instrument for studying nearby galaxies. ALMA will provide much better sensitivity than individual single dish telescopes combined with the ability to achieve routine angular resolutions of $1''$ or less. For galaxies within 20 Mpc, the high sensitivity of ALMA will allow us to observe many more molecular tracers (HCN, HCO^+, and others), which will allow us to begin to do real astrochemistry in galaxies other than the LMC. In these nearby galaxies, new space missions such as Herschel will also make important contributions in the study of atoms and molecules at far-infrared wavelengths which are not visible from the ground. For example, much more sensitive observations of H_2O in galaxies will be possible with the HIFI instrument on Herschel. For more distant galaxies, the low angular resolution of Herschel will limit it to more global studies. However, ALMA will have enough resolution and sensitivity to allow us to study galaxies out to distances of 200 Mpc with the same tracers (CO, [C I]) and spatial resolution that we now use to study galaxies within 10 Mpc. By allowing us both to study more distant galaxies and to study nearby galaxies in greater detail, ALMA promises to revolutionize our understanding of the physical and chemical processes at work in the interstellar medium of galaxies.

References

Aalto, S. 2005, this volume
Bayet, E., Gerin, M., Phillips, T. G., & Contursi, A. 2004, *A&A* 427, 45
Bolatto, A. D., Jackson, J. M., Israel, F. P., Zhang, X., & Kim, S. 2000, *Ap. J.* 545, 234
Bolatto, A. D., Jackson, J. M., Wilson, C. D., & Moriarty-Schieven, G. 2000, *Ap. J.* 532, 909
Brouillet, N., Muller, S., Herpin, F., Braine, J., & Jacq, T. 2005, *A&A* 429, 153
Dumke, M., Nieten, Ch., Thuma, G., Wielebinski, R., & Walsh, W. 2001, *A&A* 373, 853
Frayer, D. T. & Brown, R. L. 1997, *Ap. J. S.* 113, 221
Freedman, W. L., Wilson, C. D., & Madore, B. F. 1991, *Ap. J.* 372, 455
Goldsmith, P. F., Li, D., Bergin, E. A., Melnick, G. J., Tolls, V., Howe, J. E., Snell, R. L., & Neufeld, D. A. 2002, *Ap. J.* 576, 814
Gerin, M. & Phillips, T. G. 1998, *Ap. J.* 509, L17
Gerin, M. & Phillips, T. G. 2000, *Ap. J.* 537, 644
Helfer, T. T., Thornley, M. D., Regan, M. W., Wong, T., Sheth, K., Vogel, S. N., Blitz, L., & Bock, D. C.-J. 2003, *Ap. J. S.* 145, 259

Hjalmarson, A., *et al.* 2006, *A&A* 402, L39

Iono, D., *et al.* 2006, in preparation

Israel, F. P., Tilanus, R. P. J., & Baas, F. 1998, *A&A* 339, 398

Israel, F. P. & Baas, F. 2001, *A&A* 371, 433

Israel, F. P. & Baas, F. 2002, *A&A* 383, 82

Israel, F. P. & Baas, F. 2003, *A&A* 404, 495

Johansson, L. E. B., Olofsson, H., Hjalmarson, A., Gredel, R., & Black, J. H. 1994, *A&A* 291, 89

Kenney, J. D., Wilson, C. D., Scoville, N. Z., Devereux, N., & Young, J. S. 1992, *Ap. J.* 395, L79

Kim, S., Walsh, W., & Ziao, K. 2004, *Ap. J.* 616, 865

Langer, W. D. & Penzias, A. A. 1990, *Ap. J.* 357, 477

Liseau, R. *et al.* 2005, this volume

Pagani, L., *et al.* 2003, *A&A* 402, L77

Petitpas, G. R. & Wilson, C. D. 1998, *Ap. J.* 503, 219

Petitpas, G. R. & Wilson, C. D. 2000, *Ap. J.* 538, L117

Petitpas, G. R., *et al.* 2006, in preparation

Scoville, N. Z. & Solomon, P. M. 1974, *Ap. J.* 187, L67

Stark, A. A., Bolatto, A. D., Chamberlain, R. A., Lane, A. P., Bania, T. M., Jackson, J. M., & Lo, K.-Y. 1997, *Ap. J.* 480, L59

Taylor, C. L. & Wilson, C. D. 2000, *Ap. J.* 538, 134

Walsh, W., Beck, R., Thuma, G., Weiss, A., Wielebinski, R., & Dumke, M. 2002, *A&A* 388, 7

Whitmore, B. C., Zhang, Q., Leitherer, C., Fall, S. M., Schweizer, F., & Miller, B. W. 1999, *AJ* 118, 1551

Wilson, C. D. 1997, *Ap. J.* 487, L49

Wilson, C. D., Walker, C. E., & Thornley, M. D. 1997, *Ap. J.* 483, 210

Wilson, C. D., Scoville, N. Madden, S. C., & Charmandaris, V. 2000, *Ap. J.* 542, 120

Wilson, C. D., Scoville, N. Madden, S. C., & Charmandaris, V. 2003, *Ap. J.* 599, 1049

Wilson, C. D., *et al.* 2003, *A&A* 402, L59

Wilson, C. D., *et al.* 2005, *A&A* 433, L5

Young, J. S., *et al.* 1995, *Ap. J. S.* 98, 219

Discussion

PHILLIPS: One of your galaxies looks like M82?

WILSON: Yes, that is NGC2903, which shows a double-peaked structure like M82 in its nucleus. One of the things I didn't have time to discuss here is that, in some galaxies, as you move up the CO rotational ladder that double-peaked structure disappears and you see a single central concentration. So there are clearly some spatial variations in the physical conditions in those galaxies.

Photo: E. Herbst

Astrochemistry: Recent Successes and Current Challenges
Proceedings IAU Symposium No. 231, 2005
D.C. Lis, G.A. Blake & E. Herbst, eds.

© 2006 International Astronomical Union
doi:10.1017/S1743921306007277

Spitzer Observations of Deeply Obscured Galactic Nuclei

H. W. W. Spoon[1]†, J. V. Keane[2], J. Cami[3], F. Lahuis[4], A. G. G. M. Tielens[5], L. Armus[6], and V. Charmandaris[1,7]

[1]Cornell University, Astronomy Department, Ithaca, NY 14853, USA
email: spoon@astro.cornell.edu

[2]NASA Ames Research Center, MS 245-3, Moffett Field, CA 94035

[3]NASA Ames Research Center, MS 245-6, Moffett Field, CA 94035

[4]Leiden Observatory, P.O. Box 9513, 2300 RA Leiden, Netherlands

[5]SRON and Kapteyn Institute, P.O. Box 800, 9700 AV Groningen, The Netherlands

[6]Caltech, Spitzer Science Center, MS 220-6, Pasadena, CA 91125, USA

[7]Department of Physics, University of Greece, P.O. Box 2208, 71003 Heraklion, Greece

Abstract. We report on our first results from a mid-infrared spectroscopic study of ISM features in a sample of deeply obscured ULIRG nuclei using the InfraRed Spectrograph (IRS) on the Spitzer Space Telescope. The spectra are extremely rich and complex, revealing absorption features of both amorphous and crystalline silicates, aliphatic hydrocarbons, water ice and gas-phase bands of hot CO and warm C_2H_2, HCN and CO_2. PAH emission bands were found to be generally weak and in some cases absent. The features probe a dense and warm environment, in which crystalline silicates and water ice are able to survive but volatile ices, commonly detected in Galactic dense molecular clouds, cannot. If powered largely by star formation, the stellar density and conditions of the gas and dust have to be extreme not to give rise to the commonly detected emission features associated with starbursts.

Keywords. galaxies: ISM — galaxies: nuclei — infrared: ISM — ISM: evolution — ISM: molecules

1. Introduction

With the launch of the InfraRed Spectrograph (IRS) on the Spitzer Space Telescope, mid-infrared spectroscopists have been handed a powerful tool to study extragalactic objects at high signal-to-noise and over a wide mid-infrared wavelength range, previously only available for the study of Galactic sources and a few nearby galaxies.

The foundations for the Spitzer studies currently underway were laid by the Infrared Space Observatory (ISO), which freed mid-infrared spectroscopists from the confinements of the Earth's atmospheric windows and the atmosphere's glaring foreground emission.

Major extragalactic topics addressed early on in the ISO mission centered around the properties of dusty starbursts and how they evolve (*e.g.*, Thornley *et al.* 2000), the unification of optically classified type 1 and 2 active galaxies in relation to the properties of the Active Galactic Nucleus (AGN) (*e.g.*, Clavel *et al.* 2000; Laurent *et al.* 2000), and the dominant power source in Ultra-Luminous Infrared Galaxies (ULIRGs): massive starbursts or AGN activity (*e.g.*, Genzel *et al.* 1998; Tran *et al.* 2001)?

After the expiration of ISO, two unusual galaxy spectra provided the first mid-IR insights into the properties of gas and dust in deeply obscured galactic nuclei. The 2–5 μm

† Spitzer Fellow

281

spectrum of the nucleus of NGC 4945 revealed strong absorptions of water ice ($3\,\mu$m), CO_2 ($4.26\,\mu$m) and a blend of 'XCN' and CO ice ($4.65\,\mu$m), seen against a continuum obscuring the central massive black hole (Spoon *et al.* 2000). Ground-based follow-up spectra confirmed the $4.65\,\mu$m absorption band to consist of separate components of processed 'XCN' and CO ice and warm (35 K) CO gas. The detection of processed ices against the nuclear continuum is taken as an indication for the presence of dense star-forming molecular clouds towards the nucleus of this active galaxy (Spoon *et al.* 2003).

The second unusual spectrum is that of the nucleus of NGC 4418, originally classified by Roche *et al.* (1986) as a "very extinguished galaxy", for its exceptionally deep $10\,\mu$m silicate absorption feature. Instead of the commonly detected PAH emission features, the 5.5–$8\,\mu$m ISO spectrum of its nucleus is dominated by absorption features of water ice ($6\,\mu$m), hydrocarbons ($6.85\,\mu$m and $7.25\,\mu$m) and methane ice ($7.67\,\mu$m), reminiscent of the line of sight towards our own Galactic Center (Spoon *et al.* 2001). Supporting ground-based observations indicate that most of the infrared luminosity ($L_{\mathrm{IR}} = 10^{11}\,L_\odot$) from NGC 4418 is produced in a compact nucleus with a radius of less than $80\,$pc (Evans *et al.* 2003). If powered entirely by star formation, the conditions of the nuclear gas and dust within this environment must be exceptional not to give rise to emission features commonly associated with star formation.

A large-scale follow-up study into the presence of 5.5–$8\,\mu$m absorption features in ISO galaxy spectra resulted in the finding of $6\,\mu$m water ice absorption in 12 out of 19 ULIRGs, 2 out of 62 AGNs and 4 out of 21 starburst galaxies surveyed. Also, $6.85\,\mu$m hydrocarbon absorption was detected in three galaxies besides NGC 4418, all ULIRGs. The results are consistent with findings from other wavelength ranges that more molecular material is present in ULIRG nuclei than in other galaxy types (Spoon *et al.* 2002).

In the following sections we present an overview of the first results from an IRS spectroscopic study of the properties of gas and dust in strongly obscured ULIRG nuclei. The spectra were selected from a larger sample of ULIRG spectra obtained as part of the GTO program of the Spitzer IRS team.

2. The Diverse Nature of the ULIRG Family

Figure 1 offers a striking illustration of the diverse nature of the galaxies classified as ULIRGs. At the top of the figure we find ULIRG mid-infrared spectra dominated by AGN-heated hot dust, with IRAS 07598 displaying a weak silicate emission feature. Further down, the spectra of IRAS 20037 and IRAS 08559 show increasing contributions of PAH emission superimposed on the AGN continuum. Moving down to IRAS 18030 and IRAS 06009, PAH emission starts to dominate and silicate absorption at both 10 and $18\,\mu$m becomes apparent. In the spectrum of the next one down, IRAS 20087, the characteristic absorption edge of water ice at $5.7\,\mu$m starts to appear. The water ice feature deepens and the importance of PAH emission decreases moving down from Mrk 273 to IRAS 15250. At the same time, the depth of the 10 and $18\,\mu$m silicate features increases and hydrocarbon absorption bands at 6.85 and $7.25\,\mu$m become apparent. Further note how from IRAS 18030 to IRAS 15250 the characteristic PAH emission feature at $7.7\,\mu$m gradually gets replaced by a broad absorbed-continuum peak at $8\,\mu$m. Starburst diagnostics relying on the equivalent width or luminosity of the $7.7\,\mu$m PAH feature will have to carefully verify the nature of this peak (Spoon *et al.* 2004a). The bottom three spectra in Figure 1 differ from those directly above by the presence of a strong near-infrared continuum and by the relative weakness of the 5.5–$8\,\mu$m absorption features. Note how, due to the appreciable redshifts of IRAS 00183 and IRAS 00397, the IRS spectral coverage extends all the way down to rest frame $4\,\mu$m, facilitating the discovery of wide absorption features of CO gas in their spectra (Spoon *et al.* 2004b). Recent ground-based

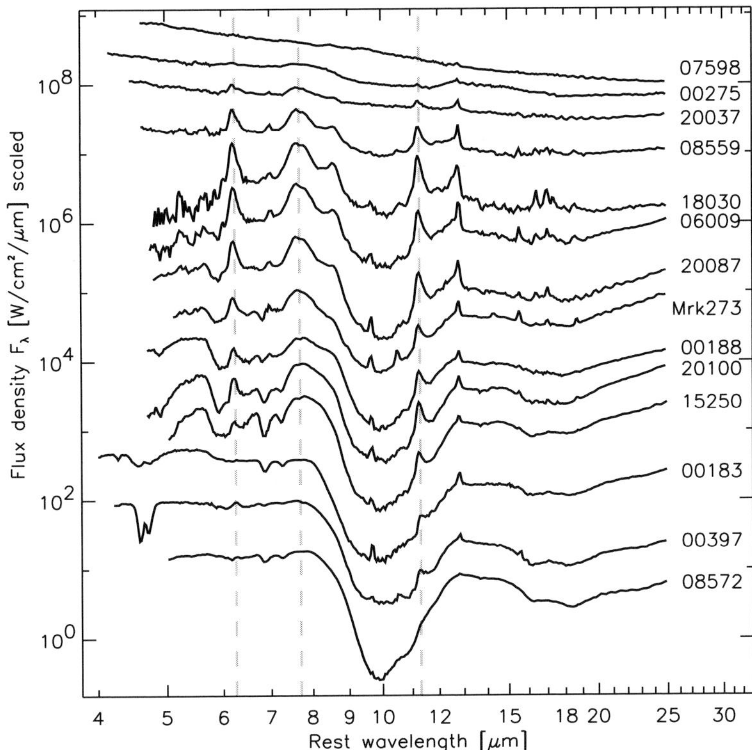

Figure 1. Spitzer IRS low-resolution spectra of ULIRGs sorted by spectral shape. The three spectra at the top are continuum-dominated (AGN-like), the next four are PAH-dominated (starburst-like) and the rest are absorption-dominated (buried nuclei). Vertical lines indicate the positions of the 6.2, 7.7 and 11.2 μm PAH emission bands.

high-resolution M-band spectroscopy reveals the CO band to be also present in the spectrum of the low-redshift ULIRG IRAS 08572 (Geballe *et al.*, in preparation).

3. Absorption Features of Crystalline Silicates

We have detected significant substructure in the deep silicate absorption features towards a sample of highly obscured ULIRG nuclei. The absorption features appear at 11, 16, 19, 23 and 28 μm and are best revealed by a spline fit to the amorphous silicate absorption component. We identify the residuals with crystalline silicates, most likely forsterite (Mg_2SiO_4). The clearest detection is offered by the spectrum of IRAS 08572, shown in detail in Figure 2. This spectrum combines the deepest silicate feature in our sample with the complete absence of all commonly present PAH emission features. In Figure 3 we show our 12 clearest crystalline silicate detections. For these we infer crystalline to amorphous silicate ratios of 7–15%, using the optical depth and opacities of the 10 μm amorphous and 16 μm crystalline band.

The detection of crystalline silicates appears to be geared to the more deeply enshrouded ULIRGs. While all 17 ULIRGs with τ(sil) > 2.9 do show the strong 16 μm feature, at lower silicate optical depth the number of detections of this band drops sharply. Between τ(sil)=2.0 and τ(sil)=2.9 only 1 out of 6 ULIRGs shows the feature and below that, only 3 out of 54. This also is apparent in the spectra shown in Figure 1, in which the 16 μm feature can only be recognized in systems with the deepest silicate features.

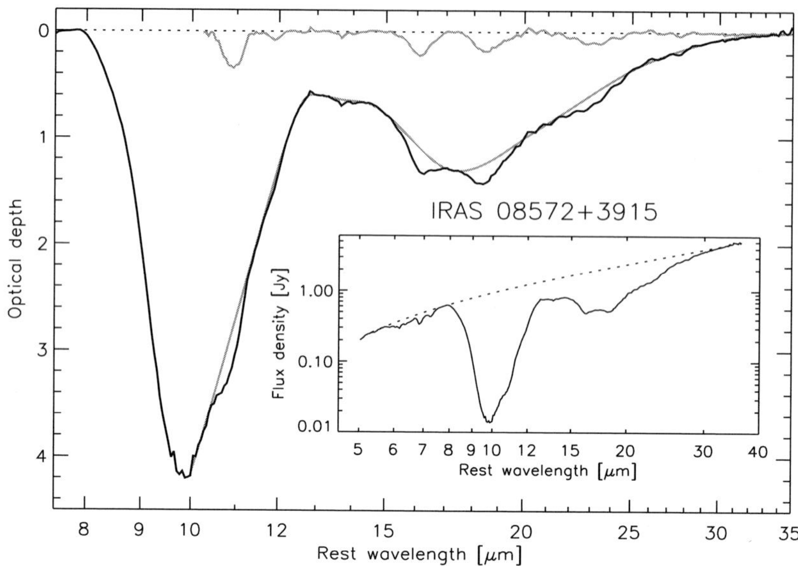

Figure 2. Silicate optical depth spectrum of IRAS 08572 (*black*) with a spline fit to the amorphous silicate component overplotted (*grey*). The residual spectrum, revealing crystalline features at 11, 16, 19 and 23 μm is shown in *grey* at the top of the plot. *Inset:* the 5–35 μm spectrum of IRAS 08572 with the adopted local continuum.

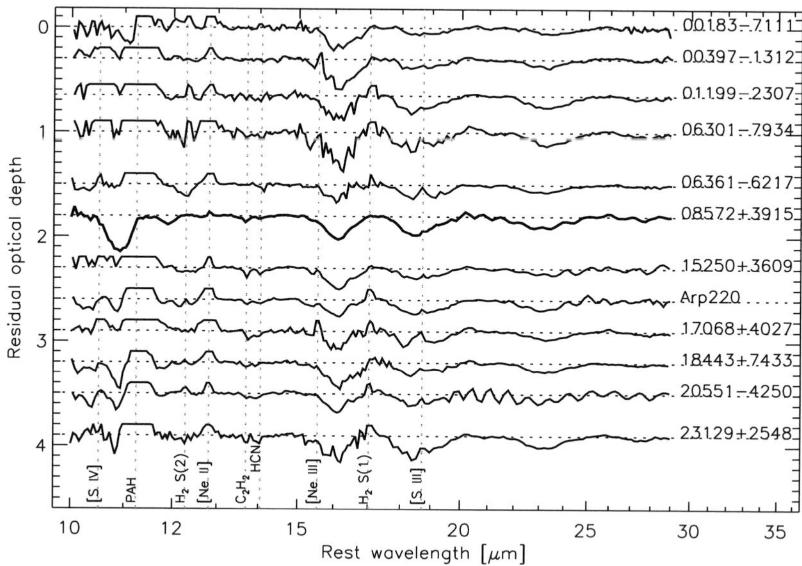

Figure 3. 10–30 μm residual optical depth spectra for 12 ULIRGs from our sample after subtraction of the amorphous silicate component. For plotting purposes, the spectra have been offset and truncated at residual optical depths of –0.1. Crystalline silicate features can be identified at 11, 16, 19, 23 and 28 μm.

The discovery of crystalline silicates in the ISM of ULIRGs stands in sharp contrast to the upper limit of 1% for the crystallinity of the Galactic ISM (Kemper *et al.* 2004, 2005). As stellar sources of silicate dust inject at least 5% of their silicates in crystalline form, the crystalline silicates arriving in the Galactic ISM must be rapidly transformed into amorphous silicates. The timescale for amorphitization is estimated to be only 70 million years (Bringa *et al.*, in preparation), considerably shorter than the timescale at which (crystalline) silicates are injected into the Galactic ISM (4 billion years; Bringa *et al.*, in preparation). In ULIRGs and other mergers these time scales may be different. ULIRGs are characterized by a high rate of star formation driven by merging events. In contrast to the Galactic ISM, the enrichment of the dusty interstellar medium will be dominated by the rapidly evolving massive stars and the contribution by the more numerous low-mass stars will lag on the timescale associated with the ULIRG-merger event ($10^8 - 10^9$ yr; Murphy *et al.* 1996). Thus, we attribute the high fraction of crystalline silicates in these ULIRGs, as compared to others, to the relative 'youth' of these systems; *e.g.*, the amorphitization process may lag the merger-triggered, star-formation-driven dust injection process. For a more detailed discussion on the discovery of crystalline silicates in ULIRG spectra we refer to Spoon *et al.* (2006).

4. Absorption Features in the 5.5–8 μm Range

The 5.5–8 μm spectral range of Galactic dense and diffuse ISM lines of sight is rich in absorption features of vibrational transitions of various molecular species, which are diagnostics of the chemical and physical states of these environments (Chiar *et al.* 2000; Keane *et al.* 2001). Toward dense clouds, the profiles and strengths of the main volatile ice absorption bands of water ice (6.0 μm), $NH_4{}^+$ ice (6.85 μm) and methane ice (7.67 μm) have been shown to vary considerably depending on the thermal history of the ice mantle. Observations of diffuse lines of sight, on the other hand, show absorption features in the 5.5–8 μm range that are associated with organic refractory dust: *e.g.*, the C–H bending modes of aliphatic hydrocarbons at 6.85 and 7.25 μm as seen towards the Galactic Center (*e.g.*, Chiar *et al.* 2000; Pendleton & Allamandola 2002). The strong detection by ISO of features of both the dense and the diffuse ISM towards several deeply enshrouded galactic nuclei (Spoon *et al.* 2001, 2002) has expanded the applicability of these diagnostics to external galaxies.

With the availability of IRS on Spitzer, both the wavelength coverage shortward of 5.7 μm and the spectral resolution in the 5.5–8 μm range have improved, enabling a better determination of the local continuum in this range, a clearer separation of the many spectral features and expansion of the study to a larger sample of galactic nuclei. Figure 4 shows the optical depth spectra for five deeply obscured merger nuclei. The top four spectra are dominated by strong absorption features centered at 6.0–6.1, 6.85 and 7.25 μm. Following earlier studies (*e.g.*, Chiar *et al.* 2000; Spoon *et al.* 2001, 2002, 2004b), we identify the main components with water ice (6.0 μm) and the bending modes of C–H in aliphatic hydrocarbons (6.85 and 7.25 μm).

Compared to the top four spectra, the optical depth spectrum of Arp 220 (bottom panel) looks very different. Strong PAH emission bands at 5.70, 6.22 and 7.7 μm and emission lines of H_2 S(5) and [Ar II] at 6.91 and 6.99 μm distort the otherwise pure absorption spectrum. While the two emission lines fill up the red wing of the 6.85 μm hydrocarbon band, the blue flank of the 7.7 μm PAH band overwhelms the 7.25 μm hydrocarbon band and the 6.22 μm PAH emission feature fills up the depth of the 6 μm water ice band. The same emission features are also weakly present in the spectrum of IRAS 15250, most notably the PAH emission feature at 6.22 μm. Evidently, the contribution of unconcealed star formation in IRAS 15250 is far smaller than in Arp 220.

At first glance, the spectral structure seen at 6.32 μm in the spectra of IRAS 08572 and IRAS 00183 appears to be the 6.22 μm PAH feature. However, neither the central wavelength nor the FWHM of the feature is consistent with an identification with interstellar

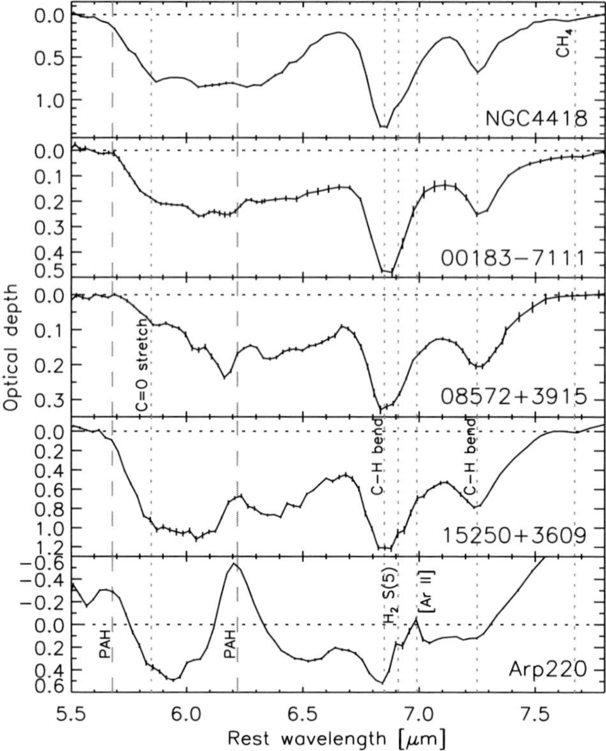

Figure 4. Optical depth spectra for five (U)LIRGs of absorption features in the $5.5\text{--}7.7\,\mu\mathrm{m}$ range. Vertical *dashed* lines denote the central wavelengths of PAH emission features at 5.7 and 6.22. Vertical *dotted* lines indicate the central positions of the C=O carbonyl stretch $(5.88\,\mu\mathrm{m})$, the C–H bending modes of hydrocarbons $(6.85$ and $7.25\,\mu\mathrm{m})$, emission lines of H_2 S(5) $(6.91\,\mu\mathrm{m})$ and [Ar II] $(6.99\,\mu\mathrm{m})$, and methane ice $(7.67\,\mu\mathrm{m})$.

$6.22\,\mu\mathrm{m}$ PAH. Instead, this feature may be 'carved out' by absorption bands of species absorbing in this range, like gas-phase water or hydrocarbons. For a detailed study of the absorption features contributing to the $6\,\mu\mathrm{m}$ absorption complex we refer to Keane *et al.* (in preparation) and for a full analysis of the origin of the hydrocarbon bands in the spectrum of IRAS 08572 to Pendleton *et al.* (in preparation).

5. Absorption Features of C_2H_2, HCN and CO_2

Using the high-resolution mode of IRS on Spitzer, we have discovered C_2H_2 $(13.7\,\mu\mathrm{m})$, HCN $(14.03\,\mu\mathrm{m})$ and CO_2 $(15.0\,\mu\mathrm{m})$ gas absorption features in the spectra of a number of deeply obscured ULIRG nuclei (Armus *et al.*, in preparation). Our four clearest detections are shown in Figure 5, together with excitation temperatures resulting from a preliminary, combined fit of the C_2H_2 and HCN profiles. The excitation temperatures found range from 150 K for Arp 220 to 400 K for IRAS 20100.

These features have previously been seen in ISO-SWS spectra of deeply embedded massive Young Stellar Objects (YSOs) (*e.g.*, Lahuis & van Dishoeck 2000) and recently in the Spitzer spectrum of a low-mass YSO, IRS 46 (Lahuis *et al.* 2006). In the massive YSOs, the features originate in the hot inner dense region of the shells surrounding the young stars. In the case of IRS 46 the features are believed to originate in the dense, hot inner region of the protostellar disk. In these cases, the chemistry is dominated by

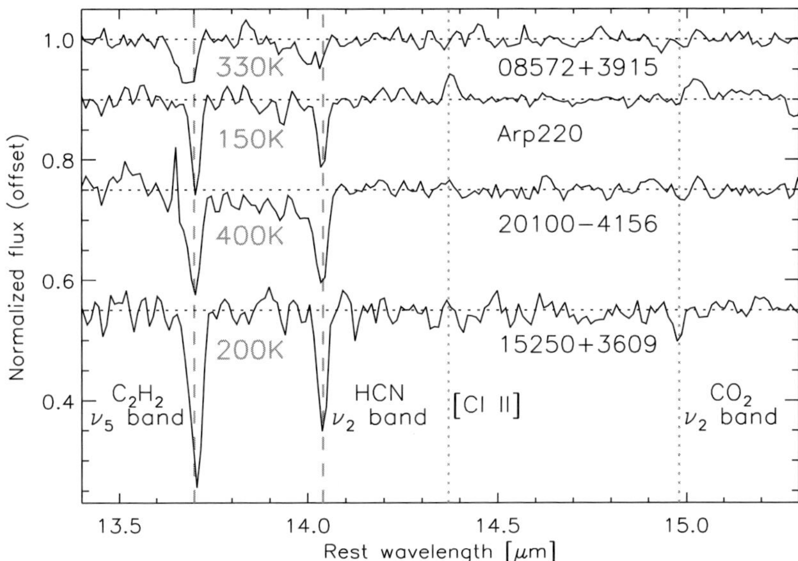

Figure 5. Continuum-normalized high-resolution IRS spectra of four deeply obscured ULIRGs. The spectra reveal the presence of gas absorption bands of C_2H_2 at $13.7\,\mu$m, HCN at $14.03\,\mu$m and CO_2 at $15.0\,\mu$m. Best fitting excitation temperatures for the C_2H_2 and HCN absorption bands are indicated on the plot. Also indicated is the [Cl II] line at $14.37\,\mu$m in Arp 220.

evaporation of the molecules from the grains with subsequent gas-phase processing, which enhances the abundances of C_2H_2 and HCN by orders of magnitude. Hence, C_2H_2 and HCN are typical tracers of high-density ($> 10^8$ cm^{-3}), high-temperature chemistry.

The detection of absorption features of C_2H_2, HCN and CO_2 in our ULIRGs might suggest the presence of a population of deeply embedded YSOs in their nuclear environments. However, an identification with embedded protostars is not supported by spectral evidence in other parts of our mid-infrared spectra (*e.g.*, by the absence of commonly detected NH_4^+ ice at $6.85\,\mu$m and CO_2 ice at $15\,\mu$m; Gibb *et al.* 2004). In the light of other evidence (*e.g.*, interferometric detection of HCN in these nuclei; extreme compactness of the ultra-luminous sources; weakness of common starburst emission features such as PAHs and fine-structure lines of neon and sulphur), it seems more likely that the features arise from highly pressure-confined star formation in a dense medium – *i.e.*, under star formation conditions more extreme than probed in ordinary starburst galaxies. This, and a more detailed analysis of the 13–$15\,\mu$m features in ULIRG spectra, will be discussed in more detail in a future paper.

6. The $4.65\,\mu$m CO Absorption Feature

Absorption spectroscopy to study features of dust and gas in the 2–$5\,\mu$m range requires the presence of a sufficiently intense 2–$5\,\mu$m background continuum source. Active galaxies, with a favorable orientation of their AGN, show AGN-heated hot dust in their spectra against which absorption features have been detected (*e.g.*, Imanishi 2000). Also some of the more deeply obscured galactic nuclei reveal 2–$5\,\mu$m absorption features against their near-infrared dust continua (*e.g.*, UGC 5101: Imanishi *et al.* 2001; NGC 4945: Spoon *et al.* 2000, 2003), while for other nuclei these are impossible to detect given the absence of a hot dust background source (*e.g.*, NGC 4418: Spoon *et al.* 2001; or IRAS 15250: Figure 1).

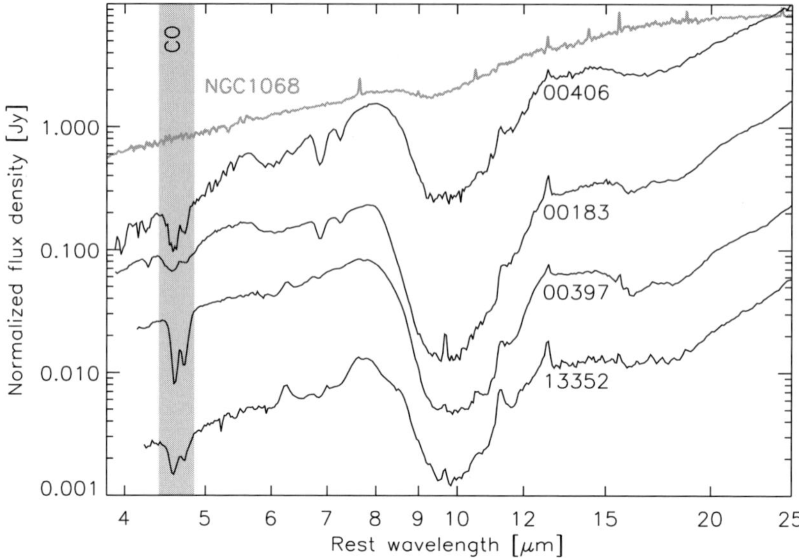

Figure 6. 4–25 μm spectra of four ULIRGs showing an absorption feature of gas-phase CO at 4.65 μm. The spectra are compared to the ISO-SWS spectrum of the nucleus of the Seyfert-2 galaxy NGC 1068 (Sturm *et al.* 2000). The spectra have been scaled for plotting purposes.

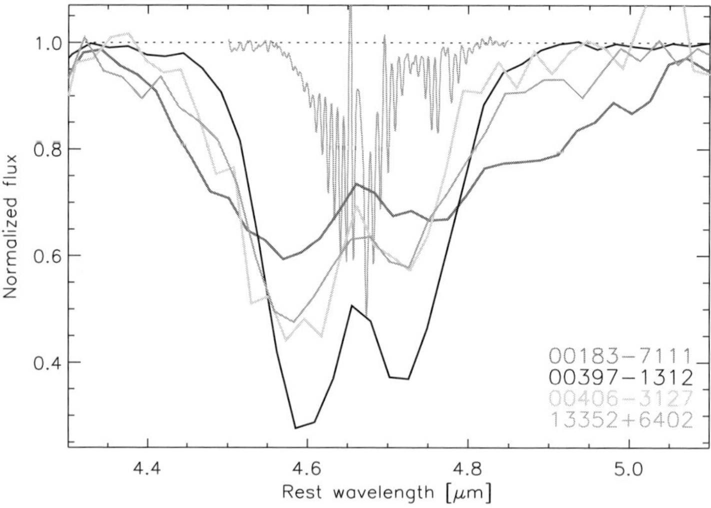

Figure 7. Comparison of the normalized flux profiles for the CO absorption features in the spectra of Sgr A* (*dark grey* narrow profile; Moneti *et al.* 2001) and the four ULIRGs shown in Figure 6. Individual lines are resolved in the high-resolution ISO-SWS spectrum of Sgr A*.

Here we report the first detection of strong absorption features of 4.65 μm gas-phase CO in the spectra of four deeply obscured ULIRG nuclei with redshifts favorable to detection by Spitzer-IRS. The 4–25 μm spectra of the sources are compared in Figure 6, their normalized flux profiles are shown in Figure 7.

We have used the isothermal plane-parallel LTE gas models of Cami (2002) to model the CO gas absorption profile of IRAS 00183. A best fit to the observed profile is found for a model with a gas temperature of 720 K, a column density of $10^{19.5}$ cm^{-2} and an intrinsic line width of $50\,\mathrm{km\,s^{-1}}$. The temperature is far higher than found toward Galactic ISM lines of sight, consistent with the clearly larger width of the feature compared to, for instance, Sgr A* (Figure 7). Note that the density in the absorbing medium has to be high ($n > 3 \times 10^6\,\mathrm{cm^{-3}}$) to give rise to high-$J$ level absorption. This limits the size of the obscuring region to less than $0.03\,\mathrm{pc}$ (Spoon *et al.* 2004b). Similar fits for the CO profiles of the other three sources in Figure 7 have not yet been made.

Following the discovery of hot CO gas in IRAS 00183, Lutz *et al.* (2004) have analyzed a sample of ISO AGN spectra for the presence of similar CO absorption features and found none to good limits. This suggests that the CO features seen in our ULIRG spectra do not arise in a typical AGN torus geometry, but rather are a signature of the special conditions in deeply obscured ULIRG nuclei – pointing to a fully covered rather than a torus geometry (Lutz *et al.* 2004).

7. Conclusions

Using the IRS spectrograph on the Spitzer Space Telescope, we have obtained mid-infrared spectra for a large sample of ULIRGs. The spectra show a great diversity in spectral shape, reflecting the diverse nature and merger evolutionary states of the sample. Especially interesting for astrochemists is the sub-sample of spectra showing signatures of strong obscuration, betraying the presence of huge amounts of dust and gas in and towards the merger nuclei. The wide spectral coverage of the IRS (5–$38\,\mu\mathrm{m}$), assisted by redshifts ranging from $z{=}0.02$ to $z{=}0.4$, allowed us to study the ISM in these sources from rest frame $\sim$$3.8\,\mu\mathrm{m}$ to $37\,\mu\mathrm{m}$, revealing absorption features of both amorphous and crystalline silicates, aliphatic hydrocarbons, water ice and gas-phase bands of hot CO and warm C_2H_2, HCN and CO_2. PAH emission bands were found to be generally weak and in some cases absent.

Our analysis of the absorption features is far from complete and requires further comparison to the results from our emission line analysis to obtain a more detailed picture of the energetic processes responsible for these rich and complicated spectra. It is clear, however, at this time that the features are probing a dense and warm environment in which crystalline silicates and water ice are able to survive but volatile ices, commonly detected in Galactic dense molecular clouds, cannot. If powered largely by star formation, the stellar density and conditions of the gas and dust have to be extreme not to give rise to the commonly detected emission features associated with starbursts.

Acknowledgements

Support for this work was provided by NASA through the Spitzer Space Telescope Fellowship Program, through a contract issued by the Jet Propulsion Laboratory, California Institute of Technology under a contract with NASA.

References

Cami, J., 2002, Ph.D. thesis, Amsterdam Univ.
Clavel, J., Schulz, B., Altieri, B., *et al.* 2000, *A&A* 357, 839
Chiar, J.E., Tielens, A.G.G.M., Whittet, D.B.C., *et al.* 2000, *Ap. J.* 537, 749
Evans, A.S., Becklin, E.E., Scoville, N., *et al.* 2003, *AJ* 125, 2341
Genzel, R., Lutz, D., Sturm, E., *et al.* 1998, *Ap. J.* 498, 579
Gibb, E.L., Whittet, D.C.B., Boogert, A.C.A., & Tielens, A.G.G.M. 2004, *Ap. J. Suppl.* 151, 35
Imanishi, M., Dudley, C.C., & Maloney, P.R. 2001, *Ap. J.* 558, 93
Imanishi, M. 2000, *MNRAS* 319, 331

Keane, J.V., Tielens, A.G.G.M., Boogert, A.C.A., *et al.* 2001, *A&A* 376, 254

Kemper, F., Vriend, W.J., & Tielens, A.G.G.M. 2004, *Ap. J.*, 609, 826

Kemper, F., Vriend, W.J., & Tielens, A.G.G.M. 2005, *Ap. J.* 633, 534, erratum

Lahuis, F., & van Dishoeck, E.F. 2000, *A&A* 355, 699

Lahuis, F. & 10 co-authors. 2006, *Ap. J.*, accepted

Laurent, O., Mirabel, I.F., Charmandaris, V., *et al.* 2000, *A&A* 359, 887

Lutz, D., Sturm, E., Genzel, R., *et al.* 2004, *A&A* 426, L5

Moneti, A., Cernicharo, J., & Pardo, J.R. 2001, *Ap. J.* 549, L203

Murphy, T.W., Armus, L., Matthews, K., *et al.* 1996, *AJ*, 111, 1025

Pendleton, Y.J., Allamandola, L.J., 2002, *Ap. J. Suppl.* 138, 75

Roche, P.F., Aitken, D.K., Smith, C.H., & James, S.D. 1986, *MNRAS* 218, 19P

Spoon, H.W.W., Koornneef, J., Moorwood, A.F.M., *et al.* 2000, *A&A* 357, 898

Spoon, H.W.W., Keane, J.V., Tielens, A.G.G.M., *et al.* 2001, *A&A* 365, L353

Spoon, H.W.W., Keane, J.V., Tielens, A.G.G.M., *et al.* 2002, *A&A* 385, 1022

Spoon, H.W.W., Moorwood, A.F.M., Pontoppidan, K.M., *et al.* 2003, *A&A* 402, 499

Spoon, H.W.W., Moorwood, A.F.M., Lutz, D., *et al.* 2004a, *A&A* 414, 873

Spoon, H.W.W., Armus, L., Cami, J., *et al.* 2004b, *Ap. J. Suppl.* 154, 184

Spoon, H.W.W., Tielens, A.G.G.M., Armus, L., *et al.* 2006, *Ap. J.*, accepted (astro_ph/0509859)

Sturm, E., Lutz, D., Tran, Q.D., *et al.* 2000, *A&A* 358, 481

Tran, Q.D., Lutz, D., Genzel, R., *et al.* 2001, *Ap. J.* 552, 527

Thornley, M.D., Förster Schreiber, N.M., Lutz, D., *et al.* 2000, *Ap. J.* 539, 641

Discussion

ALI: You presented us the Spitzer Observations of crystalline silicates in the general ISM. What kind of crystalline silicates are there on chemist's definition of vibrational spectroscopy of condensed phase? Are there any Fe-content in silicates, and how are they formed.

SPOON: I don't know.

TAPPE: Does the 16 μm absorption feature you see bear any relation to the PAH 16.4 μm (tentatively attributed to C-C-C out of plane bending models) emission observed in the ISM?

SPOON: The 16.4 μm absorption feature appears both in spectra with and spectra without PAH emission features. This already makes it hard to imagine an identification with PAH. Furthermore, other PAH bands at 12.7 μm and 13.5 μm do not appear in absorption. Finally, apart from features at 11 μm and 16.4 μm, we detect features at 23 and 28 μm, consistent with an identification with forsterite.

NEUFELD: Have you considered the energy requirements needed to maintain a column density of $10^{19.5}$ cm^{-2} at a temperature of 700 K? This would seem to be very large. Do you consider the hot CO absorbing gas to be in a torus, or must it cover a larger area to block the continuum? Are X-rays a potential source of heating?

SPOON: Our toy model (see Spoon *et al.* 2004b) is simple plane-parallel isothermal model, purely intended to give an indication of the conditions in which the CO feature arises. We did not explore the energy requirements, nor did we investigate the heating source. As pointed out by Lutz *et al.* (2004), the feature is likely associated with the fully covered geometry of the nucleus, rather than with an AGN torus, since the feature is not seen towards local Seyfert nuclei.

Astrochemistry: Recent Successes and Current Challenges
Proceedings IAU Symposium No. 231, 2005
D.C. Lis, G.A. Blake & E. Herbst, eds.

© 2006 International Astronomical Union
doi:10.1017/S1743921306007289

Molecular Gas at High Redshift

Pierre Cox

IRAM, 30 rue de la piscine, F-38406 Saint-Martin-d'Hères, France
email: cox@iram.fr

Abstract. The study of the molecular gas in quasars and submillimeter galaxies at high redshift has significantly progressed during the last few years. From the current detection of CO emission in 37 sources spanning a range in redshift from $1 < z < 6.4$ with, in some cases, the measurement of a series of CO rotational transitions, it is possible to constrain the physical conditions of the massive ($\geqslant 10^{10} \, M_\odot$) reservoirs of gas in these objects. This review will present the current status of the studies of molecular gas in high-z sources, detail the physical conditions which pertain in these systems, which are scaled-up versions of the local ULIRGS, and discuss the searches in high-z sources for species other than CO, including the fine structure lines of neutral carbon and the recent detection of the redshifted [C II] emission line in the $z = 6.4$ quasar J 1148+5251. These results hold great promise for the study of galaxy formation and their evolution with redshift. This review will conclude by outlining the expected progress in the field, in particular when future instruments such as ALMA will be operational, which will enable to study the astrochemistry and its evolution in the early universe.

Keywords. galaxies: high-redshift — galaxies: starburst — galaxies: ISM — cosmology: observations

1. Introduction

During the past decade, significant progress has been made in our knowledge of high-redshift ($z > 1$) galaxies through the study of the dense and warm molecular gas. Deep surveys and targeted observations of known galaxies and quasars done at millimeter and submillimeter wavelengths have revealed at high-z a population of objects with luminosities in excess of $10^{12} \, L_\odot$, which is are scaled-up versions of the local ultraluminous infrared galaxies (ULIRGs) – see, *e.g.*, Blain *et al.* (2002), Sanders & Mirabel (1996). In these high-z galaxies and quasars, the far-infrared (far-IR) luminosity is mainly related to the warm ($40 - 60 \, \mathrm{K}$) dust, with estimated dust masses of a few $10^8 \, M_\odot$, *e.g.*, Omont *et al.* (2001), Omont *et al.* (2003). The heating of the warm dust appears to be dominated by the starburst activity (rather than by an active galactic nucleus) with implied star formation rates of $\sim 1000 \, M_\odot \, \mathrm{yr}^{-1}$. It has been suggested that this population represents large spheroidal galaxies in the making (Blain *et al.* 2002), and the high star formation rates indicate that most of the stars could be formed in about $10^8 \, \mathrm{yr}$.

In a growing number of cases, warm and dense molecular gas has been detected in these high-z infrared luminous objects. Most of the observations were done via the detection of CO emission lines revealing giant reservoirs of molecular gas, with masses in excess of $10^{10} \, M_\odot$, which provide the fuel needed to sustain the star-forming activity in these extreme objects. Recently, species other than CO, including rotational lines of hydrogen cyanide and fine-structure lines of neutral and singly-ionized carbon, have been used to probe the molecular gas in these high-z galaxies, providing additional information on the properties of the dense and warm interstellar medium in the early universe.

The detection of these often weak lines has been made possible by the increase in sensitivity of millimeter-wave telescopes, and is facilitated by a negative K-correction for the line emission and, for many of the source, by gravitational amplification. A recent review on molecules at high redshift has been given by Solomon & Vanden Bout (2005). In this paper, we review the recent advances which have been made in our understanding of the molecular gas content in high-z galaxies and summarize the prospects of these studies in view of future instruments, in particular with the Atacama Large Millimeter Array (ALMA).

2. CO Rotational Lines

IRAS F10214, an ultraluminous galaxy which was detected by IRAS, was the first high-z galaxy wherein CO line emission was discovered (Brown & Vanden Bout 1992; Solomon *et al.* 1992). This discovery prompted further searches and the successful detection of CO line emission in two high-z quasars: the Cloverleaf at $z = 2.5$ by Barvainis *et al.* (1994) (see also Barvainis *et al.* 1997) and BR 1202 at $z = 4.7$ by Omont *et al.* (1996). For both quasars, the CO emission was resolved (Fig. 1). Since then many more searches have been made and the molecular gas has now been measured in 37 high-z objects via the detection of rotational lines of CO (Table 1). The sources include far-IR ultraluminous ($> 10^{12}\,\mathrm{L_\odot}$) radio and submillimeter galaxies, quasars and one Lyman Break galaxy. The objects are in the redshift range $1 < z < 6.4$ and about half of them are gravitationally amplified (Table 1; see also Beelen *et al.* 2004; Greve *et al.* 2005; and Solomon & Vanden Bout 2005).

All the high-z galaxies detected in the CO line emission have CO luminosities in excess of $10^{10}\,\mathrm{K\,km\,s^{-1}\,pc^2}$, a factor of a few greater than for local ULIRGs, and the masses of the molecular gas reservoirs are of the order a few $10^{10}\,\mathrm{M_\odot}$. In the few cases where the sources have been resolved and where dynamical masses could be derived, the values are in the range of a few $10^{11}\,\mathrm{M_\odot}$, implying very massive systems, dominated by baryons in their central regions.

The correlation between the CO luminosity ($L'_{\mathrm{CO}(1\to0)}$) and L_{FIR} which is seen in nearby ULIRGs (*e.g.*, Solomon *et al.* 1997) is still valid for the high-z sources out to far-IR luminosities in excess of $10^{13}\,\mathrm{L_\odot}$. However, at the highest far-IR luminosities there is a clear trend in the sense of larger far-IR luminosity (or star formation rate) for a given CO luminosity (or a given mass of molecular gas) – see Figure 2. It appears therefore that for the infrared ultraluminous massive systems, which were present in the early universe, the star-formation efficiency (star-formation rate per unit gas mass) was higher than in the less luminous local ULIRGs (see, *e.g.*, Solomon *et al.* 1997).

Most of the high-z sources which were detected in the CO line emission are sources already known from optical or radio surveys. In contrast, the search for CO emission in submillimeter galaxies (SMGs) found in blind millimeter/submillimeter surveys has been less successful. The main reason is due to the fact that the objects are extremely faint in the optical making reliable redshift determinations very difficult. Until recently, only two SMGs were detected in CO (Frayer *et al.* 1998; Genzel *et al.* 2003). A major breakthrough was made by Chapman *et al.* (2003) who initiated a major observational program at the Keck Telescope to obtain spectroscopic redshifts for a large ($\sim$70) sample of SMGs whose accurate positions were derived from deep VLA 1.4 GHz radio observations. Follow-up observations using the IRAM plateau de Bure interferometer successfully detected a large sub-sample of SMGs, bringing up to 12 the number of objects with CO detections (Neri *et al.* 2003; Greve *et al.* 2005). In some cases, higher angular resolution observations were able to resolve the CO line emission of the SMGs (Genzel *et al.* 2003; Tacconi *et al.* 2006).

From these observations, it is possible to derive the bulk properties of the SMG population. The median molecular gas mass is $\langle M(\mathrm{H_2})\rangle = (3.0 \pm 1.6) \times 10^{10}\,M_\odot$ and

Table 1. High-z Quasars and Galaxies detected in the CO line emission

Source Name	z	Telescopes	CO Line line	[Jy km s^{-1}]	1.2 mm Cont. [mJy]	Ref.
IRAS 10214+4724	2.28	12-m; 30-m	3→2	4.1±0.9	9.6±1.4	[1,2]
Cloverleaf	2.56	PdB; 30-m	3→2	9.9±0.6	18±2	[3]
BR 1202−0725	4.69	PdB; NRO	5→4	2.4±0.3	12.6±2.3	[4,5]
BRI 1335−0417	4.41	PdB	5→4	2.8±0.3	10.3±1.0	[6]
53W002	2.39	OVRO; PdB	3→2	1.20±0.15	1.7±0.4	[7,8]
MG 0414+0534	2.64	PdB	3→2	2.6±0.4	40±2[†]	[9]
SMM J02399−0136	2.80	OVRO; PdB	3→2	3.1±0.4	7.0±1.2	[10,11]
APM 08279+5255	3.91	PdB	4→3	3.7±0.5	17.0±0.5	[12]
BRI 0952−0115	4.43	PdB	5→4	0.91±0.11	2.8±0.6	[13]
Q1230+1627B	2.74	PdB	3→2	0.80±0.26	2.7±0.6	[13]
SMM J14011+0252	2.57	OVRO	3→2	2.4±0.3	≈ 3	[14]
4C60.07	3.79	PdB	4→3	2.50±0.43	4.5±1.2	[15]
6C1909+722	3.53	PdB	4→3	1.62±0.30	< 3	[15]
HR 10	1.44	PdB	5→4	1.35±0.20	4.9±0.8	[16]
MG 0751+2716	3.20	PdB	4→3	5.96±0.45	6.7±1.3	[17]
PSS 2322+1944	4.12	PdB	4→3	4.21±0.40	9.6±0.5	[18]
B3 J2330+3927	3.09	PdB	4→3	1.3±0.3	4.2±0.6	[19]
TN J0121+1320	3.52	PdB	4→3	1.2±0.4		[20]
J 1409+5628	2.56	PdB	3→2	3.28±0.36	10.7±0.6	[21]
J 1148+5251	6.42	VLA & PdB	3→2	0.18±0.04	5± 0.6	[22,23]
SMM J04431+0201	2.51	PdB	3→2	1.4±0.2	1.1±0.3	[24]
SMM J09431+4700	3.34	PdB	4→3	1.1±0.1	2.3±0.4	[24]
SMM J16358+4057	2.38	PdB	3→2	2.3±1.2	2.6±0.2	[24]
cB58	2.73	PdB	3→2	0.37±0.08	1.06±0.35	[25]
Q0957+561	1.41	PdB	2→1	1.20±0.06	5.7±1.8[†]	[26,27]
RX J0911+0551	2.79	OVRO	3→2	2.9±1.1	10.2±1.8	[28]
SMM J04135+1027	2.84	OVRO	3→2	5.4±1.3	7±1	[28]
B3 J2330+3927	3.08	PdB	4→3	1.3±0.3	4.8±1.2	[29]
4C41.17	3.79	PdB	4→3	1.8±0.2	3.8±0.4	[30]
TNJ0121+1320	3.52	PdB	4→3	1.2±0.4	–	[31]
TNJ0924−2201	5.19	ATCA	1→0	0.52±0.11	–	[32]
SMM J02396−0134	1.06	PdB	2→1	3.4±0.3	–	[33]
SMM J13120+4242	3.41	PdB	4→3	1.7±0.3	–	[33]
SMM J16366+4105	2.45	PdB	3→2	1.8±0.3	–	[33]
SMM J16371+4053	2.38	PdB	3→2	1.0±0.2	–	[33]
SMM J22174+0015	3.09	PdB	3→2	0.8±0.2	–	[33]
SMM J16359+6612	2.51	PdB & OVRO	3→2	5.75±0.25	3.0±0.7	[34]

References – [1] Brown & Vanden Bout (1992); [2] Solomon *et al.* (1992); [3] Barvainis *et al.* (1994); [4] Omont *et al.* (1996); [5] Ohta *et al.* (1996); [6] Guilloteau *et al.* (1997); [7] Scoville *et al.* (1997); [8] Alloin *et al.* (2000); [9] Barvainis *et al.* (1998); [10] Frayer *et al.* (1998); [11] Genzel *et al.* (2003); [12] Downes *et al.* (1999); [13] Guilloteau *et al.* (1999); [14] Frayer *et al.* (1998); [15] Papadopoulos *et al.* (2000); Andreani *et al.* (2000); [17] Barvainis *et al.* (2002); [18] Cox *et al.* (2002); [19] deBreuck *et al.* (2003); [20] deBreuck *et al.* (2003); [21] Beelen *et al.* (2004a); [22] Walter *et al.* (2003); [23] Bertoldi *et al.* (2003b); [24] Neri *et al.* (2003); [25] Baker *et al.* (2004); [26] Planesas *et al.* (1999); [27] Krips *et al.* (2005); [28] Hainline *et al.* (2004); [29] deBreuck *et al.* (2004a); [30] de Breuck *et al.* (2005); [31] deBreuck *et al.* (2004b); [32] Klamer *et al.* (2005); [33] Greve *et al.* (2005); [34] Kneib *et al.* (2005); Sheth *et al.* (2004).
Sources in bold face are known to be lensed.
[†] Non-thermal emission.

is distributed within a ~ 2 kpc radius, with a median dynamical mass of $\langle M_{dyn} \rangle \simeq (1.2 \pm 1.5) \times 10^{11}$ M$_\odot$. The corresponding star formation rates are ~ 700 M$_\odot$ yr^{-1}. These values are about four times greater than for local ULIRGs but similar to the most

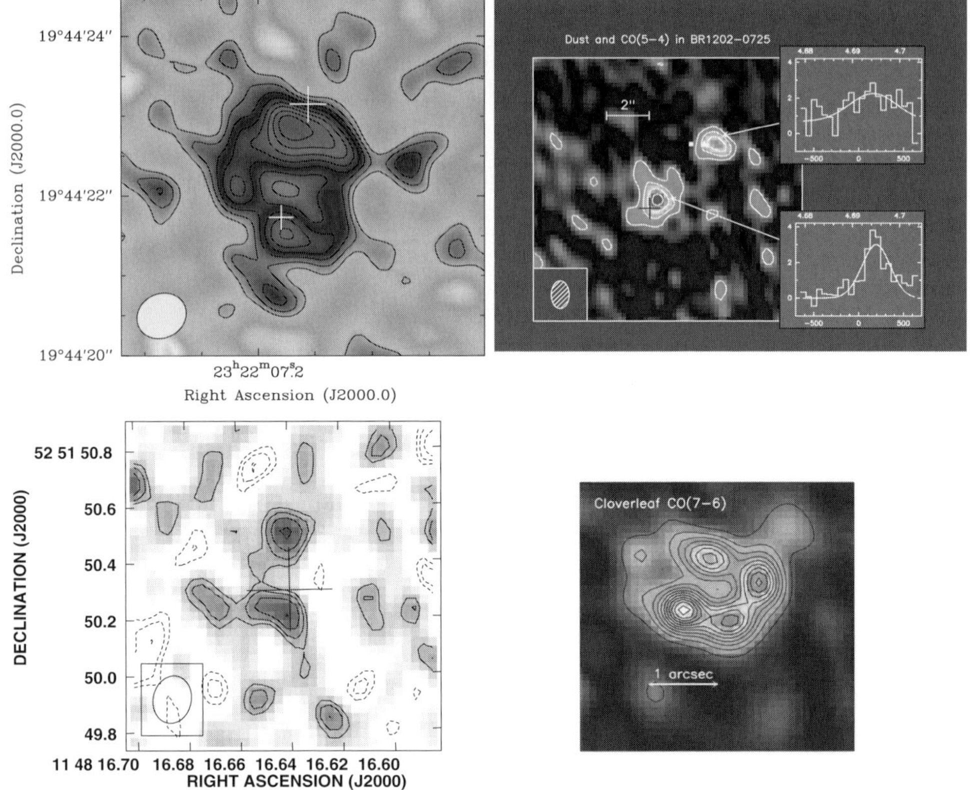

Figure 1. Examples of resolved CO emission in $z > 4$ QSOs. *From top to bottom and clockwise*: PSS 2322+1944 at $z = 4.12$ Carilli *et al.* (2003), BR 1202−0115 at $z = 4.69$ Omont *et al.* 1996; the Cloverleaf at $z = 2.6$ Alloin *et al.* (1997); J1148+5251 at $z = 6.42$ Walter *et al.* (2004).

extreme high-z radio galaxies and quasars. The CO line emission of the SMGS are broad ($\langle\text{FWHM}\rangle = 780 \pm 320\,\text{km}\,\text{s}^{-1}$) and display a high variety of profiles with, in many cases, double-peaked line profiles indicating merging systems or disks. The compactness of the sources and their properties show that the SMGs resemble scaled-up and more gas-rich versions of the ULIRGs in the local universe. Their masses, central densities and potential well depths are comparable to ellipticals galaxies or massive bulge. SMGs appear to convert most of the available initial gas into stars on typical time scales of a few $\sim\!10^8$ yr.

In a few cases, the CO line emission in high-z galaxies has been resolved and Figure 1 presents four cases of resolved CO emission in high-z quasars. The properties of these four quasars are summarized below.

(a) The gravitationally lensed quasar PSS 2322+1944 at $z = 4.12$ reveals an Einstein ring with a diameter of $1.5''$ which can be modeled as a star forming disk surrounding the QSO nucleus with a radius of 2 kpc (Carilli *et al.* 2003).

(b) The CO emission of BRI 1202−0115, a quasar at $z = 4.69$, is resolved into two sources (Omont *et al.* 1996; Carilli *et al.* 2002). The optical quasar is associated with the southern south. The CO line profiles and continuum levels are slightly different for the northern and southern sources indicating a system in the process of merging rather than gravitational lensing. Follow-up observations are still needed to confirm this result.

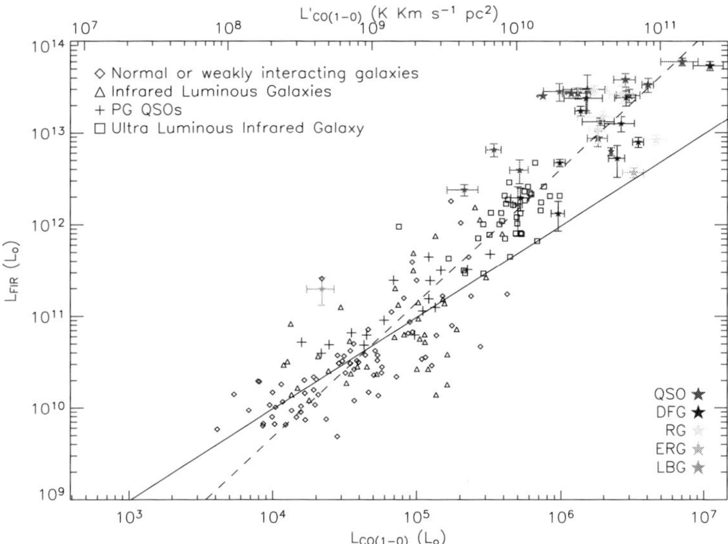

Figure 2. The correlation between the far-IR (L_{FIR}) and the CO ($L'_{CO(1\to0)}$) luminosities for high-z galaxies and quasars (shown as stars and taken from Table 1), compared to local sources. The PG quasars are from Evans *et al.* (2001), Casoli & Loinard (2001) and Scoville *et al.* (2003), and the ULIRGS from Solomon *et al.* 1997. The sample of infrared luminous galaxies is taken from Yao *et al.* (2003) and the normal and the weakly interacting galaxies sample from Solomon & Sage (1988). The solid line shows the linear fit between L_{FIR} and $L'_{CO(1\to0)}$ derived for sources with $L_{FIR} < 10^{11}\,L_\odot$. The dashed line displays the best fit to all the sources. This figure is taken from Beelen (2004b).

(c) The Cloverleaf (H1413+117), which is a strongly magnified quasar at $z = 2.5$, has the highest line flux densities of all high-z objects and hence has been observed in great detail. The high angular image (0.5″) clearly resolves the four 'leaves' of the Cloverleaf, which are similar to the optical image (Alloin *et al.* 1997). The radius of the CO emitting region is estimated to be 100 pc and the magnification is about 30.

(d) J1148+5251, the most distant quasar currently known at a redshift $z = 6.419$, is an extremely luminous quasar ($L_{\rm bol} \sim 10^{14}\,L_\odot$) powered by a super-massive ($\approx 3 \times 10^9\,M_\odot$) black hole. The detection of thermal dust emission in J 1148+5251 implies a dust mass of $\sim 7 \times 10^8\,M_\odot$, $L_{FIR} \sim 10^{13}\,L_\odot$, and a corresponding star formation rate of $\sim 2000\,M_\odot\,{\rm yr}^{-1}$ (Bertoldi *et al.* 2003a). Three transitions of CO (J=3→2, 6→5, and 7→6) were detected in J1148+5251 (Walter *et al.* 2003; Bertoldi *et al.* 2003b). The molecular gas, with an estimated mass $\approx (1-2) \times 10^{10}\,M_\odot$, is dense ($\sim 10^5\,{\rm cm}^{-3}$) and warm ($\sim 100$ K). The gas mass detected in J 1148+5251 can fuel star formation at the rate implied by the far-infrared luminosity for only 10 million years, a time comparable to the dynamical time of the region. The gas must therefore be replenished quickly, and metal and dust enrichment must occur fast.

Walter *et al.* (2004) resolved the CO emission in J1148+5251 using the VLA: the molecular gas is extended to a radius of 2.5 kpc, and two peaks are seen in the central region, separated by 0.3″ (1.7 kpc), each containing $\sim 5 \times 10^9\,M_\odot$. The dynamical mass is estimated to be $\sim 4.5 \times 10^{10}\,M_\odot$, accounting for nearly all of the molecular gas. It is much smaller than what would be expected on the basis of the local $M_{BH} - \sigma_{bulge}$ relationship (Gebhardt *et al.* 2000), which predicts a halo of mass $\sim 10^{12}\,M_\odot$, if this relation is also valid at high redshifts.

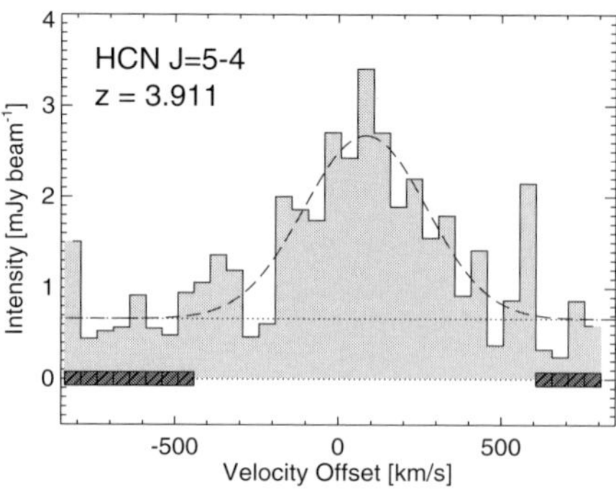

Figure 3. Spectrum of HCN(5→4) emission from the ultraluminous quasar APM08279+5255 at $z = 3.91$. The dashed line shows a Gaussian fit and the hatched regions indicate the channels sampling the continuum (from Wagg *et al.* 2006).

Finally, in a few cases it has been possible to observed in high-z galaxies various transitions of the CO rotational ladder. These measurements allow one to analyze the CO excitation and to better constrain the physical conditions of the molecular gas. A recent example of such a study is given by the observations of the strongly lensed SMG SMM 16359+6612 at $z = 2.5$ (Weiß *et al.* 2005b). Four CO lines are detected in SMM 16359+6612, namely J=3→2, 4→3, 5→4 and 6→5. The CO line spectral energy distribution (flux density *vs.* rotational quantum number) turns over at the CO(5→4) transition and is compatible with a gas density of $n_{H_2} = 10^{3.4}\,\mathrm{cm}^{-3}$ and a kinetic temperature $T_{\mathrm{kin}} = 40\,\mathrm{K}$. Other galaxies have been found to be more excited with turn overs at higher rotational quantum numbers (6→5 and even 7→6) such as in BR 1202–0725 or the Cloverleaf.

3. Hydrogen Cyanide

Hydrogen cyanide (HCN), which has a higher dipole moment than CO, traces dense gas (the critical density for excitation of the lower order transitions is $n_{H_2} \sim 10^5\,\mathrm{cm}^{-3}$) and is therefore better suited to probe the dense, warm gas directly associated with active star formation. Local ($z < 0.3$) ULIRGs show strong HCN emission (Gao & Solomon 2004) and display a tight linear relationship between the HCN luminosity (measuring the mass of dense gas available to make stars) and the far-IR luminosity (Fig. 4). This linear relationship indicates that the rate at which stars form is linearly proportional to the mass of *dense* molecular gas (Gao & Solomon 2004) as traced by HCN but not to the total molecular mass as traced by CO.

The potential to use HCN as a tracer of star formation activity in high-z sources has been demonstrated recently with the detection of HCN in four objects: the HCN(1→0) emission line is reported in the Cloverleaf at $z = 2.6$ (Solomon *et al.* 2003), in F10214 at $z = 2.3$ (Vanden Bout *et al.* 2004), and in the quasar J1409+5628 at $z = 2.6$ (Carilli *et al.* 2005); recently, the HCN(5→4) emission line was reported in the ultraluminous quasar APM08279+5255 at $z = 3.9$ by Wagg *et al.* (2006) – see Figure 3. Upper limits are available for four other high-z sources (see Carilli *et al.* 2005 and references therein). For all these source, the values of HCN/far-IR luminosity ratios are within the scatter of the relationship between HCN and far-IR emission for the $z < 0.3$ star-forming galaxies (Fig. 4) with values in between 1700 and 5000. The corresponding masses of dense molecular gas correspond to a few $10^{10}\,M_{\odot}$.

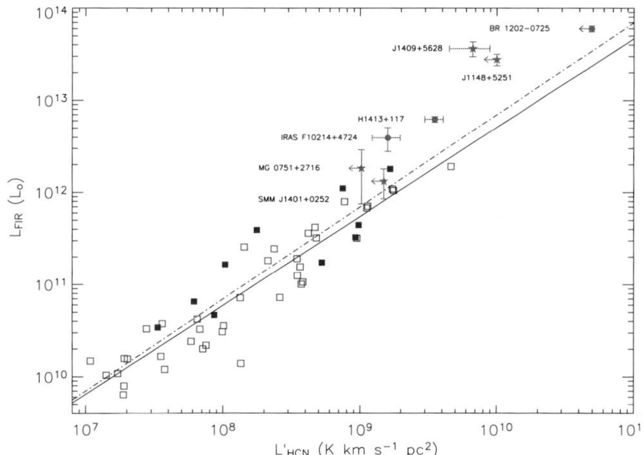

Figure 4. A comparison of the HCN line luminosity with far-IR luminosity for a sample of low-z galaxies (open squares from Gao & Solomon 2004a and filled squares from Solomon *et al.* 1992) and for high-z quasars and galaxies. The solid line is the nearly linear relationship defined by the low-z galaxies, corresponding to $\log L_{FIR} = 1.09 \log L_{HCN} + 2.0$ (from Carilli *et al.* 2005).

4. Atomic Carbon: [C I] and [C II]

Atomic carbon is an important tracer of the dense molecular gas in external galaxies and plays a central role in the cooling of the gas (Gerin & Phillips 1998, 2000). In recent years, the fine structure lines of [C I] have been reported in high-z sources providing further constraints on the physical conditions of the interstellar gas (gas column densities, thermal balance, and the UV illumination) independently of CO studies.

Five high-z sources have now been detected in [C I] emission: the Cloverleaf, which has been detected in both fine-structure transitions (Barvainis *et al.* 1997 and Weiß *et al.* 2003, 2005b); the $z = 4.12$ quasar PSS2322+1944 (Pety *et al.* 2004); F10214 and the submillimeter $z = 2.5$ galaxy SMM14011+0252 (Weiß *et al.* 2005b); and recently the $z = 6.4$ quasar J1148+5251 (Bertoldi *et al.*, in preparation). In all cases, the line widths of the atomic carbon are similar to the CO widths, indicating that both species originate in the same volume. The inferred masses of neutral carbon are of the order of a few $10^7 \, M_\odot$, and the carbon abundances are $\sim 3 - 5 \times 10^{-5}$, close to the galactic values, implying significant metal enrichment in heavy elements as early as $z \sim 2.5$.

The possibility of probing the interstellar medium and tracing star formation in galaxies at cosmological distances by using the bright C^+ emission line, when redshifted into submillimeter atmospheric windows, was proposed by Petrosian *et al.* (1969), and further discussed by Loeb (1993) and Stark (1997). The $^2P_{3/2} \rightarrow {}^2P_{1/2}$ fine-structure line of C^+ at 157.74 μm is the brightest emission line in the spectrum of galaxies, accounting for as much as ~ 0.1–1% of their total luminosity. The line is emitted predominantly by gas exposed to ultraviolet radiation in photo-dissociation regions (PDRs) associated with star forming activity, and has been extensively used to investigate the physical conditions of PDRs and to trace star formation in local galaxies, see, *e.g.*, Malhotra *et al.* (2001). Deep searches were conducted in the recent years in $z > 3$ infrared luminous galaxies and quasars known to have massive reservoirs of molecular gas but have remained so far unsuccessful. The upper limits on the line intensities are consistent with the general decrease of the $L_{[CII]}/L_{FIR}$ ratio with increasing L_{FIR} beyond $10^{11.5} \, L_\odot$ which is observed for local ULIRGs (Fig. 5).

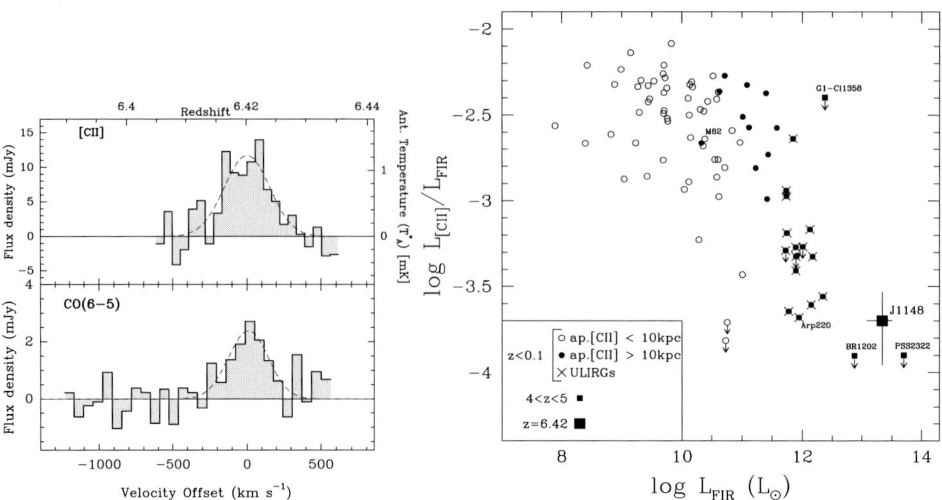

Figure 5. *Left panel*: Spectrum of the [C II] 157.74 μm emission line in the quasar J1148+5251 at $z = 6.42$ (*top panel*) compared to the CO(6→5) emission line. The dashed curves show the gaussian fits to the line profiles. *Right panel*: The $L_{[CII]}/L_{FIR}$ ratio versus L_{FIR} for normal and starburst local galaxies (circles) and for high-z sources (squares). J1148+5251 is shown with a large square. The small squares indicate the upper limits for the three other high-z sources where [C II] was searched for. Crosses indicate local ULIRGs ($L_{IR} > 10^{12}$ L$_\odot$). For reference, well-known local galaxies are also identified. From Maiolino *et al.* (2005).

Recently, Maiolino *et al.* (2005) reported the detection of the C^+ fine-structure line in the $z = 6.4$ quasar J1148+5251, using the IRAM 30-meter telescope (Fig. 5). The [C II] line has a luminosity of 4.4×10^9 L$_\odot$ and the $L_{[CII]}/L_{FIR}$ ratio is 2×10^{-4}, about an order of magnitude smaller than observed in local normal galaxies and similar to the ratio observed in local ULIRGs (Fig. 5). The detection of the [C II] line at high redshift suggests that the interstellar medium was already significantly enriched with metals at $z = 6.4$, i.e. when the universe was only 870 Myr old. This result shows the potential of using this emission line to investigate the star formation activity and the physics of the interstellar medium in the early universe.

5. Conclusions

The observations which are summarized in this review represent only the beginning of the study of the molecular gas associated with star-forming galaxies in the early universe. The data collected during the last decade have proven to be invaluable in constraining the physical characteristics (including the morphology, the masses, kinematics, densities and temperatures) of the massive reservoirs of dense and warm molecular gas from which the stars are formed when the universe was young, including the end of the reionization epoch (at $z = 5.4$.) The on-going and planned enhancements on some of the key operating facilities (such as the IRAM plateau de Bure interferometer, the Extended VLA or CARMA) will improve both on the sensitivity and the angular resolution, thereby enabling further detections and more detailed studies. It is likely that the samples of high-z for which information on the molecular gas will become available will further increase in the future and provide us with a better understanding of the processes which are involved in the formation of stars in early galaxies.

In the next decade, when ALMA will be fully operational, it will become possible to detect at submillimeter wavelengths high-z sources spanning the full range in luminosity from Milky Way type galaxies to the most massive systems in the universe. The superior sensitivity and spatial resolution of this array will enable to resolve many of the high-z sources in the CO line emission and to derive robust estimates of their dynamical masses, unhindered by extinction, thereby providing fundamental constraints to models of galaxy formation. It will also open up the possibility to detect, more routinely than today, atoms and molecules other than carbon monoxide, allowing us to probe the evolution of the chemistry and the abundances of molecular and atomic species in the early universe.

Acknowledgements

It is a pleasure to acknowledge collaborative work with F. Bertoldi, A. Omont, A. Beelen, C.L. Carilli & F. Walter which form the basis of the results presented in this paper. M. McClean is thanked for useful discussions related to molecular excitation. We are grateful to D. Lis for his nearly infinite patience in dealing with the delivery of this manuscript.

References

Alloin, D., Guilloteau, S., Barvainis, R., Antonucci, R., & Tacconi, L. 1997 *A&A* 321, 24

Alloin, D., Barvainis, R., & Guilloteau, S. 2000, *Ap. J.* 528, L81

Andreani, P., Cimatti, A., Loinard, L., & Röttgering, H. 2000, *A&A* 354, L1

Baker, A.J., Tacconi, L.J., Genzel, R., Lehnert, M.D., & Lutz, D. 2004, *Ap. J.* 604, 125

Barvainis, R., Tacconi, L., Antonucci, R., Alloin, D., & Coleman, P. 1994, *Nature* 371, 586

Barvainis, R., Maloney, P., Antonucci, R., & Alloin, D. 1997, *Ap. J.* 484, L13

Barvainis, R., Alloin, D., Guilloteau, S., & Antonucci, R. 1998, *Ap. J.* 492, L13

Barvainis, R., Alloin, D., & Bremer, M. 2002 *A&A* 385, 399

Beelen, A., Cox, P., Pety, J., Carilli, C.L., Bertoldi, F., *et al.* 2003, *A&A* 423, 441

Beelen, A. 2004, PhD thesis, Observatoire de Paris

Beelen, A., Cox, P, Benford, D., *et al.* 2005, *Ap. J.*, in press

Bertoldi, F. & Cox, P. 2002, *A&A* 884, L11

Bertoldi, F., Carilli, C.L., Cox, P., *et al.* 2003a, *A&A* 406, L55

Bertoldi, F., Cox, P., Neri, R., *et al.* 2003b, *A&A* 409, L47

Blain, A., Smail, I. Ivison, R., Kneib, J.-P., & Frayer, D. 2002, *Phys. Rep.* 369, 111

Brown, R.L. & Vanden Bout, P. 1992, *Ap. J.* 397, L11

Casoli, F. & Loinard, L. 2001, in: *Science with the Atacama Large Millimeter Array*, ed. A. Wootten (ASP Conf. Series, vol. 235), ∂. 305

Carilli, C.L., Bertoldi, F., Rupen, M.P., *et al.* 2001, *Ap. J.* 555, 625

Carilli, C.L., Kohno, K., Kawabe, R., *et al.* 2002, *A. J.* 123, 1838

Carilli, C.L., Lewis, G.F., Djorgovski, S.G., *et al.* 2003, *Science* 300, 773

Carilli, C.L., Solomon, P., Vanden Bout, P., *et al.* 2005, *Ap. J.* 618, 586

Chapman, S.C., Blain, A.W., Ivison, R.J., & Smail, I.R. 2003, *Nature* 422, 695

Cox, P. Omont, A., Djorgovski, S.G., *et al.* 2002, *A&A* 387, 406

de Breuck, C., Neri, R., Morganti, R., *et al.* 2003, *A&A* 401, 911

de Breuck, C., Bertoldi, F., Carilli, C.L., *et al.* 2004a, *A&A* 424, 1

de Breuck, C. Downes, D., Neri, R., van Breugel, W., Reuland, M., Omont, A., & Ivison, R. 2004b, *A&A*, 430, L1

De Breuck, C., Downes, D., Neri, R., van Breugel, W., Reuland, M., Omont, A., & Ivison, R. 2005, *A&A* 430, L1

Downes, D., Neri, R., Wiklind, T., Wilner, D.J., & Shaver, P.A. 1999, *Ap. J.* 513, L1

Evans, A.S., Frayer, D.T., Surace, J.A., & Sanders, D.B. 2001, *A. J.* 121, 1893

Fan, X., Hennawi, J.F., Richards, G., *et al.* 2004, *A. J.* 128, 515

Frayer, D.T., Ivison, R.J., Scoville, N.Z., Yun, M., Evans, A.S., Smail, I., Blain, A.W., & Kneib, J.-P. 1998, *Ap. J.* 506, L7

Gao, Y. & Solomon, P.M. 2004, *Ap. J.* 606, 271

Gebhardt, K., Bender, R., Dressler, A., *et al.* 2000, *Ap. J.* 539, L13

Genzel, R., Baker, A.J., Tacconi, L.J., Lutz, D., Cox, P., Guilloteau, S., & Omont, A. 2003, *Ap. J.* 584, 633

Gerin, M. & Phillips, T.G. 1998, *Ap. J.* 509, L17

Gerin, M. & Phillips, T.G. 2000, *Ap. J.* 537, 644

Guilloteau, S., Omont, A., McMahon, R.G., Cox, P., & Petitjean, P. 1997, *A&A* 328, L1

Guilloteau, S., Omont, A., Cox, P., *et al.* 1999, *A&A* 349, 363

Greve, T.R., Bertoldi, F., Smail, I. *et al.* 2005, *MNRAS* 359, 1165

Hainline, L.J., Scoville, N.Z., Yun, M.S., Hawkins, D.W., Frayer, D.T., & Isaak, K.G. 2004, *Ap. J.* 609, 61

Isaak, K.G., Priddey, R.S., McMahon, R.G. *et al.* 2002, *MNRAS* 329, 149

Kauffmann, G. & Haehnelt, M. 2000, *MNRAS* 311, 576

Kalmer, I.J., Ekers, R.D., Sadler, E.M., Weiß, A., Hunstead, R.W., & de Breuck, C. 2005, *Ap. J.* 621, L1

Kneib, J.-P., Neri, R., Smail, I., Blain, A., Sheth, K., van der Werf, P. & Knudsen, K.K. 2005, *A&A* 434, 819

Krips, M., Neri, R., Eckart, A., Downes, D., Martin-Pintado, J., & Planesas, P. 2005, *A&A* 431, 879

Loeb, A. 1993, *Ap. J.* 404, L37

Sanders, D.B. & Mirabel, I.F. 1996, *ARA&A* 34, 749

Maiolino, R., Cox, P., Caselli, P., *et al.* 2005, *A&A* 440, L51

Malhotra, S., *et al.* 2001, *Ap. J.* 561, 766

Neri, R., Genzel, R., Ivison, R.J., *et al.* 2003, *Ap. J.* 597, L113

Ohta, K., Yamada, T., Nakanishi, K., Kohno, K., Akiyama, M., & Kawabe, R. 1996, *Nature* 382, 426

Omont, A., Petitjean, P., Guilloteau, S., *et al.* 1996, *Nature* 382, 428

Omont, A., Cox, P., Bertoldi, F., *et al.* 2001, *A&A* 374, 371

Omont, A., Beelen, A., Bertoldi, F., *et al.* 2003, *A&A* 398, 857

Papadopoulos, P.P., Röttgering, H.J.A., van der Werf, P.P., Guilloteau, S., Omont, A., van Breugel, W.J.M., & Tilanus, R.P.J. 2000, *Ap. J.* 528, 626

Petrosian, V., Bahcall, J.N., & Salpeter, E.E. 1969, *Ap. J.* 155, L57

Pety, J., Beelen, A., Cox, P., *et al.* 2004, *A&A* 428, L21

Planesas, P., Martin-Pintado, J., Neri, R., & Colina, L. 1999, *Science* 286, 2493

Priddey, R.S., Isaak, K.G., McMahon, R.G., & Omont, A. 2003, *MNRAS* 339, 1183

Scoville, N.Z., Yun, M.S., Windhorst, R.A., Keel, W.C., & Armus, L. 1997, *Ap. J.* 485, L21

Scoville, N.Z., Frayer, D.T. Schinnerer, E., & Christopher, M. 2003, *Ap. J.* 585, L105

Sheth, K., Blain, A.W., Kneib, J.-P., Frayer, D.T., van der Werf, P.P., & Knudsen, K.K. 2004, *Ap. J.* 614, L5

Solomon, P.M. & Sage, L.J. 1988, *Ap. J.* 334, 613

Solomon, P.M., Downes, D., & Radford, S.J.E. 1992, *Ap. J.* 398, L29

Solomon, P.M., Downes, D., Radford, S.J.E, & Barrett, J.W. 1997, *Ap. J.* 478, 144

Solomon, P.M., Vanden Bout, P., Carilli, C.L., & Guélin, M. 2003, *Nature* 426, 636

Solomon, P.M. & Vanden Bout, P.A. 2005, *ARA&A* 43, 677

Stark, A.A. 1997, *Ap. J.* 481, 587

Tacconi, L.J., Neri, R., Chapman, S.C., Genzel, R., *et al.* 2006, *Ap. J.*, in press

Vanden Bout, P.A., Solomon, P.M., & Maddalena, R.J. 2004, *Ap. J.* 614, L97

Wagg, J., Wilner, D.J., Neri, R., Downes, D., & Wiklind, T. 2006, *Ap. J.*, in press

Walter, F., Bertoldi, F., Carilli, C., *et al.* 2003, *Nature* 424, 406

Walter, F., Carilli, C.L., Bertoldi, F., *et al.* 2004, *Ap. J.* 615, L17

Weiß, A., Henkel, C., Downes, D., & Walter, F. 2003, *A&A* 409, L41

Weiß, A., Downes, D., Henkel, C., & Walter, F. 2005a, *A&A* 249, L25

Weiß, A., Downes, D., Walter, F., & Henkel, C. 2005b, *A&A* 440, L45

Yao, L., Seaquist, E.R., Kuno, N., & Dunne, L. 2003, *Ap. J.* 597, 1271

Astrochemistry: Recent Successes and Current Challenges
Proceedings IAU Symposium No. 231, 2005
D.C. Lis, G.A. Blake & E. Herbst, eds.

© 2006 International Astronomical Union
doi:10.1017/S1743921306007290

Odin† Detection of O_2

R. Liseau and the O_2din Team

Stockholm Observatory, AlbaNova University Center, Stockholm University, Sweden
email: rene@astro.su.se

Abstract. We present the detection of molecular oxygen with Odin toward the dense molecular core ρ Oph A, which is part of a region of active star formation. The observed spectral line is the $(N_J = 1_1 - 1_0)$ ground state transition of O_2 at 119 GHz ($\lambda = 2.5$ mm). The line center is at the LSR velocity of a number of optically thin lines from other species in the region. The O_2 line also has a very similar, narrow, line width. Within the $10'$ beam, the line intensity is $\int T_A\, dv = 28$ mK km s^{-1}, which corresponds to $5\,\sigma$ of the rms noise. A standard LTE analysis results in an O_2 abundance of 5×10^{-8}, with an uncertainty of at least a factor of two. We show that standard methods, however, do not apply in this case, as the coupling of the Odin beam to the source structure needs to be accounted for. Preliminary model results indicate O_2 abundances to be higher by one order of magnitude than suggested by the standard case. This model predicts the 487 GHz line of O_2 to be easily detectable by the future Herschel-HIFI facility, but to be out of reach for observations on a shorter time scale with the Odin space observatory.

Keywords. abundances — ISM: clouds — ISM: molecules — lines and bands — stars: formation

1. Introduction

The energy balance of the interstellar medium (ISM) relies on a complex interaction of various heating and cooling processes, which of course also affect the chemistry being activated in the medium. This in turn regulates the amount of the major cooling species in the gas. Heating processes include mechanical and radiative energy input, whereas, in dense clouds, the cooling is dominated by molecules and dust particles, resulting in low equilibrium temperatures on short time scales. Whereas the coolant CO is readily observable from the ground, water and oxygen molecules need to be observed from space.

Odin is thus a mission in orbit around the Earth dedicated to the observation of H_2O and O_2. The millimeter/submillimeter (mm/submm) Odin facility is shared by aeronomers and astronomers on an equal time basis, and is described by Frisk *et al.* (2003) and Olberg *et al.* (2003). Odin's astronomy is reviewed by Hjalmarson *et al.* (2003). These articles are all collected in the special Odin edition of Astronomy & Astrophysics, which of course also contains numerous papers on initial results obtained with Odin. Energy level and excitation diagrams for the principal observable molecules can be found in Liseau (2001).

Prior to the launch of Odin in February 2001, it was already clear from the observations by the Submillimeter Wave Astronomy Satellite (SWAS) that H_2O was significantly less abundant in the ISM than was previously thought. Even more intriguing, perhaps, was the apparently total absence of O_2 (Goldsmith *et al.* 2000). However, a couple of years later, the tentative detection of the 487 GHz line (see Fig. 1) toward the dense molecular core ρ Oph A was announced by Goldsmith *et al.* (2002).

† Odin is a Swedish-led satellite project funded jointly by the Swedish National Space Board (SNSB), the Canadian Space Agency (CSA), the National Technology Agency of Finland (Tekes) and Centre National d'Etude Spatiale (CNES).

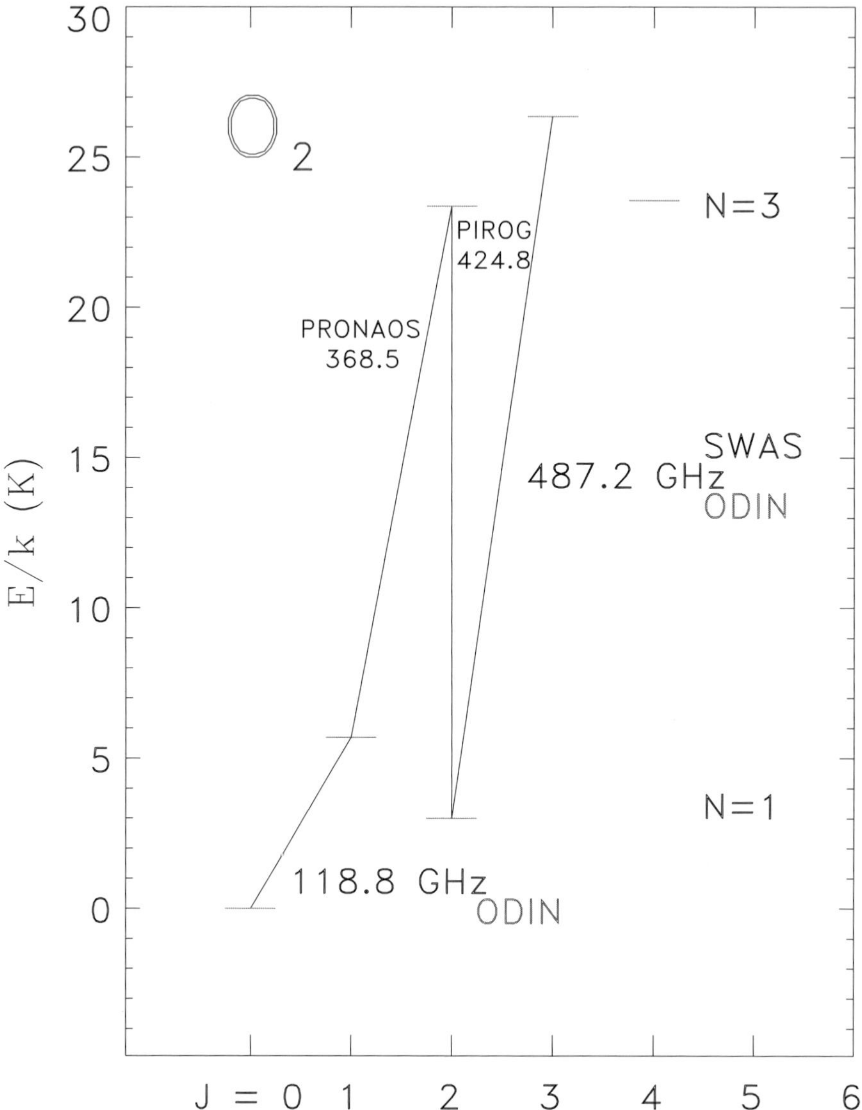

Figure 1. Rotational energy level diagram of O_2 for $E_{upper}<30$ K. Transitions accessible to various missions discussed in the text are also indicated (adopted from Liseau 2001).

The SWAS result was challenged by Pagani *et al.* (2003), who were unable to confirm the presence of oxygen in ρ Oph A based on observations with Odin in the ground state transition of O_2 at 119 GHz (Fig. 1). Taking into account the fact that this involved a different line at considerably lower angular resolution toward a somewhat different position, the status of the SWAS result was rendered inconclusive.

Besides the negative result for the dense cloud ρ Oph A, Pagani *et al.* (2003) also reported on unsuccessful observations of nearly a dozen galactic sources in the O_2 ($N_J = 1_1 - 1_0$) line at 119 GHz. Furthermore, Wilson *et al.* (2005) recently published an upper limit in this line toward an extragalactic object, *viz.* the Small Magellanic Cloud (SMC). Contrasting with these negative results, we report here the detection with Odin, at the $5\,\sigma$ level, of the 119 GHz line of O_2 toward the dense cloud ρ Oph A.

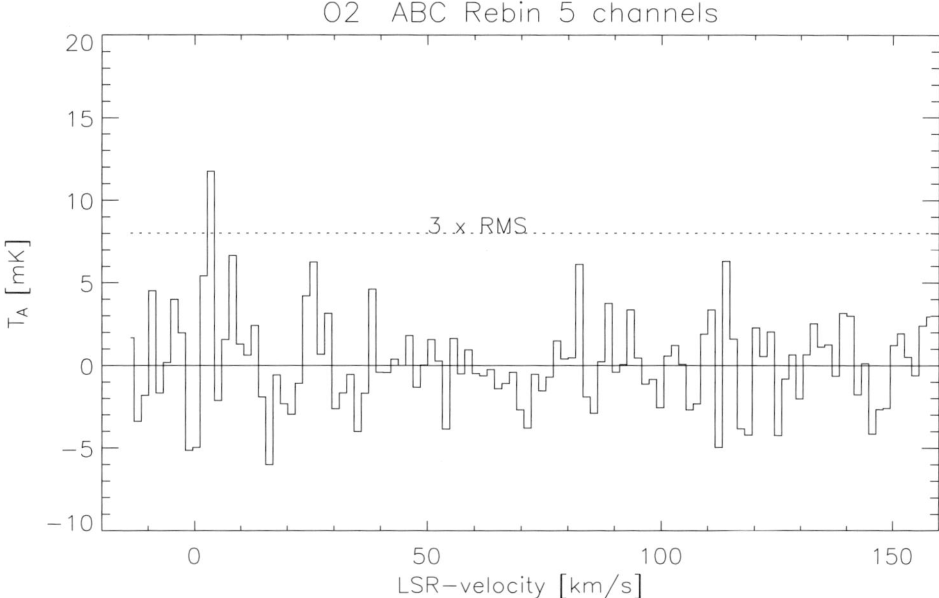

Figure 2. The 5 channels of the O_2 feature have been binned into 1 channel, corresponding to the integrated line intensity $\int T_A\, dv$ at the $5\,\sigma$ level. All other features in the entire useful spectral range over $180\,\mathrm{km\,s^{-1}}$ are below the $3\,\sigma$ leve, shown as the dashed horizontal line.

2. Odin O_2 Observations

The Odin observations of ρ Oph A were performed during 2002 and 2003 and the total on-source integration time was 78 hours. At $119\,\mathrm{GHz}$, the half-power-beam-width of the $1.1\,\mathrm{m}$ telescope is $10'$. A digital autocorrelator of $100\,\mathrm{MHz}$ bandwidth and $125\,\mathrm{kHz}$ resolution was used as the backend. In Doppler-velocity units, this corresponds to $250\,\mathrm{km\,s^{-1}}$ and $0.3\,\mathrm{km\,s^{-1}}$, respectively. More exhaustive details regarding these observations and the subsequent complex data reductions are provided by Larsson *et al.* (2005).

In Figure 2, showing the final stage of this data reduction, the Odin $119\,\mathrm{GHz}$ spectrum is presented. At full spectral resolution, the O_2 line occupies five autocorrelator channels, which in the figure have been binned into one channel. This is equivalent to the integrated line intensity, $\int T_A\, dv_z = (28 \pm 5.5)\,\mathrm{mK\,km\,s^{-1}}$.

Other Gauss-fit parameters are: peak $T_A = (17.5 \pm 3.5)\,\mathrm{mK}$, line center velocity $v_{\mathrm{LSR}} = (3.0 \pm 1.0)\,\mathrm{km\,s^{-1}}$ and line width $\Delta v_{\mathrm{FWHM}} = (1.53 \pm 0.03)\,\mathrm{km\,s^{-1}}$. The velocity and width of the O_2 line are in agreement with that of the PDR recombination line emission observed by Pankonin & Walmsley (1978) with a $7\rlap{.}'8$ beam, *i.e.* comparable to the spatial resolution of Odin. As also indicated by the $C^{18}O$ observations of Frerking *et al.* (1989) with a $2\rlap{.}'3$ beam, the line parameters center velocity and width of optically thin species appears constant on various angular scales. The values of both parameters are significantly different for the $487\,\mathrm{GHz}$ line of Goldsmith *et al.* (2002).

The Odin observations at full spectral resolution, and the comparison with other molecular line data, will be presented by Larsson *et al.* (2005).

3. O_2 Abundance

Of general interest is the relative concentration of various forms of oxygen in the interstellar medium and in particular that of the molecular constituents. Below, we shall examine the derivation of the O_2 abundance in some detail, as widely used standard methods will encounter difficulties in this particular case, leading to erroneous results.

3.1. *Column Density of Oxygen*

The data for radiative ($A_{\rm ul}$) and collisional ($q_{\rm ul}$) de-excitation have been provided by Marechal *et al.* (1997) and by Bergman (1995 and private communication). The collisional excitation $N_J = 1_0$ to $N = J$, such as ($1_0 - 1_1$), is forbidden and we assume that the collisional population of the 1_1 level occurs mainly via 3_2 and hence, the critical density for O_2 ($N_J = 1_1 - 1_0$), *viz.*

$$n_{\rm crit} = A_{11,10}/q_{32,11}(T_{\rm k}) \tag{3.1}$$

becomes $> 10^3\,{\rm cm}^{-3}$ but $< 10^4\,{\rm cm}^{-3}$ for $T_{\rm k} < 100\,{\rm K}$ and is likely lower than densities in the cloud. Local Thermodynamic Equilibrium (LTE) provides thus a reasonable assumption for the population of the lower levels. The weak O_2 ($N_J = 1_1 - 1_0$) line is most likely optically thin and can be expected to exhibit a nearly Gaussian shape, reflecting small scale stochastic motions in the cloud.

The observed integrated intensity of the optically thin 119 GHz line, formed in LTE, is then given by

$$
\begin{aligned}
I_{119}^{\rm obs} &= \int T_{\rm A}\, dv_{\rm z} \approx f_{\rm b}\, \eta_{\rm mb} \int T_{\rm R}\, dv_{\rm z} \\
&= f_{\rm b}\, \eta_{\rm mb} \left(\frac{hc}{2\pi^{1/3}k}\right)^3 \frac{A_{11,10}}{T_{11,10}^2} \left[1 - \frac{J_\nu(T_{\rm bg})}{J_\nu(T_{\rm k})}\right] \frac{g_{11}\, e^{-T_{11,10}/T_{\rm k}}}{Q(T_{\rm k})} X(O_2)\, N(H_2)
\end{aligned}
\tag{3.2}
$$

where $f_{\rm b}$ is the beam filling factor, $\eta_{\rm mb}$ is the main beam efficiency and $T_{\rm R}$ is the radiation temperature of the source. In LTE, the only dependence is on the temperature, which for our present purposes is well represented by a power law, *i.e* for the 119 GHz transition, the function $\Phi(T_{\rm k})$ is given by

$$
\begin{aligned}
\Phi(T_{\rm k}) &= \left(\frac{hc}{2\pi^{1/3}k}\right)^3 \frac{A_{11,10}}{T_{11,10}^2} \left[1 - \frac{J_\nu(T_{\rm bg})}{J_\nu(T_{\rm k})}\right] \frac{g_{11}\, e^{-T_{11,10}/T_{\rm k}}}{Q(T_{\rm k})} \\
&= 2.5 \times 10^{-11}\, T_{\rm k}^{-0.67}, \qquad T_{\rm k} > T_{11,10}
\end{aligned}
\tag{3.3}
$$

in units of $K\,{\rm cm}^3\,{\rm s}^{-1}$. In the above equations, the temperature of the background radiation field is $T_{\rm bg} = 2.734\,{\rm K}$, the statistical weight of the upper level is $g_{11} = 2J + 1 = 3$, the temperature of the transition is $T_{11,10} = 5.70\,{\rm K}$, the quasi-Planck function is $J_\nu(T) = T_{11,10}/[\exp(T_{11,10}) - 1]$ and the partition function is approximated by $Q(T_{\rm k}) \approx kT_{\rm k}/hB$, where the rotational constant is $B = 43100.460\,{\rm MHz}$.

Expressing the line intensity in $K\,{\rm km\,s}^{-1}$, the column density of O_2 is then simply found from

$$N(O_2) \geqslant 4 \times 10^{15}\, T_{\rm k}^{0.67}\, I_{119}^{\rm obs} \qquad ({\rm cm}^{-2}) \tag{3.4}$$

On the scale of the Odin beam, temperatures have variously been estimated from gas tracers (*e.g.*, CO; Loren *et al.* 1983) and multi-wavelength measurements of the dust emission (*e.g.*, Ristorcelli *et al.* 2005), yielding an average of about 30 K and, hence, $N(O_2) \geqslant 1 \times 10^{15}\,{\rm cm}^{-2}$. As can be seen from Eq. 3.4, the column density of O_2 is not very sensitive to the temperature and, *e.g.*, an uncertainty of $\pm 10\,{\rm K}$ would amount to a change in $N(O_2)$ by no more than 20%. The critical parameter for the determination of $X(O_2)$ is obviously $N(H_2)$.

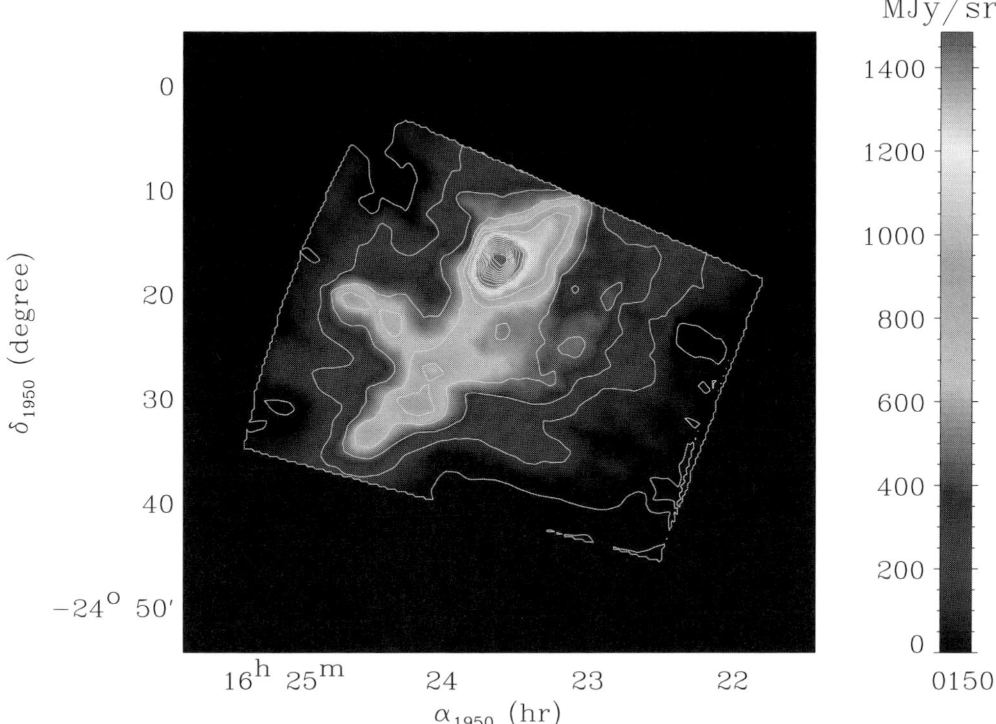

Figure 3. Continuum observations of the ρ Oph cloud at $200\,\mu$m with the $2'$ beam of the balloon-borne submm-telescope PRONAOS (Ristorcelli *et al.* 2005). ρ Oph A is the hot spot above the center of the figure. Intensities, I_ν, are given in MJy sr^{-1} (see the scale bar to the right). In addition to the $200\,\mu$m-map shown here, full $40' \times 40'$ maps have also been obtained at 260, 360 and $580\,\mu$m.

3.2. *Column Density of Hydrogen*

To find the oxygen abundance, $X(O_2) = N(O_2)/N(H_2)$, we need to specify the column density of hydrogen, H_2 (see: Eq. 3.2). Based on the $C^{18}O$ maps of Tachihara *et al.* (2000), Pagani *et al.* (2003) estimated for Odin O_2 observations with a $10'$ beam the H_2-column to be $N(H_2) = 4 \times 10^{22}$ cm^{-2}. This is higher than values recently obtained from [C I] observations by Kulesa *et al.* (2005) on a much smaller, *i.e.* $3\overset{'}{.}5$, scale, *viz.* $(2-3) \times 10^{22}$ cm^{-2} and those derived from submm dust continuum observations by Ristorcelli *et al.* (2005), *viz.* 1×10^{22} cm^{-2} for the $10'$ Odin beam. The earlier map of the LTE column density of $C^{18}O$ by Wilking & Lada (1983) would indicate a one order of magnitude higher $N(H_2)$. This is also the value obtained, on smaller angular scales, by Kamegai *et al.* (2003) from a large [C I] map of the ρ Oph cloud.

Various assumptions come into these estimates, such as the abundance of $C^{18}O$ or the opacity of the dust grains, leading to a considerable spread in $N(H_2)$. For the moment, we shall adopt a weighted average of 2×10^{22} cm^{-2}, a value which is associated with an uncertainty of at least a factor of two.

3.3. *O_2 Abundance: Extended Source*

Applying the estimates of temperature and H_2-column density to the O_2 observations with Odin yield an oxygen abundance $X(O_2) \approx 5 \times 10^{-8}$, if we assume that both η_{mb} and f_{b} are close to unity. The corresponding rotational O_2 spectrum between $100\,\mu$m and 1 cm is displayed in Figure 4. From this we infer that Odin is not capable of detecting

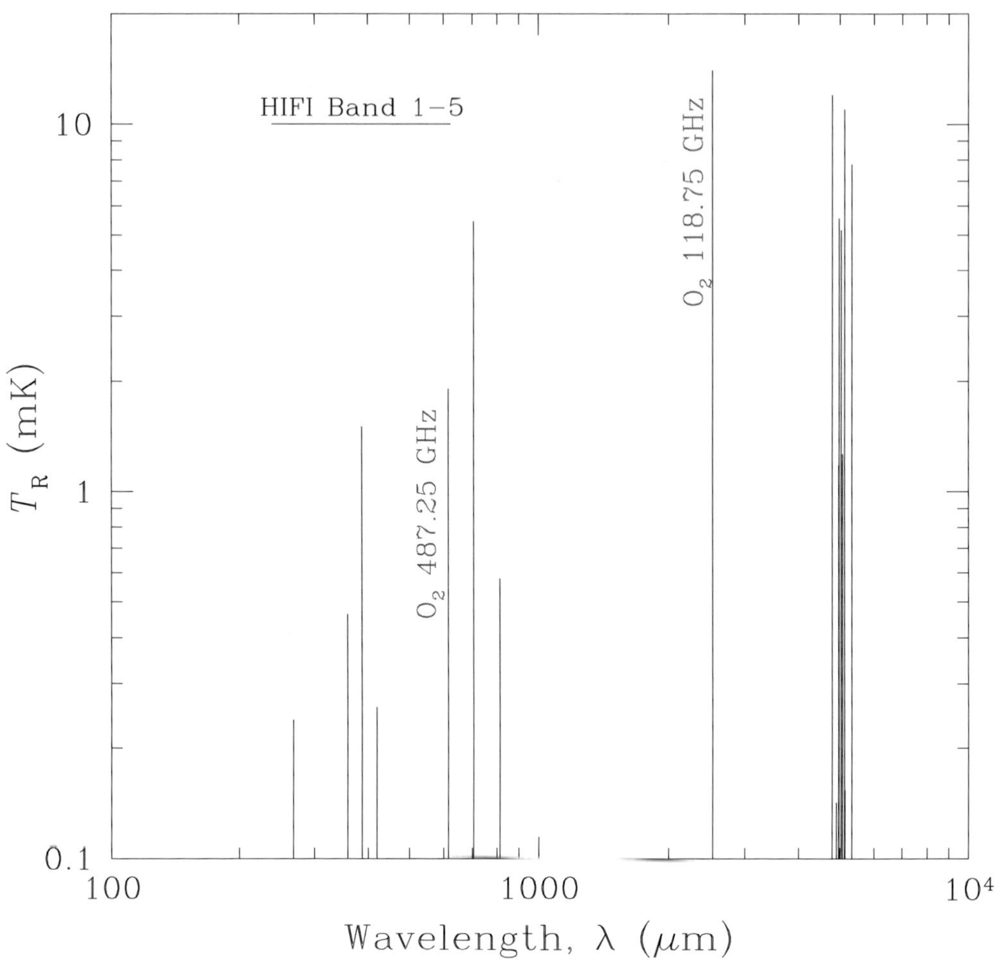

Figure 4. Theoretical (LVG) O_2 spectrum between 100 μm and 1 cm for ρ Oph A, based on the Odin observations in the O_2 ($N_J = 1_1 - 1_0$) line of ρ Oph A. The uniform source is assumed to fill the 10′ beam. The bar shows the frequency coverage of Herschel-HIFI, admitting the 487 GHz ($3_3 - 1_2$) line at its low-frequency end (see also Fig. 1), whereas the stronger 425 GHz ($3_2 - 1_2$) line falls unfortunately outside the HIFI band.

from ρ Oph A the other, in principle accessible, O_2 line, *i.e.* the ($3_3 - 1_2$) line at 487 GHz (2′3 beam). It is also clear that the detection of the 487 GHz line with Herschel-HIFI† (45″ beam) would require the O_2 source to be significantly more compact than 10′.

3.4. O_2 Abundance: Compact Source

The value of $X(O_2)$ derived above is based on an analytical model of the line excitation, where one has made the implicit assumption that the emitting region is an isothermal and homogeneous plane-parallel slab of gas. These idealized conditions would have to apply on an angular scale of 10′, corresponding to a physical length scale of half a parsec at the distance of the ρ Oph A cloud. This picture is inconsistent, however, with observations on smaller scales of the gas and the dust (see, *e.g.*, Ashby *et al.* 2000 and Johnstone *et al.* 2000, respectively), which show ρ Oph A to have a centrally condensed core structure. In

† http://www.sron.nl/divisions/lea/hifi/

other words, strong gradients in temperature, density and velocity field are known to be present inside the Odin O_2 beam. In addition, UV-radiation illuminates and heats the core preferentially only from one side (Liseau *et al.* 1999).

Modeling the complex structure of this core is in progress. Preliminary results suggest that the line formation region of O_2 119 GHz is relatively compact, with a size of 22″. Consequently, the implied abundance is relatively high, $X(O_2) = 6 \times 10^{-7}$, and the observed weakness of the line is due to substantial beam dilution (roughly $\theta^2_{\mathrm{source}}/\theta^2_{\mathrm{beam}} = 10^{-3}$). Predictions of this model for the O_2 487 GHz line are for Odin (2′.3 beam) a peak antenna temperature of 22 mK and for Herschel-HIFI (45″ beam) 0.2 K. Whereas the observation of the former would imply prohibitive integration times, the latter would be a matter of only minutes. We look impatiently forward to the launch of Herschel in the 2007/08 time frame.

4. Possible Implications

The weak detection of O_2 by Odin will likely not add to the present discussion of the energy balance of the ISM, as the role of a coolant of importance had already been altered before. For instance, the total cooling in the O_2 lines of the spectrum of Figure 4 amounts to a mere $7 \times 10^{-5}\,L_\odot$.

It is conceivable, though, that this detection eventually will lead to an increased understanding of the O_2 chemistry of the dense and cold ISM and that chemistry models may benefit from the Odin achievement.

5. Conclusions

In summary, we briefly conclude the following:

(i) Very deep integrations (3.5 mK rms) with Odin in the 119 GHz ground state line of O_2 toward the dense cloud core ρ Oph A resulted in a $5\,\sigma$ detection of the line.

(ii) Standard analysis of an optically thin line formed in LTE leads to the beam-averaged column density of molecular oxygen for ρ Oph A, *viz.* $N(O_2) = 1 \times 10^{15}\,\mathrm{cm}^{-2}$.

(iii) Estimates of the oxygen abundance need to rely on assumptions regarding the source size and structure with respect to the 10′ beam of Odin at 119 GHz. This results in $X(O_2) = 5 \times 10^{-8}$ for a homogeneous extended source and $X(O_2) = 6 \times 10^{-7}$ for a more realistic source structure, respectively.

(iv) This issue will most likely become resolved by observations with the future Herschel-HIFI, for which we predict an O_2 intensity of a few tenths of a Kelvin at 487 GHz.

Acknowledgements

The author wishes to thank the organizers for their kind invitation to this highly stimulating conference. He also wishes to express his high appreciation of the interesting and intensive discussions with many colleagues of the Odin consortium. However, Bengt Larsson deserves special mentioning, without whom the present article would not have been possible. The full list of names and affiliations of the O_2din Team will appear in Larsson *et al.* (2005).

References

Ashby, M.L.N., Bergin, E.A., Plume, R., *et al.* 2000, *Ap. J.* 539, L 119
Bergman, P., 1995, *Ap. J.* 445, L 167
Frisk, U., Hagström, M., Ala-Laurinaho, J., *et al.* 2003, *A&A* 402, L 27

Frerking, M.A., Keene, J., Blake, G.A., & Phillips, T.G., 1989, *Ap. J.* 344, 311
Goldsmith, P.F., Melnick, G.J., Bergin, E.A., *et al.* 2000, *Ap. J.* 539, L 123
Goldsmith, P.F., Li, D., Bergin, E.A., *et al.* 2002, *Ap. J.* 576, 814
Hjalmarson, Å., Frisk, U., Olberg, M., *et al.* 2003, *A&A* 402, L 39
Johnstone, D., Wilson, C.D., Moriarty-Schievens, G., *et al.* 2000, *Ap. J.* 545, 327
Kamegai, K., Ikeda, M., Maezawa, H., *et al.* 2003, *Ap. J.* 589, 378
Kulesa, C.A., Hungerford, A.L., Walker, C.K., *et al.* 2005, *Ap. J.* 625, 194
Larsson, B., *et al.* 2005, in preparation
Liseau, R., White, G.J., Larsson, B., *et al.* 1999, *A&A* 344, 342
Liseau, R., 2001, *ESA-SP* 460, p. 313
Loren, R.B., Sandqvist, Aa., & Wootten, A., 1983, *Ap. J.* 70, 620
Loren, R.B., Wootten, A., & Wilking, B.A., 1990, *Ap. J.* 365, 269
Maréchal, P., Viala, Y.P., & Benayoun, J.J., 1997, *A&A* 324, 221
Olberg, M., Frisk, U., Lecacheux, A., *et al.* 2003, *A&A* 402, L 35
Pagani, L., Olofsson, A.O.H., Bergman, P., *et al.* 2003, *A&A* 402, L 77
Pankonin, V. & Walmsley, C.M., 1978, *A&A* 64, 333
Ristorcelli, I., *et al.* 2005, in preparation
Tachihara, K., Mizuno, A., & Fukui, Y., 2000, *Ap. J.* 336, 519
Wilking, B.A. & Lada, C.J., 1983, *Ap. J.* 274, 698

Discussion

VAN DISHOECK: (1) For the high O_2 abundance of 6×10^{-7}, would the $^{16}O^{18}O$ 234 GHz line be strong enough to detect from the ground? Large single dish telescopes have beam sizes comparable to, or smaller than 22″. (2) Have you looked at any other positions in ρ Oph with Odin?

LISEAU: (1) This would seem to be a good suggestion (A rough estimate for APEX yields a few mK, *i.e.* very tough to unfeasible indeed). (2) Odin is at this very moment mapping cores in the ρ Oph main cloud.

CERNICHARO: The O_2 line at 119 GHz is a magnetic dipole transition. Hence, we could expect to have $T_{ex} \sim T_k$. Why, then, there is a difference of factor 7 between LTE and LVG calculations? Why do you need LVG? What collisional rates to you use?

LISEAU: The critical density of the transition is of the order of the average density of the core ρ Oph A. Thermalization normally occurs at densities beyond an order of magnitude higher than n_{crit}. We therefore solved the statistical equilibrium equations, checking the assumption of LTE. The difference in $X(O_2)$ is dominated by the alternate choice of $N(H)$, however.

Astrochemistry: Recent Successes and Current Challenges
Proceedings IAU Symposium No. 231, 2005
D.C. Lis, G.A. Blake & E. Herbst, eds.

© 2006 International Astronomical Union
doi:10.1017/S1743921306007307

What Have We Learned from SWAS?

Edwin A. Bergin[1] and Gary J. Melnick[2]

[1]University of Michigan, Department of Astronomy, 825 Dennison,
500 Church Street, Ann Arbor, MI 48109, USA
email: ebergin@umich.edu

[2]Harvard-Smithsonian Center for Astrophysics,
60 Garden Street, Cambridge, MA 02138, USA
email: gmelnick@cfa.harvard.edu

Abstract. The *Submillimeter Wave Astronomy Satellite* (*SWAS*) has recently completed 5.5 years of successful operation. Among the legacies of the mission has been a greater understanding of the abundance and spatial distribution of H_2O and O_2 within molecular clouds. We summarize SWAS results and discuss how the measured low abundance of water vapor and non-detections of molecular oxygen are suggestive of a general lack of atomic oxygen in the dense centers of molecular clouds. We also present a new more comprehensive model for the oxygen chemistry in the interstellar medium.

Keywords. astrobiology — astrochemistry — ISM: molecules — molecular processes — radiative transfer

1. Introduction

The *Submillimeter Wave Astronomy Satellite* (*SWAS*) was a mission primarily dedicated to the study of star formation, molecular cloud composition, and the structure of clouds. To carry out these investigations, *SWAS* was designed to observe $H_2^{16}O$, $H_2^{18}O$, O_2, C I, and ^{13}CO in transitions that are either difficult or impossible to detect from within the Earth's atmosphere. *SWAS* is a complete radio telescope in space, capable of observing the transitions listed in Table 1 with a spectral resolution of 1 km s^{-1} (Melnick *et al.* 2000). At the time of launch, the primary mission goals were to answer three important questions:

(a) Where is all of the oxygen in the dense interstellar medium (and, in particular, are H_2O and O_2 major reservoirs of oxygen)?

(b) Are H_2O and O_2 significant gas coolants?

(c) What is the large-scale distribution of C I and warm ^{13}CO in molecular clouds?

During its 5.5-year mission, *SWAS* conducted pointed observations toward more than 300 distinct galactic sources and along more than 6800 lines of sight. In so doing, *SWAS* data have been used to answer these questions. Page limitations prevent a comprehensive review of *SWAS*'s contribution to each of these topics. Instead, this paper will focus on the greater understanding of H_2O and O_2 in quiescent gas within molecular cloud cores enabled by the *SWAS* data.

2. The Observations of H_2O and O_2 in Interstellar Clouds

One of the primary scientific goals of *SWAS* was to determine the abundance of water vapor in various components of the interstellar medium by observing the lowest rotational transitions of ortho-$H_2^{16}O$ and its isotopomer ortho-$H_2^{18}O$. The study of water vapor has several motivations.

309

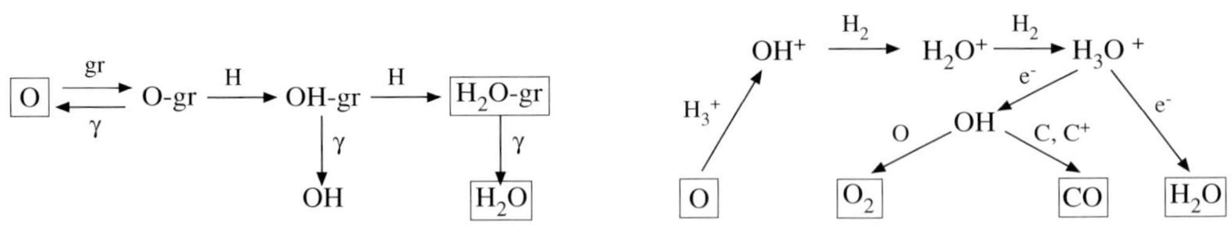

Figure 1. Key elements of the oxygen chemistry believed to occur on grain surfaces and in the cold gas phase. Photodesorption is denoted by γ. The boxed species are the most abundant and are of greatest interest in our study. O-gr, OH-gr, and H$_2$O-gr denote O, OH, H$_2$O, respectively, on grains.

Measurements of the water vapor abundance provide crucial constraints on astrochemical models that seek to elucidate the gas-phase chemical reactions and gas-grain interactions that determine the composition of the interstellar medium. In particular, chemical models (see Elitzur & de Jong 1978; Neufeld *et al.* 1995) predict that for $T \gtrsim 300$ K, water vapor will account for most of the gas-phase oxygen that is not bound as CO, as a result of the neutral-neutral reactions: H$_2$ + O $\rightarrow$ OH + H and H$_2$ + OH $\rightarrow$ H$_2$O + H. At temperatures less than 300 K, however, these reactions are negligibly slow because they possess significant activation energy barriers. Gas-phase water is then produced either by means of reactions on grain surfaces with subsequent sublimation or via cosmic-ray driven ion-neutral chemistry, as shown in Figure 1. Water formed on grains will remain frozen on the grain until either the grain temperature exceeds ~ 100 K or the water molecule is photodesorbed by a UV photon. Gas-phase ion-neutral reactions (Herbst & Klemperer 1973) proceed more slowly than the neutral-neutral reactions – requiring approximately 10^6 years to reach chemical equilibrium in well-shielded gas with $n(\text{H}_2) \sim 10^5$ cm^{-3} – and result in a maximum H$_2$O abundance more than 100 times lower than in warmer gas.

2.1. *Water and Molecular Oxygen Abundances*

SWAS has detected water vapor towards a diverse sample of astronomical objects. These objects include both dense and diffuse interstellar gas clouds, circumstellar envelopes, planetary atmospheres, and comets. As described below, the water abundances inferred from these *SWAS* observations vary by more than a factor of 10^5 and confirm some

Table 1. Spectral Lines Observed by *SWAS*

Species	Transition	Energy Above Ground State (E/k)	Frequency (GHz)	Wavelength (μm)	Critical Density (cm^{-3})
O$_2$	$3,3-1,2$	26 K	487.249	615.276	10^3
C I	$^3P_1 - {}^3P_0$ [†]	24 K	492.161	609.134	10^4
H$_2^{18}$O	$1_{10} - 1_{01}$ [†]	26 K	547.676	547.390	10^9 [‡]
^{13}CO	$J = 5 - 4$	79 K	550.926	544.161	2×10^5
H$_2^{16}$O	$1_{10} - 1_{01}$ [†]	27 K	556.936	538.289	10^9 [‡]

[†] Ground-state transition.

[‡] The critical density for H$_2^{16}$O will likely be less than this value by a factor of $10^2 - 10^4$ due to significant radiation trapping in this line. The critical density for H$_2^{18}$O could be reduced by a factor of $1.5 - 50$ due to the same effect.

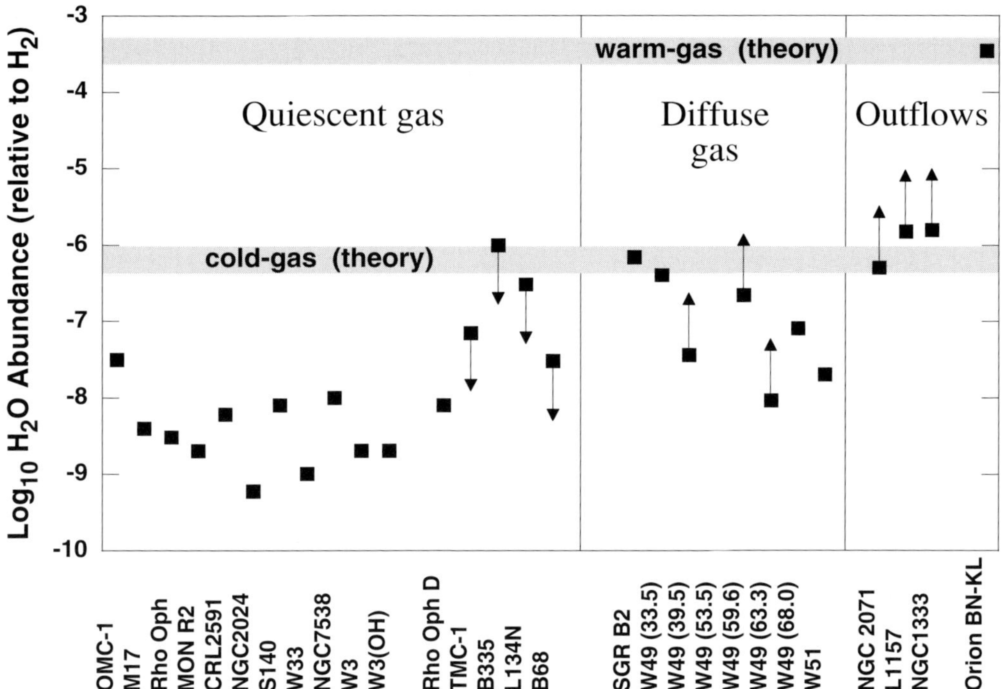

Figure 2. Measurements of the abundance of ortho-water vapor (relative to H_2) towards molecular cloud cores labeled on the abscissa. Quiescent gas refers to abundances derived from water emission in dense ambient regions (taken from Melnick *et al.* 2000; Snell *et al.* 2000a,b,c). Diffuse gas refers to abundances derived from detections of o-H_2O in absorption (Neufeld *et al.* 2000; Neufeld *et al.* 2002; Plume *et al.* 2004). Outflows abundances are from Melnick *et al.* (2000) and Neufeld *et al.* (2000).

aspects of previous theoretical predictions but also turn up some surprises. Figure 2 provides a graphical representation of the *SWAS* water abundance determinations. In particular:

– In warm ($T \gtrsim 300$ K), dense ($n(H_2) \gtrsim 10^3$ cm^{-3}) gas, the gaseous water abundance is 10^{-5} to $> 10^{-4}$ relative to H_2, in agreement with warm-cloud chemical models. Within this warm gas, H_2O is a significant repository of oxygen as well as a major gas coolant, in accord with model predictions.

– In cold ($T \lesssim 30$ K), dense clouds the abundance of gaseous H_2O is ~ 100 to 1000 times below the predictions of cold-cloud gas-phase chemical models and the O_2 abundance is at least 100 times below predictions. As a result, gaseous H_2O and O_2 are neither significant reservoirs of oxygen in cold gas nor are they important gas coolants in such clouds.

– In cold diffuse ($n(H_2) < 10^3$ cm^{-3}) gas there is limited evidence that the H_2O abundance is higher than in the cold dense gas, but O_2 remains undetected.

2.2. *Water Radiation Transfer in the Effectively Thin Limit*

It is important to note that both *SWAS* and Odin observed the ground state transition of ortho-H_2O. Water is a highly polar molecule and even small abundances can result in emission lines with high opacity. As such it is worth a discussion of the derivation of water vapor abundances under these conditions. Under normal circumstances large optical depth would make it highly difficult to extract reliable abundances. However, the emission from the o-H$_2$O ground state line is optically thick, but subthermal (given that densities are well below the critical density given in Table 1). Thus every photon that is created will scatter, but eventually escape the cloud (i.e. collisional de-excitations are rare). This so-called optically thick, but effectively thin limit is discussed by Linke *et al.* (1977) and Snell *et al.* (2000).

In the effectively thin limit the relation between the integrated emission and water vapor column density can be expressed analytically as:

$$\int T_R\, dv = C n_{H_2} \frac{c^3}{2\nu^3 k} N(\text{o-H}_2\text{O}) \frac{h\nu}{4\pi} exp(-h\nu/kT_k).$$

where C is the collisional de-excitation rate. With analysis of the source physical structure (H$_2$ volume density, total column density, and temperature) line-of-sight average abundances can be derived, which were presented in Figure 2 and summarized in § 2.1.

Two additional methods can be used to constrain the water abundance: (1) Observations of the lesser abundant isotopologue, H$_2^{18}$O, can be used to reduce concerns regarding the opacity. This line has been detected by Odin in ambient gas in Orion by Olofsson *et al.* (2003) with resulting abundances that are in agreement with the earlier analysis using H$_2^{16}$O (see also the discussion by Snell *et al.* 2000a); (2) One can use a-priori knowledge of the line-of-sight physical structure (density and temperature) and sophisticated Monte-Carlo radiation transfer models to constrain the water abundance. In this fashion average abundances for several well known objects were derived by Ashby *et al.* (2000), which again are in agreement with abundances derived assuming effectively thin emission. In addition, Bergin & Snell (2002) linked chemistry to a radiation transfer model to derive sensitive abundance limits in two pre-stellar cores (B68 and ρ Oph D) with well characterized density profiles. In these objects the average water abundance is $x(\text{o-H}_2\text{O}) < 3 \times 10^{-8}$.

3. Gas-Grain Chemistry in the ISM

In the initial interpretation of *SWAS* results Bergin *et al.* (2000) examined the various factors that will influence the oxygen chemistry in quiescent gas. In the ISM it is very difficult to stop water production. In the gas phase all of the reactions on the sequence to make water and molecular oxygen have been measured in the lab (see Millar, this volume; Sims, this volume). Thus, if oxygen is present in the gas it will react with cosmic-ray-produced H$_3^+$ and quickly create both H$_2$O and O$_2$ with abundances much greater than the measured values of $\sim 10^{-9}$ and $< 3 \times 10^{-7}$, respectively. Thus either water formation must be stunted, or our understanding of water destruction is incorrect.

Two potential solutions were isolated. Increased destruction could result if sufficient turbulent mixing were to expose water vapor to layers where photodissociation dominates or reactions with UV-produced C$^+$ and C are possible (Chièze & Pineau Des Forêts 1989; Willacy, Langer, & Allen 2002). While this solution may be relevant for some objects, it would not account for a lack of water vapor in objects dominated by thermal motions, such as B68 or other low-mass condensed cores (Bergin & Snell 2002).

On the formation side, a solution examined by several groups to varying degrees (Bergin *et al.* 2000; Viti *et al.* 2001; Charnley, Rodgers, & Ehrenfreund 2001; Roberts & Herbst 2002) is to include the formation of water ice on grains (or depletion of oxygen from the gas). In this model, as presented by Bergin *et al.* in Figure 3, the fuel for the chemistry, oxygen atoms, is frozen onto grains in the form of water ice. This ice will not evaporate unless temperatures exceed 110–120 K (Fraser *et al.* 2001) and hence most oxygen is

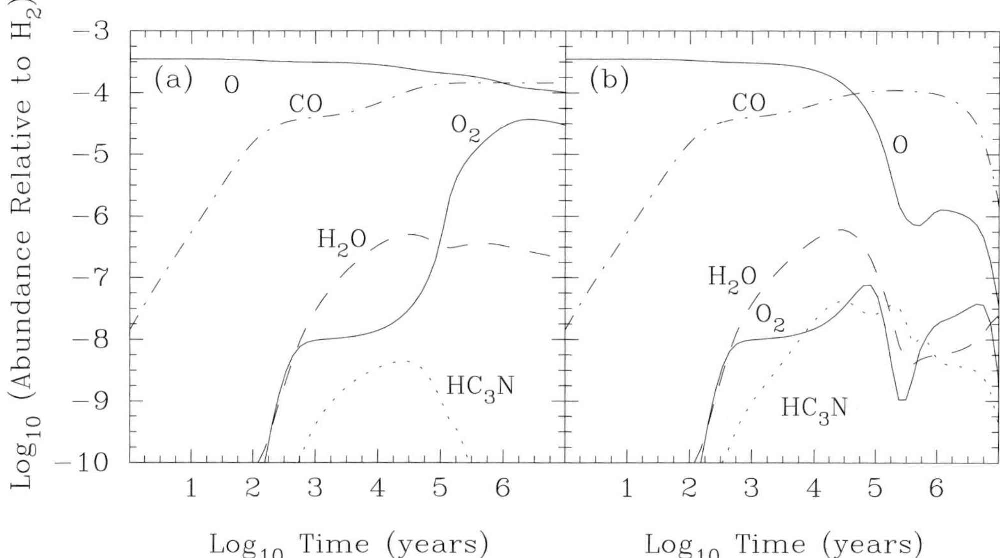

Figure 3. Abundances of CO, O_2, H_2O, and HC_3N as a function of time for two models of the gas-grain chemistry. (a) Model including depletion onto and desorption from grain surfaces, but no surface chemistry. (b) Model including depletion and desorption onto grains but with simple surface chemistry. Both models assume $n_{H_2} = 10^5$ cm^{-3}, $T_{gas} = T_{dust} = 30$ K, and $A_V = 20^m$. The second model does not include any modifications to account for the discrete nature of grains, but is nonetheless illustrative of the effects of the process. Taken from Bergin *et al.* (2000).

unavailable to make water vapor or molecular oxygen in the gas. Thus, the low abundance of water vapor hints at a lack of gaseous atomic oxygen in the densest regions of molecular cloud cores. Even small amounts of oxygen in the gas would create water vapor that could be detected by *SWAS* and Odin. In principle, this is in conflict with large atomic oxygen columns estimated from ISO [O I] observations towards many sources (Caux *et al.* 1999; Lis *et al.* 2001 and references therein). To account for the low abundances of water and molecular oxygen, this [O I] emission/absorption must be probing the outer, low-density, layers of the cloud. This is supported by an analysis of [O I] emission in ρ Ophiuchi by Liseau *et al.* (1999), who find no enhancement of oxygen emission in dense cold clumps.

4. A New Look at SWAS Observations

With the perspective of more than 5 years of observations a number of factors become evident. First, despite a wide range of conditions probed (geometry, radiation field, temperature, density) water vapor observations of GMC cores showed, in general, line intensities of $T_A(557$ GHz$) \sim 0.1 - 0.3$ K (Orion A is one of the exceptions to this statement). This seems quite puzzling and hints at some general formation mechanism that is somehow insensitive to environment. Second, maps of emission over large spatial scales began to reveal compelling clues to the origin of water vapor emission. These maps readily demonstrate that water vapor is a surface tracer inside dense photodissociation regions (PDR's) and is not tracing the volume of the cloud.

An example of this is given in Figure 4 where we show *SWAS* observations of NGC1333 presented in Bergin *et al.* (2003). The water emission in NGC1333 is divided into broad spectral lines that trace water entrained in molecular outflows (SS4, 6, 7, 8) and narrow

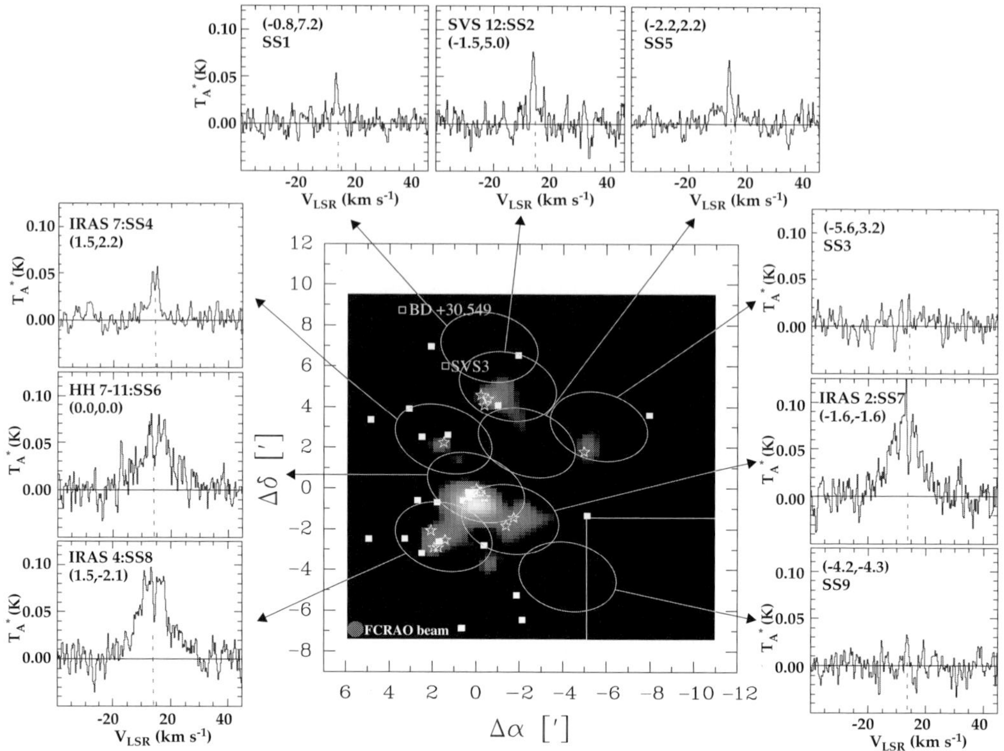

Figure 4. Spectra of the $1_{10} - 1_{01}$ 556.936 GHz transition of *ortho*-water vapor in NGC 1333. Placements of the elliptical *SWAS* beam are shown superposed on a map of integrated N_2H^+ (J=1–0) emission. The stars show the location of IRAS sources, while the solid squares are positions of known Herbig-Haro objects. Wide spectral lines are water emission arising from shocked gas, while narrow lines probe ambient material. Taken from Bergin *et al.* (2003).

($\Delta v \sim 2 - 3$ km s^{-1}) lines that probe ambient material (SS1, 2, 5). What is notable about the ambient water emission is that it does not appear to track the emission of N_2H^+ (which is effectively a probe of the dense star-forming condensations). Rather the emission clusters to the north and is correlated with high-J ^{13}CO emission (see Bergin *et al.* 2003) that is directly associated with the luminous external heating sources (SVS3 and BD+30.549) which power the reflection nebula and the PDR. This was interpreted by Bergin *et al.* as undepleted water vapor present in low-density gas that gains emissivity due to heating in the PDR (see also Spaans & van Dishoeck 2001). Extensive *SWAS* observations of the Orion A molecular cloud also suggest water is present in the PDR. A comparison of CN, C^{18}O, and water emission by Melnick *et al.* (in preparation) shows that water emission is highly correlated with CN, a known tracer of UV dominated gas in Orion (Rodríguez-Franco *et al.* 1998), and is uncorrelated with C^{18}O emission, a tracer of total volume.

That water can be a tracer of regions with substantial UV radiation presents a puzzle because water is quite easily dissociated by UV photons (Yoshino *et al.* 1996). Hollenbach *et al.* (in preparation) present an appealing model, which asserts that toward cold clouds gaseous H_2O exists only within a relatively narrow range of visual magnitudes near the cloud surface (see Figure 5). Closer to the surface than an A_V of a few, both H_2O and O_2 are photo-destroyed by the ambient galactic UV field. Deeper into the cloud than an A_V of ~ 4–8 ($\Leftrightarrow N(H+H^+ +H_2) \simeq 0.8$–$1.6 \times 10^{22}$ cm^{-2}; c.f., Bohlin *et al.* 1978) H_2O, along

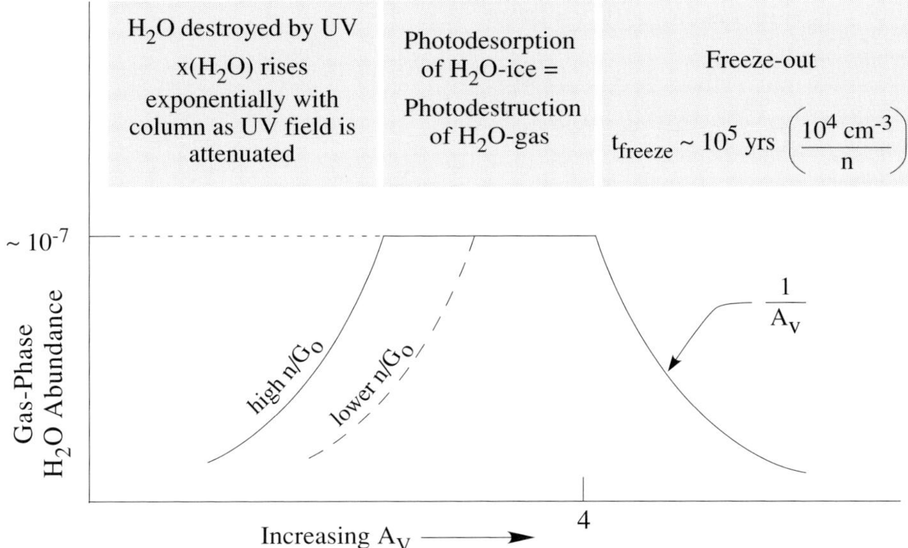

Figure 5. Schematic depiction of water vapor zone within molecular clouds.

with many other species, depletes onto grains. Between an A_V of 2 and 4–8 gas-phase abundances are set by the balance between photodesorption and photodestruction. The preferential depletion of several of the more abundant oxygen-bearing species serves to increase the gas-phase C/O ratio deep within the cloud, further suppressing the gas-phase production of H_2O and O_2 (e.g. Bergin *et al.* 2000). This picture is attractive since it naturally accounts for the low water vapor abundance inferred toward high column density lines of sight, the relatively higher water vapor abundances inferred toward lower column density lines of sight, and the high water-ice abundance derived toward many clouds.

By way of illustration, we consider a cloud of constant hydrogen density $n = 10^4$ cm^{-3} illuminated by a somewhat enhanced ultraviolet radiation field $G_0 = 10$, or about 10 times the average local interstellar field. The model results are shown in Figure 6. As discussed above, the gas-phase oxygen is atomic at the surface because the ice is photodesorbed and O-bearing compounds in the gas phase are photodissociated to atomic form. However, the oxygen is mainly incorporated as water-ice deep in the cloud. The right panel in Figure 6 shows the threshold effect for H_2O ice in the Taurus cloud. Plotted is the strength of the H_2O ice absorption feature, measured using background sources that penetrate the cloud along a number of lines of sight, as a function of cloud column density or extinction. One sees a threshold column density ($A_V \sim 4$ in Taurus) below which no ice feature is seen. Above this threshold, the absorption strength increases linearly with additional column density, and is consistent with all the elemental oxygen not in CO (or in refractory grain material like silicates) residing as water-ice (i.e., $x(H_2O\text{-ice})/x(H$ nuclei) $\sim 10^{-4}$).

The gas-phase O_2 and H_2O abundances rise steeply from $A_V = 0$ (the neutral cloud surface) to A_V of a few as the photodissociation rates decline due to dust shielding from the UV field, and then the abundances decline as A_V increases and all oxygen not in CO freezes out. In the deep interior, time-dependent chemistry must be considered since in steady state all the gas-phase oxygen, even that originally in CO, is liberated and eventually incorporated into water-ice on grains. However, this process takes more than 3×10^6 years, or comparable to cloud lifetimes, so that CO generally is abundant

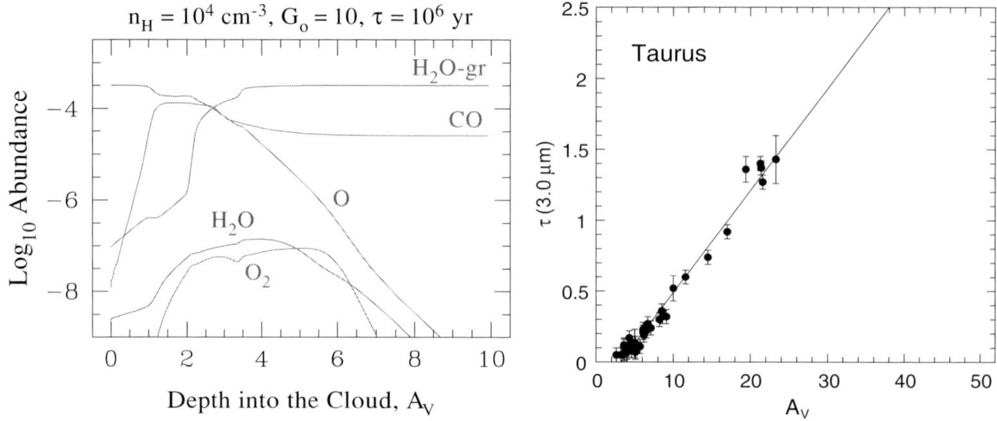

Figure 6. (*Left*) Chemical model predictions of the abundance of key oxygen-bearing species after 10^6 years as a function of distance into a molecular cloud as measured in visual magnitudes of extinction, A_v. (*Right*) The optical depth at 3 μm due to H_2O-ice absorption versus the inferred depth into the Taurus (*top*) molecular cloud as measured in visual magnitudes of extinction, A_v (after Whittet *et al.* 2001).

in the gas phase as long as the grains are warmer than about 15–20 K, the sublimation temperature for CO.

Thus, to understand the composition, this model requires that the depth dependence of the chemistry be treated. One must also include ice-forming grain surface chemistry, the change in grain surface area caused by ice formation, the ice desorption mechanisms – especially photodesorption, and time-dependent chemistry. Hopefully, this picture will motivate further laboratory studies of photodesorption yields of various ice species as a function of UV wavelength.

5. Conclusions

Water vapor and molecular oxygen are key species in the interstellar medium because their abundances are sensitive to the chemistry that occurs in both the gas and solid state. During its 5.5-year mission, *SWAS* set clear limits on the cycle that creates water in interstellar space. We eagerly await Herschel observation of numerous transitions of both ortho- and para-H_2O, along with SOFIA observations of $H_2^{18}O$ that will further define the chemistry of oxygen, the third most abundant element in the Universe.

Acknowledgements

For the duration of the *SWAS* mission this work was supported by NASA's *SWAS* Grant NAS5-30702. We are grateful to all members of the *SWAS* team who are too numerous to be named, but deserve thanks for supporting the proposal, design, development, software for the data pipeline and scheduling, data analysis tools, and the analysis summarized here.

References

Ashby, M.L.N., Bergin, E.A, Plume, R. *et al.* 2000, *Ap. J.* 539, L119
Bergin, E. A. & Snell, R. L. 2002, *Ap. J.* 581, L105
Bergin, E. A., Kaufman, M. J., Melnick, G. J., Snell, R. L., & Howe, J. E. 2003, *Ap. J.* 582, 830
Bergin, E.A., Melnick, G.J., Stauffer, J.R. *et al.* 2000, *Ap. J.* 539, L129

Bohlin, R.C., Savage, B.D., & Drake, J.F. 1978, *Ap. J.* 224, 132

Caux, E., *et al.* 1999, *A&A* 347, L1

Charnley, S. B., Rodgers, S. D., & Ehrenfreund, P. 2001, *A&A* 378, 1024

Chieze, J. P. & Pineau Des Forets, G. 1989, *A&A* 221, 89

Elitzur, M. & de Jong, T. 1978, *A&A* 67, 323

Fraser, H. J., Collings, M. P., McCoustra, M. R. S., & Williams, D. A. 2001, *M.N.R.A.S.* 327, 1165

Herbst, E. & Klemperer, W. 1973, *Ap. J.* 185, 505

Linke, R. A., Goldsmith, P. F., Wannier, P. G., Wilson, R. W., & Penzias, A. A. 1977, *Ap. J.* 214, 50

Lis, D. C., Keene, J., Phillips, T. G., Schilke, P., Werner, M. W., & Zmuidzinas, J. 2001, *Ap. J.* 561, 823

Liseau, R., *et al.* 1999, *A&A* 344, 342

Melnick, G.J., Stauffer, J.R., Ashby, M.L.N. *et al.* 2000, *Ap. J.* 539, L77

Neufeld, D.A., Lepp, S., & Melnick, G.J., 1995, *Ap. J. Suppl.* 100, 132

Neufeld, D.A., Ashby, M.L.N., Bergin, E.A. *et al.* 2000, *Ap. J.* 539, L111

Neufeld, D. A., Kaufman, M. J., Goldsmith, P. F., Hollenbach, D. J., & Plume, R. 2002, *Ap. J.* 580, 278

Olofsson, A. O. H., *et al.* 2003, *A&A* 402, L47

Plume, R., *et al.* 2004, *Ap. J.* 605, 247

Roberts, H. & Herbst, E. 2002, *Astron. & Ap.*, 395, 233

Rodriguez-Franco, A., Martin-Pintado, J., & Fuente, A. 1998, *A&A* 329, 1097

Snell, R.L., Howe, J.E., Ashby, M.L.N. *et al.* 2000a, *Ap. J.* 539, L93

Snell, R.L., Howe, J.E., Ashby, M.L.N. *et al.* 2000b, *Ap. J.* 539, L97

Snell, R.L., Howe, J.E., Ashby, M.L.N. *et al.* 2000c, *Ap. J.* 539, L101

Spaans, M. & van Dishoeck, E. F. 2001, *Ap. J.* 548, L217

Viti, S., Roueff, E., Hartquist, T. W., Pineau des Forêts, G., & Williams, D. A. 2001, *A&A* 370, 557

Whittet, D. C. B., Gerakines, P. A., Hough, J. H., & Shenoy, S. S. 2001, *Ap. J.* 547, 872

Willacy, K., Langer, W. D., & Allen, M. 2002, *Ap. J.* 573, L119

Yoshino, K., Esmond, J. R., Parkinson, W. H., Ito, K., & Matsui, T. 1996, *Chemical Physics* 211, 387

Discussion

HENNING: Did you detect H_2O in debris disks?

BERGIN: We observed several debris disks (including β Pic) for tens of hours and did not detect H_2O or [C I].

FRASER: Given that the *SWAS* results indicate a lack of atomic O in the ISM gas, what are the implications of the low atomic abundance for subsequent complex chemistry especially on grains. I presume that hydrogenation of O will dominate, and a lot of O is locked up in $CO + SiO_2$ (in grains) – so how can we explain for e.g. CO_2 ice observation and CO_2 formation?

BERGIN: My suspicion is that many of the complex species form with water ice in early stages when atomic hydrogen and oxygen are abundant. Once water formation stops at later denser stages, complex molecule production may occur more slowly.

KROES: (1) What was the frequency of UV light in experiments showing water desorption? In Leiden, S. Andersson, I, and Ewine van Dishoeck see almost no water desorption in modeling of photodissociation of water in ice, in A-band, at 8.5 eV. (2) Why does O_2 formation not accompany ice formation or if it does, why do you not observe O_2?

BERGIN: (1) The experiment by Westley *et al.* (1995, *Nature* 373) was done using Lyα photons. (2) If water vapor and ice form predominantly at low extinction, the O_2 is created but will be rapidly destroyed by UV photons.

GLASSGOLD: What is the role of OH observations in understanding the *SWAS* and Odin results on H_2O and O_2?

BERGIN: OH has long been a suspected tracer of cloud surfaces and one model readily produces OH at the surface by UV photodissociation. It would be worthwhile to revisit this issue with direct OH observations in key sources such as Orion.

HERBST: What is the efficiency you need for H_2O photo-desorption?

BERGIN: We have assumed a photo-desorption yield of 0.001. If the water photo-desorption experiments in the lab are found to be much lower than the Westley *et al.* (1995, *Nature* 373) results then water can still be a surface tracer. In this case water would be present in lower-density envelopes surrounding the dense cores, in regions where oxygen has no time to collide with grains and react with surface H to form water ice.

CECCARELLI: I have a comment related to T. Henning's question about *SWAS* non-detection of H_2O in proto-planetary disks. We (Ceccarelli *et al.*, *Ap. J.*, in press) detected HDO in the disk of DM Tau.

Astrochemistry: Recent Successes and Current Challenges
Proceedings IAU Symposium No. 231, 2005
D.C. Lis, G.A. Blake & E. Herbst, eds.
© 2006 International Astronomical Union
doi:10.1017/S1743921306007319

The Spatial Distribution of Ices in Star-Forming Regions

K. M. Pontoppidan[1] E. F. van Dishoeck[1], E. Dartois[2], H. J. Fraser[3], Z. Banhidi[4], J. K. Jørgensen[5] and the c2d Team[6]

[1]Leiden Observatory, P.O.Box 9513, NL-2300 RA Leiden, Netherlands
email: pontoppi@strw.leidenuniv.nl

[2]IAS-CNRS, Bat 121, Université Paris Sud, 91405 Orsay Cedex, France

[3]Department of Physics, University of Strathclyde, 107 Rottenrow, Glasgow G4 0NG, Scotland

[4]Stockholm Observatory, AlbaNova University Center, SE-106 91 Stockholm, Sweden

[5]Harvard-Smithsonian Center for Astrophysics, 60 Garden Street, MS42, Cambridge, MA 02138, USA

[6]http://peggysue.as.utexas.edu/SIRTF/

Abstract. We present results from an ongoing program to map the spatial distribution of ices in dark cloud cores with the Spitzer Space Telescope and VLT-ISAAC. The ice maps are used to directly trace the freeze-out of CO and the formation of H_2O, CH_3OH and CO_2.

Keywords. infrared: ISM — ISM: abundances — dust, extinction — molecular processes

1. Introduction

A significant fraction of the molecular component in dense clouds exists in the form of condensed ices. In the densest star-forming cores, more than half of all molecules apart from H_2 can be frozen out onto dust grains. Thus, a full understanding of the chemical state of a molecular cloud requires detailed observations of the structure and abundances of common ices such as water, CO_2, CH_3OH, NH_3 and CO. The only way to directly observe ice in dense interstellar clouds is via absorption bands in the mid-infrared wavelength regime. Thus, infrared continuum sources located behind or inside the cloud of interest are required. The sensitivity of ground-based 8–10 m class telescopes and the Spitzer Space Telescope now allows for spectroscopy of fainter sources concentrated close enough in the plane of the sky to produce multiple lines of sight through the same cloud fragment. Combining multiple lines of sight will produce a spatial map of the abundances of solid state species in a given cloud.

2. Observations and Results

Several observing programs using both the Very Large Telescope and Spitzer to map the spatial variation of the abundances of different ice species in a variety of dark clouds have recently been completed. The maps are obtained by selecting cloud regions with a high density of bright background stars or low-luminosity, highly extincted T Tauri stars. Absorption spectra of the ices present along the line of sight are obtained with both ground- and space-based mid-infrared spectrometers. Combining the spectra from different lines of sight produces a map of ices with spatial resolutions of 10–60″, comparable to maps of gas-phase molecules obtained with single-dish millimeter telescopes.

These results include a Spitzer 5–40 μm ice map of the protostellar envelope of the class 0 protostar Serpens SMM 4 (see also Pontoppidan *et al.* 2004). Additionally, we

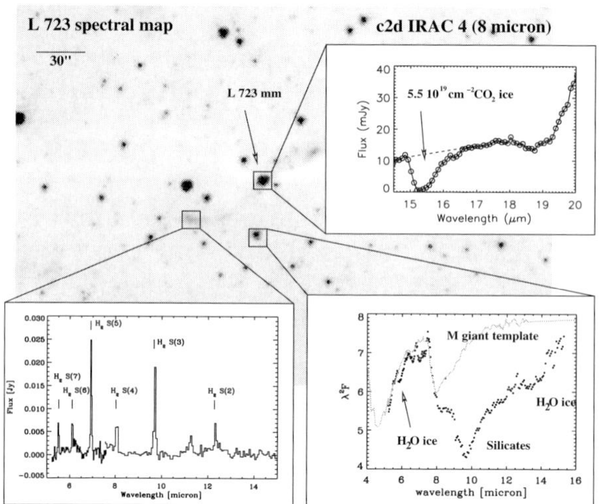

Figure 1. Examples of the Spitzer low-resolution spectra obtained for an ice map of the isolated core L723. Most of the data are spectra of background M and K giants to probe the solid state species in the cloud. We have detected a very high column density of CO_2 ice toward the central class 0 object (Dartois *et al.* 2005, *A&A*, in press).

have obtained a Spitzer map of water and CO_2 ice of the isolated core L723 as well as a combined CO/CO_2 ice map of the Oph F core in the Ophiuchus molecular complex (Pontoppidan *et al.*, in prep.) obtained using ISAAC on the VLT and Spitzer-IRS.

We generally find that the local abundance of water ice is almost constant at $5-9\times10^{-5}$ w.r.t. H_2 in most dark cloud environments, and only increases beyond this at very high densities. The abundance of CO ice is found to be highly density dependent in cold cloud environments in accordance with simple freeze-out models and CO gas-phase observations (Jørgensen *et al.* 2005 and references therein). In the Oph F core, the CO ice abundance can be traced down to a freeze-out fraction of 5% using ground-based spectroscopy of the $4.67\,\mu$m C-O stretching mode. The abundance of CO_2 ice is moderately density dependent and increases in abundance by a factor of two for the Oph F core from the outer part at 50 000 AU to the innermost 5000 AU. The profile of the CO_2 bending mode indicates that the CO_2 ice formed at higher densities is very dilute in the CO ice that dominates the ice mantle during heavy depletion. Direct comparison between the CO stretching and CO_2 bending mode profiles shows that the ratio of CO to CO_2 molecules during heavy depletion is $\sim$40. We suggest that this is a direct measure of the formation rate of CO_2 relative to the freeze-out rate of CO at densities higher than a few $10^5\,\mathrm{cm}^{-3}$.

Acknowledgements

Astrochemistry in Leiden is supported by a Spinoza grant from the Netherlands Organization of Scientific Research (NWO). This work is based in part on observations made with the Spitzer Space Telescope (GO-3336 and the Legacy Science program, under contract 1224608), which is operated by the Jet Propulsion Laboratory, California Institute of Technology under NASA contract 1407.

References

Pontoppidan, K. M., van Dishoeck, E. F., & Dartois, E. 2004, *A&A* 426, 925
Jørgensen, J. K., Schöier, F. L., & van Dishoeck, E. F. 2005, *A&A* 435, 177

Astrochemistry: Recent Successes and Current Challenges
Proceedings IAU Symposium No. 231, 2005
D.C. Lis, G.A. Blake & E. Herbst, eds.

© 2006 International Astronomical Union
doi:10.1017/S1743921306007320

Astrochemistry Results from the Spitzer c2d Project

Neal J. Evans II[1] and the c2d Team[2]

[1]Department of Astronomy, The University of Texas at Austin, Austin, TX 78712, USA
email: nje@astro.as.utexas.edu
[2]http://peggysue.as.utexas.edu/SIRTF/

Abstract. The early results on astrochemistry from the Cores to Disks (c2d) Legacy program are summarized. The c2d program focuses on the formation of low-mass stars in nearby (within about 300 pc) clouds. Spectroscopy with IRS includes the following topics: ices seen against background stars, in protostellar envelopes, and in disks; grain growth and mineralogy, PAHs, and gas in circumstellar disks.

Keywords. dust — infrared: ISM — protoplanetary disks

1. Introduction

The c2d Legacy Program (Evans *et al.* 2003) is using the instruments of the Spitzer Space Telescope to trace the evolution of molecular material from the prestellar, dense core stage through the formation of planet-forming disks. Five large clouds (covering a total of $\sim$20 sq. deg.) and about 90 smaller cores are being imaged with both IRAC (4 bands from 3.6 to 8 μm) and MIPS (3 bands from 70 to 160 μm). To study the timescale for disk dissipation, we are obtaining photometry on $\sim$190 weak-line T Tauri stars. The evolution of the physical and chemical state of the matter is being studied with spectroscopy of over 170 targets (down to sub-solar mass stars, as faint as a few mJy) with IRS. Finally, a large data base of complementary observations, from optical to millimeter wavelengths, is being assembled. Because the IRS observations bear most directly on astrochemistry, I will focus on them here.

2. Evolution of Ices

Spectroscopy of background stars can reveal the nature of the dust and ice in dense regions before any star has formed. The sensitivity of IRS on Spitzer allows study of lines of sight with very high extinction. Bergin *et al.* (2005) and Knez *et al.* (2005) have found large abundances of CO_2 ice, relative to H_2O ice, toward highly extincted background stars. The shape of the feature shows that the CO_2 is not crystallized, indicating its pristine nature. In addition, Knez *et al.* (2005) detect for the first time the 6.85 μm band toward background stars. Its strength is comparable to that previously observed toward embedded protostars, and thus its carrier (NH_4^+, tentatively) must have been formed early in the cloud history. Other, weaker features suggest the presence of HCOOH and NH_3, but these lie on the edge of stronger features and need confirmation.

Similar spectra of embedded protostars show features of solid CH_4, CH_3OH, CO, CO_2, H_2O, NH_4^+, and OCN^- (Boogert *et al.* 2004). Systematic trends in the shape of the CO and CO_2 bands indicative of increased temperatures leading to ice sublimation and changing ice structure can be seen in a range of objects, similar to what was seen toward more massive and luminous objects with ISO (*cf.* van Dishoeck 2004).

3. Evolution of Disks

Work with ISO observations also showed that disks around intermediate mass stars can have a range of profiles, ranging from flared, to self-shadowed, to disks with inner holes (van Dishoeck 2004). A similar range of SEDs is seen toward lower mass stars with Spitzer. Analysis of the silicate emission features at 10 and 20 μm shows evidence that the small end of the grain size distribution is increasing rapidly in disks (Kessler-Silacci *et al.* 2005). Considering models of rapid grain growth and settling, these observations may eventually shed light on processes such as vertical mixing and grain fragmentation.

PAH emission is detected toward at least 8% of the T Tauri stars, down to spectral type G8. The 11.3 micron feature strength suggests for some sources the presence of additional UV radiation beyond that provided by the star itself, perhaps associated with active accretion (Geers *et al.*, in preparation). Ionized PAHs around these cool stars could also be excited by visible or near-IR photons (Mattioda *et al.* 2005).

With special orientations, the disk can be probed by absorption spectroscopy. Ices in such a "grazing-incidence" disk were studied by Pontoppidan *et al.* (2005). Because the IRS instrument is not optimized for gas-phase features, the most surprising result so far is the discovery of gas-phase absorption bands of CO_2, HCN, and C_2H_2 toward one young star+disk, IRS46 in Ophiuchus (Lahuis *et al.* 2005). Large abundances of hot, dense, chemically-rich gas coming from a region less than about 10 AU in size suggest absorption in the inner regions of a disk. However, follow-up observations with high spectral resolution indicate a 25 km s^{-1} blueshift relative to the cloud velocity, raising the possibility of the footprint of a disk wind fortuitously located along the line of sight. Further observations are needed to decide between these two possible interpretations.

Acknowledgements

Support for this work, part of the Spitzer Legacy Science Program, was provided by NASA through contracts 1224608, 1230779, and 1256316 issued by the Jet Propulsion Laboratory, California Institute of Technology, under NASA contract 1407. Astrochemistry in Leiden is supported by a NWO Spinoza grant and a NOVA grant.

References

Bergin, E.A., Melnick, G.J., Gerakines, P.A., Neufeld, D.A., & Whittet, D.C.B. 2005, *Ap. J.* 627, L33

Boogert, A.C.A., *et al.* 2004, *Ap. J. Suppl* 154, 359

Evans, N.J., *et al.* 2003, *PASP* 115, 965

Kessler-Silacci, J., Augereau, J.C., Dullemond, C.P., Geers, V., Lahuis, F., Evans, N.J., II, van Dishoeck, E.F., Blake, G.A., Boogert, A.C.A., Brown, J., Jørgensen, J.,K., Knez, C., & Pontoppidan, K.M. 2005, *Ap. J.*, in press

Knez, C., Boogert, A.C.A., Pontoppidan, K.M., Kessler-Silacci, J., Evans, N.J., II, van Dishoeck, E.F., Augereau, J.C., Blake, G.A., & Lahuis, F. 2005, *Ap. J.*, in press

Lahuis, F., Boogert, A.C. A., van Dishoeck, E.F., Pontoppidan, K.M., Blake, G.A., Dullemond, C.P., Evans, N.J., II, Hogerheijde, M.R., Jørgensen, J.K., Kessler-Silacci, J., & Knez, C. 2005, *Ap. J.*, submitted

Mattioda, A.L., Allamandola, L.J., & Hudgins, D.M. 2005, *Ap. J.* 629, 1183

Pontoppidan, K.M., Dullemond, C.P., van Dishoeck, E.F., Blake, G.A., Boogert, A.C.A., Evans, N.J., Kessler-Silacci, J.E., & Lahuis, F. 2005, *Ap. J.* 622, 463

van Dishoeck, E.F. 2004, *ARAA* 42, 119

Astrochemistry: Recent Successes and Current Challenges
Proceedings IAU Symposium No. 231, 2005
D.C. Lis, G.A. Blake & E. Herbst, eds.

© 2006 International Astronomical Union
doi:10.1017/S1743921306007332

Stratospheric Observatory for Infrared Astronomy (SOFIA)

E. E. Becklin

University of California Los Angeles, 405 Hilgard Ave,
Los Angeles, CA, 90095 USA
email: becklin@astro.ucla.edu

Abstract. The joint U.S. and German SOFIA project to develop and operate a 2.5-meter infrared airborne telescope in a Boeing 747-SP is now in the final stages of development. First science flights will begin in 2008. The observatory is expected to operate for over 20 years. The sensitivity, characteristics, science instrument complement, and examples of 1-st light spectroscopic astrochemistry science are discussed.

Keywords. airborne astronomy — astrochemistry — infrared: sub-millimeter

1. Introduction

The Stratospheric Observatory For Infrared Astronomy (SOFIA) is NASA's and DLR's premier observatory for infrared and submillimeter astronomy. A Boeing 747-SP aircraft will carry a 2.5-meter telescope designed to make sensitive infrared measurements of a wide range of astronomical objects. It will fly at and above 12.5 km, where the telescope collects radiation in the wavelength range from 0.3 μm to 1.6 mm. SOFIA is being developed and operated for NASA and DLR by USRA.

The telescope and 20% of operations will be supplied by Germany through contracts with DLR (German Space Agency). The University of Stuttgart has been awarded the contract to run the Deutsches SOFIA Institut (DSI). The development of the science instruments to be attached to the SOFIA telescope will be the responsibility of the U.S. and German science communities. In the U.S., science instruments will be designed and built at universities and national centers through a USRA peer-review process.

2. SOFIA First Light Instruments

A total of nine instruments have been selected and are now under development (see Table 1). The selection includes three Facility Class (FI) Science Instruments (HAWC, FORCAST, and FLITECAM) and six Principal Investigator Class (PI). Facility Class instruments are maintained and operated by the SOFIA Science and Mission Operations Center (SSMOC) staff. PI Class instruments are operated by the PI. Two of the PI Class instruments are being developed in Germany.

3. First Light Expected in 2008

SOFIA will see first light in 2008, and is planned to make more than 120 scientific flights per year of at least 8 to 10 hours duration. SOFIA is expected to operate for at least 20 years, primarily from Moffett Field in California, but occasionally from other

Table 1. SOFIA First Light Instruments

PI	Institution	Name	Type of Instrument
E. Dunham	Lowell Observatory	HIPO	High-speed Imaging Photometer for Occultations $0.3 - 1.1$ μm
I. McLean	UCLA	FLITECAM	Near-IR Camera $1 - 5$ μm; GRISM $R = 2,000$
J. Lacy	Univ. of Texas	EXES	Echelon Spectrometer $5 - 28$ μm; $R = 10^5, 10^4$, or 3000
T. Herter	Cornell	FORCAST	Mid IR Camera 5–40 μm
D.A. Harper	Univ. of Chicago	HAWC	Far IR Bolometer Camera $50 - 240$ μm
A. Poglitsch	MPE, Garching	FIFI LS	Field Imaging Far IR Line Spectrometer $40 - 210$ μm; $R \sim 2000$
S. Moseley	NASA-GSFC	SAFIRE	Imaging Fabey-Perot Bolometer Array Spectrometer $145 - 655$ μm; $R = 1,000 - 2,000$
R. Güsten	MPIfR, KOSMA	GREAT	Heterodyne Spectrometer $60 - 200$ μm; $R = 10^4 - 10^8$
J. Zmuidzinas	Caltech	CASIMIR	Heterodyne Spectrometer $200 - 600$ μm; $R = 10^4 - 10^8$

Table 2. System characteristics of SOFIA

Nominal Operational Wavelength Range	0.3 to 1600 μm	Diffraction-Limited	$\geqslant 15$ μm
Pointing Stability	$1.''0$ rms at first light	Nominal System f-ratio	19.6
Pointing Accuracy	$0.''5$	Telescope Emissivity	10%
System Clear Aperture Diameter	2.5 meters	Recovery Temp.	240 K
Telescope's Unvignetted Elevation Range	20 to 60°		
Unvignetted Field-of-View Diameter	$8'$		
Maximum Chop Throw on Sky	$\pm 4'$ (unvignetted)		

bases around the world, especially in the Southern Hemisphere. SOFIA will fly above 12.5 km, where the typical water vapor column density is less than 10 μm.

The SSMOC, to be operated by USRA, will be located at NASA Ames Research Center at Moffett Field in the same hangar housing SOFIA. The SOFIA Program will support approximately 50 investigation teams per year.

The finished telescope has been mated into the modified aircraft and was tested in 2004. First test flights will occur in 2007, and first science in 2008.

4. Science Potential

With the parameters given under SOFIA Characteristics in Table 2, and the atmospheric transmission at flight altitudes given in Traub & Stier (1976) we calculate that SOFIA will be as sensitive as the CSO in the 350 and 450 μm windows, but for 80% of all molecular lines from 60 to 600 μm. SOFIA will be 3 times less sensitive than Herschel for heterodyne spectroscopy in the region 150 to 600 μm; advances in detectors during and after the Herschel mission should close that gap.

Using the first light heterodyne instruments, spectral surveys can be made to reveal many lines in the broad atmospheric window from 60 to 600 μm. Many new molecular lines will be observed for the first time. Of particular interest are the light molecules, such as HD at 112 μm and HF at 243 μm.

Reference

Traub, W.A., & Stier, M. 1976, *Appl. Optics* 15, 364

Astrochemistry: Recent Successes and Current Challenges
Proceedings IAU Symposium No. 231, 2005
D.C. Lis, G.A. Blake & E. Herbst, eds.
© 2006 International Astronomical Union
doi:10.1017/S1743921306007344

H$_2$ Formation on Grain Surfaces

S. M. Cazaux[1]†, P. Caselli[1], M. Walmsley[1], and A. G. G. M. Tielens[2]

[1]INAF, Largo E. Fermi, 5, 50125 Firenze, Italy
email: cazaux@arcetri.astro.it

[2]SRON-Kapteyn Astronomical Institute, 9700 AV Groningen, Netherlands
email: tielens@astro.rug.nl

Abstract. Molecular hydrogen is the most abundant molecule in the Universe and dominates the mass budget of the gas, particularly in regions of star formation. H$_2$ is also an important chemical intermediate in the formation of larger species and can be an important gas coolant when the medium lacks metals. Because of the inefficiency of gas-phase reactions to form H$_2$, this molecule is generally thought to form on grain surfaces. Observations of H$_2$ in a wide variety of objects showed that this molecule could form efficiently over a wide range of physical conditions. To understand the mechanism responsible for such an efficient formation, we developed a model for molecular hydrogen formation on grain surfaces. This model considers the interaction between atom and surface as beeing either weak (Van der Waals interaction—physisorption) or strong (covalent bound—chemisorption), as well as the mobility of the atom on a surface due to
quantum mechanical diffusion and thermal hopping. This model solves the time-dependent kinetic rate equation for the formation of molecular hydrogen and its deuterated forms. Our results have been benchmarked with laboratory experiments on silicates, carbonaceous and graphitic surfaces. This comparison allowed us to derive some characteristics of the considered surfaces. An extension of our model to astrophysical conditions gives an estimate of H$_2$ formation efficiency for a wide range of physical conditions. One of our main results is the efficient formation of molecular hydrogen for gas and grain temperatures up to several hundreds of kelvins. We also compared our predictions to observations in astrophysical objects such as photodissociation regions (PDRs). The addition of deuterium in our model for the formation of HD and D$_2$ molecules is also discussed.

Keywords. astrochemistry — ISM: molecules — molecular processes

---

1. Introduction

1.1. H$_2$ Observations in Astrophysical Environments

Molecular hydrogen is the most abundant molecule in the Universe and it constitutes the primary ingredient for astrochemistry. H$_2$ has been observed in a variety of galactic and extragalactic environments under varied physical conditions. The question of its formation is therefore of prime importance to understand the chemistry and the physics of astrophysical objects. In the Interstellar Medium (ISM), the formation of H$_2$ through gas-phase reactions is not efficient enough to explain its observed abundance in the Milky Way. Presently, in the ISM, H$_2$ forms on dust-grain surfaces and this process dominates gas-phase production by several orders of magnitude (Gould & Salpeter 1963).

Molecular hydrogen can be observed at mid- and near-infrared wavelengths in emission, but also in the far UV in emission and in absorption. A plethora of astrophysical observations testify that H$_2$ is present in environments with very different physical conditions.

† Present address: Largo E. Fermi, 5, 50125 Firenze, Italy

In diffuse interstellar clouds, with typical physical conditions of density $n(\mathrm{H})=50\ \mathrm{cm}^{-3}$, gas temperature of ~ 100 K, and dust temperature of 15 K, the FUV absorption lines of H_2 have been observed (Spitzer & Jenkins 1975, with Copernicus; Gry *et al.* 2002, with FUSE). These detections showed that H_2 formation is very efficient with a rate of 1–3 $\times 10^{-17}\ \mathrm{cm}^3\,\mathrm{s}^{-1}$ (Jura 1974; Hollenbach, Werner, & Salpeter 1971).

When molecular clouds are irradiated by nearby stars, the molecules in the region closest to the source are photodissociated. These regions are called photodissociation regions (PDRs), and are that part of the molecular cloud where photons dominate the thermal and chemical balance of the gas. In these regions, the physical conditions represent a wide range of gas and grain temperatures (100 K $\leqslant T_{gas} \leqslant$ 1000 K; 10 K $\leqslant T_{grain} \leqslant$ 100 K). Habart *et al.* (2004) derive an H_2 formation rate in the range 3×10^{-17} to $1.5 \times 10^{-16}\ \mathrm{cm}^3\,\mathrm{s}^{-1}$ for the PDRs associated with Orion Bar, NGC 2023, S140, IC 63 and Oph W.

Jets, outflows and shocks in the surroundings of star-forming regions exhibit other sets of physical conditions where H_2 is detected. Many low-mass protostars drive C-shocks into their environment (velocity $\leqslant$ 40–50 km s^{-1}). Molecular hydrogen is not dissociated in these shocks and in the warm postshock gas is collisionally excited and radiates mainly in the $v=1$–0 ro-vibrational transitions. H_2 emission is also observed associated with fast dissociative J-type shocks (velocity $\geqslant$ 40–50 km s^{-1}), where the medium in the shock front is hot ($T_{gas} \sim 10^5$ K), highly ionized, and H_2 is collisionally dissociated. Nevertheless, H_2 molecules are efficiently reformed on grain surfaces in the postshock region, where the gas is cooler ($T_{grain} \sim 75$ K and $T_{gas} \leqslant 2000$ K; Hollenbach & McKee 1989; Neufeld & Dalgarno 1989).

Supernovae remnants also exhibit H_2 emission lines. In the region where the ejecta of the supernova remnant collides with a nearby molecular cloud, excited H_2 is detected. In IC443, H_2 is seen in a slow shock (C-shock with $v_s \sim 30$ km s^{-1} and $n \sim 10^4$ cm^{-3}; Rho *et al.* 2001). In the Crab nebula, a younger supernova, H_2 is seen in emission in the complex and highly ionised filamentary structure of the nebula. H_2 has been observed in the core of the filaments, where it can be self-shielded enough to survive to the high UV flux (Graham *et al.* 1990). The physical conditions at which H_2 has been observed are $T_{grain} \sim 50$ K (Davidson & Fesen 1985), $T_{gas} \leqslant 7000$ K (Rudy *et al.* 1994), and a dust to mass ratio five to ten times higher than the normal ISM value (Sankrit *et al.* 1998). According to Graham *et al.* (1990), molecular hydrogen was formed after the explosion of the Supernova, when the density in the filaments was higher, and survived in the filamentary self shielded environments.

According to some observations discussed above, it appears that molecular hydrogen can be formed under almost all circumstances. H_2 can form on cold and warm grain surfaces, with various gas temperatures. Even when destroyed in shocked regions, it will reform again in the post-shock regions. Near strong UV or X-ray radiation fields, molecular hydrogen may even find some protected place to form and survive. The wide variety of these environments raises a key question in astronomy. *How does molecular hydrogen form in the Universe? Which physical and chemical processes can explain the presence of this molecule in such a wide range of physical parameters?*

2. H_2 Formation on Grain Surfaces

Many studies have aimed at understanding how molecular hydrogen could form on grain surfaces. Theoretically, Hollenbach & Salpeter (1970) developed a quantum mechanical model to calculate the mobility of the atoms on a grain surface. Because the interaction between atoms and grains involved in their calculations were weak, they found that H_2 formed only at low grain temperatures. To modify this result—in conflict with a variety of observations—they took the presence of lattice defects with enhanced binding into account (Hollenbach & Salpeter 1971). With this assumption, they could predict a very efficient H_2 formation for grain temperature $\leqslant 50$ K. In recent studies (Chang, Cuppen & Herbst 2005), various types of inhomogeneous and mixed surfaces are considered, allowing an efficient H_2 formation until grain temperatures of ~ 25 K.

Experimentally, Temperature Program Desorption (TPD) experiments at low temperatures revealed the weak interactions, also called physisorption, between the atoms and some surfaces of astrophysical interest. Pirronello *et al.* (1997a, 1997b, 1999) studied the formation of HD on olivine and carbonaceous surfaces at a range of surface temperatures between 5 and 25 K. These results allowed an estimate of the energy of physisorption of the atoms and molecules on the surface. Recent TPD experiments at high temperatures, performed by Zecho *et al.* (2002) on graphite, showed another type of interaction between the atoms and the surface. This interaction, also called chemisorption, is strong and allows the formation of H_2 at grain temperatures of hundreds of Kelvins.

Physisorbed atoms are weakly bound to the surface and are mobile at low grain temperatures (Ghio *et al.* 1980), whereas chemisorbed atoms are strongly bound to the surface and become mobile only at grain temperatures of a few hundred K (Barlow & Silk 1976; Aronowitch & Chang 1980; Klose 1992; Fromherz *et al.* 1993; Que *et al.* 1997; Jeloaica & Sidis 1999; Sha & Jackson 2002; Cazaux & Tielens 2002, 2004). By considering these two types of interactions between the atoms and the surface, H_2 can possibly form for a wide range of temperatures.

2.1. *Model of H_2 Formation*

We consider in our model two different interactions between the atoms and the surface, as shown in Figure 1. The formation of molecules at low surface temperatures involves atoms bound to the surface with a Van der Waals interaction, typically called physisorption (of the order of few meV), whereas formation of molecules at high surface temperature involves atoms strongly bound to the surface with an interaction called chemisorption (of the order of several eV). Therefore, these two ways of binding the atoms on a surface insure that molecules can form for a wide range of surface temperature. Another point considered in our model is the fact that atoms move on the surface by tunneling and thermal diffusion. This is essential for a comprehension of the formation of molecules at low grain temperatures where tunneling can dominate.

In order to characterise the surfaces that have been studied in the laboratory, we developed a rate equation model to follow the population of the physisorbed and chemisorbed H and D atoms, as well as the molecules. This model, benchmarked by experimental data, is used to derive the characteristics of the different surfaces. In our study, we considered in detail the experiments of Pirronello *et al.* (1997a, 1997b, 1999) made on carbonaceous and olivine surfaces at low temperatures. These TPD experiments consist of two steps: (1) in the first step the surface is set at a fixed temperature T_0, and is irradiated by H and D atoms during a time t_{irr}; (2) in a second step, the irradiation is stopped and the temperature of the surface is increased with a constant rate ($\beta = 1$ K s^{-1}).

The measurements, reported Figure 2, show the desorption peaks of HD molecules depending on the irradiation time. At high coverage ($t_{irr} \geqslant 33$ s for olivine), the peaks do not seem to shift at higher temperature with lower irradiation time. This phenomenon is called first-order desorption. When the coverage is high, the atoms find each other easily on the surface and form molecules. These molecules evaporate once an adequate surface temperature is reached and the binding energy of molecules on the surfaces is easily determined in this case (E0 is the binding energy of the molecules). At lower coverage,

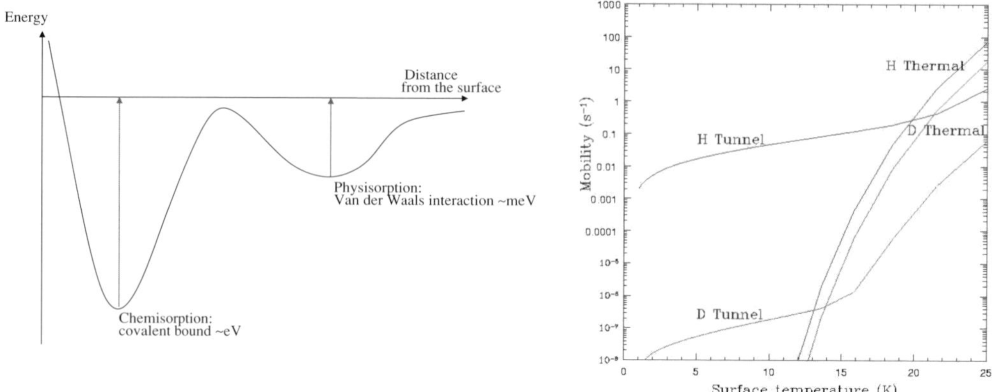

Figure 1. Two main points considered in our model. *Left:* The interaction between an atom and a surface can be either weak (physisorption) or strong (chemisorption). *Right:* The atoms move on the surface by tunneling and thermal diffusion. The mobility (in s^{-1}) to go from a physisorbed to a chemisorbed site is calculated here. Note the difference of mobility by tunneling between an H and a D atom due to the mass difference.

on the other hand, the atoms are far away from each other and need to scout the surface to associate with other atoms. The less the coverage, the more the atoms need higher mobility, and therefore higher grain temperatures to encounter another atom. This is called second-order desorption and is characterised by a shift of the desorption peaks to higher temperature with lower irradiation time. In this case, the binding energy of the atoms with the surface (E_{phys}, physisorption energy) can be determined, as well as an upper limit to the size of the barrier between two physisorbed sites (a_{pp}).

Another set of measurements performed during these TPD experiments is reported in Figure 3. These measurements represent the total HD formation efficiency during the entire experiments (irradiation + warming up). Therefore, the Y axis represents the percentage of the atoms sent on the surface that formed HD. These measurements are in fact a great tool to probe the existence of chemisorbed sites. Indeed, as reported in Figure 3, a model considering only physisorbed sites cannot reproduce the measurements. Inclusion of chemisorption sites explains the lower efficiency of HD formation because there is a leak of atoms from physisorbed sites to chemisorbed sites. Because these measurements are performed at low temperatures, the leak of the atoms from physisorbed to chemisorbed sites is governed by tunneling, and a constraint on the barrier between physisorbed and chemisorbed sites can therefore be determined. The rate of trapping in chemisorbed sites depends on the product of the width times the square root of the height of the barrier between physisorption and chemisorption ($a_{pc}\sqrt{E_a}$, where a_{pc} is the width of the barrier between physisorbed and chemisorbed sites and E_a its height). The structure of the surfaces considered here can have a physisorption-chemisorption barrier varying from a very high and thin barrier (Sha & Jackson 2002 and see discussion) to a very broad and low barrier (Parneix & Brechignac 1998). Because of these uncertainties, we consider two possible types of barrier, which we call type 1 and type 2, and assume that a realistic surface is intermediate between them.

An extension of our rate-equation model to steady state conditions is reported in Figure 4. The most important result of our model is that H_2 can form on carbonaceous and olivine dust grains for a large range of grain temperatures. This result is fundamental in order to explain the presence of H_2 molecules in almost all circumstances. The

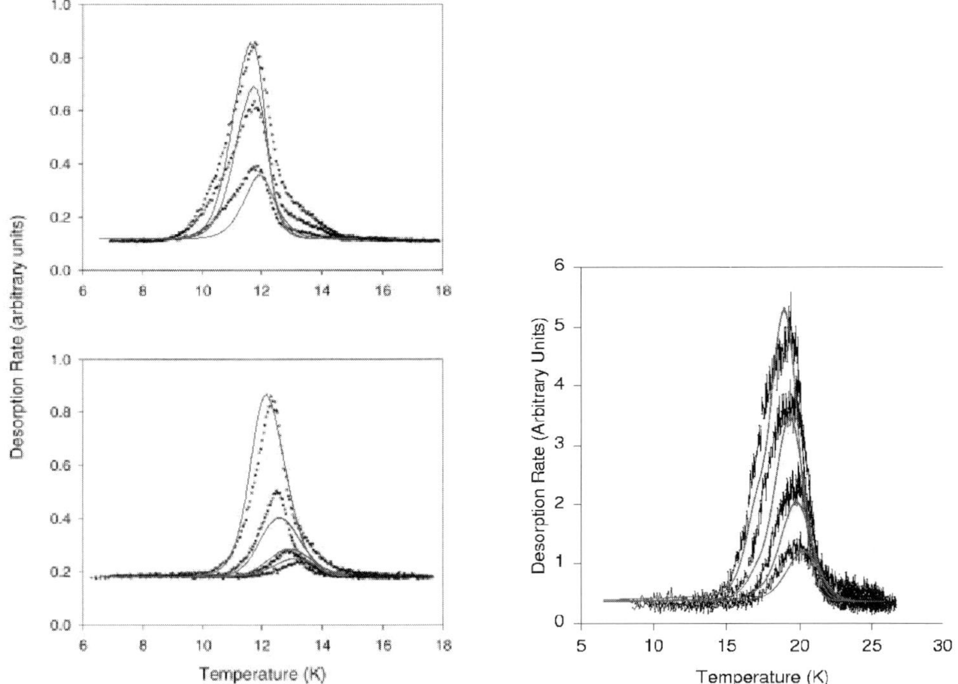

Figure 2. Temperature Program Desorption experiments from Pirronello et al (1997a,b, 1999). *Left:* The points represent the experiment performed on olivine surface for an irradition time of 480, 330 and 120 s (*top left*; from top to bottom), 33, 15, 6 and 4 s (*bottom left*; from top to bottom), benchmarked by our model (solid lines). *Right:* same experiment but on carbonaceous surfaces for an irradiation time of 192, 96, 48 and 24 s, benchmarked by our model (solid lines).

efficiencies of H_2 formation on olivine and carbonaceous grains are somewhat different but show the same behaviour. Our results show that H_2 formation efficiency is extremely sensitive to the physisorption-chemisorption barrier. Also, as discussed in previous work (Cazaux & Tielens 2004), this efficiency varies strongly with the temperature of the grains. When T_{grain} is very low ($\leqslant 10$ K), a fraction of the newly formed molecules does not spontaneously desorb or evaporate, and therefore saturates the grain surface. The formation of molecules is then suppressed since the incoming atoms cannot stick to the grain. At higher grain temperatures ($\leqslant 20$–25 K), the efficiency is extremely high, and is due to the association of physisorbed atoms. Because molecules have a lower surface binding energy than physisorbed atoms, the atoms stay on the grain and associate to form molecules, while molecules evaporate, bringing the efficiency of formation to 100 %. At higher T_{grain}, the physisorbed atoms also start to evaporate, and the formation of molecules is due to the association of physisorbed and chemisorbed atoms. Then, at high temperatures ($T_{grain} \geqslant 300$ K), molecules form through the association of chemisorbed atoms. The differences of formation efficiency between the two types of grains depends strongly on how the atoms can chemisorb. In type 1 grains, the barrier between physisorption and chemisorption is high, preventing the atoms from becoming chemisorbed. Therefore, formation of molecules becomes inefficient at temperatures higher than ~ 20 K, when the formation of molecules also involves chemisorbed atoms. In type 2 grains, on the other hand, the barrier between physisorbed and chemisorbed sites is low, and the atoms can easily become chemisorbed, allowing a more efficient formation of molecules for a large range of temperatures ($T_{grain} \sim 1000$ K).

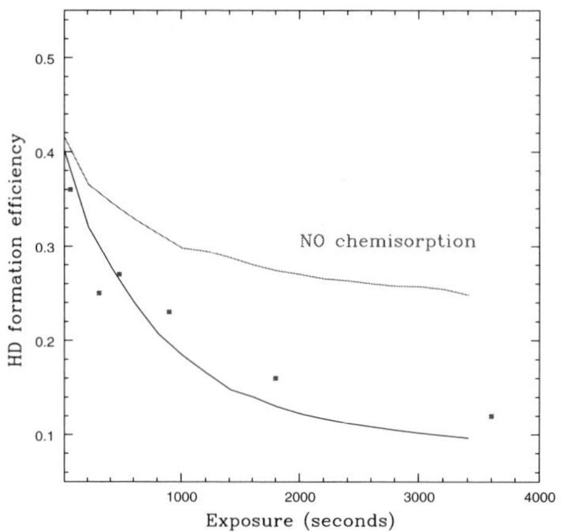
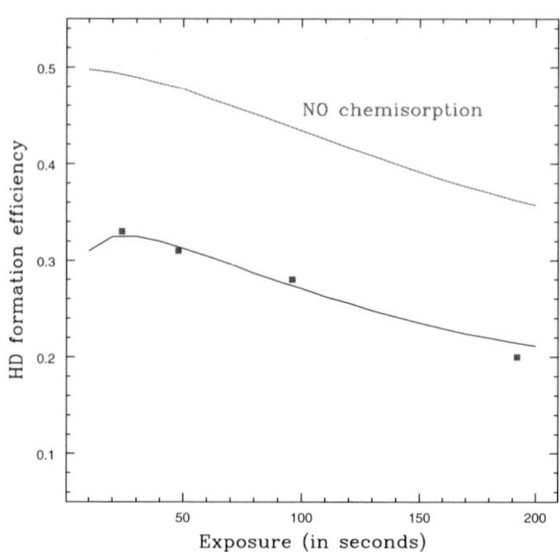

Figure 3. Total HD formation efficiency during the entire experiment (irradiation + warming up). Note that a model which doesn't take into account chemisorption cannot reproduce these measurements. *Left:* percentage of the atoms sent on olivine surface that form HD during the entire experiment (as a fonction of the irradiation time). *Right:* same but on carbonaceous surfaces.

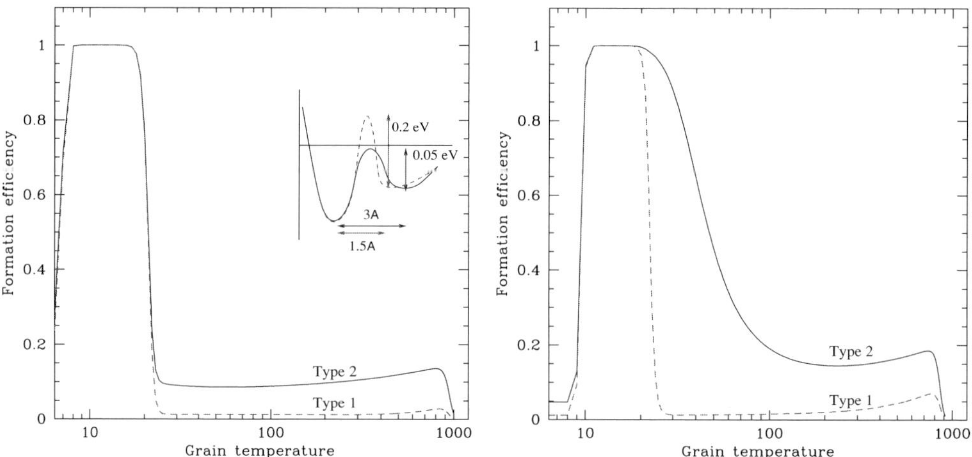

Figure 4. H_2 formation efficiency on olivine (*left*) and carbonaceous surfaces (*right*) and for type 1 (dashed lines) and type 2 (solid lines) grains.

2.2. *Formation Rate*

In astrophysical environments, the overall H_2 formation rate is written as:

$$R_d(H_2) = \frac{1}{2}n(H)v_H n_{grain}\sigma S(T)\epsilon_{H_2} \tag{2.1}$$

where $n(H)$ is the number density of H atoms, v_H the thermal velocity calculated as $\sqrt{\frac{8\pi kT}{m_H}}$ and ϵ_{H_2} the formation efficiency of H_2. We assume that the sticking coefficient, $S(T)$, decreases with T_{gas} and T_{grain} as $(1 + 0.4 \times \sqrt{(T_{gas} + T_{grain})/100})^{-1}$ (Burke & Hollenbach 1983). It is important to remember that $R_d(H_2)$ is the overall rate of H_2 formation in units of $cm^{-3}\,s^{-1}$, whereas R, the rate coefficient of H_2 formation, commonly used in the literature, is in units of $cm^3\,s^{-1}$. These two rates are linked by the expression $R_d(H_2) = R \times n \times n(H)$, where n is the abundance of H atoms in all forms. The H_2 formation rate can be indirectly measured in some astrophysical objects. In the next section we report the case of PDRs and their associate H_2 formation rate, in order to establish a comparison with our model. In the last section, we investigate the addition of deuterium in our model. We now concentrate only on carbonaceous grain surfaces, for which more studies have been dedicated to understand the nature of the barrier against chemisorption (Zecho *et al.* 2002; Sha & Jackson 2002; Klose *et al.* 1992; Aronowitz & Chang 1980, 1985; Parneix & Brechignac 1998).

3. Photodissociation Regions

The formation of H_2 has been observed in different PDRs at different physical conditions, as discussed by Habart *et al.* (2004). The grain and gas temperatures, as well as the densities and UV fluxes in these environments vary strongly from one PDR to another. For the PDRs associated with the Orion Bar, NGC 2023, S140, IC 63 and Oph W, these authors derived an H_2 formation rate in the range 3×10^{-17} to 1.5×10^{-16} $cm^3\,s^{-1}$.

In Figure 5, we compare our model with the H_2 formation rate derived for 5 PDRs (ISO observations; Habart *et al.* 2004) and a diffuse cloud (FUSE observations; Gry *et al.* 2002). In our model, we consider a grain size distribution as discribed in Weingartner & Draine (2001), which accounts for PAHs, small grains and big grains (grains with size from 5 Å to 0.8 μm). We calculate the H_2 formation rate for three different gas temperatures $T_{gas} = 100$ K, 300 K and 600 K, and a sticking coefficient decreasing with T_{gas} and T_{grain} (Burke & Hollenbach 1983). Our results show that grains with a low barrier between physisorbed and chemisorbed sites can better explain the high H_2 formation rate derived by Habart *et al.* (2004). Indeed, one needs to consider that the H atoms can "easily" cross the barrier against chemisorption in order to form H_2 in such an efficient way at high grain temperatures ($T_{grain} \geqslant 30$ K).

4. Addition of Deuterium to our Model

We reconsider our model of H_2 formation on grain surfaces, and include deuterium. We can therefore predict the formation efficiency of H_2, HD and D_2 on grain surfaces. As previously, the two different types of grains have some impact on the formation efficiency of molecules. The height and width of a barrier can segregate the H atoms from the D atoms because of their mass differences. Indeed, a very high and narrow barrier would be more easily crossed through tunneling by H atoms, whereas a low barrier, which favours mobility through thermal hopping, would not exhibit strong differences between the species. Figure 6 presents this result for the two types of grains, for a D/H of 2×10^{-5} and three different H fluxes $n(H) = 10$ atoms cm^{-3} (dashed lines), $n(H) = 100$ atoms cm^{-3} (solid lines) and $n(H) = 10^3$ atoms cm^{-3} (dotted lines). A striking result is the enhanced D_2 formation for a very narrow range of grain temperatures, while H_2 and HD form efficiently for a much broader range. Also, the formation of D_2 is more important on type 1 grains than type 2. In the case of a high and narrow barrier against chemisorption (type 1 grains) H atoms can chemisorb easily by the tunneling effect, whereas D atoms stay physisorbed. In the case of a low and broad barrier against chemisorption (type 2 grains), tunneling is much less efficient and consequently less H atoms populate the chemisorbed sites, reducing the differences between H and D atoms. Therefore the formation of D_2, which occurs through the association of two physisorbed D atoms is enhanced when the physisorbed sites are mostly occupied by D atoms. In this case,

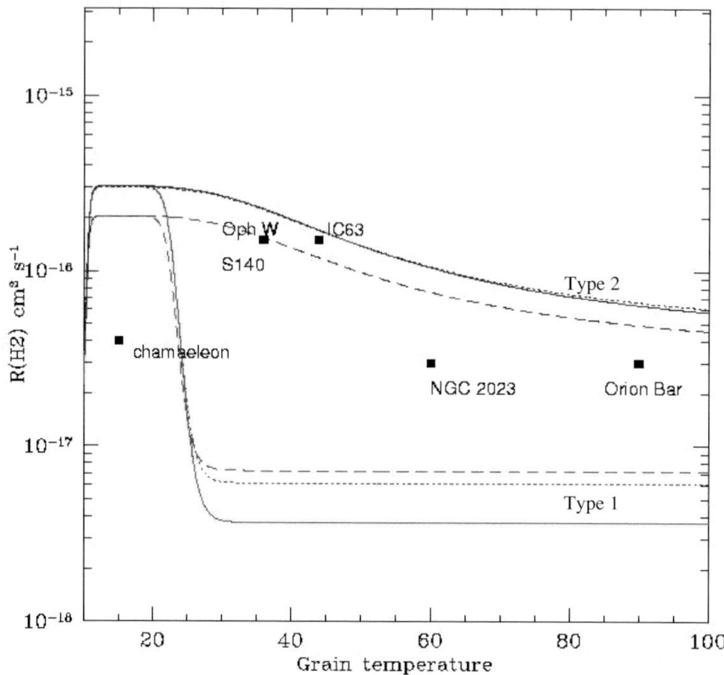

Figure 5. Comparison of our model with the H_2 formation rate derived for 5 PDRs (ISO observations, Habart *et al.* 2004) and a diffuse cloud (FUSE observations, Gry *et al.* 2002). Three different gas temperature are considered here. Solid lines: $T_{gas} = 100$ K, dashed lines: $T_{gas} = 300$ K and dotted lines: $T_{gas} = 600$ K. The three higher curves correspond to our predictions with type 2 grains, and the three lower for type 1 grains. Note the discrepancy between the predicted and the observed formation rate for the Chamaleon point. Chamaleon is a diffuse cloud and the PAHs or very small grains could be less abundant, implying a lower H_2 formation rate.

type 1 grains which jail their H atoms in chemisorbed sites and let the D atoms free to recombine in physisorbed sites, are more efficient to produce D_2.

Gas-phase chemistry has been included in our grain-surface model in order to study the formation of H_2 and its deuterated forms in diffuse clouds. A study by Le Petit *et al.* (2002) concentrates on the H/H_2 and D/HD fronts in the cloud, using the PDR model of Le Bourlot *et al.* (1993). Because this model concentrates on the gas-phase chemistry, but does not follow the evolution of the populations of the different species on the grains, we reexamine the importance of the gas-phase and grain surface chemistry and follow the population of the different species on the grain as well as in the gas phase. The H_2 formation rate derived from observations of diffuse clouds is $\sim 3 \times 10^{-17}$ cm³ s⁻¹, as discussed by Jura (1974). In order to scale our model to these observations, we adapt the mean cross section for collisions between grains and atoms as $\frac{n_{grain}}{n_H}\sigma \sim 10^{-21}$ cm². Then, we calculate the populations and formation rates of H_2 HD and D_2 until the system reaches steady state for each step towards larger Av.

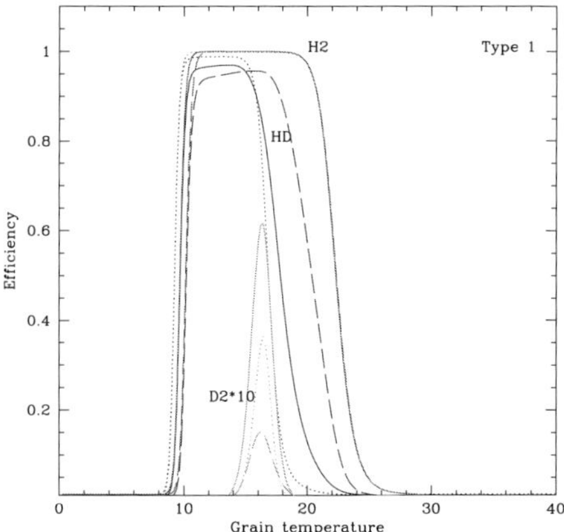
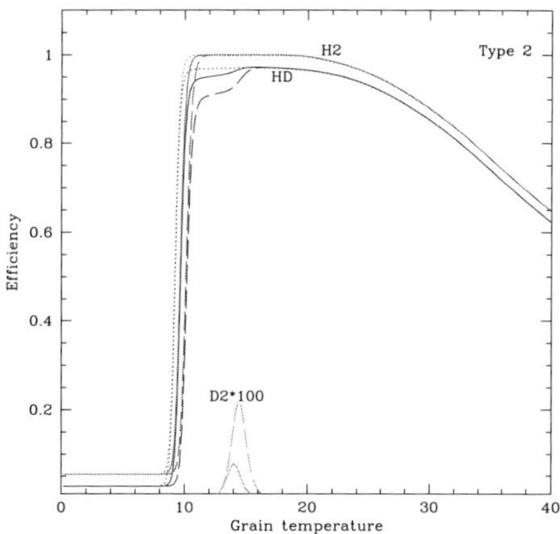

Figure 6. H$_2$, HD and D$_2$ formation efficiencies on carbonaceous grain surfaces, as a function of grain temperature, for different densities n(H). Dotted lines: n(H) $= 10$ atoms cm^{-3}, solid lines n(H) $= 100$ atoms cm^{-3} and dashed lines n(H) $= 1000$ atoms cm^{-3}. We consider in our calculation a gas temperature of 100 K and a D/H ratio is set as 2×10^{-5}. *Left:* Efficiencies for type 1 grains, when the characteristics of the surface are such that the barrier between physisorption and chemisorption is very high and narrow. *Right:* Efficiencies for type 2 grains, when the characteristics of the surface are such that the barrier between physisorption and chemisorption is low and broad.

We consider clouds with a density of 10–10^4 atoms cm^{-3}, a gas temperature of 50 K and we choose a grain temperature $\sim$15 K. We predict the H I, H$_2$, HD and D$_2$ column densities as a function of the extinction for type 1 grains in order to estimate the highest D$_2$ column density that could be observed (see Fig. 7). Our results show that D$_2$ should be observable in dense environments (n(H) $\geqslant 10^3$ atoms cm^{-3}). Also, if the density of the medium is known and D$_2$ is observed, a constraint on the grain structure could be derived. Indeed, for a type of grain considered, D$_2$ formation is more or less enhanced and therefore observations of D$_2$ could constrain which type of grains are present in the observed environment. Therefore, D$_2$ molecules could be a good probe of grain structure.

References

Aronowitz, S. & Chang, S. 1980, *Ap. J.* 242, 149

Aronowitz, S. & Chang, S. 1985, *Ap. J.* 293, 243

Barlow, M.J. & Silk, J. 1976, *Ap. J.* 207, 131

Burke, J.R. & Hollenbach, D.J. 1983, *Ap. J.* 265, 223

Cazaux, S. & Tielens, A.G.G.M. 2004, *Ap. J.* 604, 222

Cazaux, S. & Tielens, A.G.G.M. 2002, *Ap. J.* 575, L29

Chang, Q., Cuppen, H.M., & Herbst, E. 2005, *A&A* 434, 599

Davidson, K. & Fesen, R.A. 1985, *ARA&A*, 23, 119

Fromherz, T., Mendoza, C., & Ruette, F. 1993, *MNRAS* 263, 851

Ghio, E., Mattera, L., Salvo, C., Tommasini, F., & Valbusa, J.U., 1980, *Chem. Phys.* 73(1), 556

Gould, R.J. & Salpeter, E.E. 1963, *Ap. J.* 138, 393

Graham, J.R., Wright, G.S., & Longmore, A.J. 1990, *Ap. J.* 352, 172

Gry, C., Boulanger, F., Nehmé, C., Pineau des Forêts, G., Habart, E., & Falgarone, E. 2002, *A&A* 391, 675

Habart, E., Boulanger, F., Verstraete, L., Walmsley, C.M., & Pineau des Forêts, G. 2004, *A&A* 414, 531

Hollenbach, D. & McKee, C.F. 1989, *Ap. J.* 342, 306

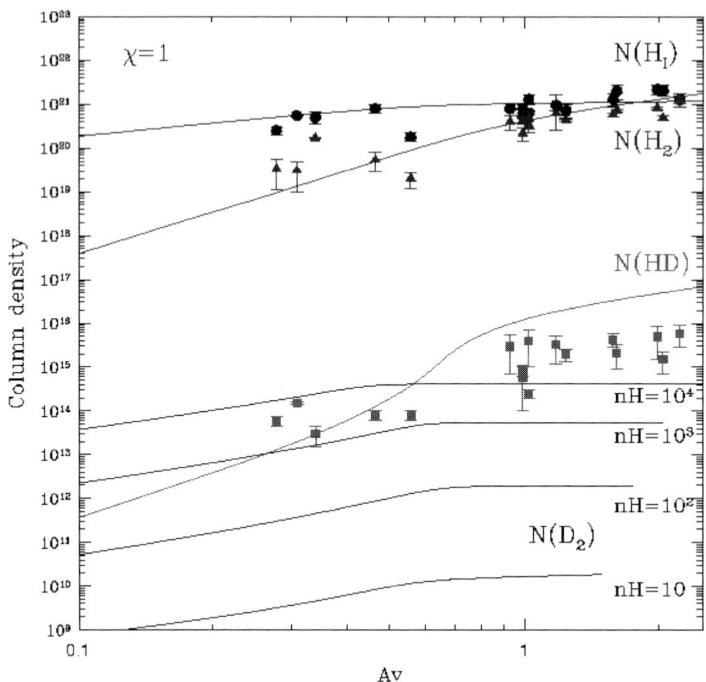

Figure 7. Column densities of H, H$_2$ and HD for a density $n(\mathrm{H}) = 10$ atoms cm^{-3} compared with the FUSE/Copernicus measurements towards translucent clouds (Lacour *et al.* 2005). D$_2$ column densities for type 1 grains are predicted for $n(\mathrm{H}) = 10$, 10^2, 10^3 and 10^4 atoms cm^{-3}. It seems that D$_2$ can be observed only in shielded environments with a density $\geqslant 10^3$ cm^{-3}. In our model, we assume a UV field of $\chi = 1$, a D/H ratio of 2×10^{-5} and a gas temperature of 50 K.

Hollenbach, D. & Salpeter, E.E. 1970, *J. Chem. Phys.* 53, 79
Hollenbach, D. & Salpeter, E.E. 1971, *Ap. J.* 163, 155
Hollenbach, D.J., Werner, M.W., & Salpeter, E.E. 1971, *Ap. J.* 163, 165
Jeloaica, L. & Sidis, V. 1999, *Chem. Phys. Lett.* 300, 157
Jura, M. 1974, *Ap. J.* 191, 375
Klose, S. 1992, *A&A* 260, 321
Lacour, S., *et al.* 2005, *A&A* 430, 967
Le Bourlot, J., Pineau Des Forets, G., Roueff, E., & Flower, D.R. 1993, *A&A* 267, 233
Le Petit, F., Roueff, E., & Le Bourlot, J. 2002, *A&A* 390, 369
Neufeld, D.A. & Dalgarno, A. 1989, *Ap. J.* 340, 869
Parneix, P. & Brechignac, P. 1998, *A&A* 334, 363
Pirronello, V., Liu, C., Roser, J.E., & Vidali, G. 1999, *A&A* 344, 681
Pirronello, V., Biham, O., Liu, C., Shen, L., & Vidali, G. 1997a, *Ap. J.* 483, L131
Pirronello, V., Liu, C., Shen, L., & Vidali, G. 1997b, *Ap. J.* 475, L69
Que, J.-Z., Radny, M.W., & Smith, P.V. 1997, *Surface Sci.* 391, 161
Rho, J., Jarrett, T.H., Cutri, R.M., & Reach, W.T. 2001, *Ap. J.* 547, 885
Rudy, R.J., Rossano, G.S., & Puetter, R.C. 1994, *Ap. J.* 426, 646
Sankrit, R., *et al.* 1998, *Ap. J.* 504, 344
Sha, X. & Jackson, B. 2002, *Surface Sci.* 496, 318

Spitzer, L. & Jenkins, E.B. 1975, *ARA&A* 13, 133
Weingartner, J.C. & Draine, B.T. 2001, *Ap. J.* 548, 296
Zecho, T., Güttler, A., Sha, X., Jackson, B., & Küppers, J. 2002, *J. Chem. Phys.* 117, 8486

Discussion

TAPPE: What happens with the formation enegy of H_2 in your model? Is it possible to include, phenomenologically, a free parameter to account for this energy and fit the experiments?

CAZAUX: We do not take into account, in our model, the energy of the newly formed molecules. In order to account for this energy when we fit the experiments, we need to include a parameter (that we call μ) which is the amount of molecules spontaneously released from the surface when formed. This corresponds to about 60–70% of the newly formed molecules.

PAPADOPOULOS: Is the H_2 formation rate of 1.5×10^{-16} to be believed for H_2 formation in PDR-type of conditions? This is 5 times higher than the canonical value of 3×10^{-17} $cm^3 \, s^{-1}$.

CAZAUX: The H_2 formation rate was estimated using the H_2 line intensity ratios as a diagnostic. Habart *et al.* (2004) used the ratio of the 0–0 S(3) to the 1–0 S(1) line, which increases strongly with R_f. (Steady state PDR models were used to examine the sensitivity of different H_2 line ratios to the H_2 formation rate R_f.) Then for each PDR, from comparison of PDR model results with ISO and ground-based data of the vibrational ground (0–0 S(3)) and excited state (1–0 S(1)) lines of H_2 a value of R_f was estimated.

BAURICHTER: I have a comment on the modelling of your type 1 grain surfaces. The calculations of Sha & Jackson you are referring to show a barrier of 0.2 eV with respect to a hydrogen atom at infinite distance to the surfaces and not in a 0.05 eV physisorption well as you assumed.

CAZAUX: I misunderstood the calculations of Sha *et al.* (2002) in this case, and this should be corrected. Our problem is that by reproducing the experiments on carbonaceous and olivine surfaces, we determined a value for the width times the square root of the height of the barrier between physisorption and chemisorption. In this case, the barrier calculated by Sha *et al.* (2002) has too large a value. This problem needs further investigation.

Astrochemistry: Recent Successes and Current Challenges
Proceedings IAU Symposium No. 231, 2005
D.C. Lis, G.A. Blake & E. Herbst, eds.

Experimental Study of H_2 Formation on Ices

**L. Hornekær[1], A. Baurichter[2], V. V. Petrunin[2], D. Field[1],
and A. C. Luntz[2]**

[1]Department of Physics and Astronomy, Aarhus University,
Ny Munkegade bygn. 520, 8000 Aarhus C, Denmark
email: liv@phys.au.dk

[2]Department of Physics, University of Southern Denmark,
Campusvej 55, 5230 Odense M, Denmark

Abstract. We present new experimental data and review previous experimental results on molecular hydrogen formation from atomic recombination on porous amorphous solid water (ASW) surfaces at temperatures from 10 K to 30 K, i.e. under conditions of relevance to cold dense interstellar clouds. We show that the desorption of molecular hydrogen formed on porous ASW surfaces is well described by a model, in which the molecules are assumed to be completely thermalized to the surface temperature and to be evenly distributed throughout the porous network of the ASW films. These results emphasize that, not only the chemical properties, but also the physical morphology of dust grain surfaces must be considered to obtain a full understanding of molecular hydrogen formation on interstellar dust grain surfaces.

1. Introduction

Molecular hydrogen is the most abundant molecule in the interstellar medium (ISM) and plays a pivotal role in the thermal and chemical development of interstellar dust and molecular clouds. There is no known effective gas-phase route for H_2 formation at the low number densities in interstellar clouds and it is the generally accepted view that H_2 is formed on the surface of interstellar dust grains. Observations indicate that H_2 formation must be a very efficient process under conditions ranging from those found in cold interstellar clouds to those found in photodissociation regions (PDRs). This entails that efficient H_2 formation mechanisms must exist on both ice-covered and bare dust grain surfaces and that these mechanisms must be efficient under grain temperatures ranging from 10 K to 30 K, or even higher (smaller dust grains might experience thermal spikes to much higher temperatures due to e.g. absorption of individual UV photons) and gas temperatures ranging from 10 K to several hundred Kelvin. In recent years several groups have embarked on experimental studies of molecular hydrogen formation under conditions of interstellar relevance. The studies include molecular hydrogen formation from weakly bound (physisorbed) states on carbonaceous and silicate surfaces (Katz *et al.* 1999), from strongly bound (chemisorbed) states on graphite (Zecho *et al.* 2002), and from physisorbed states on amorphous solid water (ASW) surfaces (Manicó *et al.* 2001; Roser *et al.* 2002; and Hornekær *et al.* 2003). In spite of all of these activities, the mechanisms behind molecular hydrogen formation in the ISM are not yet fully uncovered.

In the present paper we present new experimental data and review previous experimental results reported on in Hornekær *et al.* (2003) pertaining to molecular hydrogen formation under conditions of relevance to cold, dense interstellar clouds; e.g., molecular hydrogen formation on amorphous solid water (ASW) surfaces at temperatures from

10 K to 30 K. We show that the desorption of molecular hydrogen formed on porous ASW surfaces is well described by a model, in which the molecules are assumed to be completely thermalized to the surface temperature and to be evenly distributed throughout the porous network of the ASW films. These results emphasize that, not only the chemical properties, but also the physical morphology of dust grain surfaces must be considered to obtain a full understanding of molecular hydrogen formation on interstellar dust grain surfaces.

2. Amorphous Solid Water

Amorphous Solid Water (ASW) ice films furnish an ideal system for studying the influence of surface morphology on molecular hydrogen formation in the ISM. First, ASW ice is a well characterized porous system, which makes it possible to study molecular hydrogen formation on surfaces of controlled and varying porosity. Second, according to present models, the icy mantles on interstellar dust grains are composed predominantly of water ice with other molecular species, such as CO, CH_3OH, CO_2, NH_3, CH_4, O_2 and N_2, added in varying amounts (Tielens & Hagen 1982; D'Hendecourt *et al.* 1985). The majority of the water ice is formed in situ, by recombination of hydrogen and oxygen atoms to H_2O molecules on the grain surface, rather than by condensation of H_2O molecules from the gas phase (Jones *et al.* 1984). However, the absorption features of interstellar ice are well reproduced by the absorption features of ASW ice formed by vapor deposition of H_2O and kept at a temperature below 30 K (Hagen *et al.* 1981). Hence, we believe that ASW ice films formed by low-temperature vapour deposition constitute the best simple one-component model for icy mantles on interstellar grains.

The morphology of ASW films grown by vapor deposition on low-temperature substrates under vacuum at low deposition rates (a few $ML\,s^{-1}$ or lower) is known to depend on deposition temperature, deposition rate, angular distribution of the incoming water molecules and the thermal history of the ASW film after formation (Narten *et al.* 1976; Mayer & Pletzer 1986; Jenniskens & Blake 1994; Stevenson *et al.* 1999; Kimmel *et al.* 2001a,b; Parent *et al.* 2002). For deposition on low-temperature substrates (< 90 K), normal incidence water molecules produce non-porous ASW films, while off-normal incidence (or background dosing) produces porous films consisting of open networks of nanometer sized pores (Stevenson *et al.* 1999; Kimmel *et al.* 2001a,b). The ASW film porosity for non-normal or background dosing also depends on the growth temperature or subsequent annealing temperature of the film (Mayer & Pletzer 1986; Jenniskens & Blake 1994; Stevenson *et al.* 1999; Parent *et al.* 2002; Hornekær *et al.* 2005). For example, dosing or annealing at substrate temperatures above 90 K reduces the porosity strongly (Stevenson *et al.* 1999) and at temperatures above 120 K, fully non-porous ASW structures result (Rowland & Devlin 1991; Horimoto *et al.* 2002). At temperatures above ~136 K, crystallization into a cubic crystalline structure sets in and at even higher temperatures, the normal hexagonal crystalline ice structure is produced (Jenniskens & Blake 1994; Dohnalek *et al.* 2000). The porosity of ASW films can also be influenced by co-deposition of other gases (Bar-Nun *et al.* 1987).

Apart from changes in ASW film porosity with growth temperature, X-ray (Narten *et al.* 1976) and electron diffraction (Jenniskens & Blake 1994) experiments imply that ASW films grown at temperatures below 30 K arrange in a "local" chemical structure of higher density than ASW films grown at higher temperatures. This high-density state is referred to as the high-density amorphous (HDA) phase, while the low-density state obtained at temperatures above 30 K is referred to as the low-density amorphous (LDA) phase (Jenniskens & Blake 1994). Annealing ASW films in the HDA phase to temperatures above 30 K results in a slow and irreversible phase transition to the LDA phase (Jenniskens & Blake 1994). It is important to distinguish between the high-density (HDA) and low-density (LDA) nomenclature introduced by Jenniskens *et al.* and the actual physical density of the ice. The physical density of the ice is determined both by its local chemical structure (HDA or LDA) and, even more importantly, by its morphology/porosity. Hence a porous HDA ice film can easily have a lower physical density than a non-porous LDA ice film.

In the following, we will restrict our characterization of the ices to their porosity, since experiments indicate that porosity, rather than local chemical bulk structure, is the most important factor for understanding chemical reactions and molecular desorption from ice surfaces (see e.g. Bar-Nun *et al.* 1985; Mayer & Pletzer 1986; Rowland *et al.* 1991; Takahashi *et al.* 1999; Watanabe *et al.* 2000; Sadtchenko *et al.* 2000; Ayotte *et al.* 2001; Horimoto *et al.* 2002; Hornekær *et al.* 2003; Collings *et al.* 2003; Hornekær *et al.* 2005).

3. Experimental Setup and Results

The experimental setup has been described in detail in Hornekær *et al.* (2005). All experiments are performed under ultrahigh vacuum (UHV) with a base operating pressure below 1×10^{-10} Torr. Porous amorphous solid water (ASW) films are grown by depositing deionized H_2O, purified by a series of freeze-thaw cycles, onto a 10 K Cu substrate. H_2O deposition is done from a capillary array doser positioned 5 mm from the sample surface with a $\sim 45°$ HWHM angular spread. Typical dose rates are 0.3–3 monolayers sec^{-1} (ML s^{-1}). Under these H_2O deposition conditions and a surface temperature of $T_s \leqslant 10$ K, modestly porous ASW films are grown. Based on the angular distribution and the temperature of the sample we estimate a porosity of $\xi_{por} \simeq 0.1$ (where ξ_{por} is defined as the percentage increase in internal surface area per monolayer of ASW relative to the external surface area of the film; Kimmel *et al.* 2001), however, this estimate is somewhat uncertain. Although we have not made any independent studies of the nature of the porous distribution, we certainly anticipate that this forms a 3D interconnected network, as reported on previously in Stevenson *et al.* (1999) and predicted by ballistic growth models (Kimmel *et al.* 2001a,b).

H and D atomic beams are produced in separate microwave discharges and thermalized to 300 K by passing through phosphoric acid coated glass tubes. The beam dissociation probabilities range from 65–80 %. The beams pass through three differential pump stages before being admitted into the UHV chamber. The H beam has a diameter of 3.5 mm and is directed onto the center of the ASW film at normal incidence. The D beam has an incidence angle of 4° off normal and a diameter of 1.5 mm at the sample where it is overlapped with the H beam. The typical flux employed is 10^{13} cm^{-2} s^{-1} or $\simeq 0.01$ ML s^{-1}.

Temperature programmed desorption (TPD) measurements are performed by heating the sample with a 0.5 K s^{-1} linear ramp and detecting the desorbing species in a differentially pumped movable quadrupole mass spectrometer (QMS) fitted with a cone cap with a 3 mm aperture. During the TPD measurement the opening in the cone is moved to within 2 mm of the sample surface. Desorption of H_2, HD and D_2 can be monitored simultaneously. Molecular hydrogen formation is detected by monitoring the production of HD, which is not present in the atomic beams and can only be produced via surface reactions.

4. Model for TPD Spectra from Molecular Hydrogen Formation on Porous ASW Surfaces

Based on previous experimental results reported on in Hornekær *et al.* (2003), molecular hydrogen formation on ASW films is expected to be well described by a Langmuir-Hinshelwood, or hot-atom mechanism, in which H and D atoms are mobile at temperatures as low as 8 K, move into the pore-structure and recombine in the pores. The majority of the formed molecules are retained in the porous ASW film and thermalized to the ASW surface temperature. Subsequent desorption of the formed H_2 molecules during TPD experiments is therefore expected to follow the normal desorption kinetics for molecular hydrogen on ASW surfaces.

The desorption rate for hydrogen molecules adsorbed on a non-porous surface with binding energy E_b is:

$$\frac{d\Theta}{dt} = -k_0 e^{-E_b/k_B T_s} \Theta \tag{4.1}$$

where k_0 is the pre-exponential factor (typically of the order 10^{12}–10^{13} s^{-1}), k_B is Boltzmann's constant, T_s is the surface temperature and Θ is the H_2 coverage on the surface.

To take the effect of porosity into account, the desorption rate can be modified by considering that only molecules in the top layer of the porous film can desorb directly into the gas phase. Molecules adsorbed in the porous structure might desorb locally, but will subsequently collide with the pore walls resulting in a high probability for re-adsorption. The net result is a series of desorption-re-adsorption events, in which the molecules perform a random walk in the porous structure until they finally reach the upper layers and are able to desorb directly into the gas phase. In this picture, the measured desorption from the sample is not proportional to the total H_2 coverage, but instead to the H_2 coverage in only the top layer at the vacuum interface. If we assume that the molecules diffuse sufficiently fast to be evenly distributed throughout the ASW film, then the coverage in the top layer, Θ_s, for a N_l monolayer thick ASW film with porosity ξ_{por} is given by: $\Theta/(1 + N_l\xi_{por})$, where Θ is the total H_2 dose adsorbed on the surface. The coverage in the pores is in turn given by: $\Theta N_L \xi_{por}/(1 + N_l\xi_{por})$. Assuming fast diffusion and a uniform H_2 distribution throughout the film then leads to the following modification of the expression for the H_2 desorption rate for a single fixed binding energy E_b of H_2 on the surface:

$$\frac{d\Theta}{dt} = -k_0 e^{-E_b/k_B T_s} \Theta_s = -k_0 e^{-E_b/k_B T_s} \frac{1}{1 + N_l\xi_{por}} \Theta \tag{4.2}$$

This simple model predicts an upwards shift in the TPD peak with increasing ASW film thickness, since the pre-factor scales as $\propto 1/(1 + N_l\xi_{por})$.

One limitation of the model is the assumption of a single binding energy for H_2 on the ASW surface. Experiments show that in reality a wide distribution of binding energies exists (Hornekær *et al.* 2005). However, in the low-coverage limit the molecules preferentially occupy the high energy binding sites and the assumption of a single binding energy is more realistic.

5. Experimental Results and Comparison with Model TPD Spectra

Figure 1(a) shows background subtracted HD TPD spectra from a 1500 ML porous ASW film, after deposition of H and D atoms at a surface temperature of 10 K. Deposition times vary from 1 to 10 min. At a flux of 0.01 ML s^{-1} and a porosity of $\xi_{por} = 0.1$, a

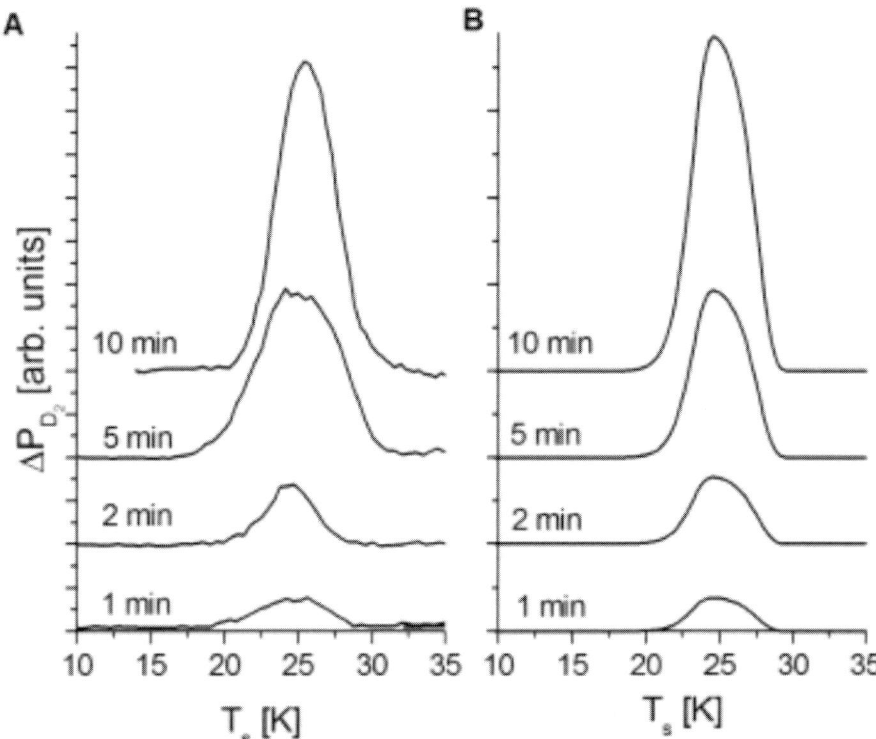

Figure 1. (*a*) TPD spectra after increasing doses of H and D atoms on a 1500 ML porous amorphous solid water film cooled to 10 K. The employed ramp rate is 0.5 K s^{-1}. The TPD traces have been offset for clarity. (*b*) Modelled TPD spectra assuming molecular desorption as described in eq. 4.2. The TPD traces have been offset for clarity.

10 min dose on a 1500 ML ASW film corresponds to a coverage of 0.04. Hence, even at the longest dose time employed we are in the low-coverage limit. A single peak at 25 K is visible in all the spectra. The fact that the peak temperature is independent of coverage indicates that the desorption is a first-order process; i.e., that the recombination of H and D atoms takes place prior to molecular desorption and that the formed molecules have been thermalized to the surface temperature before desorbing. As discussed in § 3 the desorption kinetics of HD molecules formed on the surface should therefore be well described by the same desorption kinetics as H$_2$ molecules deposited unto the ASW surface as molecules, i.e. the desorption kinetics described in § 3.

The binding energy for a D$_2$ molecule on a porous ASW surface in the low coverage limit is around 59 meV (Hornekær *et al.* 2005). A zero-point corrected binding energy for an HD molecule on a porous ASW film is then approximately 56 meV.

Figure 1(b) shows simulated TPD spectra based on the model described in Eq. 4.2. The molecules are assumed to distribute evenly between sites with binding energies ranging from 52–60 meV. The only free parameter is the detection efficiency. This has been normalized to obtain the same total peak area in theory and experiment for the 10 min. dose. As can be seen the simple model provides a reasonably good fit to the experimental data.

The fact that desorption of HD molecules formed on the ASW surface is a first-order process and can be described by the same kinetics as desorption of H$_2$ molecules deposited unto the surface as molecules (adjusting for the difference in zero-point energy) rebuts earlier models of molecular hydrogen formation on porous ASW surfaces, according to which HD desorption was described as a second-order process with thermally activated atomic diffusion as the rate limiting step (Manicó *et al.* 2001).

6. Conclusion

We have presented new experimental data on molecular hydrogen formation on porous amorphous solid water films at low temperature. The data show that the desorption of HD molecules formed on the ASW surface is a first-order process, which is well described by molecular desorption kinetics from porous ASW films. This indicates that the recombination of H and D atoms takes place prior to molecular desorption and that the formed HD molecules have been thermalized to the surface temperature before desorbing. Additional experimental data reported on in Hornekær *et al.* (2003) indicate that H and D atoms are mobile and recombine on ASW surfaces at temperatures as low as 8 K. After recombination, the molecules are retained in the porous structure and thermalize to the surface temperature. The rate limiting step in desorption of the formed HD molecules is therefore not atomic diffusion and recombination, but molecular desorption. The porosity of the ASW film plays a double role, by both facilitating the retainment of newly formed molecules and by reducing the desorption rate of the molecules. Even though the experiments have only been performed on porous ASW surfaces, we expect that the reported results can be extended to other materials exhibiting nanometer scale porosity, and that they therefore are potentially relevant for molecular hydrogen formation and desorption in diffuse clouds, where molecular hydrogen is formed on bare carbonaceous and silicate grains. Laboratory experiments on interstellar carbon dust grain analogues indicate that these could very well exhibit porosity on the nanometer scale (Schnaiter *et al.* 1999). Hence, to correctly describe both dust grain chemistry and desorption from dust grain surfaces, further knowledge of interstellar dust grain morphology is needed.

References

Ayotte, P., Smith, R.S., Stevenson, K.P., Dohnalek, Z., Kimmel, G.A., & Kay, B.D. 2001, *J. Geoph. Res.* 106, 33387

Bar-Nun, A., Herman, G., Laufer, D., & Rappaport, M.L. 1985, *Icarus* 63, 317

Bar-Nun, A., Dror, J., Kochavi, E., & Laufer, D. 1987, *Phys. Rev. B* 35, 2427

Collings, M.P., Dever, J.W., Fraser, H.J., McCoustra, M.R.S., & Williams, D.A. 2003, *Ap. J.* 583, 1058

D'Hendecourt, L.B., Allamandola, L.J., & Greenberg, J.M. 1985, *A&A* 152, 130

Dohnalek, Z., Kimmel, G.A., Ciolli, R.L., Stevenson, K.P., Smith, R.S., & Kay, B.D. 2000, *J. Chem. Phys.* 112, 5932

Hagen, W., Tielens, A.G.G.M., & Greenberg, J.M. 1981, *Chem. Phys.* 56, 367

Horimoto, N., Kato, H.S., & Kawai, M. 2002, *J. Chem. Phys.* 116, 4375

Hornekær, L., Baurichter, A., Petrunin, V., Field, D., & Luntz, A.C., 2003, *Science* 302, 1943

Hornekær, L., Baurichter, A., Petrunin, V., Luntz, A., Kay, B.D., & Al-Halabi, A. 2005, *J. Chem. Phys.* 122, 124701

Jenniskens, P. & Blake, D.F. 1994, *Science* 265, 753

Jones, A. & Williams, D. *MNRAS* 1984, 209, 955

Katz, N., Furman, I., Biham, O., Pironello, V., & Vidali, G. 1999, *Ap. J.* 522, 305

Kimmel, G.A., Stevenson, K.P., Dohnalek, Z., Smith, R.S., & Kay, B.D. 2001a, *J. Chem. Phys.* 114, 5284

Kimmel, G.A., Dohnalek, Z., Stevenson, K.P., Smith, R.S., & Kay, B.D. 2001b, *J. Chem. Phys.* 114, 5295

Manicó, G., Raguni, G., Pironello, V., Roser, J., & Vidali, G. 2001, *Ap. J.* 548, L253

Mayer, E. & Pletzer, R. 1986, *Nature* 319, 298

Narten, A.H., Venkatesh, C.G., & Rice, S.A. 1976, *J. Chem. Phys.* 64, 1106

Parent, P., Laffon, C., Mangeney, C., Bournel, F., & Tronc, M. 2002, *J. Chem. Phys.* 117, 10842

Roser, J., Manico, G., Pironello, V., & Vidali, G. 2002, *Ap. J.* 581, 276

Rowland, B. & Devlin, J.P. 1991, *J. Chem. Phys.* 94, 812

Rowland, B., Fisher, M., & Devlin, J.P. 1991, *J. Chem. Phys.* 95, 1378

Sadtchenko, V., Knutsen, K., Giese, C.F., & Gentry, W.R. 2000, *J. Phys. Chem. B* 104, 2511

Sadtchenko, V., Knutsen, K., Giese, C.F., & Gentry, W.R. 2000, *J. Phys. Chem. B* 104, 4894

Schnaiter, M., Henning, T., Mutschke, H., Kohn, B., Ehbrecht, M., & Huisken, F. 1999, *Ap. J.* 519, 687

Stevenson, K.P., Kimmel, G.A., Dohnalek, Z., Smith, R.S., & Kay, B.D. 1999, *Science* 283, 1505

Takahashi, J., Masuda, K., & Nagaoka, M. 1999, *Ap. J.* 520, 724

Tielens, A.G.G.M. & Hagen, W. 1982, *A&A* 114, 245

Watanabe, N., Horii, T., & Kouchi, A. 2000, *Ap. J.* 541, 772

Zecho, T., Güttler, A., Sha, X., Jackson, B., & Küppers, J. 2002, *J. Chem. Phys.* 117, 8486

Discussion

KROES: Have you performed, or will you perform, experiments on non-porous ice to see if H₂ then desorbs immediately upon its formation?

HORNEKÆR: We have performed experiments showing that if any HD molecules form on non-porous ASW surfaces, then they are not retained on the surface after formation. However, with our experimental setup we cannot detect HD molecules, which form during atom deposition and immediately desorb. Hence, we cannot say for sure whether any HD forms on these surfaces. Fortunately this question will probably be answered in the coming years by the group of Stephen Price at UCL.

Photo: E. Herbst

Astrochemistry: Recent Successes and Current Challenges
Proceedings IAU Symposium No. 231, 2005
D.C. Lis, G.A. Blake & E. Herbst, eds.
© 2006 International Astronomical Union
doi:10.1017/S1743921306007368

The Formation of H_2 and HD with the Master Equation Approach

Ofer Biham[1], Azi Lipshtat[1]†, and Hagai B. Perets[1]‡

[1]Racah Institute of Physics, The Hebrew University, Jerusalem 91904, Israel
email: biham@phys.huji.ac.il

Abstract. The formation of H_2 and HD molecules on interstellar dust grains is studied using rate equation and master equation models. Rate equations are used in the analysis of laboratory experiments, which examine the formation of molecular hydrogen on astrophysically relevant surfaces. However, under interstellar conditions, rate equations are not suitable for the calculation of reaction rates on dust-grain surfaces. Due to the low flux and the sub-micron size of the grains, the populations of H and D atoms on a single grain are likely to be small. In this case the reaction rates are dominated by fluctuations and should be calculated using stochastic methods. The rate of molecular hydrogen formation in interstellar clouds is evaluated using the master equation, taking into account the distribution of grain sizes.

Keywords. ISM: abundances — ISM: molecules — molecular processes

---

1. Introduction

The formation of molecular hydrogen in the interstellar medium is a process of fundamental importance in astrophysics. H_2 molecules contribute to the cooling of clouds during gravitational collapse and participate in reactions that produce more complex molecules (Duley & Williams 1984; Williams 1998). H_2 cannot form in the gas phase efficiently enough to account for its observed abundance (Gould & Salpeter 1963). It was thus proposed that H_2 formation takes place on dust grains that act as catalysts. The process of H_2 formation on grains can be broken up into several steps as follows. An H atom approaching the surface of a grain has a probability ξ to stick and become adsorbed. The adsorbed H atom (adatom) resides on the surface for an average time t_H (residence time) before it desorbs. In the Langmuir-Hinshelwood mechanism, the adsorbed H atoms quickly equilibrate and diffuse on the surface of the grain either by thermal activation or by tunneling. When two adsorbed H atoms encounter each other, an H_2 molecule may form (Williams 1968; Hollenbach & Salpeter 1970; Hollenbach & Salpeter 1971; Hollenbach *et al.* 1971; Smoluchowski 1981; Aronowitz & Chang 1985; Duley & Williams 1986; Pirronello & Averna 1988; Sandford & Allamandolla 1993; Takahashi *et al.* 1999; Farebrother *et al.* 2000). The steady-state production rate of molecular hydrogen, R_{H_2} $(\text{cm}^{-3}\,\text{s}^{-1})$ can be expressed by

$$R_{H_2} = \frac{1}{2} n_H v_H \sigma \gamma n_g, \tag{1.1}$$

where n_H (cm^{-3}) and v_H $(\text{cm}\,\text{s}^{-1})$ are the number density and the speed of H atoms in the gas phase, respectively, σ (cm^2) is the average cross-sectional area of a grain and n_g (cm^{-3}) is the number density of dust grains. The parameter γ is the fraction of H

---

† Present address: Department of Pharmacology and Biological Chemistry, Mount Sinai School of Medicine, New York, NY 10029, USA.
‡ Present address: Faculty of Physics, Weizmann Institute of Science, Rehovot 76100, Israel.

atoms striking the grain that eventually form a molecule, namely $\gamma = \xi\eta$, where η is the probability that an H adatom on the surface will recombine with another H atom to form H_2. The probability ξ for an H atom to become adsorbed on a grain surface covered by an ice mantle was calculated by Buch & Zhang (1991) and Masuda *et al.* (1998). They found that ξ depends on the surface temperature and on the energy of the irradiation beam. For a surface at 10 K and beam temperature of 350 K, Masuda *et al.* (1998) obtained a sticking coefficient around 0.5.

A simplified version of Eq. (1.1) is often used for the evaluation of the H_2 formation rate in models of interstellar chemistry. It takes the form

$$R_{H_2} = R \cdot n_H n, \tag{1.2}$$

where R $(cm^3\,s^{-1})$ is the rate coefficient and $n = n_H + 2n_{H_2}$ (cm^{-3}) is the total density of hydrogen atoms in both atomic and molecular (n_{H_2}) form. In previous studies it was assumed that the total grain mass is about 1% of the hydrogen mass in the cloud, with a single grain size of $r = 0.17$ μm and a grain mass density of $\rho_g = 2$ $(g\,cm^{-3})$ (Hollenbach *et al.* 1971). Under these assumptions, for gas temperature of 100 K and $\gamma = 0.3$, one obtains $R \simeq 10^{-17}$ $(cm^3\,s^{-1})$, in agreement with observations for diffuse clouds (Jura 1975). More recent observations in photon dominated regions indicate that in these regions the rate coefficient should be $R \simeq 10^{-16}$ $(cm^3\,s^{-1})$, in order to account for the observed H_2 abundance (Habart *et al.* 2003).

2. Laboratory Experiments

In the last few years, the formation of molecular hydrogen on interstellar dust analogues was studied in a series of laboratory experiments (Pirronello *et al.* 1997a; Pirronello *et al.* 1997b; Pirronello *et al.* 1999; Roser *et al.* 2002; Roser *et al.* 2003). In these experiments, beams of H and D atoms were irradiated on the surface. The production of HD molecules was measured both during irradiation and during a subsequent temperature programmed desorption (TPD) experiment. In this experiment, the temperature of the sample is raised quickly to either desorb particles (atoms and molecules) that got trapped on the surface or to enhance their diffusion and subsequent reaction and desorption. The formation of molecular hydrogen was studied on samples of polycrystalline olivine (Pirronello *et al.* 1997a; Pirronello *et al.* 1997b), amorphous carbon (Pirronello *et al.* 1999) and various types of amorphous ice samples, namely high density ice (HDI) and low density ice (LDI) (Manicó *et al.* 2001; Roser *et al.* 2002; Hornekær *et al.* 2003). The results of the TPD experiments, particularly on polycrystalline olivine and HDI, show second-order kinetics for low coverage of H and D atoms on the surface. In second-order kinetics the peak shifts towards higher temperature as the coverage is reduced. This is due to the larger distances that atoms need to diffuse before they encounter each other and recombine. This indicates that on these samples the surface mobility of H and D atoms is dominated by thermal hopping rather than tunneling. If tunneling were dominant, molecules could form efficiently during irradiation, resulting in first-order kinetics.

The results of the TPD experiments were analyzed using rate equation models (Katz *et al.* 1999; Cazaux & Tielens 2002; Cazaux & Tielens 2004; Perets *et al.* 2005). The analysis involves fitting the parameters of the rate equations to the experimental TPD curves. The most important parameters are the activation energy barriers for diffusion and desorption of hydrogen atoms on the surface. The models used by Katz *et al.* (1999) and Perets *et al.* (2005) take into account only physisorption sites and thermal hopping between them. The models of Cazaux & Tielens (2002, 2004) also take into account tunneling and chemisorption sites. However, the two analyses are in agreement on the fact that tunneling between physisorption sites appears to be too slow to play an important role in the TPD experiments (Vidali *et al.* 2005). Furthermore, recent experiments and calculations indicate that there are energy barriers, which are likely to prevent adsorbed H and D atoms from entering chemisorption sites at these low temperatures.

Using the parameters obtained in the laboratory experiments, the rate equations enabled the extrapolation from laboratory time scales to astrophysical time scales. Using rate equations, the efficiency of molecular hydrogen formation on these surfaces under interstellar conditions and its dependence on the flux and grain temperature were calculated. It was found that the recombination efficiency strongly depends on the grain temperature. Each sample exhibits a narrow window of high efficiency along the temperature axis. For the polycrystalline olivine sample, this window is between 7–9 K, while for amorphous carbon it is between 11–17 K (Vidali *et al.* 2005).

3. The Rate-Equation Model

Consider a diffuse interstellar cloud with a density n_H (cm^{-3}) of H atoms, and a density n_g of dust grains. The typical velocity v_H (cm s^{-1}) of H atoms in the gas phase is given by

$$v_H = \sqrt{\frac{8}{\pi} \frac{k_B T_{gas}}{m_H}}, \qquad (3.1)$$

where $m_H = 1.67 \times 10^{-24}$ (gram) is the mass of an H atom and T_{gas} is the gas temperature. To evaluate the flux of atoms onto grain surfaces we will assume, for simplicity, that the grains are spherical with a radius r (cm). The cross section of a grain is $\sigma = \pi r^2$ and its surface area is 4σ. The density of adsorption sites on the surface is s (sites cm^{-2}) and the number of adsorption sites on a grain is $S = 4\pi r^2 s$. The flux F_H (atoms s^{-1}) of H atoms onto the surface of a single grain is given by $F_H = n_H v_H \sigma$.

H atoms that collide and stick to the surface hop as random walkers between adjacent sites until they either desorb or recombine into molecules. The desorption rate of an H atom is

$$W_H = \nu \cdot \exp(-E_1/k_B T), \qquad (3.2)$$

where ν is the attempt rate (typically assumed to be 10^{12} s^{-1}), E_1 is the activation energy barrier for desorption of an H atom and T is the surface temperature. The hopping rate of an H atom is

$$a_H = \nu \cdot \exp(-E_0/k_B T), \qquad (3.3)$$

where E_0 is the activation energy barrier for H diffusion. Throughout this paper we use the parameters obtained experimentally for amorphous carbon, namely the activation energies $E_0 = 44.0$ meV and $E_1 = 56.7$ meV (Katz *et al.* 1999), and the density of adsorption sites on the surface $s \simeq 5 \times 10^{13}$ (sites cm^{-2}) (Biham *et al.* 2001). For the density of hydrogen atoms in the gas phase we take $n_H = 10$ (atoms cm^{-3}). The gas temperature is taken as $T_{gas} = 90$ K, thus $v_H = 1.37 \times 10^5$ (cm s^{-1}). These are typical values for diffuse interstellar clouds.

The number of H atoms on the grain is denoted by N_H and its expectation value is denoted by $\langle N_H \rangle$. The rate $A_H = a_H/S$ is approximately the inverse of the time t_s required for an atom to visit nearly all the adsorption sites on the grain surface. This is due to the fact that in two dimensions the number of distinct sites visited by a random walker is linearly proportional to the number of steps, up to a logarithmic correction (Montroll & Weiss 1965).

The formation of H_2 molecules on interstellar dust grains can be described by the rate equation

$$\frac{d\langle N_H\rangle}{dt} = F_H - W_H\langle N_H\rangle - 2A_H\langle N_H\rangle^2. \tag{3.4}$$

The first term on the right-hand side describes the flux of H atoms, the second term describes the desorption of H atoms and the third term describes the diffusion and recombination. Here we assume, for simplicity, that all H_2 molecules desorb from the surface upon formation. The production rate R_{H_2} (cm^{-3} s^{-1}) of H_2 molecules is given by

$$R_{H_2} = A_H\langle N_H\rangle^2 n_g. \tag{3.5}$$

By solving the rate equations under steady-state conditions for astrophysically relevant values of the flux and temperature, it was found (Biham & Lipshtat 2002) that molecular hydrogen formation is efficient only within a narrow window of grain temperatures, $T_0 < T < T_1$, where

$$T_0 = \frac{E_0}{k_B \ln(\nu S/F_H)} \tag{3.6}$$

and

$$T_1 = \frac{2E_1 - E_0}{k_B \ln(\nu S/F_H)}. \tag{3.7}$$

At temperatures below T_0, the mobility of H atoms on the surface is very low, sharply reducing the production rate. At temperatures above T_1 most atoms quickly desorb before they encounter each other and form molecules.

Rate equations are an ideal tool for the simulation of surface reactions, due to their simplicity and high computational efficiency. In particular, they account correctly for the temperature dependence of the reaction rates. However, in the limit of small grains under low flux they become unsuitable. This is because they ignore the fluctuations as well as the discrete nature of the populations of atoms on the grain (Charnley et $al.$ 1997; Caselli et $al.$ 1998; Shalabiea et $al.$ 1998; Stantcheva et $al.$ 2001). For example, as the number of H atoms on a grain fluctuates in the range of 0, 1 or 2, the H_2 formation rate cannot be obtained from the average number alone. This can be easily understood, since the recombination process requires at least two H atoms simultaneously on the surface. In this limit, the master equation approach is needed in order to evaluate the H_2 formation rate.

4. The Master Equation

Consider a grain exposed to a flux of H atoms. The probability that there are N_H hydrogen atoms on its surface is given by $P_H(N_H)$, where $N_H = 0, 1, 2, \ldots, S$. The master equation provides the time derivatives of these probabilities. It is a set of coupled linear differential equations of the form

$$\dot{P}_H(N_H) = F_H\left[P_H(N_H - 1) - P_H(N_H)\right] + W_H\left[(N_H + 1)P_H(N_H + 1) - N_H P_H(N_H)\right]$$
$$+ A_H\left[(N_H + 2)(N_H + 1)P_H(N_H + 2) - N_H(N_H - 1)P_H(N_H)\right]. \tag{4.1}$$

Each equation includes three terms. The first term describes the effect of the incoming flux F_H. The probability $P_H(N_H)$ increases when an H atom is adsorbed on a grain that already has N_H-1 adsorbed atoms [at a rate of $F_H P_H(N_H-1)$], and decreases when a new atom is adsorbed on a grain with N_H atoms on it [at a rate of $F_H P_H(N_H)$]. The second term includes the effect of desorption. An H atom that is desorbed from a grain with N_H atoms decreases the probability $P_H(N_H)$ [at a rate of $N_H W_H P_H(N_H)$], and increases the probability $P_H(N_H - 1)$ at the same rate. The third term describes the effect of H_2 formation. The production of one molecule reduces the number of H atoms on the surface from N_H to $N_H - 2$. For one pair of H atoms the recombination rate is proportional to the sweeping rate A_H multiplied by 2, since both atoms are mobile simultaneously. This rate is multiplied by the number of possible pairs of atoms, namely $N_H(N_H - 1)/2$. Note that the equations for $\dot{P}_H(0)$ and $\dot{P}_H(1)$ do not include all the terms, because at least one H atom is required for desorption to occur and at least two for recombination. The rate of formation of H_2 molecules on a single grain, $R_{H_2}^{\mathrm{grain}}$ ($\mathrm{cm}^{-3}\ \mathrm{s}^{-1}$), can be obtained from summation of master equation terms to yield

$$R_{H_2}^{\mathrm{grain}} = A_H[\langle N_H^2 \rangle - \langle N_H \rangle] \tag{4.2}$$

where

$$\langle N_H^k \rangle = \sum_{N_H=0}^{\infty} N_H^k P_H(N_H). \tag{4.3}$$

is the k-th moment of the distribution $P_H(N_H)$.

The time dependence of $R_{H_2}^{\mathrm{grain}}$ can be obtained by numerically integrating Eqs. (4.1) using a standard Runge-Kutta stepper. For the case of steady state, namely $\dot{P}_H(N_H) = 0$ for all N_H, an analytical solution for $P_H(N_H)$ was obtained, in terms of A_H/W_H and W_H/F_H (Green *et al.* 2001; Biham & Lipshtat 2002). The rate of formation of H_2 molecules on a single grain with S adsorption sites is given by (Green *et al.* 2001; Biham & Lipshtat 2002)

$$R_{H_2}^{\mathrm{grain}} = \frac{F_H}{2} \frac{I_{W_H/A_H+1}\left(2\sqrt{2F_H/A_H}\right)}{I_{W_H/A_H-1}\left(2\sqrt{2F_H/A_H}\right)}, \tag{4.4}$$

where $I_x(y)$ is the modified Bessel function.

In Figure 1, we present the H_2 production rate on a single grain of amorphous carbon vs. the grain size, which is parameterized by the number of adsorption sites, S. The conditions, specified in § 3 are typical in diffuse clouds and the grain temperature is $T = 18$ K. It is observed that for large grains the rate equation results coincide with those of the master equation. However, for small grains the rate equation over-estimates the production. The master equation accounts correctly for the rate of H_2 formation on grains of all sizes.

5. Effects of the Grain Size Distribution

Interstellar dust grains exhibit a broad size distribution between several nanometers and fractions of a micron. This distribution can be approximated by a power law of the form (Mathis *et al.* 1977; Weingartner & Draine 2001)

$$n_g(r) = cr^{-\alpha} \tag{5.1}$$

where $n_g(r)dr$ (cm^{-3}) is the density of grains with radii in the range $(r, r + dr)$ in the cloud. This distribution is bounded between the upper cutoff $r_{\max} = 0.25$ μm and the lower cutoff $r_{\min} = 5$ nm. The exponent is $\alpha \simeq 3.5$ (Draine & Lee 1984). Clearly, this distribution is dominated by the small grains. This fact has important implications on the formation rate of molecular hydrogen. The total surface area is much larger than in the case of a homogeneous population of grains with radii of the order of 0.25 μm, with the same total mass density of about 1% of the hydrogen mass density. This tends to enhance the H_2 formation rate. However, below some grain size the efficiency of H_2

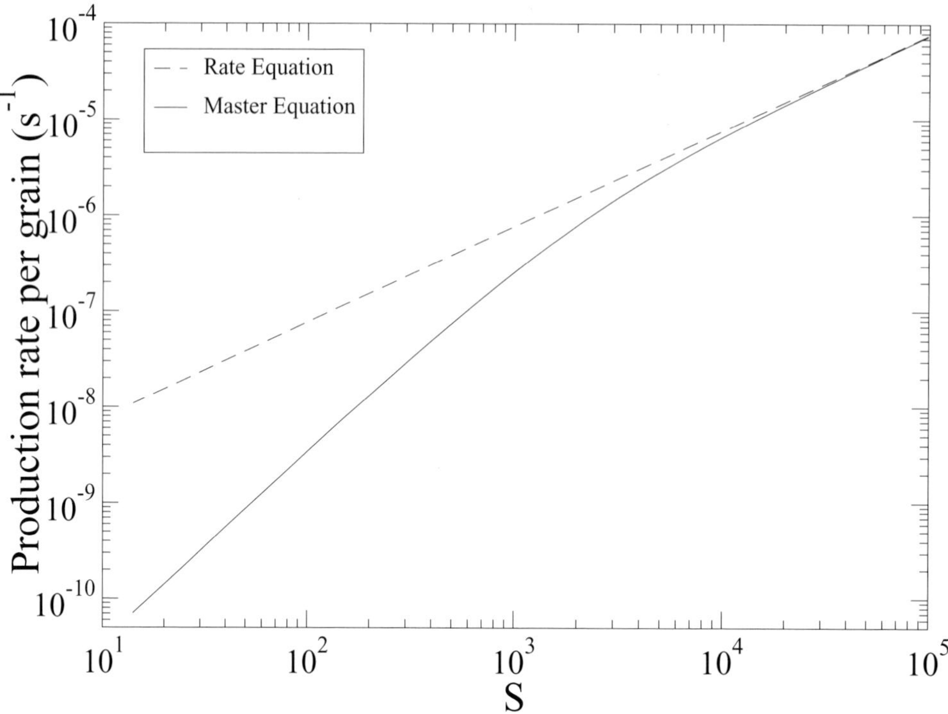

Figure 1. The production rate of H_2 molecules on a single grain of amorphous carbon with S adsorption sites. For large grains the rate equation results (dashed line) are in agreement with the master equation (solid line), while for small grains the rate equation over-estimates the production. The parameters are specified in § 3.

formation is sharply reduced. Due to these competing effects the production of molecular hydrogen should be integrated over the whole range of grain sizes.

In Figure 2 we present the rate coefficient R $(cm^3 s^{-1})$ vs. grain temperature for the case of a power-law distribution of grain sizes (solid line) and for the case, in which all grains are of radius $r = 0.17$ μm, with the same total mass of grains. It turns out that, within the temperature window in which H_2 formation is efficient, the increase in the total surface area due to the broad size distribution, enhances the production rate by about an order of magnitude compared to a homogeneous population of sub-micron-size grains (Lipshtat & Biham 2005). This is in spite of the reduced efficiency of H_2 formation on small grains.

6. Effects of Grain Porosity

The laboratory experiments on interstellar dust analogues have shown that H_2 formation on dust grains is efficient within a range of grain temperatures below 20 K. However, high abundances of H_2 have also been observed in warmer clouds, including photon-dominated regions (PDRs), where grain temperatures may reach 50 K. The possible effect of grain porosity on H_2 formation was recently examined using a rate equation model (Perets & Biham 2005). It is found that porosity extends the temperature range in which H_2 formation is efficient towards higher temperatures. This is because H atoms that desorb from the internal surfaces of the pores may re-adsorb many times and thus stay longer on the surface. However, this extension is expected to be of only several

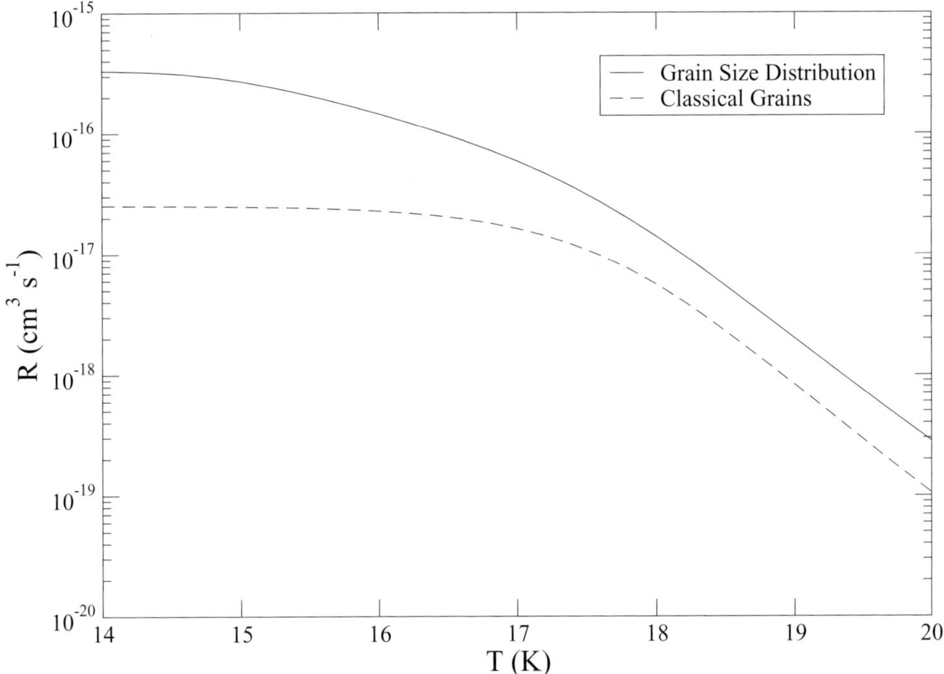

Figure 2. The rate coefficient R (cm^3 s^{-1}) of H$_2$ formation vs. grain temperature, obtained from the master equation for the case of a power-law distribution of grain sizes (solid line). The dashed line shows the rate coefficient for the case in which the same mass of grains consists of a homogeneous population of "classical grains" of radius $r = 0.17$ μm.

degrees (K) and by itself cannot account for the formation of molecular hydrogen in PDRs.

Another effect of porosity that may be relevant to molecular hydrogen formation is that porous grains may be significantly colder than predicted by models which use compact, spherical grains (Blanco & Bussoletti 1980; Voshchinnikov *et al.* 1999). While models of spherical grains predict grain temperatures in the range of $30-50$ K in PDRs, the actual temperatures of porous interstellar grains may thus be significantly lower. In the most favorable cases, the grain temperatures may go down to the range of $15-30$ K for large grains and to $10-20$ K for small grains (Blanco & Bussoletti 1980).

7. HD Formation on Grains

Unlike the H$_2$ molecules in the interstellar medium which do not form efficiently in the gas phase, HD molecules have an efficient gas-phase formation mechanism. It consists of a sequence of reactions in which an ionized D atom reacts with an H$_2$ molecule to form HD. The possibility that in diffuse clouds some HD molecules are also formed on grain surfaces was recently examined (Lipshtat *et al.* 2004). Due to the lower abundance of D atoms vs. H atoms (by a factor of about 10^{-5}), the master equation is required for such calculations even on relatively large grains. In case the D atoms are bound more strongly to the surface than H atoms, the temperature window in which HD formation is efficient is extended towards higher temperatures, compared to the range of efficient H$_2$ formation. The assumption that D atoms are more strongly bound to the grains is plausible, but so far has not been confirmed experimentally.

8. Complex Reaction Networks of Multiple Species

The master equation can be generalized to describe complex reaction networks of multiple species on grain surfaces. Recently, it was applied to the study of the reaction network that lead to the formation of H_2CO and CH_3OH on grains (Stantcheva *et al.* 2002; Stantcheva & Herbst 2003). Such surface reaction networks are dominated by H atoms because hydrogen is the most abundant specie and also because it reacts with most of the other atomic and molecular species adsorbed on grain surfaces. Furthermore, hydrogen atoms hop on the surface much faster than other adsorbed species and thus quickly react with other species even at very low coverages. For such reaction networks, the dynamical variables in the master equation are the probabilities $P(N_H, N_O, N_{OH}, \dots)$ where there are N_X copies of specie X on the grain. Setting suitable cutoffs, each population of each specie is limited to $N_X = 0, 1, \dots, N_X^{\max}$. The problem is that as the number of reactive species increases, the number of coupled equations in the master equation increases exponentially, limiting the feasibility of this approach. Using the fact that the reaction networks are usually sparse, it is possible to dramatically reduce the number of equations (Lipshtat & Biham 2004). This makes the master equation applicable even for complex surface reaction networks with a large number of reactive species.

The master equation can be solved either by direct integration, as described above, or by Monte Carlo (MC) methods (Charnley 2001). A significant advantage of direct integration is that the equations can be easily coupled to the rate equations of gas-phase chemistry. Furthermore, a single run of the direct integration method provides the entire probability distribution of populations of adsorbed species for the simulated conditions as well as the reaction rates. This is in contrast with the MC methods, which typically require large computational resources due to the need to accumulate much statistical information.

9. Summary

We have presented the master equation approach for the evaluation of molecular hydrogen formation rates on interstellar dust grains. This approach is required in the limit of small grains and low flux where the population of hydrogen atoms on each grain is small. In this limit, the reaction rate is dominated by fluctuations. Thus, the rate equation approach, which incorporates the mean-field approximation, fails. Applications of the master equation approach for the evaluation of the formation rates of H_2, HD as well as more complex molecules are discussed.

Acknowledgements

We thank E. Herbst, V. Pironello and G. Vidali for helpful discussions. This work was supported by the Israel Science Foundation and the Adler Foundation for Space Research.

References

Aronowitz, S. & Chang, S. 1985, *Ap. J.* 293, 243
Biham, O., Furman, I., Pirronello, V., & Vidali, G. 2001, *Ap. J.* 553, 595.
Biham, O. & Lipshtat A. 2002, *Phys. Rev. E* 66, 055103
Blanco, A. & Bussoletti, B. 1980, *Ap. Space Sci.* 67, 105
Buch, V. & Zhang, Q. 1991, *Ap. J.* 379, 647
Cazaux, S. & Tielens, A.G.G.M. 2002, *Ap. J.* 575, L29
Cazaux, S. & Tielens, A.G.G.M. 2004, *Ap. J.* 604, 222

Charnley, S.B., Tielens, A.G.G.M., & Rodgers, S.D. 1997, *Ap. J.* 482, L203

Charnley, S.B. 2001, *Ap. J.* 562, L99

Caselli, P., Hasegawa, T.I., & Herbst, E. 1998, *Ap. J.* 495, 309

Draine B.T. & Lee, H.M. 1984, *Ap. J.* 285, 89

Duley, W.W. & Williams, D.A. 1984, *Interstellar Chemistry* (Academic Press, London)

Duley, W.W. & Williams, D.A. 1986, *MNRAS* 223, 177

Farebrother, A.J., Meijer, A.J.H.M., Clary, D.C., & Fisher, A.J. 2000, *Chem. Phys. Lett.* 319, 303

Gould, R.J. & Salpeter, E.E. 1963, *Ap. J.* 138, 393

Green, N.J.B., Toniazzo, T., Pilling, M.J. Ruffle, D.P. Bell, N., & Hartquist, T.W. 2001, *A&A* 375, 1111

Habart, E., Boulanger, F., Verstraete, L., Pineau des Forêts, G., Falgarone, E., & Abergel, A. 2003, *A&A* 397, 623

Hollenbach, D. & Salpeter, E.E. 1970, *J. Chem. Phys.* 53, 79

Hollenbach, D. & Salpeter, E.E. 1971, *Ap. J.* 163, 155

Hollenbach, D., Werner M.W., & Salpeter, E.E. 1971, *Ap. J.* 163, 165

Hornekær, L., Baurichter, A., Petrunin, V.V., Field, D., & Luntz, A.C. 2003, *Science* 302, 1943

Jura, M. 1975, *Ap. J.* 197, 575

Katz, N., Furman, I., Biham, O., Pirronello V., & Vidali, G. 1999, *Ap. J.* 522, 305

Lipshtat, A. & Biham, O. 2004, *Phys. Rev. Lett.* 93, 170601

Lipshtat, A., Biham, O., & Herbst E., 2004, *MNRAS* 348, 1055

Lipshtat, A. & Biham, O. 2005, *MNRAS* 362, 666

Manicó, G., Raguni, G., Pirronello, V., Roser, J.E., & Vidali G. 2001, *Ap. J.* 548, L253

Masuda, K., Takahashi, J., & Mukai, T. 1998, *A&A* 330, 773

Mathis, J.S., Rumpl, W., & Nordsieck, K.H. 1977, *Ap. J.* 217, 425

Montroll, E.W. & Weiss, G.H. 1965, *J. Math. Phys.* 6, 167

Perets, H.B., Biham, O., Manicó, G., Pirronello, V., Roser, J., Swords, S., & Vidali, G. 2005, *Ap. J.* 627, 850

Perets, H.B. & Biham, O. 2005, preprint

Pirronello, V. & Averna D. 1988, *A&A* 201, 196

Pirronello, V., Liu, C., Shen L., & Vidali, G. 1997a, *Ap. J.* 475, L69

Pirronello, V., Biham, O., Liu, C., Shen L., & Vidali, G. 1997b, *Ap. J.* 483, L131

Pirronello, V., Liu, C., Roser J.E., & Vidali, G. 1999, *A&A* 344, 681

Roser, J.E., Manicó, G., Pirronello, V., & Vidali, G. 2002, *Ap. J.* 581, 276

Roser, J.E., Swords S., & Vidali, G. 2003, *Ap. J.* 596, L55

Sandford, S.A. & Allamandolla, L.J. 1993, *Ap. J.* 409, L65

Shalabiea, O.M., Caselli, P., & Herbst, E. 1998, *Ap. J.* 502, 652

Smoluchowski, R. 1981, *Ap. Space Sci.* 75, 353

Stantcheva, T., Caselli, P., & Herbst, E. 2001, *A&A* 375, 673

Stantcheva, T., Shematovich, V.I., & Herbst, E., 2002, *A&A* 291, 1069

Stantcheva, T. & Herbst, E. 2003, *MNRAS* 340, 983

Takahashi, J., Masuda, K., & Nagaoka, M. 1999, *MNRAS* 306, 22

Vidali, G., Roser, J., Manicó, G., Pirronello, V., Perets, H.B., & Biham, O. 2005, *Journal of Physics: Conference Series* 6, 36

Vidali, G., Roser, J., Manicó, G., & Pirronello, V. 2005, *IAU Sympusium 231*, this volume

Voshchinnikov, N.V., Semenov, D.A., & Henning, T. 1999, *A&A* 349, L25

Weingartner, J.C. & Draine, B.T. 2001, *Ap. J.* 548, 296

Williams, D.A. 1968, *Ap. J.* 151, 935

Williams, D.A. 1998, *Faraday Discussions* 109, 1

Discussion

HORNEKÆR: Will the inclusion of chemisorption sites on dust grains turn even small interstellar grains into efficient contributors to H_2 formation in photon dominated regions (PDRs) because thermal spikes could activate recombination of chemisorbed H atoms?

BIHAM: In our analysis so far we have not included fluctuations in grain temperatures. Such fluctuations are expected to be significant only for grains of radii smaller than about 10 nm. Chemisorbed H atoms are expected to contribute to H_2 formation when grain temperatures reach several hundred degrees (K). However, a careful analysis is required in order to quantify the contribution of such mechanism.

AIKAWA: You concluded that H migrates by thermal hopping, while Cazaux fits the same experimental data assuming tunneling. Would you tell me why you need to exclude tunneling and why you and Cazaux reached different conclusions on H-migration?

BIHAM: The temperature programmed desorption (TPD) curves obtained for the polycrystalline olivine sample (Katz *et al.* 1999) exhibit second-order kinetics at lower coverage, namely, as the coverage decreases, the TPD peak shifts to higher temperatures. This indicates that the HD molecules are formed by thermal hopping during the TPD run. If tunneling alone provided sufficient mobility for H atoms to diffuse and recombine during irradiation, when the surface temperature is around 6 K, all TPD curves would be of first-order kinetics. For other samples, such as amorphous carbon, the evidence for second order kinetics is not as strong and tunneling is not ruled out completely.

Astrochemistry: Recent Successes and Current Challenges
Proceedings IAU Symposium No. 231, 2005
D.C. Lis, G.A. Blake & E. Herbst, eds.

© 2006 International Astronomical Union
doi:10.1017/S174392130600737X

A Summary of Experimental Results on Molecular Hydrogen Formation on Dust Grain Analogues

Gianfranco Vidali[1]†, J. E. Roser[1], G. Manicó[2], and V. Pirronello[2]

[1]Syracuse University, 201 Physics Bldg., Syracuse, NY 13244-1130 USA
email: gvidali@syr.edu, jeroser@syr.edu

[2]DMFCI, University of Catania
Viale A.Doria 6, Catania (Sicily, Italy)
email: gmanico@dmfci.unict.it, pirronello@dmfci.unict.it

Abstract. We review the main laboratory results of investigations of processes of molecular hydrogen formation on surfaces. The problem of the formation of molecular hydrogen is a fundamental issue in astrophysics/astrochemistry, because of the great importance that molecular hydrogen has for the structure and evolution of our Universe. Such experiments are done using ultra-high vacuum, low temperature, and atomic/molecular beam techniques to study the formation of molecular hydrogen on dust grain analogues in conditions as close as technically feasible to the ones present in relevant ISM environments. In experiments conducted at Syracuse University, we studied H_2 formation on the three most ISM-relevant classes of surfaces: silicates, carbonaceous materials and amorphous water ice. Our experimental investigations range from the evaluation of the catalytic efficiency of the studied surfaces to the energetics of the reaction, i.e. the partition of the formation energy between the grain and the nascent molecule. Such measurements have been done by changing various parameters such as: the temperature of the interstellar dust analogue, the kinetic temperature of the atoms, the morphology of the surface and, to be completed soon, the composition of the solid. Quantitative and qualitative information on the processes of H_2 formation is then fed in theoretical models to extract results that pertain to desired ISM environments.

Keywords. astrochemistry — dust — ISM: molecules — methods: laboratory

---

1. Introduction

The problem of the formation of molecular hydrogen is a fundamental issue in astrophysics/astrochemistry because of the great importance that molecular hydrogen has for the structure and evolution of the Universe. Even though great advances have taken place in the last decade from both the experimental and theoretical standpoint, this problem is not yet fully understood.

Molecular hydrogen is the most abundant molecule in space. However, since ultraviolet photons, cosmic rays, shocks and chemical reactions continuously destroy this molecule, there must be an efficient way to form it in a variety of ISM (interstellar medium) environments (Herbst 2001). As was pointed out four decades ago by Gould & Salpeter (1963), gas-phase formation of molecular hydrogen by two-body association is extraordinarily inefficient because strongly forbidden, except in those environments—such as the early Universe—where there are enough charged particles for appreciable rates of reactions of the type: $H + e \rightarrow H^- + h\nu$ and $H^- + H \rightarrow H_2 + e$. Hollenbach & Salpeter (1970, 1971a,

† Corresponding author

355

1971b) in a series of papers, laid out a theory for the formation of molecular hydrogen. The formation occurs on the surface of dust grains, with a rate R (cm^{-3} sec^{-1}) for H+H $\rightarrow$ H$_2$ given by:

$$R_{\mathrm{H}_2} = \frac{1}{2} n_{\mathrm{H}} v_{\mathrm{H}} \sigma \xi \eta n_{\mathrm{g}}.$$ (1.1)

In this formula, n_{H} is the number density of hydrogen atoms, v_{H} their speed, σ the cross section of the grain, ξ the sticking coefficient, η the probability of bond formation once two atoms interact and n_{g} is the number density of dust grains in the ISM. For the sticking coefficient, they used the result of their calculation of the interaction of a hydrogen atom with an ice surface, $\xi = 0.3$. Assuming the probability of recombination of hydrogen atoms when on the surface to be $\eta = 1$, they found that they could recover the deduced rate.

In their calculations, Salpeter and others assumed that the surface of a dust grain was covered with polycrystalline ice. The diffusion of H atoms on such surfaces turned out to be very fast due to tunneling. Indeed, for an average grain of 0.1 μm, if two hydrogen atoms are present, they calculated that the H atoms would find each other before evaporating.

Smoluchowski (1981, 1983) considered the problem of hydrogen atoms diffusing on an amorphous ice surface. Since tunneling is very sensitive, on an atomic scale, to the details of the environment through which a hydrogen atom diffuses, it is not surprising that in an amorphous medium the mobility is greatly reduced. Unfortunately, his calculations gave the result that the mobility of hydrogen atoms on an ice surface would be much too small to yield hydrogen molecules with the rate necessary to match observations.

With the availability of ultra-high vacuum techniques beginning in the late seventies, great strides were made in the study and understanding of properties of and processes at pristine surfaces. Since then, different mechanisms of reaction at surfaces have been identified. For example, in the Langmuir-Hinshelwood reaction, the partners of the reaction on a surface go through the following steps: sticking to the surface, thermalization, diffusion on the surface, formation of a bond between two reactants, and (possible) ejection of the product-molecule from the surface. This was thought to be the predominant mechanism for reactions at surfaces until recently. Alternatively, and the experimental verification of such mechanism has been realized only recently, an incoming atom can directly react with a partner on the surface without prior accommodation on it (this is called the Eley-Rideal mechanism). Another mechanism, the hot-atom mechanism, calls for the incoming atom to retain on the surface part of its original kinetic energy (or to gain some energy from the surface) in order to move across it at superthermal speed. For a review of such mechanisms as applied to reactions of interest in the ISM, see Vidali *et al.* (2005) and references cited therein.

The strength of the bond between an adatom (i.e., an atom adsorbed on a surface) and the surface itself is of course of primary importance for reactions to proceed. In typical diffuse cloud ISM conditions, the dust grain is bare with few H atoms on it; thus, in this case, the Eley-Rideal mechanism doesn't seem to be the dominating process because the chance of an atom from the gas-phase to hit another on the surface is slim. Therefore, an incoming H atom must be able to diffuse on the surface in order to find another H atom and react with it. Assuming that the dust grain temperature is in the 15–20 K range, and that the grain (a silicate or a carbonaceous grain, or a composite particle of the two) is amorphous—as per observations (see Whittet 2003 for a review), an H atom will diffuse on the surface of the grain only if it is held by weak binding forces. In the end, successful production of molecular hydrogen in ISM conditions will occur when the H-surface bond strength is weak enough to allow mobility and strong enough to keep the H atom long enough on the surface to find another partner (the activation energy barrier for diffusion is roughly one third of the binding site energy for an atomically ordered surface, but can vary considerably due to chemical bonding and morphology). It is expected that chemical and morphological heterogeneity increase the chance of successful reactions by providing a rich landscape of energy sites where the atoms can reside long enough.

In the process of the formation of a molecule, energy is released. How this energy is partitioned between the molecule and the solid, and within the molecule, is of obvious astrophysical importance, both because the kinetic energy of H_2 can contribute to the energy budget of the cloud, and because there is the possibility of observing the formation of molecules on actual dust grains through the detection of vibrationally hot molecules in the ISM.

At the end of the '90s, we began a laboratory program to investigate how these competing processes would play out in the formation of H_2 under ISM conditions; we also wanted to learn what kind of dust grain surfaces would be most efficient in the formation of molecular hydrogen. A study of the formation of H_2 should also give us clues on the formation of more complex molecules on dust grain surfaces via hydrogenation reactions.

The first experiments were done on the polished surface of polycrystalline olivine (Pirronello *et al.* 1997a; Pirronello *et al.* 1997b.) They were followed by measurements of H_2 formation on amorphous carbon (Pironello *et al.* 1999), and amorphous water-ice layers (Manicó *et al.* 2001; Roser *et al.* 2002). In 2003, Roser *et al.* (2003) and, independently, another group, Hornekær *et al.* (2003), measured the time-of-flight of molecules that were just formed on the surface and were being ejected from it. The next step, which is already under way (Perry *et al.* 2002), is to study the distribution of the energy of the molecule within the ro-vibrational degrees of freedom. There are efforts under way to measure the excitation of the molecule in actual ISM environments; preliminary studies have not been successful, however, in singling out excitation of the molecules that can be ascribed to molecule formation events and not to pumping mechanisms due to shocks and photon absorption (Takahashi & Uehara 2001; Tiné *et al.* 2003).

In the next section, we give a brief overview of the experimental methods; the reader is referred to a recent publication (Vidali *et al.* 2005) for technical details. In § 3, we summarize the results; a discussion of the significance of these results and how they can be applied to ISM environments is given in § 4, together with comments about experiments conducted in other laboratories.

2. Experimental Methods

The experiments conducted in our laboratory at Syracuse University, N.Y. were designed to study the processes leading to the formation of molecular hydrogen on surfaces under conditions that come as close as technically feasible to the ones in the most relevant ISM environments. The main requirements are low ($\sim$10 K) sample temperature, low background pressure ($\sim$$10^{-10}$ torr, equivalent to the typical number density in dense clouds), use of non contaminating materials and techniques to preserve the integrity of samples during the the experiments, and low fluxes of H atoms. Except for the last condition, the others can be easily met with the employment of ultra-high vacuum and standard low-temperature techniques. We used atomic beam techniques to deliver an angular resolved (10^{-6} steradian), low flux (10^{12} atoms sec^{-1} cm^{-2}), low energy (200–300 K) beam of H atoms to the sample. To avoid that the signal of successful events (i.e., the detection of H_2) be overwhelmed by the contribution from the undissociated fraction of the beam (about $10-15\%$) or background gas (mostly H_2 in a UHV system), we used two beams simultaneously, one of H and the other of D atoms. In this case, HD can form only on the sample surface and we obtain a large gain in sensitivity.

The apparatus consists of two triply differentially pumped beam lines, each equipped with a radiofrequency dissociation source, a scattering chamber housing the sample and a mass-discriminating detector that can be rotated around it. There is also a time-of-flight appendix connected to the sample chamber and equipped with its own detector. The background pressure in the main chamber is in the 10^{-10} torr range and the sample can be cooled down to $\sim$ 4.5 K. The experiment consists of sending H and D atoms onto the sample for a fixed amount of time (exposure); at the same time, the products of reaction are detected using a mass-discriminating detector. At the end of the exposure, the sample temperature is quickly increased to either evaporate molecules that formed on the surface, or cause the diffusion of atoms on the surface with consequent reaction and ejection from the surface. The purpose of raising the temperature of the sample, called TPD or temperature programmed desorption, is to obtain information on the kinetics of the reaction and on its yield. After much experimentation, we find that two key features are essential for a successful experiment designed to obtain useful information about similar processes taking place in the ISM: the use of two isotopes, and of low flux. Still, our flux is several orders of magnitude higher than the ones in the ISM. Therefore, models are necessary to make predictions of H_2 formation utilizing laboratory data. For details, see Biham *et al.* (2005).

3. Results

Over the years, we have studied the formation of H_2 on the following solids that are representative (within the technical capability at that time) of the ones present in key ISM environments: a polished polycrystalline telluric sample of olivine, an amorphous carbon sample, and amorphous water ice layers. The information that we obtained can be summarized as follow:

• There is evidence that the reaction kinetics on the polycrystalline sample is of the second order at the lowest coverage (a small fraction of a layer). This means that the recombination rate depends quadratically on the surface density of the reactants. In the TPD experiments, second order kinetics is characterized by the gradual shift of the maximum of the peak to lower temperature as the coverage is increased (Pironello *et al.* 1997a; Chan *et al.* 1978). For other samples, amorphous carbon and amorphous ice, the features of the signal are broader and it is harder to separate effects due to the morphology of the surface from the signature of the kinetics. Nonetheless, in the case of desorption from low-densisty amorphous ice, at least one peak of the differential yield (dN_{HD}/dT) of HD shows a clear indication of second order kinetics (Perets *et al.* 2005).

• Thermal activation is necessary to get the reaction going, indicating that tunneling alone is not sufficient to assure the necessary diffusion at these exposures and sample temperature (see Fig. 1 and the discussion below).

• The range of temperature over which there is significant H_2 formation is rather narrow, and, in any case, is below $\sim$ 20 K, as shown in Figure 2.

From the study of the shape and position of the desorption rate *vs.* time (or temperature) —see Figure 1 for an example, one can obtain, with the help of sets of rate equations, activation energy barriers for desorption and diffusion of atoms and molecules (Katz *et al.* 1999; Peretz *et al.* 2005).

Further evidence that the formation of HD is initiated by thermal energy is given in Figure 1 which compares the differential HD yield (dN_{HD}/dT), as a function of the sample temperature T during a TPD, obtained in two different experiments on the same

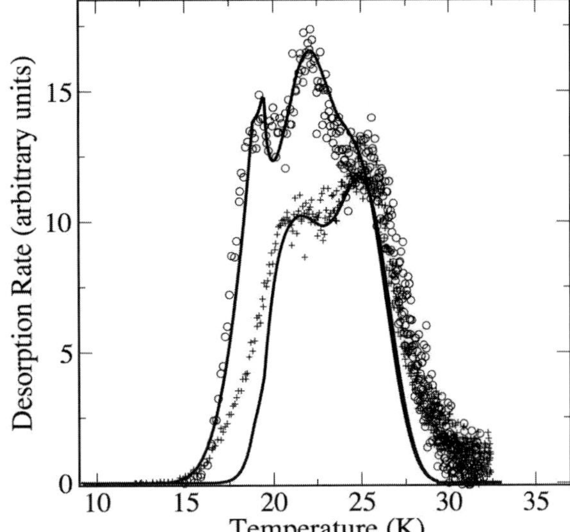

Figure 1. TPD traces of HD desorption after irradiation with HD molecules (○) and H+D atoms (+) on amorphous low-density water ice. The solid lines are fits obtained by a rate equations model. See Perets *et al.* (2005) for details.

surface of low density amorphous ice (Perets *et al.* 2005). In one, the sample is exposed to the H and D beams and then the sample temperature is raised quickly, and the HD molecules ejected from the surface are detected. In the other, the sample is dosed with HD molecules and then the sample is heated as before. If H and D were very mobile and reacted quickly making HD, then the heat pulse should give the same pattern in both experiments, since these would be experiments of desorbing HD molecules already present on the surface. On the other hand, if, in the first experiment, H and D atoms are not mobile and heat is necessary to make them diffuse and react, the desorption pattern is expected to be different, because in this case it probes the end result of three different processes: diffusion of H and D, reaction and desorption of HD; in the other, it probes the desorption of HD only. Indeed, the former is what is observed.

Experimentally, from measurements such as the ones described above, and using no assumption, one can get the recombination efficiency (defined as the probability to form a molecule for each two atoms sent onto the surface) at different sample temperature, as shown in Figure 2. To obtain it, the beam fluxes are measured first. Then TPD traces, such as those in Figure 1 are integrated to obtain the total number of molecules detected. The ratio of these two quantities (corrected for instrumental effects and for the fact that other reaction pathways are possible) gives the recombination efficiency, a number between 0 and 1. Strictly speaking, there is also a contribution to the HD yield due to H and D atoms reacting during the time of the exposure and forming HD molecules that are released in the gas phase. These molecules are found to contribute little to the yield coming from the subsequent TPD experiment.

To obtain the recombination efficiency under ISM conditions, one has to obtain the probability to form H_2 at ISM H fluxes on a grain of a given size; then one has to integrate over the distribution of grain sizes in the ISM. This procedure is described elsewhere in these proceedings (Biham *et al.* 2005) and by Vidali *et al.* (2005), Perets *et al.* (2005), Katz *et al.* (1999).

Recently, Hornekær *et al.* (2003) presented interesting results on H+D recombination on porous and non-porous amorphous solid water (ASW). Their experiments on the formation of HD on amorphous water ice are similar to ours. Their atom beams had

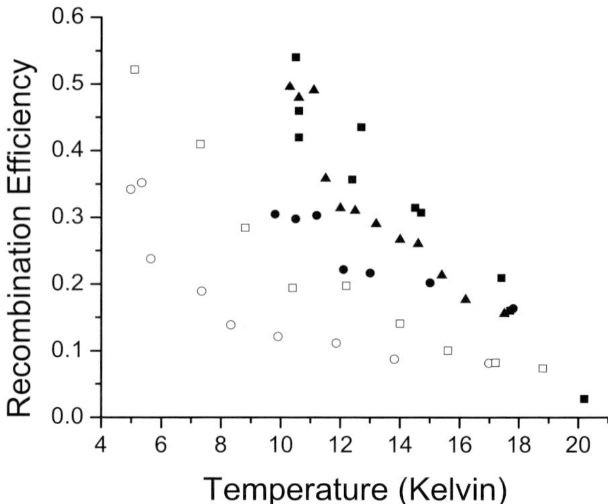

Figure 2. Recombination efficiency of molecular hydrogen *vs.* temperature of the sample. Filled circles are for high-density amorphous water ice, filled squares are for low-density amorphous water ice prepared by heating high-density amorphous ice. Filled triangles are for water vapor-deposited low-density amorphous ice. The scatter in the data points reflects the variability in the ice preparation methods. Empty circles are for polycrystalline olivine and empty squares for amorphous carbon. From Vidali *et al.* (2005). (Please note that the figure caption in Vidali *et al.* is incorrect.)

fluxes $\sim 10^{13}$ (atoms cm^{-2} sec^{-1}) about one order of magnitude higher than the ones used in our experiments, while the range of exposure times was comparable.

Hornekær *et al.* performed TPD experiments in which they irradiated H and D atoms either simultaneously or sequentially after waiting a delay time interval before dosing the other isotope. On porous ASW, their results are consistent with a recombination reaction occurring quickly during atom dosing due to the high mobility of the adsorbed atoms even at temperatures as low as 8 K. This high mobility was attributed either to quantum mechanical diffusion or to the "hot-atom" mechanism. Thermal activation is not likely to play a significant role at this temperature. This conclusion is sensible in light of the high coverages of H and D atoms used in these experiments, which required the adsorbed atoms to diffuse only for very short distances before encountering each other. As already suggested by Pirronello *et al.* (2004a, 2004b), the hot-atom mechanism may be able to provide the required mobility. In this case H and D atoms retain a good fraction of their gas phase kinetic energy during the accommodation process, enabling them to travel on the ice surface and inside its pores for several tens of Angstroms (Buch & Zhang 1991; Takahashi & Uehara 2001), exploring several adsorption sites and recombining upon encountering atoms already adsorbed on the surface.

In our experiments, we used low fluxes of atoms and short irradiation times to have a low coverage of H atoms on the surface. Therefore, in our situation the H atom that has just landed on the surface (the "hot atom") and is bouncing around and losing energy, has a far smaller probability of finding another H atom on the surface than in the Hornekær experiments because of the much lower coverage in our case. This has been confirmed by doing a calculation using rate equations in which the probability of reaction to form H$_2$ is proportional to the region of surface spanned by the hot atom. By using values of the fluxes and exposure times employed in their experiments and in ours, as well as estimates from molecular dynamics simulations of the distance traveled by an H atom landing on an ice surface (Buch & Zhang 1991; Takahashi & Uehara 2001) we do find that in our case the probability of making H$_2$ on the cold surface without thermal activation is very small, but large in the case of Hornekær's experiment.

As observed experimentally by Hornekær *et al.*, the molecules remain trapped in the ice until the TPD is initiated. In our experiment, the atoms are too far apart from each

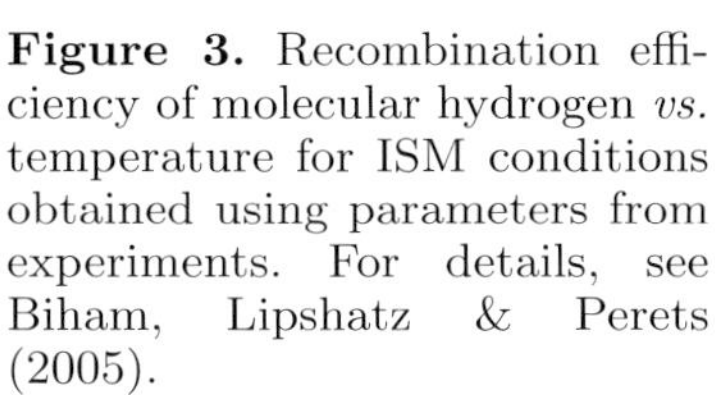

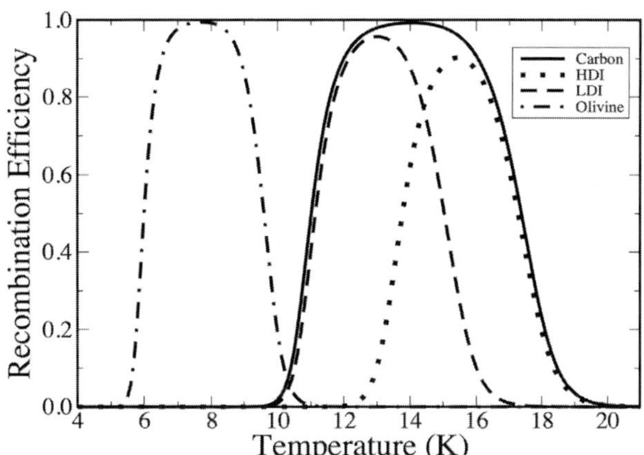

Figure 3. Recombination efficiency of molecular hydrogen *vs.* temperature for ISM conditions obtained using parameters from experiments. For details, see Biham, Lipshatz & Perets (2005).

other during the deposition phase, and the theoretical analysis confirms our original interpretation (Roser *et al.* 2002 for H_2 formation on ice; Pirronello *et al.* 1997a,b, and Katz *et al.* 1999 for H_2 formation on olivine and amorphous carbon), that atoms remain confined to their sites until the TPD is initiated. At higher coverage, but still less than one layer, first order kinetics is also observed on olivine (Pirronello *et al.* 1997a).

Finally, we did an experiment to determine the distribution of kinetic energies of hydrogen molecules forming and desorbing from the sample surface. An amorphous water ice sample was exposed to H and D atoms; the sample was turned towards the time-of-flight (t-o-f) line and a TPD was done. The molecules entering the t-o-f line were chopped by a high-speed slotted wheel. The particles emerging from the surface had Maxwell-Boltzmann distribution of velocities with an average kinetic energy corresponding to a surface temperature of ~ 24 K. A similar result was obtained independently by Hornekær *et al.* (2003). The interpretation of this result is that HD forms in the porous structure of amorphous ice and the emerging HD molecule makes enough collisions leaving the surface that it loses most of its kinetic energy (Vidali *et al.* 2004).

4. Discussion and Astrophysical Implications

We have presented a summary of experimental results on H_2 formation obtained by exposing atomic hydrogen to surfaces belonging to the three classes that can be considered good analogues of interstellar grains (silicates, carbonaceous particles, and ices). The data and theoretical modeling show that, depending on the experimental conditions, the L-H mechanism or the hot-atom/Eley-Rideal mechanisms are behind the formation of molecular hydrogen in several astrophysically relevant situations. Most of the experiments performed at low temperature (5–30 K) show that when physisorption sites are involved, hydrogen atoms that are accommodated on the surface recombine with high efficiency (close to 1) but only in a narrow temperature range through the L-H mechanism, the only one that can work when H adatoms are on the average well separated from each other and when a relatively high mobility is required for them to encounter and react. Experimental evidence (such as the detection of second order kinetics in the H + D $\rightarrow$ HD reaction and the difference between experiments done with H and D atoms and with HD molecules; see Figure 1) suggests that the mobility of adatoms that is necessary for an efficient reaction must be given by either thermal hopping or by what we call "thermally assisted tunneling", while the unassisted temperature-independent tunneling seems to be ruled out because experiments performed at liquid helium temperature show the need of thermal activation to obtain the mobility of adatoms (see Pirronello *et al.* 1997a). The drop in efficiency of this mechanism at "high temperature" (which is ~ 20 K for the studied surfaces so far) is due to the increasing importance of evaporation of the adatoms, which limits the residence time on the surface.

Whenever the "coverage" of H adatoms is high enough, the hot-atom/Eley-Rideal mechanism becomes efficient. In this case, an atom coming from the gas phase during the accommodation process (and hence during the multiple collisions of atoms with the surface) explores several adsorption sites and when it hits an H adatom it reacts with it forming H_2. At "high temperature" the efficiency of this mechanism also drops due to the shortening of the residence time of atoms on the surface which makes difficult to maintain a high enough coverage. Data presented by different groups on the formation of H_2 on amorphous water ice surfaces are in substantial agreement, but the interpretation differs considerably. Essentially, Hornekær *et al.* (2003) claim that there is fast diffusion even at low surface temperature (8 K) with the consequence that HD forms during irradiation. We proposed (Pirronello *et al.* 1997a) that this fast recombination can be explained by the hot-atom mechanism. This mechanism is expected to be efficient in the case of high coverage, which is the experimental condition used by Hornekær *et al.* for amorphous water ice and by Pirronello *et al.* for the measurements on olivine at high coverage (Pirronello *et al.* 1997a).

The implications of the reported results in the astrophysical environments can be summarized in the following way. Formation of molecular hydrogen in dense clouds or in clumps is well explained by the L-H mechanism, because the temperature of grains is well inside the range of values in which mobility of adatoms is high. In very cold regions (with temperature below 9–10 K) the hot-atom/Eley-Rideal mechanism allows efficient formation of molecular hydrogen, because on grains the coverage of H atoms reaches high values due to the lack of evaporation of adatoms and the low efficiency of the L-H mechanism in producing H_2 that could leave the grain.

Ongoing experiments on amorphous olivines could verify whether the observed trend (that recombination is higher on low-temperature amorphous surfaces than on polycrystalline ones) is correct or not.

In the case of astrophysical environments where the temperature of grains is higher yet, for example in PDRs, the possibility of formation of H_2 relies on the presence of chemisorption sites. Most of the discussion given above applies also in the case of chemisorption sites, both regarding the L-H mechanism (in an appropriate range of grain temperature), and the hot-atom/Eley-Rideal mechanism. The most relevant merit of chemisorption sites is the high binding energy (of the order of an eV) which extends the residence time of H adatoms against evaporation even when the grain temperature is in the hundreds of degree K. Of course it comes with drawbacks. In this case an energy barrier to enter the chemisorption site (which could be of the order of a few tenths of eV) might be present. Such a barrier would impede low-energy atoms to chemisorb on the surface. This is not a problem in PDRs, but obviously it gives a prominent role to physisorption sites at low temperature.

In summary, the importance of dust grain surfaces in the catalytic formation of H_2 has long been recognized, but only recently, beginning at the end of the 1990s, has it been possible to do experiments to better understand this important process. As data from different surfaces and under different conditions were obtained, it became clear that different mechanisms operate depending on the conditions chosen for the experiments. In the end, a clear picture has emerged. The formation of H_2 at low temperature on dust grain analogues requires thermal activation, and is dominated by the L-H mechanism at low coverage and the hot-atom mechanism at higher coverage, close to monolayer completion. This can explain both ours and Hornekær *et al.*'s data. We also learnt that theoretical tools such as rate equations (see Biham *et al.* 2005) are necessary to extract physical parameters from the experimental data, which are then used to predict how the reaction will go under truly ISM conditions.

5. Acknowledgments

This work was supported by NASA through grants NAG5-9093 and NAG5-11438 (G.V) and by the Italian Ministry for University and Scientific Research through grant 21043088 (V.P). We thank A. Lipshatz and H. Perets for the preparation of Figure 3.

References

Biham, O., Lipshatz, A., & Perets, H.B. 2005, in this volume
Buch, V. & Zhang, Q. 1991, *Ap. J.* 379, 647
Chan, C.M., Aris, R., & Weinberg, W.H. 1978, *Appl.Surf.Sci.* 1, 360
Gould, R. J. & Salpeter, E. E. 1963, *Ap. J.* 53, 79
Herbst, E. 2001, *Chem. Soc. Rev.* 30, 168
Hollenbach, D. & Salpeter, E.E. 1970, *J. Chem. Phys.* 53, 79
Hollenbach, D. & Salpeter, E.E. 1971a, *Ap. J.* 163, 155
Hollenbach, D., Werner, M.W., & Salpeter, E.E. 1971b, *Ap. J.* 163, 165
Hornekær, L., Baurichter, A., Petrunin, V.V., Field, D., & Luntz, A.C. 2003, *Science* 302, 1943
Katz, N., Furman, I., Biham, O., Pirronello, V., & Vidali, G. 1999, *Ap. J.* 522, 305
Manicó, G., Raguni, G., Pirronello, V., Roser, J.E., & Vidali, G. 2001, *Ap. J.* 548, L253
Perets, H.B., Biham, O., Manicó, G., Pirronello, V., Roser, J., Swords, S., & Vidali, G. 2005, *Ap. J.*, in press
Perry, J.S.A., Gingell, J.M., Newson, K.A., To, J., Watanabe, N., & Price, S.D. 2002, *Meas. Sci. Tech.* 106, 8998
Pirronello, V., Liu, C., Shen, L., & Vidali, G. 1997a, *Ap. J.* 475, L69
Pirronello, V., Biham, O., Liu, C., Shen, L., & Vidali, G. 1997b, *Ap. J.* 483, L131
Pirronello, V., Liu, C., Roser, J.E., & Vidali, G. 1999, *A&A* 344, 681
Pirronello, V., Manicó, G., Roser, J.E., & Vidali, G. 2004a, in *Astrophysics of Dust*, ed. A. Witt *et al.* (ASP Series, Sheridan Books Ann Arbor, MI, USA) p. 529
Pirronello, V., Manicó, G., Roser, J.E., & Vidali, G. 2004b, *The Dense Interstellar Medium in Galaxies*, ed. S. Pfalzner *et al.* (Springer-Verlag, Berlin) p. 525
Roser, J.E., Manicò, G., Pirronello, V., & Vidali, G. 2002, *Ap. J.* 581, 276
Roser, J.E., Swords, S., Vidali, G., Manicó, G., & Pirronello, V. 2003, *Ap. J.* 596, L55
Smoluchowski, R. 1981, *Ap. and Space Sci.* 75, 353
Smoluchowski, R. 1983, *J. Phys. Chem.* 87, 4229
Takahashi, J., & Uehara, H. 2001, *Ap. J.* 561, 843
Tine, S., Williams, D.A., Clary, D.C., Farebrother, A.J., Fisher, A.J., & Meijer, A.J.H.M. 2003, *Ap. Space Sci.* 288, 377
Vidali, G., Roser, J.E., Manicó, G., Pirronello, V., Perets, H.B., & Biham, O. 2005, *J. of Phys.: Conf. Series* 6, 36
Vidali, G., Roser, J.E., Manicó, G. & Pirronello, V. 2004, *J. Geophys. Res.* 109, E07S14
Whittet, D.C.B. 2003, *Dust in the Galactic Environment* (Cambridge University Press)

Discussion

HORNEKÆR: Since the LDA ice is also less porous (due to annealing) than the high-density ice, a difference between the TPD spectra for HD from recombination and HD from molecular deposition could be due to the immediate desorption of newly formed HD from low-energy binding sites. The case is different for high-density, porous ices. In this case I would expect the TPD spectra from directly deposited and atomically recombined HD to be identical. Is this the case in your data?

VIDALI: As to the first part of the question, the fact that there is evidence of second-order kinetics, plus the shape of the TPDs, strongly suggest that the formation of HD requires thermal activation. As to the second part of the question, we have not studied the high-density ice with the same detail as the low-density one. The reason is that the high-density ice is prepared at 10 K and the ice is never taken above ~32–34 K. We suspect that there is a dependence of the morphology of the ice on subtle changes in the preparation methods, and thus it is harder to obtain a "characteristic" high-density ice. Therefore, we concentrated most of our attention on the study of recombination processes on low-density ice.

ALLAMANDOLA: These are beautiful experiments and very hard to do. Do you have an idea of what will happen if more realistic ices are used, for example containing CO?

VIDALI: In the experiments we have done so far, we tried to keep it simple in order to get the physics of the processes right. It is of great interest to study more realistic ices. I suspect that the presence of CO would complicate matters considerably from an experimental point of view, because CO sublimates easily, but it would be definitely an interesting ice mixture to study.

Astrochemistry: Recent Successes and Current Challenges
Proceedings IAU Symposium No. 231, 2005
D.C. Lis, G.A. Blake & E. Herbst, eds.

© 2006 International Astronomical Union
doi:10.1017/S1743921306007381

Millimeter-wave Observations of Gaseous Species in Disks

Geoffrey A. Blake

Division of Geological & Planetary Sciences, California Institute of Technology,
Mail Stop 150–21, Pasadena, CA 91125, USA
email: gab@gps.caltech.edu

Abstract. The role of high spectral and spatial resolution spectroscopy in understanding the evolution of the gaseous component of circumstellar accretion disks is described. Millimeter-wave emission lines from trace constituents such as CO, CN, HCO^+, and HCN can be used to probe the kinematic and physico-chemical properties in the near-surface regions of disks beyond 50–100 AU, but, thanks to extensive depletion in the midplane, they are not a reliable proxy for the disk mass. For the special case of nearly edge-on circumstellar disks, the resulting ices can be directly observed through mid-infrared spectroscopy, using ground based facilities or spacecraft (*cf.* Najita, this volume). Emerging and planned millimeter-wave $\rightarrow$ THz arrays will possess sufficient sensitivity and resolution to probe much closer to the central star and to search for prebiotic compounds such as those detected in comets, meteorites and interplanetary dust particles.

Keywords. astrochemistry — solar system: formation — stars: planetary systems: formation — stars: planetary systems: protoplanetary disks

---

1. Introduction

As the means by which matter and angular momentum are exchanged between molecular clouds and young stars and as the birth sites of planets, circumstellar disks provide the pivotal observational link between star formation and (exo)-planetary science. Though disks must be established very early in the star formation process, this overview is concerned primarily with those systems in which the vast majority of accretion has been completed. As the accretion process wanes, reprocessing of the central stellar radiation becomes the dominant source of energy release, and in these so-called passive circumstellar disks an examination of the re-radiated longer-wavelength photons can be used to learn a great deal about the physical properties of the disk.

The basic geometry is outlined in Figure 1. Material near the star and especially at the disk surface radiates strongly in the infrared, while the more distant and interior portions of the disk can best be probed at longer millimeter wavelengths where the bulk of the dust is optically thin. Near- and mid-infrared imaging can efficiently search for the presence or absence of such excess emission, and has been used to infer the presence of disks around hundreds of young stars. Infrared surveys also yield time scale estimates of order a few million years for the disappearance of small dust grains in the inner regions of circumstellar disks (Skrutskie *et al.* 1990; Robberto *et al.* 1999) – a span similar to that measured for the creation of differentiated planetesimals in the early solar nebula from the absolute ages of chondritic meteorites (Amelin *et al.* 2002). The time scale and associated mechanism(s) by which individual optically thick disks dissipate to become the optically thin, or debris, disks are not yet quantitatively known, but are exceedingly important to nebular chemistry and to the formation of planetary systems. The more tenuous debris disks themselves are estimated to survive for several tens, perhaps hundreds, of millions of

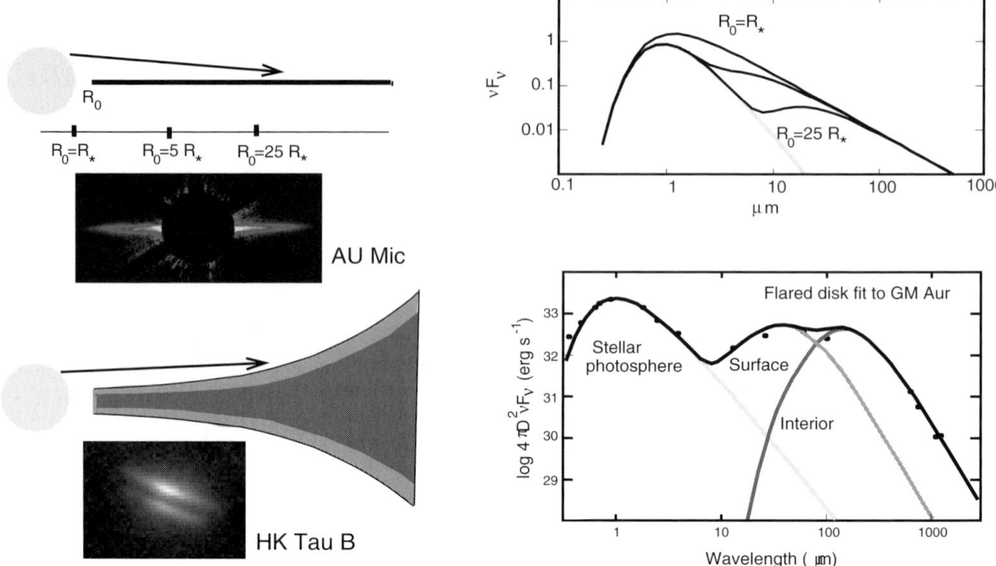

Figure 1. Schematic disk structure cross sections for passive flat *vs.* flaring (hydrostatic equilibrium) disks, and the spectral energy distributions (SEDs) that result. Conceptual models of the latter predict that the infrared portion of the SED is dominated by optically thin radiation from the disk surface layers exposed to the stellar radiation field, while the millimeter-wave emission is dominated by the cooler disk interior (Chiang & Goldreich 1997). Adapted from van Zadelhoff (2002), with inset images of AU Mic and HK Tau C from the Keck telescope (Liu 2004) and HST (Stapelfeldt *et al.* 1998), respectively. The top right panel illustrates the changes in the SED for successively larger inner holes in the dust distribution.

years in at least 15% of young main sequence stars, as judged from mid- and far-infrared continuum excesses of nearby stars (Rieke *et al.* 2005).

The spatio-temporal properties of disks as inferred from SEDs, namely masses of 0.001–0.1 $M_\odot$ and sizes of 10's to 100's of AU, are similar to those inferred for the Minimum Mass Solar Nebula (MMSN; Hayashi 1981); and have been verified for a *handful* of objects by detailed imaging, with millimeter-wave CO studies providing among the earliest and most conclusive evidence for the expected ∼Keplerian velocity fields (see Koerner *et al.* 1993; Dutrey *et al.* 1994). Direct optical/IR imaging and, especially, interferometry (Monnier & Millan-Gabet 2002) has begun to probe closer to the central stars, but cannot directly constrain critical disk properties such as the surface mass density or accretion flows, while at present millimeter-wave studies can only probe the outer regions of disks. Similarly, disk dissipation studies based on infrared excesses tell us little about the gas content of such objects. The discovery of extrasolar 'hot Jupiters' some 0.05 AU from their parent stars (Marcy, Cochran & Mayor 2000) has highlighted the important role of star-disk-protoplanet interactions and time scales, and demands new tools that can investigate the critical planet-forming zone of disks. As described by Najita (this volume), mid-infrared spectroscopy provides one current route to studies of gas in the inner disk. Below I briefly outline future possibilities with the millimeter-wave arrays presently under construction or expansion.

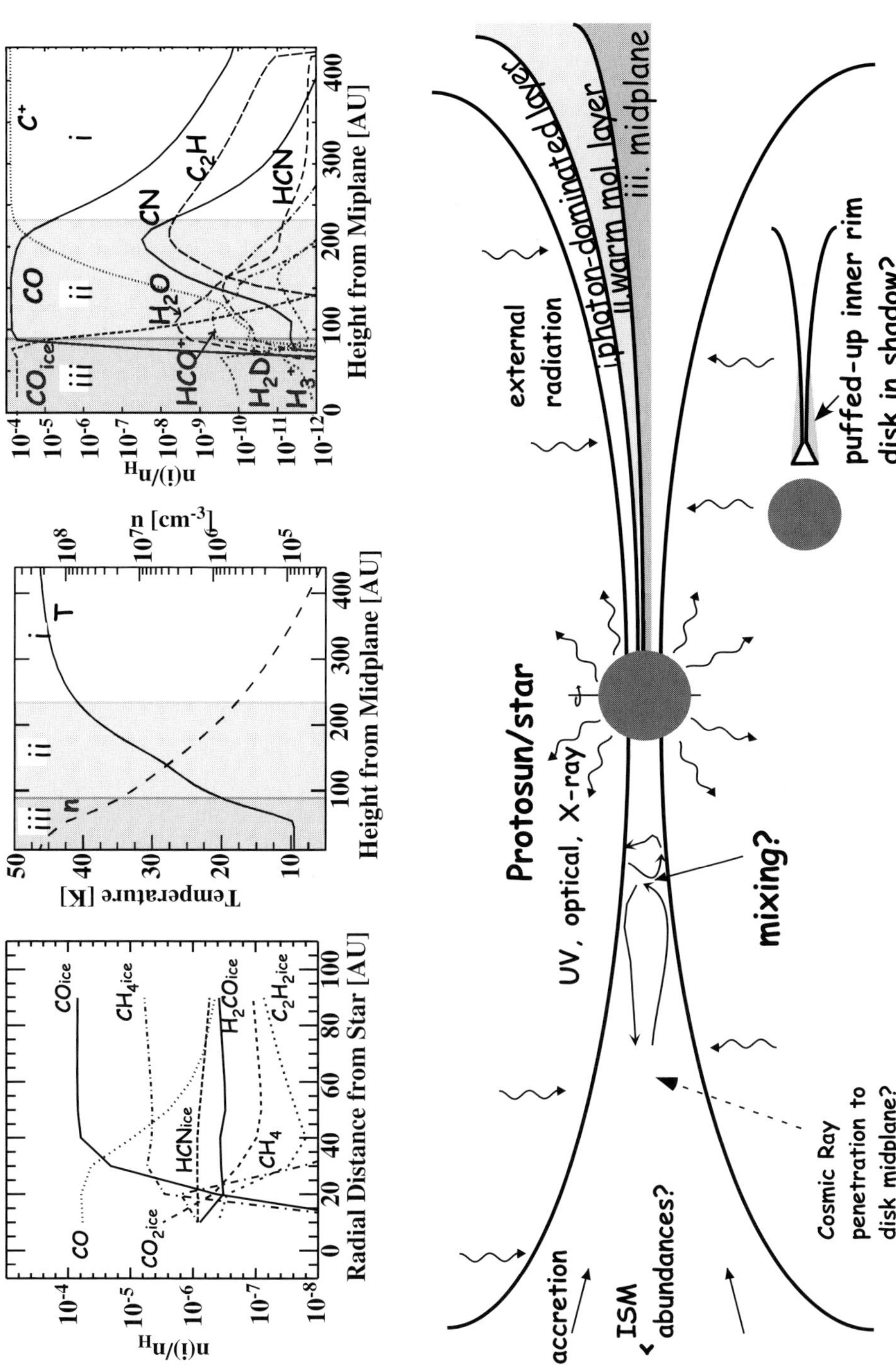

Figure 2. Chemical structure of protoplanetary disks. The disk is schematically divided into three zones: a photon-dominated layer, a warm molecular layer, and a midplane freeze-out layer. The CO freeze-out layer disappears at $r \lesssim 100$ AU as the mid-plane temperature increases. Various non-thermal inputs, cosmic ray, UV, and X-ray, drive chemical reactions. Viscous accretion and turbulence transport the disk material both vertically and radially. The *upper panel* shows the vertical distribution of molecular abundances at $r \sim 300$ AU in a typical disk model (van Zadelhoff *et al.* 2003) and a sample of the hydrogen density and *dust* temperature at the same distance (D'Alessio *et al.* 1999). In upper layers ($\gtrsim 150$ AU) the *gas* temperature will exceed the dust temperature by $\gtrsim 25$ K (Jonkheid *et al.* 2004).

2. The Basic Physical and Chemical Structure of Disks

What do disk structures such as those outlined in Figure 1 imply for chemistry? Molecular abundances are determined by both the physical conditions, such as density and temperature, and transport time scales (see Kamp *et al.*, Markwick, this volume). Recent years have seen significant progress in characterizing disk physical structure, which aids in understanding disk chemical processes. As shown below, disks that are well isolated from intense radiation fields can be quite extended with $R_{out} \sim$ few hundred AU (Simon, Dutrey, & Guilloteau 2000), much larger than expected from comparison with the MMSN. The radial distributions of midplane column density and temperature have been estimated by observing thermal emission of dust; they are fitted by a power law $\Sigma(r) \propto r^{-p}$ and $T(r) \propto r^{-q}$, with $p = 0 - 1$ and $q = 0.5 - 0.75$. At a distance of 1 AU, T(1 AU)$\sim$100–200 K, while Σ(100 AU)$\sim$0.1–10 g cm^{-2} (e.g., Beckwith *et al.* 1990).

The vertical physical structure is estimated by calculating the hydrostatic equilibrium for the density and radiation transfer for the temperature (see Kamp *et al.*, this volume). Beyond several AU the disk is mainly heated by the central star. The stellar radiation is absorbed by grains at the disk surface, which then emit thermal radiation to heat the disk interior (Calvet *et al.* 1992; Chiang & Goldreich 1997). Hence the temperature decreases towards the midplane, as shown in Figure 2. For small radii ($r <$ few AU), however, heating by mass accretion is not negligible and the midplane can be warmer than the disk surface. At $r = 1$ AU, for example, the midplane temperature can be as high as 1000 K, if the accretion rate is large (D'Alessio *et al.* 1999). The density distribution is basically Gaussian, $\exp[-(Z/H)^2]$, with some deviation due to vertical temperature variations (see Figure 2). As a whole, a gas-rich disk has a flared structure, with a geometrical thickness that increases with radius (Kenyon & Hartmann 1987). Gas-poor, or debris disks, may well have a much more flattened geometry.

Based on such physical models, the current picture of the general disk chemical structure is schematically shown in Figure 2. At $r \gtrsim 100$ AU, the disk can be divided into three layers: the photon dominated region (PDR), the warm molecular layer, and the midplane freeze-out layer. The disk is irradiated by UV radiation from the central star and ISM that drives ionization and dissociation in the surface layer. In the midplane the temperature is mostly lower than the freeze-out temperature of CO (~ 20 K), one of the most abundant and volatile molecules in the ISM. Since the timescale of adsorption onto grains is short at high density ($\sim 10\,(10^9 \mathrm{cm}^{-3}/n_\mathrm{H})$ yr), heavy-element species are significantly depleted onto grains. At intermediate heights, the temperature is several tens of K, and the density is sufficiently high ($\gtrsim 10^6$ cm^{-3}) to ensure the existence of molecules even if the UV radiation is not completely attenuated by the upper layer (Aikawa & Herbst 1999; Aikawa *et al.* 2002). Here water is still frozen onto grains, trapping much of the oxygen in ices. Thus, the warm CO-rich gas layers will have C/O ~ 1, leading to a rich and extensive carbon-based chemistry (Willacy & Langer 2000).

Do such models provide a good match to observed abundances? The observational challenges of answering this question are highlighted in Figure 3, in which it can be seen that even the strongest lines from the most optically thick species are strongly beam diluted by the resolution achievable with single-dish telescopes that can operate over a wide frequency range. The combination of millimeter-wave imaging, described next, with submillimeter-wave spectroscopy forms a powerful means of constraining the temperature, density, and molecular abundances in disks (van Zadelhoff *et al.* 2001). Work in this area began with the pioneering observations of DM Tau and GG Tau with the IRAM 30-m telescope (Dutrey *et al.* 1997) and TW Hya with the JCMT (Kastner *et al.* 1997). Further detections of the higher-J lines of high dipole moment species such

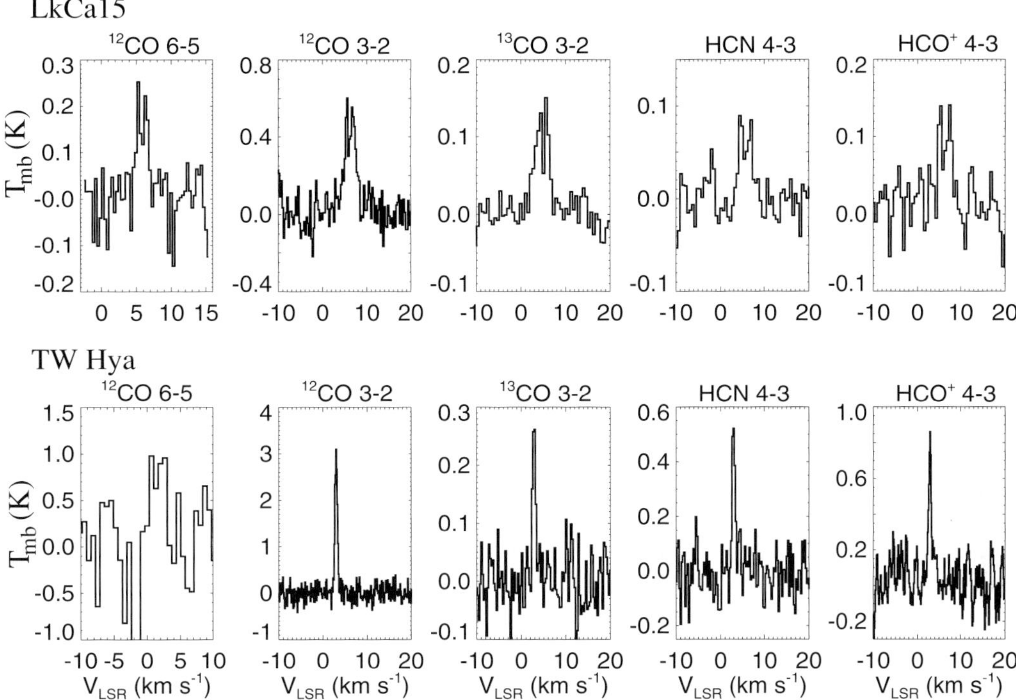

Figure 3. Submillimeter-wave emission lines from the LkCa 15 and TW Hydra circumstellar disks, observed with the JCMT and CSO (van Zadelhoff *et al.* 2001). The double-peaked nature of the emission lines from LkCa 15 is consistent with Keplerian rotation in an inclined disk ($i \sim 60°$); while TW Hydra is thought to be viewed nearly face on ($i \sim 7°$), hence the narrow lines.

as HCN and HCO$^+$ along with statistical equilibrium analyses demonstrated that the line emission indeed arises from the warm molecular layer with $n_{\mathrm{H_2}} \approx 10^6$–$10^8$ cm^{-3}, $T \gtrsim 30$ K (van Zadelhoff *et al.* 2001). In agreement with theory, the gas-phase abundances are generally lower than in dense clouds, but the CN/HCN ratio is higher (see also Thi *et al.* 2004). The low molecular abundance is caused by depletion in the midplane, and the high CN/HCN ratio originates in the surface PDR (*cf.* Figure 2), as seen in PDR's in the ISM (Rodríguez-Franco *et al.* 1998).

At $r \lesssim 100$ AU, the midplane temperature is high enough to sublimate various ice materials that formed originally in the outer disk radius or parental cloud core (e.g., Markwick, this volume). This sublimation will be species dependent with the "frost line" for a given species appearing at different radii. For, example in the solar nebula the water ice frost line appeared near 3–5 AU, while the CO snow line would appear at greater distances where the midplane dust temperatures drop below $\sim$20 K. Within these species-selective gaseous zones, sublimated molecules will be destroyed and transformed to other molecules by gas-phase reactions. In this fashion, the chemistry is similar to the so-called "hot core" chemistry, which appears in star-forming cores surrounding protostars (e.g., Hatchell *et al.* 1998; Liu, Hollis contributions to this volume). For example, sublimated CH$_4$ is transformed to larger and less volatile carbon-chain species, which can then accumulate onto grains (Aikawa *et al.* 1999). The very deepest single-dish integrations have begun to reveal more complex species (Thi *et al.* 2004), though the larger organics often seen toward hot cores/corinos remain out of reach of existing single-dish telescopes.

3. (Sub)Millimeter-wave Imaging Spectroscopy of Disks

From studies such as those outlined above, a suite of disks have been identified that are many arcseconds in diameter, either because they are nearby (TW Hya) or are intrinsically large (AB Aur). The evolutionary status of the individual objects can be difficult to obtain, but is pivotal to a proper interpretation of the molecular observations. As first argued by Hogerheijde (2002) and described further in the posters by Brinch *et al.*, Guilloteau *et al.*, and Semenov *et al.* at this meeting, the very large structures seen in dust and molecular line emission around L1489 and AB Aur (see Figure 4) may provide examples of sources in which the dynamic accretion that characterizes embedded protostars is giving way to the centrifugal support that dominates more mature star+disk systems. Such systems would not only provide estimates of the 'initial conditions' for studies of disk chemistry, at high spatial resolution they would provide stringent constraints on the mechanisms of angular momentum redistribution in star and planet formation.

For stars that are closer to the main sequence (GM Aur, LkCa 15, DM Tau), the age, large size, and masses of their surrounding disks make them important for further study since they may represent an important transitional phase, in which viscous disk spreading and dispersal competes with planetary formation processes. At present, aperture synthesis observations can only sense the line emission from the outer disk ($r > 30 - 50$ AU; see Dutrey & Guilloteau 2004; Dutrey *et al.* 2005) for stars in the nearest molecular clouds. Thus, the *chemical* imaging of disks is rarer still, with studies concentrating on a few of the best characterized T Tauri and Herbig Ae stars. Imaging studies of LkCa 15, for example, have detected a number of isotopologues of CO along with the molecular ions HCO^+ and NNH^+ and the more complex organics formaldehyde and methanol (Duvert *et al.* 2000; Qi *et al.* 2003). For this disk at least, molecular depletion of molecules onto the icy mantles of dust grains near the disk midplane is found to be extensive, but the fractional abundances and ionization in the warm molecular layer are in line with those seen toward dense PDRs. While the lines from the less abundant species can be detected (*cf.* Figure 4), they were too weak to image with good signal-to-noise. Thus, while millimeter-wave rotational line emission is a good tracer of the outer disk velocity field it is not a robust tracer of the mass unless the chemistry is very well understood.

The detection of ions such as HCO^+ is of particular interest thanks to the appreciation of the Magneto-Rotational Instability (MRI) as a potential mechanism for disk angular momentum transport (e.g., Stone *et al.* 2000). Chemistry should therefore be linked to disk dynamics as the presence of ions is necessary to couple the gas to the magnetic field. As shown by Figures 1 and 2, since most of the ionization processes are active on the surface, there exists the potential that accretion may only be active near the surface (Gammie 1996).

A central question, much discussed at this meeting, is whether cosmic rays penetrate through the possibly intense winds from young stars to reach the inner disk. Within our own planetary system, the Solar wind excludes ionizing cosmic rays. If cosmic rays do penetrate, the primary charge carriers range from metal ions or grains at small radii (~ 1 AU) to molecular ions for $r > $ a few AU with $x_e \sim 10^{-13}$ near the midplane. If ionizing cosmic rays are excluded, radionuclides can produce $x_e \sim 1.3 \times 10^{-8} (T/20 \text{ K})^{-0.5}/\sqrt{n_H}$ (assuming H_3^+ as the dominant ion). In dense protostellar cores, models and observations now suggest near total freeze-out of volatiles that results in D_3^+ and other forms of deuterated H_3^+ becoming important charge carriers (Roberts *et al.* 2004; Walmsley *et al.* 2004). The disk midplane should present a similar environment. The detection of H_2D^+ by Ceccarelli *et al.* (2004) in the outer disks of TW Hya and DM Tau supports this view, and is consistent with cosmic ray ionization.

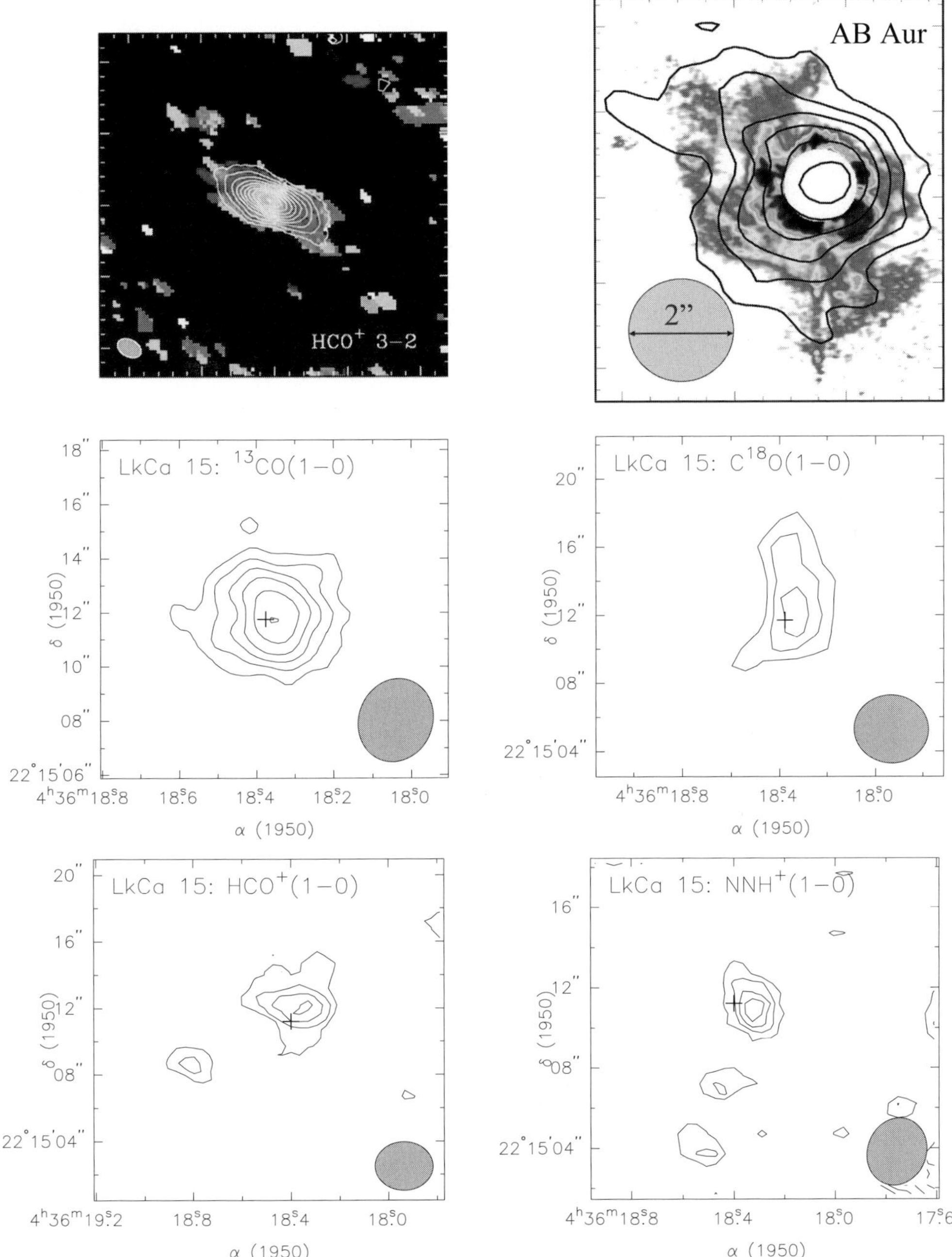

Figure 4. Integrated intensity millimeter-wave images of several disk systems. *Top:* BIMA HCO$^+$ J=3–2 image of L1489, and an OVRO image of the 3 mm dust continuum from AB Aur (contours) overlaid on the H-band scattered light image of Fukagawa *et al.* (2004). Figures adapted from Hogerheidje (2002) and Corder *et al.* (2005), respectively. *Bottom:* OVRO Millimeter Array aperture synthesis images of the ^{13}CO, C^{18}O, HCO$^+$, and N$_2$H$^+$ J=1–0 emission from the disk encircling LkCa 15 (Qi *et al.* 2003). Hatched ellipses at lower right present the synthesized beams.

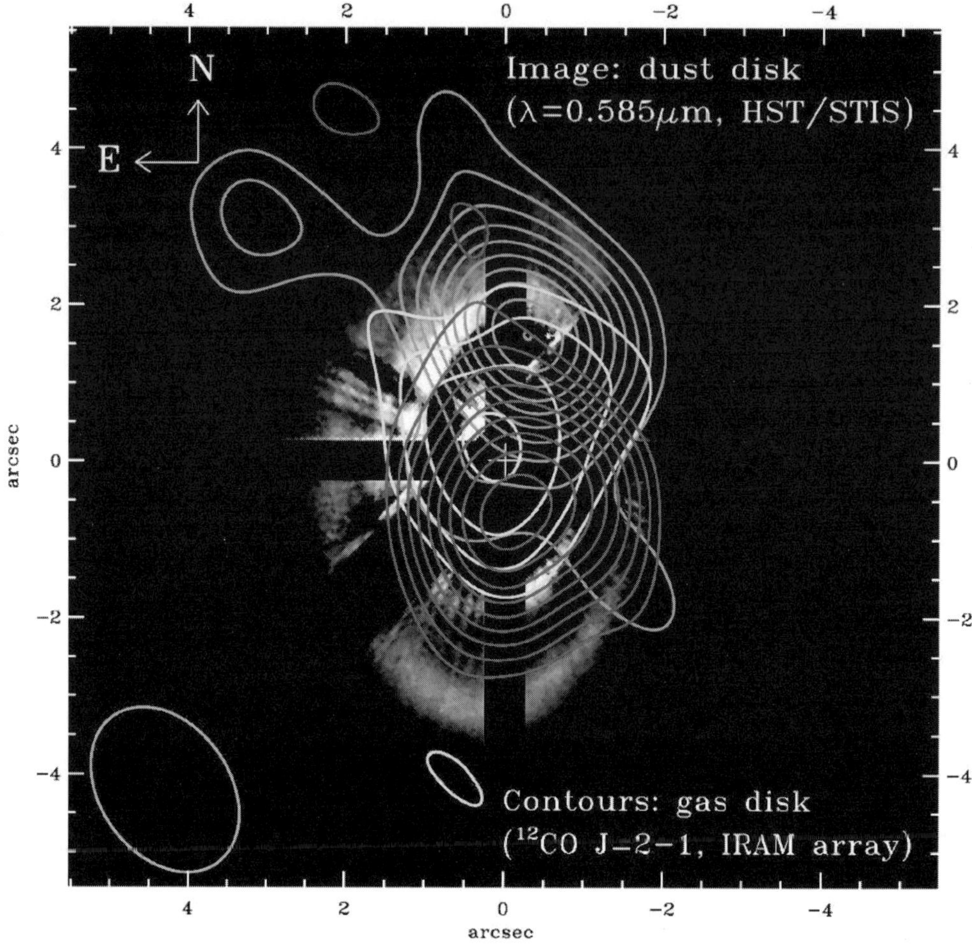

Figure 5. Plateau de Bure image of the CO emission from the transitional disk candidate HD 141569A (greyscale image from HST). Yellow contours trace gas at the systematic velocity, those in red and blue trace the expected offsets along the major axis for a disk in Keplerian rotation (from Dutrey *et al.* 2005).

At still later evolutionary stages, disk dissipation will eventually leave a young star surrounded by an optically thin, dust dominated disk. Just how this process occurs, and over what timescale, is pivotal to the development of planetary systems. The inner disk can be drained by accretion, while photoevaporation is more likely important in the outer disk – especially for stars that are members of young clusters (Hollenbach *et al.* 2000). If holes are opened in the inner disk due to the combination of planet formation and accretion onto the central star, the near- through mid-IR SED can modified substantially. A number of these potentially transitional disks are known, and more are being discovered through extensive surveys being undertaken with the Spitzer Space Telescope. High-resolution imaging and spectroscopy of these putative transitional disks is pivotal, and Figure 5 illustrates the power of millimeter-wave aperture synthesis studies through Plateau de Bure observations of one such Herbig Ae star, HD 141569A (Dutrey

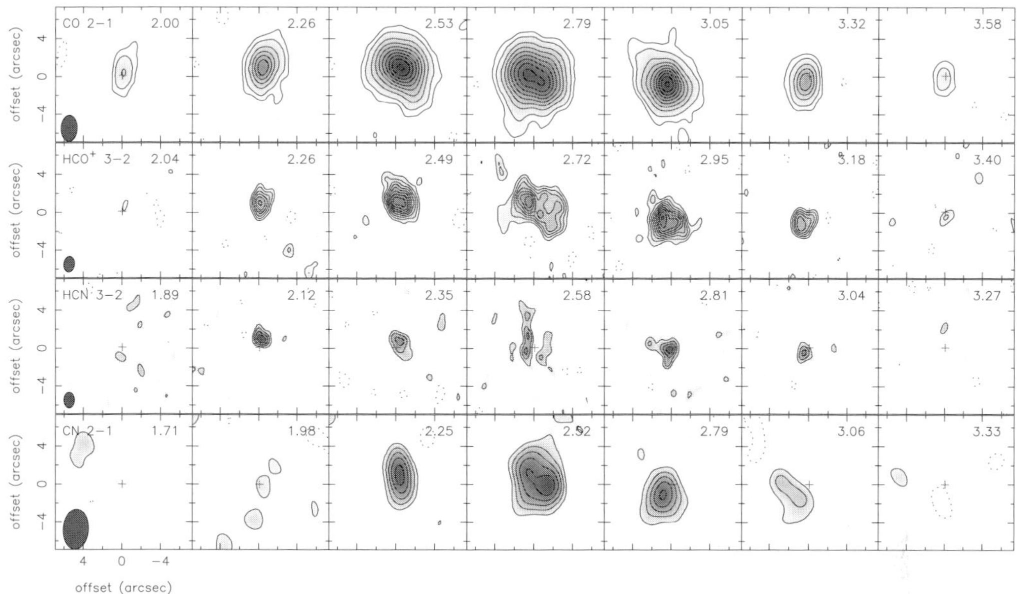

Figure 6. SMA channel maps of the millimeter and submillimeter spectral line emission from the TW Hya circumstellar disk. The ellipses at lower left in each series of panels display the synthesized beam, which for the HCN $J=3-2$ observations achieves an effective spatial resolution of $\sim$60 AU. Kindly provided by C. Qi in advance of publication (Qi *et al.* 2006, in preparation).

et al. 2005). For most such systems, however, the holes predicted by analyses of the SEDs lie on angular scales that cannot be studied with current arrays.

3.1. *Observations with New and Upgraded Arrays*

New observational facilities are poised to change our ability to probe the chemistry in disks dramatically, as illustrated by the recent SMA results on the TW Hya disk presented in Figure 6 (Qi *et al.* 2006, in preparation). At a distance of only 56 pc, observations of this source provide nearly 2–3 times the effective linear resolution of disk studies in Taurus. Thus, channel maps such as those presented can be used to derive a great deal about the physical and chemical structure of the disk – its size and inclination (Qi *et al.* 2004), the run of surface density and temperature with radius, the chemical abundance ratios with radius in the outer disk, etc. Ongoing improvements to existing arrays (SMA, PdBI, and CARMA in the north) will soon enable similar studies for a large number of disks, and will push the radii over which chemical studies can be pursued down to 10–20 AU. For transitional disks with large inner holes (or at least zones where dust grains have grown sufficiently large that this part of the disk becomes optically thin), such as that proposed to exist around GM Aur (Calvet *et al.* 2005), sub-arcsecond observations with CARMA should be able to verify the presence of an optically thin inner disk, as Figure 7 demonstrates. Resolving the vertical chemical gradients shown in Figure 2 and studying the chemistry in the 1–10 AU zone of active planet formation will require even greater sensitivity and spatial resolution, and awaits the development of ALMA. Studies of the critical water frost line that complement ongoing IR surveys (see Najita, this volume) are extremely challenging from the ground, but will be possible once SOFIA and Herschel are deployed over the next few years.

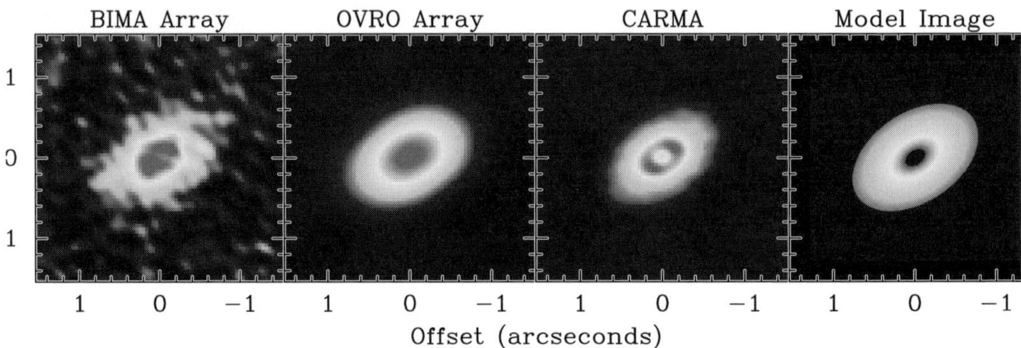

Figure 7. *Top:* Image of the fifteen telescopes of the CARMA array at Cedar Flat, CA in August 2005. The eight 3.5-m telescopes of the University of Chicago Sunyaev-Zel'dovich Array (SZA) will move to the CARMA site in 2007. *Bottom:* Simulated ν=230 GHz dust continuum imaging of a circumstellar disk at 150 pc distance using the BIMA, OVRO, and CARMA arrays in their two highest-resolution configurations, kindly provided by L.G. Mundy. The images are reconstructions of simulated observations of the model disk shown in the right panel, whose mass is 0.01 $M_\odot$. The outer radius is 120 AU, and there is an inner hole 20 AU in radius. The simulations included appropriate Gaussian noise levels for each array.

4. Conclusions

High angular resolution (sub)millimeter-wave spectroscopy of circumstellar disks provides a uniquely powerful probe of the chemistry and physics therein. While present instruments can only probe gas beyond the radius of our own Kuiper Belt in the nearest star+disk systems, the ongoing improvements to existing arrays (the SMA, CARMA, and PdBI) and particularly the deployment of ALMA herald a new epoch in which stringent observational tests of chemical and physical models will be possible. Further, the sensitivity of these arrays will be sufficient to trace the gas well into the era of disk dissipation, and will thereby provide an observational context for our emerging understanding of the processes that shape the diverse properties of (extrasolar) planetary systems.

References

Aikawa, Y. & Herbst, E. 1999, *Ap. J.* 526, 314
Aikawa, Y., Umebayashi, T., Nakano, T., & Miyama, S. M. 1999, *Ap. J.* 519, 705
Aikawa, Y., van Zadelhoff, G.J., van Dishoeck, E.F., & Herbst, E. 2002, *A&A* 386, 622
Amelin, Y., Krot, A.N., Hutcheon, I.D., & Ulyanov, A.A. 2002, *Science*, 297, 1678
Beckwith, S.V.W., Sargent, A.I., Chini, R.S., & Guesten, R. 1990, *A. J.* 99, 924
Calvet, N., Magris, G.C., Patino, A., & D'Alessio, P. 1992, *Rev. Mex. Astron. Astro.* 24, 27

Calvet, N., D'Alessio, P., Watson, D.M., Franco-Hernández, R., Furlan, E. *et al.* 2005, *Ap. J.* 630, L185

Ceccarelli, C., Dominik, C., Lefloch, B., Caselli, P., & Caux, E. 2004, *Ap. J.* 607, L51

Chiang, E.I. & Goldreich, P. 1997, *Ap. J.* 490, 368

Corder, S., Eisner, J., & Sargent, A.I. 2005, *Ap. J.* 633, L133

D'Alessio, P., Calvet, N., Hartmann, L., Lizano, S., & Cantó, J. 1999, *Ap. J.* 527, 893

Dutrey, A., Guilloteau, S., & Simon, M. 1994, *A&A* 291, L23

Dutrey, A., Guilloteau, S., & Guelin, M. 1997, *A&A* 317, L55

Dutrey, A. & Guilloteau, S. 2004, *Ap. Space Sci.* 292, 407

Dutrey, A., Lecavelier des Etangs, A., & Augereau, J.C. 2005, in Comets II, eds. M.C. Festou, H.U. Keller, H.A. Weaver (Tucson: Univ. Arizona), in press

Duvert, G., Guilloteau, S., Ménard, F., Simon, M., & Dutrey, A. 2000, *A&A* 355, 165

Fukagawa, M., Hayashi, M., Tamura, M., Itoh, Y., Hayashi, S.S. *et al.* 2004, *Ap. J.* 605, L53

Gammie, C.F. 1996, *Ap. J.* 457, 355

Hatchell, J., Thompson, M A., Millar, T.J., & MacDonald, G.H. 1998, *A&A. Supp. Ser.* 133, 29

Hayashi, C. 1981, *Prog. Th. Phys. Suppl.* 70, 35

Hogerheijde, M.R. 2002, *Ap. J.* 553, 618

Hollenbach, D.J., Yorke, H. W., & Johnstone, D. 2000, in Protostars & Planets IV, eds. V. Mannings, A. P. Boss, & S. S. Russell (Tucson: Univ. Arizona), 401

Jonkheid, B., Faas, F.G.A., van Zadelhoff, G.-J., & van Dishoeck, E.F. 2004, *A&A* 428, 511

Kastner, J.H., Zuckerman, B., Weintraub, D.A., & Forveille, T. 1997, *Science* 277, 67

Kenyon, S.J. & Hartmann, L. 1987, *Ap. J.* 323, 714

Koerner, D.M., Sargent, A.I., & Beckwith, S.V.W. 1993. *Icarus* 106, 2

Liu, M. 2004, *Science* 305, 1442

Marcy, G.W., Cochran, W.D., & Mayor, M. 2000, in Protostars & Planets IV, eds. V. Mannings, A.P. Boss, & S.S. Russell (Tucson: Univ. Arizona), 457

Monnier, J.D. & Millan-Gabet, R. 2002, *Ap. J.* 579, 694

Qi, C., Kessler, J.E., Koerner, D.W., Sargent, A.I., & Blake, G.A. 2003, *Ap. J.*, 597, 986

Qi., C., Ho, P.T.P., Wilner, D., Takakuwa, S., Hirano, N., *et al.* 2004, *Ap. J.* 616, L7

Rieke, G.H., Su, K.Y.L., Stansberry, J.A., Trilling, D., Bryden, G., *et al.* 2005, *Ap. J.* 620, 1010

Robberto, M., Meyer, M.R., Natta, A., & Beckwith, S.V.W. 1999, in The Universe as Seen by ISO, ESA SP-427, 195

Roberts, H., Herbst, E., & Millar, T.J. 2004, *A&A* 424, 905

Rodríguez-Franco, A., Martín-Pintado, J., & Fuente, A. 1998, *A&A* 329, 1097

Simon, M., Dutrey, A., & Guilloteau, S. 2000,. *Ap. J.* 545, 1034

Skrutskie, M.F., Dutkevitch, D., Strom, S., Edwards, S., Strom, K., & Shure, M. 1990, *Ap. J.* 99, 1187

Stapelfeldt, K.R., Krist, J.E., Menard, F., Bouvier, J., Padgett, D.L., & Burrows, C.J. 1998, *Ap. J.*, 502, L65

Stone, J.M., Gammie, C.F., Balbus, S.A., & Hawley, J.F. 2000, in Protostars & Planets IV, eds. V. Mannings, A.P. Boss, & S.S. Russell (Tucson: Univ. Arizona), 589

Thi, W.-F., van Zadelhoff, G.-J., & van Dishoeck, E.F. 2004, *A&A* 425, 955

van Zadelhoff, G.J., van Dishoeck, E.F., Thi, W.F., & Blake, G.A. 2001, *A&A* 377, 566

van Zadelhoff, G.-J. 2002, Ph.D. thesis, University of Leiden

van Zadelhoff, G.-J., Aikawa, Y., Hogerheijde, M.R., & van Dishoeck, E.F. 2003, *A&A* 397, 789

Walmsley, C.M., Flower, D.R., & Pineau des Forêts, G. 2004, *A&A* 418, 1035

Willacy, K. & Langer, W.D. 2000, *Ap. J.* 544, 903

Discussion

LISEAU: The results you describe pertain mostly to T Tauri stars and Herbig Ae stars with fairly low accretion rates. What is known about the chemical and physical properties of disks in FU Orionis systems for which flow onto the star is much larger?

BLAKE: Unfortunately, little is known at present, for a variety of reasons. For example, many such stars lie at distances several times further than for the nearest disk systems in Taurus and Ophiuchus, and so much improved spatial resolution will be needed to characterize FU Orionis disks to the level of detail currently available for T Tauri and Herbig Ae star disks. Some dust and line data are presented for FU Orionis and potential FU Orionis precursor stars in McMuldroch *et al.* (1993, *A. J.* 106, 2477; 1995, *A. J.* 110, 354), but at fairly low spatial resolution. Finally, the linear distances over which the dissipation of accretional energy becomes dominant are sufficiently small that only ALMA can image such regions directly at millimeter wavelengths.

Astrochemistry: Recent Successes and Current Challenges
Proceedings IAU Symposium No. 231, 2005
D.C. Lis, G.A. Blake & E. Herbst, eds.
© 2006 International Astronomical Union
doi:10.1017/S1743921306007393

Chemistry and Line Emission of Outer Protoplanetary Disks

Inga Kamp[1], Cornelis P. Dullemond[2], Michiel Hogerheijde[3], and Jesus Emilio Enriquez[4]

[1]Space Telescope Division of ESA, STScI, 3700 San Martin Drive, Baltimore, MD 21218, USA
email: kamp@stsci.edu

[2]Max-Planck Institute for Astronomy, Königstuhl 17, D-69117 Heidelberg, Germany

[3]Leiden Observatory, PO Box 9513, NL-3200 RA Leiden, The Netherlands

[4]University of Texas at El Paso, PSCI 210, 500 W. University Ave., El Paso, TX 79968, USA

Abstract. The structure and chemistry of protoplanetary disks depends strongly on the nature of the central star around which it has formed. The dust temperature is mainly set by the stellar luminosity, while the chemistry of the upper disk layers depends on the amount of intercepted UV and X-ray flux. We discuss here the differences in chemistry, thermal structure and line emission around Herbig Ae/Be, T Tauri stars and low-mass M dwarfs. Predictions will be made for future observations with SOFIA and Herschel.

Keywords. astrochemistry — planetary systems: protoplanetary disks — radiative transfer — solar system: formation — X-rays: stars — ultraviolet: stars

1. Introduction

The physical and chemical conditions in protoplanetary disks set the boundary conditions for planet formation. Although the dust component of these disks has been studied in considerable detail (grain composition, size distribution and mineralogy), very little is known about the gas phase. Over the past decade various disk models have been developed to study the chemical composition of protoplanetary disks themselves as well as the impact that it has on the disk physical structure (Aikawa *et al.* 2002; van Zadelhoff *et al.* 2003; Kamp & Dullemond 2004; Nomura & Millar 2005; Ceccarelli & Dominik 2005).

Current submm observations can study the chemical composition of a few nearby young disks; radio interferometry in particular resolves the spatial distribution of molecules and the rotation pattern of the disk for a handful of the largest or closest objects. We use these observations to understand the chemistry in the outer regions of protoplanetary disks. However, these observations are also crucial for determining the mass of these disks. So far, the dust has been used as an indirect tracer for the total disk mass. Recent publications (Shuping *et al.* 2003; Brittain *et al.* 2005; Dullemond & Dominik 2005) have shown that dust grain growth affects these disks already in very early stages; thus deriving the disk mass from the thermal emission of the small dust grains seems inadequate. A more direct approach is the detection of suitable gas tracers and the proper conversion of emission fluxes into total disk mass. The caveat here is the need for an adequate physical *and* chemical model of the disk.

2. Disk Models

Our modeling approach is described in Kamp & Dullemond (2004) for optically thick disks and Kamp & Bertoldi (2000) and Kamp & van Zadelhoff (2001) in the optically thin regime. The standard disk parameters for all models are a surface density profile $\Sigma(r) = \Sigma_0(r/\text{AU})^{-1}$ with $\Sigma_0 = 50$ g cm^{-3}, an inner radius $R_{\text{in}} = 0.1$ AU and an outer radius of $R_{\text{out}} = 300$ AU. The parameters of the central stars are listed in Table 1. The A-dwarf models are computed for a late phase of protoplanetary disks, when the disk starts to become optically thin (see Kamp & Bertoldi 2000 for details). The disk masses for these models range from 1.5×10^{-4} to 1.5×10^{-7} M$_\odot$ ($50 - 0.05$ M$_{\text{Earth}}$) and the inner and outer radius are 40 and 500 AU, respectively. Such inner hole sizes are frequently observed in later stages of disk evolution and often attributed to dust dynamics or planet formation.

Table 1. Stellar Parameters Used in the Disk Modeling

Parameter	PMS M Star	T Tauri Star	Herbig Ae Star	A Dwarf
T_{eff} [K]	3500	4000	9500	8200
$R_*/R_\odot$	0.84	2.50	2.40	1.70
$M_*/M_\odot$	0.5	0.5	2.50	2.0
$L_*/L_\odot$	0.1	1.5	13.2	12.0

The dust grains in the optically thick models are 0.1 μm astronomical silicate grains (see Dullemond *et al.* 2002 for more details). The later phases use 3 μm black body grains (see Kamp & Bertoldi 2000 for more details).

3. Stellar Models

The dust temperature is mainly set by the total luminosity of the star; hence the optically thick disk models use a blackbody radiation field with the corresponding effective temperature of the star. On the other hand, the chemistry and heating of the gas is mainly driven by UV photons and X-rays. Line blanketing in stellar atmospheres and chromospheric/accretion activity can cause the UV fluxes to differ by several orders of magnitude from those of a blackbody with the respective effective temperature. Thus, Kurucz stellar atmosphere models have been used throughout this study together with scaled UV fluxes (according to the age/activity level of the star) and observed X-ray fluxes. For the chromospheric UV excess, we used solar observations scaled with the inverse age t^{-1} (see Kamp & Sammar 2004 for details). At an age of 2 Myr, the level of UV irradiation produced by chromospheric activity would roughly correspond to that generated by an accretion rate of

$$\dot{M} = 8\,\pi R^2/v_s^2 F f \;\; \text{g s}^{-1} \tag{3.1}$$
$$\approx 2 \times 10^{-11} \;\; \text{M}_\odot \;\text{yr}^{-1}$$

where R is the inner radius of the disk, v_s is the free-fall velocity at the stellar surface, F is the accretion flux and f is the filling factor, which is assumed to be 1% (Calvet & Gullbring 1998). To mimic the UV irradiation by a T Tauri star with a typical accretion rate of 10^{-9} M$_\odot$ yr^{-1}, the chromospheric fluxes were simply scaled by a factor 50. AU Mic was used as a template for the M star UV and X-ray fluxes. This active M dwarf has been monitored for decades and the respective data were extracted from the HST and Chandra archives. In Figure 1, the spectra of the three stars, Herbig Ae, T Tauri and

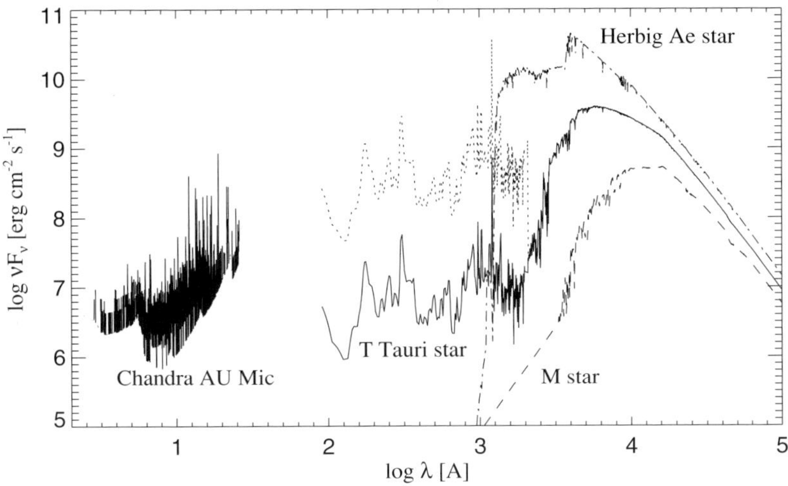

Figure 1. Full spectrum of all template stars: T Tauri star with chromosphere (solid line), T Tauri star with an accretion rate of 10^{-9} $M_\odot$ yr^{-1} (scaled chromosphere; dotted line), M star (dashed line) and Herbig Ae star (dash-dotted line); the Chandra spectrum of AU Mic consists only of emission lines.

M star, have been plotted. The UV spectrum in the M star has been approximated by a linear fit.

Figure 1 shows the Chandra fluxes of AU Mic. In our models we have used the Chandra data, because they cover a larger wavelength region than do the XMM spectra. All of these measurements are emission lines and there is no continuum detected. The X-ray heating rate is derived as an integral over the emission spectrum given a simple mean X-ray cross section (Gorti & Hollenbach 2004)

$$\sigma(E) = 1.2 \times 10^{-22} \left(\frac{E}{1 \text{ keV}} \right)^{-2.594} \text{cm}^2 ,\qquad(3.2)$$

where E is the energy in eV. This cross section is derived for elemental abundances measured through the diffuse cloud toward the star ζ Oph.

We calculated eight protoplanetary disk models to assess the influence of various parameters such as the spectral type of the central star, PAHs, and X-rays on the physical and chemical structure of the protoplanetary disks. Model 1 is the Herbig Ae star case, Model 2a–d are various T Tauri star models, and 3a/b are the M star models. The properties of these disk models are summarized in Table 2.

Table 2. Overview of the Massive Disk Models: 1—Herbig Ae Star, 2—T Tauri Star and 3—M Dwarf

Number	Star	PAH Heating	UV Flux	X-rays
1	Herbig Ae	on	Kurucz	none
2a	T Tauri	on	chromosphere	none
2b	T Tauri	off	chromosphere	none
2c	T Tauri	on	accretion	none
2d	T Tauri	off	accretion	none
3a	M dwarf	on	AU Mic fit	AU Mic observation
3b	M dwarf	on	AU Mic fit	AU Mic observation ×0.01

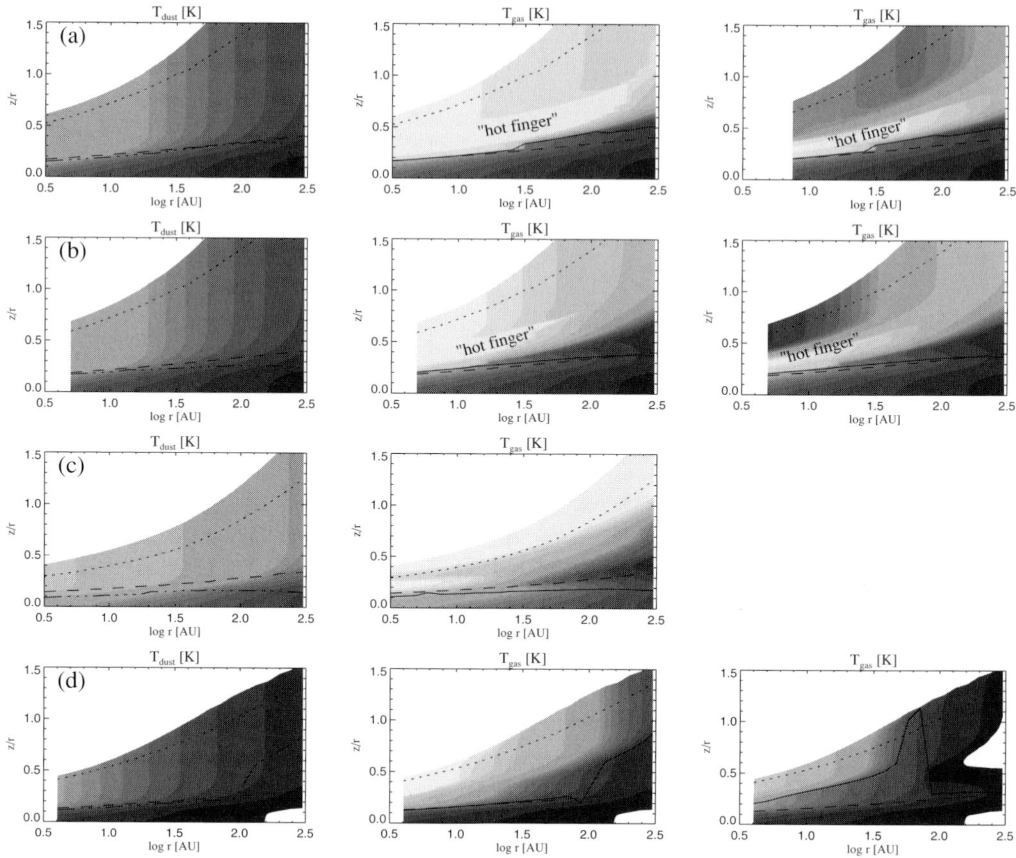

Figure 2. Dust and gas temperatures in the massive disk models: (*a*) model 2c/d: T Tauri star with an accretion rate of 10^{-9} M$_\odot$ yr^{-1}, (*b*) model 2a/b: T Tauri star with chromosphere, (*c*) model 1: Herbig Ae star, (*d*) model 3a/b: M star with and without X-rays. The left column shows the dust temperature, the middle column the gas temperature in the models with PAHs and the right column the gas temperature without PAHs; in the case of the M star model, the right column shows the gas temperature with PAHs as well, but with a hundred times weaker X-ray fluxes. The solid line indicates the location, below which gas and dust temperatures agree to within 10% (middle and right column). If over an interval $\delta z/z \sim 0.3$ both temperatures agree within 5%, complete coupling, that is $T_{\rm gas} = T_{\rm dust}$, is adopted below this depth (dashed-dotted line in left column). The 10% agreement occurs at or slightly above the $\tau = 1$ surface layer (dashed line). The dotted line marks the transition between the surrounding cloud and the disk. The grey levels correspond to the following temperatures: 10, 20, 30, 40, 50, 60, 70, 80, 90, 100, 200, 500, 1000, 2000, and 5000 K.

4. Gas Temperatures in Protoplanetary Disks

Figure 2 shows the dust and gas temperatures for the seven massive disk models. Each has the same surface density and hence the same disk mass of 0.01 M$_\odot$. The only difference between them is the central stellar radiation field. Our modeling makes clear that UV radiation is the most important heating source for the disk. X-ray heating is only important if the UV flux of the central star is low (such as is the case for M stars) or if very small grains like polycyclic aromatic hydrocarbons (PAHs) are absent.

The hot finger in the T Tauri model was first described in Kamp & Dullemond (2004) as the hot surface layer just above $\tau = 1$ (see the indication in Figure 2). The very low-density top layers are again cooler than the hot surface. The feature becomes even more prominent in the model with accretion-dervied UV irradiation. In the latter models, the hot finger extends even out to 200 AU. An external UV radiation field from a nearby O or B star will most likely have the same effect. Since such an external radiation field does not depend on the radial distance to the central star, we can speculate that it might turn this hot finger into a closed hot surface layer: a hot skin of several thousand K around the protoplanetary disk.

This hot finger is less pronounced in the Herbig Ae model, even though PAHs are included. This is due to the much lower UV flux from the central star. In the T Tauri model, the remnant material above the disk (from the parental molecular cloud) had a constant density of 3×10^3 cm^{-3}. For the Herbig Ae star, we considered a spherical cocoon with a power law $3 \times 10^3 (r/R_\mathrm{i})^{-1.3}$ cm^{-3}. This surrounding material was mainly introduced to avoid boundary effects on the disk surface. Since the density of the cocoon material decreases with distance for the Herbig Ae star, the gas stays hot even though the UV from the central star is diluted. Figure 2 shows that this gas is predicted to be very hot (5000–10,000 K).

The M star model illustrates the effects of X-ray heating. The UV fluxes of the M star are several orders of magnitude lower than those of T Tauri stars. Figure 2(d) compares the gas temperature found with the X-ray fluxes for AU Mic and with 100 times lower fluxes. Gas temperatures are typically a factor 2–10 lower in the reduced X-ray case.

5. Chemistry of Protoplanetary Disks

The chemical network used in this study is a small one with 48 species and 268 reactions. It is well suited for calculating the abundances of the dominant cooling species and contains the photochemistry important in the upper layers of protoplanetary disks. However, X-ray chemistry is missing so far, hence the results for the M star should be considered with caution. In the following, we briefly discuss the most important features of these chemical calculations.

Comparing Figure 3(c) and (d) reveals the impact of the stellar UV irradiation on the key molecules H_2 and CO. In the case of pure chromospheric UV flux, the H_2/H fraction in the outer disk upper layers is 1, meaning that all hydrogen is in molecular form. Also, the C/CO transition occurs at lower optical depth than in the T Tauri accretion model. As discussed above, the hot finger extends to much larger radii in the accretion model and this is also reflected in the OH and HCO$^+$ abundance structure. Both species arise from the fact that H_2 is collisionally destroyed by H and O in these layers (Kamp & Dullemond 2004). OH is subsequently photodissociated and this process leaves the O-atom in the ^{1}D excited state in ~ 50 % of the cases (Störzer & Hollenbach 1998). This level gives rise to the [O I] 6300 Å line, which can also be thermally excited at temperatures larger than 3000 K. The hot finger thus presents an ideal reservoir of gas for the 6300 Å line emission.

In conditions where the disk is exposed to an external UV radiation field, such as in the Orion nebula, this hot finger might turn into a hot surface of the disk. Actually, observations of the proplyds in Orion with a narrow band filter centered on the 6300 Å line have revealed that this emission comes from the "skin" of the disk (Bally *et al.* 1998). The same line emission has also been observed in Herbig Ae/Be stars by Acke *et al.* (2005), who explain the observed spectra using an OH abundance of $10^{-6} - 10^{-7}$ and non-thermal emission. Our model, however, suggests that Herbig Ae stars show a hot finger similar to T Tauri stars; thus thermal emission might contribute significantly to the total emission and help to explain the [O I] 6300 Å line flux with smaller OH abundances that are closer to those derived in our modeling ($10^{-8} - 10^{-9}$).

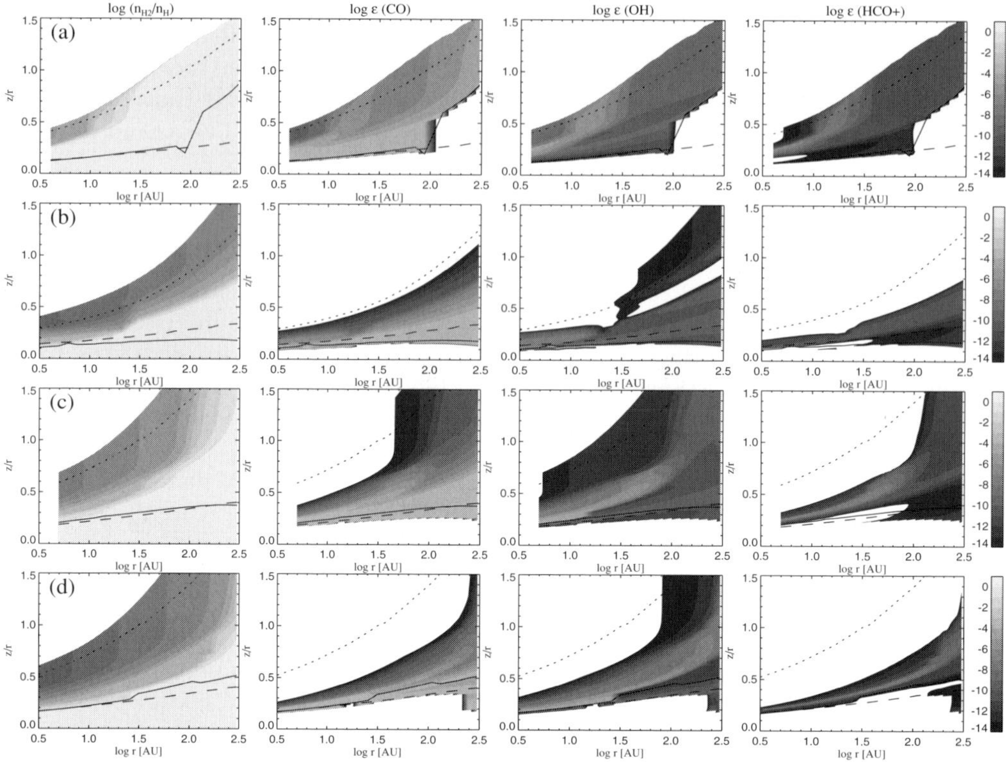

Figure 3. Selected chemical species in four of the seven disk models: (a) M star with X-rays, (b) Herbig Ae star, (c) T Tauri star with chromosphere and PAHs, (d) T Tauri star with an accretion rate of 10^{-9} $M_\odot\,\mathrm{yr}^{-1}$ and PAHs. The first column shows the H_2/H fraction, the other columns show the abundances ($\log \epsilon = \log n(\mathrm{species})/n_\mathrm{tot}$) of CO, OH and HCO^+ respectively.

Aikawa *et al.* (1998) have shown that X-ray chemistry mainly enhances the presence of ions and thus may affect species that are formed via ion-molecule reactions. However, species such as CO seem to be rather unaffected and their abundance is mainly determined by the level of UV irradiation. We abstain from a detailed discussion of the observational consequences arising from the chemistry displayed in Figure 3, and restrict ourselves instead to a purely theoretical interpretation. It is interesting to note the rather large abundance of molecular hydrogen and CO in this disk model. This is due to the extremely low UV luminosity in the M star model. Even though we used an active star, namely AU Mic, as a template for this model, the total luminosity is much smaller than in T Tauri stars of similar activity level. The maximum of the OH abundance distribution coincides with the end of the hot X-ray heated surface. Apparently, the chemistry there is similar to that noted for the hot finger.

6. Chemistry of More Evolved Disks

As protoplanetary disks evolve, the dust grows into large grains and proto-planetesimals; the total mass in large grains (micron-sized) is much smaller than the mass in the original dust disk and the entire disk becomes optically thin to continuum radiation. However, H_2 and CO photodissociation are not continuum processes and thus significant amounts of these molecules can lead to an efficient shielding of the UV radiation within the disks.

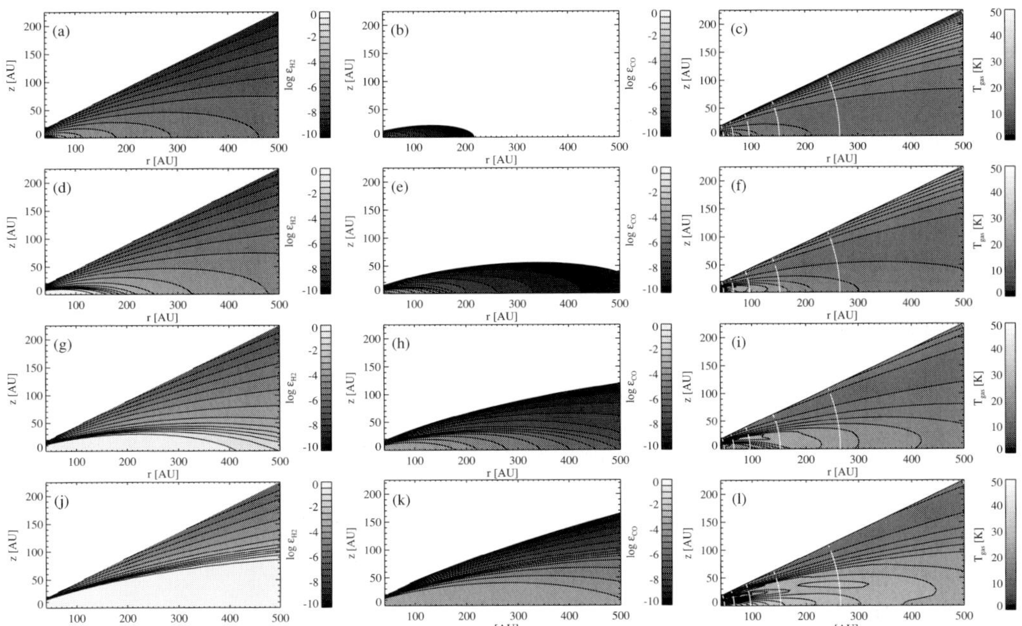

Figure 4. H$_2$ and CO abundances in the optically thin disk models and gas and dust temperatures: (a–c) 1.5×10^{-7} M$_\odot$, (d–f) 1.5×10^{-6} M$_\odot$, (g–i) 1.5×10^{-5} M$_\odot$, (j–l) 1.5×10^{-4} M$_\odot$. Contour lines show abundances $\log \epsilon = n_{\mathrm{species}}/n_{\mathrm{tot}}$ in logarithmic steps. The temperature plots show the gas temperature as filled contours and the dust temperature overplotted with white contour lines (80, 70, 60, 50, 40 K from inside out).

Figure 4 illustrates the abundances of H$_2$ and CO in four disk models whose masses range from 1.5×10^{-4} to 1.5×10^{-7} M$_\odot$. The stellar radiation field in each case is that of an A dwarf (8200 K, $\log g = 4.0$), and the interstellar radiation field corresponds to that in the solar neighbourhood ($\chi = 1$). Even though the disk masses span four orders of magnitude, the CO densities in these disks span a much larger range due to the strong photodissociation in the lowest disk mass models. As the disk mass increases, shielding by H$_2$ and CO becomes more efficient until all carbon is in the form of CO (1.5×10^{-4} M$_\odot$ model). While the CO/H$_2$ conversion factor in the most massive disks is close to the canonical value of 10^{-4} derived for molecular clouds, it can be as low as 10^{-7} in the lowest mass model.

7. Pushing Future Observations to the Limits

We used the 2D Monte Carlo code from Hogerheijde & van der Tak (2000) to calculate the line emission from the optically thin disk models. These fluxes are especially interesting in the context of the upcoming observing facilities such as APEX, ALMA, Herschel and SOFIA. We placed the disks at a typical distance of 100 pc and inclined them by 45° to the line-of-sight.

For the lowest four CO rotational lines J=1–0, 2–1, 3–2, 4–3, we assumed typical beam widths of 43, 22, 14, and 11″ respectively. At a distance of 100 pc, this means that the beam contains the entire disk. Figure 5 shows that the higher CO rotational lines are more readily detected, because the corresponding levels are more populated at the typical disk gas temperatures of 30–50 K. This figure also illustrates that the slope of the line

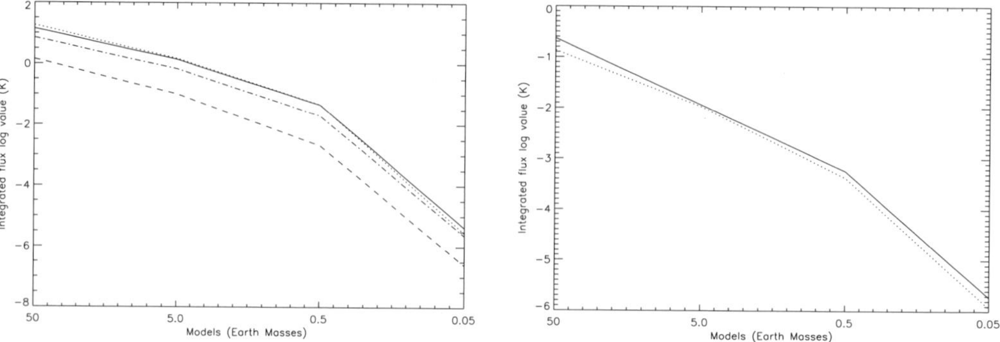

Figure 5. Integrated CO (left) and O (right) line fluxes from the optically thin disk models: the distance is assumed to be 100 pc and the inclination is 45°. Shown are the lowest four rotational lines of CO, J=1–0 (dashed), J=2–1 (dash-dotted), J=3–2 (dotted), J=4–3 (solid) and the two fine structure lines of neutral oxygen, 63 μm (solid) and 145 μm (dotted).

fluxes as a function of disk mass is much steeper than 1. As the disk masses become lower than 0.5 M_{Earth}, the gas becomes rapidly undetectable in CO even with ALMA.

For the two fine structure lines of neutral oxygen at 63 μm and 145 μm, we assumed beam widths of 7″ (SOFIA/ GREAT) and 9″ (Herschel/PACS), respectively. Assuming a typical detection limit of 5 mK in 5 hours of integration time, we can expect to detect disk masses down to a few Earth masses. The [O I] fine structure lines are thus not as sensitive as the CO lines with ALMA. However, we expect the [C II] 158 μm line to be even stronger than [O I] for these tenuous disks.

8. Conclusion

The interpretation of current near-infrared and submm observations needs self-consistent disk models, like the ones illustrated in the previous sections. Disk structure and chemistry together can be used to derive quantities such as disk gas mass and inclination. The power of such disk models as a tool for observations has been recently illustrated in various papers (Semenov *et al.* 2005; Jonkheid *et al.* 2005; Nomura & Millar 2005).

We showed the importance of a proper inclusion of all radiation sources—stellar or boundary layer UV, X-rays and external irradiation—for the gas temperature and chemical structure of the disk surface. The chemistry of the outer protoplanetary disk layers is driven by irradiation; thus it is important to take into account the shape of the radiation field and not only its strength. The wavelength distribution of UV photons in a T Tauri star is very different from that in a mean interstellar radiation field. Instead of using the UMIST photorates that were derived for a Draine field, we conclude that it is necessary to integrate the respective photoionization and dissociation cross sections over the stellar spectrum.

An important step in understanding disk evolution, and in setting the boundary conditions for planet formation, is the measurement of the disk gas mass as a function of time and type of central star. Using our disk models, we conclude that future instrumentation like ALMA will allow the detection of transition disks down to 0.5 M_{Earth} of gas.

9. Outlook

Disks span a wide range in parameter space—density, temperature, irradiation—and as a result the chemical conditions we encounter in them range from extreme PDRs, diffuse clouds to dense clouds, and even dark cores. However, disks become much denser in the midplane than molecular clouds or dark cores and hence three-body reactions will play a role there as well. Any model that wants to treat the full chemistry needs to take into account photochemistry, X-ray chemistry, gas-grain chemistry and three-body collisions as well as grain surface chemistry. Progress in the future has to be made especially in the field of gas-grain chemistry (a better understanding of desorption processes) and grain-surface chemistry. The latter is important in forming very large molecules and more complex organic species, the precursors of life.

In terms of disk modeling, the next generation models should be 2D hydrodynamical models that account for the proper treatment of mixing in the inner disk regions, where diffusion timescales are shorter than or comparable to the chemical timescales. These models would need a realistic energy equation for the gas, because the surface and intermediate layers are not dominated by accretion heating. As outlined above, treating the full chemistry (gas chemistry, gas-grain and grain-surface chemistry) is very expensive computationally, and it might be necessary to optimize the chemical networks for the various parameter regimes in the disk.

Acknowledgements

We would like to thank Jan-Uwe Ness for reducing the AU Mic Chandra data and providing the X-ray fluxes for AU Mic. Floris van der Tak helped with the Monte Carlo radiative transfer code. We also like to thank the STScI for its summer internship program during which part of this work was carried out.

References

Acke, B., van den Ancker, M. E., & Dullemond, C. P. 2005, *A&A* 436, 209
Aikawa, Y., van Zadelhoff, G. J., van Dishoeck, E. F., & Herbst, E. 2002, *A&A* 386, 622
Bally, J., Sutherland, R.S., Devine, D., & Johnstone, D. 1998, *A. J.* 116, 293
Brittain, S. D., Rettig, T. W., Simon, T., & Kulesa, C. 2005, *Ap. J.* 626, 283
Calvet, N. & Gullbring, E. 1998, *Ap. J.* 509, 802
Ceccarelli, C. & Dominik, C. 2005, *A&A* in press
Dullemond, C. P., van Zadelhoff, G. J., & Natta, A. 2002, *A&A* 389, 464
Dullemond, C.P. & Domink, C. 2005, *A&A* 434, 971
Gorti, U. & Hollenbach, D. 2004, *Ap. J.* 613, 424
Jonkheid, B., Kamp, I., Augereau, J.-C., & van Dishoeck, E.F. 2005, *A&A* submitted
Hogerheijde, M. R. & van der Tak, F. F. S. 2000, *A&A* 362, 697
Kamp, I. & Bertoldi, F. 2000, *A&A* 353, 276
Kamp, I. & van Zadelhoff, G.J. 2001, *A&A* 373, 641
Kamp, I. & Dullemond, C.P. 2004, *Ap. J.* 615, 991
Kamp, I. & Sammar, F. 2004, *A&A* 427, 561
Nomura, H. & Millar, T.J. 2005, *A&A* 438, 923
Semenov, D., Pavlyuchenkov, Ya., Schreyer, K., Henning, Th., Dullemond, C., & Bacmann, A. 2005, *Ap. J.* 621, 853
Shuping, R. Y., Bally, J., Morris, M., & Throop, H. 2003, *Ap. J.* 587, L109
Störzer, H. & Hollenbach, D. 1998, *Ap. J.* 502, L71
van Zadelhoff, G.-J., Aikawa, Y., Hogerheijde, M. R., & van Dishoeck, E. F. 2003, *A&A* 397, 789

Discussion

BERGIN: You see evidence for hot fingers with higher temperatures in layers with colder temperatures above. Photoelectric heating is important for the upper cold layers so is the temperature rise due to an increase in photoelectric heating efficiency or lower cooling.

KAMP: This temperature inversion occurs in very low-density regimes, where the main cooling ([O I] fine structure lines) is in NLTE and thus scales linearly with density. The photoelectric heating scales as $n^{1.8}$ in this regime. For a vertical slice of the disk, the density slowly increases towards the midplane and thus the heating rises faster than the cooling.

GLASSGOLD: There is some question about whether cosmic rays penetrate the entire volume of T Tauri disks due to the fact that they drive winds which sweep away galactic cosmic rays.

KAMP: Cosmic rays are mostly important in the midplane layers of the outer disk, where UV irradiation cannot reach. There, they drive the deuterium chemistry and deuterated species are actually observed in protoplanetary disks. However, additional ionization may also come from the stellar X-rays (secondary UV photons) and thus external cosmic rays may be less important in explaining the observations. Cosmic rays do not affect the thermal structure of the surface layers in protoplanetary disks.

FORREST: Comment: "Cosmic rays" for T Tauri stars are produced locally – the magnetic accretion that produces keV X-rays also produces MeV protons – like solar flares.

Astrochemistry: Recent Successes and Current Challenges
Proceedings IAU Symposium No. 231, 2005
D.C. Lis, G.A. Blake & E. Herbst, eds.
© 2006 International Astronomical Union
doi:10.1017/S174392130600740X

Infrared Spectroscopy of Molecules in Disks

Joan Najita

National Optical Astronomy Observatory, Tucson, AZ 85719, USA
email: najita@noao.edu

Abstract. Disks surrounding young stars play a fundamental role in the formation of stars and planets. Accretion through disks is believed to be responsible for the build up of stellar masses, and the gas and dust in disks is a reservoir for the potential formation of planets. As a result, one of the motivations for observing the inner regions of disks (i.e., the region within 10 AU) is to obtain clues to the processes that govern how stars and planets form. Significant progress has been made over the last decade in probing the inner regions of gaseous disks through the use of infrared molecular transitions. I discuss the observational tools that are currently available to study the gaseous component. These tools can be used to explore the evolution of gas in the inner disk and thereby help us to understand the processes of giant and terrestrial planet formation. These same tools may also be used to place constraints on the physical mechanisms that drive the disk accretion process.

Keywords. stars: pre-main-sequence — accretion disks — planetary systems: protoplanetary disks — planetary systems: formation

1. Introduction

Circumstellar disks are believed to form as a natural consequence of the collapse of molecular cloud cores. Due to the finite angular momentum of cloud cores, all of the infalling matter cannot fall directly onto the star; instead a large fraction must accrete through a disk before reaching the star. A long-standing problem regarding this phase of star formation is the nature of the physical mechanism that mediates the redistribution of angular momentum, thereby allowing accretion to occur. Several accretion mechanisms have been proposed, perhaps the foremost of which is the magnetorotational instability (Balbus & Hawley 1991; Stone *et al.* 2000), although other possibilities exist (e.g., the global baroclinic instability—Klahr & Bodenheimer 2003). Despite the significant theoretical progress that has been made in identifying plausible accretion mechanisms, observational evidence that any of these processes are actually active in disks has been difficult to come by. In this contribution, I suggest how new opportunities to study the inner region of disks may allow us to probe the nature of the accretion process.

Similarly, observations of the inner disk region may provide insight into the physical processes that govern planet formation and the resulting planetary architectures (i.e., planetary masses, orbital radii, and eccentricities). These clues are likely to be valuable given that the planet formation process appears to be more complex than was once believed. In the 10 years since their discovery, the extrasolar planets have come to span an ever increasing range in orbital properties, demonstrating a remarkable diversity that has, in turn, inspired multiple theories of their origins. The possibility that multiple theories may account for the orbital properties of the extrasolar planets drives us to search beyond planetary architectures for clues to their origins, clues that may be contained in the formation environments of planets, i.e., disks surrounding young stars.

For example, we might attempt to measure the gas dissipation timescale in the inner disk in order to constrain pathways for the formation of giant planets. Measurements of

the gas dissipation timescale would also constrain the efficiency of orbital migration and
eccentricity evolution for both terrestrial and giant planets. In this contribution, I discuss
some of the molecular probes currently available to probe the inner disk as well as current
thermal-chemical modeling of the atmospheres of inner disks. As a specific application, I
illustrate how these tools could be used to measure the gas dissipation timescale in inner
disks and thereby place constraints on the outcome of terrestrial planet formation.

2. Molecular Probes of Gaseous Inner Disks

We might first wonder what situations might give rise to emission lines from the inner
disk, since disks are typically expected to be very optically thick in the continuum due to
the high column density of gas and dust. (For example, the total column density of the
minimum mass solar nebula at 1 AU is $\sim 1500\,\mathrm{g\,cm^{-2}}$.) We can nevertheless expect to
detect emission lines in several important cases. At early times, irradiation by the central
star, or other heating processes, may heat the surface of the disk, creating a warm "disk
chromosphere" that will produce emission lines. Emission lines are also expected at late
times, once the inner disk has evolved to either low total column density (e.g., due to disk
dissipation) or low continuum optical depth (e.g., due to grain growth or the formation
of rocky bodies).

Given that line emission can plausibly arise from inner disks, we might ask what spectral features are to be expected? The temperatures involved range from several 1000 K
at the inner disk edge to ~ 100 K in the giant planet region of the disk (5–10 AU). In
addition, the disks are massive so they should be characterized by high densities. These
conditions suggest that molecules should be good probes of disks since they should be
abundant in the gas phase at these temperatures, and at high density they should be
excited enough to produce a ro-vibrational spectrum in the infrared.

This expectation is borne out in that there are now multiple molecular probes of the
inner disk region. These include the CO overtone lines, which probe the region from the
inner disk edge at ~ 0.05 AU out to a few 0.1 AU (e.g., Najita $et\ al.$ 2000 for a review)
and the CO fundamental lines, which probe the region from the inner disk edge out to
larger distances of a few AU, i.e., the terrestrial planet region of the disk (e.g., Najita
$et\ al.$ 2003). Tracers that probe somewhat larger radii than the CO overtone lines include
water ro-vibrational lines in the K-band (e.g., Carr $et\ al.$ 2004) and OH fundamental
vibrational lines in the L-band (Carr $et\ al.$, in preparation). In addition, transitions of
molecular hydrogen at UV, NIR, and MIR wavelengths (e.g., Herczeg $et\ al.$ 2002; Bary
$et\ al.$ 2003; Richter $et\ al.$ 2002) are together expected to probe both the terrestrial and
giant planet regions of the disk.

Among these, CO overtone emission (e.g., Carr 1989, Scoville $et\ al.$ 1979) is perhaps
the diagnostic first recognized to probe the inner disk region. High-resolution spectra of
$v = 2$–0 bandhead emission from young stars at 2.3 μm typically shows the characteristic shape expected for bandhead emission from a rotating disk (e.g., Carr $et\ al.$ 1993;
Chandler $et\ al.$ 1993; Najita $et\ al.$ 1996). Such bandhead profiles are seen in young stars
ranging in mass from ~ 1 M$_\odot$ to ~ 10 M$_\odot$ (e.g., Najita $et\ al.$ 2000; Blum $et\ al.$ 2004).
These results provide good evidence for rotating inner disks around young stars of low
to high mass.

If we look in more detail at the shape of the underlying rotationally broadened line,
we find that the emission is broad (typically > 100 km s^{-1}). This suggests that the inner
radius for the emission arises very close to the star, at ~ 0.05 AU. The double-peaked
profiles indicate that the emission arises over a limited range of radii, out to ~ 0.3 AU

in low-mass stars. These small radii are consistent with the high deduced excitation temperature of the emission over this range of radii ($\sim 1500 - 5000$ K).

An interesting aspect of the overtone bandhead is that it is made up of closely spaced lines with varying inter-line spacing and optical depth. This makes it possible to distinguish the intrinsic line broadening of the emitting gas from macroscopic motions such as rotation. Using spectral synthesis modeling, we are therefore able to deduce that the lines are suprathermally broadened, with line widths approximately Mach 2 (Carr *et al.* 2004; Najita *et al.* 1996). This suggests that disk atmospheres are turbulent. This may be a consequence of turbulent angular momentum transport in disks, as is expected in the case of both the magnetorotational instability and the global baroclinic instability.

Another interesting molecular probe of the inner disk is the ro-vibrational transitions of water. In the K-band, CO overtone emission is often accompanied by individual emission lines of water blueward of the bandhead (Carr *et al.* 2004; Najita *et al.* 2000; Thi & Bik 2005). Velocity resolved spectra show that the width of the CO emission is always broader than the width of the water emission. In addition, spectral synthesis modeling shows that the excitation temperature of the CO emission is larger than that of the water emission. These properties are consistent with a common origin for the water and CO emission in a differentially rotating disk with an outwardly decreasing temperature profile. This is because the lower dissociation temperature of water (~ 2500 K in inner disks) allows the water emission to extend in only to a modest radius (> 0.05 AU). In contrast, the much higher dissociation temperature of CO (~ 5000 K in inner disks) allows the CO emission to extend in to much smaller radii (~ 0.05 AU), to higher velocities and higher temperatures.

Recent spectral synthesis modeling that fits both the CO and water emission finds that the relative abundance of water to CO is low, a factor of 2–10 below the chemical equilibrium value in low-mass stars (Carr *et al.* 2004; see Thi & Bik 2005 in the context

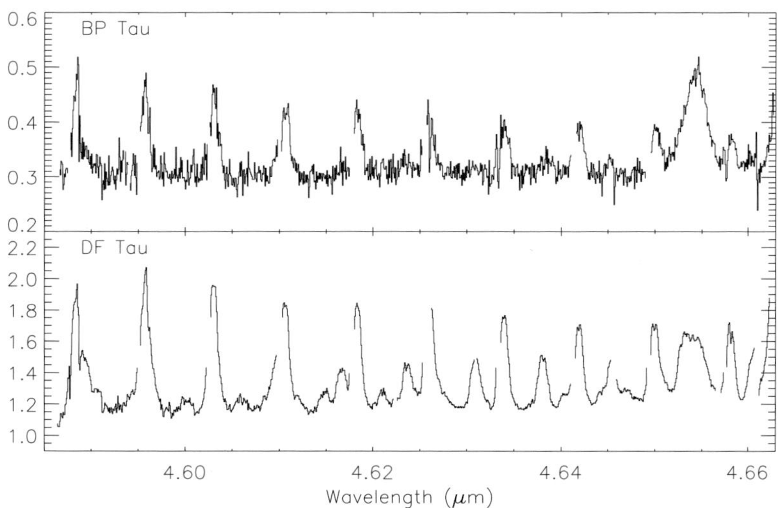

Figure 1. CO fundamental emission from accreting young stars. Regions of strong telluric absorption have been excised from the plot. The centrally peaked line profiles show that the fundamental transitions probe a wide range of disk radii, from inner disk edge at ~ 0.05 AU out to 1–2 AU, i.e., the terrestrial planet region of the disk. The emission spectrum can be rich (e.g., DF Tau), displaying lines from a range of vibrational levels (e.g., $v = 1$–0, 2–1, 3–2) as well as lines of ^{13}CO.

of high-mass stars). The low ratio may reflect strong vertical abundance gradients that result from the external irradiation of the disk (see § 3).

CO fundamental emission at 4.6 μm is interesting as a disk diagnostic because it is common: it is detected in almost all accreting young stars (e.g., Najita *et al.* 2003). This is perhaps a consequence of the much larger A-values for the fundamental lines compared to the CO overtone lines. In contrast to the overtone lines which are double-peaked, the line profiles for the fundamental emission are broad and centrally peaked. This suggests that the CO fundamental transitions probe a wide range of radii, from the inner disk edge (at ~ 0.05 AU) out to 1–2 AU, i.e., the terrestrial planet region of the disk.

The emission spectra typically show emission lines from a range of vibrational states; ^{13}CO can also be detected when the emission is bright (Fig. 1). The ability to study a wide range of vibrational states as well as isotopic species within a limited spectral range suggests the utility of CO fundamental emission in probing a wide range of gas temperatures and column densities. Indeed, an excitation analysis of the emission provides evidence for surprisingly warm gas ($\gtrsim 1000$ K) in the terrestrial planet region of disks surrounding low-mass stars. These temperatures are much higher than the dust temperatures expected for the same region of the disk (e.g., D'Alessio *et al.* 1998).

3. Models of Gaseous Atmospheres of Inner Disks

These warm temperatures can be understood in the context of current thermal-chemical models for the inner disk region. For example, in the models of Glassgold, Igea & Najita (2004), we find that both the gas and dust in the disk atmosphere experience temperature inversions, but with the gas achieving much higher surface temperatures than the dust. In these models, the dust is heated by stellar (primarily optical) photons, whereas the gas is heating by stellar X-rays at the disk surface and by accretion-related heating processes at intermediate heights. Possible accretion-related processes include *in situ* accretional heating and the dissipation of turbulence induced by a stellar wind blowing over the disk surface. These processes produce a temperature inversion in the gaseous atmosphere over a column density of $\sim 10^{22}$ cm^{-2}. At larger column densities (and higher densities), the gas temperature is driven down to the dust temperature due to efficient gas-grain coupling (Fig. 2).

If we look at the chemistry that takes place within this vertical temperature structure, we find that a small abundance of H$_2$ forms high in the atmosphere through the H$^-$ and 3-body pathways. This allows the formation of CO at a column density of $\sim 10^{21}$ cm^{-2}. In contrast, the full conversion from atomic H to molecular H$_2$ is made further down in the atmosphere, at a column density of $\sim 6 \times 10^{21}$ cm^{-2}, with more complex molecules such as water forming deeper in the disk, primarily through neutral reactions. This produces a disk atmosphere with layered molecular abundances. Similar descriptions of warm disk atmospheres with layered molecular abundances have been made for the outer disk region (e.g., Kamp & Dullemond 2004; Jonkheid *et al.* 2004; see Kamp, this volume).

Importantly, the region of the disk atmosphere over which the gas and dust temperatures differ includes the region currently accessible to observations (e.g., using the CO transitions). Thus, using models of disk atmospheres that consider the potential thermal decoupling of gas and dust can be important in interpreting observations. Indeed, looking at the temperature at which CO forms, the temperature at 1 AU is ~ 1000 K, which is similar to the observations. Perhaps interestingly, the CO is abundant in a region in which hydrogen is predominantly in atomic form. This is favorable for exciting the CO transitions since the collisional cross-section for exciting CO with atomic hydrogen is larger than the cross-section for exciting CO with H$_2$.

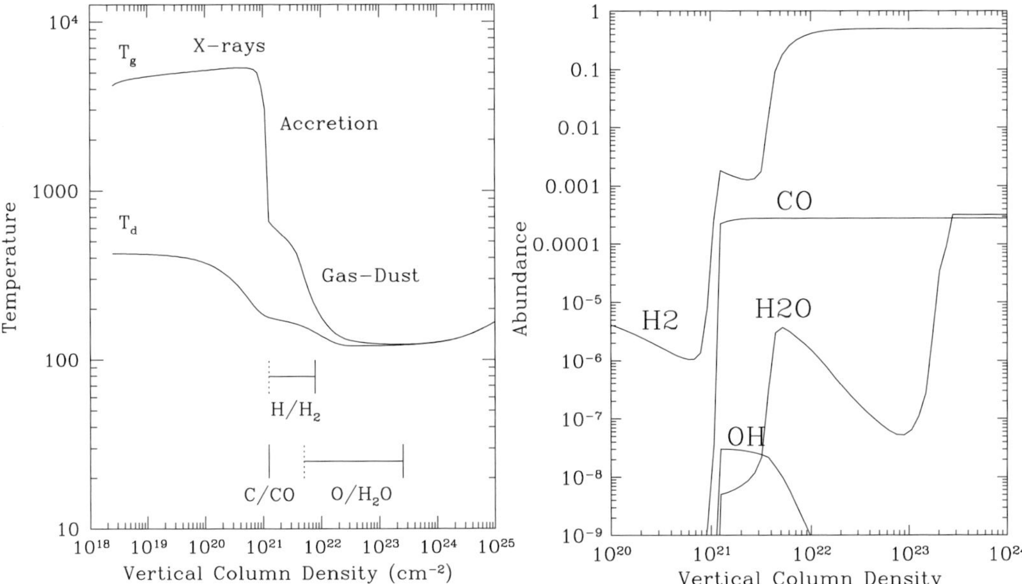

Figure 2. *Left:* Vertical temperature structure in the atmosphere of a disk surrounding an accreting young star at a radial distance of 1 AU (Glassgold *et al.* 2004). The gas and dust components both experience vertical temperature inversions, with the gas achieving much higher temperatures than the dust. *Right:* In the accompanying vertical chemical structure, a modest abundance of H_2 $(x(H_2) = 10^{-3})$ is achieved at a vertical column density of $\sim 10^{21}\,\mathrm{cm}^{-2}$, which allows CO to form at a similar column density. Full conversion to H_2 is achieved at a much larger column density $\sim 10^{22}\,\mathrm{cm}^{-2}$. Similarly, a modest water abundance $(x(H_2O) = 3 \times 10^{-6})$ is achieved at $\sim 5 \times 10^{21}\,\mathrm{cm}^{-2}$, with an asymptotic abundance $(x(H_2O) = 3 \times 10^{-4})$ achieved at much higher column densities $(> 10^{23}\,\mathrm{cm}^{-2})$.

There is additional evidence for warm molecular gas in the inner disk that comes from from UV observations of fluorescent molecular hydrogen emission. As shown by Herczeg *et al.* (2002) in the case of the young star TW Hya, Ly α is observed to pump H_2 from excited energy levels $\sim$ 2000 K above ground, showing that the excited gas is warm, with a characteristic H_2 column density of $\sim 10^{18}\,\mathrm{cm}^{-2}$. It is interesting to think about the implications of this column density in the context of the thermal-chemical models discussed in the previous section.

In the Glassgold *et al.* models, we predict that the H_2 abundance at the disk surface is small $\sim 10^{-3}$ (Fig. 2). This implies that the measured H_2 column density of $\sim 10^{18}\,\mathrm{cm}^{-2}$ for TW Hya corresponds to a total hydrogen column density of $\sim 10^{21}\,\mathrm{cm}^{-2}$. Since this is the column density over which Ly α is found to excite the gas, the dust in the disk atmosphere must be depleted relative to the gas in order for Ly α to penetrate this far. Thus, observations of this kind, when combined with gaseous disk atmosphere models, can provide one of the few ways of constraining the dust-to-gas ratio in disk atmospheres. The low dust-to-gas ratio may be taken as evidence for significant grain growth or settling in the disk atmosphere of TW Hya. This is in agreement with the low dust column density in the inner disk that is inferred from the spectral energy distribution for this source (Calvet *et al.* 2002). The gas observations show that the gas may be less depleted than the small dust grains in the inner region of the disk.

It is also interesting to look at the relative abundances of water and CO in these models. At 1 AU, we find that water is at least 100 times less abundant than CO in the disk atmosphere (Fig. 2). This is much less than the ratio inferred from spectroscopy at smaller disk radii, where water is ~ 3 times less abundant than CO (see § 2). It would therefore be interesting to model the disk atmosphere at smaller radii to see if larger water abundances are predicted there. If not, perhaps the discrepancy between the observations and the thermal-chemical predictions for static disk atmospheres is evidence for strong vertical mixing which dredges up abundant water from deeper in the disk to the disk surface.

4. Gas in the Terrestrial Planet Region of Disks

Finally, I wanted to describe how we might use gas-phase diagnostics of the inner disk region to probe the lifetime of gas in the terrestrial planet region. This issue is of interest because residual gas in the disk can affect the outcome of terrestrial planet formation, i.e., the resulting masses and eccentricities of planets and their consequent habitability. For example, in the picture of terrestrial planet formation described by Kominami & Ida (2002), only gas column densities in a narrow range around $\sim 1\,\mathrm{g\,cm^{-2}}$ are likely to produce planets with Earth-like masses and eccentricities. With gas column densities much larger than this, protoplanets (the building blocks of terrestrial planets) experience significant gravitational gas drag and are not able to acquire the eccentricities that are needed for them to collide and form more massive objects. Conversely, if the gas column density is much lower than $1\,\mathrm{g\,cm^{-2}}$, protoplanets can easily acquire the eccentricities that are needed for them to collide, producing a massive eccentric object. But then the gravitational gas drag is insufficient to re-circularize the orbit. As a result, there is only a narrow range of gas column densities around $1\,\mathrm{g\,cm^{-2}}$ that can produce planets with Earth-like masses (i.e., massive enough to hold on to life-sustaining atmospheres) and the low eccentricities that we associate with habitability on Earth.

To understand how often this situation occurs during the period of terrestrial planet formation, John Carr, Bob Mathieu, and I have been carrying out a survey for CO

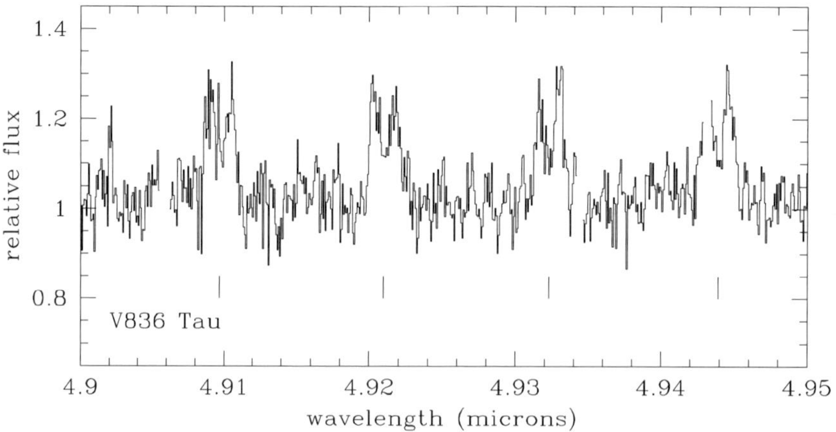

Figure 3. CO fundamental emission from the transitional T Tauri star V836 Tau, a system that is developing an optically thin inner disk. The spectra show $v=1\text{--}0$ emission from high-J P-branch lines at 4.9 μm. Regions of strong telluric absorption have been excised from the plot. The vertical lines mark the approximate CO line centers at the velocity of the star. The velocity widths of the lines indicate that the emitting gas is located within a few AU of the star, and the relative strengths of the lines suggest optically thick emission. Thus a large reservoir of gas may be present in the inner disk despite the weak infrared excess from this portion of the disk.

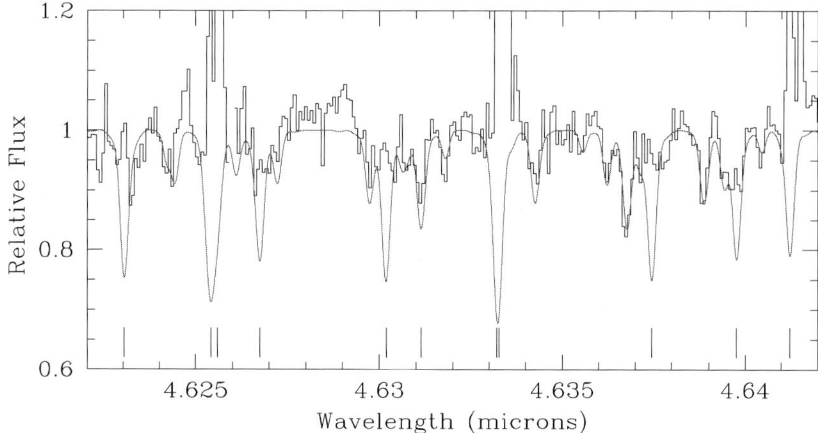

Figure 4. Spectrum of the transitional T Tauri star TW Hya in the 4.6 μm region (histogram), focusing on the structure in the continuum. The strong emission lines that extend above the plotted region are $v = 1$–0 low-J R-branch CO fundamental lines. For comparison, the predicted stellar spectrum for TW Hya is also shown (gray line). The model stellar spectrum fits many of the weaker features seen in the continuum. However, it predicts stronger stellar CO absorption in the low vibrational transitions (indicated by the lower vertical lines) than is seen in the measured spectrum. This suggests that the stellar photospheric spectrum is veiled by CO emission from warm disk gas.

fundamental emisison from young stars (see Najita 2004). CO is attractive for this purpose because it probes small column densities of gas ($10^{-4} - 1\,\mathrm{g\,cm}^{-2}$) covering the range of interest. We are targeting young stars with optically thin inner disks, because we want to know how much gas is left once disks become optically thin as a result of the formation of planetesimals, protoplanets, or larger bodies. Thus, our sample is made up of weak-lined T Tauri stars (stars with no evidence for disks or stellar accretion) and transitional T Tauri stars (stars with evidence for outer dust disks and a range of accretion rates).

Obtaining an accurate answer to our problem is challenging for several reasons. Firstly, we of course need reliable thermal-chemical models in order to convert any observed emission or upper limit into a total gas column density. Secondly, the strength of CO emission is known to correlate with the accretion rate in sources with optically thick inner disks (Najita *et al.* 2003). This suggests that the lower accretion rates that are known to characterize our sample may result in less heating of the upper disk atmosphere (see § 3 and Fig. 2) and therefore intrinsically weaker CO emission. Care may therefore be needed to detect such weak emission.

Nevertheless, thus far we have a few obvious detections of fairly strong emission. This shows that gas may survive in inner disks even after the continuum becomes optically thin. The detections include the T Tauri star V836 Tau (Fig. 3), a ~ 3 Myr old system (Siess *et al.* 1999), with an Hα equivalent width of 9 Å (Herbig & Bell 1988), and an estimated accretion rate of $\sim 4 \times 10^{-10}$ M$_\odot$yr^{-1} (Hartigan *et al.* 1995; Gullbring *et al.* 1998). CO emission is also detected from TW Hya (Fig. 4; see also Rettig *et al.* 2004, and G. Blake personal communication), an ~ 8 Myr old T Tauri star with a mass accretion rate of $4 \times 10^{-10} - 2 \times 10^{-9}$ M$_\odot$yr^{-1} (Muzerolle *et al.* 2000; Alencar & Basri 2000).

Detecting weaker emission from or placing stringent limits on the gas content in the remainder of the sample will require attention to issues such as correcting for absorption in stellar photospheres. For example, in the case of TW Hya (Fig. 4), comparison of the observed spectrum with a model stellar atmosphere appropriate for its spectral type indicates that CO absorption in the stellar atmosphere is "veiled" by CO emission from the disk. Correcting for CO absorption in the stellar photosphere would uncover emission in higher vibrational transitions of CO. The warmer gas temperatures implied by the detection of emission from higher vibrational levels is consistent with the warm molecular gas temperature that is inferred for this system from UV fluorescent H_2 emission (Herczeg *et al.* 2002).

Since we have yet to reach a conclusion about the lifetime of gas in the terrestrial planet region of disks based on spectroscopic studies, I wanted to try to guess at an answer using other data. A useful data set for this purpose is stellar accretion rates, which have been measured for many young stars (e.g., Gullbring *et al.* 1998; Hartmann *et al.* 1998; Muzerolle *et al.* 1998, 2000; Sicilia-Aguilar *et al.* 2005). If we assume that the accretion onto the star is fed by steady accretion in the disk, then the disk column density $\Sigma \propto \dot{M}_*/\alpha T$ where $\dot{M}_*$ is the accretion rate onto the star, T is the disk temperature, and the viscosity is parameterized as $\nu = \alpha c_s H$ where c_s is the sound speed and H is the disk scale height. Using a relation of this form, we can infer the disk column density given the stellar accretion rate and a value for the viscosity parameter α.

Many of the sources in our sample have stellar accretion rates of a few $\times 10^{-10}$ $M_\odot yr^{-1}$ (e.g., V836 Tau) or upper limits on the accretion rate that are in this range. For $\alpha = 0.01$, a value that is commonly adopted in the literature, accretion rates in this range correspond to column densities at 1 AU of a few $g\,cm^{-2}$, i.e., a dynamically significant column density. Since sources with this level of accretion are detected at ages up to 10 Myr or older, it seems plausible that dynamically significant reservoirs of gas can survive for long enough to influence the outcome of terrestrial planet formation.

Actually measuring total disk gas column densities using the approach described above can confirm this speculation. Once a measurement is in hand, we would know both the accretion rate and the disk column density. This would imply a value of α, which would be another way of constraining the physical mechanism that drives disk accretion.

5. Summary and Future Directions

In summary, infrared spectroscopy of the gas in disks has provided several clues to the processes that govern star and planet formation. It provides evidence for differentially rotating inner disks around young stars. It also suggests that the atmospheres of these disks are turbulent. This may indicate that the disk accretion process is itself turbulent.

The measured excitation temperature of molecular tracers in inner disks demonstrates that the gas is typically hotter than the dust in the same region of the disk. In addition, spectroscopy of multiple gas-phase species provides evidence for non-equilibrium molecular abundances in disk atmospheres. These characteristics are in rough agreement with current thermal-chemical models of inner disk atmospheres.

These results suggest several possible future directions. To explore whether the disk accretion process is turbulent, one might try to extend the techniques described in § 2 to other tracers in order to measure the turbulent line width as a function of disk radius and column density. The results could be compared with theoretical predictions for the turbulence in disk atmospheres that is produced as a consequence of, e.g., the MRI instability. Since turbulence at disk surfaces might also be produced as a consequence of a stellar wind blowing over the disk, it would be of interest to see whether these two scenarios can be distinguished on the basis of this kind of observational data.

Another approach to exploring the extent of turbulence in disk atmospheres would be to look in more detail at the chemical abundances in disks. As described in § 3, if relative abundances cannot be explained by static atmosphere models, this might suggest that vertical mixing (produced by turbulence) is playing a role. Yet another approach is to measure total gas column densities in dissipating disks that show signs of ongoing accretion. As described in § 4, knowing both the mass accretion rate and the disk column density can provide a constraint on the viscosity parameter α.

On the topic of the gas dissipation timescale in disks, there is evidence that gaseous reservoirs in the terrestrial planet region of disks can survive beyond the timescale for the inner disk to become optically thin in the continuum. Stellar accretion rates suggest that these gaseous reservoirs may be dynamically significant and may survive long enough to influence the outcome of terrestrial planet formation. Understanding whether or not this is in fact true requires improvements in data analysis techniques to optimize for the detection of weak emission as well as improvements in models of disk atmospheres in order to convert detections or upper limits into disk gas column densities. These same concerns are relevant to similar studies, now ongoing, to probe the gas dissipation timescale at larger disk radii, in the giant planet region of disks.

Acknowledgements

I would like to thank my colleagues John Carr and Al Glassgold, with whom I have discussed many of the issues reviewed in this contribution.

References

Alencar, S. H. P. & Basri, G. 2000, *Ap. J.* 119, 1881
Balbus, S. A. & Hawley, J. F. 1991, *Ap. J.* 376, 214
Bary, J. S., Weintraub, D. A., & Kastner, J. H. 2003, *Ap. J.* 586, 1138
Beckwith, S. V. W., Sargent, A. I., Chini, R. S., & Guesten, R. 1990, *A. J.* 99, 924
Bergin, E. *et al.* 2004, *Ap. J.* 614, L133
Blake, G. A. & Boogert, A. C. A. 2004, *Ap. J.* 606, L73
Blum, R. D., Barbosa, C. L., Damineli, A., Conti, P. S., & Ridgway, S. 2004, *Ap. J.* 617, 1167
Bodenheimer, P. & Lin, D. N. C. 2002, *Annual Review of Earth and Planetary Sciences* 30, 113
Boss, A. P. 1995, *Science* 276, 1836
Brittain, S. D., Rettig, T. W., Simon, T., Kulesa, C., DiSanti, M. A., & Dello Russo, N. 2003, *Ap. J.* 588, 535
Calvet, N., D'Alessio, P., Hartmann, L, Wilner, D., Walsh, A., & Sitko, M. 2002, *Ap. J.* 568, 1008
Calvet, N. *et al.* 2005, *A. J.* 129, 935
Carr, J. S. 1989, *Ap. J.* 345, 522
Carr, J. S., Tokunaga, A. T., Najita, J., Shu, F. H., & Glassgold, A. E. 1993, *Ap. J.* 411, L37
Carr, J. S., Tokunaga, A. T., & Najita, J. 2004, *Ap. J.* 603, 213
Chandler, C. J., Carlstrom, J. E., Scoville, N. Z., Dent, W. R. F., & Geballe, T. R. 1993, *Ap. J.* 412, L71
D'Alessio, P., Canto, J., Calvet, N., & Lizano, S. 1998, *Ap. J.* 500, 411
Glassgold, A. E., Najita, J., & Igea, J. 2004, *Ap. J.* 615, 972
Gorti, U. & Hollenbach, D. 2004, *Ap. J.* 613, 424
Gullbring, E., Hartmann, L., Briceño, C., & Calvet, N. 1998, *Ap. J.* 492, 323
Hartigan, P., Edwards, S., & Ghandour, L. 1995, *Ap. J.* 452, 736
Hartmann, L., Calvet, N., Gullbring, E., & D'Alessio, P. 1998, *Ap. J.* 495, 385
Herbig, G. H. & Bell, K. R. 1988, Lick Obs. Publ. 1111.
Herczeg, G. J., Linsky, J. L., Valenti, J. A., Johns-Krull, C. M., & Wood, B. E. 2002, *Ap. J.* 572, 310
Jonkheid, B., Faas, F. G. A., van Zadelhoff, G.-J., & van Dishoeck, E. F. 2004, *A&A* 428, 511
Kamp, I. & Dullemond, C. P. 2004, *Ap. J.* 615, 991
Klahr, H. H. & Bodenheimer, P. 2003, *Ap. J.* 582, 869
Kominami, J. & Ida, S. 2002, *Icarus* 157, 43

Mayer, L., Quinn, T., Wadsley, J., & Stadel, J. 2002, *Science* 298, 1756

Muzerolle, J., Hartmann, L., & Calvet, N. 1998, *A. J.* 116, 2965

Muzerolle, J., Calvet, N., Briceño, C., Hartmann, L., & Hillenbrand, L. 2000, *Ap. J.* 535, L47

Najita, J., Carr, J. S., Glassgold, A. E., Shu, F. H., & Tokunaga, A. T. 1996, *Ap. J.* 462, 919

Najita, J., Edwards, S., Basri, G., & Carr, J. 2000, in *Protostars and Planets IV.* ed. V. Mannings, A. P. Boss, & S. S. Russell (University of Arizona Press), p. 457

Najita, J., Carr, J. S., & Mathieu, R. D. 2003, *Ap. J.* 589, 931

Najita, J. 2004, in *Star Formation in the Interstellar Medium: In Honor of David Hollenbach, Chris McKee and Frank Shu* ASP Conference Proceedings 323, ed. Johnstone, F.C. Adams, D.N.C. Lin, D.A. Neufeld, & E.C. Ostriker (Astronomical Society of the Pacific), p. 271.

Rettig, T. W., Haywood, J., Simon, T., Brittain, S. D., & Gibb, E. 2004, *Ap. J.* 616, L163

Richter, M. J., Jaffe, D. T., Blake, G. A., & Lacy, J. H. 2002, *Ap. J.* 572, L161

Sako, S., *et al.* 2005, *Ap. J.* 620, 347

Scoville, N. Z., Hall, D. N. B., Ridgway, S. T., & Kleinmann, S. G. 1979, *Ap. J.* 232, L121

Sheret, I., Ramsay Howat, S. K., & Dent, W. R. F. 2003, *MNRAS* 343, L65

Sicilia-Aguilar, A., Hartmann, L. W., Hernández, J., Briceño, C., & Calvet, N. 2005, *A. J.* 130, 188

Siess, L., Forestini, M., & Bertout, C. 1999, *AA* 342, 480

Stone, J. M., Gammie, C. F., Balbus, S. A., & Hawley, J. F. 2000, in *Protostars and Planets IV..* ed. V. Mannings, A. P. Boss, & S. S. Russell (University of Arizona Press), p. 589

Thi, W.-F. & Bik, A. 2005, *A&A* 438, 557

Discussion

WILSON: Have you thought about whether these techniques would be useful for searching for gas in debris disks? Because it would be very interesting to know if those systems have gas an how much gas they have.

NAJITA: That's a good point. I did not have time to mention that indeed several different groups are currently carrying out such searches, not so much with the specific diagnostics that I have been describing as with atomic and molecular transitions in the UV and mid-infrared. As examples of current work in this field, a recent calculation of the expected mid-infrared emission from residual gas in debris disks has been carried out by Gorti & Hollenbach (2004), and searches for these emission features are underway using Spitzer/IRS (e.g., the "Formation and Evolution of Planetary Systems" Legacy project; M. Meyer, PI) and ground-based telescopes (e.g., Richter *et al.* 2002; Sheret *et al.* 2003; Sako *et al.* 2005).

FORREST: The depletion of dust in the upper layers of T Tauri disks is also indicated by modelling of the infrared $5 - 35$ μm spectra by Calvet *et al.* (2005) and Furlan *et al.* (2005, in preparation).

NAJITA: Those data certainly provide additional insight. Specifically, the low-resolution spectra tell us about the dust in the disk. Sometimes, as in the case of TW Hya, the spectral energy distribution (SED) indicates a deficit of continuum emission that can be interpreted as an inner hole to the dust opacity. In addition, using a model for how the gas and dust are distributed relative to each other (e.g., with or without settling), one might also be able to use the SED to diagnose the extent of grain depletion relative to the gas. The UV H_2 transitions that I discussed provide a complementary view. They more directly trace the gas component and constrain the gas-to-dust ratio along a specific line of sight.

Astrochemistry: Recent Successes and Current Challenges
Proceedings IAU Symposium No. 231, 2005
D.C. Lis, G.A. Blake & E. Herbst, eds.

© 2006 International Astronomical Union
doi:10.1017/S1743921306007411

Chemical Models of Inner Disks

A. J. Markwick-Kemper

University of Virginia, Department of Astronomy, Charlottesville VA 22904, USA
email: ajmk@virginia.edu

Abstract. In this paper I describe some of the recent advances made in the modelling of chemistry of protoplanetary disks, and in particular, of the inner regions where $r < 100$ AU. These advances include the treatment of mass-transport processes, the interaction of radiation with chemistry, the augmentation of chemical networks to include isotopic species and their fractionation, and the attempt to connect disk chemistry with solar system bodies like comets. In the spirit of the title of this volume, I also briefly describe what in my opinion are the current challenges facing models of disk chemistry. These include the unification of chemistry, radiation and dynamics, the treatment of gas-grain interaction and the usefulness of observations to discriminate the different models in the literature.

Keywords. astrochemistry — comets: general — solar system: formation — stars: planetary systems: formation — stars: planetary systems: protoplanetary disks

1. Motivation

Chemistry in protoplanetary disks has been studied by various authors in recent years (for a review, see Markwick & Charnley 2004a). It is important to study these objects theoretically for reasons related to observations – both current and forthcoming – for example to determine which molecules make good tracers of the disk temperature, density, ionization fraction and radiation field, and to learn how to discriminate between the different models of disk structure currently available. To date, observations have generally been limited to the outer regions of disks, but that is changing. Infrared observations already trace the inner 10 AU, and ALMA will be able to image disks on AU scales, at least in Taurus. Hence the need for good physical/chemical models of the inner disk.

There are other reasons however, for modelling chemistry in disks, which are not so obviously related to their direct observation. We can ask simply what processes affect the chemical structure of the disk – is the coupling between chemistry and dynamics or between chemistry and radiation important? Some recent studies have addressed these questions directly. It turns out, of course, that chemistry, dynamics and radiation are intimately linked in a "triangle of pain", which almost certainly means that no simple back-of-the-envelope calculation is going to be sufficient for accurately describing the chemical composition of disks.

We can realize that, as their name implies, these disks are the precursors of solar systems. We have a great deal of observational data for our solar system, at least in its current state, but also good indications of what it was like primordially from observations of comets and meteorites. There is little argument that the chemical composition of comets is representative of the chemical state of the solar system at some time in its past – although how far back is not so clear cut. Protoplanetary disks form from collapsed interstellar gas and comets form between 5 and 40 AU in the disk. They are perhaps the most pristine objects in solar systems, but are they unprocessed interstellar material? This question is important since it relates directly to what material could have been brought to the early Earth by comets.

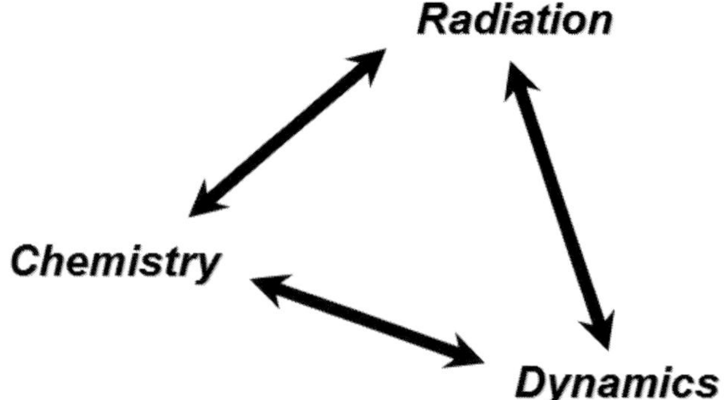

Figure 1. Disk chemistry's "triangle of pain". Chemistry, radiation and dynamics interact, rendering their consistent solution necessary. Unfortunately, this is difficult.

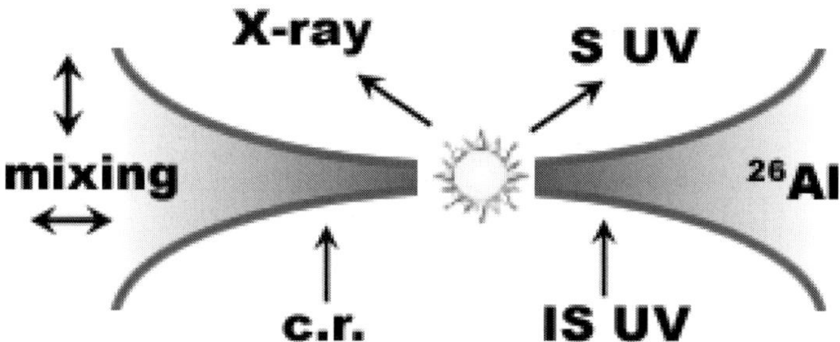

Figure 2. Schematic diagram of the protosolar nebula showing physical processes which affect chemistry. These include sources of ionization (stellar X-rays, stellar UV, interstellar UV, cosmic rays, radionuclides), heating (viscous dissipation, stellar radiation) and transport (*e.g.*, radial mixing, diffusion).

2. Recent Successes

Many physical processes should be included in a model of a protoplanetary disk, such as sources of ionization (stellar X-rays, stellar UV, interstellar UV, cosmic rays, radionuclides), heating (viscous dissipation, stellar radiation) and transport (*e.g.*, radial mixing, diffusion; Figure 2). In a model which only considers the midplane, some of these processes can be safely ignored. The temperature and density structure, however, remain crucial. Of the many different models in the literature (see Markwick & Charnley 2004a), most of them use a different disk structure.

2.1. *Structure*

The temperature profile is important because it controls the return of material into the gas phase from grain surfaces. To illustrate the difference the temperature profile can make on the molecular distributions in the inner disk, Figure 3 compares the distribution of H_2CO between the models of Markwick *et al.* (2002) and Millar *et al.* (2003). The *only* difference between these two models is that a different physical description of the disk was used in each case. In the former, the hydrostatic disk is heated only by viscous dissipation, while in the latter, the disk is heated both by viscous dissipation and stellar irradiation. The abundance of gas-phase formaldehyde follows the temperature structure of the disk very closely, due to the molecule's thermal desorption from grain surfaces.

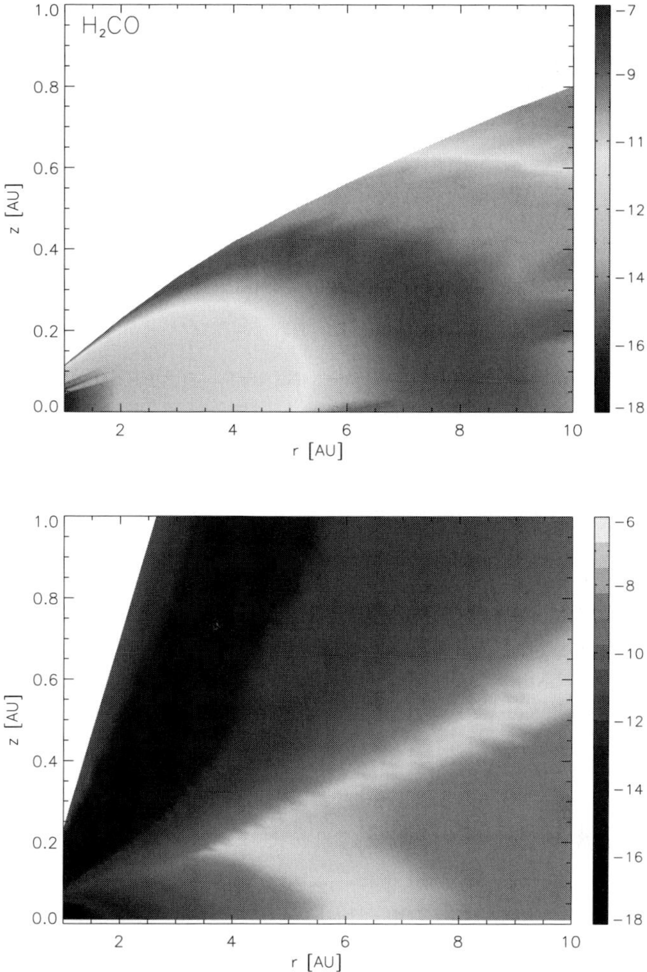

Figure 3. Comparison of the gas-phase H_2CO distribution in the inner disk produced with models using different physical structures for the disk, but with exactly the same chemistry. *Top:* Disk heated only by viscous dissipation (Markwick *et al.* 2002); *Bottom:* Disk heated by viscous dissipation and stellar radiation (Millar *et al.* 2003).

2.2. *Transport*

Other works have attempted to identify the differences that including dynamical processes makes on the chemical structure of the disk. Ilgner *et al.* (2004) compared a model with and without diffusive mass transport. The results were striking, especially for sulphur species like CS (Figure 4), indicating that dynamical processes should certainly be taken into account when modelling chemistry in the inner disk. Semenov *et al.* (this meeting), presented a 2D mixing model for a disk, in which they found the species HCN, HCN and HCO^+ were most affected by transport processes, and that in fact the HCN/HNC line ratio may be a tracer of disk diffusion. Their presentation is available on the conference website at `http://asilomar.caltech.edu`.

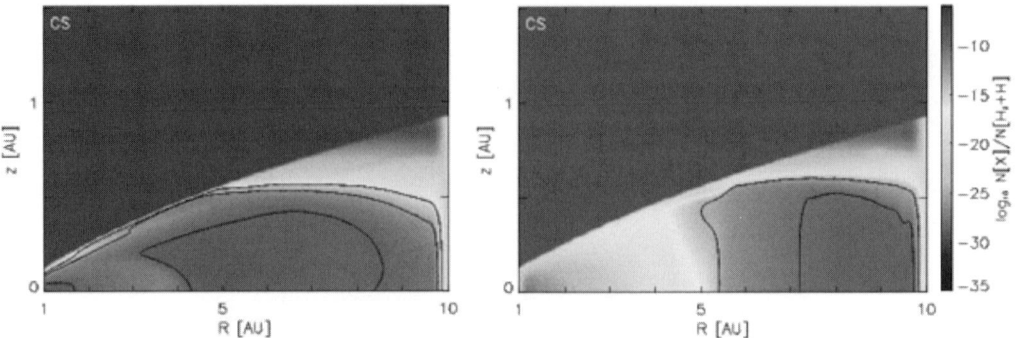

Figure 4. Comparison of the gas-phase CS distribution (log fractional abundance) in the inner disk produced with models with (*right*) and without (*left*) diffusive mass transport. (Ilgner *et al.* 2004).

2.3. *Radiation*

Considering the affect of radiation on chemistry is also important, and van Zadelhoff *et al.* (2003) used radiative transfer to model UV photo-processes in the disk. They calculated their model with different stellar radiation fields, and found in particular that the CN/HCN ratio depends strongly on the stellar spectrum. On the other hand, the HCO^+/CO ratio does not. Willacy & Langer (2000) described a detailed photo-processing calculation. They included photodesorption, finding that significant quantities of molecules can be kept in the gas phase even at visual extinctions in excess of 4 magnitudes.

2.4. *Isotopes*

Aikawa & Herbst (1999) presented the first disk model to include deuterium fractionation reactions, and therefore were able to draw conclusions between the fractionation in disks and comets. Isotope fractionation consistently proves a valuable tool in astrochemistry, and in the study of disks things are no different. The level of deuterium fractionation in molecules is very sensitive to temperature, and, subject to certain assumptions, can be used to trace the thermal processing history of interstellar material as it passes through the protoplanetary disk stage and becomes incorporated in solar system bodies. One can hope to use carbon and oxygen isotope fractionation in this way too.

To attempt to further investigate the connection between cometary and interstellar material through disks, we constructed a model in which the chemistry is, as usual, based on the UMIST Database for Astrochemistry (`http://udfa.net`; Le Teuff *et al.* 2000; Millar *et al.* 1995), but now augmented to include the deuterium chemistry of Roberts *et al.* (2003). The initial condition for the chemistry at 100 AU is the output from a dense-core model (using the same chemistry of course), in which the temperature is 10 K (Markwick-Kemper, unpublished). In these models, the solid HDO/H_2O ratio is 0.1, much higher than the observed ratio in comets. From this initial condition, we calculated two models – one assuming there is no thermal processing of material as it moves inward along the midplane, and another simulating this effect by having the material encounter a region where the temperature rises to $\sim$ 30 K. These models are described in more detail in Markwick & Charnley (2004b). Table 1 compares the results so obtained with observations of deuterated molecules in comets. It is clear that the ratios are better matched by the model with thermal processing. The ratios are decreased from their initial values because the heating removes the species from dust grains, the gas-phase

Table 1. D/H observations and upper limits reported in comets.

Ratio	Observed	Unprocessed	Processed
HDO/H_2O	0.0006	0.07	0.0004
DCN/HCN	0.002	0.01	0.002
CH_3D/CH_4	<0.3	0.1	0.14
$HDCO/H_2CO$	<0.1	>0.06	0.015
CH_2DOH/CH_3OH	<0.04	>0.1	0.0003
CH_3OD/CH_3OH	<0.01	>0.05	0.0004
C_2HD/C_2H_2	–	0.003	0.09
DC_3N/HC_3N	–	0.001	0.04
accreting D/H	–	0.6	0.005

Notes – D/H observations and upper limits reported in comets, compared with numbers from the two models discussed in the text. Also shown is the accreting gas-phase D/H ratio in the two models, and predictions for C_2HD and DC_3N. For the unprocessed model, the HDCO and methanol ratios will be lower limits due to grain-surface fractionation.

D/H ratios subsequently adjust to the warmer temperature, and then cooling freezes the molecules out onto dust grains at a temperature of around 30 K.

This temperature is supported by other observational results for comets. For example, Kawakita *et al.* (2004) measured the spin temperatures of ammonia and water molecules in comets. They found that the molecules equilibrated at 26–35 K. In addition, the discovery of crystalline silicates in comet Hale-Bopp indicates that pre-cometary amorphous silicates experienced a significant degree of heating (Wooden 2002).

Table 1 gives predicted values for C_2HD/C_2H_2 and DC_3N/HC_3N. These species were chosen firstly because the main isotopologue has been observed in comets, and secondly because the ratio shows a marked difference between the two models and therefore provides a further observational test. The difference is possible for certain species (like CH_4 as well) because there are competing routes to fractionation. At 30 K for example, the main H_3^+ route is inefficient, but the $C_2H_2^+$ and CH_3^+ routes, which have different temperature dependencies are not, so species can be fractionated through these ions at higher temperatures.

We now present for the first time results akin to those described above for carbon and oxygen isotope fractionation. The chemistry was augmented again to include ^{13}C-, ^{17}O- and ^{18}O-bearing species, for carbon chains up to C_3. This is a major modification of the chemistry, requiring manifest assumptions about the chemistry of carbon chains, and will be described in detail elsewhere (Markwick-Kemper, in prep.) Table 2 presents results from the same models with and without thermal processing, for comparison with cometary observations. Also presented are some predictions for species for which the main ^{12}C isotopologue has been observed. Of these, CS appears to provide a good test of the model.

2.5. *Fractional Ionization*

Semenov, Wiebe & Henning (2004) presented a detailed analysis of the source of fractional ionization at various locations in a protoplanetary disk. They successfully used chemical network reduction techniques to isolate the chemistries responsible for providing the ionization, in some cases finding networks with as few as 10 species. Recent work including deuterated species has shown that deuterated forms of H_3^+ will be the dominant ions in the disk midplane (Ceccarelli & Dominik 2005; Markwick & Charnley 2004b). Indeed, in the model described above, D_3^+ is by far the most abundant ion in the midplane outside of 40 AU. Although the direct detection of D_3^+ in a disk is difficult, because it requires a

Table 2. $^{12}C/^{13}C$ observations reported in comets.

Ratio	Observed	Unprocessed	Processed
$CO/C^{18}O$		498.3	492.2
$CO/^{13}CO$		75.0	72.0
$^{13}CO/^{13}C^{18}O$		495.7	490.7
$CS/^{13}CS$		74.9	62.2
$CN/^{13}CN$	$95 \pm 12, 90 \pm 10$	77.7	86.3
$HCN/H^{13}CN$	$90 \pm 15, 109 \pm 22, 111 \pm 12$	76.1	86.8
$C_2/C^{13}C$	93 ± 10	37.6	43.1
$H_2O/H_2^{18}O$	$518 \pm 45, 470 \pm 40$	502.2	502.9
$CO_2/^{13}CO_2$		80.1	73.0
$C_2H_2/C^{13}CH_2$		46.5	80.9

Notes – $^{12}C/^{13}C$ observations reported in comets, compared with numbers from the two models discussed in the text. The input ratios were $^{12}C/^{13}C = 75$, $O/^{18}O = 500$. The terrestrial ratios are $^{12}C/^{13}C = 89$, $O/^{18}O = 498$.

background source against which to see the absorption, this does at least opens up the possibility of using observations of H_2D^+ and/or HD_2^+, together with models, to measure the ionization fraction of the midplane, which is of course a parameter of consequence to MHD studies of disks.

3. Current Challenges

3.1. *Coupling*

Figure 2 shows some of the processes which ought to be included in a model of the inner disk. However, there are currently no models that include all these processes – no models that survive the "triangle of pain". This is a major current challenge of protoplanetary disk modelling. The models described in the previous section have done well in investigating individually the influence of dynamics and radiation on the chemical structure, and indeed have shown that such interactions are pertinent and should be modelled consistently. What is still required, though, is a true coupling of 2D radiation hydrodynamics and chemistry.

3.2. *Discrimination*

The many models presented in the literature include different processes and sometimes in different ways. However, it is important to calculate whether or not these differences are actually observable or not. With ALMA, for example, we will be able to image disks on solar system scales, and ideally we would like to be able to discriminate between the various models using these observations. It is not clear that we will be able to. What is really required is that a radiative transfer calculation of the observable line profiles be performed in the same way for each of the different chemical model result sets.

3.3. *Gas-Grain Interaction and Grain-Surface Chemistry*

If the temperature profile is so important to the chemical structure of the inner disk, as model results show, then the binding energies of species to grain surfaces, grain-surface chemical processes, and the exact thermal desorption mechanism are of crucial importance. Unfortunately, it is by no means clear that the values for the binding energies used in models, or the means in which thermal desorption is included, are meaningful. More laboratory work like that described by McCoustra (this volume) is required to better characterize gas-grain interactions, and more experiments and calculations like those presented by Watanabe and Charnley, respectively (both this volume), are needed to constrain our expectations of grain-surface chemistry. Finally, all these experimental and theoretical data needs to be incorporated into models.

3.4. *Completeness*

The case of multiple deuterium fractionation in general highlights another of the problems facing astrochemical models. That is, the models cannot make predictions for species which are not included in them. This, above all, is an argument for making the models as chemically complete as possible, and is an argument against the reduction of chemical networks.

4. Conclusion

Modelling inner disk chemistry is a complicated affair. There have been many improvements and advances in our understanding of the chemistry occurring in the planet and comet forming region of protoplanetary disks since the last IAU Astrochemistry Symposium in Korea and in this paper I have described some of them. It is my hope that some of the challenges facing these models will be met and we will read about them in the proceedings of the next of these meetings.

References

Aikawa, Y. & Herbst, E. 1999, *Ap. J.* 526, 314
Ceccarelli, C. & Dominik, C. 2005, *A&A* 440, 583
Ilgner, M., Henning, Th., Markwick, A.J., & Millar, T.J. 2004, *A&A* 415, 643
Kawakita, H., Watanabe, J.-I., Furusho, R., Fuse, T., & Cremonese, G. 2004, *Ap. J.* 601, 1152
Le Teuff, Y.H., Millar, T.J., & Markwick, A.J. 2000, *A&AS* 146, 157
Markwick, A.J. & Charnley, S.B., 2004a, in "Astrobiology: Future Perspectives" ed. P. Ehrenfreund *et al.*, proceedings of ISSI Workshop (Kluwer), in press
Markwick, A.J. & Charnley, S.B. 2004b, in *Formation of Cometary Material*, 25th Meeting of the IAU, Joint Discussion 14, proceedings, in press
Markwick, A.J., Ilgner, M., Millar, T.J., & Henning, Th. 2002, *A&A* 381, 632
Millar, T.J., Nomura, H., & Markwick, A.J. 2003, *Astron. Space Sci.* 285, 761
Millar, T.J., Farquhar, P.R.A., & Willacy, K. 1995, *A&AS* 121, 139
Roberts, H., Herbst, E., & Millar, T.J. 2003, *Ap. J.* 591, L41
Semenov, D.A., Wiebe, D.S., Henning, Th., & Pavlyuchenkov, Y.N. 2005, *IAUS 231 Poster*, `http://asilomar.caltech.edu`
Semenov, D.A., Wiebe, D.S., & Henning, Th. 2004, *A&A* 417, 93
van Zadelhoff, G.-J., Aikawa, Y., Hogerheijde, M.R., & van Dishoeck, E.F. 2003, *A&A* 397, 789
Willacy, K. & Langer, W.D. 2000, *Ap. J.* 544, 903
Wooden, D.H. 2002, *Earth, Moon & Planets* 89, 247

Discussion

MILLAR: Is grain-surface recombination of D_3^+ included in the model?

MARKWICK-KEMPER: Yes, it is. In fact, all molecular ions in the model are assumed to recombine on grain surfaces.

RAWLINGS: You are right to highlight the importance of knowing the correct values for the binding energies, but desorption is not a binary process. Rather, the temperature-dependence of desorption may be very complex. Does this have significant implications for your models?

MARKWICK-KEMPER: Yes, the exact details of the desorption process are very important. I referred to desorption as a 'switch' because of the way the temperature changes in the disk. The temperature gradient is so steep that material is thermally desorbed from the grains over a short distance. Of course, the way we build the gas-grain interaction into the models allows this sort of thing to happen. Our model of gas-grain interaction is probably too simplistic.

NEUFELD: Since you have waxed lyrical about the role of theory, I wanted to comment that the fractional ionization in the disk midplane, and the mass of the charge carriers, is a critical parameter in fundamental studies of accretion. Both at your and my institutions, our colleagues have substantial efforts underway to understand the magnetorotational instability; thus the ionization in disks is a crucial point of contact between astrochemistry and the wider astrophysics community.

MARKWICK-KEMPER: I agree.

Astrochemistry: Recent Successes and Current Challenges
Proceedings IAU Symposium No. 231, 2005
D.C. Lis, G.A. Blake & E. Herbst, eds.
© 2006 International Astronomical Union
doi:10.1017/S1743921306007423

Desorption of Molecules from Grain Mantles

Mark P. Collings and Martin R. S. McCoustra†

School of Chemistry, University of Nottingham, University Park,
Nottingham, NG7 2RD, UK.

Abstract. As a prerequisite to returning their molecular inventory to the gas phase, icy grain mantles must desorb from the refractory core material of the grain. This desorption process can be instigated through interactions with photons or cosmic rays but can equally well be thermally driven. In this invited paper, the application of thermal desorption techniques borrowed from ultrahigh vacuum surface science to problems of astronomical interest will be discussed. The experimental methods employed by surface scientists to probe thermal desorption processes will be described and the analysis of the resulting empirical data outlined. The results of recent laboratory measurements from a number of groups will be highlighted.

Keywords. astrochemistry — ISM: molecules — methods: laboratory — molecular processes

1. Introduction

Icy grain mantles formed from a combination of reactive accretion and condensation from the gas phase are observed in the infrared along many lines of sight toward dense molecular clouds (Whittet *et al.* 1996; Gibb *et al.* 2000). These icy mantles act as a reservoir of volatile small molecules that are returned to the gas phase in the early stage of pre-stellar core collapse and aid in radiative cooling (Williams 1998). The mantles also act as chemical nanofactories wherein these simple molecules are subject to energetic processing induced through interaction with photons and cosmic rays, converting them into more complex chemical species (for example Bernstein *et al.* 2002; Muñoz Caro *et al.* 2002; Gerakines, Moore & Hudson 2004; Loeffler *et al.* 2005; Holtom *et al.* 2005). These in turn may be returned to the gas phase as core collapse proceeds. Once there, such complex molecules could be detected if there is a sufficiently high concentration (Kuan *et al.* 2003). Alternatively, grain-grain collisions can consolidate the material into larger grain clusters that may eventually evolve into the comets capable of seeding a nascent planet with a watery ice rich in potentially pre-biotic materials (Wang *et al.* 2005).

As return of material from the icy grain mantle to the gas phase is crucial to their astronomical role, it becomes important to understand this process. In quiescent, dark regions only desorption processes associated with the interaction of cosmic rays and penetrating high-energy photons with grain mantles are likely to be important. Laboratory studies and theoretical models of the sputtering process associated with direct collisions between cosmic ray nuclei and mantle molecules are well developed (Baragiola *et al.* 2003; Vidal, Teolis & Baragiola 2005). Less so is our understanding of desorption induced by the secondary electron cascade associated with the passage of such high-energy particles through an ice film. This is also likely to be crucial in quantifying desorption induced by the high-energy photons (X-rays and γ-rays) that can penetrate optically dense regions. In optically less opaque regions, visible, ultraviolet and vacuum ultraviolet photon-stimulated desorption becomes feasible. Laboratory investigations of such photon-stimulated desorption are also well developed (Westley *et al.* 1995; Bergeld *et al.*

† Corresponding author, email: martin.mccoustra@nottingham.ac.uk

2004). However, in star-forming regions simple thermally-promoted desorption probably dominates as pre-stellar cores collapse and warm. Laboratory studies of the thermal desorption process have been reported in the astrochemical literature for a number of years (Sandford & Allamandola 1988; Hudson & Donn 1991; Notesco & Bar-Nun 2005). These have generally utilised high vacuum ($\sim 10^{-7}$ mbar) techniques coupled with, for example, transmission infrared spectroscopic measurements of the icy film itself, total pressure measurements using an ion gauge or low-quality mass spectrometric studies of the desorbing mantle materials. However, as we have demonstrated previously, high-vacuum techniques are insufficient if the purpose of the experiment is to probe the interactions occurring at the grain surface (Collings et $al.$ 2002). In high vacuum, a surface at low temperature will be completely covered by adsorption of background species (mostly water) in roughly 10 s. However, in ultrahigh vacuum ($\sim 10^{-10}$ mbar), saturation of a cold surface takes in the order of 10^4 s. Therefore, experiments can be completed before surface contamination becomes an overwhelming problem. To this end, we describe the application of ultrahigh vacuum (UHV) temperature-programmed desorption (TPD) techniques to the study of the thermally promoted desorption of icy grain mantles. Specifically we will review desorption of single-component ices and report on the issue of ice morphology as highlighted in our own recent work.

2. Experimental

The technique of temperature programmed desorption is thoroughly described in the surface science literature (Woodruff & Delchar 1994, Chapter 5; Kolasinski 2002, pp. 193–201; King 1975). The basic apparatus is shown schematically in Figure 1(a), while a practical realisation of the ideal can be found in our own work (Fraser, Collings & McCoustra 2002). In the experiment itself, the sample is heated uniformly at a constant rate. The gases evolved from the surface are detected mass spectrometrically and displayed as a plot of mass spectrometer signal $versus$ temperature. At this stage, we must realise that the efficiency with which the desorbing gas is removed from the UHV chamber is crucial to the quality of the resulting TPD traces. This is illustrated in Figure 2(a) which presents simulated TPD data for carbon monoxide chemisorbed on a platinum surface. At low pumping speeds, the desorbate accumulates in the UHV chamber. This is the so-called flash desorption limit, when the pumping speed is small compared to the rate of desorption. As the pumping speed is increased, in Figure 2(a) by a factor of 10 at each step, we see a transformation from the sigmoidal curve characteristic of flash desorption to the peaks characteristic of true TPD experiments. We should also note that within the TPD regime, increasing the pumping speed further simply reduces the detected yield of desorbate requiring that we increase the sensitivity of the mass spectrometer.

Likewise, the sample heating can play a significant role in determining the quality of the recorded TPD data. Figure 2(b) illustrates the effect of increasing the heating rate on the same system as considered in Figure 2(a). At low heating rates, good quality TPD peaks can be observed if highly-sensitive mass spectrometric detection is employed. As the heating rate increases, the TPD peaks broaden and move towards a shape more characteristic of flash desorption. Of course, this behaviour assumes that the substrate is heated evenly. In real systems, uneven sample heating as a consequence of sample size, sample mounting or heating method may give rise to peak asymmetry or even multiple TPD peaks for a system that is chemically simple. Reducing the sample size, careful design of sample mounts and optimising the heating rate to achieve uniform heating across the sample can reduce these problems. However, to eliminate them entirely,

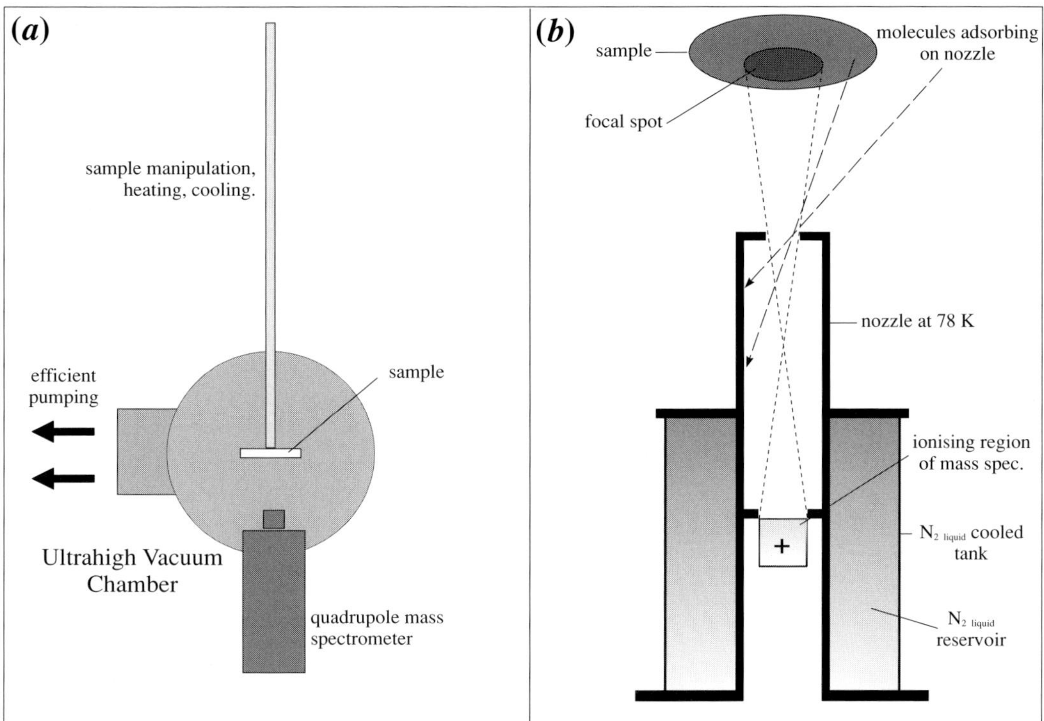

Figure 1. Schematic diagrams showing the typical layouts of (*a*) a TPD experiment in UHV, and (*b*) a line of sight TPD apparatus.

line-of-sight TPD methods as championed by Jones and co-workers can be adopted (Jones & Turton 1997; Jones & Fisher 1999), as illustrated in Figure 1(*b*).

3. Results and Discussion

Temperature programmed desorption is a kinetic technique. There is no simple relationship between the positions of peaks in a TPD trace and the energetics of the desorption processes under investigation other than a simple statement that the more strongly a species is bound to a surface, the higher the temperature at which it will desorb. Analysis of TPD data requires that we develop a kinetic model for the desorption process. This is based on the Polanyi-Wigner equation,

$$-\frac{\mathrm{d}C_s}{\mathrm{d}t} = \nu_{des}C_s{}^{n}\mathrm{e}^{-E_{des}/RT} \tag{3.1}$$

where the rate of desorption $\left(-\frac{\mathrm{d}C_s}{\mathrm{d}t}\right)$ is related to the surface concentration of adsorbate, C_s, and surface temperature, T, through the order of desorption, n, the activation energy for desorption (the surface binding energy), E_{des}, and the pre-exponential factor for desorption, ν_{des}. The activation energy and pre-exponential factor for desorption potentially can depend on the adsorbate surface concentration; *i.e* $E_{des}(C_s)$ and $\nu_{des}(C_s)$. The order of desorption, activation energy for desorption and the pre-exponential factor for desorption are all experimentally determined and can, at least in principle, be recovered by appropriate analysis of TPD data as a function of the initial surface concentration of the desorbate. To illustrate this we will look at two examples: our own work on desorption of

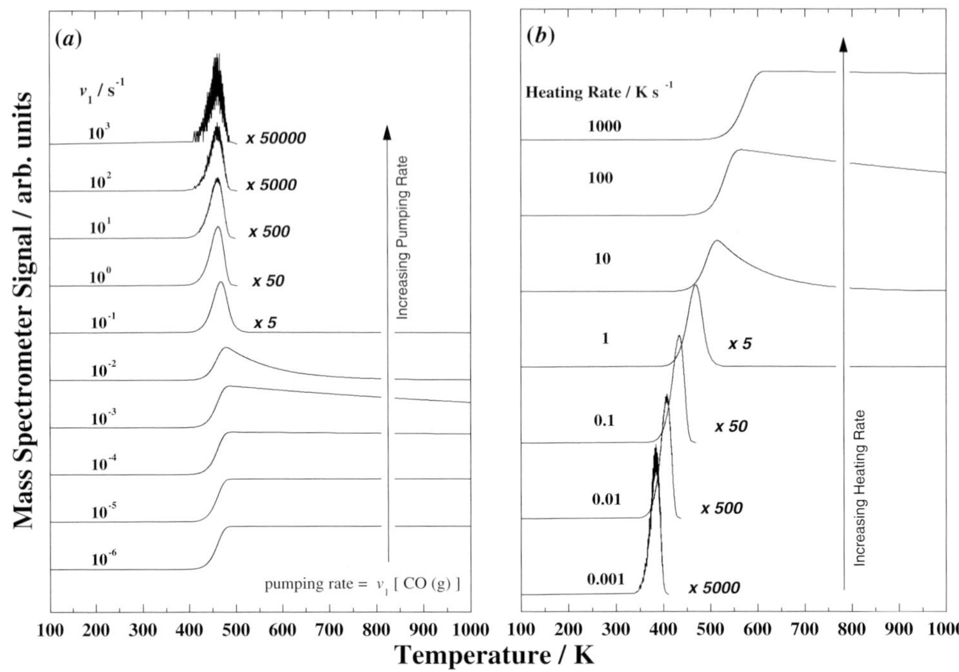

Figure 2. Model TPD spectra for a monolayer of CO desorbing from a platinum substrate illustrating the effects of (*a*) increasing the relative speed at which the desorbing CO is removed by pumping from the UHV chamber, and (*b*) increasing the heating rate. The binding energy of CO on the Pt surface is assumed to be 125 kJ mol^{-1} and a pre-exponential factor of 10^{13} s^{-1} is assumed for the first order desorption of CO from the Pt surface. The heating rate, β, is fixed at 1 K s^{-1} in (*a*), and the pre-exponential factor for pumping, ν_1, is fixed at 0.1 s^{-1} in (*b*).

water ice (Fraser *et al.* 2001) and recent work from the group of Brown and co-workers on desorption of methanol ice (Bolina 2005; Bolina, Wolff & Brown 2005a).

Figure 3(*a*) illustrates the TPD of water ice films of increasingly thickness in the range up to *ca.* 25 nm reproduced from Fraser *et al.* (2001). The TPD traces show a characteristic overlap of their exponential-like leading edges and monotonic increases in peak desorption temperature with film thickness. These observations are consistent with the water desorbing from the gold substrate in our apparatus with zeroth order kinetics, *i.e.*,

$$-\frac{\mathrm{d}C_s}{\mathrm{d}t} = \nu_{des}\mathrm{e}^{-E_{des}/RT} = k_{des,ice} \tag{3.2}$$

The desorption rate is independent of the amount of water ice on the substrate. The TPD of bulk solid materials is generally observed to follow zeroth order kinetics. The desorption of monolayers of intact molecules in contrast is generally taken to be first order, while the recombinative desorption of dissociated molecules is second order.

As a starting point, we then assumed the binding energy for water to a water ice multilayer from the work of Kay and co-workers (Speedy *et al.* 1996) and equivalent to the latent heat of sublimation of water ice. We also assumed that this value is independent of the water ice film thickness. The only unknown in our rate equation was therefore

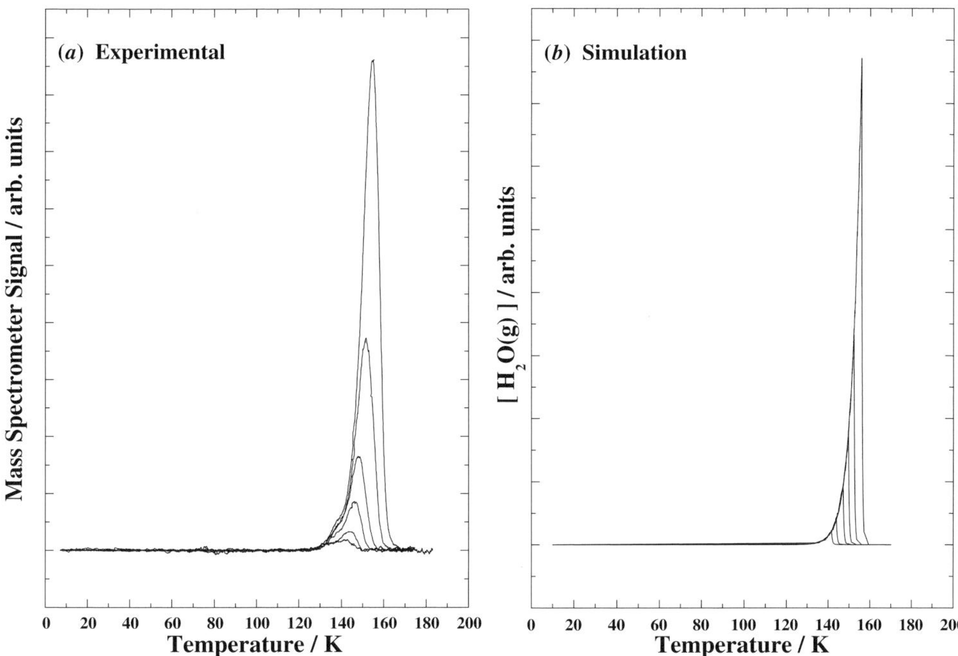

Figure 3. (*a*) A sequence of TPD spectra for water ice films for exposures of the cold substrate to 1.2, 2.4, 6.1, 12, 24 and 61 Langmuirs (L) of water vapour, $\beta = 0.02$ K s^{-1}. 1 L (10^{-6} mbar s) is approximately equivalent to 1 monolayer. (*b*) Kinetic simulations of the TPD of water ice films of thickness equivalent to Figure 3(*a*), using the simple model described in the text.

the pre-exponential factor for the desorption process. A simple kinetic model was then constructed:

$$H_2O(ice) \rightarrow H_2O(g) \quad \text{for which} \quad -\frac{dC_{H_2O(ice)}}{dt} = k_{des,ice}$$

$$H_2O(ads) \rightarrow H_2O(g) \quad \text{for which} \quad -\frac{dC_{H_2O(ads)}}{dt} = k_{des,ads}C_{H_2O(ads)}$$

$$H_2O(g) \rightarrow H_2O(pumped) \quad \text{for which} \quad -\frac{dC_{H_2O(g)}}{dt} = k_{pumping}$$

where $H_2O(ice)$ represents water in the ice multilayer, $H_2O(ads)$ represents water adsorbed directly on the gold substrate, $H_2O(g)$ is water vapour and $H_2O(pumped)$ is the water pumped out of the UHV chamber. Kinetic simulations were then run and fitted to our observed TPD data as function of film thickness allowing recovery of optimal values of $E_{des,ice}$ and $\nu_{des,ice}$ for the zeroth order desorption of water ice multilayers and $E_{des,ads}$ and $\nu_{des,ads}$ for the desorption of the water monolayer from our gold substrate. The kinetic parameters for the zeroth order desorption of water ice are summarised in Table 1, while Figure 3(*b*) illustrates the quality of the comparison between our experimental and modelled TPD data.

More recently, Brown and co-workers have adopted an analysis that avoids the assumptions of order and desorption energy used in our own work (Bolina 2005; Bolina, Wolff & Brown 2005a; Bolina, Wolff & Brown 2005b; Bolina & Brown 2005). Their analysis depends on recognising that the mass spectrometer signal as a function of temperature,

Table 1. Summary of the thermal desorption data for astronomically relevant ices recovered from analyses of UHV TPD experiments.

Species	n	E_{des} / kJ mol^{-1}	ν / cm^{-2} s^{-1}
$H_2O^{a,b,c}$	$0, 0, 0.26(0.02)$	$49.8, 48.0(0.5), 39.9(0.8)$	$4.58 \times 10^{30}, 10^{30}, 5.5(4.5) \times 10^{31}$
CH_3OH^d	$0.35(0.21)$	$31 \sim 40$	$-$
NH_3^e	$0.25(0.05)$	$23.3(1.2)$	$8(3) \times 10^{29}$
$CO^{f,g}$	0	$6.87(0.2), 7.11(0.21)$	$7(2) \times 10^{26}, 10^{30(2)}$
N_2^g	0	$6.57(0.21)$	$10^{30(2)}$

Notes: Error values, where available, are shown bracketed. Data from (a) Speedy et $al.$ (1996), (b) Fraser et $al.$ (2001), (c) Bolina, Wolff & Brown (2005b), (d) Bolina, Wolff & Brown (2005a), (e) Bolina & Brown (2005), (f) Collings et $al.$ (2003b), (g) Örberg et $al.$ (2005). Citations to other examples where the kinetic parameters for desorption of these species are published can be found within these references. *For simplicity, the units given for ν_{des} relate to a zeroth order process; the actual units can be fractional and depend on the (non-zero) value of n.

$I(T)$ say, is proportional to the rate of change of surface concentration with temperature,

$$I(T) \propto \frac{dC_s}{dT} \tag{3.3}$$

where

$$-\frac{dC_s}{dT} = \frac{\nu_{des}}{\beta} C_s{}^n e^{-E_{des}/RT} \tag{3.4}$$

and β is the heating rate in K s^{-1}. Hence,

$$I(T) \propto \frac{\nu_{des}}{\beta} C_s{}^n e^{-E_{des}/RT} \tag{3.5}$$

In contrast to our own work on water ice, Brown and co-workers cannot determine the absolute surface concentration of their desorbate and therefore this expression must be re-cast in terms of relative coverage, θ_{rel}, on the surface,

$$I(T) \propto \frac{\nu_{des}}{\beta} \theta_{rel}{}^n e^{-E_{des}/RT} \tag{3.6}$$

The relative coverage can be simply obtained by integrating the area under the TPD peaks of interest. Then at a fixed temperature, T_f say, the order of the desorption process can be recovered by plotting $\ln[I(T_f)]$ $versus$ $\ln[\theta_{rel}]_{T_x}$, while a plot of $\ln[I(T)] - n\ln\theta_{rel}$ $versus$ $\frac{1}{T}$ can be used to obtain the activation energy for the desorption process, assuming that the value is independent of coverage. Using such an approach, Brown and co-workers have been able to determine orders and activation energies for desorption of water (Bolina 2005; Bolina, Wolff & Brown 2005b), methanol (Bolina 2005; Bolina, Wolff & Brown 2005a) and ammonia (Bolina 2005; Bolina & Brown 2005) from a graphite substrate.

The key limitation of the analysis presented by Brown and co-workers is the inability of their analysis to recover the pre-exponential factor for desorption. This is a consequence of their lack of information on the absolute surface concentration of the adsorbed films. However, ices deposited at temperatures in excess of $ca.$ 50 K are typically compact amorphous materials. This allows for an intuitive estimate of these pre-exponential factors to be obtained by considering the molecular packing density in the ice and the surface area of the substrate relative to that of a single adsorbate molecule. Brown and co-workers have on this basis estimated the relevant pre-exponential factors (Bolina 2005) and these are tabulated in Table 1 in comparison with other recent literature data relating to the TPD of astronomically relevant ices.

As is clear, neither of these methods is truly assumption free. Ideally, in performing TPD experiments, we would have the necessary surface concentration data to obviate the need for estimation of the pre-exponential factor. In that regime, a full kinetic analysis based on global fitting of TPD data from multiple initial surface concentrations made with no assumption as to desorption order, activation energy for desorption and pre-exponential factor (and any possible coverage dependence thereof) would permit recovery of these quantities.

While it is important to consider the single-component ices, there is little doubt from the observational data that the ice *in situ* is a mixture of components. This adds significantly to the problem of understanding the thermal desorption of icy mantles, for in addition to considering the issue of interaction between the ice components, we must also now consider the key issue of ice morphology. This is perhaps best illustrated by our work on the CO–H_2O ice system (Collings *et al.* 2003a,b). Water ice film growth at temperatures around 10 K results in a material with high porosity (Stevenson *et al.* 1999). CO solid will spread into these pores covering the surface of the water as the temperature is raised towards and above 20 K. Strong binding of the CO to the water ice surface (E_{des} for CO from water is *ca.* 10.6 kJ mol^{-1} as opposed to 6.8 kJ mol^{-1} for a pure CO multilayer (Collings *et al.* 2003a; Collings *et al.* 2003b) and pore collapse at temperatures in the range 30 to 80 K, ensures that a significant amount of CO is trapped in the water ice matrix to elevated temperatures. Similar observations have been reported for other components of the interstellar ice inventory interacting with water ice (Collings *et al.* 2004), and for CO interacting with methanol ice (Chen *et al.* 2005). In fact for many species, release of trapped molecules is the dominant thermal desorption process, such that the binding strength of the adsorbate molecules to the water surface has an insignificant bearing on the desorption kinetics (Collings *et al.* 2004). The implications of this trapping process on gas-phase chemical networks in hot cores have been (Viti *et al.* 2004), and will continue to be the subject of investigation by astrochemical modellers.

Such systems can, of course, be interpreted using phenomenological kinetic models as we have demonstrated with the CO–H_2O system (Collings *et al.* 2003a,b). However, this simply represents a starting point. There is much more to do in order to fully understand the microscopic processes occurring in these complex systems coupling thermal desorption kinetics, diffusion kinetics and morphological change.

4. Conclusions

Temperature programmed desorption in UHV has been described. Certain experimental limitations relating to removal of the gaseous desorbate from the vacuum chamber and heating of the substrate have been outlined. The recent literature relating to the application of TPD to single-component molecular ices is summarised and methods of analysis described in detail. Orders of reaction, activation energies and pre-exponential factors for desorption of a number of species are tabulated. The problems associated with TPD of ice mixtures are introduced.

Acknowledgements

The authors thank PPARC, EPSRC, the University of Nottingham and the Leverhulme Trust for financial support for this work. We thank Dr. Helen Fraser, John Dever, Rui Chen, Simon Green and various final year undergraduate project students for their assistance and invaluable scientific contributions to our work discussed in this paper, Serena Viti and David A. Williams for helpful discussions, and Nico Bovet for assistance with graphics in this paper.

References

Baragiola, R.A., Vidal, R.A., Svendsen, W., Schou, J., Shi, M., Bahr, D.A., & Atteberry, C.L. 2003, *Nucl. Instrum. Meth.* B 209, 294

Bergeld, J., Gleeson, M., Kasemo, B., & Chakarov, D. 2004, *J. Geophys. Res.* E 109, E07s11/1.

Bernstein, M.P., Dworkin, J.P., Sandford, S.A., Cooper, G.W., & Allamandola, L.J. 2002, *Nature* 416, 401

Bolina, A.S. 2005, Ph.D. Thesis, University College London

Bolina, A.S. & Brown, W.A 2005, *Surf. Sci.*, in print

Bolina, A.S., Wolff, A.J., & Brown, W.A 2005a, *J. Chem. Phys.* 122, 44713

Bolina, A.S., Wolff, A.J., & Brown, W.A 2005b, *J. Phys. Chem.* B 109, 16836

Chen, R., Collings, M.P., McCoustra, M.R.S., Bolina, A.S., & Brown, W.A. 2005, in preparation

Collings, M.P., Dever, J.W., Fraser, H.J., & McCoustra, M.R.S. 2002, *Proc. NASA Laboratory Astrophysics Workshop* NASA/CP-2002-211863, 192

Collings, M.P., Dever, J.W., Fraser, H.J., McCoustra, M.R. S., & Williams, D.A. 2003a, *Ap. J.* 583, 1058

Collings, M.P., Dever, J.W., Fraser, H.J., & McCoustra, M.R.S. 2003b, *A&SS* 285, 633

Collings, M.P., Anderson, M.A., Chen, R., Dever, J.W., Viti, S., Williams, D.A., & McCoustra, M.R.S. 2004, *MNRAS* 354, 1133

Fraser, H.J., Collings, M.P., McCoustra, M.R.S., & Williams, D.A. 2001, *MNRAS* 327, 1165

Fraser, H.J., Collings, M.P., & McCoustra, M.R.S. 2002, *Rev. Sci. Instrum.* 73, 2161

Gerakines, P.A., Moore, M.H., & Hudson, R.L. 2004, *Icarus* 170, 202

Gibb, E.L., Whittet, D.C.B., Schutte, W.A,. Boogert, A.C.A., Chiar, J.E., Ehrenfreund, P., Gerakines, P.A., Keane, J.V., Tielens, A.G.G.M., van Dishoeck, E.F., & Kerkhof, O. 2000, *Ap. J.* 536, 347

Holtom, P.D., Bennett, C.J., Osamura, Y., Mason, N.J., & Kaiser, R.I. 2005, *Ap. J.* 626, 940

Hudson, R.L. & Donn, B. 1991, *Icarus* 94, 326

Jones, R.G. & Turton, S. 1997, *Surf. Sci.* 377–379, 719

Jones, R.G. & Fisher, C.J. 1999, *Surf. Sci.* 424, 127

King, D.A. 1975, *Surf. Sci.* 147, 384

Kolasinski, K.W. 2002, *Surface Science. Foundations of Catalysis and Nanoscience* Wiley, Chichester

Kuan, Y.-J., Charnley, S.B., Huang, H.-C., Tseng, W.-L., & Kisiel, Z. 2003, *Ap. J.* 593, 848

Loeffler, M.J., Baratta, G.A., Palumbo, M.E., Strazzulla, G., & Baragiola, R.A. 2005, *A&A* 435, 587

Muñoz Caro, G.M., Meierhenrich, U.J., Schutte, W.A., Barbier, B., Arconnes Segovia, A., Rosenbauer, H., Thiemann, W.H.-P., Brack, A., & Greenberg, J.M. 2002, *Nature* 416, 403

Notesco, G. & Bar-Nun, A.2005, *Icarus* 175, 546

Örberg, K. I., van Broekhuizen, F., Fraser, H. J., Bisschop, S. E., van Dishoeck, E. F., & Schlemmer, S. 2005, *Ap. J.* 621, L33

Sandford, S.A. & Allamandola, L.J. 1988, *Icarus* 76, 201

Speedy, R.J., Debenedetti, P.G., Smith, R.S., Huang, C., & Kay, B.D. 1996, *J. Chem. Phys.* 105, 240

Stevenson, K.P., Kimmel, G.A., Dohnálek, Z., Smith, R.S., & Kay, B.D. 1999, *Science* 283, 1505

Vidal, R.A., Teolis, B.D., & Baragiola, R.A. 2005, *Surf. Sci.* 588, 1

Viti, S., Collings, M.P., Dever, J.W., McCoustra, M.R.S., & Williams, D.A. 2004, *MNRAS* 354, 1141

Wang, H., Bell, R.C., Iedema, M.J., Tsekouras, A.A., & Cowin, J.P. 2005, *Ap. J.* 620, 1027

Whittet, D.C.B., Schutte, W.A, Tielens, A.G.G.M., Boogert, A.C.A., de Graauw, Th., Ehrenfreund, P., Gerakines, P.A., Helmich, F.P., Prusti, T., & van Dishoeck, E.F. 1996, *A&A* 315, L357

Williams, D.A. 1998, *Faraday Discus.* 109, 1

Westley, M.S., Baragiola, R.A., Johnson, R.E., & Baratta, G.A. 1995, *Nature* 373, 405

Woodruff, D.P. & Delchar, T.A. 1994 *Modern Techniques of Surface Science* Cambridge University Press, Cambridge, 2nd edt.

Discussion

D'HENDECOURT: Why not complement your studies on TPD by IR spectroscopy which, is more easily applicable to astrophysical problems? In particular, ice porosity does show up in the IR (around 3700 cm^{-1}), a feature which is not observed, for the moment, in protostellar spectra.

McCOUSTRA: TPD has, since the 1960's, provided the surface science community with a relatively simple means of determining the binding energies of molecules on surfaces. Vibrational spectroscopy, on the other hand, provides us with useful complementary information in allowing us to characterize binding sites, in particular when we use good-quality computational chemistry to support both studies. Our work published in *Ap. J.* (2003, 583, 1058) and *A&SS* (2003, 285, 633) details how we have used this combination to probe the CO–H_2O ice spectrum in some detail.

Photo: E. Herbst

Astrochemistry: Recent Successes and Current Challenges
Proceedings IAU Symposium No. 231, 2005
D.C. Lis, G.A. Blake & E. Herbst, eds.
© 2006 International Astronomical Union
doi:10.1017/S1743921306007435

Formation and Deuterium Fractionation of Organic Molecules on Grain Surfaces

Naoki Watanabe

Institute of Low Temperature Science, Hokkaido University, Sapporo, JAPAN
email: watanabe@lowtem.hokudai.ac.jp

Abstract. A series of experiments on the surface reactions of hydrogen and deuterium atoms with solid CO, formaldehyde (H_2CO), and methanol (CH_3OH) has been performed. Successive hydrogenation of CO on surfaces at ~ 10 K was found to proceed efficiently via tunneling to produce H_2CO and CH_3OH on dust grains under the typical conditions of molecular clouds. Formation rates are strongly dependent on the surface temperature and composition. The role of surface reactions in the formation of deuterated formaldehyde and methanol was investigated. The deuterium fractionation of methanol observed in molecular clouds was reproduced experimentally via H-D substitution in solid methanol at an accreting atomic D/H ratio of 0.05-0.1. This is the first evidence that grain-surface reactions can be responsible for fractionation. We have determined several effective rate constants for hydrogenation, deuteration, and H-D substitution to construct the surface reaction network for CO, H_2CO, CH_3OH, deuterated formaldehyde, and deuterated methanol.

Keywords. astrochemistry — dust, extinction — ISM: molecules — molecular processes

---

1. Introduction

In observations, large numbers of solid-phase molecules have been discovered on interstellar dust in molecular clouds. Among the observed solid-phase molecules, primordial species such as CO must first be produced in the gas phase and subsequently freeze on a cold dust surface, while it is widely accepted that the formation of the dominant H_2O molecules and more complex species requires surface chemical processes (for a review see Ehrenfreund & Charnley 2000; Boogert & Ehrenfreund 2004). Many experiments have been performed to simulate grain-surface processes and to analyze the observed infrared absorption spectra. These experiments were successful to some extent in revealing the importance of the photolysis and ion-bombardment of ice and in enabling the molecular assignment of the observed spectra. However, quantitative information on surface processes such as reaction channels, reaction rates, and activation energies, has yet to be obtained. In particular, details of nonenergetic processes, namely, surface reactions of light atoms, are as yet not well known despite their importance in the cold quiescent region.

Formaldehyde (H_2CO) and methanol (CH_3OH) have been found abundantly in not only interstellar ice (e.g., Keane *et al.* 2001) and but also in comets (e.g., Crovisier & Bockelée-Morvan 1999). It has been reported that pure gas-phase reactions (e.g., Shalabiea & Greenberg 1994) and UV photolysis of H_2O-CO ice (Allamandola *et al.* 1988; Schutte *et al.* 1996) are insufficient for explaining the observed abundances of these organic molecules in interstellar ice. In proton bombardment experiments, Hudson and Moore (1999) demonstrated that, over 4.6 billion years in space, yields of H_2CO and CH_3OH from H_2O-CO ice would be up to 7% and 12%, respectively. This process is effective for explaining the production of methanol in a comet's nucleus only. Theoretical models suggest that the successive addition of hydrogen atoms (hydrogenation) to CO:

$$CO \xrightarrow{k_1} HCO \xrightarrow{k_2} H_2CO \xrightarrow{k_3} CH_3O, CH_2OH \xrightarrow{k_4} CH_3OH \tag{1}$$

where k_n is the reaction rate, takes place on the dust surface via tunneling reactions through activation barriers as large as 1000–2000 K and produce H_2CO and CH_3OH even at a temperature of approximately 10 K (Tielens & Allamandola 1987; Tielens & Whittet 1997). These models stimulated experimental works on the hydrogenation of CO. Hiraoka *et al.* (1994; 2004) experimentally investigated this reaction system with an H atom spray apparatus. They observed only a small amount of H_2CO and no CH_3OH products upon exposure of pure solid CO to H atoms and concluded that the hydrogenation of CO would be too slow to produce H_2CO and CH_3OH on a grain surface. Unfortunately, however, they did not experimentally estimate the flux of hydrogen atoms that they used. The estimation of the H flux or fluence is essential to determine the efficiency of reactions in molecular clouds. The comparison of our experimental results described below and theirs enabled us to estimate their flux to have been 10^{12} cm^{-2} s^{-1} or less. This low flux most likely explains their results (Hidaka *et al.* 2004; Watanabe *et al.* 2005).

Recent observations have revealed that the abundances of some deuterated interstellar molecules are markedly larger than the cosmic D/H ratio of 10^{-5}. In particular, deuterium fractionation in methanol was found in not only molecular clouds (Parise *et al.* 2003; 2004) but also in comets (Crovisier *et al.* 2004), and the ratio of deuterated methanol to normal CH_3OH in column density observed toward a low-mass protostar is up to approximately 0.4. Hereafter, the term X-d_n will either denote an isotopomer with a specific number n deuterium atoms (e.g. X-d_2) or, if a general n appears, will refer to all deuterated isotopomers of a molecule. It is difficult to reproduce the observed D-enrichments with pure gas-phase models. Although some gas-grain models have been proposed that achieve the observed level of D-enrichments of interstellar methanol, they are rather ambiguous due to the lack of information about grain-surface reactions such as reaction channels and the rates of hydrogenation and deuteration. It is therefore highly desirable to experimentally clarify the role of grain-surface reactions in D-enrichments.

Here, we present the results of our series of recent experiments regarding hydrogenation and deuteration of solid CO, H_2CO, D_2CO, and CH_3OH.

2. Experimental

Two pieces of apparatus, LAboratory Setup for Surface reactions in Interstellar Environment (LASSIE) and Apparatus for SUrface Reaction in Astrophysics (ASURA), were constructed on the basis of the same concept for experiments on the surface reactions of atoms. Both systems are of basically the same design and described elsewhere (Hidaka *et al.* 2004; Watanabe *et al.* 2004, 2005). The experiments were performed with LASSIE or ASURA. The reactions of H and D atoms with solid CO, H_2CO, and D_2CO were measured mainly using LASSIE and those with methanol using ASURA. In addition, we sometimes also measured the same reaction system using both LASSIE and ASURA to crosscheck the validity of the experiments. The composition of the samples investigated were CO-H_2O mixed ice, CO (1 eML) on H_2O ice, pure solid CO, H_2CO (1 eML) on H_2O ice, D_2CO (1 eML) on H_2O ice, and pure solid methanol(-d_n). Here, the term "1 eML" represents the equivalent amount of molecules for one monolayer of a flat surface. Since the surface of amorphous H_2O ice is very rough, 1 eML of molecules cannot cover the entire surface area and some H_2O molecules will be exposed on the surface. That is, 1 eML does not represent a coverage of unity.

Solid samples were produced by vapor deposition on an aluminum substrate cooled by a closed-type He refrigerator. In the present work, the surface temperature was varied from 8 to 20 K. H_2CO and D_2CO gases were produced by the thermal cracking of paraformaldehyde (purity 99.8%, Merck) and paraformaldehyde-d_2 (purity 99%, ACROS) powders, respectively. The powders were heated to 57 °C in a glass vacuum tube. It was necessary to maintain the gas line and the variable leak valve at approximately 60 °C in order to prevent condensation of H_2CO inside the gas line. The composition of the solid samples were measured using a Fourier transform infrared spectrometer at resolutions of 0.5 cm^{-1} for pure solid CO and CO-H_2O mixed ice and 4 cm^{-1} for H_2CO, CO-H_2O mixed ice, and methanol.

Details of the atomic source have been described elsewhere (Hidaka *et al.* 2004; Watanabe *et al.* 2005). Briefly, H and D atoms were produced by the dissociation of H_2 and D_2 molecules in a microwave-induced plasma in a Pyrex glass tube (25 mm in diameter). Atoms were transferred via a series of two polytetrafluoroethylene (PTFE) tubes to the solid target. The downstream PTFE tube was tightly covered with an oxygen-free copper pipe connected to the cold head. This simple arrangement produces very low temperature atomic hydrogen (Walraven & Silvera 1982). In the present study, the beam line was cooled to 30–50 K, as measured using Au/Fe–0.07% chromel thermocouples attached to the PTFE tube at the most downstream end. We carefully eliminated the contamination in the atomic source.

The measurement of atomic flux is essential for the determination of the reaction rates and the efficiency of reactions in space. In our experiments, the fluxes of atoms were measured using a quadrupole mass spectrometer (QMS) with a Faraday cup, calibrated with an ion gauge. During the flux measurements, the substrate was de-installed and the QMS head was set at the original position of the substrate in alignment with the beam line. The typical dissociation fraction was approximately 15% at least. The H atom fluxes were in the range of $5\times10^{14}-1\times10^{15}$ cm^{-2} s^{-1}.

3. Results and Discussion

3.1. *H + CO-H_2O Mixed Ice and H + CO (1 eML) on H_2O Ice*

CO-H_2O mixed ice (CO/$H_2O \sim 0.25$) was exposed to the cold H beam. Molecules of H_2CO and, subsequently, CH_3OH were significantly produced with consumption of CO below 20 K. The sudden decrease in the reaction rates at 20 K is considered to have arisen from the decrease in sticking coefficient of hydrogen atoms on H_2O ice rather than from the decrease in the reaction rate constant (Watanabe, Shiraki, & Kouchi 2003). This surface-temperature dependence implies that the hydrogen atoms react with CO after adsorption and diffusion on the surface: the so-called Langmuir-Hinshelwood process. The formation of these molecules is not observed when a sample is exposed to just H_2 molecules. We also carefully checked for undesirable experimental factors such as contamination by, for example, carrying out a blank test. No such effects were found (Hidaka *et al.* 2004). Furthermore, since reactions of H_2 with HCO and CH_3O are endothermic, we conclude that H_2CO and CH_3OH are produced by the successive reaction (1). The intermediates, HCO and CH_3O (CH_2OH) radicals, were not observed, indicating the relationships among reaction rates to be $k_1 \ll k_2$ and $k_3 \ll k_4$.

Since the column densities of reacted CO were of the order of $10^{15}-10^{16}$ cm^{-2} over the fluence of 10^{18} atoms cm^{-2}, only 1% or less of the total number of injected H atoms were used for the hydrogenation. A significant number of H atoms would be consumed by H-H recombination to produce H_2 molecules. When the atomic flux decreases, the relative

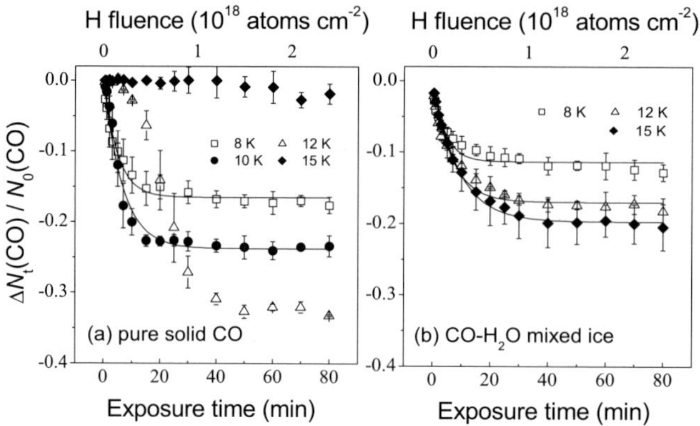

Figure 1. Attenuation of CO-column density upon H exposure for (*a*) pure solid CO and (*b*) CO-H$_2$O mixed ice. The *y*-axis is normalized to the initial column density. Solid lines are least-square fits using eq. (2).

fraction of H atoms used for recombination compared with those used for hydrogenation becomes smaller because the probability of H-H recombination is proportional to the square of the surface coverage. The depletion of CO and the formation of H$_2$CO and CH$_3$OH are saturated at long exposure times. It was found that the reason for the saturation is not due to the balance with the reverse process of reaction (1) (H-atom abstraction by another H atom) but the limit of H atom diffusion in bulk (Watanabe *et al.* 2003, 2004). Therefore, the rates of CO decay correspond only to the rates of the first part of reaction (1), namely CO + H → HCO. We also exposed CO (1 eML) on H$_2$O ice to H atoms. The results were essentially the same as those obtained for mixed ice.

We estimated the reaction rate constants for CO + H → HCO using the fits of CO data (Figure 1) to

$$\frac{\Delta N(\mathrm{CO})}{N_0(\mathrm{CO})} = \alpha \left(1 - \exp(-k n_\mathrm{H} t)\right) \tag{2}$$

and

$$\frac{dn_\mathrm{H}}{dt} = P f_\mathrm{H} - k_\mathrm{H-H} n_\mathrm{H}^2 - k_\mathrm{H-CO} n_\mathrm{H} n_\mathrm{CO} \tag{3}$$

where α is a saturation value, k the rate constants, P sticking coefficient of atom, f_H the flux of H atoms, and n_H and n_CO the surface densities of H atoms and CO, respectively. In equation (3), we neglect the terms for the reactions of H with HCO, H$_2$CO, and CH$_3$O and the desorption of H atoms, because the rates of H-atom loss at the surface by those processes must be very slow. Chigai *et al.* (2005, in preparation) found that, under our experimental conditions, the time variation of n_H is governed by the atomic flux (the first term of right side in eq. 3) and the loss due to H-H recombination (the second term) and n_H becomes constant immediately. By setting the left side of eq. (3) to zero, we obtained $n_\mathrm{H} = 53.9 \times (f_\mathrm{H} P)^{1/2}$ cm^{-2}. Here, the value of $k_\mathrm{H-H}$ was estimated by using activation energy of 103 K (Ivliev *et al.* 1982) and a square potential barrier of 1-Å width for the recombination. Consequently, the values of α and the product $k_\mathrm{H-CO} P^{1/2}$ (effective rate constants) are determined by the fittings, as shown by the solid lines in Figure 1. The effective rate constants at 15 K are 4.7×10^{-12} and 1.5×10^{-12} cm^2 s^{-1} for H + CO (1 eML) on H$_2$O and H + CO-H$_2$O mixed ice, respectively.

3.2. *H + Pure Solid CO*

Recent investigations of the observed spectra of interstellar ice have revealed the presence of pure solid CO segregated from H$_2$O ice (Teixeira *et al.* 1998; Pontoppidan *et al.* 2003).

To simulate hydrogenation on such an ice and clarify the role of H_2O ice, 10 ML of pure solid CO was exposed to H atoms. The formation of H_2CO and CH_3OH and nondetection of radicals were observed again. However, as shown in Figure 1, the dependence of the reactions on temperature is stronger than that for CO-H_2O mixed ice. The reactions become very slow at 15 K and the behavior of the CO plot at 12 K is not like a single exponential decay. The smaller sticking coefficient of H atoms on pure solid CO than that on H_2O ice would be responsible for these features. The sticking coefficient on CO may drop at around 12–15 K. The initial stage of the CO-decay curve at 12 K implies that the sticking coefficient gradually increases with exposure time due to the production of H_2CO. The H_2CO produced on CO would enhance the sticking coefficient. In fact, when we deposited 0.3 eML of H_2CO on solid CO in advance and subsequently exposed it to H atoms at 12 K, the shape of the CO decay curve became like that of a single-exponential decay curve similar to that at 10 K (Watanabe *et al.* 2004). We fitted the CO data at 8 and 10 K to eq. (2) and estimated the effective rate constants, which are shown in § 5.

3.3. *D + CO (1 eML) on H_2O Ice*

In order to compare the deuteration of CO with the hydrogenation of CO, CO (1 eML) on H_2O was exposed to cold deuterium atoms. The flux of deuterium atoms was the same as that of hydrogen atoms for H + CO (1 eML) on H_2O ice. The results will be described in detail elsewhere (Hidaka, Kouchi, & Watanabe 2005, in preparation). Briefly, the products D_2CO and CD_3OD were observed and the temperature dependence of CO decay was similar to that observed for hydrogenation. However, the formation rates of D_2CO and CD_3OD were much slower than those of H_2CO and CH_3OH for hydrogenation. Since the sticking coefficient of D atoms must be very similar to that of H atoms at the same surface temperature, this difference is considered to result from the difference in reaction rate constants. The value of $k_D P^{1/2}$ at 15 K for $D + CO \rightarrow DCO$ was determined by fitting, using eq. (2). The ratio of deuteration rate constant k_D to hydrogenation rate constant k_H, k_D/k_H, is approximately 0.1 at 15 K. This isotope effect is ascribed to the mass effect of the tunneling reaction.

3.4. *H + H_2CO (1 eML) on H_2O Ice*

In order to estimate the reaction rates for the hydrogenation of H_2CO directly from the decay of H_2CO, solid H_2CO (1 eML) was produced on H_2O ice and exposed to cold H atoms. CH_3OH was gradually produced with the consumption of H_2CO. We confirmed that H_2CO does not react with H_2. The reaction rate constant k at 15 K relative to that for CO hydrogenation on H_2O ice was obtained by the fitting of H_2CO decay curve to be approximately 0.5 on the assumption that the sticking coefficients are the same because atoms mainly adsorb first on H_2O ice which is a main component at the surface (Hidaka *et al.* 2004).

3.5. *H + D (D/H~0.1) + Pure Solid CO (10 ML) and D + Solid CH_3OH (4 ML)*

To simulate the formation of deuterated formaldehyde and methanol on grains, pure solid CO was exposed to H and D atoms simultaneously at an atomic flux ratio of D/H~0.1, at which accretion of atoms can be achieved in the gas phase (Roberts *et al.* 2003). Variations in IR spectra observed upon H and D exposure are shown in Figure 2. Peaks corresponding to the products of H_2CO (1729, 1499, 1247, and 1177 cm^{-1}), HDCO (1700 and 1401 cm^{-1}), D_2CO (1684 cm^{-1}), CH_3OH (1039 and 1126 cm^{-1}), CH_2DOH (1330, 1290, and 1047 cm^{-1}), CHD_2OH (1331, 1301, 1091, and 950 cm^{-1}), and CD_3OH (1128 and 982 cm^{-1}) are clearly observed. Radicals and Me–OD (Me; methyl group) were not detected. This seems to be consistent with the low abundance of Me–OD observed in

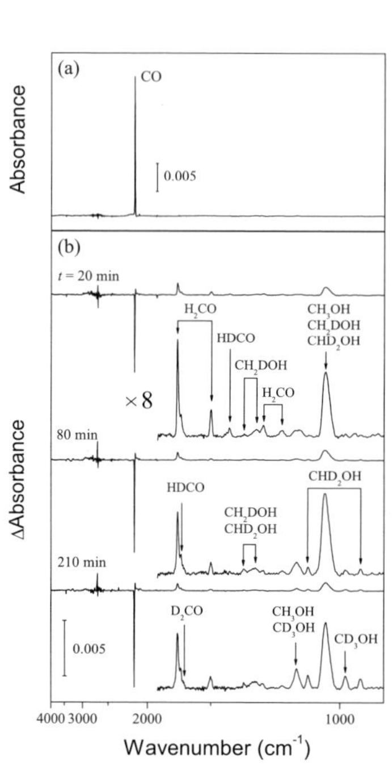

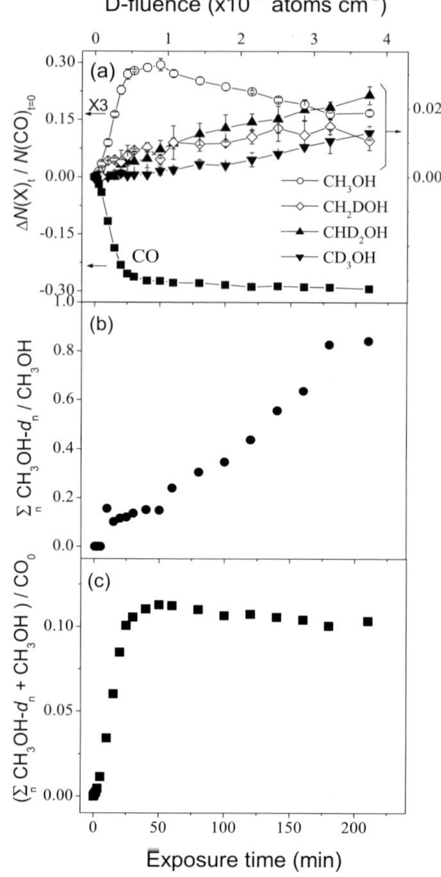

Figure 2. Variation in infrared absorption spectra for simultaneous exposure of solid CO to H and D atoms at 10 K. (a) Initial spectrum prior to exposure. (b) Spectra obtained by subtraction of the initial one from those obtained after the exposure time t.

Figure 3. Variation in column densities as a function of exposure time and D fluence. (a) column densities for CO and methanol-d_n. The y-axis is normalized to the initial column density of CO. Note that the left and right y-axes correspond to CO and normal methanol, and methanol-d_n, respectively. (b) Sum of column densities of methanol-d_n relative to column density of CH_3OH. (c) Sum of column densities for methanol-d_n and CH_3OH.

molecular clouds (Parise *et al.* 2004). The variation in column densities for parent CO, CH_3OH, and deuterated methanol and the sum of column densities of methanol-d_n are plotted in

Figure 3(a) and (b), respectively. Figure 3(c) shows a plot of the sum of the column densities for CH_3OH and methanol-d_n normalized to the initial column density of CO. Here, we concentrate on the formation of methanol-d_n and thus do not show the variation in column densities for formaldehyde. From Figure 3(b), one can see that deuterium fractionation progresses with exposure time. By examining Figure 3(a), information on the formation process of methanol-d_n is obtained. The yield of CH_3OH first increased quickly and subsequently reached a maximum. CH_2DOH, CHD_2OH, and CD_3OH were

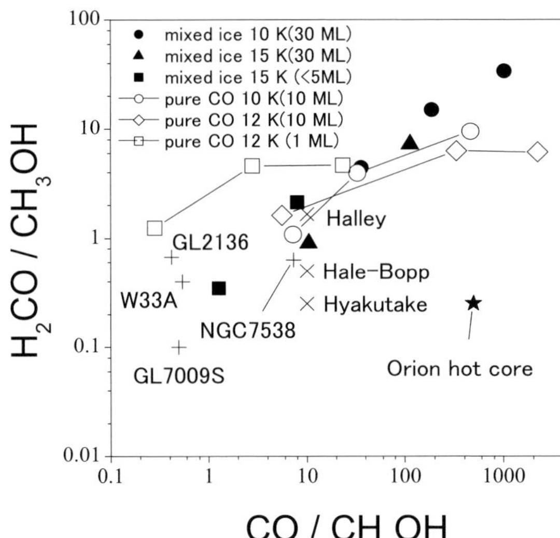

Figure 4. Column density ratios of H_2CO and CO relative to CH_3OH for the experimental and observational results. Open symbols represent results for pure solid CO, solid symbols indicate mixed-ice results, and observations are denoted by $+$ for interstellar ice (Kean *et al.* 2002), $\times$ for comets, and $\star$ for the Orion compact ridge (Ehrenfreund & Charnley 2000).

produced in sequence with decreasing amount of CH_3OH. These features cannot be explained by the successive addition of H and D to CO but indicate the occurrence of H-D substitution in methanol after the formation of CH_3OH. In Figure 3(c), the sum of CH_3OH and total deuterated methanol becomes constant after the exposure of 25 min, indicating that H-D substitution in methanol is dominant after that period of time; successive addition and substitution compete until then. As described in the next section, addition of D atoms to formaldehyde($-d_n$) is very slow. Therefore, substitution would be dominant for the formation of methanol-d_n even before 25 min.

To confirm whether H-D substitution in methanol occurs, solid CH_3OH was exposed to D atoms only. The yields of CH_2DOH, CHD_2OH, and CD_3OH were obtained with the consumption of CH_3OH. The time variation of column densities again clearly reveals successive H-D substitution in methanol (see Fig. 4 in Nagaoka, Watanabe, & Kouchi 2005). When parent solids of CH_3OD, CD_3OH, CHD_2OH, CH_2DOH, and CD_3OD were exposed to H atoms, no reaction was observed. That is, once deuterated methanol (methanol-d_n) is produced, it never returns to its predeuterated state (methanol-d_{n-1}) by reaction with H atoms. Therefore, H-D substitution in methanol on the grain surface can act as the process of deuterium fractionation and thus should be included in the theoretical models. A detailed discussion of this is described elsewhere (Nagaoka *et al.* 2005).

3.6. *D + Solid H_2CO and H + Solid D_2CO*

The substitution reactions in formaldehyde were also investigated. Detailed results will be reported elsewhere (Watanabe, Hidaka, & Kouchi 2005, in preparation). Briefly, for D + solid H_2CO, although there is competition between H-D substitution in formaldehyde producing HDCO and D_2CO and the addition of D atoms leading to deuterated methanol, the dominant process is the evolution of $H_2CO \rightarrow HDCO \rightarrow D_2CO$ by substitution. It is not clear in the present work whether the substitution proceeds via the sequence of H abstraction and D addition or via H-D direct exchange. The yields of CD_3OD and the other deuterated methanol were small even at the D fluence of $\sim 10^{18}$ cm^{-2}, indicating that the additions of D to D_2CO, HDCO and H_2CO are slower than substitution. H-D substitution in H_2CO would be an important process to produce deuterated formaldehyde in molecular clouds as much as successive H/D addition to CO.

For H + solid D_2CO, the substitution reaction by H atom ($D_2CO \rightarrow HDCO \rightarrow H_2CO$) proceeded significantly unlike the case of H + solid methanol-d_n. The yield of CHD_2OH produced by successive H addition to D_2CO is comparable to that of CH_3OH produced through the route of $D_2CO \rightarrow HDCO \rightarrow H_2CO \rightarrow CH_3OH$ (D-H substitution and subsequent H addition) throughout most of the exposure time. This implies that the rate of H addition to D_2CO is similar to that of D-H substitution (reduction of deuteration level) in D_2CO. The D-H substitution must proceed via sequence of D abstraction and H addition because D-H direct exchange in formaldehyde($-d_n$) is endothermic.

3.7. D + Solid CD_3OH, CHD_2OH, and CH_2DOH

To determine the relative rates of H-D substitution in methanol-d_n, solid CH_3OH, CH_2DOH, and CHD_2OH were exposed to D atoms independently (Nagaoka, Watanabe, & Kouchi 2005, in preparation). Briefly, from the decay curves of parent solids, the relative rates were estimated to be 4, 2, and 1 for $CH_3OH \rightarrow CH_2DOH$, $CH_2DOH \rightarrow CHD_2OH$, and $CHD_2OH \rightarrow CD_3OH$, respectively. The number of H atoms on the methyl side would govern the reactivity.

4. Implications for Astrochemistry

A comparison of the experimental results obtained for H + pure solid CO and CO-H_2O mixed ice with the observational results for molecular clouds is attempted in § 4.1. The potential contribution of the surface reactions of D atoms to the formation of deuterated formaldehyde and methanol is discussed in § 4.2.

4.1. H_2CO and CH_3OH

We compare our results for the hydrogenation of CO-H_2O mixed ice and pure solid CO with the observational results. For instance, assuming that the number density of hydrogen atoms is 1 cm^{-3} and the temperature of the atoms is the same as that of grains in a molecular cloud, the fluences for 10 K are 1.3×10^{16}, 1.3×10^{17}, and 1.3×10^{18} cm^{-2} over 10^4, 10^5, and 10^6 yrs, respectively. In the present experiments, those fluences approximately correspond to 1, 4, and 40 minutes, respectively, at 10 K. The obtained ratios of CO/CH_3OH and H_2CO/CH_3OH are plotted in Figure 4 with the observational abundances. The plots vary from left to right with the decrease in H fluence from 10^6 to 10^4 yrs. The leftmost plots correspond to 10^6 yrs. The results for the higher fluences (10^6 yrs) are reasonably consistent with the observations of molecular clouds and comets. For pure CO, the data at 12 K is closest to the observations, while the 15 K results are closest for the mixed ice. For both, thinner ice produces higher abundances of CH_3OH relative to the amounts of CO and H_2CO. For thinner ice, the fraction of reactive CO exposed on the surface relative to the fraction of unreactive CO buried in bulk is larger and thus, a ratio of CO/CH_3OH lower than that of thick ice is achieved. The ratio of H_2CO/CH_3OH for pure CO tends to be larger than that for mixed ice at lower ratios of CO/CH_3OH, and this tendency becomes stronger for thinner ice. This feature can be explained by an onion-like structure, as described below. When CO molecules are deposited on the surface, the CO molecules can readily aggregate into bulk (solid) CO even below 1 eML coverage. However, once H_2CO layers have been produced on the solid CO, only the upper layer of the H_2CO can be converted to CH_3OH, resulting in a significant fraction of remnant H_2CO in the bulk. That is, solid CO finally becomes a CO-H_2CO-CH_3OH trilayer structure (Watanabe *et al.* 2004). In the case of mixed ice, CO molecules tend to be isolated in cracks or boundaries of H_2O ice and most of the CO molecules can be finally converted to CH_3OH. The experimental conditions realized for thinner ice may simulate higher H/CO ratios of accretion onto grains in molecular clouds, where CO always has a chance to react with accreted H atoms at the surface before being buried in bulk.

Our results reveal that the hydrogenation of CO on the grains in a dense core is efficient for the production of H_2CO and CH_3OH regardless of the type of mantle − pure solid CO (apolar ice) or CO-H_2O mixed (polar) ice − as long as the abundance of CO is significant. When CO is abundant in the ice, the competition of H-atom reactions with less abundant species like C_2H_2 and C_2H_4 has little effect on the hydrogenation of CO.

It should be noted that the efficiency of the hydrogenation of CO is strongly dependent on the flux of H atoms; in other words, the number density and kinetic temperature of H atoms and the temperature of the grain. Furthermore, when the photon field of molecular clouds becomes strong, the contribution of photoinduced reactions should be considered. Strictly speaking, these factors should be taken into account.

4.2. H_2CO-d_n and CH_3OH-d_n

Assuming an H number density of 1 cm^{-3}, an atomic D/H ratio of 0.1 in the accreting gas, and that the kinetic temperature of the atoms is the same as that of the grain, the D (H) atom fluences in a 10 K molecular cloud will be 1.3×10^{16} (1.3×10^{17}), 6.5×10^{16} (6.5×10^{17}), and 1.3×10^{17} (1.3×10^{18}) cm^{-2} over 10^5, 5×10^5, and 10^6 yrs, respectively. In the experiment on H + D + solid CO, these fluences approximately correspond to exposure times of 10, 35, and 70 min, respectively. The ratios of deuterated methanol to the amount of remaining CH_3OH, denoted in the form (CH_2DOH/CH_3OH, CHD_2OH/CH_3OH, CD_3OH/CH_3OH), after exposure times of 10, 35, and 70 min are (0.16, 0, 0), (0.09, 0.05, 0.01) and (0.11, 0.13, 0.03), respectively. These values are markedly consistent with the observations (Parise *et al.* 2004) of (0.3 ± 0.2, 0.06 ± 0.05, 0.014 ± 0.014), indicating that observed deuterium fractionation can be achieved by grain-surface reactions once the atomic D/H ratio of 0.1 in the accretion gas is obtained as suggested by the models. The nondetection of Me–OD is consistent with the low abundance of Me–OD observed toward a low-mass protostar. For formaldehyde, our obtained raios of ($HDCO/H_2CO$, D_2CO/H_2CO) are (0.44, 0.034) which are also consistent with the observation of (0.14, 0.05) (Loinard *et al.* 2000). We also performed an experiment with a D/H ratio of 0.05 and found that the results were very similar to those obtained for the ratio of 0.1.

5. Surface Reaction Network for Formaldehyde and Methanol

To date, we have investigated the reactions of hydrogenation, deuteration, and substitution for the systems of solid CO, formaldehyde, and methanol. In Figure 5, a diagram of the surface reaction network for CO, formaldehyde, and methanol is shown with the relative reaction rates estimated so far. Generally speaking, hydrogen addition occurs at a faster rate than deuterium addition due to the effect of the tunneling reaction. The rate constants of hydrogenation and deuteration of CO are higher than those of H_2CO and D_2CO, respectively. An important finding is that *the reaction rates strongly depend on not only the surface temperature but also on the composition and structure of the surface.* For instance, the hydrogenation rate of pure solid CO is different from that of CO in H_2O ice, and H_2CO on solid CO enhances the hydrogenation of CO (Watanabe *et al.* 2005). H-D substitution proceeds for both formaldehyde (H_2CO-$d_n \xrightarrow{D} H_2CO$-$d_{n+1}$; n=0–1) and methanol ($CH_3OH$-$d_n \xrightarrow{D} CH_3OH$-$d_{n+1}$; n=0–2). There are two possible elementary processes for H-D substitution: direct H-D exchange (e.g., $CH_3OH + D \rightarrow CH_2DOH +$

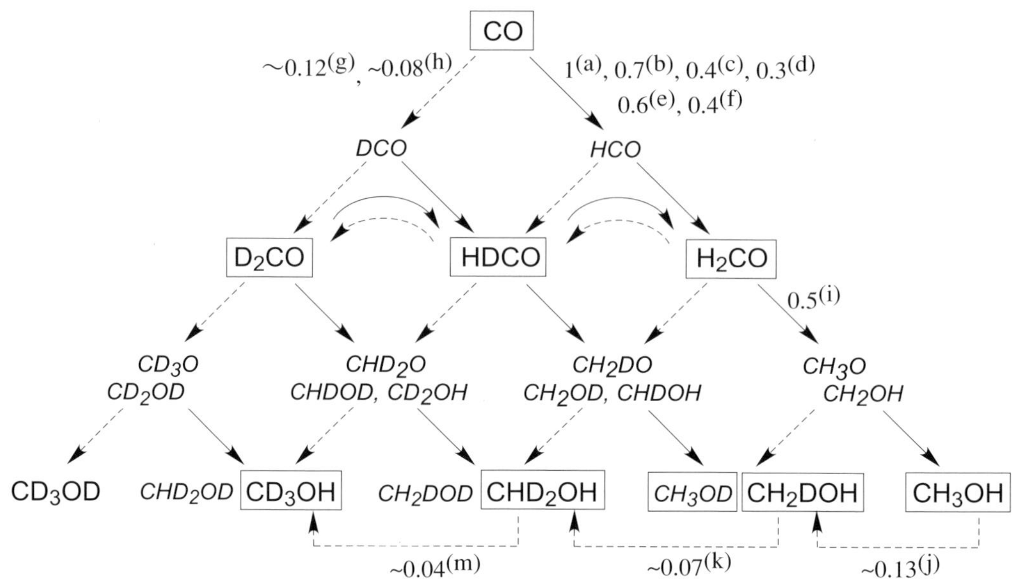

Figure 5. Surface reaction network for CO, formaldehyde, and methanol. Block and italic letters are used to denote products that were detected and not detected in the experiments, respectively. Square frames denote molecules that have been observed in molecular clouds. Solid and dashed arrows represent reactions of H and D atoms, respectively. Although it is likely that abstraction reactions proceed, they are not shown. The values shown are effective rate constants relative to that for H + CO (1 eML) on H_2O (see text) estimated experimentally for the processes represented by the arrows. It should be noted that the effective rate constant strongly depends on the surface composition, structure, and temperature. The values were determined from experiments: (a) H + CO (1 eML) on H_2O at 15 K; (b)–(d) H + CO-H_2O mixed ice at 8, 12, and 15K, respectively; (e), (f) H + pure CO at 8 and 10 K, respectively; (g) D + CO (1 eML) on H_2O at 15 K; (h) D + pure CO at 10 K. (i) H + H_2CO (1 eML) on H_2O at 15 K; (j) D + pure solid CH_3OH (20 ML) at 10 K; (k) D + pure solid CH_2DOH (20 ML) at 10 K; and (m) D + pure solid CHD_2OH (20 ML) at 10 K.

H) and the sequence of atomic abstraction and addition (e.g., $CH_3OH + D \rightarrow CH_2OH + HD$ and subsequently $CH_2OH + D \rightarrow CH_2DOH$). Although it is not clear which process, the direct exchange or the abstraction-addition, was dominant in our experiments, the activation energies for the latter process is lower than those for the former one (Blowers, Ford, & Masel 1998). For the latter case, H-abstraction/D-addition, the nondetection of Me–OD in the present experiments would be reasonable because H-abstraction in the methyl-side by either H or D has a much lower barrier than in –OH (Kerkeni & Clary 2004).

In contrast, D-H substitution proceeds in formaldehyde (H_2CO-$d_{n+1} \xrightarrow{H} H_2CO$-$d_n$) proceeds, while that in methanol (CH_3OH-$d_{n+1} \xrightarrow{H} CH_3OH$-$d_n$) was not observed. For formaldehyde, since the direct D-H exchange (e.g., $D_2CO + H \rightarrow HDCO + D$) is endothermic, the combination of D-atom abstraction and H-atom addition is responsible for D-H substitution. For methanol, the direct D-H exchange is also endothermic and, furthermore, an *ab-initio* calculation by Lendvay *et al.* (1997) indicates that the reaction of H-abstraction by H atoms from CH_3OH-d_n (e.g., $CH_2DOH + H \rightarrow CHDOH + H_2$) is much faster than that for D-abstraction by H atoms from CH_3OH-d_n (e.g., $CH_2DOH + H \rightarrow CH_2OH + HD$). This would be the reason why D-H substitution was not observed.

6. Summary

We have revealed that the hydrogenation of solid CO on grain surfaces can be a major process in the production of H_2CO and CH_3OH in the dense core of molecular clouds where the photon field is very weak. The hydrogenation rates strongly depend on not only the surface temperature but also the composition of the surface. Hydrogenation of CO at 12 K and above is found to be promoted by the coexistence of H_2CO and H_2O molecules on the surface.

The measurements of the deuterium atom exposure of solid CO, formaldehyde, and methanol demonstrated for the first time that deuteration of CO is much slower than hydrogenation of CO and that H-D substitution in solid methanol (methyl side) occurs and acts as a channel for the deuterium fractionation of methanol observed in molecular clouds.

Several effective rate constants for hydrogenation, deuteration, and H-D substitution have been determined to construct the surface reaction network for CO, H_2CO, $H_2CO\text{-}d_n$, CH_3OH, and $CH_3OH\text{-}d_n$.

References

Allamandola, L.J., Sandford, S.A., & Valero, G.J. 1988, *Icarus* 76, 225

Blowers, P., Ford, L., & Masel, R. 1998, *J. Phys. Chem. A* 102, 9267

Boogert, A. C. A. & Ehrenfreund, P. 2004, in *Astrophysics of Dust*, eds. A.N. Witt, G.C. Clayton, & B.T. Draine (APS Conference Series), Vol. 309, p. 547

Crovisier, J. & Bockelée-Morvan, D. 1999, *Space Sci. Rev.* 90, 19

Crovisier, J. *et al.* 2004, *A&A* 418, 1141

Ehrenfreund, P. & Charnley, S.B. 2000, *ARAA* 38, 427

Hidaka, H., Watanabe, N., Shiraki, T., Nagaoka, A., & Kouchi, 2004, *Ap. J.* 614, 1124

Hiraoka, K., Ohashi, N., Kihara, Y., Yamamoto, K., Sato, T., & Yamashita, A. 1994, *Chem. Phys. Lett.* 229, 408

Hiraoka, K., Sato, T., Sato, S., Sogoshi, N., Yokoyama, T., Takashima, H., & Kitagawa, S. 2004, *Ap. J.* 577, 265

Ivliev, A. V. *et al.* 1982, *JETP Lett.* 36, 472

Hudson, R.L. & Moore, M.H. 1999, *Icarus* 140, 451

Kerkeni, B. & Clary, D. C. 2004, *J. Phys. Chem. A* 108, 8966

Keane, J.V., Tielens, A.G.G.M., Boogert, A.C.A., Schutte, W.A., & Whittet, D.C.B. 2001, *A&A* 376, 254

Lendvay, G., Bérces, T., & Márta, F 1997, *J. Phys. Chem. A* 101, 1588

Loinard, L. *et al.* 2000, *A&A* 359, 1169

Nagaoka, A., Watanabe, N., & Kouchi, A. 2005, *Ap. J.* 624, L265

Parise, B. *et al.* 2003, *A&A* 410, 897

Parise, B. *et al.* 2004, *A&A* 416, 159

Pontoppidan, K. M., Fraser, H. J., Dartois, E., Thi, W.-F., van Dishoeck, E.F., Boogert, A.C.A., d'Hendecourt, L., Tielens, A.G.G.M., & Bisschop, S.E. 2003, *A&A* 408, 981

Roberts, H., Herbst, E., & Millar, T.J. 2003, *Ap. J.* 591, L41

Schutte, W.A., Gerakines, P.A., Geballe, T.R., van Dishoeck, E.F., & Greenberg, J.M. 1996, *A&A* 309, 633

Shalabiea, O.M. & Greenberg, J.M. 1994, *A&A* 290, 266

Teixeira, T.C, Emerson, J.P., & Palumbo, M.E. 1998, *A&A* 330, 711

Tielens, A.G.G.M. & Allamandola, L.J. 1987, in *Interstellar Processes*, eds. D. Hollenbach & H. Thronson (Kluwer, Dordrecht), p. 397

Tielens, A.G.G.M. & Whittet, D.C.B. 1997, in *Molecules in Astrophysics: Probes & Processes*, ed. E.F. van Dishoeck (Kluwer, Dordrecht), p. 45

Walraven J.T.M. & Silvera, I.F. 1982, *Rev. Sci. Instrum.* 53, 1167

Watanabe, N. & Kouchi, A. 2002, *Ap. J.* 571, L173

Watanabe, N., Shiraki, T., & Kouchi, A. 2003, *Ap. J.* 588, L121

Watanabe, N., Nagaoka, A., Shiraki, T., & Kouchi, A. 2004, *Ap. J.* 616, 638

Watanabe, N., Nagaoka, A., Hidaka, H., Shiraki, T., Chigai, T., & Kouchi, A. 2005, *Planet. Space Sci.*, submitted

Discussion

CASELLI: During the experiment, when methanol and deuterated methanol form, is there any evidence of evaporation of a fraction of the formed molecules? This is very important because even a small fraction may solve the problem of methanol formation in the gas phase of dark cold clouds.

WATANABE: I agree about the importance of desorption during the reactions. Unfortunately, we did not measure the desorbed molecules. We can say that more than 90% of the yields remain on the surface, but we do not have clear evidence of desorption in the present experiment. Measurements of desorption caused by reaction will be studied in the future after the modification of our apparatus.

HERBST: What is the fraction of product CH_3OH and CH_3OD that might escape into the gas in your experiments?

WATANABE: We could not detect the CH_3OD product in our experiment on the exposure of solid CH_3OH to D atoms at 10 K. It would be under the detection limit. Activation energies for the abstraction of H atom from the –OH side by either H and D atoms are much higher than those from methyl group (Kerkeni and Clary 2004). Therefore, the production rates of methyl-OD must be much slower. For desorption, we did not measure the desorbed molecules. So, we do not have direct evidence of the desorption during reactions. Measurements of infrared absorption spectra tell us that the most of products remain in the solid but we do not deny that a small fraction (several percent) of the yield may desorb by the heat of reaction.

Astrochemistry: Recent Successes and Current Challenges
Proceedings IAU Symposium No. 231, 2005
D.C. Lis, G.A. Blake & E. Herbst, eds.

© 2006 International Astronomical Union
doi:10.1017/S1743921306007447

Theory of Molecular Scattering from and Photochemistry at Ice Surfaces

G. J. Kroes[1] and S. Andersson[1,2]

[1]Leiden Institute of Chemistry, Gorlaeus Laboratories, Leiden University,
P.O. Box 9502, 2300 RA Leiden, The Netherlands
email: g.j.kroes@chem.leidenuniv.nl

[2]Leiden Observatory, Leiden University, P.O. Box 9513, 2300 RA Leiden, The Netherlands

Abstract. The classical trajectory methodology for studying scattering of ions, atoms, or molecules from ice surfaces, and photodissociation of water at or in the surface of ice, is presented. The forces between the collider and the water molecules, or between the fragments of a dissociating molecule and the surrounding water molecules, are based on pair potentials taken either from *ab initio* calculations or derived empirically. Dynamical observables like sticking probabilities and kinetic energy distributions of desorbing photo-fragments are computed by solving Newton's equations of motion, starting from representative initial conditions. Four studies with relevance to astrochemistry are considered: the sticking of H atoms to ice Ih, the sticking of CO to ice Ih and ice I_a, the sticking of protons to ice Ih, and the photodissociation of water in ice Ih and ice I_a, also with a view to the subsequent chemistry of the H- and OH-products with co-adsorbed molecules. Where possible, the theoretical results are compared with experiments.

Keywords. astrochemistry — ISM: molecules — methods: numerical — molecular processes

1. Introduction

The formation of molecules in the interstellar medium (ISM) can proceed through several classes of reactions (van Dishoeck 1998). An important class consists of surface reactions, which take place on dust particles. In dense clouds, these dust particles, can be covered by icy mantles (Ehrenfreund & Schutte 2000). The ice can consist of H_2O molecules, but molecules like CO, CO_2, NH_3, and CH_4 may also be present. The icy mantles are generally believed to have an amorphous structure (Hagen *et al.* 1981). However, crystalline ice (ice Ih) can also be relevant to astrochemistry. For instance, amorphous ice (ice I_a) deposited on amorphous silicate particles can be phase transformed to crystalline ice at temperatures close to 110 K (Maldoni *et al.* 1999). On dust particles in the ISM, this temperature can be achieved through the impact of cosmic rays. In this paper, we will therefore consider processes occurring on both amorphous and crystalline pure water ice.

An important example of a surface reaction taking place on dust particles is the formation of molecular hydrogen, the most abundant molecule in the ISM. It is generally assumed that the H_2 in the ISM is formed through surface reactions occurring on grains (Hollenbach & Salpeter 1971; Herbst 1995). Because the mechanism of hydrogen recombination always involves the trapping of at least one hydrogen atom at the surface prior to reaction, the study of trapping (sticking) of H to ice is relevant to the astrochemistry of dense clouds. We will present results (Al-Halabi *et al.* 2002) of calculations on the trapping of H on ice Ih, and compare these with calculations as well as experiments on the trapping of H on ice I_a.

Solid CO has been observed along many lines of sight towards the dense regions in the ISM, the Galactic center, and high- and low-mass young stellar objects (Tanaka *et al.* 1994; Pontoppidan *et al.* 2003). Observations suggest that CO resides in two distinct environments; *i.e.*, in van der Waals bonded ice (also called apolar ice or H_2O-poor ice), and in hydrogen bonded ice (also called polar ice or H_2O-rich ice). One question concerning the hydrogen bonded ice is related to the mixing of CO and H_2O: does an intimate mixture form, or is CO present in multilayer form on the surface of H_2O ice? Clever experiments using coadsorbates which either bond or do not bond to surface dangling OH-groups of H_2O ice (Devlin 1992; Graham *et al.* 1999) strongly suggest that intimately mixed ice should exhibit a 2152 cm^{-1} feature in the infrared due to the interaction of CO with dangling OH bonds, while a 2139 cm^{-1} feature should be due to CO-CO interactions in solid CO ice, or interactions of CO with "bonded OH". These experiments also suggest that the dangling OH-bonds should be the preferred adsorption sites of isolated CO-molecules adsorbed to H_2O ice. Molecular dynamics (MD) calculations employing accurate pair potentials can provide support for the above interpretation of laboratory experiments and observations, and yield insight into the dynamics of trapping of CO on ice. Such calculations (Al-Halabi *et al.* 2003; Al-Halabi, van Dishoeck, & Kroes 2004; Al-Halabi *et al.* 2004) are also reported here.

Interactions of protons with water ice are also relevant to astrochemistry. Cosmic rays (CRs) consist predominantly of high-energy protons. The CR-ice interaction is the predominant agent leading to mass loss of water-ice grains (Mukai & Schwehm 1981). Impact of CRs on mixed ices can promote the formation of molecules like carbonic acid (H_2CO_3) (Gerakines *et al.* 2000) and amino acids (Kobayashi *et al.* 1995). We will therefore also present new, surprising results for the sticking of H^+ ions to ice Ih (Cabrera Sanfelix *et al.* 2005). The results are for a range of energies (0.05–4.0 eV) which is much lower than the kinetic energy of CR protons. The results are nevertheless of interest. First, they show that a spin-off of research on systems of astrochemical interest may be that new discoveries in chemical physics are made, although some of these discoveries might apply to conditions that are not of direct interest to astrochemistry. Second, we expect that many of the conclusions derived from our study (Cabrera Sanfelix *et al.* 2005) should also be applicable to low-energy collisions of H^+-containing molecular ions (such as $H_3{}^+$) with ice.

Photodissociation of water in ice, which can be induced by UV light of energy exceeding 7.6 eV (Kobayashi 1983), is also of interest to astrochemistry. The products of the reaction (H and OH, and possibly other products as well) can go on to react with co-adsorbed or co-absorbed molecules. It has been postulated that CO_2 could be formed in this way, *i.e.*, by reaction of OH with CO present on or in H_2O rich ice (Allamandola *et al.* 1988), and experiments on UV irradiation of mixed $CO:H_2O$ ice have indeed shown efficient CO_2 formation (Watanabe & Kouchi 2002). We here present results on the dynamics of photodissociation of H_2O in ice, with much attention paid to the dynamics of the H and OH fragments, also with a view to their possible subsequent chemistry with co-adsorbates (Andersson *et al.* 2005; Andersson *et al.*, in prep.) The calculations to be presented may be viewed as a first step in the modeling of photoinitiated reactions of products of water photolysis with co-adsorbed molecules.

We first discuss the molecular dynamics (MD) method we use in our calculations on sticking and photodissociation of H_2O in ice, in § 2. Results of our MD calculations are presented in § 3, for H + ice Ih (§ 3.1), CO + ice Ih and ice I_a (§ 3.2), H^+ + ice Ih (§ 3.3), and photodissociation of water in ice Ih and I_a (3.4). We conclude with a summary (§ 4).

2. Method

In this section, we discuss how an MD simulation of molecular processes on or in ice is performed. Whichever problem is studied, such simulations always share common features. First of all, the initial state of the system is prepared, according to the problem of interest. This includes the setting up of a dynamic ice surface (§ 2.1), and the choice of other initial conditions, such as the selection of the coordinates and momenta of the collider (for calculations on sticking) (§ 2.4). Subsequently, Newton's equations of motion have to be solved for a long enough period of time, and relevant observables have to be extracted (§ 2.2). This can only be done if the forces between the molecules are defined, which we do through pair potentials (§ 2.3).

2.1. *The Ice Surface*

The crystalline and amorphous ice surfaces were modeled using the MD method (Allen & Tildesley 1987). Ice Ih is always modeled as proton disordered (see Kroes 1992). The lower part of the ice Ih surface is usually modeled by 2 bilayers of H_2O molecules (120 molecules) which are held fixed. The upper part of the surface is modeled by 4 or 6 bilayers of H_2O molecules (240 or 360 molecules) which are allowed to move according to Newton's equations of motion. The H_2O molecules are modeled as rigid rotors, with only one exception: if the photodissociation of water in ice is modeled, the photoexcited molecule's intramolecular motion following photoexcitation is also modeled in order to follow the dissociation dynamics and to enable the computation of the absorption spectrum. The H_2O molecules are put in a box which is replicated in two directions according to periodic boundary conditions, to create an infinite ice surface. The TIP4P pair potential (Jorgensen *et al.* 1983) is used to model the forces between the water molecules, because its use yields stable ice Ih at the relevant temperatures (Kroes 1992; Karim & Haymet 1988). For ice Ih, the initial ice configuration obeys the famous ice rules (Bernal & Fowler 1933), and has a zero dipole moment (Kroes 1992). The surface temperature (T_s) of interest is imposed on the ice surface through the use of the computational analogue of a thermostat (Berendsen *et al.* 1984). For details and for a more technical discussion of the setting up of a surface of ice Ih, the reader is referred to Kroes (1992).

In Figure 1, top (1*a*) and side views (1*c*) of ice Ih are shown, where it is the (0001) basal plane face of ice that is exposed to the vacuum. Ice Ih has a bilayer structure, the bilayers being about 3.5 Å thick. Another important observation is that the water molecules in the upper monolayer can either have one H atom pointing away from the surface (dangling OH), or both H atoms pointing obliquely down to the surface. This has important consequences for adsorption of CO and protons (see § 3).

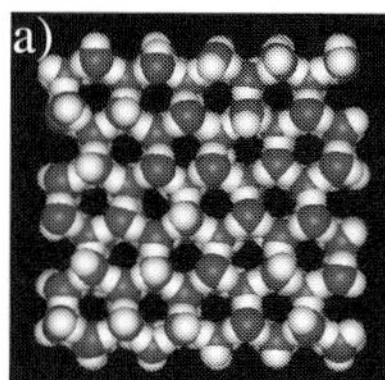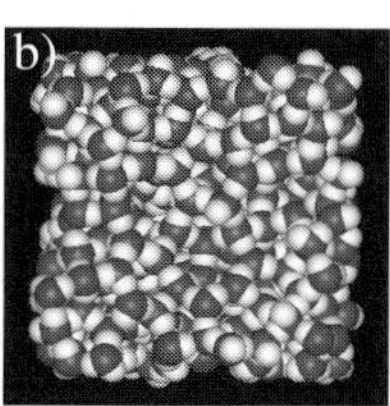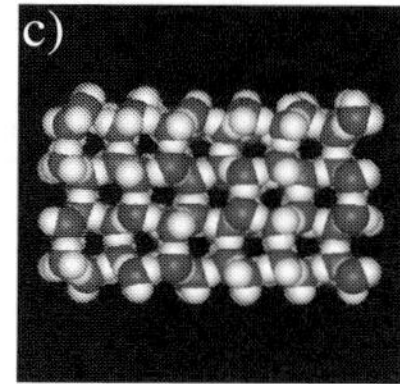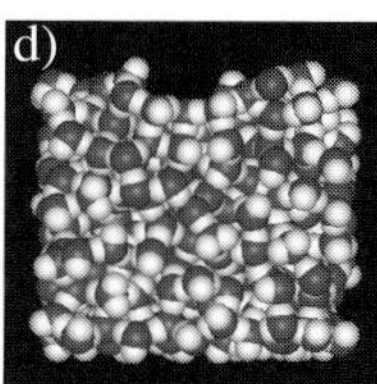

Figure 1. Top views (*a* and *b*) and side views (*c* and *d*) are shown of simulated ice Ih (*a* and *c*) and ice I$_a$ (*b* and *d*), at $T_s = 90$ K. In the figures, the dark atoms are the oxygen atoms, and the light atoms the hydrogen atoms.

Interstellar H_2O ice is believed to be predominantly amorphous in structure (Hagen *et al.* 1981). The specific morphology of amorphous ice remains an open question. The H_2O ice is thought to form through chemical reactions of H and O atoms on dust grains, but it is not known to exactly which structure this would lead. One usually assumes that the structure closely resembles that of ice grown by vapor deposition in the laboratory, known as amorphous solid water (ASW), rather than ice phases formed under high-pressure condistions, such as high-density amorphous (hda) ice, or low-density amorphous (lda) ice (Petrenko & Whitworth 1999; Ehrenfreund *et al.* 2003). We make amorphous ice (Al-Halabi, van Dishoeck, & Kroes 2004; Al-Halabi *et al.* 2004) by first heating hexagonal ice to room temperature, resulting in melting, and subsequent fast cooling (hyperquenching) to the desired temperature, using the computational analogue of a thermostat (Berendsen *et al.* 1984). The calculated average density of our amorphous ice is about 0.93 g/cm^3, which is in the range of measured densities (0.85–1.05 g/cm^3) of dense amorphous solid water deposited at $T = 22$ K for a wide range of deposition angles (Dohnálek *et al.* 2003). Just as for the case of ice Ih, the surface consists of a box of molecules which is replicated in two directions.

In Figure 1, top ($1b$) and side views ($1d$) of the dense ice I_a we made are shown. The amorphous ice completely lacks the regularity seen in ice Ih. The surface of ice I_a is irregular, displaying holes, into which CO can move and interact with many H_2O molecules simultaneously. Most of the water molecules hydrogen bond to 4 neighbouring water molecules, while a substantial fraction hydrogen bond to only 3 neighbouring water molecules (Al-Halabi *et al.* 2004).

2.2. *Classical Dynamics and Calculation of Absorption Spectrum*

To simulate the dynamics of a collision or photodissociation event, Newton's equations of motion are integrated in time. The integration is started from coordinates and momenta that are representative of the initial conditions (§ 2.4). Sticking probabilities and dynamical quantities related to photodissociation are determined by following the trajectory in time. For instance, to determine a sticking probability (P_s), first an operational definition is made of sticking. A requirement for sticking to occur in the MD is that a trajectory exhibit multiple turning points in the curve showing the molecule-surface distance as a function of time. An additional criterion may be that the final energy of the adsorbed molecule is less than a predefined threshold value. For the actual definition adopted in individual cases, the reader is referred to Al-Halabi *et al.* (2002), Al-Halabi *et al.* (2003), Al-Halabi, van Dishoeck, & Kroes (2004), Al-Halabi *et al.* (2004), and Cabrera Sanfelix *et al.* (2005). The calculation of dynamical quantities associated with photodissociation of H_2O in ice is discussed in Andersson *et al.* (2005) and Andersson *et al.*, in prep.

The absorption spectrum is calculated according to Schinke (1993):

$$\sigma_{tot}^{(i)}(\omega) \propto E_{photon} \int d\tau P_W^i(\tau)\mu_{if}^2\delta[H_f(\tau) - E_f] \qquad (2.1)$$

Here, $P_W^i(\tau)$ is the probability distribution function associated with the initial (i) vibrational state of the molecule in its ground electronic state, and $H_f(\tau)$ the Hamilton function in the upper electronic state (f). Furthermore, μ_{if} is the transition dipole function, E_i the energy of the initial state, and $E_f = E_i + E_{photon}$. $P_W^i(\tau)$ is calculated from a Wigner (semi-classical) distribution function fitted to the ground state vibrational wave function of the water molecule to be photodissociated, as discussed in van Harrevelt *et al.* (2001). The integration is over the entire phase space. The transition dipole moment function was taken from van Harrevelt & van Hemert (2000). The spectrum is calculated by determining the energy difference between the excited state potential and the ground state potential for a large number of phase space points, and the absorption spectrum is calculated as the number of times excitation energies occur within an energy

Table 1. Essential features of the pair potentials for the interaction of the collider X with an individual water molecule. Below, R_{min} refers to the distance between a defined atom pair, in the potential minimum geometry.

X	Method of derivation	Potential minimum (kJ mol^{-1})	R_{min} (Å)	Atom pair R_{min} refers to	References
CO	HF + MP2	−5.1	2.47	C-H$_w$	Al-Halabi *et al.* (2003)
H	UHF + MP4	−0.63	3.4	H-O$_w$	Al-Halabi *et al.* (2002) Zhang *et al.* 1991
H$^+$	empirical	−705	1.04	H$^+$-O$_w$	Cabrera Sanfelix *et al.* (2005) Kozack & Jordan (1992)

interval of a predetermined width. The approach outlined above has been shown to give good agreement between classical and quantum treatments of gas phase photodissociation (Schinke 1993). For more details on how we apply this procedure, see also Andersson *et al.* (2005) and Andersson *et al.*, in prep.

2.3. *Pair Potentials*

For the integration of the equations of motion, it is necessary to specify the forces that act between the water molecules belonging to the ice, and between the water molecules and the collider (sticking) or between the water molecules and the water molecule that was (or is to be) excited. The potential from which these forces are derived is always based on pair potentials acting between the collider (or the water molecule singled out for photoexcitation) and an H_2O molecule. The total interaction of the collider (or the special water molecule) with the ice surface is then taken as a sum over all (other) H_2O molecules.

The pair potentials used in the calculations on sticking are either taken from *ab initio* calculations or derived empirically. The validity of the pair potential is usually checked by comparison to experimental results. For instance, the CO-H_2O pair potential that was used (Al-Halabi *et al.* 2003; Al-Halabi, van Dishoeck, & Kroes 2004; Al-Halabi *et al.* 2004) was based on *ab initio* HF + MP2 calculations, and the minimum geometry obtained from the pair potential was checked to compare favorably with Fourier transform microwave absorption spectroscopy experiments (Yaron *et al.* 1990). Details on the pair potentials used in the calculations for H, CO, and H$^+$ + H_2O can be found in Al-Halabi *et al.* (2002), Al-Halabi *et al.* (2003), and Cabrera Sanfelix *et al.* (2005), respectively. Some essential features of the pair potentials are summarized in Table 1.

In the calculations on photodissociation of water in ice, the potential energy of the water molecule singled out for photoexcitation consists of an intramolecular and an intermolecular part. In the computation of the photoabsorption spectrum, the intramolecular potentials of the photoexcited H_2O molecule in its ground electronic and first excited states are taken from the work of Dobbyn & Knowles (1997).

For the molecule that is photoexcited, the intermolecular interaction with the surrounding water molecules is taken from the TIP3P model (Jorgensen *et al.* 1983), to avoid the complication, arising with the TIP4P model, that the negative charge is not on the (moving) O atom. In the calculation of the absorption spectrum and in the MD calculation on the dynamics following photoexcitation, the excited water molecule interacts with the surrounding water molecules as an excited state water molecule would. In the first potential model we used, charges were put on the O and H atoms of the excited molecule (Andersson *et al.* 2005) to mimic the dipole moment exhibited by gas phase water in its first excited electronic state (Klein *et al.* 1996). In our second potential model, smaller charges ($-0.2e$ on O and $0.1e$ on H) were used such that the theoretical spectrum was in good agreement with the experimental spectrum (Kobayashi 1983). For a detailed discussion of the pair interaction between excited H_2O and the surrounding H_2O molecules, the reader is referred to Andersson *et al.*, in prep.

2.4. *Initial Conditions*

The initial coordinates and momenta from which trajectories are started are selected to represent the physical situation of interest, using a Monte Carlo procedure (Porter & Raff 1976). The initial conditions are obtained by equilibrating the ice surface in a separate MD run, after a T_s has been imposed (§ 2.1). For simulating the collision of an atom, ion, or molecule with the surface, the collider's initial momentum is selected according to the collision energy and incidence angle of interest. If the collider is a molecule, its initial orientation is taken at random from uniform distributions of $\cos\theta$ and ϕ, where θ and ϕ are the polar angles of orientation of the molecule's bond axis. The molecule's rotational angular momentum J_c is taken as $J_c = \hbar\sqrt{J(J+1)}$, where J is the molecule's rotational quantum number. The direction of the angular momentum is taken from a random distribution. The collider's impact site on the surface is likewise selected at random.

In the calculations on the photodissociation of water in ice, the initial coordinates and momenta associated with the centre-of-mass motion and rotation of all H_2O molecules are obtained using a separate MD run. The water molecule to be excited is selected to be one of the water molecules present in a given layer of the simulated ice. Its initial intramolecular coordinates and vibrational momenta are selected at random from the probability distribution obtained from a semi-classical wave function – the Wigner distribution (van Harrevelt *et al.* 2001). Therefore, the initial phase-space distribution of the dissociating molecule is based on quantum mechanics.

3. Results and Discussion

3.1. *H + Ice*

The sticking probability (P_s) computed for scattering of atomic hydrogen from ice Ih at normal incidence over a range of collision energies E_i is shown for two different T_s in Figure 2 (Al-Halabi *et al.* 2002). The sticking probability could be fitted quite well to an exponential function of E_i, i.e., $P_s = \alpha\exp(-E_i/\beta)$, with $\beta = 175$ K (here and in Fig. 2, K is used as a unit of energy, 1 K = 8.31 J mol^{-1}). The values extracted for α were 1.5 for $T_s = 10$ K and 0.85 for $T_s = 70$ K (the fit for $T_s=10$ K is not valid at collision energies significantly lower than 100 K). The strong dependence of α on T_s illustrates the strong dependence of the sticking probability on T_s for H. As a result of its weak interaction with the surface, the H atom is sensitive to the fact that the water molecules are more dynamic and less packed in the warmer surface. As a result, it is more likely that an upward moving water molecule imparts momentum to the impinging H atom, and that the interaction of H with the surface is further weakened because it interacts with fewer water molecules, at higher T_s.

In Figure 2, the results for ice Ih are also compared with results of computational studies for ice I_a that used the same H-H_2O potential, so that differences should in principle be due to differences between ice Ih and ice I_a. For $T_s = 10$ K, the results for ice Ih fall between the results of ice I_a of Buch & Zhang (1991) and of Masuda *et al.* (1998). The discrepancy with the results of Masuda *et al.* is due to the use of an

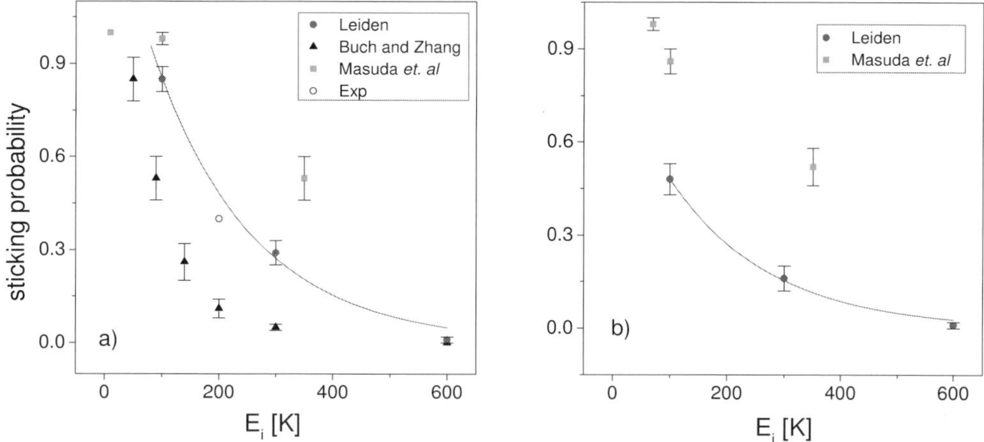

Figure 2. The sticking probability of hydrogen atoms to ice Ih is shown as a function of the collision energy E_i for (a) $T_s = 10$ K and (b) $T_s = 70$ K (Al-Halabi *et al.* 2002), together with previous computational results on sticking of hydrogen to ice I_a (Buch & Zhang 1991; Masuda *et al.* 1998). A sticking probability extracted by modeling from a TPD experiment using a 200 K atomic hydrogen beam impinging on an amorphous ice surface at low T_s is also shown (Manicó *et al.* 2001).

erroneous expression of the spherical harmonic, Y_{00}, in the equation describing the H-H_2O pair potential (Takahashi, private comm.) This expression was printed incorrectly in several papers using the H-H_2O interaction potential (Buch & Zhang 1991; Zhang *et al.* 1991).

The differences between the sticking probability for ice Ih and ice I_a computed by Buch & Zhang (1991) is probably due to differences in regularity of the ice surfaces. The amorphous ice cluster studied by Buch & Zhang is a small cluster consisting of 115 molecules. It was made by successively raining water molecules on to a small nucleus, and the surface of the cluster is quite irregular, with many sites where the impinging H may experience a significant attractive interaction to only a few water molecules. In contrast, the surface of crystalline ice (Fig. 1*a*) is quite regular, so that the impinging H atom can experience a significant attractive interaction with as many as six H_2O molecules, if it hits the center of a hexagonal ring. As a result of the stronger interaction, H exhibits a larger sticking probability on ice Ih than on the ice I_a modeled by Buch & Zhang. The large difference between trapping of H on ice Ih and on ice I_a here observed shows how importantly the ice structure may affect processes that are crucial to astrochemistry: after all, trapping of H on ice is important to the formation of molecular hydrogen, which is the most abundant molecule in the ISM, in dense clouds.

The difference we found between sticking of H on ice Ih and ice I_a also illustrates another important point: it is important to determine how the structure of interstellar amorphous ice is related to the conditions under which it is grown. It is believed that interstellar ice originates from chemical reactions between H and O on dust grains; as a result, the surface structure of the ice may be different than obtained by successively raining water molecules onto a pre-existing water nucleus, which could be an appropriate procedure for simulating vapor deposition of H_2O on ice.

The sticking probability we computed for $T_s = 10$ K is in good agreement with the sticking probability extracted by modeling thermal desorption experiments on sticking of H and D to amorphous ice, using a 200 K atomic hydrogen beam (Manicó *et al.* 2001). In the analysis of the experiments, the recombination efficiency γ measured was defined through $\gamma = P_s \eta$, and the assumption was made that the recombination probability η can be taken as one, so that $P_s = \gamma$. The resulting P_s for ice I_a is in good agreement with the theory for ice Ih. Of course, this is not consistent with the difference between the theoretical results for ice Ih and ice I_a as modeled by Buch & Zhang, $i.e.$, the experimental result for P_s on ice I_a significantly differs from the theoretical results of Buch & Zhang (1991). An open question is whether the agreement between theory and experiment for ice I_a will improve if the ice I_a is set up in a different way.

3.2. $CO + Ice$

The P_s of CO to ice Ih and ice I_a is plotted as a function of E_i in Figure 3, for $T_s = 90$ K (Al-Halabi, van Dishoeck, & Kroes 2004). The computed P_s can be fitted quite well to an exponential function of E_i, with $\alpha = 1$ and $\beta = 90$ kJ mol^{-1}, respectively, for both ice Ih and ice I_a. Interestingly, for CO the trapping probability appears to be the same for ice Ih and ice I_a. To within the statistical uncertainty implied by the number of classical trajectories run (100 for each energy), the probabilities are the same. In this respect, the trapping of CO to ice is different from the trapping of H to ice (Figure 2, § 3.1). At present, it is unclear whether this difference is due to the difference in the collider, or to the differences in the ice I_a model used in the calculations. In the calculations on H + ice I_a the ice I_a is modeled as a cluster of H$_2$O molecules successively rained on to a pre-existing H$_2$O nucleus (Buch & Zhang 1991), while the ice I_a used in CO + ice was obtained by rapid quenching of liquid water simulated at room temperature (Al-Halabi, van Dishoeck, & Kroes 2004).

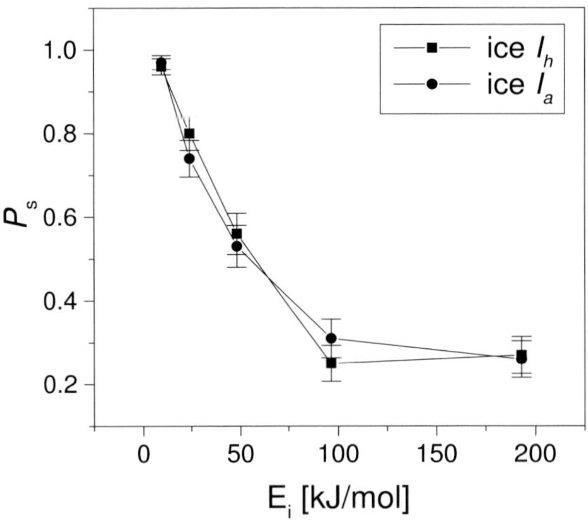

Figure 3. The P_s of CO to ice Ih and ice I_a is shown as a function of E_i, for scattering at normal incidence, and $T_s = 90$ K (Al-Halabi, van Dishoeck, & Kroes 2004).

For CO + ice, substantial P_s values were obtained for E_i (200 kJ mol^{-1}), much larger than the average adsorption of CO to ice, which is about 10 kJ mol^{-1}. This is because the energy with which the collider hits the surface can be very rapidly and efficiently dissipated through the hydrogen bonding network of ice, as also found in previous calculations on sticking of HCl to ice (Al-Halabi et $al.$ 1999; Al-Halabi et $al.$ 2001). In no case was a water molecule found to desorb from the surface, even though the highest E_i used was much higher than the energy with which surface water molecules are bound to the bulk of H$_2$O. An interesting question is whether MD simulations on sticking of atoms and molecules to CO-ice would find a similarly efficient energy dissipation and a similarly stable surface as obtained for H$_2$O ice.

The P_s for CO on ice is smaller than that found for HCl + ice (> 0.9 over the entire energy range depicted in Figure 3 (Al-Halabi *et al.* 1999; Al-Halabi *et al.* 2001) and effectively larger than that found for H + ice: the P_s of CO exceeds 0.8 at a value of E_i (10 kJ mol^{-1}) which is twice as large as the one (600 K $\approx$ 5 kJ mol^{-1}) where the trapping probability of H on ice is less than 0.1 (Fig. 2). This order in the efficiency of trapping reflects the interaction strength with the surface, which is correlated to the collider-H$_2$O pair interaction (Table 1): the stronger the interaction with the surface, the smaller the part of the sum of the incidence energy and the interaction energy that has to be transferred from incident motion to other motions, for trapping to occur. Therefore, the larger the interaction energy with the surface is, the larger the trapping probability is, for the neutral atoms and molecules we have studied (Al-Halabi *et al.* 2002; Al-Halabi *et al.* 2003; Al-Halabi, van Dishoeck, & Kroes 2004; Al-Halabi *et al.* 1999; Al-Halabi *et al.* 2001).

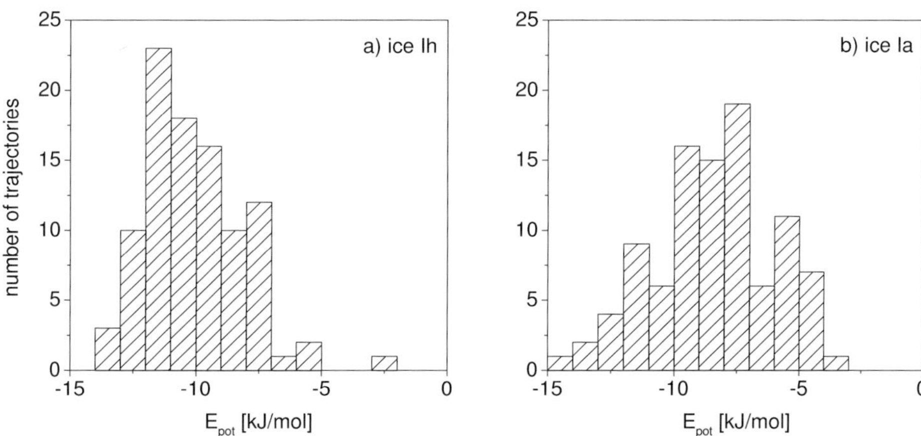

Figure 4. Using histograms, the number of sticking trajectories is shown as a function of the final value of the potential interaction energy of CO with ice, for E_i=9.6 kJ mol^{-1} and T_s=90 K, for the case of ice Ih (*a*) and ice I$_a$ (*b*) (Al-Halabi, van Dishoeck, & Kroes 2004).

From the MD simulations, it is also possible to compute the final potential interaction energy of the collider (CO) with the surface. Distributions of the adsorption energy (E_{ad}) are shown in Figure 4 for ice Ih and ice I$_a$ (Al-Halabi, van Dishoeck, & Kroes 2004). The distribution is broader for ice I$_a$ than for ice Ih. This reflects the greater variety of adsorption sites on ice I$_a$: the weakest interactions are due to CO interacting with just a few H$_2$O molecules protruding from the irregular surface, and similarly strong interactions can result from CO interacting with many water molecules, as can happen when CO moves into a hole in the irregular surface of ice I$_a$ (Fig. 1*d*). The average adsorption energies extracted from the distributions differ somewhat; *i.e.*, E_{ad} computed for ice I$_a$ was –8.4 $\pm$ 0.24 kJ mol^{-1}, in reasonably good agreement with the result (about 10 kJ mol^{-1}) of CO + ice I$_a$ FT-IR experiments (Allouche *et al.* 1998; Manca *et al.* 2000; Martin *et al.* 2002), and E_{ad} computed for ice Ih was –10.1 $\pm$ 0.20 kJ mol^{-1}.

The adsorption geometry of CO on ice I$_a$ has also been considered (Al-Halabi *et al.* 2004), in geometry optimisations in which the ice configuration (from an MD simulation) was kept fixed. The calculations suggest that CO interacts most strongly with dangling OH's (Figs. 5*a* and 5*b*), and more weakly with the bonded OH's (Figs. 5*c*–5*e*). This is in agreement with IR experiments which find the signal attributed to the interaction with dangling OH, which is observed at 2152 cm^{-1}, to be particularly strong in case of other

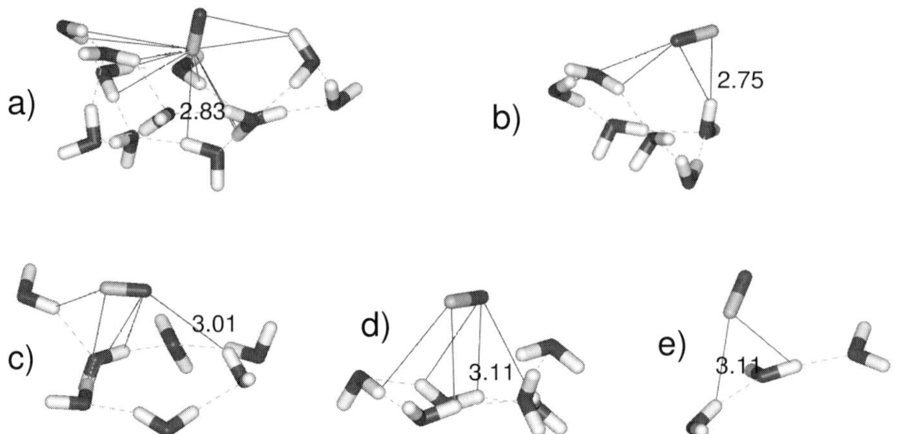

Figure 5. Examples of CO-static ice I_a configurations obtained from geometry optimisations (Al-Halabi *et al.* 2004). The five examples correspond to five different values of the potential interaction energy of CO with ice I_a, *i.e.*, (*a*) –0.155, (*b*) –0.129, (*c*) –0.111, (*d*) –0.090, and (*e*) –0.067 eV. Dotted lines illustrate hydrogen bonds between surface water molecules. Solid lines connect CO with the H-atoms of neighbouring water molecules that are closer to CO than 3.5 Å, the shortest H-CO distance being given in Å. Black cylinders represent O atoms, grey cylinders the C atom, and light grey spheres the H atoms. Bonding of CO to dangling OH bonds is seen in (*a*) and (*b*), while CO bonds more to "bonded OH" in (*c*), (*d*), and (*e*).

molecules co-adsorbed at the centres of the hexagonal rings (Devlin 1992; Martin *et al.* 2002), and absent in case of other molecules co-adsorbed at the dangling OH bonds (Devlin 1992). The absence of the 2152 cm^{-1} feature in astronomical spectra (Pontoppidan *et al.* 2003) otherwise clearly showing the presence of CO can be used as evidence that most of the CO is not present in intimately mixed CO-H$_2$O ice (Pontoppidan *et al.* 2003).

3.3. $H^+ + Ice$

The P_s computed (Cabrera Sanfelix *et al.* 2005) for H$^+$ + ice Ih is shown as a function of E_i in Figure 6 for collisions at normal incidence, at $T_s = 80$ K. The figure shows a surprising result: for low E_i (= 0.2 eV), P_s is significantly smaller than 1, and also significantly smaller than P_s for HCl, which exceeds 0.9 for these low E_i (Al-Halabi *et al.* 1999; Al-Halabi *et al.* 2001). This is surprising because the proton's interaction energy with the ice surface (computed to be about 10 eV on average (Cabrera Sanfelix *et al.* 2005) is much larger (by about a factor 50) than the interaction energy of HCl with ice (Al-Halabi *et al.* 1999; Al-Halabi *et al.* 2001). The explanation for the significant reflection (40% at $E_i = 0.05$ eV) is that in the surface monolayer of ice Ih 50% of the water molecules have an electropositively charged H atom pointing away from the surface. The impinging proton experiences electrostatic repulsion when impacting close to these molecules, causing it to be reflected at low E_i. The calculations show that the rule observed for collisions of neutrals with ice (the larger the E_{ad} of the collider on ice, the larger the trapping probability) does not hold when comparing trapping of neutrals with trapping of ions.

The calculations on H$^+$ + ice Ih also yielded another interesting prediction (Cabrera Sanfelix *et al.* 2005). Figure 7*a* shows that there is a significant probability (4–16% for E$_i$ between 0.05 to 3 eV) that a water molecule desorbs upon impact of the proton (collision induced desorption). The water desorption is not due to "sputtering"; the desorption

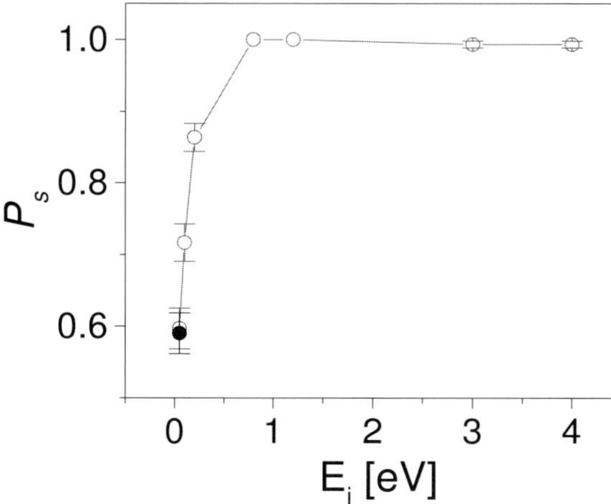

Figure 6. The sticking probability of H^+ to ice Ih is shown as a function of E_i, for normal incidence, at $T_s = 80$ K (open circles). The solid circle is a result of a control calculation (see Cabrera Sanfelix *et al.* 2005 for more details).

also occurs at values of E_i which are much lower (0.05 eV) than the energy with which surface H_2O molecules are bound to

the bulk of ice (> 0.3 eV). Visualisation of the trajectories show that the water desorption occurs because the stuck proton disrupts the hydrogen bonding network between surface water molecules. The stuck proton can bury itself between the first and the second, or the second and the third bilayer of the ice surface (Fig. 7*b*). Because the proton's interaction with a neighbouring water molecule (≈ 700 kJ mol^{-1}, Table 1) can be much stronger than a hydrogen bond between neighbouring water molecules (≈ 15–20 kJ mol^{-1}), H^+ can act as a "Coulomb bomb" or "hydrogen bond chain saw", completely disrupting the hydrogen bond network close to it. As a result, a nearby surface water molecule can all of a sudden find itself in a repulsive environment, and desorb.

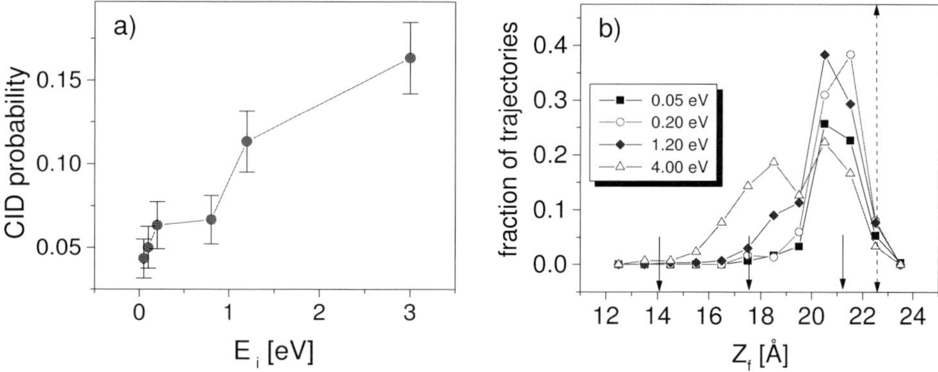

Figure 7. (*a*) The probability that one H_2O molecule desorbs upon impact of a proton is shown as a function of E_i, for $T_s = 80$ K (Cabrera Sanfelix *et al.* 2005). (*b*) Fraction of sticking trajectories plotted for four different E_i, according to the final value of the co-ordinate for motion normal to the surface, Z_f (Cabrera Sanfelix *et al.* 2005). The positions of the surface bilayers are shown by the arrows at the bottom of the figure. The surface-vacuum interface, indicated by the dashed long double arrow, is located at 22.5 Å.

3.4. *Photodissociation of H_2O Ice*

In Figure 8 the simulated spectrum of the first absorption band of ice Ih for the old (Andersson *et al.* 2005) and new (Andersson *et al.*, in prep.) potential models are shown together with the experimental spectrum (Kobayashi 1983) and the calculated first absorption band of gas-phase H_2O. The ice spectra refer to molecules in the third bilayer, which gives a good representation of bulk spectra. The previous model gave a blueshift of 2 eV compared to the gas-phase spectrum, while the new model gives a blueshift of little over 1 eV, in good agreement with experiment (Kobayashi 1983). The experimental peak (8.6 eV) and the threshold energy (7.6 eV) are reproduced by the new potential. The calculated peaks of amorphous ice (8.6 eV) and liquid water (8.2 eV) coincide with experimental values (Kobayashi 1983; Heller *et al.* 1974). These results strongly suggest that the amount of kinetic energy released into the system is also correct.

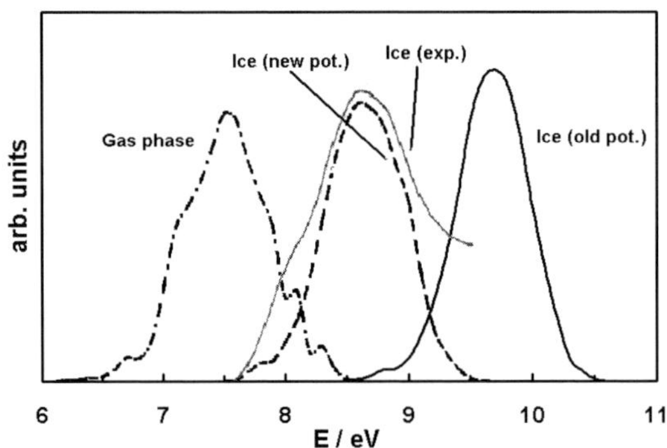

Figure 8. Simulated spectra for the first UV absorption band for ice Ih with the old (Andersson *et al.* 2005) and new (Andersson *et al.*, in prep.) potential models. Also shown are the calculated spectrum of the first band of gas-phase water UV absorption (Andersson *et al.* 2005) and the experimental spectrum for ice Ih (Kobayashi 1983).

Figure 9 shows the most important basic outcomes for ice Ih and ice I_a for photodissociation of molecules in the top three bilayers. The case where H desorbs and OH becomes trapped is the dominant outcome (60–80%) in the first bilayer. Already in the second bilayer, trapping of H and OH or of recombined water become of roughly equal importance to H atom desorption. Photodissociation in the third bilayer leads mainly to trapping of H and OH or recombined water and only in 10–15% of the cases to desorption of H. Results for the two types of ice are quite similar with a few noticeable differences. The most prominent difference is that, especially in the first two bilayers, the probability of H atom desorption is higher for ice I_a than for ice Ih (1st bilayer: 80% *vs.* 65%; 2nd bilayer: 40% *vs.* 25%). This is explained by the low density of the first bilayer in ice I_a, due to its irregular structure. H atoms moving towards the surface therefore experience fewer collisions in ice I_a than in ice Ih and have a higher probability of making it into the gas phase.

The H atoms that become trapped have on average moved about 10 Å from their original position, but atoms travelling over 60 Å have been observed. The trapped OH radicals on average only move 1–2 Å, but in cases where the photodissociation occurs at the surface, the OH has been found to move much longer distances (up to 85 Å). This is further than what is found for H atoms moving over the surface and can be explained by the OH-ice interaction being much stronger than the H-ice interaction. While H atoms moving at the surface have a high probability of desorbing, the more strongly bound OH can in most cases not desorb, but it can keep moving parallel to the surface.

Desorption of recombined water molecules is found in 0.2% of the cases for both types of ice. Indirect water desorption, where one of the molecules not initially photodissociated desorbs, has a probability of 0.2% for ice Ih and 0.04% for ice I_a. Note that these

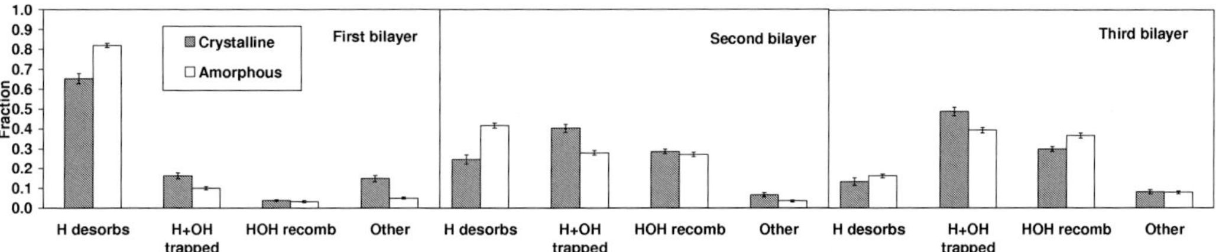

Figure 9. Probabilities (per absorbed UV photon) for the most important basic outcomes of photodissociation of water molecules in the top three bilayers in ice Ih and I_a (using the new potential model (Andersson *et al.*, in prep.). "H desorbs" is when H desorbs and OH remains trapped, "H+OH trapped" is when H and OH become trapped, and "HOH recomb" contains the cases where a recombined water molecule remains trapped. "Other" contains all other outcomes (Andersson *et al.*, in prep.)

probabilities are per absorbed UV photon. The mechanism behind this desorption is through the transfer of momentum of an energetic H atom to a water molecule, which either desorbs or initiates a further chain of momentum transfer until a molecule at the surface desorbs. The photodesorption of water is found to be quite small, but, in apparent contradiction to our results, Westley *et al.* (1995) did not detect *any* desorption of water at low temperatures ($T \leqslant 50$ K) in the limit of single-photon absorption. They however used Lyman-α (10.2 eV) photons, which probably leads to a different excited state than the lower excitation energies considered by us.

4. Conclusions

We have discussed MD calculations on sticking of atomic hydrogen, CO, and H^+ to ice Ih and ice I_a; results for HCl + ice were also briefly mentioned. A general rule that emerged for sticking/trapping of neutrals to ice is that P_s increases with the adsorption energy of the collider on ice. This rule does not hold when the sticking of neutrals and ions is compared. For H and CO, we found that, for normal incidence, the computed sticking probability could be fitted quite well to an exponential function of the collision energy.

For H and CO, it was possible to compare sticking probabilities computed for interaction with ice Ih and ice I_a, in calculations using identical pair potentials for the interaction of the collider with the ice. For CO, only a small difference was found between sticking on ice Ih and ice I_a; for H, the difference was much greater. At present, it is unclear whether this discrepancy is due to differences between the colliders, or to the different ways in which the simulated ice I_a was prepared. The observed discrepancy underlines the need for understanding how the structure of interstellar ice is related to the way it comes about in the ISM, which is a fundamental question of high importance to astrochemistry: water (mostly in ice) is the third abundant molecule in the ISM, and in dense clouds ice surfaces effectively serve as the factory for producing molecular hydrogen, which is the most abundant molecule in the ISM.

Calculations on collisions of neutrals with ice surfaces never show water desorption, even for E_i (2 eV) which significantly exceed the binding energies of surface water molecules (a few tenth's of an eV). In contrast, calculations here presented show collision induced desorption of water molecules upon impact of protons even at E_i (0.05 eV) much smaller than the binding energies of surface water molecules. In the mechanism, the proton, the interaction of which with a neighbouring water molecule can be much stronger than a hydrogen bond between neighbouring water molecules, disrupts the hydrogen bonding network, acting like a "Coulomb bomb".

We have also presented results of new calculations on photodissociation of water in ice Ih and ice I$_a$, in the first absorption band. The emphasis was on the dynamical outcome of photoexcitation of a water molecule in the top 10 Å of the ice surface (first 3 bilayers), because photoinitiated chemistry with co-adsorbed molecules, which we are interested in studying next, is most likely to occur at the surface. By adjusting the dipole moment on the excited water molecule, which affects the interaction of that water molecule with the surrounding water molecules, it was possible to reproduce the position and the shape of the first absorption band in the experimental spectrum in a semi-classical calculation.

Subsequent MD calculations show that the most important dynamical outcomes following photoexcitation in the top surface layer are dissociation followed by H-atom desorption (dominant following photoexcitation in the top bilayer) and dissociation followed by trapping of both the H and OH photofragments (dominant following photoexcitation in the third bilayer). In contrast, water desorption was found to be a very unlikely outcome ($< 0.5\%$). The dynamics is usually complete in around 1 ps, which is much shorter than the average time-interval between subsequent collisions of UV photons with a typical ice-covered grain in dense clouds (about a week). While there are subtle differences between photodissociation of water in ice Ih and ice I$_a$, the similarities in the spectrum and the dynamical outcomes are more striking than the differences.

Acknowledgements

The authors like to thank E.F. van Dishoeck, who contributed importantly to the work here presented. We also thank A. Al-Halabi, who performed the calculations on sticking of H and CO to ice, and importantly contributed to the calculations on H$^+$ + ice. We also thank A.W. Kleyn, M.C. van Hemert, H.J. Fraser, P. Cabrera Sanfelix, G.R. Darling, and S. Holloway, whose co-authorship on one or more of the papers discussed here reflects the importance of the contributions they made to the work presented here. We also thank R. van Harrevelt for help with the initial conditions for the calculations on photodissociation. The work of S. Andersson was funded by a NWO/Spinoza grant (for E.F. van Dishoeck), and a CW programme grant. The work was also funded by FOM, and through grants of computer time by the Stichting Nationale computerfaciliteiten (NCF).

References

Al-Halabi, A., Kleyn, A.W., & Kroes, G.J. 1999, *Chem. Phys. Lett.* 307, 505

Al-Halabi, A., Kleyn, A.W., & Kroes, G.J. 2001, *J. Chem. Phys.* 115, 482

Al-Halabi, A., Kleyn, A.W., van Dishoeck, E.F., & Kroes, G.J. 2002, *J. Phys. Chem. B* 106, 6515

Al-Halabi, A., Kleyn, A.W., van Hemert, M.C., van Dishoeck, E.F., & Kroes, G.J. 2003, *J. Phys. Chem. A* 107, 10615

Al-Halabi, A., Fraser, H.J., Kroes, G.J., & Van Dishoeck, E.F. 2004, *A&A* 422, 777

Al-Halabi, A., van Dishoeck, E.F., & Kroes, G.J. 2004, *J. Chem. Phys.* 120, 3358

Allamandola, L.J., Sandford, S.A., & Valero, G.J. 1988, *Icarus* 76, 225

Allen, M.P. & Tildesley, D.J. 1987, *Computer Simulations of Liquids* (Clarendon: Oxford)

Allouche, A., Verlaque, P., & Pourcin, J. 1998, *J. Phys. Chem. B* 102, 89

Andersson, S., Kroes, G.J., & van Dishoeck, E.F. 2005, *Chem. Phys. Lett.* 408, 415

Andersson, S., Kroes, G.J., & van Dishoeck, E.F., in preparation

Berendsen, H.J.C., Postma, J.P.M., van Gunsteren, V.F., Dinola, A., & Haak, J.R. 1984, *J. Chem. Phys.* 81, 3684

Bernal, J.D. & Fowler, R.H. 1933, *J. Chem. Phys.* 1, 515

Buch, V. & Zhang, Q. 1991, *Ap. J.* 379, 647

Cabrera Sanfelix, P., Al-Halabi, A., Darling, G.R., Holloway, S., & Kroes, G.J. 2005, *J. Am. Chem. Soc.* 127, 3944

Devlin, J.P. 1992, *J. Phys. Chem.* 96, 6185

Dobbyn, A.J. & Knowles, P.J. 1997, *Mol. Phys.* 91, 1107

Dohnálek, Z., Kimmel, G.A., Ayotte, P., Smith, R.S., & Kay, B.D. 2003, *J. Chem. Phys.* 118, 364

Ehrenfreund, P. & Schutte, W.A. 2000, in *Astrochemistry: From Molecular Clouds to Planetary Systems*, eds. Y.C. Minh & E.F. van Dishoeck (ASP: San Francisco), vol. 197, p. 135

Ehrenfreund, P., Fraser, H.J., Blum, J., Cartwright, J.M.E., Garcia-Ruiz, J.M., Hadamcik, E., Levasseur-Regourd, A.C., Price, S., Prodi, F., & Sarkissian, A. 2003, *Plan. Space. Sci.* 51, 473

Gerakines, P.A., Moore, M.H., & Hudson, R.L. 2000, *A&A* 357, 793

Graham, A.P., Menzel, A., & Toennies, J.P. 1999, *J. Chem. Phys.* 111, 1169

Hagen, W., Tielens, A.G.G.M., & Greenberg, J.M. 1981, *Chem. Phys.* 56, 257

Heller, J.M., Jr., Hamm, R.N., Birkhoff, R.D., & Painter, L.R. 1974, *J. Chem. Phys.* 60, 3483

Herbst, E. 1995, *Annu. Rev. Phys. Chem.* 46, 27

Hollenbach, D. & Salpeter, E.E. 1971, *Ap. J.* 163, 155

Jorgensen, W.L., Chandrasekhar, J., Madura, J.D., Impey, R.W., & Klein, M.L. 1983, *J. Chem. Phys.* 79, 926

Karim, O.A. & Haymet, A.D.J. 1988, *J. Chem. Phys.* 89, 6889

Klein, S., Kochanski, E., & Strich, A. 1996, *Theor. Chim. Acta* 94, 75

Kobayashi, K. 1983, *J. Phys. Chem.* 87, 4317

Kobayashi, K., Kasamatsu, T., Kaneko, T., Koike, J., Oshima, T., Saito, T., Yamamoto, T., & Yanagawa, H. 1995, *Adv. Space Res.* 16, 21

Kozack, R.E. & Jordan, P.C. 1992, *J. Chem. Phys.* 96, 3131

Kroes, G.J. 1992, *Surf. Sci* 275, 365

Maldoni, M.M., Robinson, G., Smith, R.G., Duley, W.W., & Scott, A. 1999, *MNRAS* 309, 325

Manca, C, Roubin, P., & Martin, C. 2000, *Chem. Phys. Lett.* 330, 21

Manicó, G., Raguni, G., Pironello, V., Roser, J.E., & Vidali, G. 2001, *Ap. J.* 548, L253

Martin, C, Manca, C., & Roubin, P. 2002, *Surf. Sci.* 502–503, 280

Masuda, K., Takahashi, J., & Mukai, T. 1998, *A&A* 330, 773

Mukai, T. & Schwehm, G. 1981, *A,&A* 95, 373

Petrenko, V.F. & Whitworth, R.W. 1999, *Physics of Ice* (University Press: Oxford)

Pontoppidan, K.M., Fraser, H.J., Dartois, E., Thi, W.F., van Dishoeck, E.F., Boogert, A.C.A., d'Hendecourt, L., Tielens, A.G.G.M., & Bisschop, S.E. 2003, *A&A* 408, 981

Porter, R.N. & Raff, L.M. 1976, in *Dynamics of Molecular Collisions, Part B*, ed. W.H. Miller (Plenum: New York), p. 1

Schinke, R. 1993, *Photodissociation Dynamics* (Cambridge University Press: Cambridge)

Takahashi, J., private communication

Tanaka, M., Nagata, T., Sato, S., Yamamoto, T. 1994, *Ap. J.* 430, 779

van Dishoeck, E.F. 1998, in *The Molecular Astrophysics of Stars and Galaxies*, eds. T.W. Hartquist & D.A. Williams (Clarendon Press: Oxford), p. 53

van Harrevelt, R. & van Hemert, M.C. 2000, *J. Chem. Phys.* 112, 5777

van Harrevelt, R., van Hemert, M.C., & Schatz, G.C. 2001, *J. Phys. Chem. A* 105, 11480

Watanabe, N. & Kouchi, A. 2002, *Ap. J.* 567, 651

Westley, M.S., Baragiola, R.A., Johnson, R.E., & Baratta, G.A. 1995, *Nature* 373, 405

Yaron, D., Peterson, K.I., Zolandz, D., Klemperer, W., Lovas, F.J., & Suenram, R.D. 1990, *J. Chem. Phys.* 92, 7095

Zhang, Q., Sabelli, N., & Buch, V. 1991, *J. Chem. Phys.* 95, 1080

Discussion

HERBST: What do you suspect happens when high energy ions strike crystalline and amorphous ice?

KROES: Brown *et al.* have studied scattering of high energy protons (0.5 MeV and 1.5 MeV, respectively) from amorphous ice (1978, *Phys. Rev. Lett.* 40, 1027). Their most important finding was that the sputtering yield (number of water molecules sputtered per incident ion) was 0.4 for a collision energy of 0.5 MeV, and 0.2 for 1.5 MeV. Extrapolating to the most usual energies of cosmic rays (approximately 100 MeV), one would think that at such high incidence energies no sputtering would be observed at all. I know of no experiments that have been done at these high energies. A recent review discussing sputtering of ice by highly energetic positive ions is that of Baragiola *et al.* (2003, *Nucl. Instruym. Methods Phys. Res. B* 209, 294), who mention that the experiments of Brown *et al.* (at 0.5 and 1.5 MeV) suggest that sputtering is an important phenomenon on icy satellites and grains subject to irradiation by energetic ions from solar flares and by magnetospheric ions.

Astrochemistry: Recent Successes and Current Challenges
Proceedings IAU Symposium No. 231, 2005
D.C. Lis, G.A. Blake & E. Herbst, eds.
© 2006 International Astronomical Union
doi:10.1017/S1743921306007459

Steps toward Identifying PAHs:
A Summary of Some Recent Results

Douglas M. Hudgins† and L. J. Allamandola

Astrophysics Branch, NASA Ames Research Center, Mountain View, CA, USA

Abstract. Based on over two decades of experimental, observational, and theoretical studies by scientists around the world, it is now widely accepted that the composite emission of mixtures of vibrationally-excited PAHs and PAH ions can accommodate the general pattern of band positions, intensities, and profiles observed in the discrete IR emission features of carbon-rich interstellar dust, as well as the variations in those characteristics. These variations provide insight into the detailed nature of the emitting PAH population and reflect conditions within the emitting regions giving this population enormous potential as probes of astrophysical environments. Moreover, the ubiquity and abundance of this material has impacts that extend well beyond the IR.

In this paper we will examine recent, combined experimental, theoretical, and observational studies that indicate that nitrogen-substituted PAHs represent an important component of the interstellar dust population, and we will go on to explore some of the ramifications of this result. We will also explore the results of recent experimental studies of the strong, low-lying electronic transitions of ionized PAH ions in the near-IR (0.7–2.5 μm) and explore the role that these transitions might play in pumping the PAH IR emission in regions of low excitation.

Keywords. astrochemistry — infrared: ISM — ISM: molecules — ISM: dust — ISM: lines and bands — methods: laboratory — molecular processes — techniques: spectroscopic

1. Introduction

Most interstellar objects with associated gas and dust exhibit a series of strong infrared emission features near 3.3, 6.2, 7.7, 8.6, and 11.2 μm which are generally attributed to polycyclic aromatic hydrocarbons (PAHs) and related molecular materials (e.g., Allamandola, Tielens, & Barker 1989; Puget & Leger 1989; Cox & Kessler 1999). Within the framework of this PAH model, the interstellar features arise from the combined emission of a complex mixture of PAH species that together comprise the molecular component of the carbon-rich interstellar dust population. The wide acceptance currently enjoyed by the interstellar PAH model is based on more than a decade of laboratory measurements and theoretical calculations (e.g., Vala *et al.* 1994; Hudgins & Allamandola 1995; Szczepanski *et al.* 1995; Langhoff 1996; Bauschlicher & Langhoff 1997; Pauzat & Ellinger 2001; Kim & Saykally 2002; Mattioda *et al.* 2003; Oomans *et al.* 2003; and references therein), which have shown that models based on the composite spectra of PAHs and PAH cations can accommodate the general pattern of band positions, intensities, and profiles observed in the interstellar IR emission spectra as well as the variations in those characteristics (e.g., Allamandola, Hudgins, & Sandford 1999; Van Kerckhoven *et al.* 2000; Hony *et al.* 2001; Verstraete *et al.* 2001; Bakes, Tielens, & Bauschlicher 2001; Bakes *et al.* 2001; Draine & Li 2001; Li & Draine 2001; Pech, Joblin, & Boissel 2002). Overall, that work indicates that the interstellar 3.3 μm band and, to a lesser extent, the

---

† Present address: Universe Division, Mail Suite 3W39, NASA Headquarters, 300 E St. SW, Washington DC 20546, USA; email: Douglas.M.Hudgins@nasa.gov

11.2 μm band are dominated by the emission of neutral PAHs, whose most prominent bands fall in these regions. Conversely, the interstellar bands in the 6–9 μm are dominated by the emission of ionized PAHs, whose most prominent bands fall in this region and are an order of magnitude more intense than those of the analogous neutral species.

Nevertheless, the agreement between the interstellar emission spectra and model spectra based on the latest experimental and theoretical data is not perfect. However, the discrepancies between observations and models have the potential to provide as much insight into the astrophysical problem as the agreements. For example, one issue that has emerged from the most recent analyses concerns the enigmatic position of the nominal 6.2 μm emission band, attributed to the CC stretching vibrations of interstellar PAH cations. Laboratory measurements and theoretical calculations of the IR spectra of a wide range of PAH cations have shown that the dominant CC stretching features of those species consistently fall at somewhat longer wavelengths than does the interstellar feature. This discrepancy is illustrated in Figure 1, which compares a typical interstellar emission spectrum to a simple PAH model in the 6.2 μm region. Hudgins & Allamandola (1999) noted this discrepancy, but also found that, for the small PAH cations ($N_C \leqslant 24$ C atoms) in their dataset, the dominant CC stretching bands shifted towards shorter wavelengths as molecular size increased, suggesting that the position of the interstellar emission band might be indicative of the size of the dominant emitters. However, as illustrated by the plot of band position verses molecular size shown in Figure 1(b), more recent experimental and theoretical studies of larger species show that this trend does not hold for larger PAHs. Instead, the positions of the CC stretching features of PAHs larger than about 30 C atoms are insensitive to molecular size and tend to cluster around 6.3 μm. Thus, molecular size alone is not sufficient to explain the 6.2 μm position of the interstellar band.

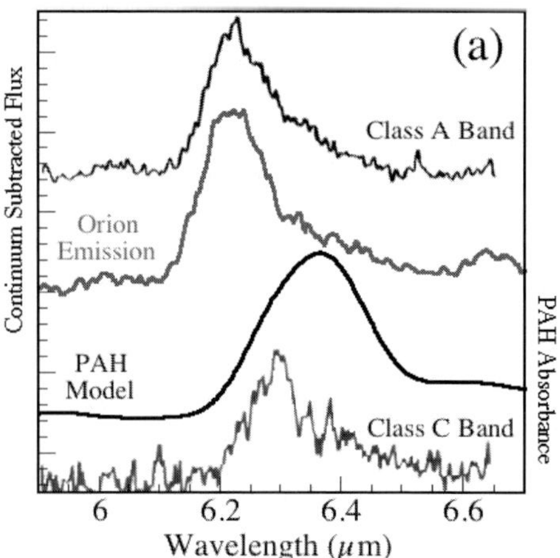

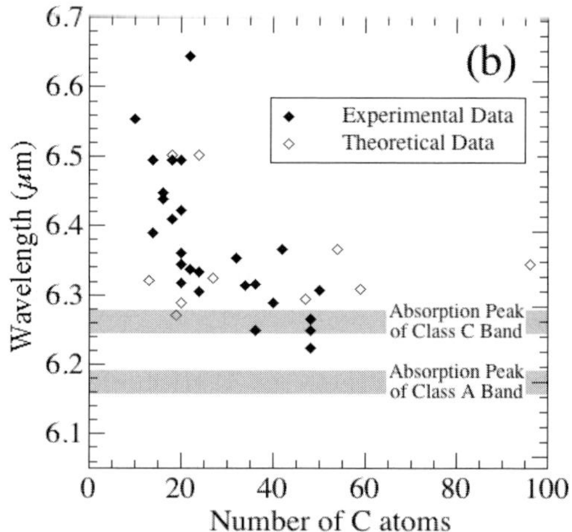

Figure 1. (a) A comparison of a typical 6.2 μm interstellar emission band (Orion) to a simple PAH model illustrating that the CC stretching features of PAH cations consistently fall at somewhat longer wavelengths than the interstellar feature. The canonical Class A and Class C emission bands identified by Peeters *et al.* (see text) are also included (plot adapted from Peeters *et al.* 2002). (b) A plot illustrating the variation in the position of the dominant CC stretching absorption feature as a function of molecular size for 3 dozen PAH cations. Filled diamonds indicate experimental data; open diamonds represent theoretical data. The bars indicate the *expected absorption* positions (see text) of the Class A and Class C emission components (plot adapted from Hudgins, Bauschlicher, & Allamandola 2005).

Further light was shed on the precise nature of this discrepancy by the detailed observational work of Peeters *et al.* (2002). In that study, the authors showed that the nominal interstellar 6.2 μm band is actually a composite of two bands, one centered near 6.2 μm and one centered near 6.3 μm. Variations in the relative contributions of these two components give rise to variations in both the position and profile of the composite feature. Bands dominated by the shorter wavelength component (by far the most common case) fall at an average position of 6.215 μm and are designated "Class A" bands. Bands that are dominated by the longer wavelength component fall at an average position of 6.295 μm and are designated "Class C" bands. "Class B" bands are a composite of Classes A and C and exhibit intermediate positions and compound profiles. The canonical Class A and Class C bands isolated by Peeters *et al.* (2002) are also shown in Figure 1(*a*) for reference. For the purpose of direct comparison, the positions of the Class A and Class C bands are also indicated in Figure 1(*b*), adjusted to compensate for the expected *ca.* 10 cm^{-1} redshift between emission and absorption peak frequencies (Joblin *et al.* 1995). Based on these data, Peeters *et al.* concluded that while moderately-sized PAH cations ($N_C \geqslant 40$ C atoms) might accommodate the position of 6.3 μm Class C component, the origin of the 6.2 μm Class A component remained anomalous within the framework of the PAH model.

2. The Effects of N Substitution on the IR Spectrum of PAHs

In an effort to explore the origin of this and other discrepancies in detail between the interstellar emission and the spectroscopic properties of PAHs, we have continued to expand our studies to encompass new and different PAH-related species – species with structural or compositional characteristics that distinguish them from the other aromatic species in the database and which might give rise to distinctive spectroscopic properties. Among these are polycyclic aromatic nitrogen heterocycles (PANHs), compounds that incorporate one or more atoms of nitrogen within their aromatic carbon skeleton. Detailed accounts of these studies and their astrophysical implications are available in the literature (Mattioda *et al.* 2003; Hudgins, Bauschlicher, & Allamandola 2005, hereafter HBA 05) and will only be summarized here.

The structures of some representative examples of this class of aromatic compound that have been considered in our studies are shown in Figure 2. To specify the position of N-substitution within the C skeletons of the various molecules, we have adopted the following convention: position 1 corresponds to replacement of a CH group on the periphery of the molecule with an N atom. Since this substitution places the N atom on the edge (or exterior) of the C skeleton, we refer to this as an *exoskeletal* PANH. Substitution at positions designated 2, 3, 4, etc. corresponds to replacing a C atom that is 1, 2, 3, etc. bonds removed from the edge (*i.e.* the nearest CH group), respectively. Such an N atom is necessarily buried within the carbon skeleton of the molecule, and the molecule is referred to as an *endoskeletal* PANH.

The studies of PANH cations summarized here are based on a combination of laboratory work and theoretical calculations. Since an N atom is isoelectronic with a CH group, neutral exoskeletal PANHs have a closed-shell electronic structure (all electrons are paired). Samples of many such molecules are stable and can be obtained for experimental work. Conversely, an N atom contains one more electron than does a C atom, so neutral endoskeletal PANHs have a radical (open shell) electronic structure. Such species are not stable and are inaccessible by current experimental techniques. Thus, the same Density Functional Theory techniques that have previously been applied to the analyses

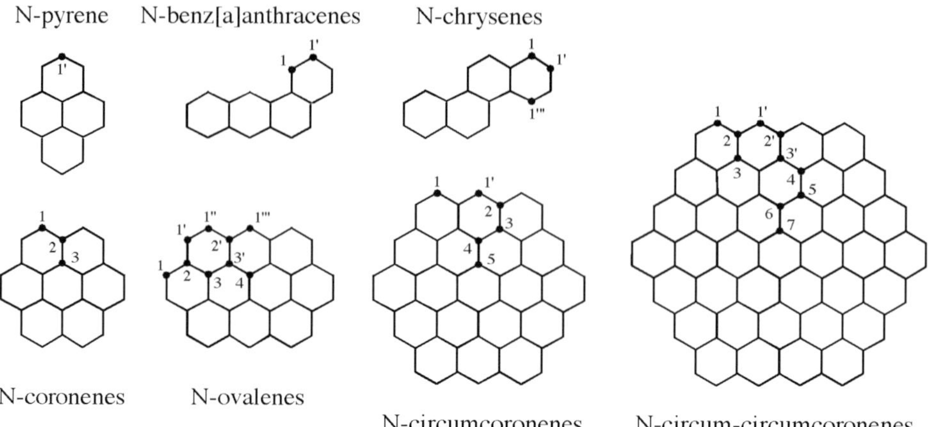

Figure 2. Generalized structures of the PANH species discussed in this manuscript. The numbers identify the possible unique positions for N atom substitution in the structure and are assigned as described in the text.

of the IR spectroscopic PAHs and their cations, have been particularly crucial to the study of endoskeletal PANHs and their cations.

Table 1 lists the positions of the dominant bands in the CC stretching region for the PANH cations shown in Figure 2 as well as their shifts from the analogous unsubstituted PAH cations. Inspection of the data for the exoskeletal PANH cations (those with a "1N" prefix) reveals that, although N-substitution is sometimes accompanied by a small blueshift, that blueshift is too modest to bring the dominant band in the region into agreement with the 6.215 μm Class A component of the interstellar spectra. However, the story changes as we move the N atom into the interior of the carbon skeleton. Consider, for example, the N-substituted coronenes shown in Table 1. Incorporation of an N atom at the endoskeletal 2 or 3 positions shifts the dominant CC stretching feature to 6.177 μm and 6.184 μm, respectively. This is precisely the absorption position required to accommodate the Class A emission component when one takes into account the 10 cm^{-1} (*ca.* 0.04 μm at 6.2 μm) redshift between emission and absorption peak frequencies (Joblin *et al.* 1995).

The effect of N substitution on the position of the dominant CC stretching band is illustrated graphically in Figure 3(*a*) which shows a spectral rendering of the calculated spectrum of the series of N-circumcoronenes in the 6.2 μm region compared to the canonical Class A and C bands of Peeters *et al.* (2002). That figure shows that as the substitutional position of the N atom moves deeper into the C skeleton of the molecule, the dominant CC stretching feature shifts to shorter wavelengths, agreeing well with the position of the Class A interstellar feature for the innermost endoskeletal positions. The data for the N-ovalene and N-circum-circumcoronene cations in Table 1 shows that a significant blueshift of the CC stretching band is a general characteristic of endoskeletal PANH cations. This effect is quantified in Figure 3(*b*) which shows a plot of the position of the dominant CC stretching band as a function of N substitution position for the four large, condensed cations in Table 1. It is interesting to note that the impact of N atom substitution (*i.e.* the slope of each curve in Figure 3*b*) is dependent on molecular size – the larger the molecule, the more modest the effect. Additional studies detailed elsewhere (HBA 05) have shown that, for species larger than 30 or so C atoms, 2 or more endoskeletal N atoms are required to achieve the full measure of the blueshift needed to accommodate the interstellar Class A band. Based on that result, it is estimated that

Table 1. Molecular characteristics and positions of the dominant 6.2 μm bands for a range of singly-substituted PANHs compared to their parent hydrocarbons.

Species	Formula	Exo-/endo-skeletal N	λ (μm)	$\Delta\lambda$ (μm)	$\tilde{\nu}$ (cm^{-1})	$\Delta\tilde{\nu}$ (cm^{-1})
		Experimental Measurements[a]				
Pyrene Cation	$C_{16}H_{10}{}^{+}$	-	6.439		1553	
1'N-pyrene^{+}	$C_{15}H_{9}N^{+}$	exo	6.456	+0.017	1549	-4
Benz[a]anthracene cation	$C_{18}H_{12}{}^{+}$	-	6.494		1540	
1N-benzanthracene^{+}	$C_{17}H_{11}N^{+}$	exo	6.502	+0.008	1538	-2
1'N-benzanthracene^{+}	↓	exo	6.532	+0.038	1531	-9
Chrysene Cation	$C_{18}H_{12}{}^{+}$	-	6.410		1560	
1N-chrysene^{+}	$C_{17}H_{11}N^{+}$	exo	6.414	+0.004	1559	-1
1'N-chrysene^{+}		exo	6.439	+0.029	1553	-7
1'''N -chrysene^{+}	↓	exo	6.394	-0.016	1564	+4
		Theoretical Calculations[a]				
Coronene cation	$C_{24}H_{12}{}^{+}$	-	6.441	-	1553	-
1N-coronene$^{+,c}$	$C_{23}H_{12}N^{+}$	exo	6.254	-0.187	1599	+35
2N-coronene^{+}		endo	6.177[b]	-0.222	1619[a]	+55
3N-coronene^{+}	↓	endo	6.184	-0.257	1617	+64
Ovalene cation	$C_{32}H_{14}{}^{+}$	-	6.415	-	1559	-
1N-ovalene$^{+,c}$	$C_{31}H_{14}N^{+}$	exo	6.320	-0.095	1582	+23
1'N-ovalene$^{+,c}$		exo	6.332	-0.083	1579	+20
1''N-ovalene$^{+,c}$		exo	6.324	-0.091	1581	+22
1'''N-ovalene$^{+,c}$		exo	6.268	-0.147	1595	+36
2N-ovalene^{+}		endo	6.267	-0.148	1596	+37
2'N-ovalene^{+}		endo	6.277	-0.138	1593	+34
3N-ovalene^{+}		endo	6.243	-0.172	1602	+43
3'N-ovalene^{+}		endo	6.231	-0.184	1605	+46
4N-ovalene^{+}		endo	6.202	-0.213	1612	+53
Circumcoronene cation	$C_{54}H_{18}{}^{+}$	-	6.365	-	1571	-
1N-circumcoronene$^{+,c}$	$C_{53}H_{14}N^{+}$	exo	6.252[d]	-0.113	1600[c]	+29
1'N-circumcoronene$^{+,c}$		exo	6.309	-0.056	1585	+14
2N-circumcoronene^{+}		endo	6.250	-0.115	1600	+29
3N-circumcoronene^{+}		endo	6.234	-0.131	1604	+33
4N-circumcoronene^{+}		endo	6.215	-0.15	1609	+38
5N-circumcoronene^{+}	↓	endo	6.211	-0.154	1610	+39
Circum-circumcoronene cation	$C_{96}H_{24}{}^{+}$	-	6.345	-	1576	-
2N-circumcircumcor^{+}	$C_{95}H_{24}N^{+}$	endo	6.262	-0.083	1597	+21
2'N-circumcircumcor^{+}		endo	6.266	-0.079	1596	+20
3N-circumcircumcor^{+}		endo	6.305	-0.040	1586	+10
3'N-circumcircumcor^{+}		endo	6.281	-0.064	1592	+16
4N-circumcircumcor^{+}		endo	6.277	-0.068	1593	+17
5N-circumcircumcor^{+}		endo	6.256	-0.089	1599	+23
6N-circumcircumcor^{+}		endo	6.258	-0.087	1598	+22
7N-circumcircumcor^{+}	↓	endo	6.242	-0.103	1602	+26

[a] — experimental data correspond to argon matrix-isolation measurements from Mattioda et al. 2003; theoretical data were computed using density functional theory at the B3LYP/4-31G and are taken from Hudgins, Bauschlicher, and Allamandola 2005.

[b] — shorter wavelength component of a doublet; a longer wavelength component with similar intensity falls at a position of 6.274 μm (1594 cm^{-1}).

[c] — closed-shell, protonated form of the exoskeletal PANH cation. See discussion in text (Hudgins, Bauschlicher, and Allamandola 2005).

[d] — shorter wavelength component of a doublet; a longer wavelength component with similar intensity falls at a position of 6.452 μm (1550 cm^{-1}).

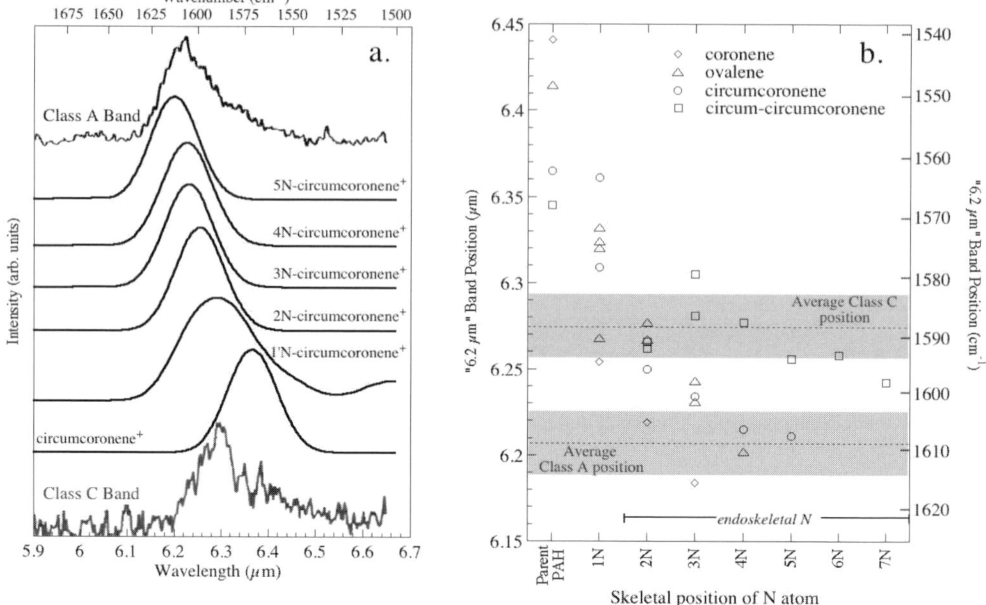

Figure 3. This figure illustrates the spectroscopic effects of endoskeletal N atom substitution on the position of the dominant CC stretching modes of PANH cations (HBA 05). The structures of the species are shown in Figure 2 above. (*a*) A comparison of the Class A and Class C 6.2 μm emission components to a spectral rendering of the theoretical results for various possible N-circumcoronene cations. The plot is adapted from HBA 05 and details can be found therein. (*b*) The position of the CC stretching band plotted as a function of N substitution position. All data calculated using DFT at the B3LYP/4-31G level.

the C/N ratio in the interstellar PANH population can be no more than ~ 30, and that, consequently, approximately 1%–5% of the available cosmic N is tied up in PAHs.

Finally, not only are endoskeletal PANH cations the first species found that can accommodate the 6.215 m Class A emission band identified by Peeters *et al.*, a strong case can be made that they are also by far the most likely origin of this feature. That case is discussed in detail in HBA 05 and is beyond the scope of this paper. However, in a nutshell, focused theoretical analyses of key representative species in conjunction with the chemical and astrophysical constraints of the problem can be used to rule out or render unlikely such alternatives as larger and or different PAH species, dehydrogenated/superhydrogenated PAHs, aromatic heterocycles involving elements other than N, and PAH-metal atom complexes (*i.e.* "metallocenes") as the origin of the anomalous Class A emission band. Consequently, all the available evidence indicates that endoskeletal PANHs are unique in their capacity to accommodate the Class A 6.2 μm emission component and that this feature provides a tracer of N in the interstellar PAH population.

3. The Rotational Spectroscopy of PANHs

Although the last two decades have seen a renaissance in our understanding of the IR spectroscopic properties of PAHs, the spectroscopic properties of PAHs in that other great realm of astronomical molecular spectroscopy – the radio region – have received relatively little attention. There are several significant challenges to detecting the radio emission from specific PAH species in astronomical environments. First, although PAHs as a family are the most abundant interstellar molecules, the population consists of an enormous number of discrete molecular species. It is unclear whether any one particular PAH species will be present in sufficient abundance to permit unambiguous identification. The problem is further compounded by the extreme complexity of the rotational spectra of all but the most symmetric PAH species. Because of this complexity, the total energy emitted at radio frequencies will, in general, be distributed over a large number of weak lines rather than concentrated into a few stronger ones. Happily, with advances in observational technologies and techniques at radio wavelengths, the radio frequency spectroscopy of PAHs is receiving increased attention as evidenced by the work of Thorwirth and others highlighted elsewhere in this volume.

Another obstacle to studies of the pure rotational spectroscopy of interstellar PAHs has been the fact that pure PAHs tend to have relatively small dipole moments and, consequently, weak rotational transition. PANHs, on the other hand, possess significant dipole moments and would be expected to exhibit intrinsically stronger rotational transitions. Given the evidence indicating that N-substituted aromatic compounds play an important role in the interstellar PAH population, these species may represent attractive targets for an astronomical search. Toward this end, the dipole moments and rotational constants have been calculated at the B3LYP/4–31G level of theory and are tabulated in Tables 2 and 3, respectively. In those tables, the rotational axes of each molecule are identified as a, b, and c on the basis of the moment of inertia, I, about each axis and according to the convention $I_a < I_b < I_c$. Calibration calculations at the same level of theory for the benzene molecule yield rotational constants that are accurate to within 1%. Thus, while rotational constants such as those listed in Table 2 are insufficient to calculate precise rotational line positions, they are sufficient to constrain the strengths and wavelength regions in which PANH rotational transitions are expected.

Given the enormous number of PANH structures possible and the wide range of rotational constants they encompass, interstellar PANHs should produce a very dense forest of lines over a very broad spectral range. For example, a mixture of PANHs up to a size of circum-circumcoronene at 100 K are expected to emit in a pseudo-continuum between about 0 and 40 GHz. On the other hand, given the large dipole moments associated with PANHs, it may also be possible to distinguish the rotational emission lines of interstellar PANHs amongst the forest of molecular rotational lines observed in cold, dark molecular clouds where the low ambient temperatures ($T_{\mathrm{gas}} \simeq 10$ K) collapse the rotational energy into a relatively small number of transitions. Furthermore, it may be that there are some PANH species which possess large dipole moments and have a high enough degree of symmetry that the number of substitutional isomers is sufficiently small that these species could have unusually intense radio lines.

Table 2. Calculated rotational constants for the cations of several singly-substituted PANHs.

Cations	Rotational Constants (GHz)			Cations	Rotational Constants (GHz)		
	R_a	R_b	R_c		R_a	R_b	R_c
1'N-pyrene	1.027	0.551	0.359	N-coronenes	0.334–	0.331–	0.166–
1N-benz[a]anthracene	1.166	0.262	0.214		0.337	0.336	0.168
1'N-benz[a]anthracene	1.173	0.256	0.210	N-ovalenes	0.238	0.148	0.091
1N-chrysene	1.279	0.261	0.217	N-circumcoronenes	0.066	0.066	0.033
1'N-chrysene	1.283	0.259	0.216	N-circum-	0.021	0.021	0.011
1'''N -chrysene	1.270	0.265	0.220	circumcoronenes			

Table 3. Calculated dipole moments for the cations of several singly-substituted PANHs.

Species	Dipole Moments (Debye) μ_a	μ_b	μ	Species	Dipole Moments (Debye) μ_a	μ_b	μ
N-pyrene cation, $C_{15}H_9N^+$				**N-ovalene cations, $C_{31}H_{14}N^+$**			
1'N	3.49	0.00	**3.49**	1N	7.10	0.98	**7.17**
				1'N	5.38	4.81	**7.21**
N-benz[a]anthracene cations, $C_{17}H_{11}N^+$				1"N	4.92	4.26	**6.51**
1N	1.63	1.48	**2.20**	1'''N	0.00	3.47	**3.47**
1'N	2.09	4.92	**5.35**	2N	5.25	1.19	**5.38**
				2'N	1.59	3.65	**3.98**
N-chrysene cations, $C_{17}H_{11}N^+$				3N	4.32	1.02	**4.44**
1N	3.72	0.04	**3.72**	3'N	1.29	1.99	**2.37**
1'N	3.61	2.22	**4.24**	4N	0.00	1.56	**1.56**
1'''N	0.08	2.17	**2.17**				
				N-circum-circumcoronene cations, $C_{95}H_{24}N^+$			
N-coronene cations, $C_{23}H_{11}N^+$				2N	10.12	0.33	**10.13**
1N	5.48	0.19	**5.49**	2'N	9.09	0.00	**9.09**
2N	3.69	0.00	**3.69**	3N	7.47	1.94	**7.72**
3N	2.67	0.00	**2.67**	3'N	8.31	0.00	**8.31**
				4N	7.33	0.63	**7.72**
N-circumcoronene cations, $C_{53}H_{14}N^+$				5N	4.75	0.62	**4.79**
1N	9.23	0.23	**9.23**	6N	3.06	0.00	**3.06**
1'N	6.99	0.00	**6.99**	7N	2.54	0.00	**2.54**
2N	6.77	0.47	**6.79**				
3N	5.30	1.20	**5.43**				
4N	4.55	0.00	**4.55**				
5N	1.32	0.00	**1.32**				

4. The Near-IR Spectroscopy of PAHs and PANHs

Broadening the scope of our discussion to encompass the whole population of interstellar PAHs (N-bearing or otherwise) that dominate the mid-IR emission and turning our attention to shorter wavelengths, there is another aspect of the astrophysical problem that has emerged in recent years that warrants consideration here. While significant effort has been made to understand the global infrared spectral properties of PAHs under conditions appropriate to the emission zones, significantly less attention has been directed toward understanding their overall UV, Visible, and Near-IR (UV/Vis/NIR) spectral properties. Until recently, most experimental and theoretical studies of the electronic spectroscopy of PAHs and PAH ions in the UV/Vis/NIR regions have focused on specific bands in narrow wavelength regions to test the hypothesis that PAHs are responsible for some of the discrete diffuse interstellar bands (DIBs, e.g., Salama *et al.* 1996; Brechignac and Pino 1999; Biennier *et al.* 2003; Hirata *et al.* 2003 – references 22 through 49).

To remedy this situation, Mattioda *et al.* (2005a,b) measured the electronic spectra of many PAH ions, focusing on transitions in the NIR, the region least studied. This work stressed the importance of the oft-neglected point that PAH ions indeed have NIR transitions that warrant consideration when evaluating the interstellar radiation field. Mattioda *et al.* show that the PAH IR emission features can be pumped by NIR photons and that PAHs should impose broadband structure on the NIR portion of the extinction curve. Due to space limitations here we only briefly describe the role NIR photons play in pumping the IR emission features.

Since the IR emission bands were initially found to be associated with UV rich objects, and PAHs were known to absorb most strongly in the UV, it was tacitly assumed that they were pumped primarily by UV photons. However subsequent measurements showed that UV flux alone was insufficient to pump the measured intensities of the IR emission bands in certain objects, seeming to pose a problem for the PAH model (Aitken & Roche 1983; Uchida, Sellgren, & Werner 1998; Uchida *et al.* 2000). While it is well known that small neutral PAHs have a steep and sharp absorption cutoff in the UV with much weaker absorption extending into the visible, this cutoff smoothly moves to longer wavelength with increasing PAH size (e.g., Birks 1970; Malloci, Mulas, & Joblin 2004). Since this absorption characteristic of neutral (closed shell) species was incorporated in most PAH emission models (Schutte *et al.* 1993; Draine and Li, 2001; Li & Draine 2001; Bakes *et al.* 2001) these models could not account for observations of the IR bands from UV poor objects. However, upon ionization the neutral PAH electronic configuration changes from closed- to open-shell (radical) and moderate-to-strong absorptions occur at much longer wavelengths.

Taking this into account, Li & Draine (2002) extended their model and showed that ionized PAHs could indeed account for the IR emission features in these objects. Since the number of near-IR PAH ion experimental spectra available upon which to extend their model was severely limited, Li & Draine adopted the Desert *et al.* (1990) cutoff. However, the experimental measurements of Mattioda *et al.* (2005a,b) showed that this cutoff greatly underestimated PAH ion NIR absorption strengths and band widths. To properly take the role of NIR radiation into account, Mattioda *et al.* (2005b) extended Draine & Li's PAH UV/Visible formalism with their NIR experimental data and derived a model for the PAH ion absorption cross section spanning the FUV through the NIR.

With this model, Mattioda *et al.* (2005b) assess the amount of radiant energy PAHs absorb in the UV, Vis, and NIR wavelength regions individually and as function of stellar type. Figure 4 summarizes these results. The figure presents the UV through NIR blackbody stellar output for a variety of different star types, the model PAH ion absorption function and the convolution of the two. The integrated amount of radiant energy absorbed by these PAHs is indicated by the shaded areas in Figure 4 labeled: UV, Vis, and NIR. While the convolved function represents the amount of energy absorbed per carbon atom per wavelength, the area gives the total amount of radiant energy absorbed per C atom.

From the convolved curves in Figure 4 one can clearly see that as stellar temperature decreases, the amount of energy absorbed by PAH ions in the NIR relative to the amount absorbed in the UV and Visible increases markedly. Analysis of the curves shows the power absorbed in the NIR becomes nearly equivalent to the power absorbed in the UV and Visible for stellar temperatures around 4,000 K and 3,000 K respectively. For stars with lower stellar temperatures the NIR power absorbed exceeds that in the UV and Vis. These crossover temperatures are lower limits as the stellar radiation fields were derived asuming a blackbody radiation field. Using Kuruscz models reduces the UV output substantially with respect to the Vis and NIR, especially for early type stars.

Thus, while the availability of UV-Vis photons to pump the PAH emission features drops with decreasing star type, the contribution by the stellar NIR photons remains steady and actually becomes comparable to that of the higher energy photons for the later type stars. Inspection of the integrated power curves in Figure 4 shows that the net effect of decreasing stellar temperature is a steady but slow and gradual decrease in the integrated radiative power pumped into the PAH ions and a concomitantly slow decrease in the power emitted in the mid-IR by the PAH bands, not a precipitous drop as previously expected. This predicts that, if present in the vicinity of even very late

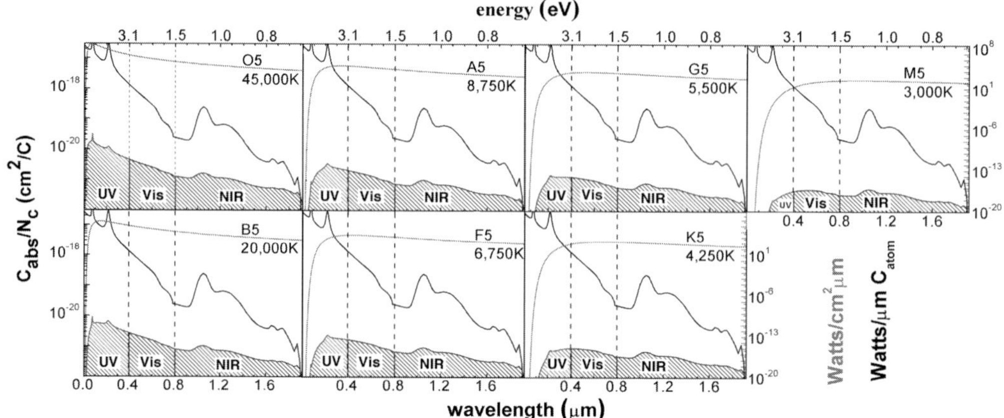

Figure 4. Stellar radiative energy absorbed by PAHs as a function of star type. The UV through NIR stellar (black body) output for a variety of different star types (smooth gray curve, right axis units: Watts/cm^2/μm) is plotted in conjunction with PAH ion absorption cross-section (structured curve, left axis units). The integrated amount of radiant power absorbed by an optically thick cloud of PAH ions is indicated by the shaded areas labeled UV, Vis, and NIR (right axis units: Watts/μm/Catom). PAH ion UV-Vis absorption is modeled using the Li & Draine (2002) formalism and NIR absorption is modeled using the formalism described in Mattioda et al. 2005b. Figure adapted from that reference. The vertical lines at 0.4 μm and 0.78 μm separate the spectral domains.

type stars, open shell PAH species (PAH ions and PANHs with an odd number of N atoms) can be vibrationally excited by the stellar radiation field. A much more detailed discussion can be found in Mattioda et al. (2005a).

Acknowledgements

We acknowledge the National Research Council (NRC) and NASAs Long Term Space Astrophysics (UPN #399-20-40), Astrobiology (UPN #344-53-92), and Exobiology (UPN #344-58-21) programs for supporting this work. In addition, we are indebted to Bob Walker for his outstanding technical support of all phases of the experimental work; Jan Cami for detailed discussions on stellar radiation fields; Charlie Bauschlicher for his computational efforts, expertise and keen insight without which the crucial computational components of the foregoing work would not have been possible; Els Peeters and Xander Tielens for bringing the variations in the CC stretching feature to our attention and many subsequent discussions; and Andy Mattioda for helpful discussions concerning laboratory measured PANH spectra.

References

Aitken, D.K. & Roche P.F. 1983, *MNRAS* 202, 1233

Allamandola, L.J., Hudgins, D.M., & Sandford, S.A. 1999, *Ap. J.* 511, L115

Allamandola, L.J., Tielens, A.G.G.M., & Barker, J.R. 1989, *Ap. J. Supp.* 71, 733

Bakes, E.L.O., Tielens, A.G.G.M., & Bauschlicher, C.W. 2001, *Ap. J.* 556, 501

Bakes, E.L.O., Tielens, A.G.G.M., Bauschlicher, C.W., Hudgins, D.M., & Allamandola, L.J. 2001, *Ap. J.* 560, 261.

Bauschlicher, C.W. & Langhoff, S.R. 1997, *Spectrochim. Acta A* 53, 1225

Biennier, L., Salama, F., Allamandola, L.J., & Scherer, J.J. 2003, *J. Chem. Phys.* 118, 7863

Birks, J.B. 1970, *Photophysics of Aromatic Molecules* (London: Wiley-Interscience)

Brechignac, P. & Pino, T. 1999, *A&A* 343, L49

Cox, P. & Kessler, M.F. (eds.), 1999 *The Universe as Seen by ISO, vol. I, II* (ESTEC, Noordwijk, The Netherlands: ESA Pub. Div.)

Desert, F.X., Boulanger, F., & Puget, J.L. 1990, *A&A* 237, 215

Draine, B.T. & Li, A. 2001, *Ap. J.* 551, 807

Hirata, S., Head-Gordon, M., Szczepanski, J., & Vala, M. 2003, *J. Phys. Chem. A* 107, 4940

Hony, S. *et al.* 2001, *A&A* 370, 1030

Hudgins, D.M. & Allamandola, L.J. 1995, *J. Phys. Chem.* 99, 3033

Hudgins, D.M. & Allamandola, L.J. 1999, *Ap. J.* 513, L69

Hudgins, D.M., Bauschlicher, C.W., & Allamandola, L.J. 2005, *Ap. J.* 632, 316

Joblin, C., Boissel, P., Leger, A., d'Hendecourt, L.B., & Defourneau, D. 1995, *A&A*, 299, 835

Kim, H.-S. & Saykally, R.J. 2002, *Ap. J. Supp.* 143, 455

Langhoff, S.R. 1996, *J. Phys. Chem.* 100, 2819

Li, A. & Draine, B.T. 2001, *Ap. J.* 554, 778

Li, A. & Draine, B.T. 2002, *Ap. J.* 572, 232

Malloci, G., Mulas, G., & Joblin, C. 2004, *A&A* 426, 105

Mattioda, A.L., Hudgins, D.M., Bauschlicher, C.W., Rosi, M., & Allamandola, L.J. 2003, *J. Phys. Chem. A* 107, 1486

Mattioda, A.L., Allamandola, L.J., & Hudgins, D.M. 2005a, *Ap. J.* 629, 1183

Mattioda, A. L., Hudgins, D.M., & Allamandola, L.J. 2005b, *Ap. J.* 629, 1188

Oomens, J., Tielens, A.G.G.M., Sartakov, B.G., von Helden, G., & Meijer, G. 2003, *Ap. J.* 591, 968

Pauzat, F. & Ellinger, Y. 2001, *MNRAS*, 324, 355

Pech, C., Joblin, C., & Boissel, P. 2002, *A&A* 388, 639

Peeters, E. *et al.* 2002, *A&A* 390, 1089

Puget, J.L. & Leger, A. 1989, *ARAA* 27, 161

Salama, F., Bakes, E., Allamandola, L.J., & Tielens, A.G.G.M. 1996, *Ap. J.* 458, 621

Schutte, W.A., Tielens, A.G.G.M., & Allamandola, L.J. 1993, *Ap. J.* 415, 397

Szczepanski, J., Drawdy, J., Wehlburg, C., & Vala, M. 1995, *Chem. Phys. Lett.* 245, 539

Uchida, K.I., Sellgren, K., & Werner M.W. 1998, *Ap. J.* 493, L109

Uchida,K.I., Sellgren,K., Werner,M.W., & Houdashelt, M.L. 2000, *Ap. J.* 530, 817

Vala, M., Szczepanski, J., Pauzat, F., Parisel, O., Talbi, D, & Ellinger, Y.J. 1994, *J. Phys. Chem.* 98, 9187

Van Kerckhoven, C. *et al.* 2000, *A&A* 357, 1013

Verstraete, L. *et al.* 2001, *A&A* 372, 981

Discussion

MCCALL: You have suggested that a very large number of species and a very large number of isomers are necessary to explain the unidentified infrared bands, and expressed some pessimism about the identification of individual PAHs. Does this imply that PAHs are poor candidates for the carries of the Diffuse Interstellar Bands, which are relatively few in number?

HUDGINS: I am surprised by the comment that the DIBs are "relatively few in number." It is my understanding that in excess of 300 DIBs of all shapes and sizes have been cataloged. Far from being a challenge to the PAH/DIB theory, I think that the richness of the DIP spectrum implies a large and diverse population of molecules-entirely consistent with the PAH/DIB theory. Also, while the interstellar PAH (PANH?) population is undoubtedly made up of a distribution of countless discrete molecular species, they will certainly not all be present in equal abundance. Therefore, not every single species in the population need be represented among the DIBs – only those members of the population that are present in sufficient abundance to produce a measurable absorption.

SNOW: You have shown that the 6.2 μm UIR is matched best by a PANH mix, with an interior nitrogen atom substitution in place of a carbon atom. I find this, including the implications for the cosmic nitrogen budget, fascinating. But how does this nitrogen substitution affect the other UIRs?

HUDGINS: Since the N-involved vibrations of PANH molecules blend effectively with the other vibrations of the carbon skeleton (CC stretches, bends) PANHs do not have any distinctive spectral features that would otherwise distinguish them from ordinary PAHs. The impact of N-substitution in PAHs is actually quite subtle. The small but significant blue shift of the dominant CC stretching band is at this point the only consistent, measurable effect that distinguishes the spectra of PAHs and PANHs.

Astrochemistry: Recent Successes and Current Challenges
Proceedings IAU Symposium No. 231, 2005
D.C. Lis, G.A. Blake & E. Herbst, eds.

© 2006 International Astronomical Union
doi:10.1017/S1743921306007460

Rotational Spectroscopy of PAHs: Acenaphthene, Acenaphthylene and Fluorene

S. Thorwirth†, P. Theulé‡, C. A. Gottlieb, M. C. McCarthy, and P. Thaddeus

Department of Engineering and Applied Sciences, Harvard University,
Pierce Hall, 29 Oxford Street, Cambridge, MA 02138, U.S.A.

Harvard-Smithsonian Center for Astrophysics,
60 Garden Street, Cambridge, MA 02138, U.S.A.
email: sthorwirth@mpifr-bonn.mpg.de

Abstract. Pure rotational spectra of three polycyclic aromatic hydrocarbons – acenaphthene, acenaphthylene and fluorene – have been obtained by Fourier transform microwave spectroscopy of a molecular beam and subsequently by millimeter wave absorption spectroscopy for acenaphthene and fluorene. The data presented here will be useful for deep radio astronomical searches for PAHs employing large radio telecopes.

Keywords. astrochemistry — ISM: molecules — molecular data

Polycyclic aromatic hydrocarbons (PAHs) have been studied extensively in the laboratory over the last 20 years (see Salama 1999 and Tielens & Peeters 2004 for reviews) owing to their astronomical significance as possible carriers of the unidentified infrared bands (UIRs, e.g. Allamandola *et al.* 1989). These studies have been performed almost exclusively in the uv, optical, and infrared regions of the electromagnetic spectrum. Very little is known, however, about the rotational spectra of small and polar PAHs, since microwave studies have been reported so far only for azulene ($C_{10}H_8$, Huber *et al.* 2005 and references therein) and corannulene ($C_{20}H_{10}$, Lovas *et al.* 2005).

In the present study, we have investigated the rotational spectra of selected small PAHs (see Fig. 1) employing Fourier transform microwave (FTM) spectroscopy (Balle & Flygare 1981) using the spectrometer at Harvard (McCarthy *et al.* 1997, 2000) in combination with a heated nozzle recently developed for studies of low-volatility compounds

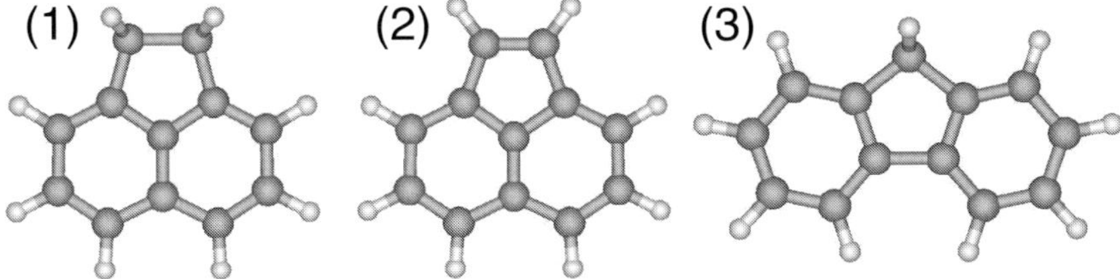

Figure 1. Molecular structures of acenaphthene ($C_{12}H_{10}$, **1**), acenaphthylene ($C_{12}H_8$, **2**) and fluorene ($C_{13}H_{10}$, **3**).

† Present address: Max-Planck-Institut für Radioastronomie, Auf dem Hügel 69, 53121 Bonn, Germany
‡ Present address: Physique des interactions ioniques et moléculaires, Université de Provence, Centre de Saint Jérôme, 13397 Marseille Cedex 20, France

(Thorwirth *et al.* 2005). Initial searches were guided by rotational constants obtained from quantum chemical calculations performed at the B3LYP/cc-pVTZ level of theory (see Table 1) using the program package Gaussian03 (Frisch *et al.* 2003). All three molecules exhibit *b*-type rotational spectra and are calculated to be moderately polar,

Table 1. Rotational constants (in MHz) and dipole moments μ (in D) for (**1**), (**2**) and (**3**) as determined in the present study.

Molecule	B3LYP/cc-pVTZ				Experiment		
	A_e	B_e	C_e	μ	A_0	B_0	C_0
Acenaphthene (**1**)	1416.5	1200.6	655.0	0.9	1410.3	1193.9	652.1
Acenaphthylene (**2**)	1520.7	1228.2	679.4	0.3	1511.8	1220.6	675.5
Fluorene (**3**)	2195.1	588.2	465.2	0.5	2176.2	586.7	463.6

with dipole moments of order 0.3 to 0.9 D. Rotational transitions were found readily for all three molecules and based on improved predictions several tens of lines could be measured for each one. Rotational constants obtained from least-squares analyses of the experimental data are shown in Table 1. As can be seen, the calculated equilibrium values and experimentally obtained ground state rotational constants agree very well, to within 1%. Subsequently, selected rotational transitions of acenaphthene and fluorene could also be measured by standard millimeter wave absorption spectroscopy at 90 GHz.

The present investigation highlights the potential of FTM spectroscopy for the characterization of polar PAHs, including the nitrogen variants (PANHs; e.g., see Hudgins, this volume).

A detailed account of the present study will be given elsewhere.

Acknowledgements

This work was supported in part by NASA grant NAG5 9379 and NSF grant CHE-0353693. S. Thorwirth is grateful to the Alexander von Humboldt-Foundation for a Feodor Lynen research fellowship. P. Theulé would like to thank the Swiss National Science Foundation for a research fellowship.

References

Allamandola, L.J., Tielens, A.G.G.M., & Barker, J.R. 1989, *Ap. J. Suppl.* 71, 733

Balle, T.J. & Flygare, W.H. 1981, *Rev. Sci. Instrum.* 51, 33

Frisch, M.J., Trucks, G. W., Schlegel, H.B. *et al.* 2003, Gaussian03, Revision B.04, Gaussian, Inc., Wallingford CT

Huber, S., Grassi, G., & Bauder A. 2005, *Mol. Phys.* 103, 1395

Lovas, F.J., McMahon, R.J., Grabow, J.-U., Schnell, M., Mack, J., Scott, L.T., & Kuczkowski, R. L. 2005, *J. Am. Chem. Soc.* 127, 4345

McCarthy, M.C., Travers, M.J., Kovács, A., Gottlieb, C.A., & Thaddeus, P. 1997, *Ap. J. Suppl.* 113, 105

McCarthy, M.C., Chen, W., Travers, M.J., & Thaddeus, P. 2000, *Ap. J. Suppl.* 129, 611

Salama, F. 1999, in *Solid Interstellar Matter: The ISO Revolution*, eds. L. d'Hendecourt, C. Joblin, & A. Jones (Springer-Verlag, New York), 65

Thorwirth, S., McCarthy, M.C., Gottlieb, C.A., Thaddeus, P., Gupta, H., & Stanton, J.F. 2005, *J. Chem. Phys.* 123, 054326

Tielens, A.G.G.M. & Peeters, E. 2004, in *The Dense Interstellar Medium in Galaxies*, eds. S. Pfalzner, C. Kramer, C. Staubmeier, & A. Heithausen, Springer Proceedings in Physics, Vol. 91 (Springer-Verlag, Berlin), 497

Astrochemistry: Recent Successes and Current Challenges
Proceedings IAU Symposium No. 231, 2005
D.C. Lis, G.A. Blake & E. Herbst, eds.

© 2006 International Astronomical Union
doi:10.1017/S1743921306007472

Silicates—Space and Laboratory

Thomas Henning[1], Harald Mutschke[2], and Cornelia Jäger[2]

[1]Max Planck Institute for Astronomy, Königstuhl 17, D-69117 Heidelberg, Germany
email: henning@mpia.de

[2]Astrophysical Institute and University Observatory, Schillergässchen 2, D-07745 Jena, Germany
email: mutschke@astro.uni-jena.de, conny@astro.uni-jena.de

Abstract. In this paper, we will review recent progress made in the understanding of cosmic silicates and the experimental characterization of relevant analogue materials. We will introduce the main structural properties of silicates. Then we will discuss recent infrared observations with a special emphasis on protoplanetary disks and AGNs. In the experimental section we will describe the optical properties of silicate grains. Here we will concentrate on fundamental optical data and the dependence of the optical behavior on particle shape and temperature.

Keywords. astrochemistry — cosmic dust — planetary systems: protoplanetary disks — galaxies: quasars

1. Introduction

Apart from carbonaceous grains, silicates are the major refractory component of interstellar dust. Silicates regulate the thermal structure of regions in space where dust grains dominate the opacity. The various spectral features of silicate particles in the infrared spectral domain are an unique diagnostic tool for the determination of the optical depth of the medium, its temperature, and chemical state as well as for dynamical processes such as turbulent mixing in protoplanetary disks. Silicates provide the surface for chemical reactions and the condensation of volatile components in the cold environments of molecular clouds and the outer regions of protoplanetary disks. Finally, we should stress that silicates are an interesting class of materials with a large variety of physical, chemical, and optical properties.

Due to the large amount of data and the limited space available, we will concentrate on major new developments in the observational characterization of cosmic silicates and the laboratory study of their optical behavior relevant to astronomy. We refer the reader to Henning (2003a) for an extensive review of earlier results and to Colangeli *et al.* (2003) for a detailed discussion of the properties of silicon-based materials. Data from the Infrared Space Observatory ISO and the Spitzer Space Telescope led to the characterization of crystalline silicates in space and, together with data from meteorites and interplanetary dust particles, opened up the new field of astromineralogy (Henning 2003b).

2. Structural Properties of Silicates

The most interesting cosmic silicates belong to the mineral groups of pyroxenes and olivines. The olivine group consists of silicates with the general sum formula $A^{2+}B^{2+}SiO_4$. Here A and B are divalent cations, the most abundant of which in astrophysical environments are Mg and Fe. Mg^{2+} and Fe^{2+} can replace each other in the crystal structure. This means that olivines can be considered as solid solutions of Mg_2SiO_4 and Fe_2SiO_4; the

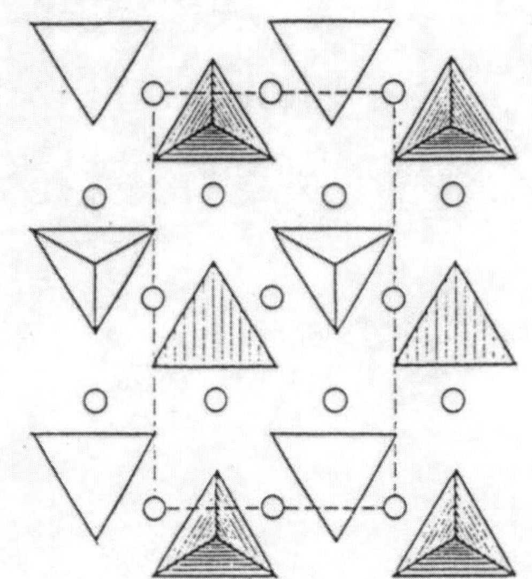

Figure 1. The crystalline structure of olivine. The dashed line marks the border of the elemental cell.

mixture ratio can be expressed by the subscript x in the general formula $Mg_{2x}Fe_{2-2x}SiO_4$ with $1 \leqslant x \leqslant 0$. The two end members of the isomorphous crystalline compounds are Mg_2SiO_4 ($x = 1$) and Fe_2SiO_4 ($x = 0$), named forsterite and fayalite, respectively. The mineral "olivine" contains 8 to 20 mass percent FeO. Minerals with higher FeO contents are called hortonolite and ferrohortonolite. The olivines are neso-silicates (island silicates) because they consist of isolated SiO_4 tetrahedra that are linked by divalent cations (see Fig. 1). The metal ions are coordinated by six oxygens. There are two nonequivalent six-coordinate positions in the olivine structure which are both distorted. According to the lattice symmetry, the olivines belong to the rhombic crystal system.

The pyroxenes form a large group of minerals belonging to the inosilicates (chain silicates). The basic structural units of the silicates, the SiO_4 tetrahedra, form $[Si_2O_6]_\infty$ chains, i.e. each tetrahedron shares two of its oxygens with the adjacent two tetrahedra. The direction of the chains is the crystallographic c-axis. There are two types of SiO_4 tetrahedral chains with different Si–O distances. Pyroxenes have the symbolic formula $Mg_xFe_{1-x}SiO_3$. The metal ions, which link the SiO_4 chains, are coordinated by six oxygens (see Fig. 2). There are also two different types of distorted MeO_6 octahedra. Pyroxenes also form solid solution series. The Fe-free and Mg-free end members are enstatite ($x=1$) and ferrosilite ($x=0$), respectively. Enstatite can contain up to 10% of Fe by weight. The minerals with higher Fe^{2+} content are bronzite and hypersthene. In contrast to olivines, pyroxenes can occur in two different main crystallographic systems. The lattice can be of rhombic or monoclinic structure. The rhombic crystal structure is produced by twinning of the unit cell of clinopyroxene by operation of a b-glide parallel to the (100)-plane. This group is called orthopyroxene and represents the most common Mg–Fe pyroxenes. Pyroxenes containing cations with radii considerably larger than that of Mg^{2+}, e.g. Ca^{2+}, belong to the monoclinic system. Under extreme formation conditions, monoclinic Mg-Fe pyroxenes can also be produced, e.g., clinoenstatite. Synthetic Mg pyroxenes produced via melting usually are of this type.

Other groups of crystalline silicates considered as possible dust analogues are hydrous silicates mostly crystallizing in the form of layer silicates (phyllosilicates). In these materials the silicate tetrahedra form a layer with the general formula $(Si_2O_5^{2-})_x$ which is combined with an adjacent OH layer. Both belong to an octahedral layer with

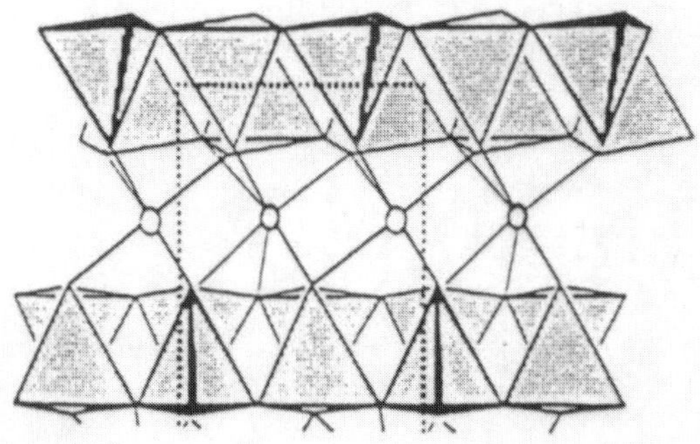

Figure 2. The crystalline chain structure of orthopyroxene.

octahedrally coordinated cations. For example, in the hydrous silicate talc with the chemical formula $Mg_3(OH)_2[Si_2O_5]_2$ brucite $(Mg(OH)_2)$ layers are inserted between the silicate planes. The number of bridging oxygen atoms in these silicates amounts to three. AlO_4 tetrahedra can easily replace the SiO_4 groups and big cations like K^+, Na^+, Ca^+ and water molecules can additionally be incorporated in the silicate structure. Besides talc, serpentine $(Mg_3(OH)_4[Si_2O_5])$, chlorite and montmorillonite belong to this phyllosilicate group and represent minerals of certain astrophysical interest. However, the chemical composition of these minerals is rather uncertain and depends on the varying content of water.

Amorphous silicates are built of the same basic structural units, the SiO_4 tetrahedra, but they do not show periodic structures like crystalline silicates. The SiO_4 tetrahedra form a three-dimensional disordered network. The incorporation of metal cations like Fe^{2+} or Mg^{2+} leads to a partial destruction of the oxygen bridges and the formation of non-bridging oxygen (NBO). In the produced hollow space the cations are octahedrally coordinated by such NBOs. However, small cations like Mg^{2+} can act as network formers and can replace the silicon in the SiO_4 groups. The overall number of bridging oxygen atoms depends of course on the general stoichiometry of the silicate. For example in an amorphous $MgSiO_3$ silicate the average number of bridging oxygen per silicate tetrahedron is 2. However, the resulting amorphous structure can be visualized by mixing of different structural SiO_4 arrangements, like islands, small chains, rings and and sheets, on a very small scale (Mysen *et al.* 1982). It is generally known that a phase transition between crystalline and amorphous silicates leads to a dramatic change of their spectral properties from the ultraviolet to the far-infrared region.

Amorphous silicates of pyroxene or olivine composition typically show two broad IR bands at about 10 and 18 μm corresponding to Si–O stretching and bending vibrations. The large width of both bands results from a distribution of bond lengths and angles within the amorphous structure. The 18 μm band is additionally broadened due to the coupling of the Si–O bending to the Me–O stretching vibration in this spectral region (Gervais *et al.* 1987). The position of the Si–O stretching vibration depends on the level of SiO_4 polymerization. For example, the band is shifted from 9 μm for pure SiO_2 to about 10.5 μm for $Mg_{2.4}SiO_{4.4}$ (Jäger *et al.* 2003a). Simultaneously, the O–Si–O bending mode is shifted towards shorter wavelengths.

In contrast to amorphous silicates crystalline pyroxenes and olivines produce a wealth of narrow bands from the mid-infrared to the far-infrared range due to metal-oxygen vibrations. Generally, the spectrum can be divided into three main regions. The bands between 1100 and 800 cm^{-1} correspond to different asymmetric and symmetric stretching vibrations of the SiO_4 tetrahedra. The group of modes in the region between 700 and $470\,cm^{-1}$ can be attributed to bending vibrations of the SiO_4 tetrahedra. The far-infrared bands in the low-frequency region beyond $470\,cm^{-1}$ are caused by translational motions of the metal cations within the oxygen cage and complex translations involving metal and Si atoms. However, more accurate band assignments can be found in the literature (see, e.g., Servoin & Piriou 1973; Hofmeister 1997).

The study of crystalline pyroxenes and olivines with different Mg/Fe ratios has demonstrated that the majority of the IR peaks are shifted to longer wavelengths with growing iron content. The observed shift is caused by an increase of the bond lengths between the metal cations and oxygen when Mg^{2+} is substituted by Fe^{2+} (Cameron & Papike 1980). The wavenumber shift $\Delta\nu$ is closely correlated with the Fe content and promises at least the possibility to derive values of the Mg/Fe ratio of crystalline silicates from the band positions (Jäger $et\ al.$ 1998).

The study of phase conversions from the amorphous into the crystalline state and vice versa are essential for the understanding of astronomical spectra. The conversion from the amorphous into the crystalline state can be caused by annealing at sufficiently high temperatures. Crystallization is a complex process including nucleation and crystal growth. Infrared spectroscopy has been used for monitoring the progress of annealing (see Hallenbeck $et\ al.$ 1998; Fabian $et\ al.$ 2000). The activation energy of crystallization for different amorphous silicate materials has been calculated using the equation $\tau^{-1} = \nu_0 e^{-E_a/kT}$, wherein E_a is the activation energy and ν_0 is a constant proportional to the mean vibrational frequency of the silicate lattice ($\nu_0 = 2.0 \times 10^{13}\ s^{-1}$). The estimation of the required annealing time at a given temperature is then based on the calculated activation energy. Since the crystallization is related to a preliminary diffusion of metal and oxygen atoms, the activation energy is strongly related to the viscosity behavior of the silicate during annealing. Since OH^- ions are able to reduce the viscosity, the annealing temperature as well as the annealing time can be considerably reduced by the presence of OH in the silicate network (Jäger $et\ al.$ 2003a). Apart from annealing of amorphous silicates, crystalline silicates can be formed directly from the gas phase. A gas-phase condensation route for the formation of crystalline silicates in the solar nebula and other protoplanetary disks has been recently discussed by Scott & Krot (2005).

Since the discovery that a significant fraction of stardust silicates leave their stellar environment in crystalline form, the question has arisen of why the interstellar silicate dust component does not show any indications of crystallinity. Amorphization of crystalline silicates due to irradiation by cosmic rays is one possible scenario. Interaction between the solid particles and ions may be expected in an energy range of a few eV up to a few MeV. In the astrophysical context, a series of irradiation experiments on crystalline olivines and pyroxenes were performed (see Bradley 1994; Dukes $et\ al.$ 1999; Demyk $et\ al.$ 2001). An efficient amorphization by light ions like H^+ and He^+ could only be observed for low energies ($\leqslant 50$ keV). At very low energies (20 keV H^+, 4–10 keV He^+) the amorphization process can be accompanied by a selective sputtering of Mg and Si relative to oxygen. Irradiation experiments of submicrometre-sized clinoenstatite ($MgSiO_3$) particles with 400 keV Ar^+ and 50 keV He^+ ions lead to an amorphization process without a change of the chemical composition but with morphological consequences (Jäger $et\ al.$ 2003c).

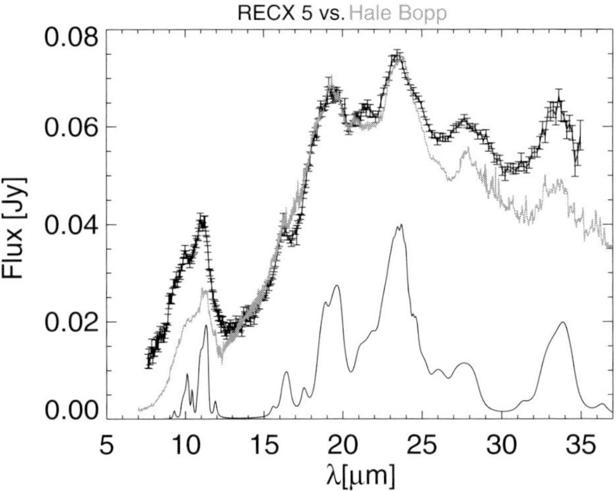

Figure 3. A comparison between the spectra of RECX5 and Hale-Bopp. The lower line is an emission spectrum for a distribution of hollow forsterite spheres (compact equivalent radius of 0.1 μm) at 200 K. After Bouwman *et al.* (2005).

3. Progress in Space

In this section, we will exclusively concentrate on protoplanetary disks and AGNs and quasars because of the tremendous progress made in the silicate spectroscopy of these objects.

Earlier studies of T Tauri disks showed that the 10 μm silicate features occur predominantly in emission (see, e.g., Natta *et al.* 2000). This fact together with models of protoplanetary disks (Menshchikov & Henning 1997; Chiang & Goldreich 1997) clearly suggested the presence of an optically thin surface layer in flaring disks. Here we should note that the feature goes from emission to absorption in the rare cases of protoplanetary disks seen edge-on.

ISO provided the first clear evidence for the presence of crystalline silicates in the spectra of relatively bright Herbig Ae/Be stars (Meeus *et al.* 2001) with HD 100546 (Malfait *et al.* 1998) and HD 163296 (van den Ancker *et al.* 2000) being good examples. Silicates in the diffuse interstellar medium and in molecular clouds are predominantly in the amorphous state. The detection of crystalline silicates in disks implies that they are either produced by annealing of amorphous silicates at higher temperatures in inner regions of disks or form directly from the gas phase. We refer the reader to a discussion of thermal processing of silicate dust in the solar nebula and clues coming from the analysis of primitive chondritic meteorites by Scott & Krot (2005). The amazingly perfect match of the "crystalline" features of HD 100546 with those of comet Hale-Bopp (Crovisier *et al.* 1997) establishes a bridge between circumstellar and cometary material. Most of the features in these two sources can be identified with crystalline forsterite. With Spitzer we are now in the position to measure fainter sources in the mass range of T Tauri stars (see, e.g., Forrest *et al.* 2004). However, we should note that in the 8–13 μm atmospheric window such data are also becoming available from ground-based observatories (see, e.g., Przygodda *et al.* 2003; Kessler-Silacci *et al.* 2005). The disadvantage of these data is obviously the limited wavelength range and the thermal background from the atmosphere. Figure 3 shows a comparison between the Spitzer spectrum of a disk around the M-type star RECX5 (age about 9 Myr) and the ISO spectrum of Hale-Bopp (Bouwman *et al.* 2005), again with a remarkable agreement between these two spectra. Another very interesting Spitzer result is the finding that even spectra of disks around brown dwarfs show evidence for the presence of crystalline silicates (Apai *et al.* 2005). Presently, we are trying to find out if the fraction of crystalline silicates is a function of stellar or disk mass. There is some indication that disks around more massive stars have a higher fraction of crystalline silicates (van Boekel *et al.* 2005).

Spitzer spectroscopy of older stars with debris disks has usually not led to the detection of features due to silicates. An explanation for this finding may be the fact that the particles have already grown to such sizes that their features are no longer visible. An exception is the remarkable spectrum of the 2 Gyr old star HD 69830, which shows clear evidence for the presence of crystalline silicates (Beichman *et al.* 2005).

Long-baseline interferometry at mid-infrared wavelengths is now opening a new chapter in the study of silicates. For the first time, we are able to determine the radial distribution of silicates in protoplanetary disks. The very first observations of disks around Herbig Ae stars with the MIDI instrument at the Very Large Telescope Interferometer showed that dust in the inner regions of these disks is highly crystallized, more so than any other dust observed in protoplanetary disks so far (van Boekel *et al.* 2004). Figure 4, taken from this paper, clearly shows that the inner-disk spectrum is dominated by crystalline olivine and pyroxene particles. One can exclude a significant contribution from hydrous silicates, clearly pointing to the fact that the dust has not been modified by hydrous alteration. For HD 142527 we even see evidence that the innermost disk regions are dominated by forsterite particles and that the enstatite/forsterite fraction increases with radius. This is in agreement with chemical equilibrium models as well as the radial mixing of material. These models predict that forsterite, present in the innermost disk regions, is converted into enstatite at slightly lower temperatures by a reaction with gas-phase silicon (Gail 2004).

Another important result is the accumulating evidence of the growth of silicate particles in the disk surface layer, from sizes of about 0.1 μm to particle sizes of few microns (see, e.g., Bouwman *et al.* 2001; van Boekel *et al.* 2003; van Boekel *et al.* 2005). An interesting recent finding by van Boekel *et al.* (2005) is the observation that all sources with crystalline silicates are dominated by large grains. This issue needs certainly more studies, investigating what happens at longer wavelengths and how much this conclusion is influenced by the increasing contrast between "crystalline" features and the flatter profiles of larger silicate grains.

We will now discuss new observational results concerning silicate features in AGNs and quasars. The unified models for AGNs assume that the central black hole is surrounded by dusty material that absorbs light from the central source. Many models assume the presence of a somewhat clumpy circumnuclear torus. In the unified picture of AGNs the different types of AGNs are produced by a different viewing angle. In the case of AGNs of Type 2, the torus is seen edge-on and the 10 and 18 μm silicate features should be seen in absorption. In the case of Type 1 AGNs (face-on view) the features should occur in emission. As discussed before, this behavior of the silicate feature has been found in protoplanetary disks of different orientation. Furthermore, the mid-infrared spectra of many Type 2 AGNs show silicate absorption features. However, Type 1 AGNs have never shown clear evidence for any silicate feature. A variety of models have been developed to explain this behaviour ranging from tori with larger grains, to anisotropic radiation coming from the accretion disk, to clumpy tori. However, recent observations with the IRS spectrometer on board the Spitzer Space Telescope led to the detection of 10 and 18 μm emission in a number of radio-quiet and radio-loud quasars (Hao *et al.* 2005; Siebenmorgen *et al.* 2005). Meanwhile, silicate emission features have also been detected in the spectra of AGNs of lower luminosity (Sturm *et al.* 2005). Here, we should note that the inferred dust temperatures are lower than what we would expect for silicates located in the hot torus walls or their surfaces. This may actually indicate that the silicates are

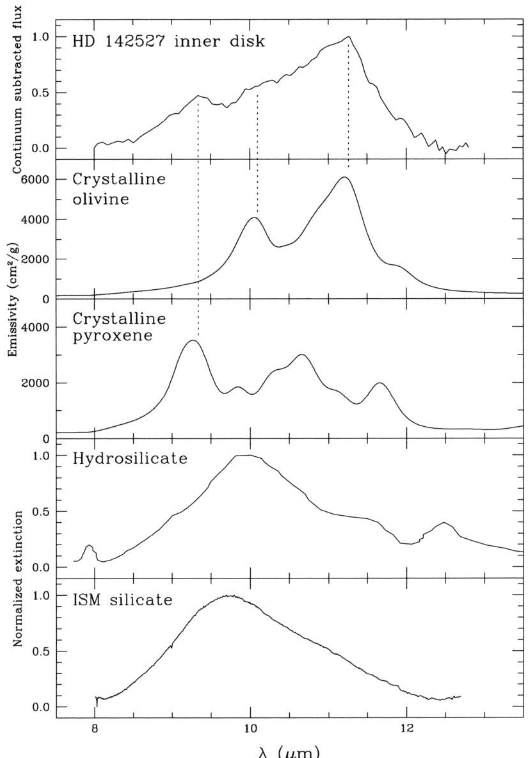

Figure 4. The spectrum of the innermost regions of HD 142527 compared with the laboratory spectra of typical dust species. After van Boekel *et al.* (2004)

not directly associated with the torus, but located in its optically thin environment. In addition, the silicate profiles are different from galactic interstellar medium dust, pointing to the presence of larger grains and differences in the chemical composition.

4. Progress in the Laboratory

Following the recent observational discoveries, especially with ISO, the spectroscopic data of relevant silicates have been considerably refined in collaboration between observers and laboratory astrophysics groups. Especially in the case of crystalline silicates, new optical constants of the most important pyroxene and olivine crystals have been derived (Jäger *et al.* 1998; Fabian *et al.* 2001; Suto *et al.* 2002), and the dependence of the IR band positions on the chemical composition has now been studied in great detail (Chihara *et al.* 2002; Koike *et al.* 2003). For amorphous silicates, the database so far consisting mainly of the optical constants of glasses produced from melts (Dorschner *et al.* 1995; Mutschke *et al.* 1998) and of silicate films (Scott & Duley 1996) has been broadened by new optical constants of sol-gel products with different Mg/Si ratios (Jäger *et al.* 2003a; see Fig. 5). The optical constants of these materials, which often comprise not only the mid- and far-infrared but also the UV and the important near-infrared spectral ranges, are available for observers and theoreticians from WWW databases such as the ones found at `http://www.kyoto-phu.ac.jp/labo/butsuri` and `http://www.astro.uni-jena.de/Laboratory/OCDB` (Henning *et al.* 1999; Jäger *et al.* 2003b). Furthermore, efforts have been undertaken in order to study the dependence of silicate spectral properties on morphological factors, i.e. grain size and shape, as well as on the grain temperature. We will review some of the latest results on these topics in the following discussion.

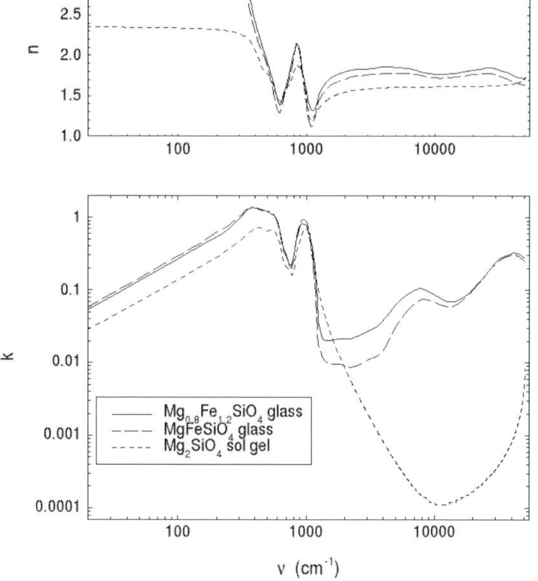

Figure 5. Optical constants of amorphous silicates with olivine stoichiometry: Mg-Fe silicates (glasses) by Dorschner *et al.* (1995), Mg silicate (sol-gel) by Jäger *et al.* (2003).

The influence of the grain morphology (size, shape, agglomeration, porosity) on both the IR band profiles and the far-IR continuum emission or absorption was noted a long time ago. For both cases, theoretical investigations so far dominate over experimental studies, because it is generally very difficult and expensive to control agglomeration, grain shape, and often even grain size in laboratory experiments. Whereas numerical simulations of morphological effects in the far-infrared continuum have led to a good understanding of the effects of porosity, fluffiness, or even metallic inclusions (e.g. Henning & Stognienko 1996), in the mid-infrared bands theoretical approaches often encounter problems with convergence because of the relatively large optical constants. Currently, this seems to be overcome by the development of efficient statistical approaches, which are able to simulate the extinction and emission cross sections of inhomogeneous grains by layered- or hollow-sphere models (Voshchinnikov *et al.* 2005; Min *et al.* 2005). Figure 6 demonstrates how these new theoretical models can be used together with the available optical constants to derive band profiles of non-spherical silicate grains for a mineralogical analysis of the 10 μm band of Herbig Ae/Be stars (from van Boekel *et al.* 2005). Laboratory measurements would be extremely important for the validation of these theoretical results. So far, shape effects on laboratory-measured crystalline silicate bands have been noted in a couple of papers (e.g. Fabian *et al.* 2001; Koike *et al.* 2005), but systematic measurements are still lacking. Recently, first attempts have been made to avoid the influence of embedding media on IR spectroscopy of particulate samples by dispersing particles in air (Tamanai *et al.* 2005), which is a first step to disentangle morphological effects.

The profiles of vibrational bands generally depend on the temperature of emitting or absorbing grains, because of the anharmonicity of the vibrational potentials (Henning & Mutschke 1997). The question of whether this effect can be used as a thermometer has become especially important after the discovery of crystalline silicates in many environments (see, e.g., Bowey *et al.* 2002; Molster *et al.* 2002). Crystalline olivines, e.g., possess promising sharp and isolated far-infrared bands at about 49 μm and 69 μm wavelength, which broaden and shift to larger wavelengths with increasing temperature (Bowey *et al.*

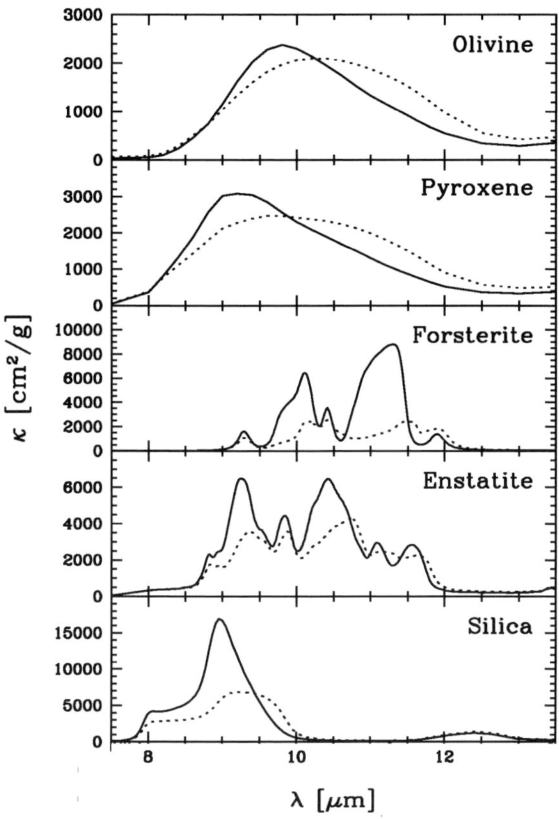

Figure 6. Mass absorption coefficients of amorphous and crystalline silicate grains with volume equivalent radii of 0.1 μm (solid lines) and 1.5 μm (dotted lines) calculated using the hollow-sphere model (from van Boekel *et al.* 2005).

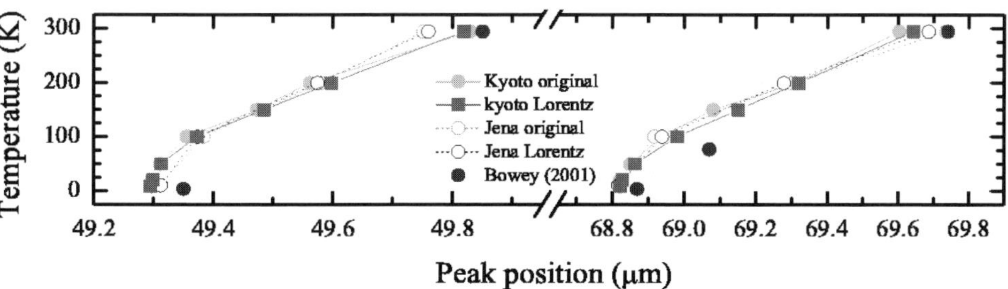

Figure 7. Temperature dependence of the far-infrared peak positions measured in the Kyoto and Jena laboratories on two different synthetic forsterites (Mg_2SiO_4) (from Koike *et al.* 2005). Labels "original" and "Lorentz" refer to the determination of the peak position just by reading by eye and by fitting of a Lorentzian band profile, respectively. Previous data by Bowey *et al.* (2001) are shown for comparison.

2001; Chihara *et al.* 2001). The exact temperature dependence of the peak position of this and other bands has recently been independently measured by two laboratories and has turned out to be nonlinear with stronger changes at higher temperatures (Fig. 7; Koike *et al.* 2005).

The far-infrared continuum absorption of crystalline silicates is known to be strongly temperature-dependent. However, the far-infrared emission of astronomical sources will often be dominated by absorption by amorphous grains, the temperature-dependence of which is less well known. First experimental studies on this subject for astronomically

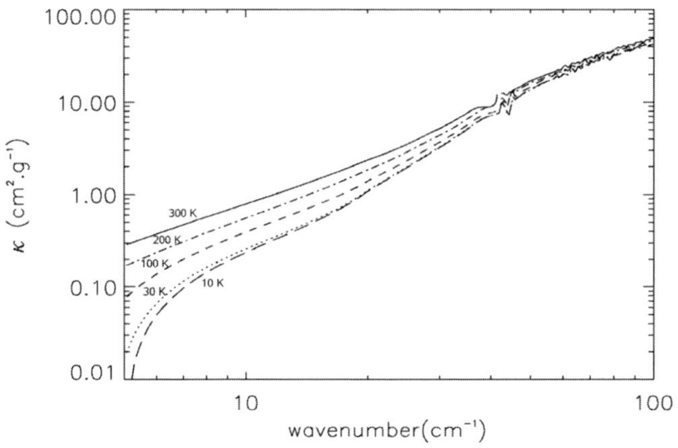

Figure 8. Mass absorption coefficient of $MgSiO_3$ glass for temperatures between 10 K (long-dashed line) and 300 K (solid line) (from Boudet *et al.* 2005).

relevant amorphous silicates have been carried out by Agladze *et al.* (1996) and Mennella *et al.* (1998), which recently have been extended by Boudet *et al.* (2005). It has been confirmed by these studies that the absorption/emission cross section of amorphous silicates at submillimeter and millimeter wavelengths is significantly lower (up to more than an order of magnitude) at 10 K compared with room temperature (Fig. 8). The spectral power-law index of the opacity's slope at these wavelengths may increase from about unity to values between two and three. This behavior, which is due to transitions in two-level systems probably formed by network-modifier cations, apparently correlates with an observed change of the power-law index in the submillimeter emission of ISM dust at different temperatures (Dupac *et al.* 2003).

Recently, the information about cosmic silicates obtained from infrared spectroscopy has been extended by the analysis of presolar silicate grains in primitive meteorites (Mostefaoui & Hoppe 2004; Nguyen & Zinner 2004), antarctic micrometeorites (Yada *et al.* 2005), and interplanetary dust particles (Messenger *et al.* 2003). These grains are found to be olivines, pyroxenes, and amorphous silicates originating from AGB stars and supernovae, in accordance (apart from the relative abundances) with the spectroscopic information from evolved objects.

5. Summary and Outlook

Over the last several years, we have seen a tremendous progress in our understanding of cosmic silicates, mainly coming from infrared spectroscopy and the chemical and isotopic analysis of primitive meteorites and interplanetary dust. This progress would not have been possible without the efforts of laboratory astrophysics groups, which provided the necessary experimental data for an adequate interpretation of astronomical spectra. We now have available a fundamental database of the optical properties of the most important silicate components. Experiments to clarify the influence of shape and agglomeration on the silicate profiles are lagging behind theoretical computations of this effect. Such studies may open a way to discriminate between the origin of crystalline silicates from annealing or direct condensation from the gas phase. The latter process is poorly understood and is presently lacking a firm experimental basis. We can expect to see a flood of new observational data coming from Spitzer, especially for sources which are too faint to have been observed with ISO. In addition, Herschel and SOFIA will soon provide more information on the far-infrared properties of cosmic silicates.

Acknowledgements

We would like to thank our collaborators J. Bouwman, J. Dorschner, C. Koike, R. van Boekel, and L.B.F.M. Waters for many interesting discussions and for providing material for this review.

References

Agladze, N.I., Sievers, A.J., Jones, S.A., Burlitch, J.M., & Beckwith, S.V.W. 1996, *Ap. J.* 462, 1026

Apai, D., Pascucci, I., Bouwman, J., Natta, A., Henning, Th., & Dullemond, C.P. 2005, *Science* 310, 834.

Beichman, C.A., Bryden, G., Gautier, T.N., Stapelfeldt, K.R., Werner, M.W., Misselt, K., Rieke, G., Stansberry, J., & Trilling, D. 2005, *Ap. J.* 626, 1061

Boudet, N., Mutschke, H., Nayral, C., Jäger, C., Bernard, J.-P., Henning, Th., & Meny, C. 2005, *Ap. J.* 633, 272

Bouwman, J., Meeus, G., de Koter, A., Hony, S., Dominik, C., & Waters, L.B.F.M. 2001, *A&A* 375, 950

Bouwman, J., *et al.*, 2005, in preparation

Bowey, J.E., Lee, C., Tucker, C., Hofmeister, A.M., Ade, P.A.R., & Barlow, M.J. 2001, *MNRAS* 325, 886

Bowey, J.E., Barlow, M.J., Molster, F.J., Hofmeister, A.M., Lee, C., Tucker, C., Lim, T., Ade, P.A.R., & Waters, L.B.F.M. 2002, *MNRAS* 331, L1

Bradley, J. 1994, *Science* 265, 925

Cameron, M. & Papike, J.J. 1980 in *Reviews in Mineralogy Vol. 7; Pyroxenes*, ed. Ch.T. Prewitt (BookCrafters, Inc.: Chelsea), p. 5

Colangeli, L., Henning, Th., Brucato, J.R., Henning, Th., Brucato, J.R., Clément, D., Fabian, D., Guillois, O., Huisken, F., Jäger, C., Jessberger, E.K., Jones, A., Ledoux, G., Manicò, G., Mennella, V., Molster, F.J., Mutschke, H., Pirronello, V., Reynaud, C., Roser, J., Vidali, G., & Waters, L.B.F.M. 2003, *Astron. Ap. Rev.* 11, 97

Crovisier, J., Leech, K., Bockelée-Morvan, Brooke, T.Y., Hanner, M.S., Altieri, B., Keller, H.U., & Lellouch, E. 1997, *Science* 275, 1904

Chiang, E.I. & Goldreich, P. 1997, *Ap. J.* 490, 368

Chihara, H., Koike, C., & Tsuchiyama, A. 2001, *Publ. Astron. Soc. Japan* 53, 243

Chihara, H., Koike, C., Tsuchiyama, A., Tachibana, S., & Sakamoto, D. 2002, *A&A* 391, 267

Demyk, K., Carrez, P., Leroux, H., *et al.* 2001, *A&A* 368, L38

Dorschner, J., Begemann, B., Henning, Th., Jäger, C., & Mutschke, H. 1995, *A&A* 300, 503

Dupac, X., Bernard, J.-P., Boudet, N., Giard, M., Lamarre, J.-M., M'eny, C., Pajot, F., Ristorcelli, I., Serra, G., Stepnik, B., & Torre, J.-P. 2003, *A&A* 404, L11

Dukes, C, Baragiola, R., & McFadden, L. 1999, *Geophys. Res.* 104, 1865

Fabian, D., Jäger, C., Henning, Th., Dorschner, J., & Mutschke, H. 2000, *A&A* 364, 282

Fabian, D., Henning, Th., Jäger, C., Mutschke, H., Dorschner, J., & Wehrhan, O. 2001, *A&A* 378, 228

Forrest, W.J., Sargent, B., Furlan, E., D'Alessio, P., Calvet, N., Hartmann, L., Uchida, K.I., Green, J.D., Watson, D.M., Chen, C.H., Kemper, F., Keller, L.D., Sloan, G.C., Herter, T.L., Brandl, B.R., Houck, J.R., Barry, D.J., Hall, P., Morris, P.W., Najita, J., & Myers, P.C. 2004, *Ap. J. Suppl.* 154, 443

Gail, H.-P. 2004, *A&A* 413, 571

Gervais, F., Blin, A., Massiot, D., Coutures, J.P., Chopinet, M.H., & Naudin, F. 1987, *J. Non-Cryst. Sol.* 89, 384

Hallenbeck, S.L., Nuth, J.A., & Daukantas, P.L. 1998, *Icarus* 131, 198

Hao, L., Spoon, H.W.W., Sloan, G.C., Marshall, J.A., Armus, L., Tielens, A.G.G.M., Sargent, B., van Bemmel, I.M., Charmandaris, D.W., Weedman, D.W., & Houck, J.R. 2005, *Ap. J.* 625, L75

Henning, Th. 2003a, in *Solid State Astrochemistry*, eds. V. Pirronello, Krelowski, J. & G. Manicò (Kluwer: Dordrecht), p. 85

Henning, Th., ed. 2003b, *Astromineralogy* (Springer:Berlin)

Henning, Th. & Stognienko, R. 1996, *A&A* 311, 191

Henning, Th. & Mutschke, H. 1997, *A&A* 327, 743

Henning, Th., Il'in, V.B., Krivova, N.A., Michel, B., & Voshchinnikov, N.V. 1999, *A&AS* 136, 405

Hofmeister, A.M. 1997, *Phys. Chem. Minerals* 24, 535

Jäger, C., Molster, F.J., Dorschner, J., Henning, Th., Mutschke, H., & Waters, L.B.F.M. 1998, *A&A* 339, 904

Jäger, C., Dorschner, J., Mutschke, H., Posch, Th., & Henning, Th. 2003a, *A&A* 408, 193

Jäger, C., Il'In, V.B., Henning, Th., Mutschke, H., Fabian, D., Semenov, D.A., & Voshchinnikov, N.V. 2003b, *J. Quant. Spectr. Rad. Transfer* 79-80, 765

Jäger, D. Fabian, C., Schrempel, F., Dorschner, J., Henning, Th., & Wesch, W. 2003c, *A&A* 401, 57

Kessler-Silacci, J.E., Hillenbrand, L.A., Blake, G.A., & Meyer, M.R. 2005, *Ap. J.* 622, 404

Koike, C., Chihara, H., Tsuchiyama, A., Suto, H., Sogawa, H., & Okuda, H. 2003, *A&A* 399, 1101

Koike, C., Mutschke, H., Suto, H., Naoi, T., Chihara, H., Henning, Th., Jäger, C., Tsuchiyama, A., Dorschner, J., & Okuda, H. 2005, *A&A*, in press

Malfait, K., Waelkens, C., Waters, L.B.F.M., Vandenbussche, B., Huygen, E., & de Graauw, M.S. 1998, *A&A* 332, 25

Menshchikov, A.B. & Henning, Th. 1997, *A&A* 318, 879

Meeus, G., Waters, L.B.F.M., Bouwman, J., van den Ancker, M.E., Waelkens, C., & Malfait, K. 2001, *A&A* 365, 476

Mennella, V., Brucato, J.R., Colangeli, L., Palumbo, P., Rotundi, A., & Bussoletti, E. 1998, *Ap. J.* 496, 1058

Messenger, S., Keller, L.P., Stadermann, F.J., Walker, R.M., & Zinner, E. 2003, *Science* 300, 105

Min, M., Hovenier, J.W., & de Koter, A. 2005, *A&A* 432, 909

Molster, F.J., Waters, L.B.F.M., Tielens, A.G.G.M., Koike, C., & Chihara, H. 2002, *A&A* 382, 441

Mostefaoui, S. & Hoppe, P. 2004, *Ap. J.* 613, L149

Mutschke, H., Degemann, B., Dorschner, J., Gürtler, J., Gustafson, B., Henning, Th., & Stognienko, R. 1998, *A&A* 333, 188

Mysen, B.O., Virgo, D., & Seifert, F.A. 1982, *Rev. Geophys. Space Phys.* 20, 353

Natta, A., Meyer, M.R., & Beckwith, S.V.W. 2000, *Ap. J.* 534, 838

Nguyen, A.N. & Zinner, E. 2004, *Science* 303, 1496

Przygodda, F., van Boekel, R., Àbràham, P., Melnikov, S.Y., Waters, L.B.F.M., & Leinert, C. 2003, *A&A* 412, L43

Servoin, J.L. & Piriou, B. 1973, *Phys. Stat. Sol. (b)* 55, 677

Scott, A. & Duley, W.W. 1996, *Ap. J. Suppl.* 105, 401

Scott, E.R. & Krot, A.N. 2005, *Ap. J.* 623, 571

Siebenmorgen, R., Haas, M., Krügel, E., & Schulz, B. 2005, *A&A* 436, L5

Sturm, E., Schweitzer, M., Lutz, D., Contursi, A., Genzel, *et al.* 2005, *Ap. J.* 629, L21

Suto, H., Koike, C., Sogawa, H., Tsuchiyama, A., Chihara, H., & Mizutani, K. 2002, *A&A* 389, 568

Tamanai, A., *et al.*, 2005, in preparation

van den Ancker, M.E., Bouwman, J., & Wesselius, P.R. 2000, *A&A* 357, 325

van Boekel, R., Waters, L.B.F.M., Dominik, C., Bouwman, J., de Koter, A., Dullemond, C.P., & Paresce, F. 2003, *A&A* 400, L21.

van Boekel, R., Min, M., Leinert, Ch., Waters, L.B.F.M., Richichi, *et al.* 2004, *Nature* 432, 479

van Boekel, R., Min, M., Waters, L.B.F.M., de Koter, A., Dominik, C., van den Ancker, M.E., & Bouwman, J. 2005, *A&A*, 437, 189

Voshchinnikov, N.V., Il'in, V.B., & Henning, Th. 2005, *A&A* 429, 371

Yada, T., Stadermann, F.J., Floss, C., Zinner, E., Olinger, C.T., Graham, G.A., Bradley, J.P., Dai, Z.R., Nakamura, T., Noguchi, T., & Bernas, M. 2005 *Lunar Planet. Sci.* 36, 1227

Astrochemistry: Recent Successes and Current Challenges
Proceedings IAU Symposium No. 231, 2005
D.C. Lis, G.A. Blake & E. Herbst, eds.

© 2006 International Astronomical Union
doi:10.1017/S1743921306007484

Observations of Molecules in Comets

D. Despois[1], N. Biver[2], D. Bockelée-Morvan[2], and J. Crovisier[2]

[1]Obs. de Bordeaux, OASU/L3AB, BP89, F-33270 Floirac, France
email: despois@obs.u-bordeaux1.fr

[2]LESIA, Obs. de Paris, 5 pl. Janssen, F-92195 Meudon, France

Abstract. Comets are among the most primitive bodies of the solar system, and their chemical composition is rich in information on the protosolar nebula and its possible connection with interstellar cloud chemistry. Comets are also a source of light atoms and, possibly, prebiotic organic molecules for the early Earth. We know better and better the abundances of cometary volatiles through spectroscopy, mainly at infrared and radio wavelengths. Another crucial component of cometary matter – organic refractories – is still poorly characterized, however.

We summarize here the $\sim$30 abundances and $\sim$20 upper limits obtained on cometary volatiles and highlight a few species and problems: ethylene glycol, NS, HNC/HCN, ^{14}N/^{15}N, the origin of CN, CS_2, PAHs, and H_2O. Comet-to-comet variations and comet internal heliocentric variations can now be studied, and cometary comas can be mapped with a variety of techniques. We list a number of temperature indicators, for they can help understand the relation between IS and cometary matters, which present both a global similarity but also marked differences, such as the high ethylene glycol content of comets.

We conclude by outlining a few key problems to be addressed by future ground-based and space instruments or by cometary sample analysis. For many species, we stress that laboratory data are missing on spectroscopy, photodissociation and collisions with H_2O.

Keywords. astrochemistry — comet: general — planetary systems: protoplanetary disks — solar system: formation

---

1. Introduction

On the 4th of July, 2005, comets were again a hot topic, as the Deep Impact projectile hit the 5×7 km nucleus of comet 9P/Tempel 1. The experiment has brought a wealth of information on the physics of the nucleus, surface details and strength of cometary matter (A'Hearn *et al.* 2005); it may also have provided new insights in the chemical composition: beside known species, Spitzer IR observations (Lisse *et al.* 2005) may have detected Al_2O_3, and PAHs (already tentatively detected in 1P/Halley, see below). More results are given in Meech *et al.* (2005) and Keller *et al.* (2005).

Comets are the most pristine bodies in the solar system. Their astrochemical interest is at least fourfold: (i) they hold a wealth of information on the physics and chemistry of the protosolar nebula (PSN), (ii) they offer a link, at least partial, to the chemistry of the parent cloud from which the solar system has formed, (iii) they are, through violent impacts as well as through smooth delivery of dust particles (meteors, micrometeorites, IDPs) one source of carbon and prebiotic molecules for the early Earth; (iv) finally, as all astronomical objects presented in this Symposium, they are a remarkable laboratory for molecular physics and chemistry. Kinetic temperatures in the coma first decrease from $\sim$200 K (water ice sublimation) to a few 10 K or even less, due to adiabatic expansion, then warm up again due to heating by photodissociation. Rotational populations, first at LTE, progressively decouple from the gas translational energy, due to the rapid decrease in density and collisions far from the nucleus. Solar radiation then takes over as the main source of excitation. For vibrational and electronic states, even very close to the nucleus solar excitation dominates, provided the coma is not optically thick. The coma density can be as high as $10^{12}\,\mathrm{cm}^{-3}$ close to the surface of the nucleus, and decreases rapidly to the interplanetary level ($\sim 1\,\mathrm{cm}^{-3}$). Gas expansion velocities are on the order of $1\,\mathrm{km\,s}^{-1}$. The dominant coma species is usually water, except far ($>4\,\mathrm{AU}$) from the Sun, where CO takes over this role. Ions and radicals are photo-produced which means that ion-molecule and radical reactions are possible; ions finally interact with the solar wind, building up the plasma tail.

The time scales for chemistry are diverse. Icy planetesimals can form over periods as short as 2.5×10^5 yr according to models, and their formation must take place when a gaseous disk still encircles the Sun, i.e. during the first $\sim 10\,\mathrm{Myr}$ of the solar system. Note that some ingredients, especially among dust particles, are prestellar, and thus older. The formation is followed by a 4.5 Gyr mostly dormant phase, in one of the two comet reservoirs: the Oort Cloud or the Kuiper Belt. Despite the generally very low temperature ($<30\,\mathrm{K}$?) of the nucleus, it is not certain that such a long period is devoid of chemical transformation (an episode of liquid water is even considered as possible in the core of the largest – a few $100\,\mathrm{km}$ – KB objects). The active phase during which the coma is observed typically lasts months, whereas the lifetimes of sublimated molecules range from 10^3 to 10^6 s (at $1\,\mathrm{AU}$).

For physicists and chemists unfamiliar with comet science, a general introduction is given in Despois & Cottin (2006), and Crovisier (2004), with special emphasis on the connection to astrobiology. A very recent, deep and complete summary on cometary science is provided by the book *Comets II* (eds. Festou *et al.* 2004), where the chapters by Bockelée-Morvan *et al.* (parent species) and Feldman *et al.* (secondary species) are the most directly relevant to astrochemistry. The possible connection between ISM and comets is addressed in detail by Irvine *et al.* (2000) and Ehrenfreund *et al.* (2004). We will discuss in the following mainly cometary volatiles; when there is no reference to other authors, the original data are from Bockelée-Morvan *et al.* (2000) and Crovisier *et al.* (2004a,b; detections and upper limits in comet Hale-Bopp), and Biver *et al.* (2002a, 2005a) and Crovisier *et al.* (2005; inter and intra comet variations).

2.   The Composition of Cometary Volatiles

Comets, if the coma composition reflects that of the nucleus, are made of roughly equal masses of dust and ices (within a factor of 10, depending on the comet). Dust itself is half rocky material (e.g., silicates), half organic refractories ("CHON" particles). We will mainly discuss here the volatile composition. One should, however, keep in mind that: (i) many complex molecules may be hidden in the organic refractories, and are not yet accessible to precise identification; (ii) macromolecular carbon (e.g., heterogeneous copolymers, *cf.* Fomenkova 1999) may be the dominant form of carbon in these refractories as it is in meteorites (Binet *et al.* 2004).

Figure 1 gives "typical" cometary volatile abundances, together with the range observed in various comets, and the number of comets observed/detected in a given species. "Typical abundances" means in practice that most coma abundances are from C/1995 O1 (Hale-Bopp) observed near $r_h = 1\,\mathrm{AU}$; however, a few species have been measured under more favorable circumstances (CO_2) or only (S_2, C_4H_2) in other comets . The comet-to-comet dispersion, comet internal variation, and coma *vs.* nucleus abundance discussions (including differential sublimation and coma chemistry) should not be forgotten when using these data.

category	Molecule	Hale-Bopp Abundance (H2O= 100)	Intercomet variation	comets detected +upperlimits (a)	Notes	
H	H2O	100				water
	H2O2	<0.03				hydrogen peroxide
C,O	CO	23	<1.4-23	9+8	variable, extended source	carbon monoxide
	CO2	6	2.5-12	4	IR, UV from various comets	carbon dioxide
C,H	CH4	1.5	0.14-1.5	8	IR	methane
	C2H6	0.6	0.1-0.7	8	IR	ethane
	C2H2	0.2	<0.1-0.5	5	IR	acetylene
	C4H2	0.05?			IR From Ikeya-Zhang(prelim.val.)	diacetylene
	CH3C2H	<0.045				propyne
C,O,H	CH3OH	2.4	<0.9-6.2	25+2		methanol
	H2CO	1.1	0.13-1.3	18+3	extended source	formaldehyde
	CH2OHCH2OH	0.25				ethylene glycol
	HCOOH	0.09	<0.05-0.09	3+2		formic acid
	HCOOCH3	0.08				methyl formate
	CH3CHO	0.025				acetaldehyde
C,O,H upper limits	H2CCO	<0.032				ketene
	c-C2H4O	<0.20				oxirane
	C2H5OH	<0.1			trans	ethanol
	CH2OHCHO	<0.04				glycolaldehyde
	CH3OCH3	<0.45				dimethyl ether
	CH3COOH	<0.06				acetic acid
N	NH3	0.7	<0.2-1	4		ammonia
	HCN	0.25	0.08-0.25	32+0		hydrogen cyanide
	HNCO	0.1	0.02-0.1	4+2		isocyanic acid
	HNC	0.04	<0.003-0.035	12+3	Variable with rh	hydrogen isocyanide
	CH3CN	0.02	0.013-0.035	9+2		methyl cyanide
	HC3N	0.02	<0.003-0.03	3+7		cyanoacetylene
	NH2CHO	0.015				formamide
N upper limits	NH2OH	<0.25				hydroxylamine
	HCNO	<0.0016				fulminic acid
	CH2NH	<0.032				methanimine
	NH2CN	<0.004				cyanamide
	N2O	<0.23				nitrous oxide
	NH2-CH2-COOH	<0.15			conformer I	glycine I
	C2H5CN	<0.01				cyanoethane
	HC5N	<0.003				cyanobutadiyne
S	H2S	1.5	0.13-1.5	15+5		hydrogen sulfide
	OCS	0.4	<0.09-0.4	2+5	Extended source	carbonyl sulfide
	SO	0.3	<0.05-0.3	4+1	Extended source	sulfur monoxide
	SO2	0.2				sulfur dioxide
	CS2	0.17	0.05-0.17	15+3	from CS	carbon disulfide
	H2CS	0.02				thioformaldehyde
	S2	0.005	0.001-0.25	5	UV; from Hyakutake	disulfur
	CH3SH	<0.05				methanethiol
	NS	0.02	<0.02-0.02	1+1	Parent molecule ??	
P	PH3	<0.16				phosphine
Metals	NaOH	<0.0003				sodium hydroxide
	NaCl	<0.0008				sodium chloride

Figure 1. Cometary volatiles: detections and upper limits. Abundances are in % by number relative to water. (a) refers to radio data unless otherwise noted.

It is of prime importance, when comparing these abundances to laboratory experiments on interstellar/cometary ice analogs, or to chemical models of the ISM or protosolar disk/nebula, to take into account both detections and upper limits. One should especially keep in mind that the modelling used to derive upper limits is less elaborate than that used for the major species, e.g., H_2O or HCN: LTE is typically assumed, and photodissociation rates are in several cases only educated guesses – due to the lack of precise laboratory data. However, radio single dish observations of comet Hale-Bopp by Crovisier *et al.* (2004a) did not suffer too much from these approximations, as the inner coma inside the beam was dense and little affected by photodissociation.

3. A Few Highlights

3.1. *Ethylene Glycol CH_2OHCH_2OH*

Ethylene glycol (EG) has been securely identified in comet Hale-Bopp from IRAM 30m, PdBI and CSO observations (Crovisier *et al.* 2004b). Ten transitions with S/N of 3 to 10, spanning the range 20 to 100 K in upper level energy, were detected. The derived abundance is 0.29% if one takes the rotational temperature of 77 K derived from the rotational plot, or 0.17% if one adopts the 110 K average rotational temperature derived from other coma species. The photodissociation rate was assumed to have the value $2.10^{-5}\,s^{-1}$ at AU, as no data were available; but the small size of the beam for these observations makes this uncertainty not too critical.

Ethylene glycol is a known IS species, observed for the first time by Hollis *et al.* (2002 and this volume) in Sgr B2, but some striking differences are apparent when comparisons are made with other molecules. The EG/CH_3OH ratio is $\sim$0.1 in comets, whereas it is $\sim$0.001 towards Sgr B2; glycolaldehyde, CH_2OHCHO, a molecule very similar to ethylene glycol, has not yet been detected in comets ($x < 0.16$ EG) whereas it has comparable abundances in Sgr B2. More remarkably, ethanol, a simple alcool (CH_3CH_2OH) has not yet been detected in comets, while ethylene glycol, a dialcohol, has: CH_3CH_2OH/EG $<$ 0.4. All this seems to indicate a specific production mechanism in comets. Considering the high relative abundance of methanol in comets (2.5% in Hale-Bopp, up to 6.2% in other comets) it is tempting to invoke a direct formation from methanol, e.g., by ice irradiation. This has been experimentally performed with success (e.g., Moore, this volume, on H_2O–CO_2 and H_2O–CH_3OH ices). One should note, however, that HCOOH is also produced in large quantities in these experiments, whereas it is even less abundant in comets ($\sim$0.3 EG) than in Sgr B2 ($\sim$EG).

3.2. *NS*

The NS radical has been detected in comet Hale-Bopp by Irvine *et al.* (2000) with the JCMT through its J=15/2–13/2 a and f transitions, with an abundance $x_{NS} > 0.02\%$. NS is a well known IS molecule present in both warm and cold clouds. Searches through the $J = 11/2$–$9/2$ e and f transitions at 253 GHz in comet Hale-Bopp (Crovisier *et al.* 2004a) and 153P/Ikeya-Zhang (Biver *et al.* 2005a) led to upper limits of 0.01% and 0.02% respectively, assuming (?) a parent molecule distribution.

The origin of NS is not yet known. Could it be present in the ice? If it is a photodissociation product, which is the parent molecule? NH_4SH (predicted as a cometary species by Fegley *et al.* 1993) is a possibility, but its photodissociation behavior is unknown. NH_2SH is another candidate. No map exists, so we have no constraint on the NS parent lifetime, which also prevents a good estimate of its production rate. The $x_{NS} > 0.02\%$ abundance in comet Hale-Bopp is a lower limit and assumes a NS photodissociation rate $\beta \sim 0.1$–$2 \times 10^{-4}\,s^{-1}$. A chemical model has been proposed to produce NS by coma reactions, but it cannot account quantitatively for the NS abundance (Canaves *et al.* 2002).

3.3. *HNC vs. HCN*

The HNC/HCN ratio is seen to decrease in the coma with increasing heliocentric distance r_h. Irvine *et al.* (1998) proposed a chemical conversion in the inner coma by ion-molecule reactions and electronic recombination of the molecular ion $HCNH^+$ as in cold dark clouds. Reactions of HCN with suprathermal hydrogen atoms issued from photodissociation were shown to be more efficient, and capable of quantitatively explaining the comet Hale-Bopp results, but still not the presence of HNC in less dense comas (Rodgers and Charnley 2001, Charnley *et al.* 2002). To match the HNC/HCN *vs.* r_h data (Biver *et al.* 2002b, 2005a; Irvine *et al.* 2003), thermo-desorption from grains is also proposed. Comet Hale-Bopp interferometric observations (Bockelée-Morvan *et al.* 2005a,b) and new observations (Biver *et al.* 2005a; Crovisier *et al.* 2005), which show a plateau in HNC/HCN around 0.2 for $r_h = 0.1$–0.5 AU, favor thermo-desorption and do not seem to be compatible with HNC production by chemical reaction of a daughter species (H). Finally, it is not yet excluded that HNC could be stable in ice at very low temperature; but observations limit the HNC fraction from the nucleus to 6% of HCN.

3.4. *The ^{15}N Problem and the Origin of CN*

A $^{14}N/^{15}N$ isotopic ratio of 140 (± 30) in the CN radical has been systematically derived (Arpigny *et al.* 2003; Manfroid *et al.* 2005) from visible observations of Hale-Bopp and other comets, including the Deep Impact target 9P/Tempel 1 (Meech *et al.* 2005); this is half the solar value (270) and contrasts with the $^{14}N/^{15}N$ ratio in HCN from radio observations (323 ± 46 Jewitt *et al.* 1997; 330 ± 98 Ziurys *et al.* 1997). Other isotopic abundances in cometary volatiles are, within the (still large) error bars, also close to solar: ^{13}C, ^{18}O, ^{34}S (see, e.g., Table 3 of Bockelée-Morvan *et al.* 2004, and Altwegg and Bockelée-Morvan 2003). In comet Halley, strongly non-solar $^{12}C/^{13}C$ values were found in a limited number of grains, indicating their presolar nature; however, on the average, isotopic ratios in the grains appeared to be solar; the low $^{12}C/^{13}C$ ratio in HCN measured in comet C/1996 B2 (Hyakutake) by Lis *et al.* (1997) is likely due to contamination by another line. $C^{14}N/C^{15}N$ is up to now the only clearly non-solar isotopic ratio derived on the scale of the whole comet and involving heavy atom isotopes.

That HCN is not the unique parent of CN has also been questioned on other grounds: comparing HCN and CN parent scalelengths, HCN and CN production rates, or CN jet morphology (e.g., Fray *et al.* 2005). This has lead to the consideration of other parents: HC_3N, C_2N_2 or refractory nitrogenated compounds (Fray *et al.* 2004). The production of some CN from ^{15}N rich grains is not sufficient to reconcile the $^{14}N/^{15}N$ measurements in CN with the HCN data; there is currently no proposed explanation which works.

3.5. *Detection of Carbon Disulfide CS_2*

Since the detection of the daughter species CS in comet C/1975 V1 West (Smith *et al.* 1980), its parent has been sought, and various candidate parent molecules have been proposed and discussed based on photodissociation lifetime arguments. That of CS_2, proposed as a CS parent as early as 1982 by Jackson *et al.*, has been revised in the laboratory from 100 to 500 s (at 1 AU), whereas cometary data required about 1000 s (Feldman *et al.* 1999). The establishment of $S + CO$ as main photodissociation path for OCS (Huebner *et al.* 1992) ruled it out as a candidate. This exemplifies the crucial role played by laboratory data in the interpretation of cometary observations. CS_2 seems now to have been identified with confidence in 122P/de Vico, through 28 visible lines (Jackson *et al.* 2004). Laboratory experiments (in supersonic molecular beams) are in progress to confirm that the line intensities are correct. CS_2 would be the third parent molecule (along with CO and S_2) identified directly from UV/visible spectra.

3.6. *Polycyclic Aromatic Hydrocarbons (PAHs)*

The detection of PAHs, a major component of IS matter, would be an important link between the ISM and comets. The 3.28 μm band of PAHs was seen in some comets, but not in Hale-Bopp, perhaps due to the large heliocentric distance (2.7 AU) where it was searched for. This band is subject to blending with OH prompt emission and CH_4 fluorescence emission. In comet Halley, a recent reanalysis of TKS data (Clairemidi *et al.* 2004) added a tentative detection of pyrene $C_{16}H_{10}$ to that of phenanthrene $C_{14}H_{10}$ (Moreels *et al.* 1993). PAHs are seen in laser ablation of those IDPs thought to be cometary in origin (Clemett *et al.* 1993). Recently, PAH features have been seen in the IR spectrum of 9P/Tempel 1 after the impact of Deep Impact (Lisse *et al.* 2005). According to Kress *et al.* (2004), PAHs are not likely to be produced by the impact itself, as their formation, while thermodynamically allowed, is not favored kinetically. The search for PAHs in the cometary dust sample collected by Stardust will be extremely interesting (if they have survived the collection process).

3.7. *The H_2O Reference*

A new development which improves the measurement of relative abundances is the direct determination of H_2O production through space-based submillimeter observations. This complements the widely employed but indirect technique using OH maser emission (Crovisier *et al.* 2002), ground-based IR observations of hot bands of water (Dello Russo *et al.* 2004), narrow band UV OH emission (A'Hearn *et al.* 1995) and other techniques based on water photodissociation products (Feldman *et al.* 2004). SWAS (Bensch *et al.* 2004) and Odin (Lecacheux *et al.* 2003) have observed the 1_{10}–1_{01} submillimeter line of water in various comets. The $^{18}O/^{16}O$ ratio could even be determined with precision in comet 153P/Ikeya-Zhang from Odin data (Lecacheux *et al.* 2003). SWAS (Bensch, this volume), Odin (Biver *et al.* 2005b), and the smaller radio telescope MIRO aboard Rosetta all attempted to follow the 9P/Tempel 1 temporal evolution after Deep Impact. Remarkable features are the pre-impact natural variability, and the line profile changes seen in Odin spectra during and a few hours after impact.

4. Comet Variation and Comet Population Homogeneity

4.1. *Comet-to-Comet Variation*

As a "first approximation", for the sake of simplicity, and due to the lack of more complete data, comet composition was first thought of as universal. Dust-to-gas ratio variations were, however, already obvious (Rolfe & Battrick 1987) and CN was used early on (A'Hearn *et al.* 1995) to distinguish comet classes. We are now, due to the increased sensitivity in radio and IR instruments, at the beginning of a statistical investigation of the comet population based on volatiles. Biver *et al.* (2002a, 2005a) and Crovisier *et al.* (2005) have now gathered data on a handful of comets, and dispersions in the abundances are given in Figure 1. They remain generally within a factor 3 of the mean, except for the highly volatile CO molecule. No clear correlation has yet been found with dynamical parameters characterizing the orbit.

4.2. *Comet Internal Heliocentric Variation*

The study of heliocentric variation of cometary abundances is an important tool to discuss processes responsible for the production of parent molecules (Biver *et al.* 2002b; "Christmas tree plot"). Similar evolution is expected in a group of molecules if the release of a single major species (H_2O, or CO) controls their outgassing; any departure from the average behavior indicates, to the contrary, the interplay of other processes. Several have been proposed: chemistry in the coma (HNC *vs.* HCN, *cf.* above), differential sublimation (CO *vs.* H_2O) and release from grains (H_2CO and, possibly, CN). Less pronounced departures seem to affect even the production rate ratio of quite similar species such as H_2O and CH_3OH (Biver *et al.* 2002b., 2005a), and a layered nucleus model has been proposed to explain it (Rodgers & Charnley, this volume).

4.3. *Coma Maps*

Another important recent development is the ability to map molecules in the coma at radio wavelengths. This allows us to address the origin (nuclear or not) and the photodissociation lifetime of a species, and to reveal structures ("jets") in the coma. Large-scale maps (a few arcminutes, or a few 10^5 km) were produced first point-by-point (Biver *et al.* 1999; Hirota *et al.* 1999), but more homogeneous maps are produced by detector arrays (QUARRY at FCRAO; Lovell *et al.* 1999) or raster techniques (NRAO 12m; Milam *et al.*, IAUS 231 poster). High resolution interferometric maps ($\sim$ a few arcsecs or thousands of km) have been obtained on the very bright comet Hale-Bopp (Blake *et al.* 1999 at OVRO; Veal *et al.* 2000 at BIMA; Henry *et al.* 2002 at PdBI). A specific technique has been developed to analyze PdBI observation of Hale Bopp – model fitting in the (uv) plane – which allows us to reconstruct the main CO jet (Henry 2003; Bockelée-Morvan *et al.* 2005a). Interferometric studies of comets are especially complex due to the time variability of the object, which limits the use of super-synthesis. In this respect ALMA, with its large instantaneous uv coverage, will bring a noticeable improvement.

4.4. *Temperature History and Origin of Cometary Matter, Connection to ISM*

Comets are the most pristine bodies of the solar system, but just how pristine are they? How much information on IS matter will comets provide? Several indicators exist which should inform us on the past temperature history of comae. Within a simple evolutionary model of cometary nucleus formation, they could give us a "formation temperature" of the icy material, dating back to the protosolar nebula. However, radial mixing in the PSN (e.g., Bockelée-Morvan *et al.* 2002), as well as vertical mixing between mid-plane and high-z layers of the protostellar disk (*cf.* disk structures presented in this volume) may have led to a complex history, possibly different for refractories and volatiles.

Important temperature indicators are:

• The presence/absence of volatiles and supervolatiles. The observation of CO at large heliocentric distances in comet Hale-Bopp (Biver *et al.* 2002b) may indicate that cometary matter incorporated it at low temperature (22 K for pure CO ice, up to several tens of K for CO-H_2O mixture – the exact limit is still debated, and depends on the structure of the ice: amorphous, crystalline, clathrate). It may also indicate that the nucleus internal temperature remained at such low temperatures for 4.5 Gyr, and may provide a constraint for models of the comet accretion process: perhaps CO was incorporated after the build-up of the ice-dust planetesimal, as the energy produced by the accretion process itself might have released this very volatile species.

H is strongly depleted in comets (FUSE observations of H_2 in the coma are an indication of coma production by H_2O photodissociation). Rare gases, also very volatile, are crucial probes of the temperature history, but have not yet been convincingly detected (Bockelée-Morvan *et al.* 2004). An interesting case is N_2. Its detection in comets through N_2^+ is ambiguous. Nitrogen as an element is depleted in comets (*cf.* references in Iro *et al.* 2003), a fact explained generally by the volatility of N_2. However, some estimations give a very similar volatility for CO and N_2, even on H_2O ices , which is clearly a problem; clathrates have been proposed as a solution to this puzzle (Iro *et al.* 2003).

• Ortho/Para (O/P) ratios. These have been summarized by Kawakita *et al.* (2004). A recent development is the use of a daughter species, NH_2, whose study can be used to trace the ratios in NH_3. The O/P ratios generally point toward a temperature of about 30 K, except in one case – comet C/1986 P1(Wilson). No difference is yet seen between the various comets despite their different dynamical histories. But the meaning of this spin temperature is unclear: Is it the formation temperature of the molecule in the gas phase? Formation on grain surfaces? Equilibration with the comet interior during the 4.5 Gyr of its existence?

• Further constraints come from the isotopic ratios (HCN/DCN and H_2O/HDO; Mousis *et al.* 2000), the conservation of prestellar grains and from the crystalline and/or amorphous nature of silicates and water ice (still unknown).

While an overall correspondence exists between cometary and interstellar molecules found in hot cores (Bockelée-Morvan *et al.* 2000), definite differences exist; one of the most interesting being the suspected relatively high abundance of ethylene glycol in cometary ices. Another result still to be understood is the deuteration level (HCN/DCN $\sim 2.3 \times 10^{-3}$ and $D/H_{H_2O} \sim 3 \times 10^{-4}$), and the presence of crystalline silicates (Crovisier *et al.* 1998; Hanner and Bradley 2004) not observed in the ISM (but observed around Herbig AeBe stars; Malfait *et al.* 1998).

5. Conclusion: What is Important for the Future

About 30 abundances of parent species, and nearly 20 upper limits on related compounds have already been determined in at least one comet. In parallel with the increase of this list, the following questions need to be addressed with high priority in the future: (i) the homogeneity of the cometary nucleus; (ii) the homogeneity of the comet population; (iii) the nature of the organic refractories (which likely hide many complex molecules together with heterogeneous organic polymers). The resolution of all these issues would help to elucidate the origin of cometary matter, and to determine the respective roles of the interstellar matter (from the solar system parent cloud) and of processing in the Protosolar Nebula. In view of the complexity of hot cores like Orion, of the processes in the protosolar nebula, and of the possible additional complexity of cometary chemistry, the correlation previously found between interstellar and cometary abundances remains a surprising observational fact, which needs to be explained.

Chemical analysis *in situ* (with Rosetta) and on samples (dust samples, with Stardust, and – hopefully soon – ice samples with a future space mission) may enlarge considerably (up to 100 species ?) the chemical inventory of comets. It will also provide the means to carry out a chiral analysis, impossible to do from remote observing; this in turn will feed models of the early Earth for which comets may have been a source of prebiotic compounds. Ground-based and space-based tele-detection (e.g., the D/H ratio with Herschel/HIFI) will allow the statistical characterization of the comet population(s). With high spatial resolution (using large radio interferometers like ALMA and large optical telescopes, close-up maps from the Rosetta orbiter with e.g., MIRO or VIRTIS), we will study comet variability and heterogeneity (analyzing the respective chemical composition of jets versus more isotropic outgassing). Space observations will also provide the fundamental reference, Q_{H_2O}, as they have already begun to do.

The increased precision required to model comets asks for plenty of new laboratory measurements: spectroscopy (only a few unidentified lines at mm wavelength but more than 4000 in the UV/visible), photodissociation rates (most of them are now only educated guesses), and collision cross sections with H_2O.

References

A'Hearn, M.F., Millis, R.L., Schleicher, D.G., Osip, D.J., & Birch, P.V. 1995, *Icarus* 118, 223

A'Hearn, M.F., Belton, M.J.S., Delamere, W.A., *et al.* 2005, *Science* 310, 258

Altwegg, K. & Bockelée-Morvan, D. 2003, *Space Science Reviews* 106, 139

Arpigny, C., Jehin, E., Manfroid, J., Hutsemékers, D., Schulz, R., Stüwe, J.A., Zucconi, J.-M., & Ilyin, I. 2003, *Science* 301, 1522

Bensch, F., Bergin, E.A., Bockelée-Morvan, D., Melnick, G.J., & Biver, N. 2004, *Ap. J.* 609, 1164

Binet, L., *et al.* 2004, in *Astrobiology: Future Perspectives* (ASSL Vol. 305), 333

Biver, N., *et al.* 1999, *A. J.* 118, 1850

Biver, N., *et al.* 2002a, *Earth Moon and Planets* 90, 323

Biver, N., *et al.* 2002b, *Earth Moon and Planets* 90, 5

Biver, N., *et al.* 2005a, *A&A*, in press

Biver *et al.* 2005b, *Asteroids, Comets and Meteors 2005* (IAU Symp. 229), abstract

Biver, N., *et al.* 2005c, *BAAS* 37, 710

Blake, G.A., Qi, C., Hogerheijde, M.R., Gurwell, M.A., & Muhleman, D.O. 1999, *Nature* 398, 213

Bockelée-Morvan, D., *et al.* 2000, *A&A* 353, 1101

Bockelée-Morvan, D., Gautier, D., Hersant, F., Huré, J.-M., & Robert, F. 2002, *A&A* 384, 1107

Bockelée-Morvan, D., Crovisier, J., Mumma, M.J., & Weaver, H.A. 2004, in *Comets II* (Univ. Arizona Press), 391

Bockelée-Morvan, D., *et al.* 2005a, *A&A* in preparation

Bockelée-Morvan, *et al.* 2005b, Abstract, *Asteroids, Comets and Meteors 2005* (IAU Symp. 229), abstract

Canaves, M.V., De Almeida, A.A., Boice, D.C., & Sanzovo, G.C. 2002, *Earth Moon and Planets* 90, 335

Charnley, S.B., Rodgers, S.D., Butner, H.M., & Ehrenfreund, P. 2002, *Earth Moon and Planets* 90, 349

Clairemidi, J., Bréchignac, P., Moreels, G., & Pautet, D. 2004, *Plan. Sp. Sci.* 52, 761

Clemett, S.J., Maechling, C.R., Zare, R.N., Swan, P.D., & Walker, R.M. 1993, *Science* 262, 721

Crovisier, J. 2004, in *Astrobiology: Future Perspectives* (ASSL Vol. 305), 179

Crovisier, J., Colom, P., Gérard, E., Bockelée-Morvan, D., & Bourgois, G. 2002, *A&A* 393, 1053

Crovisier, J., Leech, K., Bockelée-Morvan, D., Brooke, T.Y., Hanner, M.S., Altieri, B., Keller, H.U., & Lellouch, E. 1997, *Science* 275, 1904

Crovisier, J., Bockelée-Morvan, D., Colom, P., Biver, N., Despois, D., Lis, D.C., & the Team for target-of-opportunity radio observations of comets 2004a, *A&A* 418, 1141

Crovisier, J., Bockelée-Morvan, D., Biver, N., Colom, P., Despois, D., & Lis, D.C. 2004b, *A&A* 418, L35

Crovisier, J., *et al.* 2005, *BAAS* 37, 646

Despois, D. & Cottin, H. 2005, in *Lectures in Astrobiology Vol. I*, eds. M. Gargaud, B. Barbier, H. Martin, & J. Reisse (Springer), 289

Ehrenfreund, P., Charnley, S.B., & Wooden, D. 2004, in *Comets II* (Univ. Arizona Press), 115

Fegley, B., Jr. 1993, in *The Chemistry of Life's Origins*, eds. Greenberg, Mendoza-Gomez and Pirronello (Kluwer), 75

Feldman, P.D., Weaver, H.A., A'Hearn, M.F., Festou, M.C., McPhate, J.B., & Tozzi, G.-P. 1999, *BAAS* 31, 1127

Feldman, P.D., Cochran, A.L., & Combi, M.R. 2004, in *Comets II* (Univ. Arizona Press), 425

Festou, M.C., Keller, H.U., & Weaver, H.A. (eds.) 2004, *Comets II* (Univ. Arizona Press)

Fomenkova, M.N. 1999, *Space Science Reviews* 90, 109

Fray, N., Bénilan, Y., Cottin, H., Gazeau, M.-C., Minard, R.D., & Raulin, F. 2004, *Meteoritics and Planetary Science* 39, 581

Fray, N., Bénilan, Y., Cottin, H., Gazeau, M.-C., & Crovisier, J. 2005, *Plan. Sp. Sci.* 53, 1243

Hanner, M.S. & Bradley, J.P. 2004, in *Comets II* (Univ. Arizona Press), 555

Henry, F. 2003, Ph.D. Thesis, Paris

Henry, F., Bockelée-Morvan, D., Crovisier, J., & Wink, J. 2002, *Earth Moon and Planets* 90, 57

Hirota, T., Yamamoto, S., Kawaguchi, K., Sakamoto, A., & Ukita, N. 1999, *Ap. J.* 520, 895

Hollis, J.M., Lovas, F.J., Jewell, P.R., & Coudert, L.H. 2002, *Ap. J.* 571, L59

Huebner, W.F., Keady, J.J., & Lyon, S.P. 1992, *Astr. Sp. Sci.* 195, 291

Iro, N., Gautier, D., Hersant, F., Bockelée-Morvan, D., & Lunine, J.I. 2003, *Icarus* 161, 511

Irvine, W.M., *et al.* 1998, *Nature* 393, 547

Irvine, W.M., Senay, M., Lovell, A.J., Matthews, H.E., McGonagle, D., & Meier, R. 2000, *Icarus* 143, 412

Irvine, W.M., Schloerb, F.P., Crovisier, J., Fegley, B., & Mumma, M.J. 2000, in *Protostars and Planets IV* (Univ. Arizona Press), 1159

Irvine, W.M., Bergman, P., Lowe, T.B., Matthews, H., McGonagle, D., Nummelin, A., & Owen, T. 2003, *Origins of Life and Evolution of the Biosphere* 33, 609

Jackson, W.M., Halpern, J.B., Feldman, P.D., & Rahe, J. 1982, *A&A* 107, 385

Jackson, W.M., Scodinu, A., Xu, D., & Cochran, A.L. 2004, *Ap. J.* 607, L139

Jewitt D.C., Matthews, H.E., Owen, T., & Meier, R. 1997, *Science* 278, 90

Kawakita, H., Watanabe, J.-I., Furusho, R., Fuse, T., Capria, M.T., De Sanctis, M.C., & Cremonese, G. 2004, *Ap. J.*, 601, 1152

Keller, H.U., Jorda, L., Kuppers, M., *et al.* 2005, *Science* 310, 281

Kress, M., Tielens, A., & Frenklach, M. 2004, *35th COSPAR Scientific Assembly*, 3692

Lecacheux, A., *et al.* 2003, *A&A* 402, L55

Lis, D.C., *et al.* 1997, *Icarus* 130, 355

Lisse, C.M., van Cleve, J., Fernandez, Y.R., & Meech, K.J. 2005, *IAU Circ.* 8571, 2

Lovell, A.J. 1999, Ph.D. Thesis,

Magee-Sauer, K., Dello Russo, N., DiSanti, M.A., Gibb, E., & Mumma, M.J. 2002, *BAAS* 34, 868

Malfait, K., Waelkens, C., Waters, L.B.F.M., Vandenbussche, B., Huygen, E., & de Graauw, M.S. 1998, *A&A* 332, L25

Manfroid, J., Jehin, E., Hutsemékers, D., Cochran, A., Zucconi, J.-M., Arpigny, C., Schulz, R., & Stüwe, J.A. 2005, *A&A* 432, L5

Meech, K.J., Ageorges, N., A'Hearn, M.F., *et al.* 2005, *Science* 310, 265

Moreels, G., Clairemidi, J., Hermine, P., Brechignac, P., & Rousselot, P. 1994, *A&A* 282, 643

Mousis, O., Gautier, D., Bockelée-Morvan, D., Robert, F., Dubrulle, B., & Drouart, A. 2000, *Icarus* 148, 513

Rodgers, S.D. & Charnley, S.B. 2001, *MNRAS* 323, 84

Rolfe, E.J. & Battrick, B. 1987, *Diversity and Similarity of Comets* (ESA SP-278)

Smith, A.M., Stecher, T.P., & Casswell, L. 1980, *Ap. J.* 242, 402

Veal, J.M., *et al.* 2000, *A. J.* 119, 1498

Ziurys, L.M., Savage, C., Brewster, M.A., Apponi, A.J., Pesch, T.C., & Wyckoff, S. 1999, *Ap. J.* 527, L67

Astrochemistry: Recent Successes and Current Challenges
Proceedings IAU Symposium No. 231, 2005
D.C. Lis, G.A. Blake & E. Herbst, eds.

© 2006 International Astronomical Union
doi:10.1017/S1743921306007496

Organic Chemistry in Meteorites, Comets, and the Interstellar Medium

Oliver Botta†

International Space Science Institute, 3012 Bern, Switzerland
email: obotta@pop600.gsfc.nasa.gov

Abstract. With the notable exception of those originating on the Moon and Mars, all known meteorites are pieces of objects in the asteroid belt. As such, they have recorded a succession of chemical processes, starting from reactions in the interstellar medium (ISM), followed by reactions that accompanied the formation and evolution of the early solar system, and culminated with reactions during aqueous alteration in the meteorite parent bodies. One of the challenges in meteorite research is to decipher this record and to learn about interstellar formation processes as well as to conditions in the early solar system. The rare class of carbonaceous chondrites contains up to 5% by weight of organic carbon, most of which is locked in an insoluble macromolecular material and only about 20% of it is in the form of distinct organic compound classes. The molecular and isotopic data of these organic compounds clearly show an interstellar heritage, but a fraction of these precursors were later modified. For example, the amino acids were probably formed inside the meteorite parent body during the aqueous alteration period from simple molecules such as HCN, NH_3 and carbonyl compounds. However, the CI type carbonaceous chondrites contain a significantly distinct amino acid composition, indicating that there may be other synthetic processes involved. Polycyclic aromatic hydrocarbons (PAHs) are probably the most abundant form of organic carbon in the gas phase in the ISM. PAHs are among the most abundant organic compounds in carbonaceous meteorites, and they have been shown to have a presolar origin. A fraction of these PAHs are present in an extractable form, while the rest is part of the insoluble macromolecular matter. Progress is being made in the understanding of the evolution of this material in relation to aqueous alteration and oxidation. Although the potential of cometary meteorites cannot be ruled out, no such macro-meteorite has been recognized in the meteorite collections. Therefore, the organic composition of comets has been inferred mostly from astronomical observations. Future in-situ investigation of comets with spacecraft such as Rosetta will deliver new data on their organic composition, in particular the non-volatile fraction. However, in order to understand the contributions of different formation processes in primitive solar system objects, the analysis of the organic composition of meteorites remains essential.

Keywords. astrobiology — astrochemistry — molecular data — molecular processes — methods: laboratory — comets: general — minor planets — asteroids — solar system: formation — stars: AGB and post-AGB

---

1. Introduction

Life on Earth is a consequence of increasing complexity of chemical pathways, which have lead to the formation of stable carbon compounds, themselves being the building blocks of biopolymers and cell components. This evolution started early on in the history of the Universe with the synthesis of the elements heavier than helium in the first generation of stars. During the course of several cycles of star formation and death, these

† Current address: NASA Goddard Space Flight Center, Code 699, Greenbelt, MD 20771, USA

heavy elements – including carbon, oxygen, nitrogen, sulfur and phosphorous – were redistributed into the interstellar medium (ISM) and provided a basis for the synthesis of complex organic molecules. Even heavier elements such as iron are necessary to form solid planets around stars, so the first few generations of stars are not expected to harbor any type of (terrestrial) planetary system.

2. Formation of Organic Compounds in Space

Organic molecules can be formed in different environments in space. In diffuse interstellar clouds, ions are produced via photoionization of small molecules by external UV radiation. Since the UV flux in these regions is rather high, photodissociation occurs in parallel to ionization, preventing the formation of larger molecules. Consequently, only small molecules such as HCN, C_3H_2, H_2CO or CS are present in abundances that are above the detection limit of the current infrared and mm-wave telescopes.

In cold molecular clouds, at temperatures of 10 to 40 K, most of the atoms and molecules are frozen out on grain surfaces. Since 99.9% of the atoms are hydrogen, the formation of molecular hydrogen on these grain surfaces is the most common process, the details of which are not yet completely understood. However, it has been shown that only surface chemistry can produce molecules such as methane and ammonia, and even methanol, by hydrogen addition reactions in timescales short enough to lead to abundances that are in accordance with the observations. The dust grains act as chemical catalysts for these reactions. On the other hand, more complex molecules form through a chemistry network dominated by gas-phase ion-molecule and neutral-neutral reactions (Herbst 2001), which is driven by galactic cosmic rays that can penetrate throughout the dark cloud. This chemistry results in a distinctive suite of many unsaturated carbon chain molecules, including cyanoplyynes, various cumulene carbenes and chain radicals (Ehrenfreund & Charnley 2000).

AGB and post-AGB stars are copious producers of carbonaceous dust particles. Interstellar PAHs, and those found in meteorites, probably originated in the cool circumstellar envelopes (CSEs) of carbon stars (Messanger *et al.* 1998), where their photostability allows them to survive passage out of the CSE into the ISM, ultimately to become incorporated into more complex presolar organic material (Peeters *et al.* 2003).

Finally, organic molecules could also be produced in the accretionary disk environment around a high- or low-mass star or inside the planetesimals that are being formed in a later stage of planetary system formation. The chemistry in some parts of the disk may be very similar to that in molecular hot cores, where the ices on grains evaporate and more complex molecules are formed in the gas phase driven by thermal heating and the UV radiation coming from the star. In addition, in the accretionary disk, these processes could be repeated several times as the molecules settle back down into the midplane and freeze out again. Aqueous alteration, driven by the radioactive decay of short-lived isotopes inside the meteorite parent bodies, has been proposed to be an early solar system phenomenon which transforms anhydrous minerals into hydrated ones, for example phyllosilicates. In this environment, organic compounds could be formed in aqueous solution from precursor molecules inherited from the interstellar medium.

3. Investigation Techniques

3.1. *Astronomical Searches for Molecules in the Interstellar Medium*

Searches for organic molecules in space rely on the cooperation of theoretical modeling, laboratory investigations (for example to obtain the exact infrared or radio frequencies of a certain organic molecule in the gas phase or in interstellar ice analogs), and obviously the astronomical observations. Deep searches have been carried out using single-dish telescopes such as the NRAO 12 m telescope at Kitt Peak, JCMT, CSO and the Green Bank Telescope (GBT). Recently, molecules such as propanal (Hollis *et al.* 2004), glycolaldehyde (Hollis *et al.* 2000) and ethyleneglycol (Hollis *et al.* 2002) have been detected in the interstellar medium. The detection of the simplest amino acid glycine, obtained with the NRAO 12 m telescope, was reported in 2003 (Kuan *et al.* 2003).

3.2. *Searches for Molecules in Comets*

Our knowledge of the composition of comets is predominantly based on the evaporation of volatile species and thermal emission from dust that occur when bright comets pass through their perihelion. Although no direct observations of molecules on the cometary surface can be obtained with telescopes, the source of the molecules observed in the coma can be inferred from their outgassing properties. It should be noted that although the existence of cometary meteorites cannot be ruled out, no such object has been recognized in the meteorite collections (Campins & Swindle 1998). Most of the species observed at optical and UV wavelengths are radicals, atoms, and ions that are not directly subliming from the nucleus, but are instead produced in the coma, usually by photolysis of parent molecules, but also by chemical reactions (Bockelée-Morvan *et al.* 2004). Spectral observations in the mid-IR, in particular by the Infrared Space Observatory (ISO) of comets Hale-Bopp and 103P/Hartley 2, as well as the radio ranges, with SWAS and Odin, have produced a rich database of cometary parent molecules. The first *in situ* investigation of a cometary nucleus with spacecraft was the pioneering encounter of several interplanetary probes with the active comet Halley in 1986. Only limited information on carbonaceous matter was obtained through mass-spectrometric measurements of Halley's coma. Almost 20 years later, the *Stardust* spacecraft, while collecting cometary dust particles, encountered comet Wild 2 on January 2, 2004, providing spectacular imagery and mass-spectrometric data Brownlee *et al.* (2004). Although the particles that will return with *Stardust* are only of very small mass (1000 particles of >15 μm in size), it may be possible to investigate their organic composition in the laboratory, providing a first direct insight into the refractory organic component of a cometary nucleus. More recently, to investigate the subsurface composition of a comet, the penetrator component of the *Deep Impact* spacecraft impacted on comet Tempel 1, while the fly-by component recorded the event. Observations of organic and carbonaceous components were carried out during the event in a space- and ground-based campaign. In order to obtain direct information on the refractory organic composition of cometary nuclei, either *in-situ* analysis of comet nucleus or a sample return mission are necessary. The European *Rosetta* mission, carrying a small lander that is equipped to carry out a pyrolysis/gas chromatography/mass spectrometry (py-GCMS) analysis of surface material, was launched in 2004 for an encounter with comet 67P/Churyumov-Gerasimenko in 2014.

3.3. *Investigations of Carbonaceous Chondrites*

Meteorites are pieces of asteroids, which are solar system objects most of which failed to grow large enough to ever experience 'planetary' processes like core formation, volcanism or plate tectonics. The chondrite meteorites have near-solar volatile composition and radiometric ages of 4.566 billion years, reflecting their formation during, or shortly after, the birth of the solar system. Within the chondrites, there exists a carbonaceous subclass that contains significant amounts of organic carbon, a characteristic that contributes to a generally dark appearance. The chemistry and mineralogy of carbonaceous chondrites suggests that any organic matter they contain should predate the formation of the Earth, therefore reflecting the product of prebiotic chemical evolution in our solar system and the molecular cloud from which it originated. Hence these objects, being the only "macroscopic" samples from beyond the Earth-Moon system, represent a window back in time providing valuable information about the physical and chemical conditions in the early solar system.

3.4. *The Insoluble Component*

The total carbon content of carbonaceous chondrites can be partitioned into soluble and insoluble components. The soluble fraction can be obtained by treating a crushed meteorite sample with a series of solvents of different polarity, which leads to complex mixtures of compounds in the individual extracts. This total soluble fraction of C1 and C2 chondrites was estimated to contain 30–40% of the total carbon, but this number is probably an upper limit due to the additional dissolution of inorganic salts in the polar solvents (Cronin93). The insoluble fraction of carbonaceous chondrites is composed of macromolecular matter that is commonly referred to as 'kerogen-like' material. (Kerogen is insoluble macromolecular organic matter, operationally defined as the organic residue left after acid demineralization of a rock.) A protocol using supercritical fluid extraction (SFE), which provides simplified compound mixtures, of meteoritic hydrous pyrolysates in combination with isotope ratio monitoring-GCMS in order to identify and measure the $\delta^{13}C$ values of individual fragments of macromolecular materials in several carbonaceous has been developed and applied to the Orgueil, Cold Bokkeveld and Murchison meteorites (Sephton *et al.* 1998; Sephton *et al.* 2000). The results indicate that the macromolecular materials in these meteorites are qualitatively similar, but also reveal significant quantitative differences that correlate with the aqueous alteration stage of the meteorite.

3.5. *The Soluble Fraction*

The techniques that are used to investigate the soluble organic fraction of carbonaceous chondrites were originally developed for the (unsuccessful) search for organic molecules in returned Apollo lunar samples. However, the fall of the Murchison meteorite in 1969 provided the opportunity to search for amino acids in a fresh fall, which led to the detection of the first extraterrestrial organic compound (Kvenvolden *et al.* 1970). In the meantime, more than 80 amino acids were identified in Murchison, which is a representative of the CM2 class. Similar amino acid compositions were also found in other CM2 carbonaceous chondrites such as Murray or LEW90500 (see Figure 1; Botta & Bada 2002). However, investigations of the amino acid composition of the CI carbonaceous chondrites Orgueil and Ivuna revealed a distinct pattern, indicating that the CIs originated from a parent body, or a family of parent bodies, that probably formed from different precursor material than did the CM2 parent bodies (Ehrenfreund *et al.* 2001).

Although the soluble component of carbonaceous chondrites is only a minor fraction, it contains a suite of organic compounds that has a strong structural link to the origin of life on Earth. Apart from the amino acids already mentioned, a wide variety of N-heterocyclic compounds were also found in meteorites, among them four purines and one pyrimidine (uracil) (Stoks & Schwartz 1979; Stoks & Schwartz 1981; Stoks & Schwartz 1982), compounds that are used in the replicating systems of terrestrial organisms. A persistent issue with these compounds is the lack of unambiguous evidence that they are indeed indigenous to the meteorites rather than terrestrial contamination. However, their structural diversity implies an abiotic origin and provides some means of distinguishing them from their biological counterparts. More recently, a variety of compounds that are related to sugars (sugar alcohols, sugar acids) and the simplest sugar molecule, dihydroxyacetone, were detected in Murchison (Cooper *et al.* 2001). For two compounds (glycerol

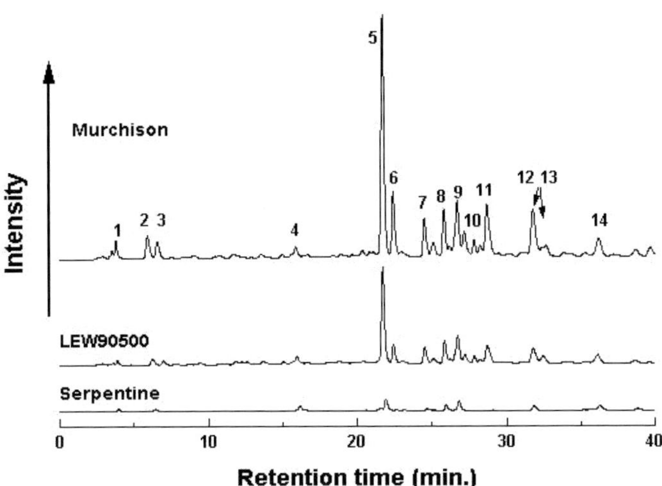

Figure 1. The 0 to 40 min region (no peaks were found outside this time period) of HPLC traces of 6 M HCl hydrolyzed hot water extracts of the CM carbonaceous chondrites LEW90500 and Murchison and a serpentine blank after pre-column derivatization with OPA/NAC. Signals: 1 DL-Asp, 2 L-Glu, 3 D-Glu, 4 DL-Ser, 5 Gly, 6 b-Ala, 7 g-ABA, 8 D-Ala, 9 L-Ala, 10 DL-b-ABA, 11 AIB, 12/13 Dl-a-ABA/DL-Isoval, 14 L-Val. Excitiation: 340 nm, Emission: 450 nm.

and glyceric acid), the abundances were determined to be on the same order of magnitude as that for the amino acid glycine. Yet the simplicity and diversity of meteoritic sugar-related compounds contrasts sharply with the high selectivity for sugars found in terrestrial organisms, where deoxyribose and ribose are present in DNA and RNA respectively, glucose is the principle energy source and the highly ordered polysaccharide cellulose is the most abundant organic molecule in the biosphere.

The most abundant class of soluble organic molecules to be found in carbonaceous meteorites are carboxylic acids, at an overall concentration of $\sim$300 ppm. Fullerenes were found in the Allende meteorite at a concentration of $>$100 ppb. Linear amides are the next most abundant class, with concentrations of $>$70 ppb, followed by amino acids, which are between 15-60 ppb for CM2 and about 5 ppb in CI1 carbonaceous chondrites. For an review on organic compounds found in meteorites see Botta & Bada (2002).

4. Formation of Amino Acids and PAHs is Space

4.1. *Amino Acids in Interstellar Ice Analogs*

Theoretical models on the formation of organic molecules predict the synthesis of compounds such as methanol inside interstellar ices. More complex molecules are formed in gas-phase reactions after these compounds desorb from the refractory core of the grains, either by collisions, by interactions with photons or cosmic rays, or by thermal desorption.

It has been reported that amino acids can form in interstellar ice analogs in the laboratory. Muñoz Caro *et al.* (2002) showed a chromatogram, obtained from a irradiated ice mixture containing $H_2O/CH_3OH/NH_3/CO/CO_2$, which was warmed up to room temperature after irradiation and the residual hydrolyzed in 6M HCl, with the identifications of a variety of amino acids, including di-amino acids. Although this mixture contained a number of amino acids that are not used in modern biochemistry, and can therefore not be contaminants of any kind, the mixture is dissimilar to amino acid compositions of any carbonaceous chondrite ever analyzed. In particular, some of the most abundant compounds, such α-aminoisobutyric acid (AIB) and isovaline, which have been identified in *non-acid hydrolyzed* meteorite extracts (Botta & Bada 2002), are not found in the hydrolyzed residue of the interstellar ice analog, indicating that a direct link between the synthetic pathways of these two sources is missing. Since no information about the presence of these amino acids *before* acid hydrolysis is given in the paper, there is no evidence that the amino acids were actually formed in the ice. However, it is more plausible that the variety of amino acids were formed during acid hydrolysis from precursors that were synthesized in the ice during irradiation. This is in agreement with experiments by Bernstein *et al.* (2002), who found in their irradiated ice mixture only amino acid precursor before acid hydolysis, which then hydrolyzed to the amino acids glycine, *rac*-alanine and *rac*-serine as well as other compounds.

4.2. *Synthesis of Amino Acids in Meteorite Parent Bodies*

The reaction of HCN, NH_3 and carbonyl compounds in the so-called Strecker-cyanohydrin synthesis leads to an equilibrium in aqueous solution between cyanohydrins and aminonitriles. These intermediates then undergo irreversible hydrolysis yielding a mixture of α-hydroxy- and α-amino acids, respectively. All other conditions being equal, the ratio of these two species in equilibrium depends on the concentration of ammonia in solution. Both α-hydroxy and α-amino acids are found in the Murchison meteorite, and in some instances, the molar ratios of structurally analogous hydroxy/amino acid pairs are consistent with equilibration of the nitrile precursor at a common ammonia concentration (Peltzer *et al.* 1984). Further support for the hypothesis that Strecker synthesis is the main pathway for the formation of α-amino acids in carbonaceous chondrites is the recent identification of iminodicarboxylicacids in Murchison, which are expected by-products of this reaction (Lerner & Cooper 2005).

One feature of Strecker synthesis is the fact that the complexity of the side chain of the resulting amino acid is inherited from the the substitution pattern of the carbonyl group. For example, if formaldehyde is the starting product, which only has two hydrogen atoms apart of the carbonyl group, the resulting amino acid glycine has the two hydrogen on the α-carbon. Analogously, acetaldehyde will form alanine and acetone will form AIB. Of course, the corresponding α-hydroxy acids will also be formed. Hypothetically, one could introduce some of the recently discovered interstellar aldehydes into a Strecker synthesis, since it can be expected that they would react the same way. Propanal, which is a three-carbon aldehyde, will form α-amino-n-butyric acid, on of the most prominent amino acids in carbonaceous chondrites, and α-hydroxy-n-butyric acid (see Figure 2). More interestingly, glycolaldehyde, assuming that it is present in the meteorite parent body in the first place, would react to form glyceric acid and serine. While the latter was also found in meteorites (how much of it is indigenous is not clear, however, due to incomplete separation of the enantiomers), the former is one of the sugar-related compounds found recently by Cooper *et al.* (2001) in Murchison. Therefore, perhaps Strecker synthesis may be a synthetic pathway to form not even amino acids, but also other astrobiologically relevant compounds in the meteorite parent bodies.

It should be noted here that the synthesis of a variety of amino acids such as β-alanine, β-amino-n-butyric acid and γ-amino-n-butyric acid cannot be explained by Strecker pathway. Alternatively, β-amino acids could be synthesized through Michael-addition of ammonia to substituted α,β-unsaturated nitriles followed by hydrolysis, a mechanism

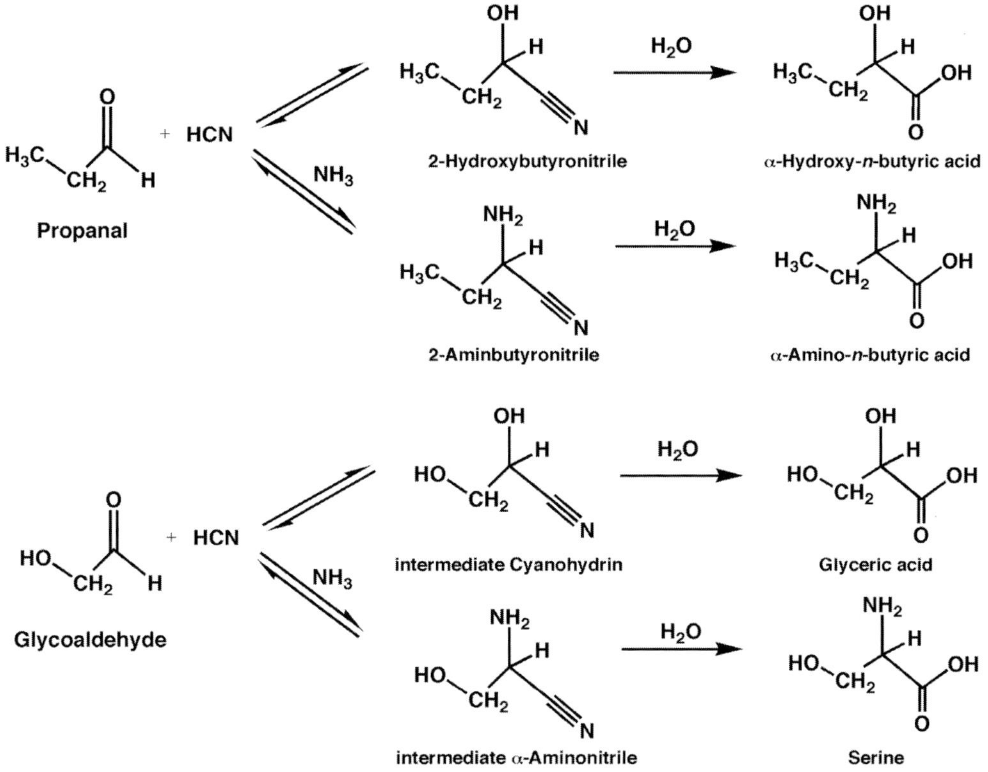

Figure 2. Theoretical synthetic reaction schemes for astrobiologically interesting compounds via Strecker-cyanohydrin pathway from aldehydes detected in the interstellar medium. *Top:* Formation of α-amino-*n*-butyric acid and the corresponding α-hydroxy acid from propanal. *Bottom:* Formation of glyceric acid and α-serine from glycolaldehyde

proposed for the synthesis of β-alanine in spark discharge experiments. The starting material for the synthesis of β-alanine for this process would be acrylonitrile, a molecule that has been detected in the ISM. Also, the recently detected di-amino acids in Murchison (Meierhenrich *et al.* 2004) are not simple products of Strecker chemistry, indicating that other synthetic pathways need to be explored to explain the diverse suite of amino acids identified in carbonaceous chondrites.

4.3. *Polycyclic Aromatic Hydrocarbons (PAHs)*

As already mentioned earlier, the major source for PAHs in space are outflows of of carbon-rich AGB, post-AGB and protoplanetary nebulae (PPNe). Due to the chemical and physical conditions in the photospheres and winds of these stars, which resemble those in sooting flames, it is speculated that the PAHs are intermediates towards the production of macromolecular assemblies of carbon and carbon dust particles (e.g., Cherchneff *et al.* 1992). Recent ISO observations have led to detection of benzene in the PPNe CRL 618 (Cernicharo *et al.* 2001), the discovery of the first aromatic compound outside the solar system. In carbonaceous chondrites, free PAHs have been investigated using two-step laser mass spectroscopy (Hahn *et al.* 1988). However, most of the aromatic compounds in meteorites are locked in the macromolecular material, which is difficult to access analytically. There are basically two ways to obtain information about the structure of this material. The first is the use of solid state ^{1}H and ^{13}C Nuclear Magnetic Resonance (NMR) spectroscopy, which allows the determination of functional group abundances in the sample. Gardinier *et al.* (2000) re-examined the Murchison and Orgueil macromolecular materials and determined the amount of aromatic carbon to be between 69 and 78% in Orgueil and 61 and 67% in Murchison. Subsequent studies by Cody and coworkers (e.g., Cody & Alexander 2005) have refined these numbers to be between 61 and 66% for both CI and CM-type meteorites, 48 to 52% for CR2, and 79 to 83% in the unusual Tagish-Lake meteorite. Generally, these results suggest that the primary differences observed in the macromolecular material of different meteorite classes most likely reflect processes that occurred on the meteorite parent body. The second analytical method to investigate the macromolecular material involves hydrous pyrolysis, where the sample remains in in contact with water during pyrolysis, followed by GCMS. Qualitatively, hydrous pyrolysates of different meteorites were found to be very similar, indicating that macromolecular materials of different classes of meteorites are made up of essentially the same units (Sephton *et al.* 1998). More recently, H_2-pyrolysis has been applied to investigate the composition of the meteorite macromolecular material (Sephton *et al.* 2004), releasing significant amounts of high molecular weight PAHs with varying degrees of alteration. The largest PAH identified by this method is the seven-ring PAH coronene ($C_{24}H_{12}$), which is about half the size expected for interstellar PAHs and components of the carbonaceous dust (Pendleton & Allamandola 2002). This result, and the deduction that even larger entities should be present in the residue, partly reconciles the apparent discrepancy between the meteoritic and interstellar aromatic components of the organic material.

5. Summary

A combination of interstellar ice processing at extremely low temperatures, circumstellar formation pathways and gas-phase chemistry results in the distribution of organic molecules observed in the interstellar medium. During formation of the solar system, some interstellar organic compounds were incorporated into growing planetesimals, but most of them were destroyed by shocks, radiative and thermal processes that occurred in the very early phases. Comets, being the most pristine, though not completely unmodified objects in the solar system, are a key to the understanding of the composition of the presolar nebula. However, astronomical observations are limited to organic volatiles. The detection of refractory organic molecules requires space missions that are capable of *in situ* analysis or sample return. The organic composition of carbonaceous chondrites is a result of a combination of interstellar synthetic processes and secondary processing of these precursors in the meteorite parent body. Therefore, in order to understand the contributions of different formation processes of organic compounds before, during and after the formation of the solar system, the analysis of the organic composition of meteorites, both in the soluble and the macromolecular fractions, remains an essential analytical tool in the future.

References

Allamandola, L.J., Tielens, A.G.G.M., & Barker, J.R. 1989, *Ap. J. Suppl.* 71, 733

Bernstein, M.P., *et al.* 2002, *Nature* 416, 401

Bockelée-Morvan D., Crovisier, J., Mumma, M., & Weaver, H.A. 2004, in *Comets II* (University of Arizona Press), p. 391

Botta, O. & Bada, J.L. 2002, *Lunar and Planetary Science XXXIII*, Abstract #1391, Lunar and Planetary Institute, Houston (CD-ROM)

Botta, O. & Bada, J.L. 2002, *Surv. Geophys.* 23, 411

Brownlee, D., *et al.* 2004, *Science* 304, 1764

Campins, H. & Swindle, T.D. 1998, *Meteorit. Planet. Sci.* 33, 1201

Cernicharo, J., *et al.* 2001, *Ap. J.* 546, L123

Cherchneff, I., Barker, J.R., & Tielens, A.G.G.M. 1992, *Ap. J.* 401, 269

Cody, G.D. & Alexander, C.M.O'D. 2005, *Geochim. Cosmochim. Acta* 69, 1085

Cooper, G., *et al.* 2001, *Nature* 414, 879

Cronin, J.R. & Chang, S. 1993, in *The Chemistry of Life's Origins* (Kluwer, Dordrecht), p. 209

Ehrenfreund, P. & Charnley, S.B. 2000, *ARAA* 38, 427

Ehrenfreund, P., *et al.* 2001, *Proc. Natl. Acad. Sci. USA* 98, 2138

Gardinier, A., *et al.* 2000, *Earth Planet. Sci. Lett.* 184, 9

Hahn, J.H., *et al.* 1988, *Science*, 239, 1523

Herbst, E. 2001, *Chem. Soc. Rev.* 30, 168

Hollis, J.M., *et al.* 2000, *Ap. J.* 540, L107

Hollis, J.M., *et al.* 2002, *Ap. J.* 571, L59

Hollis, J.M., *et al.* 2004, *Ap. J.* 610, L21

Kuan, Y.-J., *et al.* 2003, *Ap. J.* 593, 848

Kvenvolden, K., *et al.* 1970, *Nature* 228, 923

Lerner, N.R. & Cooper, G.W. 2005 *Geochim. Cosmochim. Acta* 69, 2901

Meierhenrich, U.J., *et al.* 2004 *Proc. Natl. Acad. Sci. USA* 101, 9182

Messenger, S., *et al.* 1998, *Ap. J.* 502, 284

Mumma, M.J., *et al.* 2003, *Adv. Space Res.* 31, 2563

Muñoz Caro, G.M., *et al.* 2002, *Nature* 416, 403

Peeters, Z., *et al.* 2003, *Ap. J.* 593, 129

Peltzer, E.T., *et al.* 1984, *Adv. Space Res.* 4, 69

Pendleton, Y.J. & Allamandola, L.J. 2002, *Ap. J. Suppl.* 138, 75

Sephton, M.A., Pillinger, C.T., & Gilmour I. 1998, *Geochim. Cosmochim. Acta* 62, 1821

Sephton, M.A., Pillinger, C.T., & Gilmour I. 2000, *Geochim. Cosmochim. Acta* 64, 321

Sephton, M.A., *et al.* 2004, *Geochim. Cosmochim. Acta* 68, 1385

Stoks, P.G. & Schwartz, A.W. 1979, *Nature*, 282, 709

Stoks, P.G. & Schwartz, A.W. 1981, *Geochim. Cosmochim. Acta*, 45, 563

Stoks, P.G. & Schwartz, A.W. 1982, *Geochim. Cosmochim. Acta*, 46, 309

Discussion

IRVINE: Can we constrain the length of the acqueous phase in carbonaceous chondrites?

BOTTA: Strontium and Manganese/Chromium isotopic dating established aqueous alteration as a very early process on the meteorite parent body, occurring only 10 to 20 Myr after the time of formation of the oldest known solar nebula condensates (Endress *et al.* 1996; *Nature* 379, 701).

ELLINGER: How can we be sure of the pH inside meteorites?

BOTTA: The pH in meteorite parent bodies has been estimated from the composition of the fluid (DuFresne & Anders 1962; *Geochim. Cosmochim. Acta* 26, 1085). The value is estimated to be in the range $6 < \mathrm{pH} < 8$.

Photo: E. Herbst

Astrochemistry: Recent Successes and Current Challenges
Proceedings IAU Symposium No. 231, 2005
D.C. Lis, G.A. Blake & E. Herbst, eds.

© 2006 International Astronomical Union
doi:10.1017/S1743921306007502

SWAS Observations of Comet 9P/Tempel 1 and Deep Impact

Frank Bensch[1,2], Gary J. Melnick[2], David A. Neufeld[3], Martin Harwit[4], Ronald L. Snell[5], and Brian M. Patten[2]

[1]Radioastronomisches Institut, Universität Bonn, Auf dem Hügel 71, 53121 Bonn, Germany

[2]Harvard-Smithsonian Center for Astrophysics, 60 Garden Street, Cambridge, MA 02138, USA

[3]Department of Physics and Astronomy, Johns Hopkins University, Baltimore, MD, USA

[4]511 H Street, SW, Washington, DC, USA; also Cornell University

[5]Department of Astronomy, University of Massachusetts, Amherst, MA 01003, USA

Abstract. On 4 July 2005 at 5:52 UT the Deep Impact mission successfully completed its goal to hit the nucleus of 9P/Tempel 1 with an impactor, forming a crater on the nucleus and ejecting material into the coma of the comet (A'Hearn *et al.* 2005). The 370 kg impactor collided with the sunlit side of the nucleus with a relative velocity of 10.2 $km\,s^{-1}$. NASA's Submillimeter Wave Astronomy Satellite (SWAS) observed the $1_{10} - 1_{01}$ ortho-water ground-state rotational transition in comet 9P/Tempel 1 before, during, and after the impact. No excess emission from the impact was detected by SWAS. However, the water production rate of the comet showed large natural variations of more than a factor of three during the weeks before the impact.

Keywords. comets: individual (9P/Tempel 1) — radiative transfer — radio lines — solar system

---

SWAS is a complete space-borne radio observatory (Melnick *et al.* 2000), capable of observing the 556.9 GHz transition of ortho $H_2^{16}O$, among other species, with a velocity resolution of $1\,km\,s^{-1}$ and a beam size of $3.3 \times 4.5'$. SWAS began near-daily monitoring observations of comet 9P/Tempel 1 on 5 June 2005, and the observations were continued post-impact until 1 September 2005. The total water production rate Q of the comet is determined from the velocity-integrated intensity of the emission detected in the SWAS spectra (typically being 2- to 3-day co-adds) and employing the radiative transfer model for water line emission in comets by Bensch & Bergin (2004). The left panel of Figure 1 shows the SWAS-measured water production rate for the observations made from 5 June through 9 July. The cometary activity traced by the water evaporation rate varies by more than a factor of three, from 3.8×10^{27} to 12.9×10^{27} molecules per second (115 to 385 $kg\,s^{-1}$), respectively.

No statistically significant increase in the water line emission was detected by SWAS immediately following the impact. The average velocity-integrated intensity during the three days after the impact was $0.16 \pm 0.04\,K\,km\,s^{-1}$, virtually identical to the average intensity measured for the three days before impact (Fig. 1, right panel). This corresponds to a total water production rate of $(6.6 \pm 1.5) \times 10^{27}\,s^{-1}$.

In order to derive an upper limit on the water released by the impact we extended the radiation transfer model to include comets where the water production rate is time-dependent. The total water production rate is replaced by $Q = Q_q + Q_b(t)$, where Q_q is a constant (quiescent) component. $Q_b(t)$ is the time-variable component, assumed to be a box-car function for the present simulations. In this case the water production rate is elevated (but constant) for the duration τ of the outburst and returns to the pre-outburst level afterwards. The total number of water molecules released by the outburst is

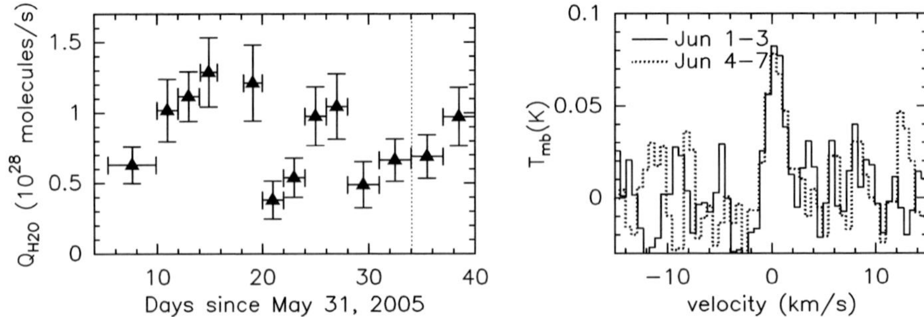

Figure 1. *Left:* Water production rate of comet 9P/Tempel 1, derived from SWAS observations of the $1_{10} - 1_{01}$ transition of o-$H_2^{16}O$. The vertical line gives the impact time. *Right:* SWAS spectrum before and after the impact. Each spectrum is a co-average of 3 days.

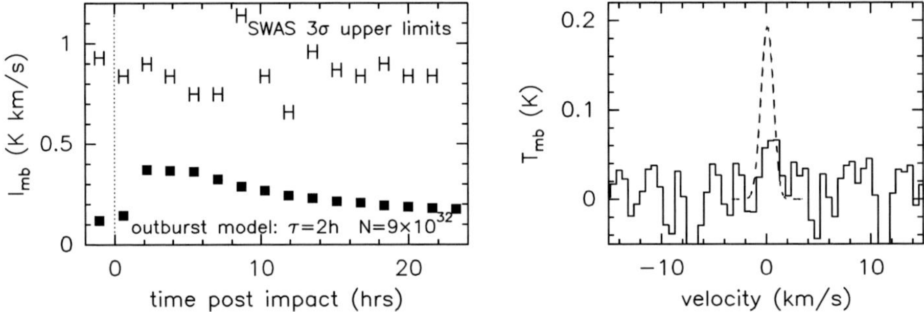

Figure 2. *Left:* The velocity-integrated intensity calculated by an outburst model with $\tau = 2\,\mathrm{hrs}$ and $N = 9 \times 10^{32}$ is compared to the 3σ upper limit derived from the SWAS spectra observed during each orbital segment of 35 min duration (error bars). No emission is detected in individual segments, consistent with the low Q_{H_2O} measured before impact. *Right:* the weighted average of line profile for this outburst model (dashed line) is compared to the SWAS-measured spectrum. The individual spectra are weighted by the square of integrated intensity predicted by the model for each orbital segment and the inverse square of the rms noise in the spectrum.

$N = Q_b \times \tau$. An example for the time evolution of the velocity-integrated intensity derived with this outburst model is shown in Figure 2. Preliminary model results for the SWAS observations up to 24 hrs after the impact give a 3σ upper limit of $\sim 9 \times 10^{32}$ molecules (2.7×10^4 tons) for the vaporized water. This result is not very sensitive to the assumed duration of the outburst, as a similar upper limit is obtained for models with $\tau = 0.5$, 2, 8 and 16 hrs. In addition, a doubling of the water production rate can be excluded at a 3σ confidence level if we assume that the impact created a new, permanently active area on the comet nucleus. Further results from the Deep Impact event are highlighted by D. Despois (this volume).

Acknowledgements

Support of the SWAS mission is provided by NASA through SWAS contract NAS5-30702. Additional support for the SWAS 9P/Tempel 1 observing campaign was obtained by the Deep Impact team. The help and cooperation of NASA HQ is acknowledged.

References

A'Hearn, M.F., *et al.* 2005, *Science* 310, 258
Bensch, F. & Bergin, E.A. 2004, *Ap. J.* 615, 531
Melnick, G.J. 2000, *Ap. J.* 539, L77

Astrochemistry: Recent Successes and Current Challenges
Proceedings IAU Symposium No. 231, 2005
D.C. Lis, G.A. Blake & E. Herbst, eds.

© 2006 International Astronomical Union
doi:10.1017/S1743921306007514

Exoplanet Atmospheres and Photochemistry

S. Seager[1], M.-C. Liang[2], C. D. Parkinson[2], and Y. L. Yung[2]

[1]Department of Terrestrial Magnetism, Carnegie Institution of Washington,
5241 Broad Branch Rd. NW, Washington, DC 20015, USA

[2]Division of Geological and Planetary Sciences, California Institute of Technology,
1201 East California Boulevard, Pasadena, CA 91125, USA

Abstract. Over 150 extrasolar planets are known to orbit sun-like stars. A growing number of them (9 to date) are transiting "hot Jupiters" whose physical characteristics can be measured. Atmospheres of two of these planets have already been detected. We summarize the atmosphere detections and useful upper limits, focusing on the MOST albedo upper limit and H exosphere detection for HD 209458b as the most relevant for photochemical models. We describe our photochemical model for hot Jupiters and present a summary explanation of the main results: a low gas-phase abundance of hydrocarbons; an absence of hydrocarbon hazes; and a large reservoir of H atoms in the upper atmospheres of hot Jupiters. We conclude by relating these model results to the relevant observational data.

Keywords. planetary systems — radiative transfer

---

1. Introduction

Over 150 extrasolar planets are now known to orbit main-sequence stars. Most of the extrasolar planets have been discovered by the Doppler technique (Marcy *et al.* 2005; Santos *et al.* 2005), which measures the star's motion about the planet-star common center of mass. The resulting minimum masses and orbital parameters are guiding planet formation and migration models. The planet transit survey technique has recently succeeded in discovering exoplanets. Although planets are known to orbit pulsars and direct imaging of young stellar systems has revealed bright, low-mass objects 100s of AU from their star, here we focus on solar-system-aged planets around main-sequence stars.

Now that the extrasolar planets' existence is firmly established, we want to learn more about their physical properties. Direct imaging of solar-system-aged planets is not yet possible, because the adjacent star outshines the planet by up to ten orders of magnitude. Fortunately, transiting planets provide us with many opportunities for observations with current technology. In particular, the transiting hot Jupiter (planets within 0.05 AU of their stars) planet atmospheres have been successfully detected. The hot Jupiter HD 209458b has been extensively observed because, until recently, it was the only one hosted by a bright star.

2. Transiting Planet Data

2.1. *Primary Transit*

When an exoplanet goes in front of its parent star, both its density and atmosphere can potentially be measured. The drop in brightness of a parent star during transit gives the planet-to-star area. If the parent star's radius is known (usually known to about 10% for main-sequence stars), so is the planet's. A mass measurement is needed to identify a transiting object as a planet, because low-mass stars, brown dwarfs (i.e., failed stars) and giant planets all have similar sizes due to the hydrogen equation of state. The mass and radius give planetary density.

During planet transit some of the starlight passes through part of the planet's atmosphere. The planet's "transmission spectrum", therefore, is hiding in the stellar spectrum during transit. Astronomers can compare the stellar spectrum before and during transit to get the planet's spectrum. Detecting the transmission spectrum is a very difficult measurement, because the planet's spectrum is on order 10,000 times fainter than the star's. The differential nature of the measurement and the on/off nature of the transit is what enables such a precise measurement. For the planet HD 209458b, the following useful observations or upper limits have been obtained:

- Na atmosphere detection (Charbonneau *et al.* 2002);
- H Lyα exosphere detection (Vidal-Madjar *et al.* 2003);
- CO atmosphere upper limit (Deming *et al.* 2005a).

The H Lyα detection is the transmission spectrum observation most relevant to photochemistry. A huge occulting area in Lyα was detected, a 15% drop in stellar brightness. This is a 10 times greater area than the planet's transit at visible wavelengths – implying an extended exosphere out to 3 or 4 Jupiter radii. Theoretical estimates for hydrogen escape from hot Jupiter atmospheres (Baraffe *et al.* 2004; Yelle 2004; Hubbar *et al.* 2005) vary over 4 orders of magnitude – translating from minimal to substantial planet mass loss over the planet's lifetime. Researchers do agree that an extremely high thermospheric temperature of 10,000 K is required for atmospheric escape to explain the Vidal-Madjar *et al.* (2003) observations.

2.2. *Secondary Eclipse*

A planet that passes in front of its star also goes behind its star. A differential measurement before and after the planet enters into "secondary eclipse" can be used to measure the thermal emission from the planet. The thermal emission measurement is easier than the transmission spectrum measurement, because the whole face of the planet contributes to the signal and not just the thin shell of the atmosphere. The detections or useful upper limits for HD 209458b and where noted for the planet TrES-1 are:

- Thermal emission at 24 microns (Deming *et al.* 2005);
- Thermal emission TrES-1 at 4.5 and 8 microns (Charbonneau *et al.* 2005);
- H$_2$O upper limit 2.2 microns (Richardson *et al.* 2003b);
- Albedo upper limit, 0.3–0.7 microns (Rowe *et al.* 2005).

The albedo upper limit is the secondary eclipse observation most relevant to photochemistry. This was measured with the Canadian Space Agency microsatellite called MOST (Microvariability and Oscillations of STars; Walker *et al.* 2003; Matthews *et al.* 2004). The upper limit on the geometric albedo is 0.25 (1σ) in MOST's single broad-band filter which spans 0.3–0.7 microns. With an albedo of 0.5 in the same bandpass (using data from Karkoschka 1994), Jupiter is much more reflective than HD 209458b.

3. Model Atmospheres and Chemical Equilibrium

Model atmospheres used for extrasolar giant planets are typically 1D, plane-parallel, LTE (local thermodynamic equilibrium), hydrostatic equilibrium and chemical equilibrium computer models (see Marley *et al.* 2005; Seager *et al.* 2005 and references therein). Hot Jupiters may range in equilibrium temperature from 900–1800 K depending on albedo. The major uncertainty for the model is clouds: their composition, particle size distribution, fraction of gas condensed and cloud vertical extent. The clouds would be composed of high-temperature condensates, including potentially silicates and iron. The second most important uncertainty is heat redistribution by atmospheric circulation. The hot Jupiters are almost certainly "tidally-locked", presenting the same face to the star at all times (just as our moon does to Earth). In an orbit almost 10 times closer to their stars as Mercury is to our sun, these planets are intensely heated. Strong winds, even approaching the sound speed, may or may not efficiently redistribute the energy (e.g., Showman & Guillot 2002; Cho *et al.* 2003; Cooper & Showman 2005). Other model uncertainties include opacities, internal luminosities, and non-equilibrium chemistry.

Considering chemical equilibrium we can specify the major chemical species in a hot Jupiter atmosphere. Figure 1 shows some of the abundant chemical species (expressed as partial pressure) for solar abundances present at 0.1 bar for different temperatures. This pressure/temperature regime can be thought of as representative of the atmospheric location where the spectral lines form. Although most hydrogen is in the form of H_2, H_2O is abundant. H_2O is a strong absorber and its absorption features are expected to sculpt the near-IR spectrum of hot Jupiters (Figure 2). For solar abundances CO is the dominant form of carbon at high temperatures, whereas CH_4 dominates at lower temperatures. This makes the potential detection of CO or CH_4 a useful temperature discriminant. The hot Jupiter atmospheres also "live" in the regime of high-temperature condensates. Two examples are shown in Figure 1: Fe vapor converting into solid Fe and Ti becoming locked into Ti condensates at low temperature. The latter is significant, since TiO is the most important absorber for low-mass (i.e., M) stars at visible wavelengths, and hence likely will not be present in the cooler hot Jupiter atmospheres. The alkali metals are expected to be in their atomic form in hot Jupiter atmospheres, due to their low ionization potentials. This is in contrast to both stars, where most of a given alkali metal is ionized, and to Jupiter where alkali metals are locked into condensates deep in the atmosphere. See the excellent review by Lodders (2005) for details on chemistry in brown dwarf atmospheres – this also applies to hot Jupiter atmospheres which are similar in temperature to brown dwarf atmospheres.

Fortney *et al.* (2005) and Seager *et al.* (2005) have made forays into hot Jupiter atmosphere models with non-solar abundances, in particular exploring the C/O ratio. With a higher C ratio, CO becomes more abundant at the expense of H_2O. CH_4 also becomes more abundant than in the solar abundance case.

Given the chemical equilibrium profile, we can understand the "big picture" spectrum of a hot Jupiter. Figure 2 shows a theoretical spectrum for a typical hot Jupiter (in this case the 51 Peg system). The stellar spectrum is computed from Kurucz model atmospheres (Kurucz 1993). In this theoretical example, the planet's visible wavelength spectrum is dominated by scattered light, showing the same features as in the star. At visible wavelengths the alkali metals Na and K are strong absorbers, but in the IR strong H_2O absorption features and some CH_4 bands shape the spectrum. See Seager *et al.* (2000, 2005) for details.

4. Photochemistry

The spectra of solar system giant planets Jupiter and Saturn are affected by photochemistry. Solar UV radiation photodissociates CH_4, and the subsequent reactions with photolytic products of CH_4 synthesize higher hydrocarbons, in turn forming hydrocarbon hazes. The hazes make the UV- and blue-wavelength albedos lower by a factor of 2 to 3 and "wash out" spectral features. See Figure 3 in Karkoschka (1994) for full disk albedo spectra of Jupiter, Saturn, Uranus, and Neptune for the effects of photochemistry.

Hot Jupiters are 100 times closer to their star than Jupiter is to the Sun and therefore receive 10^4 times as much flux. This raises the question about whether hydrocarbon hazes

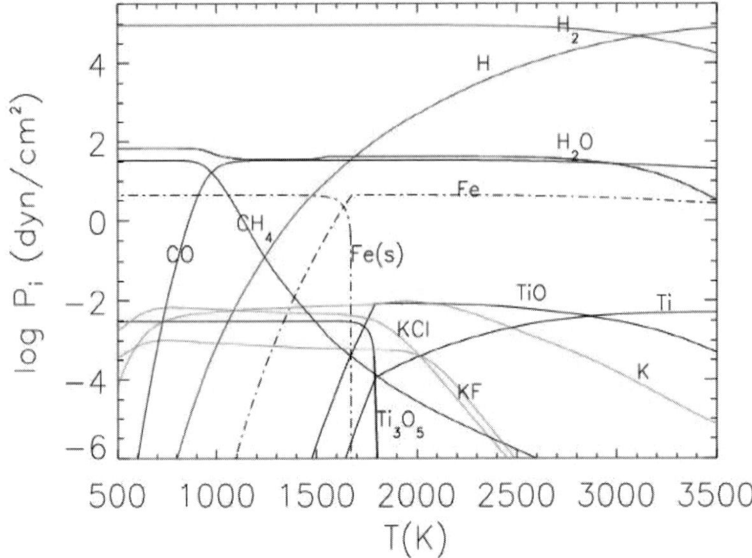

Figure 1. Partial pressures of some molecules and solids in a typical hot Jupiter atmosphere. Solar abundances are used. The total pressure is 10^5 dyne cm^{-2} (0.1 bar), representative of where spectral lines are formed. A Gibbs free energy minimization routine was used for this computation, described in Seager (1999).

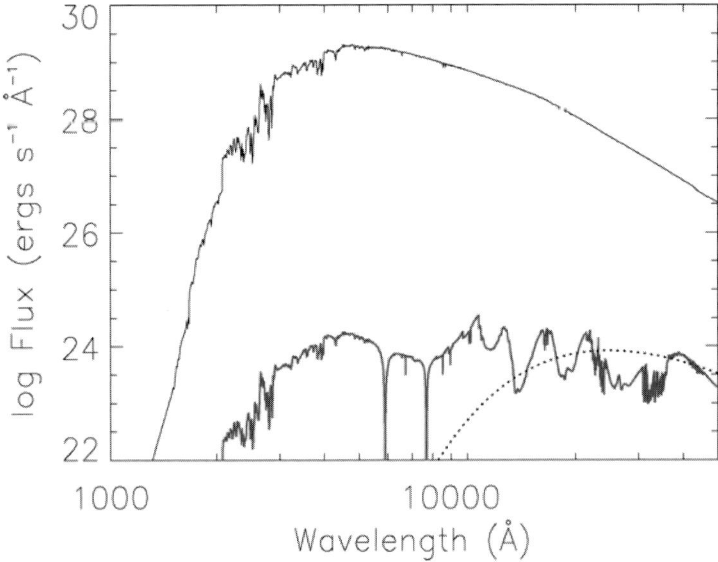

Figure 2. Generic spectrum of a hot Jupiter planet and its parent star. See text for details.

are highly abundant and in turn cause an even more extreme affect on the planetary visible-wavelength spectra.

We have modeled photochemistry on hot Jupiters using the 1D Caltech/JPL KINETICS model developed by Yuk Yung's group (Strobel 1973; Gladstone *et al.* 1996). The model reproduces hydrocarbon observations on Jupiter, as well as He 584 Å and H Lyα

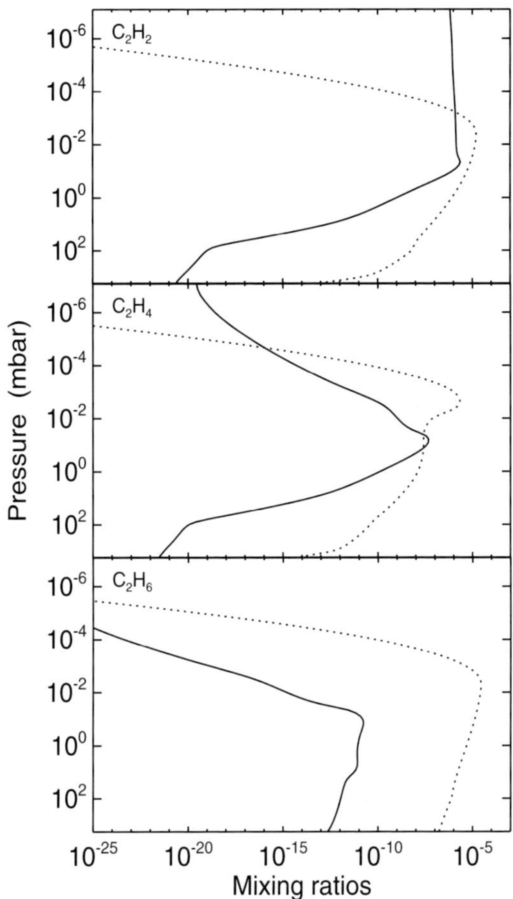

Figure 3. Mixing ratios of photochemical products in hot Jupiter atmospheres. The solid line is for a fiducial hot Jupiter atmosphere model (see Liang *et al.* 2004) and the dotted line is for Jupiter. Overall, the hot Jupiter model shows a low total gas-phase abundance of hydrocarbons. These three hydrocarbon compounds are basic ingredients for synthesizing complex hydrocarbons, such as benzene and PAHs, which are also expected to be of low abundance.

airglow on Jupiter. We used a reduced version involving four parent molecules: H_2, CO, H_2O, and CH_4, and 253 chemical reactions.

The hot Jupiter atmosphere composition and temperature/pressure profiles are unknown. To overcome these uncertainties, we used three different input atmospheres (i.e., temperature/pressure profiles) from Seager *et al.* (2000), Barman *et al.* (2002), Fortney *et al.* (2003). We also used five different initial values of the input species, H_2, CO, H_2O and CH_4, and considered a four orders of magnitude variation over the CH_4 abundance. The model and results are fully described in Liang *et al.* (2003, 2004). See those papers for a quantitative description of results.

The main results are as follows. No hydrocarbon hazes are expected on hot Jupiters and gas-phase hydrocarbons should have a very low abundance (Figure 3). These results are due to the hot Jupiters' proximity to their parent stars.

Primarily, the hot temperatures create fundamentally different chemical pathways than on Jupiter (Figure 4). Although the hydrocarbon formation rate is higher than on Jupiter,

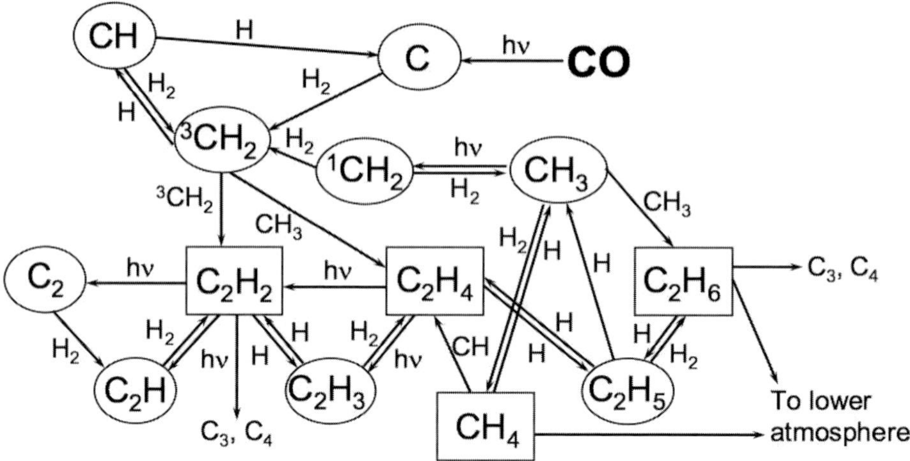

Figure 4. Photochemical pathways in hot Jupiter atmospheres.

so is the hydrocarbon destruction rate (i.e., the lifetime is decreased). This is because a key reaction in the destruction of C_2H_2 is fast at the high temperatures found on hot Jupiters, but a bottleneck at the low temperature found on Jupiter. The reaction is $(C_2H_3+H_2 \rightarrow C_2H_4 + H)$, and see Liang *et al.* (2003) for the full set of reactions. Because C_2H_2 is the main starting point for higher hydrocarbons, its low abundance leads to an overall low hydrocarbon abundance on hot Jupiters.

Secondly, CO and H_2O are expected to be abundant on hot Jupiters at the expense of CH_4, which is the dominant form of C on the colder solar system giant planets. Hydrocarbon formation on Jupiter is driven by the photodissociation of CH_4 and the subsequent reactions of the products – C_2H_2, C_2H_4, C_2H_6 – are important for forming more complex hydrocarbons and hydrocarbon aerosols. The high-UV radiation environment close to the star, together with abundant H_2O, facilitates hydrocarbon destruction. H_2O photodissociation in the upper atmosphere provides excess atomic H that partially enables the fast destruction of C_2H_2 by hydrogenation to CH_4 (with the second criteria being temperature, described above) and forms a large atomic reservoir in the upper atmosphere. This is in contrast to the equilibrium chemistry case where all H exists as H_2.

These results tie in to observations of HD 209458b described in § 2. First, the low albedo on HD 209458b is not caused by hydrocarbon hazes. Second, the extended hydrogen exosphere can form partially because of a huge H reservoir in the planet's upper atmosphere, caused by H_2O photodissociation.

5. Near Future Prospects

In the coming year, the Spitzer Space telescope will observe four transiting hot Jupiters to detect their thermal emission during secondary eclipse. One of these planets, HD 209458b, also has a growing dataset of observations and upper limits at different wavelengths and a good understanding of this planet's atmosphere should emerge. See Seager *et al.* (2005) for a discussion about future prospects. The number of transiting hot Jupiters around bright stars is growing and an increasing number will be characterized in terms of atmospheric composition, temperature and presence of an exosphere. On the computational side, photochemistry of S and N compounds needs to be further explored in their role as hazes on hot Jupiter atmospheres. The near future is bright for both observations and models of transiting hot Jupiters.

Acknowledgements

S.S. thank the conference organizers for an interesting meeting.

References

Baraffe, I., Selsis, F., Chabrier, G., Barman, T.S., Allard, F., Hauschildt, P.H., & Lammer, H. 2004, *A&A* 419, L13

Barman, T.S., Hauschildt, P.H., Schweitzer, A., Stancil, P.C., Baron, E., & Allard, F. 2002, *Ap. J.* 569, L51

Charbonneau, D., Brown, T.M., Noyes, R.W., & Gilliland, R.L. 2002, *Ap. J.* 568, 377

Charbonneau, D., *et al.* 2005, *Ap. J.* 626, 523

Cho, J.Y.-K., Menou, K., Hansen, B.M.S., & Seager, S. 2003, *Ap. J.* 587, L117

Cooper, C.S. & Showman, A.P. 2005, *Ap. J.* 629, L45

Deming, D., Brown, T.M., Charbonneau, D., Harrington, J., & Richardson, L.J. 2005, *Ap. J.* 622, 1149

Deming, D., Seager, S., Richardson, L.J., & Harrington, J. 2005, *Nature* 434, 740

Fortney, J.J., Marley, M.S., Lodders, K., Saumon, D., & Freedman, R. 2005, *Ap. J.* 627, L69

Fortney, J.J., Sudarsky, D., Hubeny, I., Cooper, C.S., Hubbard, W.B., Burrows, A., & Lunine, J.I. 2003, *Ap. J.* 589, 615

Gladstone, G.R., Allen, M., & Yung, Y.L. 1996, Icarus 119, 1

Hubbard . W.B., Hattori M.F., Burrows, A., Hubeny, I., & Sudarsky, D. *Icarus*, submitted

Karkoschka, E. 1994, *Icarus* 111, 174

Kurucz, R.L. 1993, VizieR Online Data Catalog, 6039, 0

Liang, M.-C., Seager, S., Parkinson, C.D., Lee, A.Y.-T., & Yung, Y.L. 2004, *Ap. J.* 605, L61

Liang, M.-C., Parkinson, C.D., Lee, A.Y.-T., Yung, Y.L., & Seager, S. 2003, *Ap. J.* 596, L247

Lodders, K. 2005, *Science*, in press

Marcy, G, Butler, R.P., Fischer, D., Vogt, S., Wright, J.T., Tinney, C.G., & Jones, H. 2005, *Prog. Theor. Phys Supp.*, in press

Marley, M.S., Fortney, J.J., Seager, S., & Barman, T. 2005, *PPV Conference Proceedings*, in press

Matthews, J.M., Kusching, R., Guenther, D.B., Walker, G.A.H., Moffat, A.F.J., Rucinski, S.M., Sasselov, D., & Weiss, W.W. 2004, *Nature* 430, 51

Rowe, J. *et al.* 2005, *Ap. J.*, submitted

Richardson, L.J., Deming, D., & Seager, S. 2003, *Ap. J.* 597, 581

Richardson, L.J., Deming, D., Wiedemann, G., Goukenleuque, C., Steyert, D., Harrington, J., & Esposito, L.W. 2003, *Ap. J.* 584, 1053

Santos, N.C., Benz, W., & Mayor, M. 2005, *Science* 310, 251

Seager, S., Richardson, L.J., Hansen, B.M.S., Menou, K., Cho, J.Y.-K., & Deming, D. 2005, *Ap. J.* 632, 1122

Seager, S., Whitney, B.A., & Sasselov, D.D. 2000, *Ap. J.* 540, 504

Seager, S. 1999, Ph.D. Thesis,

Shkolnik, E., Walker, G.A.H., Bohlender, D.A., Gu, P.-G., & Kuumlrster, M. 2005, *Ap. J.* 622, 1075

Showman, A.P. & Guillot, T. 2002, *A&A* 385, 166

Strobel, D.F. 1973, *Journal of Atmospheric Sciences* 30, 489

Vidal-Madjar, A., *et al.* 2004, *Ap. J.* 604, L69

Vidal-Madjar, A., Lecavelier des Etangs, A., Désert, J.-M., Ballester, G.E., Ferlet, R., Hébrard, G., & Mayor, M. 2003, *Nature* 422, 143

Walker, G., *et al.* 2003, *PASP* 115, 1023

Yelle, R.V. 2004, *Icarus* 170, 167

Discussion

BLACK: What can be said about magnetospheres and ionospheres of hot exoplanets?

SEAGER: We do not know if the hot exoplanets have magnetospheres. If they do, interaction between the stellar and planetary magnetic fields could show up observationally, because of their close proximity. Shkolnik *et al.* (2005) find suggestive evidence that two stars hosting hot Jupiters have chromospheric activity with the same period as the planet, suggesting a magnetic field interaction. No dramatic superflares from magnetic reconnection have been observed. The decametric radiation (as seen from Jupiter) due to synchrotron emission, is too faint to be detected with today's radio telescopes.

HERBST: Are your "equilibrium" chemical models truly thermodynamic ones? If so, they are probably used to too low temperatures.

SEAGER: The JANAF Thermochemical tables go down to 500K or much lower, so the calculations are appropriate.

VAN DISHOECK: What are the prospects for detecting the other H_2O photodissociation products, OH or O?

SEAGER: With the demise of HST/STIS we do not have any way of detecting narrow absorption features in the UV-visible wavelengths at this time. In addition, because the transmission spectrum measurement is so difficult, an atomic or molecular feature in an exoplanet atmosphere has to be extremely strong. However, a tentative detection of O in HD 209458b's exosphere has been reported by Vidal-Madjar *et al.* (2004).

STRELNITSKI: We have calculated the possibility of fluorescence and maser radiation in radio lines of molecules (H_2O in particular) from exoplanets. An important parameter for possible masering is the pressure of the gas. What is the typical pressure in the region where H_2O is abundant?

SEAGER: For solar abundance models we expect water vapor to be abundant throughout the planet atmosphere – this includes pressures ranging from 10 bars to $\sim 10^{-5}$ bars, and possibly even a greater range.

Astrochemistry: Recent Successes and Current Challenges
Proceedings IAU Symposium No. 231, 2005
D.C. Lis, G.A. Blake & E. Herbst, eds.

© 2006 International Astronomical Union
doi:10.1017/S1743921306007526

Molecular Abundances in AGB Circumstellar Envelopes

Hans Olofsson

Stockholm Observatory, AlbaNova, SE-10691 Stockholm, Sweden
email: hans@astro.su.se

Abstract. In this review the present status of molecules in circumstellar envelopes of AGB stars is presented. Emphasis is put on the determination of abundances, and estimates of their uncertainties, from an observational point of view. Despite an impressive number of circumstellar species detected, about 60, there remains much work before general conclusions can be drawn. In particular, sophisticated radiative transfer modelling of circumstellar line emission must be done. This requires a detailed knowledge of the stellar and circumstellar properties, as well as basic molecular physics/chemistry data.

Keywords. stars: AGB and post-AGB — stars: abundances — stars: mass loss — stars: circumstellar matter — radiative transfer

1. Introduction

The circumstellar envelopes (CSEs) of Asymptotic Giant Branch (AGB) stars, formed through extensive stellar mass loss, contain gas (primarily in the form of molecules) and dust. The molecules are formed through a sequence of processes. The high-density, high-temperature LTE chemistry in the stellar atmosphere produces the parent molecules. The result is strongly determined by the C/O ratio in the atmosphere. The mass-loss mechanism expels these molecules through the dynamical atmospheres of the pulsating AGB star. Presumably, this activates a shock chemistry that partly changes the chemical composition of the outflowing gas. The formation of bulk particles, dust grains, also takes place in the upper stellar atmosphere. This may decrease some molecular abundances because of grain adsorption, while new molecules are produced through grain surface chemistry. Eventually, the dispersion of the gas/dust CSE leads to the destruction of molecules by interstellar UV photons. In this way new molecules are formed as photodissociation products, and a photo-induced chemistry is initiated where these play a crucial role. Even if this sequence of processes may lead to some surprising results, the molecular setup of a CSE is primarily determined by the stellar C/O ratio (Millar 2003). The general elemental composition of the star shows up only in the finer details.

The ultimate goal is to understand the stellar/circumstellar chemistry on the AGB (and beyond), and from this derive important results relevant to stellar evolution as well as galactic chemical evolution. The relative simplicity of the AGB circumstellar environment lends the hope that various chemical processes can be studied in great detail. In addition, through the radial structure produced by the outflowing gas and dust, there is also the possibility of understanding 'time-resolved' chemistry.

2. The AGB Circumstellar Environment

The circumstellar envelope around an AGB star provides, at least to a first approximation, a rather well-defined environment in which physical and chemical processes can be studied. Overall, they are spherically symmetric, and expand with a constant velocity. A constant mass-loss rate (at least over the time scale probed by molecular emissions, $\leqslant 10^4\,\mathrm{yr}$) assures that the densities scale as r^{-2}. The properties of the radiation fields, due to the aging red giant and circumstellar dust, are relatively well characterized. A unique property of AGB CSEs is that they provide two distinctly different chemical environments, those that are O-rich ($C/O < 1$) and those that are C-rich ($C/O > 1$). Finally, regular luminosity variations of the central star and the expansion of the CSE allow temporal chemical evolution studies.

Basic to all circumstellar abundance estimates is a good circumstellar model, which is normally obtained in two ways. Radiative transfer modelling of CO radio line emission combined with a solution of the energy-balance equation provides the mass-loss rate (assuming a CO abundance), the terminal expansion velocity, and the kinetic temperature law (Groenewegen 1994; Schöier & Olofsson 2001). Radiative transfer modelling of the spectral energy distribution combined with a dust-driven-wind model provide an estimate of the mass-loss rate, the dust mass-loss rate, the velocity law, and the dust radiation field. In general, the mass-loss rate, which enters the abundance through the H_2 density, of an individual object has an uncertainty of at least a factor of three.

3. Circumstellar Molecules

There are presently 63 molecular species detected in AGB CSEs, Table 1, and a significant number of them are unique to the circumstellar medium (when compared to the interstellar medium). About 80% are detected at radio wavelengths, and the rest (except C_2) are detected at IR wavelengths. This is impressive considering that AGB CSEs are low-mass objects.

However, the situation is far from satisfactory. About half of the molecules have been detected in only one object, IRC+10216, and an additional 10 in less than 10 objects. Measured brightness distributions are essentially limited to a dozen species, in a single line each, observed towards IRC+10216 (with the exception of CO). The CSE of IRC+10216 has turned out to be a gold mine, since this object is probably the nearbest C-star ($\approx 120\,\mathrm{pc}$), and it happens to have a very high mass-loss rate ($\approx 2 \times 10^{-5}\,\mathrm{M_\odot\,yr^{-1}}$). If it is also representative of a C-rich CSE remains to be determined.

In the C-rich CSE of IRC+10216 a large number of relatively complicated molecules exists, mainly as the result of an efficient C-based circumstellar chemistry. Here we find sequences of long carbon chains: the cyanopolyynes HC_nN, the hydrocarbons C_nH (Cernicharo & Guélin 1996), and the carbenes H_2C_n (Cernicharo $et\ al.$ 1991). Nitrogen–carbon chains C_nN (Guélin $et\ al.$ 1998), silicon–carbon chains C_nSi (Ohishi $et\ al.$ 1989), and sulphur–carbon chains C_nS (Cernicharo $et\ al.$ 1987) are also present. Simple ring molecules (SiC_2, cyclic C_3H_2, and SiC_3) have been detected (Apponi $et\ al.$ 1999) as well, but branched molecules are notably rare. On the spectacular side, we have the 'metal' species NaCl, AlCl, AlF, KCl, MgNC, etc. (Cernicharo & Guélin 1987; Ziurys $et\ al.$ 2002). Notably, only one ion, HCO^+, has been detected. Most likely, even more complicated species exist in C-rich AGB CSEs. Their detections may come through sub-mm and far-IR observations of ro-vibrational lines of various bending and flopping modes.

The O-CSEs are apparently less rich in molecules, certainly an effect of the lack of the chemically potent carbon atoms. However, the situation may have been slightly different had an M-star equivalent in mass-loss rate and proximity to IRC+10216 existed. Here OH, H_2O, and SiO dominate (through their strong maser lines), but also S-bearing species, e.g., SO, SO_2, and H_2S, are frequently found (Omont $et\ al.$ 1993). Quite unexpectedly, detections of C-bearing species, HCN, CS, and CN, have been made (Olofsson

Table 1. Molecules detected in AGB CSEs. The (rough) number of sources detected in each species is given (Σ), as well as abundances w.r.t. H_2 (O: C/O<1; C: C/O>1; $k(l) = k \times 10^l$)

	Molecule	Σ	Chem. O	Chem. C	Molecule	Σ	Chem. O	Chem. C
2-atoms	AlCl	1		$1(-7)$	NaCl	1		$1(-9)$
	AlF	1		$4(-8)$	OH	2000	$2(-4)$	$4(-8)$
	C_2	1		$2(-6)$	PN	1		?
	CO	600	$5(-4)$	$1(-3)$	SiC	2		$4(-8)$
	CN	40	$2(-7)$	$5(-6)$	SiN	1		$2(-8)$
	CP	1		$2(-8)$	SiO	500	$5(-6)$	$1(-7)$
	CS	35	$1(-7)$	$1(-6)$	SiS	20	$7(-7)$	$2(-6)$
	KCl	1		$2(-9)$	SO	20	$2(-6)$	
3-atoms	AlNC	1		$1(-9)$	HNC	15	$1(-7)$	$1(-7)$
	C_3	1		$1(-6)$	MgCN	1		$1(-9)$
	C_2H	20		$4(-6)$	MgNC	1		$2(-8)$
	C_2S	5		$1(-6)$	NaCN	1		$2(-8)$
	CO_2	15	$3(-7)$		SiC_2	5		$3(-7)$
	HCN	120	$4(-6)$	$2(-5)$	SiCN	1		$4(-9)$
	H_2O	300	$3(-4)$	$1(-6)$	SiNC	1		$4(-9)$
	H_2S	20	$1(-5)$		SO_2	15	$2(-6)$	
4-atoms	ℓ-C_3H	2		$4(-8)$	HC_2N	1		$8(-9)$
	C_3N	5		$3(-7)$	H_2CO	1		$1(-8)$
	C_3S	1		$3(-8)$	NH_3	5	$4(-6)$	$1(-7)$
	C_2H_2	7		$5(-5)$	SiC_3	1		$3(-9)$
5-atoms	C_5	1		$1(-7)$	HC_3N	10		$1(-6)$
	C_4H	5		$3(-6)$	HC_2NC	1		$2(-9)$
	C_4Si	1		$3(-9)$	H_2C_3	1		$2(-9)$
	c-C_3H_2	5		$3(-8)$	SiH_4	1		$2(-7)$
	CH_4	1		$4(-6)$				
6-atoms	C_5H	1		$6(-8)$	CH_3CN	5		$3(-9)$
	C_5N	1		$9(-9)$	HC_4N	1		$1(-9)$
	C_2H_4	1		$1(-8)$	H_2C_4	1		$5(-9)$
$\geqslant$*7-atoms*	C_6H	1		$8(-8)$	HC_7N	2		$4(-8)$
	C_7H	1		$3(-9)$	HC_9N	1		$1(-8)$
	C_8H	1		$1(-8)$	H_2C_6	1		?
	HC_5N	5		$2(-7)$				
Ions:	HCO^+	2		$1(-9)$				

et al. 1991; Bujarrabal *et al.* 1994), showing that there exists a source of free carbon atoms that drive a carbon chemistry in these CSEs.

The increasing ultraviolet flux and the presence of shocks in post-AGB objects should have a noticeable effect on the chemistry (Woods *et al.* 2003a). This is verified by the detections of polyacetylenes in the C-rich proto-PNe AFGL618 and AFGL2688, and methylpolyynes and the benzene ring in AFGL618 (Cernicharo *et al.* 2001a; Cernicharo *et al.* 2001b). A number of ionic species have been detected in the CSE around a young PN, NGC7027 (Latter *et al.* 1993). In addition, abundant molecular hydrogen is readily detectable in PNe of the bipolar type (Cox *et al.* 2002). The molecular species detected only in post-AGB CSEs are CH, CH^+, CO^+, H_2, N_2H^+, OCS, HC_4H, HC_6H, CH_3C_2H, CH_3C_4H, and C_6H_6.

4. Abundance Estimate Methods

4.1. *Single Line*

Single-line estimates are commonly used for (simpler) species where few lines have been observed. This works only in the optically thin regime, and requires assumptions about a Boltzmann population distribution at an assumed temperature, as well as an assumption of the size of the emitting region (since this is normally not measured). This leads to order-of-magnitude estimates at best, but it is important to check that the assumption of optically thin emission is fulfilled. It should be noted that even for CO, the lower rotational transitions are very far from thermally excited at low mass-loss rates; in fact, they are inverted over substantial regions of the CSE.

4.2. *Rotational Temperature Diagram*

In particular, heavier molecules and asymmetric tops allow the measurement of line intensities from a large number of transitions (preferably sampling a large range in excitation energies). This can be used in a rotation-temperature diagram to estimate both the excitation temperature and the column density, and should, in principle, lead to an abundance with a (often significantly) higher accuracy than that obtained from a single line. However, the lines must be optically thin, and originate in the same region (its size is often assumed). The latter may not be true in an AGB CSE where the stellar radiation intensity and kinetic temperature gradients are rather steep. Yet, when comparing abundances for a single source this method is expected to provide fairly reliable results.

4.3. *Radiative Transfer Modelling*

Detailed radiative transfer modelling, including a chemical model for the species in question, is still rarely used. There are a number of reasons for this, e.g., the lack of reliable collisional cross sections (however, see Schöier *et al.* 2005), and a complicated radiative transfer where the details of the radiation field or the molecular energy level structure play an important role. Interestingly, a sophisticated model often leads to abundance estimates with rather large uncertainties.

5. Abundance Estimates

The majority of the abundance (with respect to H_2) estimates given in Table 1 are either based on single lines or using the rotation-temperature-diagram method on multi-line data. Only in very few cases have more sophisticated radiative transfer analyses been used. Therefore, the given abundances are in general order-of-magnitude estimates. In addition, for some species the abundance is an average obtained from estimates for a number of sources. We proceed to look at this in more detail.

5.1. *An Individual Source, IRC+10216*

The heavier (often linear) carbon species detected at radio wavelengths are particularly suitable for a rotation-temperature-diagram analysis. This has been used for the IRC+10216 CSE where broad band spectral scans provide the data (e.g., Cernicharo *et al.* 2000). As an example Figure 1 shows a comparison of a number of carbon-chain species abundances. Although the absolute scale is uncertain, the relative abundances are expected to be rather accurate. The immediate conclusion is that the abundance decrease with increasing 'complexity' is moderate. For the cyanopolyynes, $HC_{2n+1}N$, the abundance decreases by about a factor of four for each step in the sequence $n = 1$ to 4. The same applies to the related species $C_{2n}H$ when n goes from 1 to 4. Note also that the

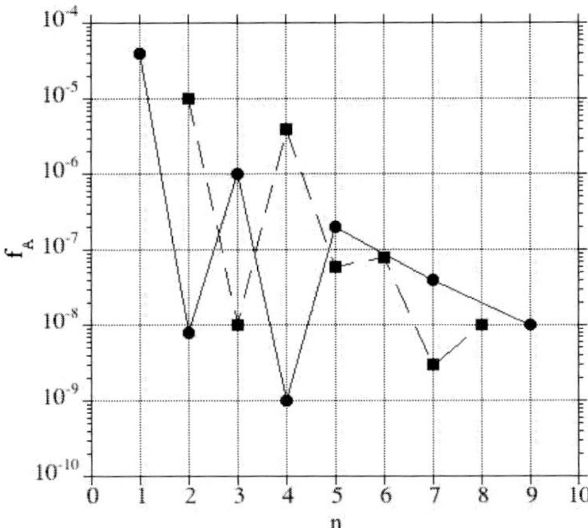

Figure 1. Abundances (with respect to H_2) of carbon-chain species, HC_nN (circles) and C_nH (squares), in the CSE of the C-star IRC+10216.

$HC_{2n}N$ and $C_{2n+1}H$ species are considerably less abundant. In addition, ratios involving isomers, e.g., $HC_2NC/HC_3N \approx 0.003$, and branched species, e.g., $CH_3CN/HCN \approx 0.0003$, provide further constraints on the chemical models.

Interesting results appear when comparisons with young post-AGB objects are done. Pardo *et al.* (2005) found that the HCN/HC_3N abundance ratio is at least an order of magnitude lower in the C-rich proto-PN AFGL618 than in IRC+10216, while the decrease in abundance beyond HC_3N is the same. This suggests a very efficient transformation of HCN into longer cyanopolyyne chains. Likewise, polyacetylenes are abundant in AFGL618 (C_4H_2 and C_6H_2 are only a factor of a few less abundant than C_2H_2), while they are not detected in IRC+10216 (Cernicharo *et al.* 2001b). The increasing UV flux and shocks produce efficient factories of organic molecules in young post-AGB objects.

Another interesting, recent development concerns O-bearing molecules in C-rich CSEs. Surprisingly, H_2O was discovered towards IRC+10216 (Melnick *et al.* 2001), and a destruction of icy Kuiper-belt-like objects was suggested as an explanation. This stimulated further searches leading to the successful detections of OH (Ford *et al.* 2003) and H_2CO (Ford *et al.* 2004), but a low upper limit to CH_3OH (Ford *et al.* 2004). Further observations are required before the chemical origin is firmly established.

It should be emphasized that whether these are generally applicable results we do not know, since data of these quality exist only towards IRC+10216.

5.2. *Comparison Between Sources*

Very few studies comparing molecular abundances in different sources have been made. Woods *et al.* (2003b) observed the same molecular species, 15 in total, in the CSEs of seven high-mass-loss-rate C-stars with similar characteristics (including IRC+10216). Using the single-line method (although crude rotation-temperature estimates are obtained for a few species), and theoretical estimates of the photodissociation radii, they concluded that in five sources (including IRC+10216) the abundances vary between the sources by no more than a factor of five (i.e., within the uncertainties). Two sources, CIT6 and, in particular, IRAS15194–5115, show a somewhat different chemical composition, but this remains to be confirmed by more sophisticated modelling.

A notable exception to the general agreement is SiO which shows a variation of more than an order of magnitude between the sources, and the abundance is, on average,

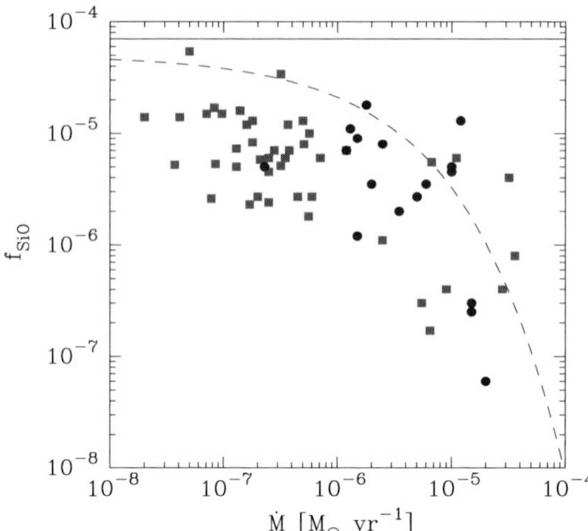

Figure 2. Circumstellar SiO abundances as a function of stellar mass-loss rate for sample of stars: M-stars (squares) and C-stars (circles). The solid line gives the maximum stellar photospheric abundance for solar abundances, and the dotted line the result of a 'simple' grain condensation model assuming an M-star photospheric equilibrium SiO abundance of 5×10^{-5}. The C-star photospheric equilibrium SiO abundance is about 5×10^{-8}.

more than $10\times$ higher than predicted from stellar atmosphere chemistry. This result is confirmed by radiative transfer modelling, see § 5.3.

5.3. *Radiative Transfer Modelling Results*

The excitation conditions in CSEs are, for most molecular species, far from LTE, and the effects of non-local radiative transfer may be substantial (not the least in the case of the stellar radiation field). Thus, collisional cross sections and the details of the radiation fields are often crucial.

CO is probably the species with the most accurate collisional cross sections, and, in addition, the sensitivity to the radiation fields is limited. However, the abundance of CO is often assumed and serves as an estimator of the H_2 density, i.e., the stellar mass-loss rate (Schöier & Olofsson 2001; Olofsson *et al.* 2002). This assumption is probably not so bad, since the stability of CO suggests that it is the C and O content that determines the CO abundance in O- and C-CSEs, respectively, but it leads to a basic uncertainty in other abundance estimates.

The first more detailed study of abundances in a larger sample of sources was performed by González Delgado *et al.* (2003), who examined SiO in about 40 M-stars. This is a more complicated problem than the radiative transfer modelling of CO: the dependence on radiative excitation is stronger, and the chemistry is expected to be more complex (due to grain formation effects as well as photodissociation). González Delgado *et al.* (2003) used multi-line data to derive the size of the emitting region, and obtained reasonable fits to both brightness sizes measured by interferometers and theoretical estimates from photodissociation models, as well as the SiO abundance. Schöier & Olofsson (2005) used the same method to estimate the SiO abundance in a sample of 17 C-rich CSEs. The SiO abundance estimates for both samples are given as a function of stellar mass-loss rate in Figure 2. There are some striking results. For the M-stars the circumstellar abundance is much lower than that expected from photospheric equilibrium chemistry (5×10^{-5}), on average by an order of magnitude, in many cases by two orders of magnitude. For the C-stars the circumstellar abundance is much higher than that expected from

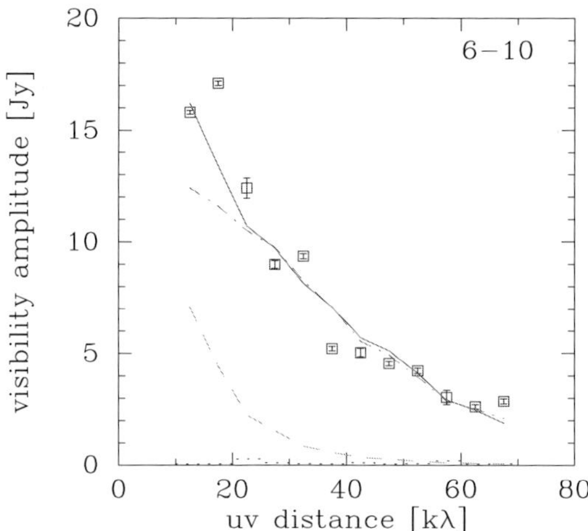

Figure 3. A radiative transfer model fit to the $SiO(v = 0, J = 2\text{-}1)$ uv data obtained towards the M-star R Dor. The solid line gives the best fit model, which is based on a two-component SiO abundance distribution (see text). The dashed-dot line gives the result for only an inner high-abundance component, and the dashed line the best-fit model using only single-dish data.

photospheric equilibrium chemistry (5×10^{-8}), on average by two orders of magnitude. In addition, there is a trend of decreasing circumstellar abundance with increasing mass-loss rate, probably due to a trend of increasing SiO depletion onto grains with increasing circumstellar density.

Recent interferometric results show that a more complicated SiO abundance distribution is required to explain the data (as is also expected due to grain formation and photodissociation). In the case of the semi-regular M-star R Dor Schöier *et al.* (2004) found that an inner component of high SiO abundance (at the level expected from stellar atmosphere chemistry), and an outer, extended component of low SiO abundance (as expected from SiO depletion onto grains), are required to fit the data, Figure 3.

The final example concerns circumstellar H_2O. Water has been known for decades through the detections of H_2O (and OH) masers. More data were provided by ISO, where the SWS and LWS instruments delivered line intensities, but no kinematical information. Recently, the SWAS and Odin satellites have provided measurements of the ortho-H_2O ground-state line at 557 GHz at high spectral resolution (Harwit & Bergin 2002; Justtanont *et al.* 2005). Modelling of circumstellar H_2O line emission is complicated by the very high optical depths, and the highly sub-thermal excitation in CSEs. The former leads to strong P Cygni profiles (when the continuum and line flux densities are comparable), which requires comparison with high-spectral-resolution data (e.g., the SWS lines of NML Cyg; Zubko *et al.* 2004). The latter leads to a dependence on collisional cross sections (where the ortho/para structures of both H_2O and H_2 produce an additional complication).

W Hya, a semiregular M-star, has been the target of a number of H_2O investigations with highly disparate results (Barlow *et al.* 1996; Neufeld *et al.* 1996; Zubko & Elitzur 2000). Recently, Justtanont *et al.* (2005) analysed Odin and ISO data. All available H_2O line intensities could be fitted within the estimated observational uncertainties (Figure 4), and the mass-loss rate and the H_2O abundance were both determined from the data (provided that the ISO SWS lines were included). The mass-loss rate is estimated to be about $2 \times 10^{-7}\,M_\odot\,yr^{-1}$, a result consistent with those obtained from CO radio line modelling and a fit to the SED combined with a dynamical model of a dust-driven

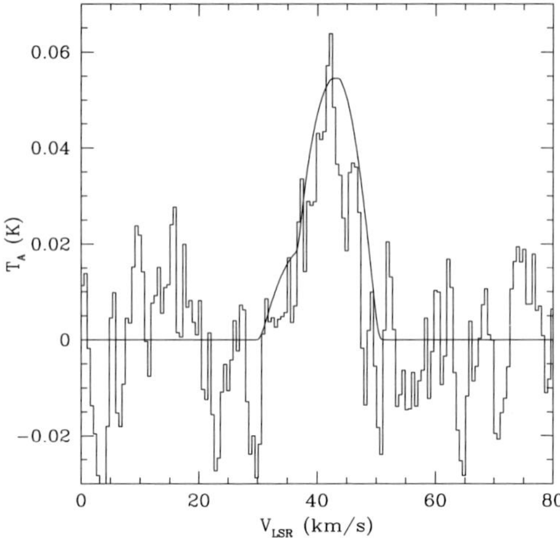

Figure 4. The $H_2O(1_{0,1}-1_{1,0})$ line obtained towards the M-star W Hya by the Odin satellite, and a radiative transfer model fit. The line is strongly affected by self absorption due to a high optical depth.

wind. The resulting circumstellar H_2O abundance (ortho + para) is 2×10^{-3}, i.e., about a factor of five higher than that expected from stellar atmosphere chemistry, and about a factor of two higher than allowed by solar abundances. This could reflect a problem with the radiative transfer modelling, e.g., optical depth effects, strong dependence on radiation fields (excitation to vibrational states was not included in the analysis), the use of erroneous collisional cross sections, or a too low mass-loss rate (although there are many indications that it is low). Nevertheless, an extra source of circumstellar H_2O (with an associated underabundance of H_2) remains an interesting possibility. Clearly, these results show that the interpretation of Herschel data on circumstellar H_2O will require careful modelling and adequate input data.

6. Summary

Our understanding of circumstellar molecules is far from satisfactory due to the limited observational basis (the low number of sources observed, the lack of imaging data, etc.), and the limited use of detailed radiative transfer modelling. The results of the latter depend crucially on the properties of the circumstellar medium (morphology, density structure, radiation fields, chemistry, etc.), and basic molecular physics data since highly non-LTE excitation is the rule. The first studies of this type on larger samples suggest that very interesting results can be obtained.

Nevertheless, the extensive studies of the C-rich CSE of IRC+10216 have led to a number of secure results that can be compared in detail with astrochemical modelling predictions. It should be emphasized that we do not know to what extent these results are applicable to C-rich CSEs in general, i.e., the concept that the IRC+10216 CSE is the prototype of C-rich CSEs still rests on rather loose grounds. Further, comparisons of the IRC+10216 results with the extensive studies of the young C-rich post-AGB object AFGL618 strongly suggest that the increased UV flux and the presence of shocks make these objects efficient factories of more complex organic molecules.

Finally, even though this review has focussed on abundance determinations based on radio line observations, we can expect considerable progress when IR spectroscopy data will become available for a larger number of sources. These data contains such a wealth of lines that very good constraints can be put on the modelling, and, in addition, provide observations of non-polar molecules.

References

Apponi, A.J., McCarthy, M.C., Gottlieb, C.A., & Thaddeus, P. 1999, *Ap. J.* 516, L103

Barlow, M.J., Nguyen-Q-Rieu, Truong-Bach, *et al.* 1996, *A&A* 315, L241

Bujarrabal, V., Fuente, A., & Omont, A. 1994, *A&A* 285, 247

Cernicharo, J., Gottlieb, C.A., Guélin, M., *et al.* 1991, *Ap. J.* 368, L43

Cernicharo, J. & Guélin, M. 1987, *A&A* 183, L10

—. 1996, *A&A* 309, L27

Cernicharo, J., Guélin, M., & Kahane, C. 2000, *A&AS* 142, 181

Cernicharo, J., Heras, A.M., Pardo, J.R., *et al.* 2001a, *Ap. J.* 546, L127

Cernicharo, J., Heras, A.M., Tielens, A.G.G.M., *et al.* 2001b, *Ap. J.* 546, L123

Cernicharo, J., Kahane, C., Guélin, M., & Hein, H. 1987, *A&A* 181, L9

Cox, P., Huggins, P.J., Maillard, J.-P., *et al.* 2002, *A&A* 384, 603

Ford, K.E.S., Neufeld, D.A., Goldsmith, P.F., & Melnick, G.J. 2003, *Ap. J.* 589, 430

Ford, K.E.S., Neufeld, D.A., Schilke, P., & Melnick, G.J. 2004, *Ap. J.* 614, 990

González Delgado, D., Olofsson, H., Schwarz, H.E., *et al.* 2003, *A&A* 411, 123

Groenewegen, M.A.T. 1994, *A&A* 290, 531

Guélin, M., Neininger, N., & Cernicharo, J. 1998, *A&A* 335, L1

Harwit, M. & Bergin, E.A. 2002, *Ap. J.* 565, L105

Justtanont, K., Bergman, P., Larsson, B., *et al.* 2005, *A&A* 439, 627

Kahane, C., Cernicharo, J., Gomez-Gonzalez, J., & Guelin, M. 1992, *A&A* 256, 235

Kahane, C., Dufour, E., Busso, M., *et al.* 2000, *A&A* 357, 669

Latter, W.B., Walker, C.K., & Maloney, P.R. 1993, *Ap. J.* 419, L97

Melnick, G.J., Neufeld, D.A., Ford, K.E.S., Hollenbach, D.J., & Ashby, M.L.N. 2001, *Nature* 412, 160

Millar, T.J. 2003, in H.J. Habing & H.Olofsson (eds), *Asymptotic giant branch stars*, Astronomy and astrophysics library (New York, Berlin: Springer), p. 247

Neufeld, D.A., Chen, W., Melnick, G.J., *et al.* 1996, *A&A*, 315, L237

Ohishi, M., Kaifu, N., Kawaguchi, K., *et al.* 1989, *Ap. J.* 345, L83

Olofsson, H. 2003, in H.J. Habing & H.Olofsson (eds), *Asymptotic giant branch stars*, Astronomy and astrophysics library (New York, Berlin: Springer), p. 325

Olofsson, H., González Delgado, D., Kerschbaum, F., & Schöier, F. 2002, *A&A* 391, 1053

Olofsson, H., Lindqvist, M., Winnberg, A., Nyman, L.-Å., & Nguyen-Q-Rieu. 1991, *A&A* 245, 611

Omont, A., Lucas, R., Morris, M., & Guilloteau, S. 1993, *A&A* 267, 490

Pardo, J.R., Cernicharo, J., & Goicoechea, J.R. 2005, *Ap. J.* 628, 275

Schöier, F.L. & Olofsson, H. 2000, *A&A* 359, 586

—. 2001, *A&A* 368, 969

—. 2005, *A&A*, submitted

Schöier, F.L., Olofsson, H., Wong, T., Lindqvist, M., & Kerschbaum, F. 2004, *A&A* 422, 651

Schöier, F.L., van der Tak, F.F.S., van Dishoeck, E.F., & Black, J.H. 2005, *A&A* 432, 369

Wannier, P.G., Andersson, B.-G., Olofsson, H., Ukita, N., & Young, K. 1991, *Ap. J.* 380, 593

Woods, P.M., Millar, T.J., Herbst, E., & Zijlstra, A.A. 2003a, *A&A* 402, 189

Woods, P.M., Schöier, F.L., Nyman, L.-Å., & Olofsson, H. 2003b, *A&A* 402, 617

Ziurys, L.M., Savage, C., Highberger, J.L., *et al.* 2002, *Ap. J.* 564, L45

Zubko, V. & Elitzur, M. 2000, *Ap. J.* 544, L137

Zubko, V., Li, D., Lim, T., Feuchtgruber, H., & Harwit, M. 2004, *Ap. J.* 610, 427

Photo: E. Herbst

Astrochemistry: Recent Successes and Current Challenges
Proceedings IAU Symposium No. 231, 2005
D.C. Lis, G.A. Blake & E. Herbst, eds.

© 2006 International Astronomical Union
doi:10.1017/S1743921306007538

Physical and Chemical Conditions in the Dust Formation Zone of IRC+10216

J. P. Fonfría Expósito[1], J. Cernicharo[1], M. J. Richter[2], and J. Lacy[3]

[1]Dept. Molecular and Infrared Astrophysics (DAMIR, IEM), Consejo Superior de Investigaciones Científicas (CSIC), C/ Serrano, 121, 28006, Madrid, Spain
email: jpablo.fonfria@damir.iem.csic.es, cerni@damir.iem.csic.es

[2]Physics Department, UC Davis, One Shield Ave., Davis, CA 95616, USA
email: richter@physics.ucdavis.edu

[3]Astronomy Department, University of Texas, Austin, TX 78712. USA
email: lacy@shrub.as.utexas.edu

Abstract. A mid-infrared high-resolution spectral survey of the source IRC+10216 (CW Leo) has been carried out between 11 and 14 μm. A large number of lines of C_2H_2 and HCN and their most abundant isotopologues, have been identified. Lines involving high-energy ro-vibrational levels allow an accurate derivation of the physical and chemical conditions in the innermost envelope. We have developed a radiative transfer model capable of fitting the observed lines satisfactorily. The fit of more than 200 ro-vibrational lines allowed us to get the kinetic, vibrational and rotational temperatures and the abundances of the C_2H_2 and HCN between 1 and 300 R_*.

Keywords. instrumentation: spectrographs — line: identification — line: profiles — radiative transfer — stars: AGB and post-AGB — stars: carbon — stars: mass loss — surveys — techniques: spectroscopic

1. Introduction

IRC+10216 is a carbon-rich AGB star surrounded by a circumstellar envelope (CSE) at $\simeq$180 pc from the Earth. The low stellar temperature, $\simeq$2300 K, and a high mass-loss rate of $\simeq 2 \times 10^{-5}$ $M_\odot$ yr^{-1} (Keady *et al.* 1988; Cernicharo *et al.* 1999), turn the CSE into an environment friendly to high molecular abundances. By now, 60 different molecular species have been detected, with CO the most abundant species with a fractional abundance of 8×10^{-4}, C_2H_2 with 8×10^{-5} and HCN with 4×10^{-5} (Keady & Ridgway 1993; Cernicharo *et al.* 1996). The dust grains, assumed to consist of amorphous graphite and refractory species (e.g., SiC), condense in two different shells at $\simeq$5 R_* and $\simeq 15 - 20$ R_* (Keady *et al.* 1988; this work). The acceleration, produced by the interaction between the dust and the stellar radiation and other phenomena, produces a complex velocity profile equal to 1–5 km s^{-1} ($1 \leqslant r/R_* \lesssim 5$), 11 km s^{-1} ($5 \lesssim r/R_* \lesssim 15 - 20$) and 14 km s^{-1} ($15 - 20 \lesssim r/R_*$) (Keady *et al.* 1988; Ridgway & Keady 1988; this work).

2. Observations, Detections and Results

The observations were obtained in 2002 December with the 3 m optimized infrared telescope IRTF in Hawaii and the TEXES spectrometer (Lacy *et al.* 2001), working between 5 and 25 μm with a power resolution $R \sim 10^5$.

In the observed spectrum we have identified many lines corresponding to the R and Q branches of ro-vibrational transitions ν_5, $\nu_4 + \nu_5 - \nu_4$, $2\nu_5 - \nu_5$, $2\nu_4 + \nu_5 - 2\nu_4$, $\nu_4 + 2\nu_5 - \nu_4 + \nu_5$ and $3\nu_5 - 2\nu_5$ for C_2H_2, ν_2 and $2\nu_2 - \nu_2$ for HCN and some of them for their isotopologues. Many lines remain still unidentified (see Figure 1).

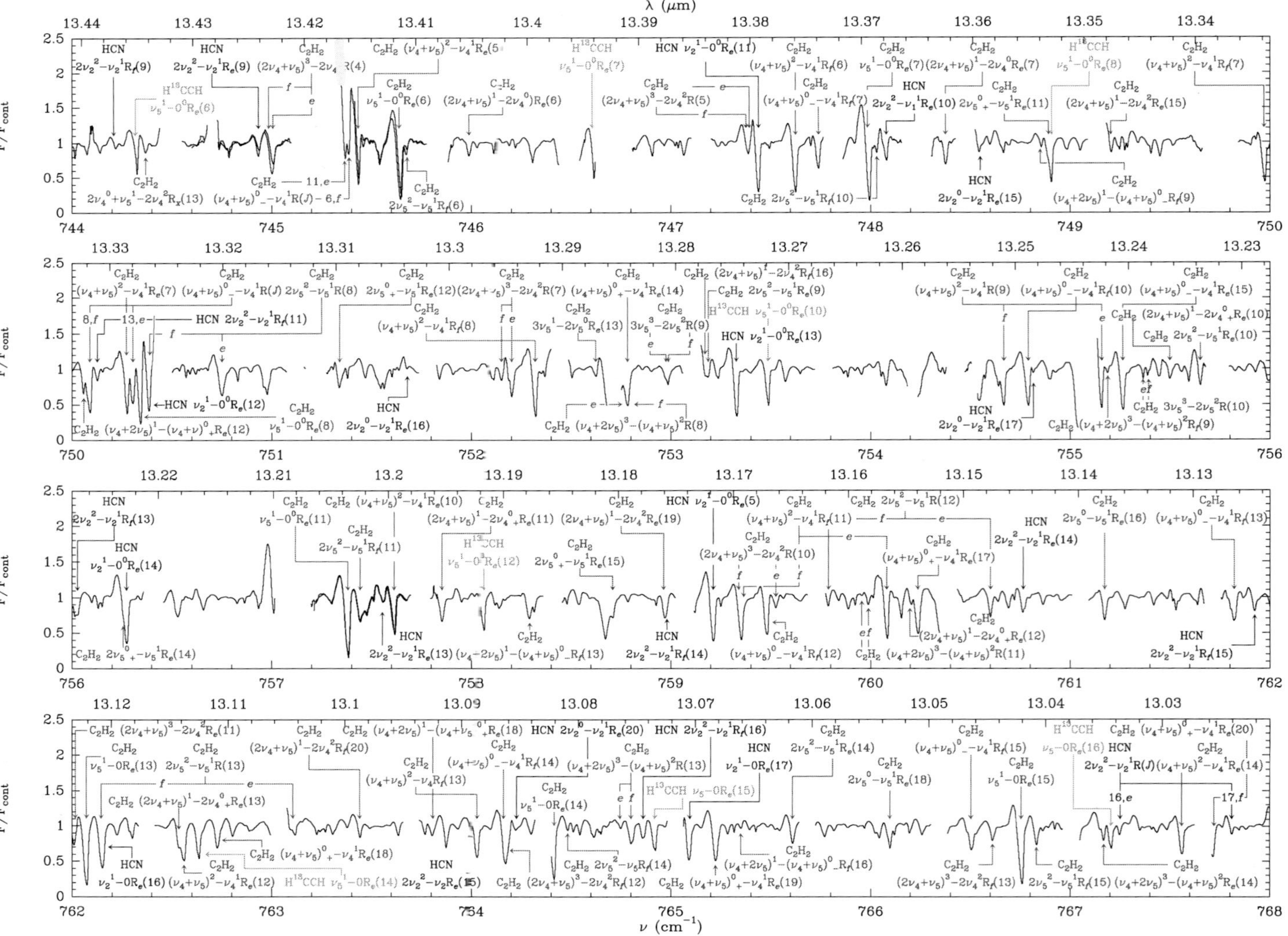

Figure 1. The 744 to 768 cm^{-1} spectrum observed toward IRC+10216.

The main results derived through the fitting of the lines show that the abundances of C_2H_2 and HCN reach their maxima in the middle region (between the two dust formation shells) and keep their values nearly constant as the kinetic chemistry models predict and the outer CSE observations suggest. The derived C_2H_2 abundances suggest a possible condensation of this species onto the dust grains beyond the first dust formation shell, explaining the last acceleration process at $15 - 20$ R_*. The vibrational levels of C_2H_2 and HCN are out of LTE, supporting the existence of a radiative pumping mechanism in the middle region related to the near-IR radiation field (Cernicharo *et al.* 1999). The deviations from LTE depend on the molecule, being very different for C_2H_2 and HCN because of their own radiative selection rules (e.g., C_2H_2 $\nu_4\,(\pi_g) \rightarrow$ G.S.(σ_g^+) is forbidden). While most of the vibrational levels of C_2H_2 are at LTE only in the innermost envelope, two of the three studied vibrational transitions of HCN present a marked non-LTE behavior, even near the star. The rotational levels seem to follow the LTE condition for $J \lesssim 20$ for C_2H_2 with good accuracy except for few lines which present non-thermal populations, probably produced by overlaps with other lines.

We derive the following velocity profile, v_e, kinetic temperature, T_K, and C_2H_2 and HCN abundances, $x\,(C_2H_2)$, $x(HCN)$:

$$v_e\,(\mathrm{km/s}) = \left\{ \begin{array}{ll} 5, & 1 \leqslant r/R_* < 5.2 \\ 11, & 5.2 \leqslant r/R_* < 21.2 \\ 14.5, & 21.2 \leqslant r/R_* < 300 \end{array} \right\}$$

$$T_K = \left\{ \begin{array}{ll} 2330 \left(\frac{1}{r}\right)^{0.58}, & 1 \leqslant r/R_* < 5.2 \\ 900 \left(\frac{5.2}{r}\right)^{0.58}, & 5.2 \leqslant r/R_* < 21.2 \\ 400 \left(\frac{21.2}{r}\right)^{1.00}, & 21.2 \leqslant r/R_* < 300 \end{array} \right\}$$

$$x\,(C_2H_2,\ HCN) = \left\{ \begin{array}{lll} 7.5 \times 10^{-6}, & 2.5 \times 10^{-5}, & 1 \leqslant r/R_* < 5.2 \\ 8.0 \times 10^{-5}, & 4.9 \times 10^{-5}, & 5.2 \leqslant r/R_* < 21.2 \\ 8.0 \times 10^{-5}, & 4.9 \times 10^{-5}, & 21.2 \leqslant r/R_* < 300 \end{array} \right\}$$

Acknowledgements

J. Cernicharo and J. P. Fonfría Expósito thank the Spanish MEC (grants AYA 2003-2785, ESP2004-00665) and the EU for funds under the FP6 program "Molecular Universe". M. J. Richter thanks the NSF (grant AST-0307497). TEXES was built with funds from the NSF. The Infrared Telescope Facility is operated by the University of Hawaii under Cooperative Agreement no. NCC 5-538 with the National Aeronautics and Space Administration, Office of Space Science, and Planetary Astronomy Program.

References

Cernicharo, J., *et al.* 1996, *A&A* 315, L201
Cernicharo, J., *et al.* 1999, *Ap. J.* 526, L41
Keady, J.J., Hall, D.N.B., & Ridgway, S.T. 1988, *Ap. J.* 326, 832
Keady, J.J. & Ridgway, S.T. 1993, *Ap. J.* 406, 199
Lacy, J.H., Richter, M.J., Greathouse, T.K., Jafee, D.T., & Zhu, Q. 2001, *PASP* 144, 153
Ridgway, S.T. & Keady, J.J. 1988, *Ap. J.* 326, 843

Photo: D. Lis

Astrochemistry: Recent Successes and Current Challenges
Proceedings IAU Symposium No. 231, 2005
D.C. Lis, G.A. Blake & E. Herbst, eds.

© 2006 International Astronomical Union
doi:10.1017/S174392130600754X

The Red Rectangle: Solid State Components of Varying Composition in the Outflow

F. Markwick-Kemper[1], **J. D. Green**[2], **and E. Peeters**[3]

[1]Department of Astronomy, University of Virginia, P.O. Box 3818,
Charlottesville, VA, 22903-0818, USA
email: ciska@virginia.edu

[2]Department of Physics and Astronomy, University of Rochester, Rochester, NY 14627, USA

[3]NASA Ames Research Center, MS 245-6, Moffett Field, CA 94035, USA

Abstract. We report the discovery of broad mid-infrared resonances in the outer regions of the Red Rectangle outflow (Markwick-Kemper *et al.* 2005). The peak position and the strength of the resonances vary spatially, but the full width at half maximum, as well as the shape of the feature, appears remarkably constant. While emission due to polycyclic aromatic hydrocarbons (PAHs) is also present at these locations, we show that PAHs cannot be the carriers of these new components. Instead, we argue that these resonances are caused by solid state components, perhaps simple Mg-Fe-oxides. The presence of such O-rich species in the otherwise C-rich outflows further complicates the picture of the formation and chemistry of the Red Rectangle nebula.

Keywords. infrared: stars — ISM: dust, extinction — ISM: individual (HD 44179; Red Rectangle) — stars: AGB and post-AGB — stars: circumstellar matter

---

The Red Rectangle is a relatively nearby system (estimates range from 300–700 pc), consisting of a post-AGB star HD 44179 with a circumstellar disk and a biconical outflow. The disk is seen almost exactly edge-on, providing superb conditions to study the emission from the outflows. The system shows a remarkable chemistry. Besides the so-called Extended Red Emission (Schmidt *et al.* 1980), for which the carrier is still unknown, there are emission features related to diffuse interstellar bands (Scarrott *et al.* 1992), and a wealth of PAHs detected in the outflow (Bregman *et al.* 1993). In contrast, the circumstellar disk contains predominantly oxygen-rich silicate dust (Waters *et al.* 1998). Optical imaging obtained with the Hubble Space telescope reveals that the biconical outflow shows a large amount of substructure, tracing variations in the physical conditions (Cohen *et al.* 2004). The question arises whether these variations in the physical conditions give rise to differences in the chemical composition of the outflows.

We have obtained infrared spectroscopy from 10–19.5 μm at a resolution of $R \sim 600$ using Infrared Spectrograph (IRS) aboard Spitzer. We looked at three different pointings, about 30$''$ away from the central star, in the northern outflow. The high-resolution slits are 5 pixels long, while the point spread function is $\sim$2–3 pixels wide. We have performed sub-slit extractions with a width of three pixels. Figure 1 shows spectra obtained at pointings inside and outside the biconical shape of the outflow. Besides the PAHs emission around 11 μm, there are two very strong features present at $\lambda > 13$ μm: one at 13–16 μm and one longwards of 17 μm. The peak position and the relative strength vary spatially, while the FWHM and the shape remain remarkably constant.

We rule out that the feature is due to PAHs (Markwick-Kemper *et al.* 2005). Instead, we argue that it is due to a solid state carrier. A possible explanation of the $\lambda > 17$ μm features could be provided by simple Mg-Fe-oxides. Figure 1 shows the comparison with laboratory spectroscopy (Henning *et al.* 1995). The change in peak position could be

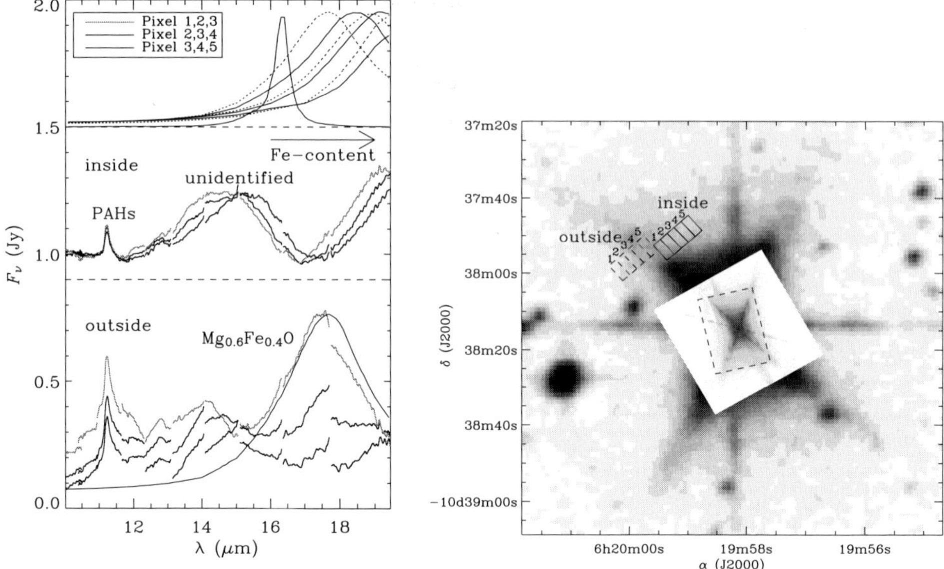

Figure 1. *Left:* IRS high resolution spectra (bottom two panels) compared with the laboratory spectroscopy of simple oxides of varying composition (top panel). The map on the *right* shows the slit and pixel positions of these observations.

explained with a change in composition. The features at 13–16 μm remain unidentified, but we are looking into mixtures of the simple oxides with either spinels or silicates.

The presence of oxides in the C-rich outflow of the Red Rectangle paints a complex picture of circumstellar chemistry, as does the suggested presence of PAHs in the disk (Vijh *et al.* 2005). The origin of the O-rich species in the outflow is unclear. It could be a relic from earlier mass loss, or it could be caused by stellar wind erosion of the circumstellar O-rich disk. Moreover, the varying composition is yet to be explained.

References

Bregman, J.D., Rank, D., Temi, P., Hudgins, D., & Kay, L. 1993, *Ap. J.* 411, 794
Cohen, M., Van Winckel, H., Bond, H.E., & Gull, T.R. 2004, *A. J.* 127, 2362
Henning, T., Begemann, B., Mutschke, H., & Dorschner, J. 1995, *A&AS* 112, 143
Markwick-Kemper, F., Green, J.D. & Peeters, E. 2005, *Ap. J.* 628, L119
Scarrott, S.M., Watkin, S., Miles, J.R., & Sarre, P.J. 1992, *MNRAS* 255, 11P
Schmidt, G.D., Cohen, M., & Margon, B. 1980, *Ap. J.* 239, L133
Vijh, U.P., Witt, A.N., & Gordon, K.D. 2005, *Ap. J.* 619, 368
Waters, L.B.F.M., Waelkens, C., van Winckel, H., *et al.* 1998, *Nature* 391, 868

Discussion

BOWEY: Have you considered temperature shifts for the origin of the finer detail changes in the 14–16 μm feature positions between slit pointings?

MARKWICK-KEMPER: Yes, we have. Besides composition, such effects as temperature changes and grain properties affect the spectral appearance. However, temperature specific optical properties are not yet available for the oxides.

Astrochemistry: Recent Successes and Current Challenges
Proceedings IAU Symposium No. 231, 2005
D.C. Lis, G.A. Blake & E. Herbst, eds.

© 2006 International Astronomical Union
doi:10.1017/S1743921306007551

Astrochemistry: A Summary

Alexander Dalgarno

60 Garden Street, Cambridge, MA 02138

Abstract. A summary is given of the presentations delivered at IAU Symposium 231.

According to Big Bang cosmology the universe began as a very hot, weakly fluctuating, not quite uniform expanding plasma. It was not a promising scenario for the formation of structures. Yet as the Universe evolved, the baryons were distributed into clouds of gas and dust, condensed into stars, stellar remnants and planets, and assembled into groups and clusters of galaxies. On at least one occasion life emerged.

Astrochemistry deals with the molecules and the molecular processes that are central to the growth of complexity in the Universe. Because of their energy level structure molecules can be detected by observing transitions at wavelengths and frequencies ranging from the ultraviolet to the gigahertz, and they provide powerful unique diagnostic probes of astronomical environments. But they do much more than provide insight into the nature of astronomical events. In their control of the ionization balance and the thermal structure of cosmic gases, molecular processes are fundamental in bringing about the physical conditions in which gravitational collapse occurs and galaxies, stars and planets are formed.

The first International Astronomical Union Symposium in Astrochemistry took place twenty years ago in December 1985. This IAU symposium on Astrochemistry is the fifth. The series has charted the development of Astrochemistry into an extensive intellectual discipline in which Astronomy and Chemistry have merged and observers, experimenters and theorists have joined in a mutually enriching interaction. The future of Astrochemistry is assured by the exciting advances that have occurred since the last symposium in August 1999 and by the promise of new telescopes with improved spatial and spectral resolution operating over an extended range of frequencies. Results of observations with a wide range of telescopes have been presented here and we have the Atacama Large Millimetre Array (ALMA), the Herschel Space Observatory and SOFIA to look forward to, though we must take care not to be engulfed by the wave of data that we will be able to accumulate.

A major emphasis of this symposium has been star formation. Progress in understanding the several stages of star formation has been rapid, much of it stimulated by the ISO satellite. For low-mass stars a plausible picture has been constructed that begins with the collapse of a pre-stellar dense core in a molecular cloud. The collapse leads to a protostar embedded in an infalling envelope of gas and dust surrounded by an accreting protoplanetary disk. The infall is accompanied by an outflow driven by the evolving protostar. The outflows give rise to shocks and the dispersal of the envelope material exposes the gas to ultraviolet radiation from the parent protostar, creating photon-dominated regions (PDRs).

The several evolutionary stages are characterized by distinct spectral energy distributions (SEDs). They have been labelled Classes 0 and I (early and late), II and III. They are assumed to reflect an ordering in time. In a remarkable application of astrochemistry, the different continuum and line spectra arising from changes in molecular composition have been interpreted to yield a quantitative description of the density and temperature distribution and variations in velocity. There is chemical evidence for a freeze-out region in the envelope producing a fall in the molecular abundances. Depletion and desorption are critical elements in the modeling. A differential depletion of CO and N_2 appears necessary to explain the relative behavior of HCO^+ and N_2H^+, though recent experiments raise questions about the magnitudes of the binding energies.

Observations of low-mass protostars were reviewed by Cecilia Ceccarelli. She showed how chemical composition and SEDs could be employed to construct two-phase models consisting of a cold envelope around a hot core. She noted that the changing chemical composition might serve as a clock.

Observations of pre-stellar cores, dense condensations of dust and gas with no apparent sign of an interval energy source, were summarized by Mario Tafalla. He also stressed the usefulness of changes in chemical composition and in tracking the evolution of different objects.

Observations place new demands on the development of adequate theoretical models with which to interpret the data and to predict further observations. One such model of pre-stellar cores was advanced by Valery Shematovich.

The most direct evidence of depletion is the existence of abundant multiply-deuterated species. Because of depletion the destruction of deuterated molecular ions is slowed and the ions have time to undergo fractionation driven by reactions with HD molecules and with D atoms. Helen Roberts gave an enlightening discussion of the deuterium chemistry. Deuterium chemistry came up time and time again in the meeting.

Because of the large disruptive effect of the formation of high-mass stars, the sequence of events is less clear. High-mass Young Stellar Objects (YSOs) reveal large variations in chemical composition. They can be generally described as composed of a hot compact core surrounding an ultra compact H II region embedded in an envelope with an inner warm and outer cold region. The complex environment of young massive stars was discussed by Stan Kurtz. He identified a hot phase in which molecules that had been frozen onto grains during the collapse undergo sublimation back into the gas phase as the newly formed star warms its environment, and he drew attention to the occurrence of high-mass starless cores.

Steven Doty gave an instructive comparison of chemical models of the composition of low- and high-mass YSOs. The chemistry of hot cores was the subject of Serena Viti's talk. She demonstrated the value of molecules as evolutionary tracers. Hot cores are the places where complex organic molecules are found.

Geoff Blake gave us a broad instructive account of observations of circumstellar disks and their relationship to planetary formation. He discussed the diagnostics of infall and rotation and pointed out the importance of the ionization distribution. Inga Kemp described how models of the disk chemistry are similar to but different from those of interstellar chemistry. She showed in more detail how chemistry serves as a control of the structure of protoplanetary disks, as did Andrew Markwick-Kemper, who drew attention to the usefulness of observations of isotopic molecules.

Joan Najita demonstrated the value of infrared observations in understanding the formation of planets from disks and the circumstances in which terrestrial-like planets could come into existence.

Hans Olofsson reviewed the molecular abundances in the circumstellar envelopes of AGB stars and stressed the need for improved models and physical data. Jose Cernicharo examined the AGB star IRC + 10216 in which 60 molecular species have been detected from which the physical parameters of the gas and dust could be derived and Ciska

Markwick-Kemper discussed the dust in the outflow of the Red Rectangle observed with the Spitzer Space Telescope.

Complex organic molecules are, by their connection to living organisms, of absorbing interest, though as yet no biological molecule has been found in interstellar space. Apparently the discovery of glycine has to be withdrawn. New mechanisms of the formation of complex molecules were proposed by Steve Charnley, who suggested that deuterated organic molecules would provide a test of the chemical sequences. Marla Moore described laboratory data on complex molecules resulting from proton bombardment of an icy surface and Mike Hollis reviewed observations and mechanisms for the production of isomers. The difficulties in spectroscopic identifications were discussed by Aldo Apponi. Their detection in star-forming regions was reviewed by Sheng-Yuan Liu who showed the value of interferometric observations and who drew attention to the large number of unidentified lines.

Shocks are a ubiquitous phenomenon in the interstellar medium and their effects on molecular gas are profound. In fast shocks the molecules are dissociated and the gas is ionized. The chemistry is that of a cooling recombining gas subject to precursor radiation. In slow shocks the gas is heated and not only is the chemistry modified by endothermic reactions, it is enhanced by materials released from grains. The chemistry is a valuable determinant of the physical parameters of the shock as shown by Malcolm Walmsley. The chemistry serves to differentiate shocks from PDRs, which are regions heated by intense ultraviolet radiation, and from X-ray dominated regions (XDRs), which are regions subjected to intense X rays. Amiel Sternberg reviewed the essential physics and chemistry of PDRs and XDRs and Maryvonne Gerin gave a more specialized talk on carbon chemistry in PDRs.

The basic data that enter into models of the composition of gases and dust were the subject of twelve presentations. Basic data are the identification and determination of spectroscopic lines and bands, calculations and measurements of cross sections for collisional excitation and ionization, charge transfer, photodissociation, photoionization and chemical processes in the gas phase and in and on grains. These studies are a vital component of astrochemistry and we must do what we can to encourage them. Remarkable progress has been made.

In the first session which was on gas-phase processes, Tom Millar summarized the present state of our knowledge, listed the available databases and identified some critical needs. Extraordinary technical skills were exhibited by four experimenters. Thomas Giessen reported measurements of spectral lines at teraherz frequencies expected from light hydrides, carbon chains and complex molecules. Experiments at low temperatures on neutral-neutral reactions were reported by Iam Sims. The variation with temperature of rate coefficients that he found was unexpected. Wolf Geppert measured rate coefficients of dissociative recombination of protonated methanol and Stephen Schlemmer measured rate coefficients of the deuterium fractionation reactions. As an example of theoretical contributions, Marie-Lise Dubernet presented her sophisticated calculations of heavy particle excitation cross sections.

The physics and chemistry of grains and of polycyclic aromatic hydrocarbons (PAHs) occupied the second session. Martin McCoustra described elegant careful measurements of desorption from grain-mantle analogues, Naoki Watanabe investigated grain reactions as a source of deuterated organic molecules and Thomas Henning gave a broad informative review of observations and experiments on silicates and their use as diagnostics, including their appearance in meteorites. Douglas Huggins reported on the Spectroscopy of PAHs (are PADs interesting?) and introduced the idea of possible contributions to the 6.2 μm emission feature of polycyclic aromatic nitrogen heterocycles (PANHs). Sven

Thorwirth gave results of measurements of rotational transitions of specific PAHs. There was an enlightening contribution from Geerd-Jan Kroes on the theory of scattering and photochemistry of ice surfaces which yielded some unexpected interesting results.

Because of the exciting discoveries of extrasolar planets, the study of the solar system, and its relationship to its parent star and to the interstellar medium out of which it was formed, assumes a new importance as a nearby example of a planetary system, though the extrasolar planets detected so far appear to have highly unusual properties, as Sara Seager told us. Actually, Geoff Blake remarked that it is the solar system planets that are unusual. Sara discussed aspects of the photochemistry of their atmospheres. She made it clear that we are at the beginning of an exciting extension of astrochemistry which leads me to suggest we include more on the atmospheres of the planetary satellites such as Titan, Triton and Europa in our next symposium. There were two posters on reactions of CH and C_4H with hydrocarbons by Bergeat *et al.* and Berteloite *et al.* that may occur in the atmospheres of Titan and Triton and in interstellar clouds.

Didier Despois reviewed a large amount of information on comets and explored the connection with the interstellar medium. He compared organic molecules and PAHs in comets and interstellar clouds with particular references to Hale-Bopp which is an abundant source of complex molecules. Oliver Botta discussed the molecular content of carbonaceous meteorites and described the procedures used to analyze them. He provided a list of amino acids found in meteorites and noted that two of them are not found on earth. He included a comparison of the cometary molecules and discussed pathways for the production of amino acids. Frank Bensch reviewed SWAS data on comets and reported new results on Tempel 1 and Deep Impact. He derived water production rates and showed they varied with time. Deep Impact was tracked. The analysis and interpretation continue. Don Brownlee gave a fascinating account of interplanetary dust particles. They offer an additional source of information of a unique character. He concluded with the idea that femto-rocks are the original building blocks of the planets. He pointed to hotspots of D/H as indicating an origin in molecular clouds.

Astrochemistry began as a discipline with the chemistry of diffuse clouds, and diffuse clouds present a continuing challenge. Ben McCall reported on optical and infrared observations, Harvey Liszt on millimeter observations and Ted Snow on ultraviolet observations. Ted spoke mostly on the FUSE data. He was not optimistic about the future of ultraviolet astronomy despite, as he showed, its considerable potential. Ben McCall showed some of his data on the diffuse interstellar bands (DIBs). Their origin is an outstanding spectroscopic mystery. Evelyne Roueff put forth new models of diffuse clouds which have three components with different densities and temperatures. They achieved general but not precise agreement with measured abundances of diatomic species. The abundances found by Harvey Liszt of species like HCO^+, HCN, HNC and C_2H in what may be translucent clouds present an interesting challenge. The chemistry may be correct; it is the physical model and perhaps the assumption of a steady state that require attention. The simple models are successful in explaining the abundances of HF, unexpected though its discovery was. David Neufeld *et al.* went on to predict the presence of CF^+ (isoelectronic with CO). David reported here its detection towards the Orion Bar.

The observation of absorption by H_3^+ by Tom Geballe, Takeshi Oka and Ben McCall was a major advance, long anticipated. Despite the simplicity of its chemistry, interpretation may not be immediate. From measurements towards ζ Persei, McCall *et al.* (2004) inferred an ionization rate of 1.2×10^{-15} s^{-1}. With a more elaborate model, Franck Le Petit, Evelyne Roueff and Eric Herbst derived the much lower rate of 2.5×10^{-16} s^{-1}. The slower rate agrees with the value presented by Stephen Lepp at the IAU Symposium in Brazil in 1991, which was based on measurements of HD and OH. Stephen Lepp was quoting Ewine van Dishoeck who had obtained 2×10^{-16} s^{-1} using the now accepted faster rate for dissociative recombination of H$_3^+$. The lower ionization rates obtained at the time were based on a slow recombination rate. It would be interesting, Ewine, to know what you predicted for the abundance of H$_3^+$ in those models. Whatever, it seems that a larger ionization rate is appropriate. These rates are larger than the value of 5×10^{-17} s^{-1} that appears to be typical of dense clouds. That the rate in dense clouds could be less than in the intercloud region was pointed out years ago.

Extragalactic molecules offer an exciting project for chemically probing the universe. Molecules in external galaxies promise to become a major source of information about the evolution and structure of galaxies and should enable distinctions to be drawn between different galaxy types. Carbon monoxide has been detected in a quasar at a red shift of 6.42 when the universe was less than a billion years old. Cooling by collisional excitation of H$_2$ and HD was crucial to the formation of the first distinct cosmological objects, as Tom Abel noted in his simulations. Not much is yet understood about where and when the first heavy elements were made and still less is understood about the production of heavy molecules. One of the first heavy elements was probably oxygen and the first heavy molecule would have been OH.

I mention OH because it could provide limits on the variation in time and space of the proton gyromagnetic ratio. Observations of molecules in external galaxies and their interpretation were described by Suzanne Aalto, Pierre Cox, Christine Wilson, and Henrik Spoon. They demonstrated the value of chemical composition as a diagnostic probe. The CO/H$_2$ ratio is an important number which will depend on galaxy type.

There were five special talks, focusing on the unexpected lack of O$_2$ in molecular clouds. The SWAS data were examined by Ted Bergin who gave a comprehensive picture of the distribution of water. The ODIN data were examined by Rene Liseau. As both speakers made clear, we learn much from the near-absence of O$_2$. Klaus Pontoppidan introduced us to the idea and reality of mapping the ice distribution in star-forming regions. Neal Evans surveyed broadly the exciting results from Spitzer, and Eric Becklin told us what we can expect from SOFIA.

The special session was followed by a Panel Discussion presented by a distinguished assembly of older and, we hope, mature astrochemists. It was chaired by David Williams, who managed to keep control of a potentially unruly group.

The consensus of the panel was that grain-gas interactions were the major uncertainty in astrochemistry and that is primarily an experimental question. Tom Phillips raised the question—where is all the deuterium? It was agreed that theorists should attempt to predict what observers measure rather than have observers derive what theorists calculate. It is better apparently to have brilliant insights than to carry out large-scale calculations. In my personal view, they are not mutually exclusive and both can be attempted at the same time, each supporting and amplifying the other.

The fundamental question of the formation of H$_2$ and HD was addressed in several challenging experiments. Measurements were reported by Stephanie Cazaux, Liv Hornekær and Gianfranco Vidali, and a theoretical analysis of H$_2$ and HD was advanced by Ofer Biham. We need to understand not only the rates of formation, chemisorption, physisorption and quantum tunneling, but also the details of the energetics. There is astrophysical evidence that the velocity dispersion increases with the level of rotational excitation and observations of PDRs appear to indicate a more rapid rate for a warmer grain-gas interaction as Stephanie Cazaux noted.

The posters, both in number and in content, were a testament to the vitality of Astrochemistry today. Over 200 posters were presented, all directly relevant and interesting and all addressing real questions. Regrettably I cannot mention them individually. There were observational, experimental and theoretical papers relating to low- and high-mass star formation and to protoplanetary discs. PAH's, neutral and ionized, received much attention as did diffuse cloud chemistry, and the Red Rectangle was the favorite object, followed by DIBs and the Early Universe. There were extensive laboratory and theoretical posters on grain and gas phase processes and chemical schemes for the formation of organic molecules were proposed.

Photo: E. van Dishoeck

Astrochemistry: Recent Successes and Current Challenges
Proceedings IAU Symposium No. 231, 2005
D.C. Lis, G.A. Blake & E. Herbst, eds.

Future Directions in Astrochemistry

David A. Williams

Department of Physics and Astronomy, University College London,
Gower Street, WC1E 6BT, UK
email: daw@star.ucl.ac.uk

Abstract. An account of a free-ranging Panel Discussion held during the Symposium.

1. Introduction

'Future Directions in Astrochemistry' was the title of a Panel Discussion held during the Symposium. The Panel members were selected to represent 'voices of experience' in the fast-moving field of astrochemistry, and were invited by the Scientific Organising Committee to discuss briefly their perceptions of the big questions facing modern astrochemistry or of the new directions, in which our subject should develop. It proved to be an interesting and entertaining evening, in which some useful scientific perspectives were gained and new prospects described. The Panellists were Lou Allamandola, John Black, Ewine van Dishoeck, Michel Guélin, Eric Herbst, Tom Phillips, Jonathan Rawlings, Xander Tielens, and Malcolm Walmsley. The proceedings were chaired by David Williams.

There was no prior collusion between the Panel members, nor were the discussions prepared or rehearsed in any way. Yet there was an interesting convergence of views as to some of the challenges facing astrochemistry today. These included:

- the high level of complexity of modern astrochemical models: are the model outputs realistic and credible?
- future observational data rates from facilities such as ALMA: will the future astrochemical community be able to cope?
- can we use the detailed mapping of entire galaxies now available to determine the implications for comprehensive understanding of galactic-scale physics and chemistry?
- are we able to develop useful treatments of non-equilibrium chemistry in weakly ionised, magnetic, and non-Maxwellian media?
- how should we account in our models for the interaction of gaseous atoms, molecules and ions with grain surfaces?
- do we properly understand the chemistry of interstellar water and the formation of water ice on dust grains?
- dirty ice chemistry; what is the role of stable ions in solids?
- can cosmic rays induce the ejection of all kinds of molecules from ices?
- can we constrain galactic astrochemistry by a comparison of interstellar regions of differing physical parameters?
- is deuterium fractionation the key to understanding the physics and chemistry of star-forming regions?
- have we made any progress in understanding long-standing pathological cases, such as interstellar CH^+?
- can astrochemistry contribute to the important astronomical question: the origin of planets?

Of course, as Mr Rumsfeld might say, such a list inevitably contains only the 'known knowns' and 'known unknowns'. There are sure to be the 'unknown unknowns', some of which will (we hope) become apparent at the next Astrochemistry Symposium. All these and other topics were discussed, but two areas were particularly emphasised and will be described in more detail here. These were, firstly, the ever-increasing complexity of the models by which we try to understand and interpret observations, and, secondly, the major gaps in our understanding of the interaction of atoms and molecules with surfaces, and the growth and processing of ices.

2. Complexity in Astrochemistry: are Astrochemical Models Reliable and do They Represent Reality in Some Limited Way?

Over the four decades in which astrochemistry has grown and matured, the models by which we try to understand and interpret astronomical observations have become vastly more complex. Initially, we simply computed steady-state chemical abundances for uniform and static regions to compare with molecular abundances deduced from observations. Now, our models may need to include time-dependence, temperature variability, gas dynamics, the role of magnetic fields, and spatial heterogeneity. The gas-phase networks we use have grown in size from a few tens to several thousands of reactions of many types. Thanks to the dedication and skill of laboratory scientists and theoreticians large numbers of these reactions have been studied experimentally and theoretically. It has been an astounding response to the needs of astrochemistry, though it is still true that many reactions (and their temperature dependences and products) are still poorly understood. New laboratory work continually reveals how deeply subject to error is our understanding of astrochemistry. Throughout this period in which our subject evolved, the importance of the gas/dust interaction has become gradually accepted, and there are many uncertainties to be associated with that interaction, as described below. In addition to these uncertainties, the improvements in observing facilities, particularly those affording high angular resolution, reveal ever-increasing spatial structure on smaller and smaller scales, with the implication of the relevance of shorter and shorter timescales. Thus, in many situations, the original concepts within which chemistry was explored are simply no longer valid. For example, the well-known object TMC-1 has been known since the beginnings of our subject to be exceptionally rich in carbon-chain molecules. The anomalous region was believed to lie in what was found to be a sub-component of TMC-1, known as Core D; but we now know that Core D is itself complex and has a very high degree of sub-structure within it (nearly 50 sub-units), with the implication that it may be evolving rapidly. Hence, the original models of polyyne formation inTMC-1 inevitably rested on a poor foundation.

Thus, complexity arises from incomplete knowledge of chemistry in the gas phase, on surfaces and in the solid state; it is compounded by a limited understanding (driven by ever-improving observations) of the physical situations we seek to explore and describe. The extent to which non-thermal processes or turbulence play a role is unclear. Can we believe the output of our models? Are they useful?

The Panel's discussion tended to stress the incremental approach. No, we could not, as we used to, simply 'put everything in the pot, stir gently, and serve' and hope to arrive at the truth. Computational complexity needs to be tested through simpler approaches. More limited chemical networks can highlight reactions that might have a crucial role, and which should therefore be examined in detail. Modellers must acknowledge that the background of chemical information is continually improving, and update their databases. But it is also important to recognise that the physical basis of any model is open to question, since densities and temperatures deduced from observations are volume-averaged quantities, where the volumes are so large that the important fine detail is lost. For example, the chemical processes determining the abundance of CH^+ molecules by gas-phase reactions in diffuse clouds appear to be well understood and and their rate coefficients well determined, but the physical description of the gas in which those processes play a part is evidently not well characterised or cannot be described sufficiently well.

So the conclusion offered by the Panellists from this part of the discussion might be: do not be disheartened by the slow emergence of the truth, nor make gross claims for one's own work. The enterprise is necessarily slow and iterative between laboratory, theory, observations, and modelling. And that's what makes it fun!

3. The Interaction of Gas and Dust: What are the Issues for Today?

The second area of extensive discussion was also related to concerns about complexity. It is now apparent that the interaction between atoms and molecules with dust grains is an important factor in astrochemistry, and one in which current knowledge is a seriously limiting factor. The scientific issues are the adsorption of atoms and molecules on grain surfaces, the mobility of surface species and their reaction on the surface, the deposition of mixed molecular ices, the chemical processing of such ices, and their removal by thermal or non-thermal means.

The present situation for our understanding of the gas/grain interaction is analogous to that of gas-phase chemistry when our subject began. While the general concepts were then broadly understood, the extent of reliable information on gas-phase reactions relevant to astrochemistry was very small indeed. For example, discussion in the literature from the 1960s and later on the formation of molecular hydrogen on the surfaces of dust grains considered the same issues as now: physisorption and chemisorption, mobility of adsorbed species, the reaction mechanism and the energy budget, and product desorption (either prompt or delayed). Those theoretical speculations have now been very greatly influenced by the results of several recent and sophisticated experiments on molecular hydrogen formation, some of which were described elswhere in the Symposium. Thus, technical advances in the laboratory and in computing power have made possible much more detailed and realistic studies of various surface and solid state processes.

Yet it is clear that we are only at the beginning of this process of learning. It is much easier to pose questions arising from astrochemistry on surfaces or in the solid state than to answer them. However, we have at least advanced from a state of denial of the importance of the gas/dust interaction (an early cause of contention). Molecular hydrogen and other species certainly form on grain surfaces in the interstellar medium, and under suitable conditions molecular ices are deposited on those surfaces. The consequent loss of species from the gas phase is measureable in dense cores in molecular clouds, and this demonstrates that time-dependent freeze-out of species occurs and is an interstellar clock that merely needs to be normalised by a laboratory determination of the relevant sticking probabilities. Detection of relatively complex species in the vicinity of star-forming regions indicates that relatively simple molecular ices may be converted to greater chemical complexity, and desorbed. Many laboratory experiments demonstrate how this chemical complexity may be induced by energy deposition from UV or fast particle irradiation. An important open question at present is desorption: what are the processes desorbing molecules from interstellar ices? Obviously, thermal desorption will be important where the ice temperature can be raised significantly above its normally low value, but even in this apparently simplest of cases recent laboratory work shows unexpected desorptive behaviour linked to the physical nature of ice itself. This is an important issue to address: what is the nature of the interstellar ice? Is it porous or compact? Is its structure dependent on its formation mechanism? Why is there at least partial separation between ice materials?

One area of current interest is the balance of reduction and oxidation reactions occurring within interstellar ices. For example, laboratory work indicates that carbon monoxide may be hydrogenated in some circumstances to formaldehyde, and ultimately to methanol. The relatively high fraction of methanol in interstellar ices indicates that such hydrogenation may be occurring. On the other hand, carbon dioxide, also detected to be abundant in interstellar ices, is surely the oxidation product of carbon monoxide. While it is evident that for gas-phase interstellar chemistry we have some understanding of how the network responds to changes in the physical or chemical conditions, for chemistry in the solid state such intuition eludes us. Nevertheless, it seems likely that simple addition and exchange reactions of neutral species are occurring. The actual situation may, however, be considerably more complex than that. Photolysis of laboratory ices has been shown to produce stabilized ions of polycyclic aromatic hydrocarbons. If such ions and ions of other species exist within interstellar ices, as seems likely, then they may certainly have important roles in cosmic chemistry and physics.

Panel members were confident that the interaction of gaseous atoms and molecules with interstellar dust grains was a very significant influence on interstellar chemistry and therefore on interstellar physics. The broad outlines of that interaction are clear, through surface reactions, ice deposition and processing, and the later ejection of ice molecules back into the gas phase. The details of the processes, however, are mostly obscure, but at least these processes are now capable of being studied with current laboratory techniques, and the next Astrochemistry Symposium should show a major step forward in our understanding. It is, however, at least as likely that new experiments will reveal gaps in our understanding of which we are currently unaware. As Charles Townes (echoing Donald Rumsfeld's words, but with greater clarity) said at the opening of this Symposium 'There is much that we don't understand; in many cases we don't even understand that we don't.' The audience of the Panel discussion seemed excited, rather than depressed, by the challenges outlined during the evening.

Acknowledgements

DAW thanks the UK Particle Physics and Astronomy Research Council for support.

Astrochemistry: Recent Successes and Current Challenges
Proceedings IAU Symposium No. 231, 2005
D.C. Lis, G.A. Blake & E. Herbst, eds.
© 2006 International Astronomical Union
doi:10.1017/S1743921306007575

Poster Presentations

Session I

I.74 *Reaction Rates for Hydrogenation and Deuteration of Solid CO at 15 K.*
Naoki Watanabe, Hiroshi Hidaka, Akihiro Nagoka, & Akira Kouchi

I.75 *Chemical Evolutions in a UV-Illuminated Magnetized Pre-Stellar Core.*
D. Wiebe, V. Shematovich, B. Shustov, Ya. Pavlyuchenkov, M. Kirsanova, R. Launhardt, Th. Henning, & Z.-Y. Li

I.76 *Mapping of CO, $J=5-4$, in Orion Using the Odin Satellite.*
E.S. Wirström, P. Bergman, U. Frisk, Å. Hjalmarson, M. Olberg, A.O.H. Olofsson, C.M. Persson, & Aa. Sandqvist

I.77 *Variations of Molecular Abundances in Regions of High Mass Star Formation.*
Igor Zinchenko, Lev Pirogov, Paola Caselli, Lars E.B. Johansson, Sergey Malafeev, & Barry Turner

Session II

II.01 *Production of Complex Bio-Molecules in Collapsing Interstellar Clouds.*
Kinsuk Acharyya, S.K. Chakrabarti, S. Chakrabarti, & Ankan Das

II.02 *From Molecules to Cosmic Silicate Grains: An Experimental Study of Cluster Intermediates and Growth Mechanisms.*
Ashraf Ali & A.W. Castleman, Jr.

II.03 *CO and C Cooling in Galaxy Nuclei, New Tracers of the Star Forming Activity?*
E. Bayet, M. Gerin, T. Phillips, & A. Contursi

II.04 *Methylidyne Radical Reactions with Hydrocarbons: Kinetics at Low Temperature and Product Branching Ratios.*
Astrid Bergeat, Philippe Caubert, Michel Costes, Nicolas Daugey, & Jean-Christophe Loison

II.05 *Rate Constants for the Reaction of C_4H Radical with Various Hydrocarbons at Very Low Temperatures Relevant to the Atmospheric Chemistry of Titan and other Astronomical Sources.*
Coralie Berteloite, Sébastien D. Le Picard, André Canosa, & Ian R. Sims

II.06 *Gas Phase Ion Chemistry of Polycyclic Aromatic Harbocadrons.*
Nicholas Betts, Veronica M. Bierbaum, & Theodore P. Snow

II.07 *From Organic Molecules to Carbon Particles: Implications for the Formation of Interstellar Dust.*
Ludovic Biennier, Matt Hammond, Jamie Elsila, Richard Zare, & Farid Salama

II.08 *Overtones of Silicate and Aluminate Minerals and the 5–8 μm Ice Bands of Deeply Embedded Objects.*
Janet E. Bowey & Anne M. Hofmeister

II.09 *From Cold Gas Phase Coronene Clusters to Hydrocarbonated Nanograins.*
Phillipe Bréchignac & Martin Schmidt

II.10 *Gas-Grain Interactions in Interstellar Chemistry Explored with PIRENEA.*
N. Bruneleau, C. Joblin, A. Simon, C. Nayral, M. Armengaud, & P. Frabel

II.11 *A Database of Synthetic Molecular Spectra for Astrophysical Applications.*
J. Cami & A.J. Markwick-Kemper

II.12 *First Results of an Unbiased Millimeter Spectral Survey of the Solar-Type Protostar IRAS 16293-2422.*
Emmanuel Caux, Aurore Bacmann, Alain Castets, Stephanie Cazaux, Cecilia Ceccarelli, Claudia Comito, Frank Helmich, Claudine Kahane, Alain Klotz, Bérengère Parise, Peter Schilke, Alexander Tielens, Ewine van Dishoeck, Valentine Wakelam, & Adam Walters

II.13 *Extra-Galactic Diffuse Interstellar Bands.*
Nick Cox, Pascale Ehrenfreund, Lex Kaper, Marco Spans, & Bernard Foing

Session III

Astrochemistry: Recent Successes and Current Challenges
Proceedings IAU Symposium No. 231, 2005
D.C. Lis, G.A. Blake & E. Herbst, eds.

© 2006 International Astronomical Union
doi:10.1017/S1743921306007587

Constants, Units, and Conversion Factors

Solar mass	$M_\odot$	1.9891×10^{33}	g
Jupiter mass	M_J	1.8987×10^{30}	g
Earth mass	$M_\oplus$	5.9742×10^{27}	g
Moon mass		7.3483×10^{25}	g
Solar radius	$R_\odot$	6.9551×10^{10}	cm
Earth radius (equatorial)	$R_\oplus$	6.378×10^{8}	cm
Astronomical Unit	AU	1.4960×10^{13}	cm
Light year	ly	9.4607×10^{17}	cm
Parsec	pc	3.0857×10^{18}	cm
		3.2616	ly
Julian year		3.1558×10^{7}	s
Solar luminosity	$L_\odot$	3.845×10^{33}	$\mathrm{erg\,s^{-1}}$
Jansky	Jy	10^{-23}	$\mathrm{erg\,s^{-1}\,cm^{-2}\,Hz^{-1}}$
Gravitational constant	G	6.67259×10^{-8}	$\mathrm{cm^{3}\,g^{-1}\,s^{-2}}$
Proton mass	m_p	1.67262×10^{-24}	g
Electron mass	m_e	9.10939×10^{-28}	g
Mass of unit atomic weight	u	1.66054×10^{-24}	g
Radius of first Bohr orbit	a_0	5.29177×10^{-9}	cm
Speed of light	c	2.99792×10^{10}	$\mathrm{cm\,s^{-1}}$
Stephan-Boltzmann constant	σ	5.67051×10^{-5}	$\mathrm{erg\,s^{-1}\,cm^{-2}\,K^{-4}}$
Boltzmann constant	k	1.38066×10^{-16}	$\mathrm{erg\,K^{-1}}$
Planck constant	h	6.62608×10^{-27}	$\mathrm{erg\,s}$
Electron-volt	eV	1.60218×10^{-12}	erg
		8065.54	$\mathrm{cm^{-1}}$
		$11{,}604.45$	K
Hartree	H	4.3597×10^{-11}	erg
		27.2114	eV
$\mathrm{kcal\,mol^{-1}}$		6.947×10^{-14}	$\mathrm{erg\,atom^{-1}}$
Electron charge	e	4.80321×10^{-10}	esu
Debye	D	10^{-18}	$\mathrm{esu\,cm}$
Avogadro constant	N_A	6.02214×10^{23}	$\mathrm{mol^{-1}}$

Based on Appendix 1 of Minh & van Dishoeck (eds., 2000; *Astrochemistry: from Molecular Clouds to Planetary System*, ASP). Numerical values from Cox (ed., 2000; *Allen's Astrophysical Quantities*, Springer). The cgs system is still most commonly used in the astrophysical literature.

Author Index

Molecule and Atom Index

Notes: For an up-to-date list of interstellar molecules see Table 1 of Gerin *et al.* and the URL in Charnley & Rodgers in this volume. Table 1 of Moore & Hudson in this volume lists organic molecules detected in interstellar and circumstellar regions. Table 1 of Despois in this volume lists cometary volatiles (detections and upper limits). Table 1 of Olofsson in this volume lists molecules detected in AGB CSEs.

^{13}CO (^{13}C-carbon monoxide) – 187, 261, 271, 309, 365
$^{15}NH_3$ (^{15}N-ammonia) – 227

Al_2O_3 (aluminum oxide) – 469
$AlCl$ (aluminum monochloride) – 499
AlF (aluminum monofluoride) – 499
ArH^+ (argon hydride ion) – 125

C (carbon atom) – 141, 187, 261, 271, 291, 309, 377
$C_{10}H_8$ (azulene) – 455
$C_{12}H_{10}$ (acenaphthene) – 455
$C_{12}H_8$ (acenaphthylene) – 455
$C_{13}H_{10}$ (fluorene) – 455
$C^{18}O$ (^{18}O-carbon monoxide) – 17, 37, 309, 365
C_2 (diatomic carbon) – 97, 165, 197, 237, 499,
$C_{20}H_{10}$ (corannulene) – 455
$C_{24}H_{12}$ (coronene) – 479
C_2Cl_4 (perchloroethylene) – 97
C_2H (ethynyl) – 77, 97, 187, 197, 237
C_2H_2 (acetylene) – 237, 247, 281, 491, 499, 509
$C_2H_2^+$ (acetylene ion) – 125
C_2H_3CN (vinyl cyanide) – 207, 217
C_2H_4 (ethylene) – 97, 247, 491
C_2H_5CN (ethyl cyanide) – 1, 207, 217
C_2H_5OH (ethanol) – 67, 247
C_2H_6 (ethane) – 97, 247, 491
C_2N_2 (cyanogen) – 469
C_3 (triatomic carbon) – 165, 197
$C^{34}S$ (^{34}S-carbon monosulfide) – 17
C_3H_2 (cyclopropenylidene) – 17, 187, 479
$C_3H_6O_4$ (glyceric acid) – 479
$C_3H_7NO_2$ (α-alanine) – 479
$C_3H_7NO_2$ (β-alanine) – 479
$C_3H_7NO_3$ (serine) – 479

C_3H_8 (n-propane) – 97
C_3O_2 (carbon suboxide) – 97
C_4 (4-cabon cluster) – 165
C_4H (butadiynl radical) – 97
C_4H_2 (diacetylene) – 97, 469, 499
C_4H_3 (vinyl acetylene) – 97
$C_4H_8O_3$ (α-hydroxy-n-butyric acid) – 479
$C_4H_9NO_2$ (α-aminoisobutyric acid; AIB) – 479
$C_4H_9NO_2$ (α-amino-n-butyric acid) – 479
$C_4H_9NO_2$ (β-amino-n-butyric acid) – 479
$C_4H_9NO_2$ (γ-amino-n-butyric acid) – 479
C_5 (5-carbon cluster) – 165
$C_5H_{11}NO_2$ (isovaline) – 479
C_6H (hexadiynal radical) – 237
$C_6H_{12}N_4$ (HMT) – 237
C_6H_2 (triacetylene) 97, 499
C_6H_6 (benzene) 499
c-C_2H_4O (ethylene oxide) – 207, 217, 237, 247
c-C_3H_2 (cyclopropenylidene) – 197, 499
CCS (thioxoethenylidene) – 37
CF^+ (fluoromethylidynium) – 163
CH (methylidyne) – 77, 141, 165, 187, 197, 499
CH^+ (methyliumylidene) – 37, 141, 165, 187, 197, 499, 521
CH_2 (methylene) – 141
CH_2^+ (methylene ion) – 125, 141
CH_2CHCHO (propenal) – 227, 247
CH_2CHCN (acrylonitrile) – 67
CH_2CHOH (ethanol; vinyl alcohol) – 207, 217, 237, 247
CH_2CO (ketene) – 67
CH_2D^+ (methylium-d) – 125
CH_2NH (mehylenimene) – 247
CH_2OH (hydroxymethyl radical) – 237, 247, 415
CH_2OHCHO (glycoaldehyde) – 207, 217, 227, 237, 247, 469, 479
CH_3 (methyl radical) – 77
CH_3^+ (methylium) – 27, 37, 77, 125, 141
CH_3C_4H (methyldiacetylene) – 499
CH_3CCH (methyl acetylene; propyne) – 1, 247, 499
CH_3CH_2CHO (propanal) – 227, 247, 479

Subject Index